The Toxicology of Aflatoxins

❖

The Toxicology of Aflatoxins
Human Health, Veterinary, and Agricultural Significance

❖

Edited by

David L. Eaton

Department of Environmental Health
School of Public Health and Community Medicine
and Institute for Environmental Studies
University of Washington
Seattle, Washington

John D. Groopman

Department of Environmental Health Sciences
School of Hygiene and Public Health
The Johns Hopkins University
Baltimore, Maryland

ACADEMIC PRESS, INC.

A Division of Harcourt Brace & Company

San Diego   New York   Boston   London   Sydney   Tokyo   Toronto

Copyright © 1994 by ACADEMIC PRESS, INC.

All Rights Reserved.
No part of this publication may be reproduced or transmitted in any form or by any means, electronic or mechanical, including photocopy, recording, or any information storage and retrieval system, without permission in writing from the publisher.

Academic Press, Inc.
1250 Sixth Avenue, San Diego, California 92101-4311

United Kingdom Edition published by
Academic Press Limited
24–28 Oval Road, London NW1 7DX

Library of Congress Cataloging-in Publication Data

The Toxicology of aflatoxins : human health, veterinary and
  agricultural significance / edited by David L. Eaton, John D.
  Groopman.
        p.   cm.
    Includes index.
    ISBN 0-12-228255-8
    1. Aflatoxins--Toxicology. I. Eaton, David L. II. Groopman,
John D.
  RA1242.A344T68   1993
  615.9'52923--dc20   93-13269
                                                                CIP

PRINTED IN THE UNITED STATES OF AMERICA
93 94 95 96 97 98 BC 9 8 7 6 5 4 3 2 1

Contents

❖

Part I
Experimental Toxicology of Aflatoxins

1 ❖ Acute Hepatotoxicity of Aflatoxins
John M. Cullen and Paul M. Newberne

2 ❖ Biochemical Mechanisms and Biological Implications of the Toxicity of Aflatoxins as Related to Aflatoxin Carcinogenesis
Bill D. Roebuck and Yulia Y. Maxuitenko

3 ❖ Biotransformation of Aflatoxins

David L. Eaton, Howard S. Ramsdell, and Gordon E. Neal

4 ❖ Pharmacokinetics and Excretion of Aflatoxins

Dennis P. H. Hsieh and Jeffrey J. Wong

5 ❖ Nonhepatic Disposition and Effects of Aflatoxin B_1
Roger A. Coulombe, Jr.

6 ❖ Carcinogenicity of Aflatoxins in Nonmammalian Organisms
Jerry D. Hendricks

7 ❖ Role of Aflatoxin DNA Adducts in the Cancer Process
George S. Bailey

8 ❖ Mechanisms by Which Aflatoxins and Other Bulky Carcinogens Induce Mutations
Edward L. Loechler

9 ❖ Aflatoxin Carcinogenesis in the Context of the Multistage Nature of Cancer
Yvonne P. Dragan and Henry C. Pitot

10 ❖ Nutritional Modulation of Aflatoxin Carcinogenesis
Adrianne E. Rogers

Part II
Human Carcinogenicity and Toxicity

11 ❖ Epidemiology of Aflatoxin-Related Disease
Andrew J. Hall and Christopher P. Wild

12 ❖ Molecular Dosimetry Methods for Assessing Human Aflatoxin Exposures
John D. Groopman

Part III
Agricultural and Veterinary Problems

18 ❖ Approaches to Reduction of Aflatoxins in Foods and Feeds

Timothy D. Phillips, Beverly A. Clement, and Douglas L. Park

Part IV
Analytical Identification of Aflatoxins

19 ❖ Recent Methods of Analysis for Aflatoxins in Foods and Feeds

Mary W. Trucksess and Garnett E. Wood

20 ❖ Problems Associated with Accurately Measuring Aflatoxin in Food and Feeds: Errors Associated with Sampling, Sample Preparation, and Analysis

Thomas B. Whitaker and Douglas L. Park

21 ❖ Development of Antibodies against Aflatoxins

Fun S. Chu

Part V
Economic and Regulatory Aspects of Aflatoxins

22 ❖ Human Risk Assessment Based on Animal Data: Inconsistencies and Alternatives
Nancy J. Gorelick, Robert D. Bruce, and Mohammad S. Hoseyni

23 ❖ Economic Issues Associated with Aflatoxins
Simon M. Shane

Contributors

❖

Numbers in parentheses indicate the pages on which the authors' contributions begin.

George S. Bailey (137), Department of Food Science and Technology, Oregon State University, Corvallis, Oregon 97331

Deepak Bhatnagar (327), United States Department of Agriculture, Agricultural Research Service, Southern Regional Research Center, New Orleans, Louisiana 70179

Mary G. Bolton (281), Department of Environmental Health Sciences, School of Hygiene and Public Health, The Johns Hopkins University, Baltimore, Maryland 21205

Robert D. Bruce (493), Human and Environmental Safety Division, The Procter & Gamble Company, Miami Valley Laboratories, Cincinnati, Ohio 45239

Fun S. Chu (451), Food Research Institute, Department of Food Microbiology and Toxicology, University of Wisconsin, Madison, Wisconsin 53706

Beverly A. Clement (383), Department of Veterinary Public Health, College of Veterinary Medicine, Texas A & M University, College Station, Texas 77843

Thomas E. Cleveland (327), United States Department of Agriculture, Agricultural Research Service, Southern Regional Research Center, New Orleans, Louisiana 70179

Peter J. Cotty (327), United States Department of Agriculture, Agricultural Research Service, Southern Regional Research Center, New Orleans, Louisiana 70179

Roger A. Coulombe, Jr. (89), Department of Animal, Dairy, and Veterinary Science, Center for Environmental Toxicology, Utah State University, Logan, Utah 84322

John M. Cullen (3), Department of Microbiology, Parasitology, and Pathology, College of Veterinary Medicine, North Carolina State University, Raleigh, North Carolina 27606

Elaine F. Davis (281), Department of Environmental Health Sciences, School of Hygiene and Public Health, The Johns Hopkins University, Baltimore, Maryland 21205

Yvonne P. Dragan (179), McArdle Laboratory for Cancer Research, University of Wisconsin, Madison, Wisconsin 53706

David L. Eaton (45), Department of Environmental Health, HSB F-561, SC-34, University of Washington, Seattle, Washington 98195

Nancy J. Gorelick (493), Human and Environmental Safety Division, The Procter & Gamble Company, Miami Valley Laboratories, Cincinnati, Ohio 45239

John D. Groopman (259), Department of Environmental Health Sciences, School of Hygiene and Public Health, The Johns Hopkins University, Baltimore, Maryland 21205

Andrew J. Hall (233), Communicable Disease Epidemiology Unit, Department of Epidemiology and Population Sciences, London School of Hygiene and Tropical Medicine, University of London, London WC1E 7HT, United Kingdom

Jerry D. Hendricks (103), Department of Food Science and Technology, and the Marine Freshwater Biomedical Center, Oregon State University, Corvallis, Oregon 97331

Mohammad S. Hoseyni (493), Biometrics and Statistical Sciences, The Procter & Gamble Company, Cincinnati, Ohio 45241

Dennis P. H. Hsieh (73), Department of Environmental Toxicology, University of California, Davis, Davis, California 95616

Thomas W. Kensler (281), Department of Environmental Health Sciences, School of Hygiene and Public Health, The Johns Hopkins University, Baltimore, Maryland 21205

Edward L. Loechler (149), Department of Biology, Boston University, Boston, Massachusetts 02215

Yulia Y. Maxuitenko (27), Department of Pharmacology and Toxicology, Dartmouth Medical School, Hanover, New Hampshire 03756

Doris M. Miller (347), Athens Veterinary Diagnostic Laboratory, University of Georgia, Athens, Georgia 30602

Gordon E. Neal (45), MRC-Toxicology Unit, Medical Research Council Laboratories, University of Leicester, Leicester LE1 9HN, United Kingdom

Paui M. Newberne (3), Department of Pathology and, Mallory Institute of Pathology, Boston University School of Medicine, Boston, Massachusetts 02118

Douglas L. Park (383, 433), Department of Nutrition and Food Science, Food Toxicology Research Laboratory, University of Arizona, Tucson, Arizona 85721

Gary A. Payne (309), Department of Plant Pathology, North Carolina State University, Raleigh, North Carolina 27695

Timothy D. Phillips (383), Department of Veterinary Public Health, College of Veterinary Medicine, Texas A & M University, College Station, Texas 77843

Henry C. Pitot (179), McArdle Laboratory for Cancer Research, and Departments of Oncology and of Pathology, and Laboratory Medicine, University of Wisconsin, Madison, Wisconsin 53706

Howard S. Ramsdell (45), Department of Environmental Health, Colorado State University, Fort Collins, Colorado 80523

Bill D. Roebuck (27), Department of Pharmacology and Toxicology, Dartmouth Medical School, Hanover, New Hampshire 03755

Adrianne E. Rogers (207), Department of Pathology, Boston University School of Medicine, and Laboratory Medicine and Mallory Institute, Boston City Hospital, Boston, Massachusetts 02118

Simon M. Shane (513), Department of Epidemiology and Community Health, School of Veterinary Medicine, Louisiana State University, Baton Rouge, Louisiana 70803

Mary W. Trucksess (409), Center for Food Safety and Applied Nutrition, U.S. Food and Drug Administration, Washington, D.C. 20204

Thomas B. Whitaker (433), United States Department of Agriculture, North Carolina State University, Raleigh, North Carolina 27695

Christopher P. Wild (233), Unit of Mechanisms of Carcinogenesis, International Agency for Research on Cancer, World Health Organization, Lyon Cedex 08, France

David M. Wilson (309, 347), Department of Plant Pathology, Coastal Plain Experiment Station, University of Georgia, Tifton, Georgia 31793

Jeffrey J. Wong (73), Department of Toxic Substances Control, California Environmental Protection Agency, Sacramento, California 95812

Garnett E. Wood (409), Center for Food Safety and Applied Nutrition, U.S. Food and Drug Administration, Washington, D.C. 20204

Hans P. van Egmond (365), Laboratory for Residue Analysis, National Institute of Public Health and Environmental Protection, 3720 BA Bilthoven, The Netherlands

Preface

❖

Over the past thirty years more than 8000 research articles describing the exposure, toxic effects, and mechanisms of action of aflatoxins have been published. The first published reports on the toxicology of aflatoxins pertained primarily to their acute effects, but by the end of the 1960s the carcinogenic potency of these agents was well established and became the focal point of much research. Indeed, the pioneering efforts of George Buchi, who first determined the structural characterization and synthesis of aflatoxins, and Gerald Wogan and Paul Newberne, who individually and collectively pioneered our understanding of the basic biochemistry, toxicology, and carcinogenicity of these agents, are outstanding examples of the application of basic science to a public health problem of global significance. Thus, by the end of the first decade of research on aflatoxins, it was suspected that these compounds were significant human and animal health hazards in various parts of the world. In 1969, Leo A. Goldblatt edited *Aflatoxin: Scientific Background, Control, and Implications,* published by Academic Press. This important book served as a source authority for many years on aflatoxins and provides many insights into the early studies of these agents.

In 1971, aflatoxins were reviewed in Volume 1 of the International Agency for Research on Cancer (IARC) Monographs on the Evaluation of Carcinogenic Risk of Chemicals to Man. That review was only eleven pages long in the monograph. It was concluded at that time that aflatoxins were possible human carcinogens, but the database was still extremely limited. Twenty years later these mycotoxins were classified as Group I, known human carcinogens, and the summary of data published in 1993 in Volume 56 of the IARC Monographs is over 150 pages long. This explosion of information reflects the vast increase in information about the mechanisms of action of the aflatoxins over this period of time and reflects the large number of research scientists studying these toxins. The specific knowledge of the chemistry, biochemistry, toxicology, and epidemiology of aflatoxins is far greater than that for any other environmentally occurring chemical carcinogen. Indeed, it is possible to consider the studies of aflatoxin as a template for researching other environmental carcinogens. Toward this end, we have attempted to bring together as comprehensive a group of scientists as possible in assembling this book.

Part I focuses on the acute toxic effects of aflatoxins (Chapters 1 and 2), their biological disposition (Chapters 3–5), and specific aspects of aflatoxin carcinogenicity (Chapters 6–12). Included in these chapters are detailed reviews of the many important mechanistic aspects of aflatoxins that dictate individual and species susceptibility to aflatoxins. The hepatic biotransformation (Chapter 3), pharmacokinetics (Chapter 4), and genotoxin actions (Chapters 7 and 8) of aflatoxins as well as effects on nonhepatic tissues (Chapter 5), nonmammalian organisms (Chapter 6), and modulation by nutritional factors (Chapter 10) are described in detail. Finally, in this section the carcinogenesis of aflatoxin in animal models (Chapter 9) is discussed in the context of the multistage nature of chemical carcinogenesis.

Part II focuses more specifically on our current level of understanding of human exposures and effects of aflatoxins. The current status of the epidemiology of human aflatoxin exposures is detailed in Chapter 11, while Chapter 12 focuses on recent advances in the application of molecular biomarkers to the study of human cancer incidence in aflatoxin-exposed populations. Chapter 13 concludes with an enlightening review and discussion of potential avenues for human therapeutic and/or dietary interventions aimed at reducing liver cancer incidence in high risk populations, derived from our mechanistic understanding of aflatoxin carcinogenesis.

Part III examines agricultural and veterinary aspects of aflatoxin contamination of food and feed, including reviews on the fungal processes and factors that influence aflatoxin production by *Aspergillus* (Chapters 14 and 15), specific veterinary problems associated with aflatoxin contamination of feeds (Chapter 16), and the unique issues and concerns that arise from the excretion of aflatoxin M_1 in milk of dairy cows (Chapter 17). Lastly, current approaches for reducing the presence of aflatoxin contamination in animal feed and human food crops are discussed in detail (Chapter 18).

Part IV focuses on the complicated and difficult, yet extremely important, aspect of analysis of aflatoxins in food, feed, and biological samples. Chapter 19 reviews current approaches to the analytical determination of aflatoxins in complex matrices; Chapter 20 discusses strategies and problems in accurate sampling, preparation, and analysis of aflatoxins in food and feed; and Chapter 21 discusses the recent development of specific antibodies toward aflatoxin—methods that have found widespread use both in biomonitoring human populations and for analytical detection of aflatoxins in food and feed.

Part V concludes the book with a discussion of the important and sometimes controversial aspect of "quantitative risk assessment" of aflatoxins, which have profound regulatory implications, using either or both animal and human data (Chapter 22), and a discussion of the economic impacts of aflatoxin contamination that impact us all (Chapter 23).

David L. Eaton
John D. Groopman

Acknowledgments

❖

We are grateful to all of the contributors for taking the time and effort to complete these chapters in a thorough yet timely manner. In addition to the individual contributions of each of our chapter authors, who are recognized leaders in their respective areas of specialization, many other scientists from around the world have contributed extensively to our understanding of aflatoxin biochemistry and toxicology. The lifelong contribution of Gerald Wogan to the study of aflatoxins should be especially noted, not only for his extensive and continuing publications in this area, but also for his role in training many of the top researchers in the field today. Indeed, many of the chapter authors in this book trained in Dr. Wogan's laboratory.

Many of us who are involved in the investigation of aflatoxins and other natural products find this work fascinating and compelling; however, we are also drawn to these compounds because of their real public health significance. Worldwide variations in daily exposure to aflatoxins are at least 5000-fold and, in some underdeveloped countries, human exposure can exceed 1 mg per day at certain times of the year. In other parts of the developed world, human exposure is much lower, less than 50 ng per day, but the veterinary and other economic consequences of aflatoxin exposure are still great. Given this situation it is hoped that the understanding of basic mechanisms of action of aflatoxins will lead to the design of effective prevention strategies for both the developed and developing world. Fortunately, the past support of many governments and their agencies have provided the finances necessary to do this research and train the scientists with the skills to tackle these problems. In the future, as new environmental contaminates are discovered, the studies on aflatoxins will provide an important model for how mechanistically driven research can be used to devise and implement appropriate safety regulations to protect the public's health.

Finally, in addition to the many scientists who have provided the scientific substance for this book, we thank Azure Skye for her extensive and excellent administrative support in the development of this book from its inception. We are also grateful to the many people at Academic Press for their assistance, patience, and understanding in working with us to complete this book in a timely manner. We would be remiss without specifically acknowledging the research support provided by the National Institutes of Health, specifically the NCI and NIEHS, for research and training grants that have fostered our own interest in completing this book.

David L. Eaton
John D. Groopman

Part I

❖

Experimental Toxicology
of Aflatoxins

1

Acute Hepatotoxicity
of Aflatoxins

❖

John M. Cullen and Paul M. Newberne

HISTORY

In approximately 1960, an acute hepatotoxic disease in turkeys focused the attention of many scientific laboratories on a common problem affecting animals in many areas of the world (Asplin and Carnaghan, 1961; Blount, 1961; Lancaster *et al.*, 1961). Shortly after the report of turkey "X" disease, reports were made of poisoned chickens and ducklings as well as turkeys, all characterized by acute hepatic necrosis, marked bile duct hyperplasia, acute loss of appetite, wing weakness, and lethargy (Blount, 1961). Similar symptoms and lesions in poultry were reproduced later (Asplin and Carnaghan, 1961) and the toxins identified as metabolites of some strains of *Aspergillus flavus*. The toxins were characterized chemically and were designated aflatoxins (Asao *et al.*, 1963). Within the same time-frame of the early 1960s, many species of animals were exposed either by contaminated feed or by direct experimental procedures (Loosmore and Harding, 1961; Loosmore and Markson 1961; Allcroft and Lewis, 1963).

The outbreak of turkey X disease, clearly described and documented, had been preceded by a number of less well-described episodes of epizootics in a number of animal species. In retrospect, and with subsequent corroborating animal data in hand, these incidents were largely attributable to aflatoxin poisoning. For example, in the mid-1940s, a diet for laboratory animals (MRC #18) that contained peanut meal as a source of part of the protein was formulated in England (Bruce and Parkes, 1947). This diet was associated by Paget (1954) and

The Toxicology of Aflatoxins:
Human Health, Veterinary, and Agricultural Significance

3

Schoental (1961) with a noninfectious disease in guinea pigs and rats that was characterized by gross subcutaneous edema, ascites, and hepatic injury, and ultimately liver cancer in rats. Other outbreaks of the same or a similar disease in guinea pigs indicated that the source of trouble was a toxic factor or a deficiency state associated with some batches of the formulated MRC #18 diet. Identical symptoms and lesions were reproduced later in guinea pigs by Butler (1966), using aflatoxin.

In 1957, abortions and deaths in a guinea pig colony at the Veterinary Laboratory at Weybridge, England, were associated with unknown dietary factor(s). Losses ceased when the diet was changed. Subcutaneous edema and ascites were constant findings at necropsy. Paterson et al. (1962) demonstrated that this disease was attributable to toxic peanut meal included in the diet and, as noted earlier, was reproduced by use of more purified forms of the toxins.

Additional epizootic events confirmed a role for mycotoxins in disease outbreaks in animals (Newberne, 1967). During the period 1945–1953, practicing veterinarians and veterinary diagnostic laboratories in the southeastern United States encountered sporadic recurring outbreaks of a noninfectious, hepatotoxic disease in swine, cattle, and dogs. The disease in hunting dog kennels was especially disastrous. These incidents initiated an intensive investigation into the cause of the malady, but results of these studies left many questions unanswered (Seibold and Bailey, 1952; Newberne et al., 1955; Bailey and Groth, 1959). The disease in swine and cattle was associated with moldy feeds, more often, badly damaged corn; in all documented outbreaks in dogs, the commercially prepared feed reportedly contained peanut meal as a source of dietary protein. At the time, little information of a definitive nature, other than morphologic characteristics of the disease, was learned beyond these facts. Later studies using similar diets from the same source produced identical symptoms and lesions in dogs. The peanut meal in these later studies was contaminated with aflatoxin (Newberne et al., 1966). Similar studies reproduced earlier observations in cattle; in these studies, pure mold cultures were used (Burnside et al., 1957).

Several veterinary diagnostic laboratories in the United States had observed occasional isolated outbreaks of a disease in turkeys, with characteristic histopathologic changes in the liver that were not attributable to previously diagnosed conditions. Newberne encountered the turkey "X" disease and trout hepatoma while working in the veterinary diagnostic laboratory at the University of Missouri in the years 1955–1957 (Newberne, 1967).

Gross and microscopic lesions observed as well as the history in the cases in poultry presented to our laboratory were identical to those described for cases in Britain (Blount, 1961; Siller and Ostler, 1961). In addition, trout hepatoma occurred in hatchery-raised fish in southeast Missouri state laboratories during this same period, examples of which were presented to the veterinary diagnostic laboratory at Missouri University. These typical liver tumors were the same as those described by Halver (1965). The diagnostic service was not involved beyond performing the morphologic diagnosis, but investigators noted that the hatchery fish were fed a commercial chow that contained cottonseed meal as a source of part of the protein source; this meal later was shown to be contaminated with aflatoxins.

Thus, aflatoxicosis appears to have existed for a considerable time prior to the epizootic outbreak in Britain in 1960. However, that dramatic outbreak of the hepatotoxic disease, which initially destroyed more than 100,000 turkeys, demonstrated the seriousness of the problem facing the food animal industry, and ultimately led to the recognition that aflatoxin is both an economic and a public health problem in many areas of the world.

BIOLOGICAL ACTIVITY—ACUTE AFLATOXICOSIS

Epidemiological Evidence

Acute structural and functional damage to the liver, the principal target organ for aflatoxins as observed in field outbreaks, has been reproduced experimentally in most laboratory animals and in several domestic animal species. However, data on clinical aflatoxicosis in humans is still limited although ample evidence exists for substantial exposure in subsets of human populations in many areas of the world (Busby and Wogan, 1984). Campbell et al. (1970) found aflatoxin M_1 (AFM) to be the major hydroxylated metabolite of aflatoxin B_1 (AFB$_1$) in urine of Filipinos exposed to peanut butter contaminated with ~0.5 mg AFB$_1$/kg. Further, extensive studies in a large population of mainland Chinese has confirmed significant exposure to aflatoxin, and a relationship of this exposure to liver cancer seems likely (Yeh et al., 1989; T. C. Campbell, personal communication; Ross et al., 1992). Such exposure is associated with a high incidence of hepatocellular carcinoma, but the extent to which concomitant hepatitis B virus (HBV) infection is involved is not known. Evidence of acute aflatoxicosis has been reported from Taiwan and Uganda (Shank, 1977, 1981) that is characterized by vomiting, abdominal pain, pulmonary edema, and fatty infiltration and necrosis of the liver.

An outbreak of putative aflatoxin poisoning in western India was described (Van Rensburg, 1977; Shank, 1981), ostensibly as a result of the consumption of heavily moldy corn. Specimens were analyzed and shown to contain 6–16 mg aflatoxin/kg corn. The contaminated corn was consumed by people in over 200 villages. Of the nearly 400 patients examined, over 100 fatalities occurred, mainly from gastrointestinal hemorrhage that was reminiscent of aflatoxin poisoning in dogs (Newberne et al., 1966). Cases were found only in households in which the contaminated corn was consumed. Liver specimens revealed marked parenchymal cell necrosis and extensive bile duct proliferation, lesions often seen in experimental animals after acute aflatoxin exposure. Concurrent presence of other mycotoxins and multiple etiology cannot, however, be ruled out.

According to convincing evidence, a disease of children in Thailand with symptoms identical to those of Reye's syndrome has been associated with human aflatoxicosis (Shank 1977, 1981). The disease in children was characterized by vomiting, convulsions, and coma in addition to cerebral edema and fat infiltration in the liver, kidney, and heart. Aflatoxin poisoning was suspected because the symptoms of Reye's syndrome in humans closely approximate those observed with acute aflatoxicosis in monkeys (Bourgeois, 1971). In one fatal case, consumption of aflatoxin-contaminated rice was documented. In addition, Shank *et al.* (1971) demonstrated AFB_1 in the liver, brain, kidney, bile, and gastrointestinal tract contents of 22 of 23 Thai Reye's syndrome fatalities. The AFB_1 content of these materials was elevated substantially relative to the low levels of AFB_1 detected in 10 of 15 patients dying from other causes. Aflatoxin residues have not been associated with incidences of Reye's syndrome in the United States.

Nonhuman Evidence

A review of the effects of acute exposure to aflatoxins reveals that a wide variety of vertebrates, invertebrates, plants, bacteria, and fungi are sensitive to these toxins but the range of sensitivity is wide. The basis for the species and strain variation in the acute toxicity of aflatoxin is not fully understood. Two important factors are (1) the proportion of AFB that is metabolized to the 8,9-epoxide relative to other metabolites that are considerably less toxic and (2) the relative activity of phase II metabolism, which forms nontoxic conjugates and inhibits cytotoxicity. The 8,9-epoxide of AFB is short lived but highly reactive and is believed to be the principal mediator of cellular injury.

The activation of AFB_1 to the reactive intermediate is carried out by the P450 enzyme system (Guengerich and Shimada, 1991; see Chapter 3). Although the specific molecular mechanisms are unknown, binding of AFB_1-epoxide to various cellular macromolecules is believed to be responsible for hepatocellular injury and death. Once bioactivated, AFB_1 can bind to DNA, RNA, and proteins, resulting in a diminution in synthesis of DNA, nuclear and nucleolar RNA, and protein (Clifford and Rees, 1967; Neal, 1973; Neal *et al.*, 1981; Yu, 1981).

Formation of DNA adducts of AFB_1-epoxide is well characterized. The primary site of adduct formation is the N7 position of the guanine nucleotide (Essigmann *et al.*, 1977). In addition to a possible increased risk of cancer, acute toxicity also may be associated with these adducts because of the greater tendency for adduct formation to occur in sites of active gene transcription, for example, at ribosomal genes (Yu, 1983; Irvin and Wogan, 1984). Unlike transcriptionally inactive DNA, which is packaged tightly into nucleosomes by histone proteins, active genes are not associated with histone proteins and are more accessible to the epoxide. Protection of DNA from adduct formation by association with nucleosomes has been demonstrated *in vitro* using cloned DNA packaged into nucleosomes, which sustained reduced adduct formation compared with cloned DNA that was not associated with histones (Moyer *et al.*, 1989). Thus, AFB_1 binding to transcriptionally active sites in DNA may be the mechanism for the dramatic decrease in nuclear and nucleolar RNA synthesis observed in AFB_1-treated animals (Yu, 1983, 1988a,b). Reduced nuclear RNA synthesis is attributed to an inhibition of function of the chromatin template and inhibition of RNA polymerase. The principle cause of diminution of ribosomal RNA synthesis is attributed to altered chromatin template function (Yu, 1981). The mechanism by which template function is altered appears to be a direct interaction of AFB_1-epoxide with DNA rather than with associated histone proteins (Yu, 1988a). Inhibition of RNA polymerase II by AFB_1-epoxide also contributes to diminished RNA transcription (Yu *et al.,* 1986). Adduct formation also occurs in mitochondrial DNA which is not associated with histone proteins and is bound preferentially by AFB_1-epoxide compared with nuclear DNA. Adduct formation occurs at guanine residues of mitochondrial DNA, as it does in nuclear DNA. Injury to mitochondrial DNA is an area of active interest with respect to carcinogenesis, but its contribution to acute injury or carcinogenesis currently is unclear.

Guanine nucleotides of RNA also can form adducts with AFB_1-epoxide (Lin *et al.,* 1977). These RNA adducts, especially those in messenger RNA, are believed to interfere with cellular protein synthesis and to inhibit protein translation at ribosomes in acute aflatoxicosis (Sarasin and Moule, 1975). RNA injury is considered the major factor in diminished cellular protein synthesis by ribosomes, since little AFB_1 binding of intrinsic or ribosome-associated proteins occurs in isolated AFB_1-treated hepatocytes (Yu *et al.,* 1988b).

AFB_1-epoxide can bind covalently to various proteins, which may affect structural and enzymatic protein functions. AFB_1 treatment has been shown to alter phosphorylation of proteins in rats (Viviers and Schabort, 1985). The significance of this effect must be explored more fully, since protein phosphorylation controls the activity of many enzymes and the assembly of some structural proteins. Cellular respiration, another critical aspect of cellular metabolism, may be inhibited by mitochondrial damage in acute aflatoxicosis. Mitochondrial injury manifested as altered electron transport has been demonstrated in AFB_1-treated rats (Doherty and Campbell, 1972). Aflatoxin also can bind to free nuclear proteins, chromatin-bound nonhistones and histones (Yu *et al.,* 1988a and c). This behavior may influence gene expression and chromosome structure, but no inhibitory effect on RNA synthesis occurs. Among the histones, preferential binding to histone H3 occurs, a result that is surprising since AFB_1 binds covalently to the amino acid lysine in serum albumin and lysine is present in relatively low amounts in this histone compared with others. However, binding may occur because of the presence of cysteine in the H3 histone. Because of the abundance of structural and enzymatic proteins in cells and their critical role in cellular homeostasis, the degree of sensitivity to acute aflatoxin toxicity may be related to the amount of protein binding by the AFB_1 8,9-epoxide (Ueno *et al.,* 1980; Neal *et al.,* 1981). Inhibition of cellular respiration, induction of lysosomal enzyme release, and blocked uptake of metabolic precursors into the cell may be very important factors in determining the level and severity of acute toxic responses. An area that should receive more experimental attention because it may be especially relevant to the acute toxic effects of the aflatoxins is the activity of these substances as membrane-active agents.

Acute effects of aflatoxins on plants and microorganisms are important components of the total effects of these mycotoxins on the environment. However, the limitations in space preclude the inclusion of all but a few of these observations here. Details of these effects can be found in a review by Busby and Wogan (1984). However, brief coverage of selected effects on microorganisms, cell cultures, and lower vertebrates, prior to further consideration of more conventional species, is justified.

Microorganisms

In terms of microorganisms and plants, growth of a number of species of *Bacillus*, a *Clostridium*, and a *Streptomyces* was inhibited by AFB$_1$ at a concentration of 7 μg/ml in the culture medium (Burmeister and Hesseltine, 1966). However, most of the 329 species of microorganisms, including bacteria, yeast, fungi, and protozoa, were not inhibited at the concentration used in the medium. No inhibition of growth was observed for selected species of common gram-positive and gram-negative organisms including *Serratia marcescens, Pseudomonas aeruginosa, Staphylococcus aureus,* and *Escherichia coli* (Arai *et al.,* 1967). Several other microorganisms, however, are sensitive to aflatoxins, including members of the genus *Bacillus* (Ueno and Kubota, 1976) such as *Bacillus thuringiensis* (Boutibonnes, 1979) and *Salmonella typhimurium,* when rat liver S9 supernatant preparations were added to the incubation media (Garner *et al.,* 1971).

In Vitro Effects

The aflatoxins have been studied extensively *in vitro* [International Agency for Research on Cancer (IARC) 1987]. All the members of the family (AFB$_1$, AFB$_2$, AFG$_1$, AFG$_2$) produce injury consistent with their characterization as genotoxic agents. AFB$_1$ is the most potent genotoxic agent *in vitro* (as it is *in vivo*) and AFG$_2$ the least potent, but all the naturally occurring aflatoxins, as well as the metabolite AFM, produce some injury to DNA. AFB$_1$ produces chromosomal aberrations, micronuclei, sister chromatid exchange, unscheduled DNA synthesis, and chromosomal strand breaks, and forms adducts in rodent and human cells (IARC, 1987). AFG$_1$ produces chromosomal aberrations in Chinese hamster bone marrow cells *in vivo* and forms adducts to DNA of kidney and liver cells of rats. This aflatoxin induces unscheduled DNA synthesis in human fibroblasts and rat hepatocytes *in vitro,* as well as chromosomal aberrations and sister chromatid exchange in Chinese hamster cells *in vitro.* AFB$_2$ binds covalently to DNA of rat hepatocytes, produces sister chromatid exchange in Chinese hamster ovary cells, and stimulates unscheduled DNA synthesis in rat hepatocytes but not in human fibroblasts. AFG$_2$ produces sister chromatid exchange in Chinese hamster cells but does not induce unscheduled DNA synthesis in human fibroblasts *in vitro.* Unscheduled DNA synthesis in rat and in hamster hepatocytes results following exposure to AFG_2 *in vitro*. AFM_1, a hydroxylated metabolite of AFB_1, causes unscheduled DNA synthesis in rat hepatocytes *in vitro*.

The aflatoxins have been examined for potency in a range of primary cultures as well as in established cell lines. Liver cell cultures from chick embryos demonstrate cytotoxicity of AFB_1 to both mesenchymal and parenchymal cells; the latter are more sensitive (Terao, 1967). Human embryo liver cells are also susceptible to AFB_1 toxicity; 50% of cells exhibit cytotoxic effects to 1 μg/ml after 24 hr (Zuckerman *et al.*, 1967). Electron microscopy reveals nucleolar capping of the chromatin, rounding of the cells, and degranulation of the endoplasmic reticulum.

Space limitations do not permit further description of cultured cell effects; the reader is referred to the review by Busby and Wogan (1984). Cells from most mammalian tissues are sensitive to cytotoxic effects; the range of potencies is similar to those observed in whole animals (Engelbrecht and Altenkirk, 1972).

Vertebrates

Fish larvae and amphibians have been used to screen for toxicity of the aflatoxins, but with variable results. A major problem with these models is a lack of relevance to human toxicity. Nevertheless, some indication for level of toxicity has been drawn from some of the studies. For example, Trucksess and Stoloff (1980) observed that 15 to 20-mm larvae of *Bufo melanostictus, Rhacophorus leucomystax,* and *Uperodon* sp. were sensitive to AFB_1 with an LC_{50} of 2.8, 1.6, and 0.5 μg/ml of aqueous solution, respectively, for the three species. Average potency (LC_{50}) of aflatoxins to zebra fish larvae (in μg/ml) was 0.51 for AFB_1, 0.79 for AFG_1, 1.0 for AFB_2, and 4.2 for AFG_2. These treatments produced nuclear changes in parenchymal cells and pleomorphism, in addition to cytotoxic cell necrosis in the liver of the larvae, changes similar to those observed in mammalian species. In addition to liver pathology, AFG_1 also induced kidney damage, primarily in the proximal tubules, as was observed in rats and ducklings (Newberne *et al.*, 1964).

Comparative species toxicity of AFB_1, the major toxic member of aflatoxin congeners, is listed in Table 1. Note that all species tested are susceptible to AFB_1 toxicity but wide variation in response exists among the many species. Potency ranges more than two orders of magnitude from a very sensitive species (i.e., the rabbit) to a relatively insensitive one (rat or mouse). Although this wide variation in acute toxicity could not be explained in the early years when the data were being generated, now the variation in response seems to depend primarily on the manner in which the individual species metabolizes the compound (by either activation or detoxification) but also on the rate and balance at which these processes take place (Busby and Wogan, 1984; Guengerich and Shimada, 1991). We now know that the metabolism of AFB_1 is primarily, if not totally, dependent on the P450 enzyme system. However, the rate of bioactivation by P450 enzymes is only one factor. Some species, such as mice and hamsters, which are resistant to acute AFB toxicity, can produce AFB-epoxide readily. The resistance is attri-

TABLE 1 Comparative Hepatotoxicity of Aflatoxin B$_1$ in Various Species of Vertebrates[a]

Species	Strain	Sex	Age or weight	Route of administration[b]	LD$_{50}$ (mg/kg)	Acute necrosis hemorrhage	Bile duct hyperplasia	Regeneration nodules
Duck	Khaki–Campbell	M, F	1 day	P.O.	0.36	+	+	+
	Pekin	M, F	1 day	P.O.	0.34	+	+	+
Chicken		M	21 days	P.O.	18.00	+	±	0
Turkey	Beltsville	M	15 days	P.O.	3.20	+	+	0
Trout	Mt. Shasta	M, F	9 months	I.P.	0.81	+	+	+
Catfish	Channel	M, F	9.3–0.5 kg	P.O.	11.5	+	+	0
Mouse	Swiss	M, F	newborn	P.O.	1.50	±	0	0
	CD-1	M	weanling	P.O.	7.30	+	0	0
Rat	Porton	M	42 days	P.O.	6.25	+	+	+
		F	42 days	P.O.	18.00	+	+	+
	Fischer	M	42 days	I.P.	4.20	+	+	+
Hamster	Syrian	F	42 days	I.P.	5.85	+	+	+
		M	30 days	P.O.	12.80	+	0	0
Guinea Pig		M	56 days	P.O.	1.00	+	0	0
		F	56 days	P.O.	1.80	+	0	0
Rabbit	Dutch breed	M, F	90 days	I.P.	0.30	+	0	0
Cat	Mixed breed	M, F	adult	P.O.	0.55	+	0	0
Dog	Mixed breed	M, F	weanling	P.O.	0.80	+	+	+
Pig	Poland China	M	weanling	P.O.	0.62	+	+	+
Sheep	Cross breed	M	2 years	P.O.	2.00	+	0	0
Baboon	Wild	M	adult	P.O.	2.2	+	0	0
Monkey Cynomolgus		M	adult	P.O.	2.2	+	+	+
Macaque		F	adult	P.O.	8.0	+	+	0

[a]Data shown in this table are from many literature sources including Butler (1964), Newberne et al. (1966), Newberne (1967), Wogan and Newberne (1967), and Busby and Wogan (1984).

[b]P.O., Per os; I.P., intraperitoneal.

buted to active detoxification of the epoxide by glutathione S-transferases (Neal and Green, 1983). Many investigators are now examining factors or conditions that can modulate metabolism in a way that provides more safety for humans and lower animals (see Chapter 13).

Inspection of Table 1 reveals an effect of age and sex on the response of some species to AFB_1. Generally, weanling rats are more sensitive than newborn and older animals (1 yr or more) to toxic effects of AFB_1 (Newberne, 1986). Male rats are more sensitive than females. In the Swiss mouse, little difference existed between sexes but the mouse was more sensitive than the rat. In carcinogenecity studies, all mouse strains tested are refractory to AFB_1. The hybrid B6C3F1, however, it the exception and shows sensitivity comparable to that of rats (Newberne *et al.*, 1982). Mechanisms for these differences appear to reflect, in part, differences in metabolism among the various strains and species; activation or deactivation with formation of covalent adducts differs among the species and strains.

Route of administration sometimes has been a determinant in the manner in which a species or strain responds to AFB_1. Oral administration of AFB_1 to Porton strain rats was not as potent as intraperitoneal (ip) administration (Butler, 1964) but the opposite result was seen in mice. These differences again appear to reflect differences in metabolism.

Zarba *et al.* (1992) have demonstrated that inhalation of AFB_1 for 20–120 min results in an increase in the amount of aflatoxin N^7-guanine adduct formed per mg DNA in a dose-dependent manner; exposure for 20, 40, 60, or 120 min results in mean adduct formation of 4.2, 15.3, 21.6, and 56.8 pmol AFB_1–N^7-guanine/mg DNA, respectively. Wilson *et al.* (1992) have shown that instillation of AFB_1 into the trachea of hamsters can result in bronchogenic carcinomas and in multiple hepatobiliary adenomas without hepatic parenchymal cell proliferation. These two instances of experimental exposure to AFB_1 via the lung indicate that this route can pose an important risk for carcinogenesis (see also Chapter 5).

Daniels and Massey (1992) have used the knowledge that the bulk of P450-related activity is located in the Clara and type II cells of the lung to identify where AFB_1 is activated in the rabbit respiratory system. Microsomes from Clara-rich cell fractions had 13–22 times the activation potency of whole lung microsomes, whereas type II cell microsomes had only minimal activity.

In comparing the relative potency of aflatoxin congeners, limited data are available that indicate the wide differences in toxicity among the various chemical members of the aflatoxin family. For example, Carnaghan *et al.* (1963) calculated relative potency in ducklings, the most sensitive of commonly used test animals, and reported $AFB_1 > AFG_1 > AFB_2 > AFG_2$ with LD_{50} values of 0.36, 0.78, 1.70, and 3.44 mg/kg, respectively. Wogan *et al.* (1971) determined single dose LD_{50} values (mg/kg body wt) in ducklings of 0.73 AFB_1, 1.18 AFG_1, 1.76 AFB_2, and 2.83 AFG_2. Oral LD_{50} values for 200-g male Fischer rats were 1.16 mg/kg for AFB_1 and 1.5–2.0 mg/kg body wt for AFG_1. The other two aflatoxins, AFB_2 and AFG_2, were nonlethal at 200 mg/kg in Fischer rats of 200-g body weight using the ip route of administration. In male Porton rats of 150–200 g, the LD_{50} for AFB_1 was 7.2 mg/kg and for AFG_1 was 14.9 mg/kg. In 9-month-old rainbow trout (Bauer *et al.,* 1969), the LD_{50} for AFB_1 was 0.81 mg/kg, compared with 1.90 mg/kg for AFG_1 when administered ip. Thus, AFB_1 is significantly more toxic than the other congeners, a characteristic that carries over into carcinogenicity, as addressed in other chapters in this volume.

In studies with aflatoxin M_1 (AFM_1), this hydroxylated metabolite of AFB_1 retained about the same toxicity as the parent compound for 1-day-old ducklings (Purchase, 1967) and rats (Pong and Wogan, 1971). For 1-day-old ducklings, the comparative oral doses were 12 μg AFB_1 and 16 μg AFM_1 per duckling. Synthetic racemic AFM_1 was lethal to rats at 1.5 mg/kg when given by ip injection. This dose is about twice that of the natural compounds. AFB_2 was about 200 times less toxic than AFB_1 in the induction of bile duct hyperplasia in 1-day-old ducklings (Wogan, Edwards, and Newberne, 1971). Assessment of bile duct hyperplasia was the primary bioassay prior to chemical or immunoassay and proved to be a very accurate assay for toxicity of parent AFB_1 as well as of a mixture of the crude substances (Newberne *et al.,* 1964).

Acute comparative AFB_1 poisoning has been reviewed by Newberne and Butler (1969) and Newberne and Rogers (1981). The principal target organ is the liver. Hemorrhagic necrosis is the major gross pathologic observation noted in all animal species given a lethal dose of AFB_1. Necrosis and hemorrhage also occur occasionally in other organs (kidney, heart, spleen, and pancreas), depending on variables such as animal species, dose, route, and treatment protocol used. Hemorrhage in the liver and other organs might result from hepatic failure and the resultant reduction in the production of blood-clotting factors. Liver regeneration, with a substantial increase in the mitotic index and DNA synthesis, is prominent in animals surviving AFB_1 treatment, but only after 1 week or more. Liver fibrosis and cirrhosis are not a posttreatment finding in rodents, but these effects have been recorded for monkeys, pigs, calves, ducklings, and turkeys (Newberne and Butler, 1969).

The histopathology of aflatoxicosis has been derived largely from studies in the rat. In these studies, thorough time-course observations have been made with light and electron microscopy. The most prominent alterations detected with light microscopy included hepatic (parenchymal cell) necrosis, fatty infiltration, and bile duct proliferation. A sequential study of livers of rats given an LD_{50} dose of AFB_1 (7 mg/kg) showed focal glycogen loss and cytoplasmic basophilia 16–24 hr after treatment (Butler, 1964). No mitoses were present, but some parenchymal cell lysis was detected. By 48 hr, hepatic necrosis and early bile duct proliferation were evident with histiocytic infiltration of the necrotic zone. Many of the surviving parenchymal cells contained pyknotic nuclei. After 72 hr, biliary proliferation was extensive; the surviving parenchymal cells exhibited fat deposits. After 1 week, the necrotic areas largely had been removed by the histiocytes; a few parenchymal cell mitoses were seen. Cytoplasmic swelling and pyknotic nuclei were induced in kidney tubule cells by 24 hr, but rapid regeneration of these cells was underway within 48 hr of treatment. The renal changes in rats and ducks were seen with crude mixtures and with AFG_1. In the duck, these changes were associated with tubule cell tumors (Newberne, 1967).

Svoboda and Higginson (1968) observed essentially the same sequence of events in rat liver, although a much lower dose of AFB_1 (0.45 mg/kg) was used to reduce the severity of necrosis. The liver nucleoli were much smaller and less numerous 48–72 hr after treatment. Electron micrographs showed separation of the fibrillar and granular components of the nucleolus, proliferation of the cytoplasmic smooth endoplasmic reticulum, dissociation of ribosomes from the rough endoplasmic reticulum, and changes in mitochondrial configuration. These changes in the smooth and rough endoplasmic reticulum were verified by Butler (1971) with an LD_{50} dose of AFB_1. Nucleolar segregation was noted as early as 15 min after a 1 mg/kg dose of AFB_1. The nucleoli returned to normal after 36 hr (Pong and Wogan, 1970). Svoboda *et al.* (1966) presented a detailed summary of the ultrastructural pathology associated with exposure to AFB_1. Similar data are available for the guinea pig (Butler, 1966), dog (Newberne *et al.*, 1966), and monkey (Deo *et al.*, 1970). Rainbow trout, a strain very sensitive to AFB_1, exhibit liver lesions similar to those seen with AFB_1 following treatment with AFG_1 (Bauer *et al.*, 1969).

Indole-3-carbinol (I3C) induces P4501A1 in mammals. This enzyme is involved in the metabolism of AFB_1 to AFM_1, a means of detoxification and protection against hepatocarcinogenesis (Takahashi *et al.*, 1992; see Chapter 5). In rainbow trout, P4501A1 induction by I3C occurs in a dose-dependent manner, but the induction is small and transient. However, a condensation product of I3C inhibits P4501A1-dependent ethoxyresorufin *O*-dethylase activity, leading to the conclusion that I3C induction of P4501A1 in rainbow trout is not a relevant mechanism in anticarcinogenesis.

Other investigators have shown that water temperature affects tumor induction in rainbow trout, increasing liver tumor incidence is associated with increasing water temperature from 11 to 18°C (Carpenter *et al.*, 1992). These same investigators (Zhang *et al.*, 1992) observed that the temperature effect was associated with more covalent binding of AFB_1 to DNA and less detoxification at the higher temperature than at the lower temperature.

Plakas *et al.* (1991) examined the absorption kinetics of tissue residues as well as renal excretion of AFB_1 and its metabolites after oral administration of ^{14}C-labeled AFB_1 to channel catfish (*Ictaburus punctatus*). After dosing with 250 μg/kg ^{14}C-labeled AFB_1 (7.5 μCi/kg body wt), peak plasma concentration (503 ppb) occurred at 4.1 hr with absorption and elimination half-lives of 1.5 and 3.7 hr, respectively. The concentrations of [14]C (in AFB_1 equivalents) in tissues were highest at 4 hr and ranged from ~600 ppb in plasma to 40 ppb in muscle. At 24 hr, the concentrations of AFB_1 residues were 32 and 5 ppb in plasma and muscle, respectively. This change represented a very rapid depletion from the tissues. Concentration in the urine was 51 ppb at 4- to 6-hr collection interval but bile concentration exceeded 2000 ppb at 24 hr postdosing. Liver concentration peaked after 4 hr at 421 ppb but sustained a level of 53 and 44 ppb at 24 and 96 hr after dosing, respectively. On the other hand, muscle peaked at 4 hr (40 ppb) and presented only trace amounts 24–96 hr after dosing. The rapid elimination half-life of 3.7 hr in catfish, compared with 15.5 hr in rainbow trout, and the low transient concentration in muscle imply that channel catfish are unlikely to contain action levels (20 ppb) of the toxin. The single oral dose (250 μg/kg body weight) administered in this study is equivalent to an assumed feed consumption rate of 3% body weight per day. In spite of this high level of exposure, AFB_1 residues in the edible flesh were below the limit of determination (<5 ppb) and well below the Food and Drug Administration (FDA) action level (20 ppb).

The pattern of metabolism of AFB_1 administered to various mammals appears to be similar. Metabolism of AFB_1 is nearly complete, with most metabolites excreted in the bile. AFM_1 excreted in the bile and urine is the major hydroxylated metabolite (Nabney, 1967; Allcroft et al., 1968; Dalezios, 1973; Chau and Marth, 1976). Monkeys excrete increased amounts of AFM_1 in urine than other mammals (Dalezios et al., 1973). Following a nontoxic ip dose of AFB_1 in rats is an initial phase of rapidly rising serum levels of AFB_1 over a 2-hr period. This increase is followed by a second phase lasting 2–12 hr during which a slow rise in serum levels is seen. Injected AFB_1 is found bound noncovalently to albumin. Hepatocyte levels reach a maximum level of AFB_1 ~2 hr postinjection. AFB_1 is associated initially with microsomes. Approximately 2 hr post injection, AFB_1 levels in the mitochondria and the nucleus peak and plateau. *In vitro* hepatocyte studies and *in vivo* studies with other strains of rats and dose levels yield similar results (Wogan et al., 1967). Approximately 1% of administered AFB_1 is bound covalently to DNA; this binding peaks at 2 hr postadministration (Croy and Wogan, 1981). AFB_1 administered ip can be visualized first in the hepatocellular cytoplasm followed by a gradual transfer to the nucleus (Stora et al., 1979). The liver attracts 15–18% of AFB_1 by 3 hr postinjection; 60% of that is covalently bound (Niranjan et al., 1982). Of AFB_1 administered orally, 1–2% can be found bound to plasma protein 24 hr after administration.

In the rat, AFG_1 causes parenchymal cell necrosis and bile duct proliferation in the liver, extensive hemorrhagic necrosis of the zone reticularis of the adrenal, and necrosis with subsequent desquamation of the kidney tubule epithelium (Newberne et al., 1966a). AFG_1 also induces a significant level of kidney tumors in the rat (Salmon and Newberne, 1963). AFM_1 and AFM_2 cause lesions that are qualitatively similar to those produced by AFB_1 in the liver and kidney of 1-day-old ducklings (Purchase, 1967); synthetic racemic AFB_1 and AFM_1 induce liver lesions in the rat that are indistinguishable from those produced by the natural parent matabolite (Pong and Wogan, 1971).

Alterations in mitosis and liver nuclei populations accompany the previously described cytomorphological observations. Rogers and Newberne (1967) and Newberne and Rogers (1968) noted that, in rats, reduced mitosis in hepatocytes was notable as early as 1 hr after a single 3 mg/kg dose of AFB_1; the mitotic index remained depressed for at least 50 hr after treatment and recovery occurred by 5 days. Significant reductions in the tetraploid hepatocyte population correlated well with the development of periportal parenchymal cell necrosis in rats given an LD_{50} dose of AFB_1 (Neal *et al.*, 1976). The corresponding increase in the diploid hepatocyte population appeared concurrently with bile duct proliferation and an increase in the infiltrating histiocytes.

Observations by Schrager *et al.* (1990) may help explain some of the changes that occur immediately after exposure to single and multiple doses of AFB_1. A single dose of 25 μg AFB_1 can cause a measurable increase in AFB_1–DNA adducts that is more pronounced following a second daily dose. These adducts and the behavior of hepatic parenchymal cells described by Newberne and Rogers (1968) appear to correlate. However, since these phenomena are magnified by dietary manipulation, they will be described in the next section.

FACTORS AFFECTING AFB_1 TOXICITY

Numerous factors can influence the magnitude of response in vertebrates to exposure to AFB_1. One of the more striking modulating effects on AFB_1 toxicity is associated with dietary components that the animals consume prior to or following exposure to AFB_1 (see also Chapters 3 and 10).

Dietary and Nutrient Effects

Amino Acids

An early study with ducklings, one of the most sensitive species to AFB_1 toxicity, determined the short-term effects of dietary modifications on the response to aflatoxin. In the presence of aflatoxin, dietary supplements of 4.0% methionine, 1.0% arginine, or 0.8% lysine as individual additions depressed weight gain but also decreased mortality (Newberne *et al.*, 1966a). The addition to the diet of 1.0% arginine and 0.8% lysine with, but not without, aflatoxin sharply decreased weight gain and increased mortality. The addition of glutathione or cysteine to the diet as sources of sulfhydryl groups had no effect on toxicity. Autoclaving aflatoxin-contaminated peanut meal that was used as the source of the toxin decreased toxicity and markedly increased weight gains of ducklings over a 9-day period. The most significant observation of this investigation was the profound increase in sensitivity of ducklings to aflatoxin when dietary concentrations of arginine and lysine were increased only slightly above normal.

Protein and vitamin content of the diet, hormonal status, and treatment with pharmacologically active compounds, among other changes, modify the response of experimental animals to acute AFB_1 toxicity. These responses—whether measured as animal mortality, histopathological alterations in the liver, or clinical evidence of liver damage—all register differences compared with untreated animals. Many of these factors also have been studied with respect to their effects on AFB_1-induced hepatocarcinogenesis, the metabolism of AFB_1, the biochemical response to AFB_1 toxicity (especially in terms of inhibition of nucleic acid and protein synthesis), and the metabolism and pharmacokinetics of AFB_1.

Low Protein

Rats fed a low-protein diet (4% casein) and given 50 µg of an undefined aflatoxin preparation exhibited a significant degree of toxic liver lesions within 3 weeks, compared with rats fed a high-protein diet (20% casein) (Madhavan and Gopalan, 1965). Similarly, a high protein diet (16% casein) protected monkeys from aflatoxin liver injury in comparison with animals on a 1% casein diet (Madhavan et al., 1965).

Choline and Methionine

No mortality was recorded 2 weeks after a single dose of 7–9 mg AFB_1/kg in male Fischer and Sprague–Dawley rats that were maintained on a high-fat, lipotrope-deficient diet for 2 weeks prior to treatment (Rogers and Newberne, 1971). Mortality in the rats fed a nutritionally adequate diet was 60–100%. In contrast, animals on the deficient diet were much more sensitive (50% mortality) to repeated daily doses of 25 µg AFB_1 than were those fed the control diet (4% mortality), suggesting that enzyme induction increased AFB_1 activation and DNA adduct formation with repeated dosing. Even diets marginally deficient in choline, given for 10 days prior to AFB_1, were sufficient to protect against the acute toxic effects of AFB_1.

Using choline or methionine deficient diets, Schrager et al. (1990) showed that a methyl group deficit has a profound effect on the formation and removal of AFB_1–DNA adducts. Several laboratories have documented modulation in sensitivity of male Fischer rats to hepatocarcinogenesis by a choline-deficient/low-methionine diet. This regimen dramatically increases hepatocarcinogenesis and reduces time to first tumors induced by AFB_1.

However, early events (acute exposure) and how they relate to cancer induction have received only minimal attention. Therefore, the effect of a choline-deficient/low-methionine diet on hepatic aflatoxin–DNA adduct burden in male Fischer rats dosed with a carcinogenic regimen of AFB_1 was examined in this model. After 3 weeks of ingestion of a choline-deficient/low-methionine diet or a control semipurified diet (sufficient time for enzyme activity to be altered), the rats were administered a carcinogenic regimen of 25 µg [³H]AFB_1 for 5 days/week over 2 weeks. Six choline-deficient and four control diet rats were killed 2 hr after each dose so liver DNA could be isolated. In addition, hepatic DNA was isolated from animals 1, 2, 3, and 11 days after the last [³H]AFB_1 administration. At all time points, high-performance liquid chromatography (HPLC) analysis of aflatoxin–DNA adducts was performed to confirm radiometric determinations of DNA binding levels. No significant quantitative differences in AFB_1–DNA adduct formation were observed among the dietary groups after the first exposure to $[^3H]AFB_1$; however, total aflatoxin–DNA adduct levels in the choline-deficient animals were increased significantly during the multiple-dose schedule. When total aflatoxin–DNA adduct levels were integrated over the 10-day dose period, a 41% increase in adduct burden was determined for the choline-deficient animals. This increase in DNA damage is consistent with the hypothesis that DNA damage is related to tumor outcome, but the biochemical basis for this effect is not known. Moreover, adduct formation correlates closely to the early onset of parenchymal cell necrosis first described by Newberne and Rogers (1968).

Other evidence for nutrient effects on acute toxicity of AFB_1 has accumulated over the past two decades; effects are sometimes variable. For example, oral treatment of rats with 300 mg/kg of ubiquinone, vitamin K_1, menadione or α-tocopherol immediately after an oral dose of 7 mg AFB_1/kg did not alter the 50–75% mortality within the various treatment groups (Rogers and Newberne, 1971). In addition, vitamin A deficiency or supplementation was not a factor in the severity of aflatoxin-produced liver damage in female rats (Reddy *et al.,* 1973). Vitamin deficiency in males, however, increased the susceptibility of the liver to aflatoxin damage; this increased susceptibility was overcome largely by vitamin A supplementation. Pretreatment of rats fed a control diet with carotene 30 min before an LD_{50} dose of AFB_1 (7 mg/kg) reduced the 7-day morality from 80% to less than 10% (Newberne *et al.,* 1974).

In an earlier study (Rogers and Newberne, 1967), the effects of AFB_1 and of dimethylsulfoxide (a frequently used carrier solvent for the toxin) on DNA synthesis and parenchymal cell mitosis were striking. Prior to this study, Butler (1964) found that a single dose of AFB caused extensive necrosis and delayed mitotic response in the liver of rats. In this study, half of an LD_{50} dose of AFB_1 was used. Since this dose is not carcinogenic, the toxic effects were measured. DNA synthesis was measured by $[^3H]$thymidine labeling of nuclei in autoradiographs. Mitotic figures were counted in the livers of 148 weanling Fischer rats who received a single intragastric dose of AFB_1 in dimethylsulfoxide and in 190 control rats who were given either dimethylsulfoxide, isotonic sodium chloride, or no treatment. A marked reduction occurred in DNA synthesis and mitosis in the parenchymal cells within 7 hr of administration of aflatoxin; this decrease persisted for approximately 50 hr, after which both parameters of cell division gradually returned to normal. DNA synthesis also was reduced in the Kupffer cells, but less severely than in the parenchymal cells. However, Kupffer cell phagocytic function was maintained, as evidenced by the presence of nuclear fragments in the cytoplasm, presumably from the demise of surrounding parenchymal cells. Histologic evidence of damage to both parenchymal and Kupffer cells was found within 3 hr of administration of aflatoxin. Controls given dimethylsulfoxide exhibited an elevation of hepatic parenchymal mitoses 6–50 hr later, indicative of a mitogenic response. After only 3 hr, an increase in cellular debris from dead and disintegrating cell nuclei in Kupffer cells was found that correlated with more recent observations on AFB_1–DNA adducts at this early time point (Schrager *et al.*, 1990).

Vitamin A

With respect to vitamin A and retinoid effects on AFB_1 exposure, we demonstrated earlier (Newberne and Rogers, 1973; Rogers and Newberne, 1975) that chronic subclinical dietary vitamin A in rats results in a significant incidence of colon tumors in addition to the usual hepatic neoplasms which, incidentally, also were increased significantly. In attempts to understand mechanisms for increased liver cancer (as well as the 29% incidence of colon cancer) we examined the acute effects of AFB_1 on liver morphology, enterohepatic recirculation, reduced glutathione (GSH) content in liver, and differing capacities for conjugation of aflatoxin to GSH (Newberne and Suphakarn, 1977; Suphakarn *et al.*, 1983). Enzyme concentrations in liver, intestinal and colon mucosa, and intestinal and colon contents suggested that AFB_1 had different metabolites in vitamin deficiency that may result in differing susceptibilities of colon mucosa to carcinogenesis. Binding studies supported this hypothesis. Previous studies had shown that colon epithelium from vitamin A-deficient rats binds more AFB_1 than colon epithelium from normal vitamin A-supplemented animals. In this acute study, vitamin A supplementation to the vitamin A-deficient rats before oral administration of [^{3}H]AFB_1 significantly decreased the bind capacity of AFB_1 to colon epithelium at 12 and 15 hr after dosing with the carcinogen, suggesting that vitamin A affects the metabolism of the carcinogen and binding of AFB_1 to cellular macromolecules, which might explain the mechanism by which vitamin A modifies toxicity of aflatoxin. This effect may be influenced in part through enzymatic mechanisms similar to the DNA–AFB_1 adducts referred to earlier. Repair of AFB_1-damaged DNA also may be defective in vitamin A deficiency.

Enzyme Inhibitors

Current interest in cell injury and necrosis, proliferation, and neoplasia have focused attention on early events that follow exposure to chemicals (Ames, 1990). Ornithine decarboxylase (OCD), a key enzyme in the biosynthesis of the polyamines putrescine, spermidine, and spermine, is important to growth and replication of cells (Sjoerdsma, 1981) and can be manipulated by enzyme inhibitors. The OCD inhibitor difluoromethylornithine (DFMO) has been used to inhibit DNA synthesis, mitosis, and tumor induction by AFB_1 (Sondergaard *et al.*, 1985). DFMO reduced [^{3}H]thymidine labeling to 50% 3 days after the last of a series of AFB_1 doses, paralleled by a similar reduction in gamma glutamyl transpeptidase (GGT)-positive foci of hepatocytes. Since these early events were associated with a later significant reduction in tumor incidence (40–50%), clearly early modulation of AFB_1-induced cell proliferation can have profound effects on chronic tissue injury and ultimately on tumor incidence.

Antioxidants and Selenium

Protective effects were observed with butylated hydroxyanisole (BHA) and AFB_1 administered to mice and with selenium and AFB_1 in the B6C3F1 mouse (Newberne *et al.*, 1987). However, [^{3}H]thymidine labeling, cell necrosis, and GGT-positive foci were increased markedly with selenium deficiency or excess (a U-shaped curve) compared with more normal dietary concentrations of 0.1–2.0 ppm. These observations followed earlier studies (Newberne and Conner, 1974) in which mortality in rats from AFB_1 exposure was reduced significantly at 1.0 ppm selenium in the diet, compared with either 0.02 or 5.0 ppm. Note that the recommended concentration of selenium in the rodent diet is 0.1 ppm. Exposure to AFB_1 increases that requirement ~10-fold, to 1.0–2.0 ppm, for optimum protective effect.

Gender Effects

The increased sensitivity of male rodents, compared with females, to the toxic and carcinogenic effects of AFB_1 has been reported by many investigators and has focused efforts by some laboratories to investigate hormonal effects on response to AFB_1 exposure. Neal and Judah (1978) demonstrated that hypophysectomy of rats decreased liver lesions induced by 2 mg AFB_1/kg. This result was consistent with the antihepatocarcinogenic effect of pituitary extirpation reported by Goodall and Butler (1969). However, interpretation of these results is complicated by the failure of the rats to grow normally. A continuous-dosing study, described by Chedid *et al.* (1980), demonstrated decreased mortality in AFB_1-treated rats if they were administered hydrocortisone, corticosterone, or adrenal corticotropic hormone (ACTH).

Castrated 4-week-old male rats were protected completely from the lethal effects of mixed aflatoxins given in the diet starting at 12 weeks of age (Righter *et al.*, 1972). This effect was obliterated by testosterone treatment prior to and during aflatoxin feeding, which resulted in 100% mortality within 1 week. Conversely, treatment of female rats with an estrogen- and progesterone-based oral contraceptive agent (Ovral-28) prevented death from a dose of 3 mg AFB_1/kg (Mgbodile and Holscher, 1976). This protective effect correlated well with conservation of liver GSH and cytochrome P450 levels, which remained at or near control levels. These results suggest that the sensitizing effect of testosterone may be mediated through these factors.

Glutathione

Conjugation with GSH is an important detoxification pathway for AFB_1. Thus, factors that alter cellular GSH levels have profound effects on the biological activity of AFB_1. Diethyl maleate pretreatment, which lowers tissue GSH concentrations, increases the mortality from 0 to 50% in rats given an ip dose of

3 mg AFB_1/kg. This treatment also accentuates the severity of liver damage (Mgbodile *et al.,* 1975). Pretreatment with cysteine, a GSH precursor, does not alter liver cytochrome P450 levels, either with or without AFB_1 treatment. Diethyl maleate treatment reduces P450 levels even more than AFB_1 does alone. However, a simplistic interpretation of the mechanism of diethyl maleate action, in terms of GSH depletion, may not be sufficient to explain its observed effects. Acute and subchronic AFB_1-induced injury can be reduced significantly in rats treated with oltipraz, a 1,2-dithiol-3-thione, which acts to elevate GSH levels and induce detoxification pathways, including glutathione *S*-transferases (Liu, 1988; see also Chapter 13). Glutathione *S*-transferases can inhibit AFB_1-epoxide binding to DNA by conjugation with GSH (see Chapter 3). Presumably, binding of the AFB-epoxide with other macromolecules is diminished also. Dithiolthiones can protect against AFB_1-induced altered foci in liver that are indicators of carcinogenesis (see Chapter 13).

Goats pretreated with cysteine and methionine, or with sodium thiosulfate, had prolonged survival time after a lethal intraruminal dose of 4 mg AFB_1/kg. Diethyl maleate reduced survival time by over 60%. Cystamine and cysteine, given 12 hr after AFB_1 treatment did not protect rats from developing liver necrosis, nor did cysteine given 8 hr after AFB_1 prolong survival time in goats.

Ethanol

Based on severity of hepatocellular necrosis and elevated serum transaminases (ALT and AST), ethanol pretreatment potentiates the toxic effects of AFB_1 in female rats (Glinsukon *et al.,* 1978). Pretreatment with phenobarbital, which induces several cytochromes P450 and glutathione *S*-transferases, prevented the induction of liver lesions in rats by AFB_1 (Mgbodile *et al.,* 1975). This result correlated with the reduced liver tumors in phenobarbital-treated rats exposed to AFB_1 (Swenson *et al.,* 1977) and *in vivo* activation of AFB_1 to form DNA covalent adducts (Garner, 1975). However, pretreatment with 3-methylcholanthrene, which induces cytochrome P450 1A1, had no influence on liver damage from AFB_1 (Garner, 1975).

Other Environmental Factors

Dietary AFB_1 (0.2 mg/kg for 6 weeks) potentiated the toxic effects of rubratoxin B, a mycotoxin produced by *Penicillium rubrum,* when given concurrently to male rats at a sublethal dose (25 mg/kg orally three times weekly for 5 weeks) (Wogan *et al.,* 1971). Single-toxin treatment groups had no mortality but mortality was 45% when the toxins were administered together.

Studies with human epidermal cells have shown that dioxin (TCDD) potentiates AFB_1 toxicity that is associated with a 20-fold increase in DNA adduct formation, suggesting that the TCDD stimulates the epoxidation of AFB_1 through induction of cytochrome P450 1A enzymes (Walsh *et al.,* 1992).

Immunocompetence

Finally, an area of considerable significance that has received very little attention is the potential for adverse effects of AFB_1 on immunocompetence. Studies in poultry, swine, guinea pigs, and cattle (Pier *et al.,* 1977) have been done, but other species have been neglected. The influence of long-term exposure to low levels of AFB_1, similar to environmental situations, should be given special emphasis. In addition, exposure to mixed mycotoxins such as AFB_1 and toxins produced by members of the *Fusarium* family of molds (tricothecenes, zearalenones, etc.) as well as to combinations of AFB_1 and other environmental contaminants (pesticides, PCBs) should be accorded appropriate attention.

REFERENCES

Allcroft, R., and Lewis, G. (1963). Groundnut toxicity in cattle. *Vet. Rec.* **75,** 487–493.

Allcroft, R., Carnaghan, B. A., Sargeant, K., and O'Kelly (1961). Toxic factor in Brazilian groundnut meal. *Vet. Rec.* **73,** 428–429.

Allcroft, R., Roberts, B. A., and Lloyd, M. K. (1968). Excretion of aflatoxin in a lactating cow. *Food Cosmet. Toxicol.* **6,** 619–625.

Ames, B. N. (1990). Carcinogenesis debate. *Science* **250,** 1498–1499.

Arai, T., Ito, T., and Koyama, Y. (1967). Antimicrobial activity of aflatoxins. *J. Bacteriol.* **93,** 59–64.

Asao, T., Buchi, G., Abdel-Kader, M. M., Chang, S. B., Wick, E. L., and Wogan, G. N. (1963). Aflatoxins B and G. *J. Am. Chem. Soc.* **85,** 1706–1707.

Asplin, F. D., and Carnaghan, R. B. A. (1961). The toxicity of certain groundnut meals for poultry with special reference to their effect on ducklings and chickens. *Vet. Rec.* **73,** 1215–1219.

Bailey, W. S., and Groth, A. H. (1959). The relationship of hepatitis X of dogs and moldly corn poisoning in swine. *J. Am. Vet. Med. Assoc.* **134,** 514–516.

Bauer, D. H., Lee, D. J., and Sinnhuber, K. O. (1969). Acute toxicity of aflatoxains B_1 and G_1 in the rainbow trout. *Toxicol. Appl. Pharmacol.* **15,** 415–419.

Blount, W. P. (1961). Turkey "X" disease. *J. Brit. Turkey Fed.* **9,** 52–54.

Bodine, A. B., Fisher, S. F., and Gangjee, S. (1984). Effect of aflatoxin B_1 and major metabolites on phytohemagglutanin-stimulated lymphoblastogenesis of bovine lymphocytes. *J. Dairy Sci.* **67,** 110–114.

Bourgeois, C. H. (1971). Acute aflatoxin B_1 toxicity in the macaque and its similarities to Reye's syndrome. *Lab. Invest.* **24,** 206–216.

Boutibonnes, P. (1979). Sensitivity of *B. thuringiensis* to aflatoxin. *Mycopathologia* **67,** 45–50.

Bruce, H. M., and Parkes, A. S. (1947). Peanut meal diets for laboratory animals. *J. Hyg. Camb.* **45,** 70–72.

Burmeister, H. R., and Hesseltine, C. W. (1966). Survey of the sensitivity of microorganisms to aflatoxin. *Appl. Microbiol* **14,** 403–404.

Burnside, J. E., Sippell, W. C., Forgacs, J., Carl, W. T., Atwood, M. B., and Doll, E. R. (1957). A disease of swine and cattle caused by eating moldy corn II. Experimental production with pure cultures of mold. *Am. J. Vet. Res.* **18,** 817–824.

Busby, W. F., Jr., and Wogan, G. N. (1984). Aflatoxins. *In* "Chemical Carcinogens" (C. E. Searle, ed.), pp. 945–1136. American Chemical Society, Washington, D.C.

Butler, W. H. (1964). Acute toxicity of aflatoxin B_1 in rats. *Brit. J. Cancer* **18,** 756–762.

Butler, W. H. (1966). Acute toxicity of aflatoxin B in guinea pigs, *J. Pathol. Bacteriol.* **91,** 277–280.

Butler, W. H. (1971). Further ultrastructural observations on injury of rat hepatic parenchymal cells induced by AFB_1. *Chem. Biol. Interact.* **4,** 49–65.

Campbell, T. C., Caedo, J. P., BulataoJayme, J., Salamat, L., and Engel, R. W. (1970). Aflatoxin M1 in human urine. *Nature (London)* **227**, 403–404.

Carnaghan, R. B. A., Hartley, R. D., and O'Kelly, J. (1963). Toxicity and fluorescence properties of the aflatoxins. *Nature (London)* **200**, 1101–1102.

Carpenter, H. M., Zhang, Q., Bailey, G. S., Hendricks, J. O., Buhler, D. R., Mirauda, C. L., and Curtis, L. R. (1992). Temperature affects initiation and promotion of AFB$_1$-induced hepatic tumors in rainbow trout. *Toxicologist* **12**, 208.

Chau, C. C., and Marth, E. H. (1976). Radioactivity in urine and feces of mink (*Mustela vison*) treated with [^{14}C] aflatoxin B$_1$. *Arch. Tox.* **35**, 75–81.

Chedid, A., Halfman, C. J., and Greenberg, S. R. (1980). Hormonal influences on chemical carcinogenesis; studies with the aflatoxin B$_1$ model in the rat. *Digest. Dis. Sci.* **25**, 869–874.

Clifford, J. I., and Rees, K. R. (1967). The action of AFB$_1$ on rat liver. *Biochem. J.* **103**, 65–75.

Croy, R. G., and Wogan, G. N. (1981). Temporal patterns of covalent DNA adducts in rat liver after single or multiple doses of aflatoxin B$_1$. *Cancer Res.* **41**, 197–203.

Dalezios, J. I., Hsieh, D. P. H., and Wogan, G. N. (1973). Excretion and metabolism or orally administered aflatoxin B$_1$ by Rhesus monkeys. *Food Cosmet. Toxicol.* **11**, 605–616.

Daniels, J. M., and Massey, T. E. (1992). Activation of AFB$_1$ in isolated rabbit lung microsomes and the effects of beta-naphthaflavone treatment. *Toxicologist* **12**, 327.

Deo, M. G., Dayal, Y., and Ramalingaswami, V. J. (1970). Aflatoxins and liver injury in the Rhesus monkey. *J. Pathol.* **101**, 47–56.

Doherty, J., and Campbell, T. C. (1972). Inhibition of rat liver mitochrondria electron transport flow by aflatoxin B$_1$. *Res. Commun. Chem. Pathol. Pharmacol.* **3**, 601–612.

Engelbrecht, J. C., and Altenkirk, B. (1972). Comparison of some biological effects of sterigmatocystin and aflatoxin analogues on primary cell cultures. *J. Natl. Cancer Inst.* **48**, 1647–1655.

Essigmann, J. M., Croy, R. G., Nadzan, A. M., Busby, W. F., Reinhokd, M., Buchi, G. B., and Wogan, G. N. (1977). Structural identification of the major DNA adduct formed by aflatoxin B$_1$ *in vitro*. *Proc. Natl. Acad. Sci. U.S.A.* **74**, 1870–1874.

Ewaskiewicz, J. I., Devlin, T. M., and Ch'ih, J. J. (1991). The *in vivo* disposition of aflatoxin B$_1$ in rat liver. *Biochem. Biophys. Res. Commun.* **179**, 1095–1100.

Fujimoto, Y., Hampton, L. L., Luo, L. D., Wirth, P. J., and Thorgeirsson, S. (1992). Low frequency of p53 gene mutation in tumors induced by aflatoxin B$_1$ in nonhuman primates. *Cancer Res.* **52**, 1044–1046.

Garner, R. C. (1975). Reduction in binding of [^{14}C] AFB$_1$ to rat liver molecules by phenobarbitone pretreatment. *Biochem. Pharmacol.* **24**, 1553–1556.

Garner, R. C., Miller, E. C., Miller, J. A., Garner, J. L., and Hanson, R. S. (1971). Formation of a factor lethal for *S typhimurium* TA 1530, TA 1531 on incubation of aflatoxin B$_1$ with rat liver microsomes. *Biochem. Biophys. Res. Commun.* **45**, 774–780.

Glinsukon, T., Taycharpipranai, S., and Toskulkao, C. (1978). Aflatoxin B$_1$ hepatotoxicity in rats pretreated with ethanol. *Experientia* **34**, 869–870.

Goodall, C. M., and Butler, W. H. (1969). Aflatoxin carciogenesis: Inhibition of liver cancer induction in hypophysectomized rats. *Int. J. Cancer* **4**, 422–429.

Groopman, J. D., Busby, W. F., and Wogan, G. N. (1980). The nuclear distribution of aflatoxin B, and its interaction with histones *in vivo*. *Cancer Res.* **40**, 4343–4351.

Guengerich, F. P., and Shimada, T. (1991). Oxidation of toxic and carcinogenic chemicals by human cytochrome P-450 enzymes. *Chem. Res. Toxicol.* **4**, 391–407.

Halver, J. E. (1965). Aflatoxicosis and rainbow trout hepatoma. *In* "Mycotoxins in Foodstuffs" (G. N. Wogan, ed.), pp. 209–234. MIT Press, Cambridge, Massachusetts.

IARC Monographs (1987). Suppl. 7. "Evaluation of Carcinogenic Risks to Humans." pg. 83–87. IARC Press, Lyon.

Irvin, T. R., and Wogan, G. N. (1984). Quantitation of aflatoxin B$_1$ adduction within the ribosomal RNA gene sequences of rat liver DNA. *Proc. Natl. Acad. Sci. U.S.A.* **81**, 664–668.

Lancaster, C. M., Jenkins, F. P., and Philip, P. (1961). Toxicity associated with certain samples of groundnuts. *Nature (London)* **192**, 1095–1096.

Lin, J. K., Miller, J. A., and Miller, E. C. (1977). 2,3-Dihydro-2-(guan-7-yl)-3-hydroxy-aflatoxin B_1, a major acid hydrolysis product of aflatoxin B_1–DNA or ribosomal RNA adducts formed in hepatic microsome-mediated reactions and in rat liver *in vitro. Cancer Res.* **37**, 4430–4438.

Liu, Y. L., Roebuck, B. D., Yager, J. D., Groopman, J. D., and Kensler, T. W. (1988). Protection by 5-(2-pyrazinyl)-4-methyl-1,2-dithiol-3-thione (oltipraz) against the hepatotoxicity of aflatoxin B_1 in rat. *Toxicol. Appl. Pharmacol.* **93**, 442–51.

Loosmore, R. M., and Harding, J. D. (1961). A toxic factor in Brazilian groundnut causing liver damage in pigs. *Vet. Rec.* **73**, 1362–1364.

Loosmore, R. M., and Markson, L. M. (1961). Poisoning of cattle by Brazilian groundnut meal. *Vet. Rec.* **73**, 813–814.

Madhavan, T. V., and Gopalan, C. (1965). Effect of dietary protein on aflatoxin liver injury in weanling rats. *Arch. Pathol.* **80**, 123–126.

Madhavan, T. V., Rao, K. S., and Tulpule, P. G. (1965). Effect of dietary protein level on suscep-tibility of monkeys to aflatoxin liver injury. *Indian J. Med. Res.* **53**, 984–989.

Mainigi, K. D., and Sorof, S. (1977). Carcinogen-protein complexes in liver during hepatocar-cinogenesis by aflatoxin B_1 (1). *Cancer Res.* **37**, 4304–4312.

Mgbodile, M. U. K., and Holscher, M. (1976). Protective effects of oral contraceptive for aflatoxin. *Fd. Cosmet. Toxicol.* **14**, 171–174.

Mgbodile, M. U. K., Holscher, M., and Neal, R. A. (1975). A possible protective role for reduced glutathione in aflatoxin B_1 toxicity: Effect of pretreatment of rats with phenobarbital and 3-methylcholanthrene on aflatoxin toxicity. *Toxicol. Appl. Pharmacol.* **34**, 128–142.

Moyer, R., Marien, K., vanHolde, K., and Bailey, G. (1989). Site-specific aflatoxin B_1 adduction of sequence-positioned nucleosome core particles. *J. Biol. Chem.* **264**, 12226–12231.

Nabney, J., Burbage, M. B., Allcroft, R., and Lewis, G. (1967). Metabolism of aflatoxin in sheep: Excretion pattern in the lactating ewe. *Fd. Cosmet. Toxicol.* **5**, 11–17.

Neal, G. E. (1973). Inhibition of rat liver RNA synthesis by aflatoxin B_1. *Nature (London)* **244**, 432–435.

Neal, G. E., and Green, J. A. (1983). Requirement for glutathione S-transferase in the conjugation of activated aflatoxin B_1 during aflatoxin hepatocarcinogenesis in the rat. *Chem. Biol. Interact.* **45**, 259–275.

Neal, G. E., and Judah, D. J. (1978). Effect of hypophysectomy and AFB_1 on rat liver. *Cancer Res.* **38**, 3460–3467.

Neal, G. E., Godoy, H. M., Judah, D. J., and Butler, W. H. (1976). Some effects of acute and chronic dosing with AFB_1 on rat liver nuclei. *Cancer Res.* **36**, 1771–1778.

Neal, G. E., Judah, D. J., Stirpe, F., and Patterson, D. S. P. (1981). The formation of 2,3-dihydroxy-2,3-dihydro-aflatoxin B_1 by the metabolism of aflatoxin B_1 by liver microsomes isolated from certain avian and mammalian species and the possible role of this metabolite on the acute toxicity of aflatoxin B_1. *Toxicol. Appl. Pharmacol.* **58**, 431–437.

Newberne, J. W., Bailey, W. S., and Seibold, H. R. (1955). Notes on a recent outbreak and experimen-tal reproduction of "Hepatitis X" in dogs. *J. Am. Vet. Med. Assoc.* **134**, 514–516.

Newberne, P. M. (1967). "Biological Activity of the Aflatoxins in Domestic and Laboratory Ani-mals." U. S. Department of Interior Research Report 70. U.S. Government Printing Office, Washington, D.C.

Newberne, P. M. (1986). Aging animal models for the study of drug-nutrient interactions. *In* "Nutri-tion of Aging" (S. Gershoff, ed.), pp. 99–115. Academic Press, New York.

Newberne, P. M., and Butler, W. H. (1969). Acute and chronic effects of aflatoxin on the liver of domestic and laboratory animals. A review. *Cancer Res.* **29**, 236–250.

Newberne, P. M., and Conner, M. W. (1974). Effect of selenium on acute response to aflatoxin B_1. *In* "Trace Substances in Environmental Health—VIII" (D. D. Hemphill, ed.), pp. 323–328. University of Missouri Press, Columbia.

Newberne, P. M., and Rogers, A. E. (1968). Carcinoma, thymidine uptake and mitosis in the livers of rats exposed to aflatoxins. *New Zeal. Med. J.* **67**, 8–17.

Newberne, P. M., and Rogers, A. E. (1973). Rat colon carcinomas associated with aflatoxin and marginal vitamin A. *J. Natl. Cancer Inst.* **50**, 439–448.

Newberne, P. M., and Rogers, A. E. (1981). Animal toxicity of major environmental mycotoxins. *In* "Environmental Risks" (R. C. Shanks, ed.), pp. 51–106. CRC Press, Boca Raton, Florida.

Newberne, P. M., and Suphakarn, V. (1977). Preventive role of vitamin A in colon carcinogenesis in rats. *Cancer* **40**, 2553–2556.

Newberne, P. M., Carlton, W. W., and Wogan, G. N. (1964a). Hepatomas in rats and hepatorenal injury in ducklings fed peanut meal or *Aspergillus flavus* extract. *Pathol. Vet.* **1**, 105–118.

Newberne, P. M., Wogan, G. N., Carlton, W. W., and Abdel-Kader, M. M. (1964b). Histopathologic lesions in ducklings caused by *Aspergillus flavus* cultures, culture extracts and crystallin aflatoxins. *Toxicol. Appl. Pharmacol.* **6**, 542–547.

Newberne, P. M., Wogan, G. N., and Hall, A. (1966a). Effects of dietary modifications on response of the duckling to aflatoxin. *J. Nutr.* **90**, 123–130.

Newberne, P. M., Russo, R., and Wogan, G. N. (1966b). Acute toxicity of aflatoxin B_1 in the dog. *Pathol. Vet.* **3**, 331–340.

Newberne, P. M., Chan, W. C., and Rogers, A. E. (1974). Influence of light, riboflavin, and carotene on the response of rats to the acute toxicity of aflatoxin and monocrotaline. *Toxicol. Appl. Pharmacol.* **28**, 200–208.

Newberne, P. M., de Camargo, J. L. V., and Clark, A. J. (1982). Choline deficiency, partial hepatectomy, and liver tumors in rats and mice. *Tox. Pathol.* **10**, 95–109.

Newberne, P. M., Punyarit, P., de Camargo, J., and Suphakarn, V. (1987). The role of necrosis in hepatocellular proliferation and liver tumors. *In* "Mouse Liver Tumors" (P. L. Chambers, D. Menschler, & F. Oesch, eds.), pp. 54–67. Springer-Verlag, Berlin.

Niranjan, B. G., Bhjat, N. K., and Avadhani, N. G. (1982). Preferential attack of mitochondrial DNA by aflatoxin B_1 during hepatocarcinogenesis. *Science* **215**, 73–75.

Paget, G. E. (1954). Exudative hepatitis in guinea pigs. *J. Pathol. Bacteriol.* **67**, 393–396.

Paterson, J. S., Crook, J. C., Shand, A., Lewis, G., and Allcroft, R. (1962). Groundnut toxicity as the cause of exudative hepatitis in guinea pigs. *Vet. Rec.* **74**, 639–643.

Pier, A. C., Fichther, R. E., and Cysewski, S. J. (1977). Effects of aflatoxin on the cellular immune system. *Ann. Nurt. Alim.* **31**, 781–787.

Plakas, S. M., Loveland, P. M., Bailey, G. S., Blazer, V. S., and Wilson, G. L. (1991). Tissue disposition and excretion of ^{14}C-labelled aflatoxin B_1 after oral administration in channel catfish. *Fd. Chem. Toxic.* **29**, 805–808.

Pohland, A. E., Cushmac, M. E., and Andrellos, P. J. (1968). Aflatoxin B_1 Hemiacetal. *J. Assoc Off. Anal Chem.* **51**, 907–910.

Pong, R. S., and Wogan, G. N. (1971). Toxicity, biochemical and fine structural effects of synthetic aflatoxin M_1 and B_1 in rat liver. *J. Natl. Cancer Inst.* **47**, 585–592.

Purchase, I. F. H. (1967). Susceptibility of vitamin A-deficient rats to aflatoxin. *Food Cosmet. Toxicol.* **5**, 339–342.

Reddy, G. S., Tilak, T. B. G., and Krshnamusthi, D. (1973). Susceptibility of vitamin A-deficient rats to aflatoxin. *Food. Cosmet. Toxicol.* **11**, 467–470.

Righter, H. F., Shalkop, W. T., Mercer, H. D., and Leffet, E. C. (1972). Influence of age and sexual status on the development of toxic effects in the male rat fed aflatoxins. *Toxicol. Appl. Pharmacol.* **21**, 435–439.

Rogers, A. E., and Newberne, P. M. (1967). Effects of AFB_1 and DMSO on ^3H-thymidine uptake and mitosis in rat liver. *Cancer Res.* **27**, 855–861.

Rogers, A. E., and Newberne, P. M. (1971). Diet and aflatoxin B_1 toxicity in rats. *Toxicol. Appl. Pharmacol.* **20**, 113–121.

Rogers, A. E., and Newberne, P. M. (1975). Dietary effects on chemical carcinogenesis in animal models for colon and liver tumors. *Cancer Res.* **35**, 3427–3431.

Ross, R. K., Yuan, J.-M., Yu, M. C., Wogan, G. N., Qian, G.-S., Tu, J.-T., Groopman, J. D., Gao, Y.-T., and Henderson, B. E. (1992). Urinary aflatoxin biomarkers and risk of hepatocellular carcinomas. *Lancet* **339**, 943–946.

Salmon, W. D., and Newberne, P. M. (1963). Occurrence of hepatomas in rats fed diets containing peanut meal as a major source of protein. *Cancer Res.* **23**, 571–577.

Sarasin, A., and Moule, Y. (1975). Translational step inhibited *in vivo* by aflatoxin B_1 in rat-liver polysomes. *Eur. J. Biochem.* **54**, 329–340.

Schoental, R. (1961). Liver changes and primary liver tumors in rats given toxic guinea pig diet (MRC Diet 18). *Br. J. Cancer* **15**, 812–815.

Schrager, T. F., Newberne, P. M., Pikul, A. H., and Groopman, J. D. (1990). Aflatoxin–DNA adduct formation in chronically dosed rats fed a choline deficient diet. *Carcinogenesis* **11**, 177–180.

Seibold, H. R., and Bailey, W. S. (1952). An epizootic of hepatitis in the dog. *J. Am. Vet. Med. Assoc.* **121**, 201–206.

Shamsuddin, A. M., Harris, C. C., and Hinzman, J. (1987). Localization of aflatoxin B_1–nucleic acid adducts in mitochondria and nuclei. *Carcinogenesis* **8**, 109–114.

Shank, R. C. (1977). Epidemiology of aflatoxin carcinogenesis. *In* "Environmental Cancer" (H. F. Kraybill and M. A. Mehlman, eds.), pp. 291–318. John Wiley and Sons, New York.

Shank, R. C. (1981). Environmental toxicoses in humans. *In* "Mycotoxins and *N*-Nitroso-Compounds: Environmental Risks" (R. C. Shank, ed.), vol. 1, pp. 107–140. CRC Press, Boca Raton, Florida.

Shank, R. C., Bourgeois, C. H., Keschamras, N., and Candavimal, P. (1971). Aflatoxins in autopsy specimens from Thai children with an acute disease of unknown etiology. *Food Cosmet. Toxicol.* **9**, 501–507.

Siller, W. G., and Ostler, D. C. (1961). The histopathology of an enterohepatic syndrome of turkey poults. *Vet. Res.* **73**, 134–138.

Sjoerdsma, A. (1981). Commentary. Suicide enzyme inhibitors as potential drugs. *Clin. Pharmacol. Ther.* **30**, 3–22.

Sondergaard, D., Taylor, F., and Newberne, P. M. (1985). Effects of the irreversible ornithine decarboxylase inhibitor, alpha-difluoromethyl-ornithine, AFB_1, and choline deficiency on hepatocarcinogenesis. *Toxicol. Path.* **13**, 36–49.

Stora, C., Aussei, C., Mayzaud, O., and Masseyeff, R. (1979). Hepatocarcinogenesis by aflatoxin B_1: Relationship between cellular localization of the carcinogen and early histological changes in rat liver *Biomedicine* **31**, 173–176.

Suphakarn, V. S., Newberne, P. M., and Goldman, M. (1983). Vitamin A and aflatoxin: Effect on liver and colon cancer. *Nutr. Cancer* **5**, 41–50.

Svoboda, D., and Higginson, J. (1968). A comparison of ultrastructural changes in rat liver due to chemical carcinogenesis. *Cancer Res.* **28**, 1703–1733.

Svoboda, D., Grady, H. J., and Higginson, J. (1966). Aflatoxin B_1 injury in rat and monkey liver. *Am. J. Pathol.* **49**, 1023–1029.

Swenson, D. H., Lin, J. K., Miller, E. C., and Miller, J. A. (1977). Aflatoxin B_1-2,3-oxide as a probable intermediate in the covalent binding of aflatoxin B_1 and B_2. *Cancer Res.* **37**, 172–181.

Takahashi, N., Stresser, D. M., Williams, D. E., and Bailey, G. S. (1992). The significance of induction of cytochrome P4501A1 by indole-3-carbinol an anticarcinogen protective against aflatoxin B_1 hepatocarcinogenesis in rainbow trout. *Toxicologist* **12**, 208.

Terao, K. (1967). The effect of aflatoxin on chick-embryo liver cells. *Exp. Cell Res.* **48**, 151–155.

Ueno, Y., and Kubota, K. (1976). DNA-attacking ability of carcinogenic mycotoxins in recombination deficient nutrient cells of *Bacillus subtilis*. *Cancer Res.* **36**, 445–451.

Ueno, Y., Freidman, L., and Stone, C. L. (1980). Species differences in the binding of AFB_1 to hepatic macaromolecules. *Toxicol. Appl. Pharmacol.* **52**, 177–180.

van Rensburg, S. J. (1977). Role of epidemiology in the elucidation of mycotoxin health risks. *In* "Mycotoxins in Human and Animal Health" (J. V. Rodricks, C. W. Hesseltine, and M. A. Mehlman, eds.), pp. 699–711. Pathotox, Forest Park South, Illinois.

Viviers, J., and Schabort, J. C. (1985). AFB_1 alters protein phosphorylation in rat livers. *Biochem. Biophys. Res. Commun.* **129,** 342–349.

Walsh, A. A., Hsieh, D. P. H., Whitehead, W. E., and Rice, R. H. (1992). Aflatoxin toxicity in cultured epidermal cells: Stimulation by TCCD. *Toxicologist* **12,** 406.

Wilson, D. W., Harris, R. A., and Coulombe, R. A. (1992). Lung and liver lesions induced by multiple intratracheal instillations of aflatoxin B_1. *Toxicologist* **12,** 207.

Wogan, G. N., and Newberne, P. M. (1967). Dose–response characteristics of aflatoxin B_1 carcinogenesis in rats. *Cancer Res.* **27,** 2370–2376.

Wogan, G. N., Edwards, E. S., and Shank, R. C. (1967). Excretion and tissue distribution of radioactivity from aflatoxin B_1-^{14}-C in rats. *Cancer Res.* **27,** 1729–1736.

Wogan, G. N., Edwards, G. S., and Newberne, P. M. (1971). Structure–activity relationships in toxicity and carcinogenecity of aflatoxin and analogs. *Cancer Res.* **31,** 1936–1942.

Yeh, F. S., Yu, M. C., Mo, C. C., Luo, S., Tong, M. J., and Henderson, B. E. (1989). Hepatitis B virus, aflatoxins and hepatocellular carcinoma in southern Guangxi, China. *Cancer Res.* **49,** 2506–2509.

Yu, F.-L. (1981). Studies on the mechanism of aflatoxin B_1 inhibition of rat liver nucleolar RNA synthesis. *J. Biol. Chem.* **256,** 3292–3297.

Yu, F.-L. (1983). Preferential binding of aflatoxin B_1 to the transcriptionally active regions of rat liver nucleolar chromatin *in vitro* and *in vivo*. *Carcinogenesis* **4,** 889–893.

Yu, F.-L., Dowe, R. J., Geronimo, I. H., and Bender, W. (1986). Evidence for an indirect mechanism of aflatoxin B_1 inhibition of rat liver nuclear RNA polymerase II activity *in vivo*. *Carcinogenesis* **7,** 253–257.

Yu, F.-L., Geronimo, I. H., Bender, W., and Permthamsin, J. (1988a). Correlation studies between the binding of aflatoxin B_1 to chromatin components and the inhibition of RNA synthesis. *Carcinogenesis* **9,** 527–532.

Yu, F.-L., Bender, W., and Geronimo, I. H. (1988b). The DNA of aflatoxin B_1 to rat liver nuclear proteins and its effect on DNA-dependent RNA synthesis. *Carcinogenesis* **9,** 533–540.

Yu, F.-L., Bender, W., Geronimo, I. H. (1988c). The binding of aflatoxin B_1 to rat liver nuclear proteins and its effect on DNA-dependent RNA synthesis. *Carcinogenesis* **9,** 533–540.

Zarba, A., Jakab, R., and Groopman, G. (1992). Genotoxic damage in liver following inhalation exposure to aflatoxin B_1 in rats. *Toxicologist* **12,** 207.

Zhang, Q., El-Zahr, C., Carpenter, H. M., Selivonchick, D., and Curtis, L. R. (1992). Temperature acclimation differentially modulates metabolic pathways for AFB_1 in rainbow trout. *Toxicologist* **12,** 326.

Zuckerman, A. J., Tsiquaye, K. N., and Fulton, F. (1967). Tissue culture of human embryo liver cells and the cytotoxicity of aflatoxins B_1. *Br. J. Exp. Pathol.* **48,** 20–27.

2

Biochemical Mechanisms and Biological Implications of the Toxicity of Aflatoxins as Related to Aflatoxin Carcinogenesis

❖

Bill D. Roebuck and Yulia Y. Maxuitenko

Aflatoxins are very toxic, particularly to the liver. Overwhelming evidence suggests that specific aflatoxins are hepatocarcinogens. The mechanisms accounting for the toxicity of the aflatoxins, including aflatoxin B_1 (AFB_1), to hepatic parenchymal cells have not been elucidated. Compared with the toxicity of aflatoxins, mechanisms regarding the causation of hepatic cancer by aflatoxins (especially AFB_1) are fairly well understood. In the field of hepatocarcinogenesis, evidence is mounting that hepatotoxicity plays a significant role in hepatocellular carcinogenesis. For the aflatoxins, particularly AFB_1, the role of hepatic toxicity in the causation of liver cancer has not been investigated thoroughly.

ACUTE AND CHRONIC TOXICITY OF AFLATOXINS

An epizootic poisoning in England in 1960 that resulted in the death of thousands of poultry rapidly led to the isolation and structural elucidation of the aflatoxins. The acute toxicity of the aflatoxins was instrumental in their discovery. This toxicity has been the subject of numerous reviews (Goldblatt, 1969; Busby and Wogan, 1984; Chapter 1). The poisoning, then termed "turkey X disease," was characterized by massive hepatic necrosis, parenchymal cell degeneration, and bile duct proliferation. Simultaneously, similar episodes of hepatic toxicity were described in ducklings, pigs, and calves in Uganda and Kenya

(Allcroft, 1969); hepatic cancer was reported in hatchery-raised trout in the United States (Halver, 1969). Eventually, both the acute toxicity and the hepatic tumors were traced to the contamination of livestock feeds by strains of the molds *Aspergillus flavus* and *Aspergillus parasiticus* that produced the aflatoxins. Although the acute hepatic toxicity led to the discovery of the aflatoxins, the chronic effects of the aflatoxins, primarily, the hepatic carcinogenicity has received the most intense scientific attention. Several pieces of evidence indicate that the acute toxicity of the aflatoxins may be an important determinant in cancer development.

In Vivo Systems

Acute Exposure

Largely, the aflatoxins were discovered because of their acute toxicity to poultry. For some representative experimental and domestic livestock species, Table 1 gives the acute LD_{50} values for the most extensively studied, naturally occurring aflatoxin, AFB_1. This list is not comprehensive; Busby and Wogan (1984) and Cullen and Newberne (Chapter 1) give additional acute LD_{50} values. Acute toxicity as assessed by LD_{50} values spans two orders of magnitude from highly susceptible species such as ducklings and rabbits to very resistant species such as chickens. Interestingly, ducks exposed to aflatoxins develop hepatocellular carcinomas (Carnaghan, 1965; Cova *et al.*, 1990) whereas chickens are appar-

TABLE 1   Acute Toxicity of Aflatoxin B₁

Species	Strain	LD_{50} (mg/kg)	Sex	Age (days)	Route[a]	Reference
Duckling	Pekin	0.34	M, F	1	p.o.	Lijinsky and Butler (1966)
Rabbit	Dutch Belted	0.3	M, F	90	p.o./i.p.	Newberne and Butler (1969)
Trout	Rainbow	0.81	M, F	270	i.p.	Bauer *et al.* (1969)
Guinea pig		1.4	M, F	250[b]	i.p.	Butler (1966)
Rat	Fischer	1.1	M, F	2	s.c.	McGuire (1969)
		8	M, F	21	i.p.	McGuire (1969)
		0.75	M	70	i.p.	McGuire (1969)
		1.3	F	70	i.p.	McGuire (1969)
	Sprague–Dawley	1.36	M, F	1	p.o.	Hayes *et al.* (1977)
Mouse	CFW Swiss	1.36	M, F	1	p.o.	McGuire (1969)
		>150	M	30	i.p.	McGuire (1969)
		40	M	58	i.p.	McGuire (1969)
		12	M	100	i.p.	McGuire (1969)
Hamster	Syrian	10.2	M	30	p.o.	Wogan (1966)
Chicken	Australia	15–18	M	21	p.o.	Bryden *et al.* (1980)

[a]p.o., Per os; i.p., intraperitoneal; s.c., subcutaneous.
[b]Guinea pig is weight in grams.

ently highly resistant, since hepatic cancer from aflatoxins has not been reported. Rats have LD_{50} values that are smaller than those of mice; generally, rats (Wogan *et al.*, 1974) are more susceptible to the carcinogenic effects of AFB_1 than mice (Wogan, 1973). Within species, the LD_{50} values may vary with strain, sex, route of administration of aflatoxin, and age of the animal. Additionally, the nutritional status of the animal or the concurrent composition of the diet may modulate the acute toxicity. For example, the Fischer (F344) rat strain is more susceptible to the acute toxicity of AFB_1 and also to its carcinogenecity (Wogan *et al.*, 1974) than is the Sprague–Dawley strain (Rogers *et al.*, 1971). Male rats have lower LD_{50} values than females, and male rats are more susceptible than females to AFB_1-induced hepatic tumors (Wogan and Newberne, 1967). Although mice are relatively more resistant to the acute toxic effects of AFB_1 than are rats, the neonatal mouse has a very low LD_{50} value and also is susceptible to AFB_1-induced cancers (Vesselinovitch *et al.*, 1972). Collectively, these data indicate that the acute toxicity to AFB_1 correlates with the susceptibility to hepatic cancer induced by AFB_1. We do not imply that the relationship between acute toxicity and cancer is direct, but only that a species that is sensitive to the acute toxic effects of AFB_1 is more susceptible to hepatic cancers by some regime of AFB_1 exposure.

Four aflatoxins commonly occur in contaminated food and feeds: AFB_1, aflatoxin B_2 (AFB_2), aflatoxin G_1 (AFG_1), and aflatoxin G_2 (AFG_2). The acute toxicity of these naturally occurring aflatoxins has been evaluated in a limited number of species. Acute LD_{50} values are presented in Table 2. AFB_1 is the most toxic and is also the most carcinogenic of the four congeners. Butler *et al.* (1969) observed that, when MRC rats are dosed with aflatoxins via the drinking water, AFB_1 leads to 6 times the cancer incidence generated by a similar level of AFG_1. In Fischer rats, Wogan *et al.* (1971) found that AFG_1 is less carcinogenic than

TABLE 2   Acute Toxicity of Naturally Occurring Aflatoxins

Species	Strain	Sex	Route[a]	Weight (g)	Aflatoxin	LD_{50} (mg/kg)	Reference
Duck	Pekin	M	i.p.	50	AFB_1	0.73	Wogan *et al.*
					AFB_2	1.76	(1971)
					AFG_1	1.18	
					AFG_2	2.83	
	Khaki–Campbell	—	p.o.	50	AFB_1	0.36	Carnaghan
					AFB_2	0.78	(1963)
					AFG_1	1.70	
					AFG_2	3.44	
Rat	Fischer	M	p.o.	200	AFB_1	1.16	Wogan *et al.*
					AFB_2	>200	(1971)
					AFG_1	1.5–2.0	
					AFG_2	>200	

[a]i.p., Intraperitoneal; p.o., per os.

**TABLE 3   Acute Toxicity of Metabolites of Aflatoxins
Relative to Acute Toxicity of AFB$_1$**

Species	Strain	Sex	Route[a]	Age or weight (g)	Aflatoxin metabolite	LD$_{50}$ relative to AFB$_1$	Reference
Mouse	C57BL/6J	M, F	i.p.	newborn	AFP$_1$	<5%	Büchi et al. (1973)
Chicken embryo	White leghorn	—	—	—	AFQ$_1$	<18×	Hsieh et al. (1974)
Duck	Pekin	—	p.o.	40–50	AFM$_1$ AFM$_2$	similar <4×	Purchase (1967)

[a]i.p., Intraperitoneal; p.o., per os.

AFB$_1$ by approximately 5-fold and AFB$_2$ is 100 times less carcinogenic than AFB$_1$.

The aflatoxins are transformed metabolically by liver enzyme systems. A few metabolic products have been evaluated for their acute toxicity and carcinogenicity. LD$_{50}$ values for selected metabolites relative to AFB$_1$ are presented in Table 3. Based on these limited data, AFM$_1$ and AFB$_1$ have similar acute toxicity in ducks (Purchase, 1967) and in rats (Pong and Wogan, 1971). In the rat, AFM$_1$ is carcinogenic, but less so than AFB$_1$ (Wogan and Paglialunga, 1974). The other hydroxylated metabolites of AFB$_1$—AFP$_1$ and AFQ$_1$—are less toxic; their carcinogenicity has not been evaluated.

Chronic Exposure

Chronic exposure to low levels of aflatoxins is likely to occur more commonly than acutely toxic exposure. Much accumulated evidence indicates that chronic exposure to aflatoxins more readily leads to cancer than does acute exposure; thus, chronic exposure almost certainly represents a more serious public health concern. Loss of weight or poor growth is one of the first and most consistent signs of intoxication (Edds, 1973). Failure to grow and reduced egg production have been observed in poultry exposed to aflatoxin-contaminated feed (Hegazy et al., 1991). van Halderen et al. (1989) reported mortality in Friesland dairy calves in South Africa. The calves had been fed maize contaminated with A. flavus and contained mixed aflatoxins totalling 11.8 ppm. The presence of aflatoxins has been postulated to contribute to the chronic effect of some common forms of malnutrition in growing children (Hendrickse, 1991).

The chronic effects of aflatoxins have been studied in greatest detail in the rat (Busby and Wogan, 1984). The histopathology resulting from chronic exposure to aflatoxins is described in a subsequent section. In this section we describe the toxic effects of chronic exposure and the magnitude of the hepatic response. Liu et al. (1988) demonstrated that chronic treatment of rats with AFB$_1$ resulted in failure to gain body weight, loss of liver weight, and loss of hepatic DNA content over a 2-week period of exposure to AFB$_1$. Additionally, Liu et al. (1988)

showed that the toxic effects of AFB_1 are ameliorated by pretreatment with 1,2-dithiole-3-thiones, of which oltipraz is representative. (Intervention in toxicity is discussed in a subsequent section.)

This treatment protocol—25 μg AFB_1 per rat for 5 days per week for 2 successive weeks—has been shown to be a carcinogenic dose (Roebuck et al., 1991). The results of this experiment are summarized graphically in Figure 1. Prior to AFB_1 treatment, the hepatic DNA was labeled. Rats 7 weeks of age were subjected to a two-thirds partial hepatectomy. The regenerating DNA was labeled with [^{3}H]thymidine, so 50% of the parenchymal cell nuclei were labeled. Loss of prelabeled DNA has been interpreted as cell death (Yager and Potter, 1975). The rats were allowed a 2-week recovery period. These 9-week-old male F344 rats were gavaged with 25 μg AFB_1/kg body weight daily for 2 successive 5-day periods over 2 weeks (Figure 1A,B,C). These AFB_1-treated rats failed to gain weight, whereas controls not treated with AFB_1 grew (Figure 1A). AFB_1 treatment resulted in a loss in liver weight that was nearly 50% of the pretreatment hepatic weight (Figure 1B). The rats not treated with AFB_1 showed an increase in hepatic weight over the 2-week treatment period; thus, the loss in hepatic weight was of an especially large magnitude compared with the non-AFB_1-treated control rats that grew during this same period (Figure 1B). Chronic AFB_1 exposure resulted in a rapid and dramatic loss of hepatic DNA. Rats not treated with AFB_1 showed no change in their hepatic DNA content; rats treated with AFB_1 lost approximately half their hepatic DNA content by the end of the first week of AFB_1 treatment. This hepatic DNA loss from AFB_1 exposure approaches the effects of the classic two-thirds partial hepatectomy that is used to induce hepatic DNA synthesis. The rate of DNA synthesis as a result of AFB_1-induced hepatic cell loss was not measured in this experiment.

Using the same AFB_1 treatment protocol, we observed repeatedly that the size of the liver at 3 months after AFB_1 treatment does not differ from hepatic weights of non-AFB_1-treated rats. Thus, significant DNA synthesis and hepatocellular regeneration occurred. However, at what stage in the treatment protocol DNA synthesis occurred and at what rates, or if DNA synthesis largely occurred after the treatment with AFB_1 had ceased, is not clear (see subsequent discussion).

In Vitro Systems

The aflatoxins have been studied in in vitro systems. This approach has been reviewed by Busby and Wogan (1984). Although in vitro approaches have been useful for studying structure–activity relationships of the aflatoxins, their use has been limited severely for several reasons. The aflatoxins are both activated and detoxified by hepatic enzyme systems. Rarely have such complex pathways been monitored or controlled in in vitro systems. Additionally, cells maintained in vitro drift over time with respect to their ability to metabolize chemicals. Finally, as with in vivo systems, the significant in vitro toxicological end points are not obvious nor are they easy to quantify.

Two examples of in vitro approaches to study toxic events induced by aflatox-

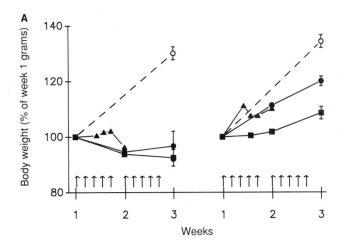

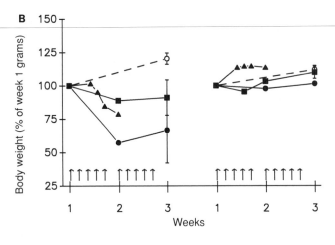

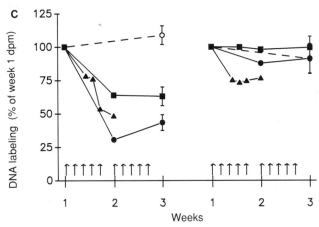

ins are discussed. In the work of Cole *et al.* (1989), primary cultures of hepatocytes from rats, mice, and humans were employed. One of the earliest and most readily quantified end points, the segregation of the fibrillar and granular components of the hepatocyte nucleolus, was used as an index of toxicity. On exposure to AFB_1 *in vivo*, nucleolar components of hepatocytes long have been known to segregate. This response represents one of the earliest effects of AFB_1 (Busby and Wogan, 1984). Cole *et al.* (1989) showed that rat and human, but not mouse, hepatocytes readily develop nucleolar segregation on exposure to AFB_1. For rat, but not mouse, nucleolar segregation is an early event of AFB_1 exposure *in vivo* (Busby and Wogan, 1984). The significance of nucleolar segregation, especially in a quantitative sense, for hepatocyte toxicity and death is not clear. Interestingly, Cole *et al.* (1989) did not study the toxicity of AFB_1 per se, but studied the carcinogenesis of AFB_1. They regarded, as many others have, that relatively direct parellels exist between toxicity and cancer development, but this idea has not been proven.

Other comparisons between rat and human liver cells have been undertaken. For example, Butterworth *et al.* (1989) used primary cultures of hepatocytes to evaluate whether DNA repair was similar in hepatocytes of humans and of rats for nearly 20 known carcinogens, including AFB_1. Clearly, several human liver specimens showed unscheduled DNA synthesis (DNA repair) similar to that seen in the rat. Again, the focus was on aspects of cancer, not on the death of hepatocytes.

The sequence of events leading to systemic or even organ-specific toxicity by an aflatoxin is complex. Any specific *in vitro* system is extremely limited in its utility. However, *in vitro* approaches used in concert with *in vivo* observations are of great utility to study mechanisms of toxicity.

MECHANISMS OF TOXICITY

The toxicity of aflatoxins has been difficult to study because of the wide range of effects following exposure to aflatoxins. Determining which effects are the most critical for hepatocyte death and which effects are secondary or perhaps

---

FIGURE 1 Hepatotoxicity resulting from chronic AFB_1 treatment and protection against the toxicity by oltipraz in young rats (Reprinted with permission from Liu *et al.,* 1988). (A) Growth of rats during treatment with multiple doses of AFB_1 (arrows). The rats were fed either a control diet (*left*) or the control diet supplemented with oltipraz (0.075%; *right*). The first data point (Week 1) represents the body weight just prior to AFB_1 treatment, which is 2 weeks following partial hepatectomy and labeling of hepatic DNA with [³H]thymidine and 1 week after dietary supplementation with oltipraz. The protocol was replicated 3 times; each replicate has a different symbol. The open circle indicates the mean weight gain of two control groups of rats not receiving AFB_1. Each symbol is the mean of 4–6 rats; representative standard error bars are shown. (B) Liver weights following multiple doses of AFB_1 (arrows) are shown for the same rats as in A. The symbols are defined in A. (C) Loss of [³H]thymidine-labeled DNA from whole liver following multiple doses of AFB_1 (arrows) is shown for the same rats as in A and B. See A for description and legend to symbols.

unrelated to the toxicity has been particularly difficult. The characteristic histopathological observations and the major biochemical events are summarized briefly here for acute exposure to AFB_1. With chronic exposure, the toxic effects are more complex. For example, gastrointestinal bleeding can be extensive and may be the cause of death from chronic exposure to aflatoxins (Edds, 1973; B. D. Roebuck and Y. Y. Maxuitenko, personal observations). Whether this bleeding is the result of a lack of clotting factors from the extensive liver damage, the anticoagulant properties of aflatoxin, or a combination of both effects is unclear.

Histopathological Observations

The sequential events following acute exposure to AFB_1 have been reviewed by Busby and Wogan (1984). Many detailed reports of the histopathological events following exposure to aflatoxins are available (e.g., Rogers and Newberne, 1967; Kalengayi and Desmet, 1975; Godoy *et al.*, 1976). Clearly, the effects of AFB_1 are dose related with respect to degree and extent of hepatic involvement. Within 1 hr of dosing with AFB_1, the liver parenchymal cells show a separation of the fibrillar and granular components of the nucleoli. Glycogen loss, RNA depletion, and proliferation of the endoplasmic reticulum all occur within the first day. Parenchymal cell mitosis is decreased dramatically within this first day. Over the next several days, the sequence of events is: periportal necrosis of parenchymal cell, fatty infiltration of the remaining parenchymal cells, and extensive bile duct cell proliferation. The parenchymal cells are considerably more susceptible to the toxic effects of AFB_1 than are the nonparenchymal Kupffer and bile duct cells. Parenchymal cell proliferation is absent for the first few days after an acute dose of AFB_1; thereafter, mitoses occur in this cell population. The regeneration of the parenchymal cell population begins after the proliferation of the bile duct cells is well underway. Death and regeneration of parenchymal cells are discussed in a later section.

Biochemical Observations

The biochemical events following an acutely toxic dose of AFB_1 parallel the histopathological observations detailed earlier. These biochemical events have been reviewed by Busby and Wogan (1984). Covalent binding of aflatoxin to nucleic acids occurs within minutes of treatment of an animal with AFB_1, and results in a precipitous decrease in both DNA and RNA synthesis rates in the liver. The degree of inhibition is even more pronounced in a liver in which parenchymal cells have been stimulated to divide as a result of partial hepatectomy (Rogers *et al.*, 1971; Neal and Cabral, 1980). Nucleic acid synthesis only begins to recover after 1–2 days. Protein synthesis is inhibited by AFB_1, but not as rapidly or as extensively as that of nucleic acids. Within 1 hr or so after AFB_1 exposure, protein synthesis declines. Polysomal disaggregation parallels this inhibition and is likely to represent the mode of inhibition of protein synthesis.

Reaggregation of ribosomes and a return of protein synthesis to normal occurs within 2–5 days of acute aflatoxin treatment. Lipid accumulation in response to aflatoxin exposure occurs in surviving parenchymal cells and returns to normal even more slowly than do nucleic acid and protein synthesis. This latter effect is probably secondary to the inhibition of synthesis of proteins for transport of lipids from the hepatocyte.

As mentioned earlier, the study of biochemical events resulting from the chronic exposure to aflatoxins is not extensive and is complicated by an inability to differentiate relevant primary toxicity from secondary or unimportant and unrelated toxicological events, as well as from the inevitable secondary effects of chronic toxicity. Godoy *et al.* (1976) have described the sequence of histological and biochemical events similar to those events described earlier, emphasizing their protracted nature through 6 weeks of chronic exposure of rats to dietary AFB_1 at the level of 4 ppm. This dose represents approximately 40 µg AFB_1/day for the rat. To emphasize one complexity of chronic administration of aflatoxins, such long-term administration of AFB_1 induces metabolic enzymes that are involved in the biotransformation of aflatoxin (Kensler *et al.,* 1992). Therefore, the metabolic fate of AFB_1 in an animal that has never been exposed to AFB_1 may be quite different from that in an animal that is chronically exposed.

Enhancement of Toxicity

By the modulation of the *in vivo* toxicity of aflatoxins, the sequence of histological and biochemical events has been defined more clearly. In this and the subsequent section, examples of enhancement and inhibition of toxicity will be presented. These examples are of cases in which mortality is modified or the histopathological status of the liver is modified. As in the rest of this chapter, we have limited the scope of toxicity to cell death or events closely associated with cell death.

Neal and Cabral (1980) found that a two-thirds partial hepatectomy of rats increased the acute toxicity of AFB_1. This result occurred despite the fact that Neal and Cabral decreased the LD_{50} dose of AFB_1 to adjust for the reduced mass of liver that resulted from the hepatectomy. The remnant liver metabolized the AFB_1 and, in fact, generated considerably more reactive species of AFB_1 than the sham-operated liver, indicating that AFB_1 metabolism was not inhibited by partial hepatectomy. The specific role of the reactive metabolites in toxicity and mortality is not known.

Madhavan and Gopalan (1968) found that rats fed a low protein (5% casein) diet were more susceptible to the hepatotoxic effects of aflatoxin than rats fed a high casein (20% by weight) diet. Not only was mortality higher, but the livers of these low-protein-fed rats showed parenchymal necrosis, bile duct proliferation, and fatty accumulation. Of the rats surviving more than 1 year, the low-protein group had significantly fewer hepatic tumors than the rats fed the high-protein diet. Appleton and Campbell (1983) have confirmed and expanded these observations. Rats fed 5% casein during treatment with 25 µg AFB_1/100 g body weight over a 2-week period developed the characteristic signs of aflatoxin toxicity—hepatomegaly, severe bile duct proliferation, and cholangiofibrosis—whereas rats fed 20% casein had only mild signs of hepatotoxicity that were limited to mild bile duct proliferation.

Inhibition of Toxicity

Protection against AFB_1 toxicity, especially in terms of mortality, was demonstrated several years ago. Mgbodile *et al.* (1975) showed that the level of glutathione played an important role in protection against AFB_1. For example, administration of cysteine prior to AFB_1 exposure protected against hepatic necrosis. Conversely, the prior administration of diethyl maleate, a compound that depletes hepatic glutathione, enhanced the toxicity as expressed by a dramatic increase in mortality. Clearly, glutathione status was demonstrated to modulate the toxicity of AFB_1. We have used the knowledge concerning glutathione modulation of AFB_1 toxicity to evaluate drugs that modulate glutathione status to protect the liver against aflatoxin toxicity (Kensler *et al.*, 1987, 1992; Liu *et al.*, 1988; Roebuck *et al.*, 1991). Oltipraz, an example of such a drug, is discussed next.

Pretreatment of rats with oltipraz reduced the toxicity of AFB_1 (Liu *et al.*, 1988). A single high dose (10 mg/kg) of AFB_1 by gavage killed 10 of 12 rats fed a control diet whereas rats pretreated for 1 week with oltipraz (0.075% in the diet) only suffered death in 4 of 11 animals. This result represented a significant reduction in mortality from 83 to 36%. At a lower acute dose of AFB_1 (4 mg/kg), oltipraz reduced the acute mortality from 33 to 0% of the rats. At 24 hr after a single dose of AFB_1 (1 or 0.25 mg/kg), serum levels of alanine amino transferase and sorbitol dehydrogenese (both markers of hepatotoxicity) were reduced significantly by previous dietary exposure to oltipraz. At all these doses of AFB_1, we measured a reduction in histological evidence of necrosis in those rats exposed to oltipraz. Figure 1 summarizes the effects of oltipraz on weight gain, liver weight, and death of hepatocytes in rats treated with 10 doses of 25 µg AFB_1. Oltipraz fed at 0.075% by weight in the diet prior to and concurrent with chronic AFB_1 treatment ameliorated the hepatotoxicity, allowed for nearly normal growth of the rat, and largely prevented cell death. With oltipraz, the bulk of the evidence indicates that AFB_1 is detoxified readily by phase II enzymes, specifically glutathione *S*-transferases (Kensler *et al.*, 1992).

ROLE OF ACUTE AND CHRONIC TOXICITY IN CARCINOGENESIS

Hepatocyte toxicity leading to cell death results in a compensatory hyperplasia that restores the liver to its original size. This classic regenerative capacity of the liver is well recognized and has been studied extensively (Michalopoulos, 1990). The acute and chronic toxicity of aflatoxins, with the accompanying regenerative hyperplasia of parenchymal cells, could contribute to the development of cancer in at least two ways. First, cell replication is essential for carcinogenesis. Second, the carcinogen-altered hepatocytes appear better able to survive and grow in the face of a general hepatotoxin.

Increased Cell Proliferation

Many liver carcinogens are hepatotoxic (Solt and Farber, 1976; Columbano *et al.*, 1987) and lead to a compensatory hyperplasia. Indeed, this event appears to occur in response to AFB_1. The toxicity of AFB_1 is well documented; its specificity for the parenchymal cell also is well established (see preceding discussion). With large doses of AFB_1, liver failure leads to death of the animal. However, with smaller doses of AFB_1, biochemical and histological evidence of cell death is found. Hepatocyte regeneration is a late and well-documented event in response to the toxicity of AFB_1. Equally well documented are the failures of AFB_1 to induce cancer by single doses (except in rainbow trout; see Chapter 6) or the extraordinary measures that must be evoked to produce cancers in mammals by a single dose of AFB_1.

Chronic exposure to AFB_1 is a most efficient method to produce hepatic cancer. For example, 2 cycles each of 5 days duration with 25 μg AFB_1 per day to young (approximately 100 g) F344 rats yield putative preneoplastic foci at 2–3 months postdosing (Appleton and Campbell, 1983; Roebuck *et al.*, 1991; Kensler *et al.*, 1992). Hepatic cancers arise by approximately 1 year and a small incidence (12%) of hepatomas by 23 months postdosing (Roebuck *et al.*, 1991). With this AFB_1 protocol, an approximate 50% loss in hepatocytes occurs (Liu *et al.*, 1988). On autopsy of the rats some months after this large hepatocyte loss, the livers are normal in weight. Complete regeneration of the hepatocyte cell mass after the termination of the AFB_1 treatment is assumed.

Cell Replication in Carcinogenesis

Cell replication in carcinogenesis long has been a phenomenon of interest. For example, Cayama *et al.* (1978) have documented the requirement for cell division as an essential and early event in carcinogenesis. These researchers used a modification of the protocol of Solt and Farber (1976). The general protocol is discussed in the next section. Specifically, rats were initiated with *N*-methyl-*N*-nitrosourea and, during weeks 2–4 postinitiation, were fed an inhibitor of mitogenesis. At week 3 (the midpoint in this 2-week stage), during the feeding of the "mito-inhibitor", the rats were given a strong stimulus of hepatocyte growth. The hepato-mitogen α-hexachlorocyclohexane and the necrogenic agent carbon tetrachloride were used in separate experimental trials. These authors showed that cell division immediately after exposure to the carcinogen was essential for the development of foci. Focal development was greatest when partial hepatectomy occurred within 24 hr of initiation. The exact role of mitogenesis is not clear, but evidence suggests that it is important to "fix" the mutations induced by a chemical carcinogen.

The type or extent of the mitogenic stimulus also appears to be very important. Columbano *et al.* (1987) accumulated evidence that chemicals that only induce mitogenesis of the liver do not stimulate carcinogenesis whereas regenerative hyperplasia after partial surgical or chemical hepatectomy supports carcinogenesis. These investigators induced hepatic proliferation by one of two general methods. Hyperplasia was induced by one of several hepatic mitogens (e.g., lead nitrate, ethylene dibromide, cyproterone acetate, or nafenopin). At the doses used, these agents did not induce cell death. A compensatory cell proliferation resulted from a two-thirds hepatectomy by surgery or from a necrogenic dose of carbon tetrachloride. At the peak of DNA replication, the [^{3}H]thymidine labeling index of hepatocytes was similar for both types of liver cell proliferation. However, only compensatory cell proliferation supported the development of putative preneoplastic foci. Why only compensatory cell proliferation led to foci and why these two types of proliferation have differing outcomes is unknown.

Selection Pressure for Focal Growth

Foci of putative preneoplastic parenchymal cells appear to be resistant to hepatic toxins. They absorb and activate some toxic chemicals less efficiently than do normal liver cells, have a decreased level of microsomal enzymes, and proliferate after partial hepatectomy or with carcinogen administration (for references, see Solt and Farber, 1976). Solt and Farber (1976) have taken these observations and unified them into a protocol that permits study of events that occur early in initiation. This protocol is worth reiterating.

As originally developed, the protocol is as follows: a single dose of the hepatocarcinogen diethylnitrosamine is administrated to rats. The animals are allowed 2 weeks to recover from the initial cell damage. Then the animals are fed a diet with a generalized growth inhibitor for the remaining 2 weeks of the protocol. Usually this growth inhibitor is 2-acetylaminofluorene (2-AAF). At the midpoint of this 2-week period of feeding 2-AAF, the liver is subjected to a strong growth stimulus. In the original protocol, this stimulus was a two-thirds partial hepatectomy. All three components are needed for the development of foci—the carcinogen, the generalized growth inhibitor, and a strong stimulus for liver growth. With this protocol, carcinogen-initiated foci are "forced" to develop by the strong stimulus for liver growth, but are selected over normal liver cells by the fact that a population of carcinogen-initiated cells is resistant to the growth inhibitory properties of 2-AAF. The application of the protocol of Solt and Farber (1976) to the chronic exposure to AFB$_1$ is discussed next.

CELL REPLICATION IN RESPONSE TO AFLATOXINS

Few experiments have evaluated the role of cell replication in carcinogenesis by aflatoxins. As discussed earlier, AFB_1 is a potent inhibitor of DNA synthesis and hepatocyte replication.

Experiments using partial hepatectomy to stimulate hepatocyte replication illustrate the complexity of the situation. Rogers *et al.* (1971) found that AFB_1 suppressed both [^{3}H]thymidine incorporation into parenchymal cells and parenchymal cell mitoses when the cells were stimulated to divide by partial hepatectomy. The time of aflatoxin treatment relative to the posthepatectomy waves of DNA synthesis did not affect the ultimate development of hepatomas. Further, hepatectomy during the latent period of tumor development did not enhance the development of tumors. Rogers *et al.* (1971) concluded that hepatectomy did not influence the development of liver cancer. These researchers emphasized that surgical procedures do not mimic chronic hyperplasia as a result of malnutrition, viral or parasitic infection, or exposure to hepatotoxic agents.

In rats treated *in vivo* with AFB_1, Neal and Cabral (1980) found inhibition of DNA synthesis as measured by [^{3}H]thymidine incorporation. The inhibition of hepatic DNA synthesis occurred in livers of control, sham-operated, and partially hepatectomized rats. Interestingly, the degree of inhibition of DNA synthesis was much greater in the livers of the partially hepatectomized rats. AFB_1 was metabolized by the remnant liver of the hepatectomized rat. Neal and Cabral (1980) found that single doses of AFB_1 in the partially hepatectomized rat resulted in the induction of putative preneoplastic foci. However, focal incidence was very low and the size and multiplicity of foci was unimpressive. These investigators hypothesized that any stimulatory effect of cell division is countered by the effects of AFB_1 as an inhibitor of DNA synthesis. Further, these researchers speculated that the effects of chronic low level AFB_1 treatment probably select foci that are resistant to the toxic DNA-inhibitory effects of AFB_1.

Other experiments investigating the effects of hepatic replication on aflatoxin carcinogenesis have met with more success. Newberne *et al.* (1987) have addressed this question directly. Selenium at very low or high dietary levels causes hepatic toxicity. In combination with AFB_1, dramatic effects on hepatocytes were seen. In these experiments, AFB_1 was given to young rats at 25 µg/day for 15 days. At low levels of dietary selenium (0.05 and 0.10 ppm), AFB_1 induced hepatic necrosis that persisted for up to 3 months whereas, at adequate dietary levels of selenium (1.00 and 2.00 ppm), significantly less necrosis occurred and did not persist. At high levels of dietary selenium, necrosis in response to AFB_1 was high and significant necrosis was still occurring 3 months later. Apparently, both low and high dietary selenium confer a state of chronic necrosis although AFB_1 exposure has ceased. In all groups, [^{3}H]thymidine labeling of hepatic nuclei was suppressed significantly 24 hr after the last aflatoxin dose; by 1 week, the labeling had recovered to slightly greater than that in the non-AFB_1-treated rat livers. These results imply that regeneration occurred in the interim. The number of putative preneoplastic foci at 3 months post-AFB$_1$ and the incidence of hepatocellular cancers at 14 months were elevated in the two aflatoxin-treated groups that had high indices of chronic necrosis, that is, in the two groups receiving either the low or the high selenium levels.

IMPLICATIONS OF AFLATOXIN TOXICITY FOR HEPATIC CARCINOGENESIS

The evidence is overwhelming that the toxicity of AFB$_1$ is important in its carcinogenic properties, yet the definitive experiments have not been undertaken. The protocol of Solt and Farber (1976) seems to be applicable to aflatoxin carcinogenicity; however, the situation is not simple because of the multiple properties of the aflatoxins. In any chronic regimen of AFB$_1$ exposure, initiation of DNA adducts and mutations occurs with the first dose of AFB$_1$. Abundant evidence suggests that AFB$_1$ can inhibit cell replication, thus fulfilling the second requirement for the Solt–Farber protocol. Finally, the necrogenic properties of AFB$_1$ fulfill the final requirement of the protocol by providing a strong stimulus for initiated cells to grow.

Several complications arise in defining the role of hepatic toxicity in aflatoxin carcinogenesis. First, AFB$_1$ does not initiate DNA damage on just day 1 of chronic exposure, but throughout the exposure period (Kensler et al., 1986; Groopman et al., 1992). From these studies, a major contribution to the DNA adduct burden is known to occur in the first several days of the chronic protocols. Additionally, the relative importance of cumulative DNA damage and mutations is unknown. Second, the proportion of any dose of AFB$_1$ that inhibits cell replication rather than inducing DNA damage is not known. In fact, the induction of DNA damage may act as a generalized inhibitor of cell growth. Third, the extent of necrosis and the subsequent regeneration rarely have been monitored carefully. For example, the two-cycle AFB$_1$ treatment protocol used by Appleton and Campbell (1983) and the investigators in this laboratory represents virtually the only protocol to attempt to define the extent of AFB$_1$-induced necrosis (Liu et al., 1988). The time-course and extent of regenerative hyperplasia remains to be defined in that protocol.

The hepatonecrotic toxicity of the aflatoxins was the feature that led to their discovery. This toxicity also appears to be important in their carcinogenicity in the liver. The enormous interest in cell proliferation (Cohen and Ellwein, 1991) and the circumstantial evidence of an important role for proliferation in aflatoxin carcinogenesis (Newberne et al., 1987; Liu et al., 1988) is encouraging. The study of the overall event following exposure to aflatoxins may shed light on the role of liver carcinogens such as AFB$_1$ in hepatic insults such as parasites, malnutrition, hepatitis viruses, and other toxins.

REFERENCES

Allcroft, R. (1969). Aflatoxicosis in farm animals. *In* "Aflatoxin: Scientific Background, Control, and Implications" (L. A. Goldblatt, ed.), pp. 237–264. Academic Press, New York.

Appleton, B. S., and Campbell, T. C. (1983). Effect of high and low dietary protein on the dosing and postdosing periods of aflatoxin B_1-induced hepatic preneoplastic lesion development in the rat. *Cancer Res.* **43,** 2150–2154.

Bauer, D. H., Lee, D. J., and Sinnhuber, R. O. (1969). Acute toxicity of aflatoxins B_1 and G_1 in rainbow trout (*Salmo gairdneri*). *Toxicol. Appl. Pharmacol.* **15,** 415–419.

Bryden, W. L., Cumming, R. B., and Lloyd, A. B. (1980). Sex and strain responses to aflatoxin B_1 in the chicken. *Avian Pathol.* **9,** 539–550.

Büchi, G., Spitzner, D., Paglialunga, S., and Wogan, G. N. (1973). Synthesis and toxicity evaluation of aflatoxin P_1. *Life Sci.* **13,** 1145.

Busby, W. E., and Wogan, G. N. (1984). Aflatoxins. *In* "Chemical Carcinogens" (C. E. Searle, ed.), 2d Ed., pp. 945–1136. Americal Chemical Society, Washington, D.C.

Butler, W. H. (1966). Acute toxicity of aflatoxin B_1 in guinea pigs. *J. Pathol. Bacteriol.* **91,** 277–280.

Butler, W. H., Greenblatt, M., and Lijinsky, W. (1969). Carcinogenesis in rats by aflatoxin B_1, G_1, and B_2. *Cancer Res.* **29,** 2206–2211.

Butterworth, B. E., Smith-Oliver, T., Erle, L., Loury, D. J., White, R. D., Doolittle, D. J., Working, P. K., Cattley, R. C., Jirtle, R., Michalopoulow, G., and Strom, S. (1989). Use of primary cultures of human hepatocytes in toxicology studies. *Cancer Res.* **49,** 1075–1084.

Carnaghan, R. B. A. (1963). Toxicity and fluorescence properties of the aflatoxins. *Nature (London)* **200,** 1101.

Carnaghan, R. B. A. (1965). Hepatic tumors in ducks fed a low level of toxic groundnut meal. *Nature (London)* **208,** 308.

Cayama, E., Tsuda, H., Sarma, D. S. R., and Farber, E. (1978). Initiation of chemical carcinogenesis requires cell proliferation. *Nature (London)* **275,** 60–62.

Cohen, S. M., and Ellwein, L. B. (1991). Genetic errors, cell proliferation, and carcinogenesis. *Cancer Res.* **51,** 6492–6505.

Cole, K. E., Jones, T. W., Lipsky, M. M., Trump, B. F., and Hsu, I.-C. (1989). Comparative effects of three carcinogens on human, rat, and mouse hepatocytes. *Carcinogenesis* **10,** 139–143.

Columbano, A., Ledda-Columbano, G. M., Lee, G., Rajalakshmi, S., and Sarma, D. S. R. (1987). Inability of mitogen-induced liver hyperplasia to support the induction of enzyme-altered islands induced by liver carcinogens. *Cancer Res.* **47,** 5557–5559.

Cova, L., Wild, C. P., Mehrotra, R., Turusov, V., Shirai, T., Lambert, V., Jacquet, C., Tomatis, L., Trepo, C., and Montesano, R. (1990). Contribution of aflatoxin B_1 and hepatitis B virus infection in the induction of liver tumors in ducks. *Cancer Res.* **50,** 2156–2163.

Edds, G. T. (1973). Acute aflatoxicosis: A review. *Am. J. Vet. Med. Assoc.* **162,** 304–309.

Godoy, H. M., Judah, D. J., Arora, H. L., Neal, G. E., and Jones, G. (1976). The effects of prolonged feeding with aflatoxin B_1 on adult rat liver. *Cancer Res.* **36,** 2399–2407.

Goldblatt, L. A., ed. (1969). "Aflatoxin: Scientific Background, Control, and Implications." Academic Press, New York.

Groopman, J. D., DeMatose, P., Egner, P. A., Love-Hunt, A., and Kensler, T. W. (1992). Molecular dosimetry of urinary aflatoxin-N^7-guanine and serum aflatoxin–albumin adducts predicts chemoprotection by 1,2-dithiole-3-thione in rats. *Carcinogenesis* **13,** 101–106.

Halver, J. E. (1969). Aflatoxicosis and trout hepatoma. *In* "Aflatoxin: Scientific Background, Control, and Implications" (L. A. Goldblatt, ed.), pp. 265–306. Academic Press, New York.

Hayes, A. W., Cain, J. A., and Moore, B. G. (1977). Effect of aflatoxin B_1, ochratoxin A, and rubratoxin B on infant rats. *Food Cosmet. Toxicol.* **15,** 23–27.

Hegazy, S. M., Azzam, A., and Gabal, M. A. (1991). Interaction of naturally occurring aflatoxins in poultry feed and immunization against fowl cholera. *Poultry Sci.* **70,** 2425–2428.

Hendrickse, R. G. (1991). Kwashiorkor: The hypothesis that incriminates aflatoxins. *Pediatrics* **88**, 376–378.

Hsieh, D. P. H., Solhab, A. S., Wong, J. J., and Yang, S. L. (1974). Toxicity of aflatoxin Q_1 as evaluated with the chicken embryo and bacterial auxotrophs. *Toxicol. Appl. Pharmacol.* **30**, 237–241.

Kalengayi, M. M. R., and Desmet, V. J. (1975). Sequential histological and histochemical study of the rat liver after single-dose aflatoxin B_1 intoxication. *Cancer Res.* **35**, 2836–2844.

Kensler, T. W., Egner, P. A., Davidson, N. E., Roebuck, B. D., Pikul, A., and Groopman, J. D. (1986). Modulation of aflatoxin metabolism, aflatoxin-N^7-guanine formation, and hepatic tumorigenesis in rats fed ethoxyquin: Role of induction of glutathione *S*-transferase. *Cancer Res.* **46**, 3924–3931.

Kensler, T. W., Egner, P. A., Dolan, P. M., Groopman, J. D., and Roebuck, B. D. (1987). Mechanism of protection against aflatoxin tumorigenicity in rats fed 5-(2-pyrazinyl)-4-methyl-1,2-dithiol-3-thione (oltipraz) and related 1,2-dithiol-3-thiones and 1,2-dithiol-3-ones. *Cancer Res.* **47**, 4271–4277.

Kensler, T. W., Groopman, J. D., Eaton, D. L., Curphey, T. J., and Roebuck, B. D. (1992). Potent inhibition of aflatoxin-induced hepatic tumorigenesis by the monofunctional enzyme inducer 1,2-dithiole-3-thione. *Carcinogenesis* **13**, 95–100.

Lijinsky, W., and Butler, W. H. (1966). Purification and toxicity of aflatoxin G_1. *Proc. Soc. Exp. Biol. Med.* **123**, 151–154.

Liu, Y.-L., Roebuck, B. D., Yager, J. D., Groopman, J. D., and Kensler, T. W. (1988). Protection by 5-(2-pyrazinyl)-4-methyl-1,2-dithiol-3-thione (oltipraz) against the hepatotoxicity of aflatoxin B_1 in the rat. *Toxicol. Appl. Pharmacol.* **93**, 442–451.

McGuire, R. A. (1969). M. S. Thesis. Factors affecting the acute toxicity of aflatoxin B_1 in the rat and mouse. Massachusetts Institute of Technology, Cambridge.

Madhavan, T. V., and Gopalan, C. (1968). The effect of dietary protein on carcinogenesis of aflatoxin. *Arch. Pathol.* **85**, 133–137.

Mgbodile, M. U. K., Holscher, M., and Neal, R. A. (1975). A possible protective role for reduced glutathione in aflatoxin B_1 toxicity: Effect of pretreatment of rats with phenobarbital and 3-methylcholanthrene on aflatoxin toxicity. *Toxicol. Appl. Pharmacol.* **43**, 128–142.

Michalopoulos, G. (1990). Liver regeneration: Molecular mechanisms of growth control. *FASEB J.* **4**, 176–187.

Neal, G. E., and Cabral, J. R. P. (1980). Effect of partial hepatectomy on the response of rat liver to aflatoxin B_1. *Cancer Res.* **40**, 4739–4743.

Newberne, P. M., and Butler, W. H. (1969). Acute and chronic effects of aflatoxin on the liver of domestic and laboratory animals: A review. *Cancer Res.* **29**, 236–250.

Newberne, P. M., Punyarit, P., de Camargo, J., and Suphakarn, V. (1987). The role of necrosis in hepatocellular proliferation and liver tumors. *Arch. Toxicol. Suppl.* **10**, 54–67.

Pong, R. S., and Wogan, G. N. (1971). Toxicity and biochemical and fine structural effects of synthetic aflatoxin M_1 and B_1 in the rat liver. *J. Natl. Cancer Inst.* **47**, 585–592.

Purchase, I. F. H. (1967). Acute toxicity of aflatoxin M_1 and M_2 in one-day-old ducklings. *Food Cosmet. Toxicol.* **5**, 339–342.

Roebuck, B. D., Liu, Y.-L., Rogers, A. E., Groopman, J. D., and Kensler, T. W. (1991). Protection against aflatoxin B_1-induced hepatocarcinogenesis in F344 rats by 5-(2-pyrazinyl)-4-methyl-1,2-dithiole-3-thione (oltipraz): Predictive role for short-term molecular dosimetry. *Cancer Res.* **51**, 5501–5506.

Rogers, A. E., and Newberne, P. M. (1967). The effects of aflatoxin B_1 and dimethylsulfoxide on thymidine-^3H uptake and mitosis in rat liver. *Cancer Res.* **27**, 855–864.

Rogers, A. E., Kula, N. S., and Newberne, P. M. (1971). Absence of an effect of partial hepatectomy on aflatoxin B_1 carcinogenesis. *Cancer Res.* **31**, 491–495.

Solt, D., and Farber, E. (1976). New principle for the analysis of chemical carcinogenesis. *Nature (London)* **263**, 701–703.

Van Halderen, A., Green, J. R., Marasas, W. F., Thiel, P. G., and Stockenstrom, S. (1989). A field outbreak of chronic aflatoxicosis in dairy calves in the Western Cape Province. *J. S. Afr. Vet. Assoc.* **60**, 210–211.

Vesselinovitch, S. D., Mihailovich, N., Wogan, G. N., Lombard, L. S., and Rao, K. V. N. (1972). Aflatoxin B_1, a hepatocarcinogen in the infant mouse. *Cancer Res.* **32**, 2289–2291.

Wogan, G. N. (1966). Chemical nature and biological effects of the aflatoxins. *Bacteriol. Rev.* **30**, 460–470.

Wogan, G. N. (1973). Aflatoxin carcinogenesis. *Methods Cancer Res.* **7**, 303–344.

Wogan, G. N., and Newberne, P. M. (1967). Dose–response characteristics of aflatoxin B_1 carcinogenesis in the rat. *Cancer Res.* **27**, 2370–2376.

Wogan, G. N., and Paglialunga, S. (1974). Carcinogenicity of synthetic aflatoxin M_1 in rats. *Food Cosmet. Toxicol.* **12**, 381–384.

Wogan, G. N., Edwards, G. S., and Newberne, P. M. (1971). Structure–activity relationship in toxicity and carcinogenecity of aflatoxins and analogs. *Cancer Res.* **31**, 1936–1942.

Wogan, G. N., Paglialunga, S., and Newberne, P. M. (1974). Carcinogenic effects of low dietary levels of aflatoxin B_1 in rats. *Food Cosmet. Toxicol.* **12**, 681–685.

Yager, J. D., and Potter, V. R. (1975). A comparison of the effects of 3′-methyl-4-dimethylamino-azobenzine, 2-methyl-4-dimethyl-aminoazobenzene, and 2-acetylaminofluorene on rat liver DNA stability and new synthesis. *Cancer Res.* **35**, 1225–1234.

3

Biotransformation
of Aflatoxins

David L. Eaton, Howard S. Ramsdell, and Gordon E. Neal

INTRODUCTION

Biotransformation and Aflatoxin Toxicity

Biotransformation plays an important role in the biological activity and disposition of aflatoxins. Bioactivation of aflatoxins has been demonstrated as a necessary step in the most dramatic of their toxic and carcinogenic effects. Garner, working in the Millers' laboratory, was the first to show that metabolic activation of aflatoxin was necessary for mutagenic activity (Garner *et al.*, 1972). Several detoxification mechanisms involving biotransformation are known also. In this chapter, pathways of aflatoxin biotransformation that have been characterized in animals, and their significance to aflatoxin toxicity, are discussed.

The majority of research on aflatoxin biotransformation has focused on the metabolic alterations of aflatoxin B_1 (AFB_1). The interest in this most potently toxic and carcinogenic of the aflatoxins is understandable. AFB_1 is also usually the aflatoxin found in the highest concentrations in contaminated food and feed. This chapter thus necessarily focuses on AFB_1; other aflatoxins are discussed whenever possible.

Figure 1 shows the biotransformation pathways for AFB_1. Note that not all metabolites have been identified in all species and that significant quantitative differences in the formation of the various products may exist.

Virtually all the toxic effects of AFB_1 are now recognized to be attributable to

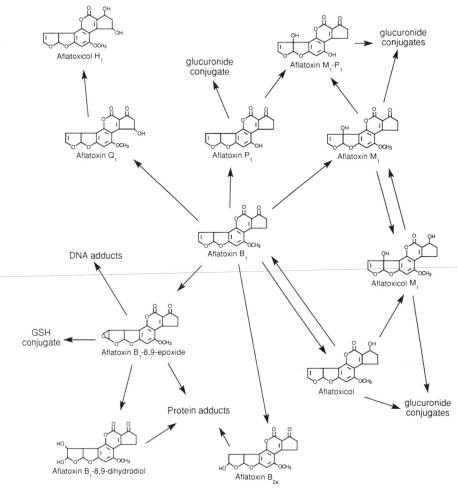

FIGURE 1   Biotransformation pathways for aflatoxin B₁.

the action of its metabolites that are capable of reacting with cellular macro-molecules. Oxidation by cytochrome P450 plays a central role in the pathways that lead to the formation of such products. Although the 8,9-epoxide of AFB long was postulated to be the ultimate reactive intermediate, this highly reactive epoxide was first synthesized only in 1988 (Baertschi *et al.*, 1988).

Detoxification Pathways

In addition to the recognized pathways of aflatoxin bioactivation, metabolic detoxification of aflatoxins and their reactive metabolites has been demonstrated. Microsomal oxidation products that are less toxic than the parent compounds may be formed. An important detoxification mechanism involves the conjugation of the reactive epoxides with glutathione (GSH). The detoxification pathways may be more crucial as determinants of the effects of aflatoxins than the bioactivation mechanisms.

Biotransformation as a Determinant of Susceptibility to Aflatoxin Toxicity

Evidence has accumulated that supports the hypothesis that the relative efficacy of competing pathways of aflatoxin biotransformation is a critical factor in the susceptibility of a given species to the toxic effects of aflatoxin. The proportion of the mycotoxin that is converted to metabolites that bind to critical cellular macromolecules determines the extent of toxicity or carcinogenicity.

Much of aflatoxin biotransformation research has focused on the formation of metabolites that are capable of reacting with DNA, processes that are related primarily to the carcinogenicity of aflatoxins. Far less attention has been given to the role of biotransformation in acute aflatoxin toxicity. A much broader range of sensitivity to the carcinogenic effects of aflatoxin exists than for acute toxicity. The acute LD_{50}s for 9 species vary over a 30-fold range (Newberne and Butler, 1969; see also Chapter 1). The dose level of AFB_1 that is required to produce hepatocellular carcinoma in 100% of the rats treated (15 ppb in diet; Wogan and Newberne, 1967) is 67 times lower than a dietary exposure level (1 ppm) that fails to cause any liver tumors in mice (Wogan, 1973). Whereas the acute LD_{50} values for AFB_1 and aflatoxin G_1 (AFG_1) differ by a factor of 2–3 in the rat (Wogan et al., 1971) and trout (Bauer et al., 1969), the mutagenic (Gurtoo et al., 1978; Coulombe et al., 1982) and carcinogenic (Butler et al., 1969; Sinnhuber et al., 1977) potencies of these compounds differ by over an order of magnitude, suggesting the existence of important differences in the mechanisms of acute aflatoxin toxicity and carcinogenesis. The roles of different biotransformation pathways in these differences remain to be elucidated.

OXIDATION

Epoxidation

Cytochrome P450-dependent epoxidation of the terminal furan ring double bond of AFB_1 and AFG_1 generates very potent electrophilic species that are capable of alkylating nucleic acids (Essigmann et al., 1977; Swenson et al., 1977; Baertschi et al., 1989). Because of its lability in aqueous solution, AFB_1-8,9-epoxide has not been isolated intact from biological systems, but its formation is inferred from the isolation of products of its reaction with biological nucleophiles, including DNA and GSH (Essigmann et al., 1977; Degen and Neumann, 1978). AFB_1-8,9-epoxide can be synthesized, but is only stable in the absence of both electrophiles and nucleophiles. When the synthetic epoxide is added to

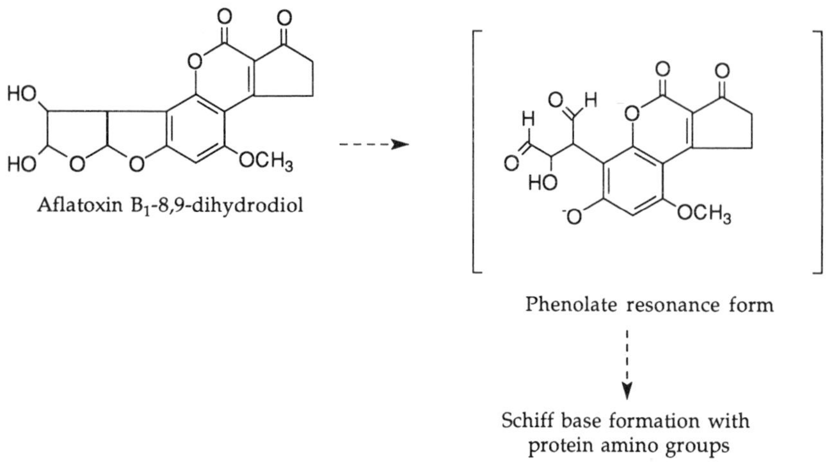

FIGURE 2   Schiff base formation between protein amino groups and aflatoxin B$_1$-8,9-dihydrodiol.

aqueous solutions, it hydrolyzes very rapidly and nonenzymatically to the dihydrodiol, disappearing completely in less than 5 sec. However, as discussed later, in the presence of GSH and glutathione S-transferase, the synthetic epoxide can be "trapped" efficiently as the GSH conjugate (Raney et al., 1992b).

Analysis of microsomal incubations with AFB$_1$ indicates the presence of AFB$_1$-8,9-dihydrodiol (and its Tris adduct, if this buffer is present) (Lin et al., 1978; Neal and Colley, 1979; Neal et al., 1981). The diol presumably is formed by the rapid hydrolysis of the 8,9-epoxide. AFB$_1$-8,9-dihydrodiol can exist in a resonance form as a phenolate ion that is capable of forming Schiff base adducts with protein amino groups, particularly lysine (Figure 2). The diol (Lin et al., 1978) is most likely to be involved in the binding of AFB$_1$ to protein that is observed in vivo (Garner et al., 1979; Appleton et al., 1982). The AFB–lysine adduct is the principal adduct found in plasma albumin after in vivo AFB exposure (Sabbioni et al., 1987). This type of reaction could be involved in the mechanisms of acute toxicity of aflatoxin. The additional reductive metabolism of AFB$_1$-diol, catalyzed by enzymes induced in rat liver by antioxidants and certain carcinogens, could play a role in the resistance to AFB-induced cytoxicity observed in some studies (G. E. Neal, D. J. Judah, and J. D. Hayes, unpublished observations).

Conflicting evidence exists regarding the role of epoxide hydrolase in the hydrolysis of AFB$_1$-8,9-epoxide. Studies with specific inhibitors of epoxide hydrolase in isolated hepatocytes (Decad et al., 1979; Ch'ih et al., 1983) have suggested that this enzyme does play a role in detoxification of AFB-8,9-epoxide, whereas addition of epoxide hydrolase inhibitors to microsomal incubations with AFB$_1$ do not result in an increase in dihydrodiol formation (Lin et al., 1978; Neal and Colley, 1978) or binding to exogenous DNA (Lotlikar et al., 1984b). One study showed that binding of AFB$_1$ to DNA catalyzed by two different purified P450 isoenzymes in a reconstituted system was reduced markedly when purified epoxide hydrolase was added (Shayiq and Avadhani, 1989). Whether epoxide hydrolase activity toward AFB_1-8,9-epoxide is relevant in all species is unclear, given the efficiency of hepatic glutathione S-transferase as a scavenger of AFB_1-8,9-epoxide: no dihydrodiol could be detected in microsomal incubations containing GSH and mouse cytosol (Monroe and Eaton, 1987).

Covalent binding to DNA is generally a property of all aflatoxin derivatives with an unsaturated terminal furan ring, presumably via epoxidation. DNA adducts of aflatoxin P_1 (AFP_1) and aflatoxin M_1 (AFM_1) epoxides were identified in liver extracts following treatment of rats with AFB_1 (Essigmann et al., 1983), implying that the hydroxylation of AFB_1 can be followed by epoxidation to form electrophilic metabolites. Aflatoxin Q_1 (AFQ_1), on the other hand, is a relatively poor substrate for epoxidation (Raney et al., 1992c) and thus would not be expected to show very high DNA binding. Treatment of rats in vivo with AFG_1 resulted in covalent binding to both liver and kidney (Garner et al., 1979). Activation of AFG_1 by liver 9000 $\times$ g supernatant in vitro also was shown to result in binding to exogenous DNA.

Biotransformation of AFB_2, which is saturated at the 8,9 position, leads to DNA-binding metabolites (Swenson et al., 1977; Roebuck et al., 1978). The adducts formed from AFB_2 in vivo are identical to those identified following AFB_1 administration (Groopman et al., 1981), consistent with the hypothesis that AFB_2 can be reduced to AFB_1 and then epoxidized (Swenson et al., 1977; Roebuck et al., 1978).

Mechanisms of AFB_1 activation that do not involve cytochrome P450-mediated oxidation have been demonstrated. Co-oxygenation of AFB_1 can occur in the presence of arachidonic acid and a source of prostaglandin H synthase, forming mutagenic metabolites with chromatographic properties identical to the N^7-guanine adducts formed in other activation systems (Battista and Marnett, 1985). Whereas the amount of prostaglandin H synthase-dependent binding of AFB_1 to DNA in guinea pig liver microsomes was 1–2% of that resulting from cytochrome P450 oxidation, the two enzyme systems had the same amount of activity in kidney microsomes (Liu et al., 1990). Prostaglandin H synthase thus appears to contribute significantly to the activation of AFB_1 in the latter organ.

Lipoxygenases from soy beans and from guinea pig tissues also are able to activate AFB to DNA-binding metabolites, most likely through co-oxidation secondary to peroxy radical generation (Liu and Massey, 1992). Although the total capacity of guinea pig lipoxygenases to activate AFB was far lower than that of P450-mediated oxidation, the hepatic half-maximal DNA binding concentration for lipoxygenases was substantially lower than that for total P450s and prostaglandin H synthase, suggesting that this pathway could be significant for activation of AFB in vivo at the low AFB concentrations encountered in the diet (Liu and Massey, 1992).

Aflatoxins (including AFB_1 and AFB_2) can be activated to DNA-binding species by the action of UV light, apparently related to the presence of the furocoumarin moiety as a photosensitizing group (Israel-Kalinsky *et al.*, 1984). Singlet oxygen formation has been suggested as the mechanism leading to the oxidation of the 8,9 double bond of AFB_1, whereas activation of AFB_2 proceeds first through hydrogen abstraction to form AFB_1 (Stark *et al.*, 1990). These studies have been conducted *in vitro;* no evidence for or against a significant role of photoactivation in the biological activities of aflatoxins has been found. The observation of preferential retention of administered AFB_1 in pigmented tissues through noncovalent interactions with melanin (Larsson *et al.*, 1988) is interesting, but the biological significance of this interaction remains uncertain.

Hydroxylation

Oxidation of aflatoxins by microsomal cytochromes P450 forms hydroxylated metabolites that have generally lower biological activity than the parent compounds. In the case of AFB_1, hydroxylation at the 3 and 9a positions forms AFQ_1 and AFM_1, respectively (Figure 1).

AFQ_1 has been observed as a major AFB_1 metabolite *in vitro*, particularly using primate liver microsomes (Buchi *et al.*, 1974; Masri *et al.*, 1974; Moss and Neal, 1985; Yourtee *et al.*, 1987; Ramsdell and Eaton, 1990b). AFQ_1 generally is regarded as a detoxification product of AFB_1, having much lower acute toxicity (Hsieh *et al.*, 1974), mutagenicity (Gurtoo *et al.*, 1978; Coulombe *et al.*, 1982), and carcinogenicity (Hendricks *et al.*, 1980), although it shows a surprising level of mutagenic activity in the absence of a metabolic activating system (Yourtee and Kirk-Yourtee, 1986). Raney *et al.* (1992c) demonstrated that human liver microsomes oxidized AFQ_1 to the epoxide very poorly, and that synthetic AFQ-epoxide yielded only low levels of DNA adducts with calf thymus DNA and had little mutational activity toward *Salmonella typhimurium* strain TA98.

AFM_1, identified as a metabolite of AFB_1 in milk and urine (Holzapfel *et al.*, 1966; Masri *et al.*, 1967; see also Chapter 17), is less biologically active than AFB_1, but nevertheless is a relatively potent carcinogen. For example, at a dietary level of 50 ppb, AFM_1 induced a 33% incidence of liver tumors in rats, compared with a 95% incidence for AFB_1 at the same concentration (Hsieh *et al.*, 1984). Similarly, trout fed AFM_1 at 4 ppb in the diet had a 40% incidence of hepatomas, compared with 60% for AFB_1 at the same level (Sinnhuber *et al.*, 1974). AFM_1 has been found to have approximately 2% of the mutagenic potency of AFB_1 (Gurtoo *et al.*, 1978; Coulombe *et al.*, 1982). The acute toxicity of AFM_1 is only slightly less than that of AFB_1, however (Purchase, 1967; Pong and Wogan, 1971). AFM_1 has been identified as a biotransformation product of AFB_1 in a number of species (Holzapfel *et al.*, 1966; Schabort and Steyn, 1969; Masri *et al.*, 1974; Moss and Neal, 1985; Loveland *et al.*, 1988; Ramsdell and Eaton, 1990b). Using rat liver-derived epithelial cell lines that lack cytochrome P450-catalyzed mixed-function monooxygenase activity, $5-10\ \mu M$ AFM_1 exhibits cytotoxicity whereas AFB requires metabolic activation for cytotoxicity (G. E.

Neal and D. Judah, unpublished observations), suggesting that AFM_1 may have cytotoxic properties distinct from those of the epoxide.

Secondary hydroxylation at the 9a position of the AFB_1 O-demethylation product AFP_1 appears to be the primary route of 4,9a-dihydroxyaflatoxin B_1 formation by mouse and rat liver microsomes (Eaton *et al.*, 1988). The glucuronide conjugate of this dihydroxy metabolite also has been identified in bile of rats after AFB_1 administration (Holeski *et al.*, 1987; Eaton *et al.*, 1988). AFB_2 apparently also undergoes 9a hydroxylation (demonstrated *in vivo*), with the formation of AFM_2 (Holzapfel *et al.*, 1966; Dann *et al.*, 1972). The analogous 10a-hydroxy derivative of AFG_1 (GM_1) also has been identified (by thin layer chromatography) in the urine of treated rats (Dann *et al.*, 1972).

O-Demethylation

O-Demethylation of AFB_1 is a route of biotransformation in some species that produces AFP_1 (Dalezios *et al.*, 1971; Wong and Hsieh, 1980). This product is much less toxic than AFB_1 (Stoloff *et al.*, 1972) and shows little mutagenic activity (Gurtoo *et al.*, 1978; Coulombe *et al.*, 1982). However, apparently AFP_1 can be converted to a reactive epoxide, as noted earlier (Essigmann *et al.*, 1983). Microsomal fractions from a series of primary liver tumors of Thai patients have been observed consistently to produce, in *in vitro* incubations, higher levels of AFP_1 from AFB microsomes than microsomes from secondary normal liver (G. E. Neal, C. Wilde, and R. Wolf, unpublished observations). This result could have some relevance to a report by Ross *et al.* (1992), which showed that AFP_1 had the highest correlation of all urinary aflatoxin metabolites with the presence of liver cancer. Note that AFP_1 production in human liver microsomal preparations is usually very low relative to other hydroxylated metabolites (Ramsdell and Eaton, 1990b), although precision human liver slices incubated under similar conditions show substantial AFP_1 formation (D. L. Eaton and J. Heinonen, unpublished observations). Thus, *in vitro* microsomal incubation methods may not accurately reflect the *in vivo* production of AFP_1 and, perhaps, of other oxidative metabolites.

Oxidative O-demethylation also appears to be possible with AFM_1 as the substrate to provide an alternative route for 4,9a-dihydroxyaflatoxin B_1 formation (Eaton *et al.*, 1988). No evidence exists for microsomal O-demethylation of AFG_1 or the 8,9-saturated aflatoxins.

AFB(G)$_{2\alpha}$ Formation

Hydration of the double bond of the terminal furan ring results in the formation of $AFB_{2\alpha}$ or $AFG_{2\alpha}$ from AFB_1 and AFG_1, respectively. This reaction can occur nonenzymatically under acidic conditions (Pohland *et al.*, 1968) and may occur in the stomach following oral aflatoxin ingestion. Microsomal oxidation also has been suggested as a source of these compounds (Schabort and Steyn, 1969; Patterson and Roberts, 1970, 1971, 1972). Microsomal hydroxylation of AFB_2 forming $AFB_{2\alpha}$ also may occur, constituting a mechanism for bioactivation of that 8,9-saturated aflatoxin (Groopman et al., 1981). The formation of $AFB_{2\alpha}$ and $AFG_{2\alpha}$ in microsomal incubations has not been proven by isolation and chemical characterization, however, possibly because of binding to protein at physiological pH. The biological formation of these compounds in vitro instead has been inferred from absorption spectral characteristics (Patterson and Roberts, 1970) or thin layer chromatography (Schabort and Steyn, 1969). Because similar spectral properties are exhibited by the dihydrodiol hydrolysis products of the aflatoxin epoxides, the microsomal formation of $AFB_{2\alpha}$ and $AFG_{2\alpha}$ has been disputed (Lin et al., 1978; Neal et al., 1981; Eaton et al., 1988). Protein binding by $AFB_{2\alpha}$ or $AFG_{2\alpha}$ could be involved in the toxicity of AFB_1 and AFG_1, but such a role remains to be proven conclusively. Synthetic $AFB_{2\alpha}$ has been shown to form Schiff bases with amines, including Tris buffer (Eaton et al., 1988).

Enzymology of Aflatoxin Oxidations

Several studies have examined the role of specific human cytochrome P450 enzymes in the activation of AFB_1. This role is of more than academic interest because a significant contribution of isoforms that are inducible or expressed polymorphically could have important implications for interindividual variations in the susceptibility of humans to AFB_1-induced carcinogenesis.

Ramsdell and Eaton (1990b) demonstrated that the various P450s that are active in the biotransformation of AFB exhibit different apparent K_ms, so metabolite profiles obtained at high substrate concentration differ considerably from those obtained at low concentrations in some species, including humans. These data suggest that multiple forms of cytochrome P450 are involved in the formation of AFB metabolites. Correlation of AFB_1 mutagenicity and AFB_1-8,9-dihydrodiol formation with cytochrome P450 isoenzyme content measured by Western blot analysis also suggests that more than one human cytochrome P450 is involved in activation of AFB_1 (Forrester et al., 1990).

Using a genotoxicity assay, an association was observed between the activation of both AFB_1 and AFG_1 and expression of cytochrome CYP3A4 (Shimada and Guengerich, 1989). However, nifedipine oxidation, which is a specific marker of CYP3A4 activity, is correlated poorly with AFB-8,9-epoxide formation at low AFB substrate concentrations. Inhibitory antibodies to CYP3A4 partially inhibited microsomal formation of AFB-8,9-epoxide only at high (124 μM) AFB concentrations; at low substrate concentrations (15.6 μM), only slight inhibition of AFB-8,9-epoxide formation occurred that was sufficient to block nifedipine oxidation (Ramsdell et al., 1991). These results suggest that activation of AFB by human liver, at concentrations of AFB that are likely to be encountered in the human diet, may occur predominantly by a form(s) of cytochrome P450 other than CYP3A4. Guengerich and co-workers (Raney et al., 1992c) disagree with this conclusion because, in similar studies, they found a positive correlation of AFB-epoxide production at all substrate concentrations in three different human liver samples. Antibody inhibition data showed substantial inhibition of epoxide production with anti-P450 3A4 but not with anti-P450 1A or anti-P450 2C. Thus, the relative importance of various cytochromes P450 in the oxidation of AFB to the epoxide and to hydroxylated detoxification products remains uncertain.

Correlations among the various rates of formation of different oxidative metabolites of AFB at low and high substrate concentrations in human liver microsomes provide some insight into the complexity of oxidative AFB metabolism. At low substrate concentrations, AFM_1 and AFB-epoxide formation were correlated almost perfectly ($r^2 = 0.976$) in 14 different human liver samples, suggesting that the same P450 (or multiple coordinately regulated P450s) was responsible for these metabolites (Fig. 3). At high substrate concentrations, the correlation became less, whereas the correlation between AFQ formation and AFB-epoxide was very poor at low substrate concentrations ($r^2 = 0.001$) but much higher ($r^2 = 0.550$) at the high AFB concentration (Fig. 3). These results suggest that the "high affinity" form of P450 that is responsible for activation of AFB at low substrate concentrations also forms AFM_1, whereas the "low affinity" form, probably CYP3A4, forms AFQ_1 as well as AFB-8,9-epoxide (Ramsdell *et al.*, 1991; Raney *et al.*, 1992c).

Expression of human cytochrome P450 cDNA has proven to be a powerful tool, avoiding the uncertainties of activity correlations, antibody inhibition experiments, and variable activity in reconstituted systems. In a study using the Ames assay as a genotoxicity end point, five different human cytochrome P450s were found to activate AFB_1 (Aoyama *et al.*, 1990). Recombinant cells expressing human cytochrome P450 1A2 were found to activate AFB_1 cytotoxicity and mutagenicity to a greater extent than those transfected with a vector containing P450 3A4 cDNA (Crespi *et al.*, 1991). Human cytochrome P450 2A3 expressed from cDNA also has been shown to activate AFB_1 (Aoyama *et al.*, 1990), as well as to catalyze the formation of AFM_1 and AFP_1 (Crespi *et al.*, 1990).

A few studies have been done of the AFB_1-biotransforming capabilities of cytochrome P450 isoenzymes of other species. Rat cytochrome P450p (CYP3A1 or CYP3A2) was shown in a series of induction studies to be active in the formation of AFQ_1 (Halvorson *et al.*, 1988). This conclusion is supported by correlation studies between human P450 expression and *in vitro* AFQ_1 formation (Forrester *et al.*, 1990). Formation of AFM_1 has been associated with the cytochrome P_3-450 (CYP1A2) of the mouse, an aryl hydrocarbon-inducible form (Faletto *et al.*, 1988; Koser *et al.*, 1988). Microsomes from 3-methylcholanthrene-treated rats demonstrate increased production of AFM_1 (Metcalfe *et al.*, 1981). In contrast with the results in the mouse, a cytochrome P450 isoenzyme induced by 3-methylcholanthrene in the hamster showed very high AFB_1 activation (Fukuhara *et al.*, 1989, 1990; Lai and Chiang, 1990). However, treatment of 3-methylcholanthrene-induced hamster liver microsomes with an inhibitory concentration of anti-P450 1A antibody stimulated mutagenicity of AFB_1 in the μmu

AFB Concentration: 124 μM

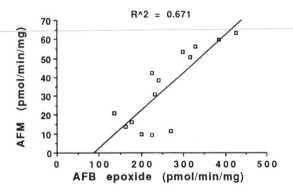

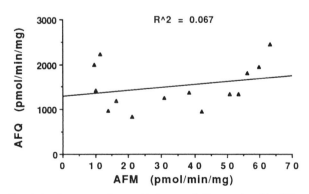

FIGURE 3 Correlation between various oxidative metabolites produced in different human liver microsomal preparations. Human liver microsomes (1.5 mg protein/ml) were incubated in the presence of an NADPH-regenerating system for 10 min with either 124 μ*M* or 15.6 μ*M* AFB₁. Reactions were stopped by addition of cold methanol, centrifuged to remove precipitate, and the supernatants

AFB Concentration: 15.6 μM

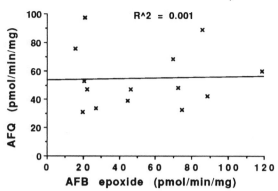

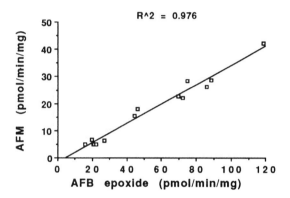

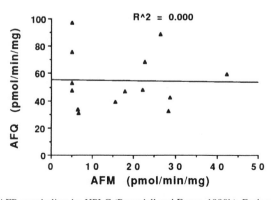

were analyzed for AFB metabolites by HPLC (Ramsdell and Eaton, 1990b). Each symbol represents one human liver sample. A total of 14 different human liver samples was assessed. Correlations were determined by least squares linear regression analysis.

system, suggesting that hamster CYP1A enzymes function primarily in a detoxification pathway for AFB_1 under the experimental conditions used (Lai and Chiang, 1990). Interestingly, precision liver slices from rats treated with the phenolic antioxidant ethoxyquin demonstrated a 5-fold increase in AFM_1 formation, a 2-fold increase in AFB–GSH formation, and a 3-fold decrease in DNA binding (J. T. Heinonen and D. L. Eaton, unpublished observations). These results are consistent with studies using microsomes from ethoxyquin-treated rats, in which a nearly 3-fold increase in AFM_1 formation was observed (Mandel *et al.*, 1987). The protective effects of ethoxyquin through induction of hepatic glutathione *S*-transferases are well described (Kensler *et al.*, 1986; Mandel *et al.*, 1987), although induction of AFM_1 formation, probably through induction of P450 1A enzymes, appears to be relatively more important.

In vitro studies with human liver microsomes demonstrated that the specific human CYP 1A2 inhibitor, furafylline, greatly decreased AFB-epoxide formation, and abolished AFM_1 formation, at low (16 μM) AFB concentrations. In parallel incubations, the specific CYP 3A4 inhibitor, troleandomycin, had little effect on AFB-epoxide production at low substrate concentrations but significantly reduced AFB-epoxide and AFQ_1 formation at high (128 μM) AFB concentrations (E. P. Gallagher, D. L. Eaton and K. Kunze, unpublished observations). Additional experiments with microsomes obtained from lymphoblastoid cell lines that had human cDNAs incorporated to express only one form of CYP were also used to evaluate the roles of CYP 1A2 and 3A4 in AFB-epoxide production. CYP 1A2-expressing cell lines produced large amounts of epoxide, with a ratio of AFB-epoxide to AFM_1 formation of about 3:1. Cell lines expressing only CYP 3A4 produced large amounts of AFQ_1 and lesser amounts of AFB-epoxide, with a ratio of AFQ_1:AFB-epoxide of about 12:1 (E. P. Gallagher, D. L. Eaton and K. Kunze, unpublished observations). These results further demonstrate the important role of CYP 1A2 in the formation of AFB-epoxide at AFB concentrations of relevance in the human diet.

A mitochondrial rat cytochrome P450 isoenzyme inducible by phenobarbital has been shown to be capable of activation of AFB_1 to DNA-binding metabolites (Shayiq and Avadhani, 1989). The toxicological significance of this activity remains unclear, however.

AFB-epoxide has been shown to exist in two stereomeric forms: the *exo*-epoxide, in which the epoxide ring is pointing "below" the plane and *cis* to the 5a and 9a protons, and the *endo*-epoxide, in which the epoxide ring is positioned "above" the plane and *trans* to the 5a and 9a protons (Raney *et al.*, 1992a). The *exo* form appears to be primarily responsible for AFB–DNA adduct formation, and the ratio of *endo* to *exo* epoxide formed may vary depending on the specific P450 involved (Raney *et al.*, 1992b). Different glutathione *S*-transferases also appear to possess different catalytic activities toward these two epoxides (Raney *et al.*, 1992b).

In summary, activation of AFB to AFB-8,9-epoxide appears to be mediated by several cytochromes P450, including CYP1A2 and CYP3A4. Substantial evidence indicates that CYP1A2 is relatively more active at low substrate concentrations, but with limited capacity, whereas at higher substrate concentrations CYP3A4 activity predominates. AFQ_1 appears to be formed predominantly, if not exclusively, by CYP3A family enzymes, whereas AFM_1 is formed by CYP1A enzymes (either or both 1A1 and 1A2) at low substrate concentrations, and also perhaps by CYP2A family enzymes. The role of phenobarbital-inducible forms of P450 (e.g., CYP2B) in oxidative metabolism of AFB remains unclear. Other CYP2 class enzymes (e.g., CYP2C8 and CYP2A1) also have been implicated in AFB oxidation (Forrester *et al.*, 1990), but their significance to overall metabolism is questionable. The human polymorphic CYP2D6 (debriso-quine hydroxylase activity) does appear not to be involved significantly in AFB metabolism (Forrester *et al.*, 1990).

REDUCTION

Reduction of the 1-keto group of AFB_1 forms the metabolite aflatoxicol (AFL) (Detroy and Hesseltine, 1970). This product has been found as a major AFB_1 metabolite in rat plasma (Wong and Hsieh, 1978). Postmitochondrial fractions of liver from rabbits, trout, and avian species were shown to have relatively high activity for AFL formation *in vitro* (Patterson and Roberts, 1971; Salhab and Edwards, 1977; Loveland *et. al.*, 1979). A cytosolic reductase that catalyzes the conversion of AFB_1 to AFL has been purified from chicken liver (Chen *et al.*, 1981).

Aflatoxicol formation does not appear to represent a significant detoxification mechanism (unless it is converted to a conjugated form; see subsequent discussion). AFL is equally potent to AFB_1 as a carcinogen (Schoenhard *et al.*, 1981) and has about 70% the mutagenicity (Coulombe *et al.*, 1982). AFL is oxidized readily back to AFB_1, a reaction that has been demonstrated using liver fractions from several species (Loveland *et al.*, 1977; Salhab and Edwards, 1977; Wong *et al.*, 1979). These observations led to the hypothesis that AFL may, in fact, serve as a "reservoir" for AFB_1 *in vivo*, prolonging the effective lifetime in the body (Wong and Hsieh, 1980) (see also Chapter 4).

AFL also can be oxidized at other carbon positions; 9a hydroxylation forms $AFL-M_1$ (Loveland *et al.*, 1984, 1988). $AFL-M_1$ apparently also can be formed by the reduction of AFM_1 (Loveland *et al.*, 1988). 1-Keto reduction of AFQ_1 produces aflatoxicol H_1 (Cole and Cox, 1981).

CONJUGATION

Glutathione Conjugation of Epoxide

The GSH conjugate of AFB_1-8,9-epoxide has been identified as the major metabolite in the bile of rats given AFB_1 (Degen and Neumann, 1978; Holeski *et al.*, 1987), as well as in *in vitro* incubations containing microsomes, cytosol, and GSH (Degen and Neumann, 1978; Moss and Neal, 1985; Monroe and Eaton,

1987; Neal et al., 1987). GSH conjugation is an important reaction in determining the susceptibility of different species to the toxic effects of AFB_1 (Degen and Neumann, 1981; O'Brien et al., 1983; Hayes et al., 1991b; Eaton and Ramsdell, 1992). This conjugation reaction is mediated by cytosolic glutathione S-transferase (Coles et al., 1985; Neal et al., 1987; Ramsdell and Eaton, 1990a). Activities vary widely among species (O'Brien et al., 1983; Monroe and Eaton, 1987). Although some apparently nonenzymatic conjugation with GSH can be observed when AFB_1-8,9-epoxide is generated in situ by liver microsomes, incubation of synthetic AFB-8,9-epoxide with GSH does not form any detectable quantity of AFB–GSH conjugate (H. S. Ramsdell, D. Slone, and D. L. Eaton, unpublished observations). These results demonstrate the critical catalytic role of glutathione S-transferases in the detoxification of AFB-8,9-epoxide.

A most striking species difference in GSH conjugation of AFB_1-8,9-epoxide has been observed between the mouse and the rat (O'Brien et al., 1983; Monroe and Eaton, 1987; Quinn et al., 1990). Although the mouse has very high microsomal AFB_1 epoxidation activity (Monroe and Eaton, 1987), the species is very resistant to AFB_1 induction of tumors. The very high level of glutathione S-transferase activity toward AFB_1-8,9-epoxide in the mouse appears to the basis of its resistance (O'Brien et al., 1983; Monroe and Eaton, 1987; Quinn et al., 1990; Hayes et al., 1991b).

The differences in cytosolic glutathione S-transferase activity toward AFB_1-8,9-epoxide in the mouse and the rat are not reflected by the activities with the standard glutathione S-transferase substrate 1-chloro-2,4-dinitrobenzene (CDNB). Mouse liver cytosolic glutathione S-transferase activity toward CDNB was 2.6 times that of the rat enzyme, but the difference with AFB_1-8,9-epoxide as the substrate was 52-fold (Monroe and Eaton, 1987). The basis for this observation has been investigated through studies of the glutathione S-transferase isoenzyme specificity for AFB_1-8,9-epoxide conjugation. Of the constitutively expressed forms, three rat isoenzymes—1-1 (YaYa), 1-2 (YaYc), and 2-2 (YcYc)—showed the highest AFB_1-8,9-epoxide-conjugating activity (Coles et al., 1985). A high pI glutathione S-transferase from mouse liver showed relatively high activity toward AFB_1-8,9-epoxide but low CDNB specific activity (Neal et al., 1987). A mouse isoenzyme fraction with similar properties has been purified and characterized (Ramsdell and Eaton, 1990a) and appears to belong to the alpha class of isoenzymes (Mannervik et al., 1985).

An alpha-class glutathione S-transferase with high specific activity toward AFB-8,9-epoxide has been cloned and sequenced from mouse tissue (Buetler and Eaton, 1992; Hayes et al., 1992). This form (called mYc) shares 85% amino acid and cDNA sequence homology with the rat Yc subunit (rat GST 2) (Buetler and Eaton, 1992; Hayes et al., 1992). When the cDNAs for both the rat and the mouse Yc forms were expressed in a bacterial expression system, the mouse Yc form had about 100-fold higher specific activity than the rat form (T. M. Buetler and D. L. Eaton, unpublished observations). Hayes et al. (1992) also demonstrated that their mouse Yc clone, which differed by only 4 nucleotides (1 in the coding region and 3 in the 3′ noncoding region) and no amino acids from the mouse Yc clone of Buetler and Eaton (1992), had high specific activity toward AFB-epoxide relative to other mouse alpha-class glutathione S-transferases. Hayes *et al.* (1991a) reported that a nonconstitutively expressed but ethoxyquin-inducible alpha-class glutathione S-transferase subunit is present in rats that appears to have high AFB-8,9-epoxide-conjugating activity. Based on partial amino acid sequence analysis, this form (called Yc2) is about 91% homologous with the mouse Yc form that has high specific activity toward AFB-8,9-epoxide. Although rat Yc2 normally is not expressed constitutively, it is expressed in relatively high levels in preneoplastic lesions induced by a variety of carcinogens (D. J. Harrison, G. E. Neal, and J. D. Hayes, unpublished observations; Z. Y. Chen and D. L. Eaton, unpublished observations).

Thus, glutathione S-transferase isoforms with high AFB detoxifying ability are likely to differ from other glutathione S-transferase isoforms with much lower activity by only one or a few amino acids. Expression of a chimeric construct of the rat Yc1 and the mouse Yc cDNAs (in which the first 56 residues are rat Yc1, the middle 85 residues are mouse Yc, and the remaining 89 residues are rat Yc1) yielded a protein with low AFB-epoxide-conjugating activity but significant CDNB activity, suggesting that one or more of the 15 nonconserved amino acids in this region may be important in conferring high AFB-epoxide conjugating activity (K. P. Van Ness, T. M. Buetler, and D. L. Eaton, unpublished observations). Site-directed mutagenesis studies are currently underway to identify the specific molecular basis for the large differences in glutathione S-transferase activity toward AFB-8,9-epoxide between different isoforms.

As noted previously, evidence indicates that the various glutathione S-transferases have differential activity toward the *endo* and *exo* forms of AFB-8,9-epoxide. Rat alpha-class glutathione S-transferases, particularly GST 1-1, had relatively high activity toward the *exo* epoxide but no detectable activity toward the *endo* form (Raney *et al.*, 1992b). Conversely, rat glutathione S-transferase mu forms, notably GST 4-4, had very high activity toward the *endo* epoxide and approximately equal activity to GST 1-1 toward the *exo* form (Raney *et al.*, 1992b). Purified human alpha-class glutathione S-transferases (GST Ha 1-1 and Ha 2-2) had very low activity (20- to 100-fold less than rat GST 1-1) toward the *exo* epoxide and no measurable activity toward the *endo* form (Raney *et al.*, 1992b). However, purified human glutathione S-transferase mu form M1a-1a showed considerable activity toward both the *exo* and the *endo* epoxide (Raney *et al.*, 1992b). Since this particular enzyme is polymorphically distributed in the human population (about 50% of Caucasians possess a gene deletion for this particular enzyme), further demonstration of the importance of this and other specific human glutathione S-transferases in detoxification of AFB-*exo*-8,9-epoxide will be of great importance.

Glucuronidation of Hydroxylated Derivatives

Some of the metabolites just described appear in the bile as glucuronide conjugates. Glucuronides of AFL and AFL-M_1 are the principal biliary metabolites of AFB_1 in trout (Loveland *et al.*, 1984). Thus, enterohepatic circulation of AFL could prolong the effective half-life of AFB_1. The glucuronide conjugate of AFP_1 is a major biliary metabolite of AFB_1 in the rat (Holeski *et al.*, 1987; Eaton *et al.*, 1988). 4,9a-Dihydroxyaflatoxin B_1 also is found in the bile as a glucuronide conjugate after treatment of rats with AFB_1 (Eaton *et al.*, 1988). The AFP_1–glucuronide conjugate also is identified as the only significant glucuronide or sulfate conjugate of hydroxylated AFB metabolites produced in isolated hepatocytes (Ch'ih *et al.*, 1983). However, enzymatic hydrolysis of nonaromatic glucuronides and/or sulfates (such as AFM_1 and AFQ_1 conjugates) with bacterial β-glucuronidase/sulfatase preparations is inefficient and may underestimate the actual amount of conjugate present (Metcalfe and Neal, 1983).

Collectively, all the data suggest that the phenolic hydroxyl group present in AFP_1 serves as a much better site for glucuronide conjugation than do either of the aliphatic hydroxyl groups present in AFM_1 and AFQ_1. An exception is aflatoxicol, which also may be conjugated in species in which it is a significant metabolite. Although the rate of conjugation of AFM_1 and AFQ_1 is probably much slower than that for AFP_1, as demonstrated in isolated hepatocytes (Metcalfe and Neal, 1983), the potential significance of conjugation of AFM_1 and AFQ_1 to enhance the rate of elimination remains uncertain.

INTEGRATION OF AFLATOXIN
BIOTRANSFORMATION PATHWAYS

Competing Pathways: Activation versus Inactivation

As shown schematically in Figure 4, the fate of AFB_1 is dependent on the relative activity of several biotransformation pathways, in addition to other factors such as DNA repair rates. The amount of the mycotoxin that is going to exert carcinogenic or toxic effects will depend on the amount converted to various metabolites as well as on the biological activity of those metabolites. With respect to carcinogenicity, AFB_1-8,9-epoxide is the key active metabolite. As indicated in Figure 4, hydroxylated metabolites of AFB_1 (AFM_1, AFP_1, AFQ_1) are assumed to represent detoxification products. Detoxification of the reactive epoxide also may occur through conjugation with GSH. Hydrolysis of AFB_1-8,9-epoxide forms a dihydrodiol that probably still is capable of causing toxic effects (via binding to protein) but presumably is a less potent carcinogenic species than the epoxide.

Activation and inactivation ratios are a convenient means of comparing the relative activities of AFB_1 biotransformation pathways in different species (Degen and Neumann, 1981). Activation is defined by Degen and Neumann as the amount of AFB_1-8,9-epoxide formed divided by the total formation of oxidative metabolites (AFM_1 + AFP_1 + AFQ_1). Inactivation is the amount of epoxide conjugated with GSH divided by the amount of epoxide formed. (The amount of

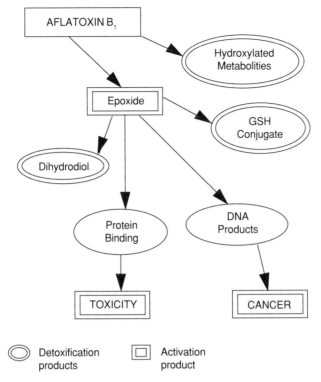

FIGURE 4 Schematic representation of the role of various biotransformation pathways in the disposition, toxicity, and carcinogenicity of aflatoxin B_1. [a]Activation product. [b]Detoxification product.

TABLE 1 Activation and Inactivation of Aflatoxin B_1 in Vitro

Species	Activation[a]	Inactivation[b]	Activation/inactivation
Rat	0.42	0.068	6.2
Mouse	0.39	2.9	0.13
Monkey	0.09	0.033	2.8
Human	0.14	0.008[c]	18

[a]Activation = (rate of AFB_1-8,9-epoxide formation)/Σ(rates of AFB_1-8,9-epoxide, AFQ_1, AFM_1, and AFP_1 formation). Calculated from data in Ramsdell and Eaton (1990b). $[AFB_1]$ = 124 µM.

[b]Inactivation = (rate of cytosolic AFB_1-8,9-epoxide conjugation with GSH)/(rate of microsomal AFB_1-8,9-epoxide formation.Epoxidation data as above; conjugation data: H. S. Ramsdell, D. H. Monroe, and D. L. Eaton, unpublished observations.

[c]Calculated from the detection limit of 2 pmol/mg protein/min that represents the upper bound of human liver cytosol/glutathione S-transferase activity toward AFB_1-8,9-epoxide.

epoxide detoxified by hydrolysis is assumed to be relatively small and equivalent for different species.) Equivalent results are obtained if rates of formation are used for these calculations. An index of species susceptibility can be calculated as activation/inactivation. This ratio was calculated for rat, mouse, monkey, and human based on data obtained *in vitro* using hepatic microsomes and cytosol (Table 1).

Such quantitative comparisons of AFB_1 biotransformation pathway activities should facilitate interspecies extrapolations with respect to susceptibility to hepatocarcinogenesis. This comparison is of particular interest for the assessment of the risk of human exposure to aflatoxin.

Kinetic Considerations

For the preceding discussion to be relevant to environmental aflatoxin exposures, comparisons must consider the kinetics of the pathways involved. This consideration is critical because the concentrations of AFB_1 that are likely to be present in the liver after ingestion of food contaminated with the mycotoxin are much lower than the concentrations typically used in *in vitro* biotransformation assays. Extrapolations of relative activities to low concentrations will not be legitimate unless all the enzymes involved have parallel variation in their substrate affinities. This situation is clearly unlikely, especially for comparisons across species. Indeed, substantial variations in the patterns of liver microsomal AFB_1 oxidation over a range of substrate concentrations has been demonstrated for four species: rat, mouse, monkey, and human (Ramsdell and Eaton, 1990b). In the case of rat and human microsomes, the proportion of AFB_1 converted to the epoxide increases with decreasing AFB_1 concentration. In other words, the activation ratio is greater at lower substrate concentrations, suggesting that humans may be relatively susceptible to AFB_1-induced hepatocarcinogenesis.

As with microsomal oxidation, affinity of glutathione *S*-transferases for AFB_1-8,9-epoxide is likely to vary among species. This question is more difficult than microsomal oxidation to approach experimentally. Use of the *in situ* epoxide generation system (Monroe and Eaton, 1987) does not allow determination of the concentration of AFB_1-8,9-epoxide during the enzymatic conjugation reaction. The use of synthetic AFB_1-8,9-epoxide, although allowing calculation of its concentration in the reaction mixture at the instant of addition, is of dubious validity because of the very rapid hydrolysis of the compound in aqueous environments.

FACTORS AFFECTING AFLATOXIN BIOTRANSFORMATION

In addition to the notable differences in biotransformation of aflatoxin B_1 among different species, modulation of biotransformation—and thus toxicity and carcinogenicity—can result from a variety of dietary and drug treatments.

Dietary Factors

Nutrient and nonnutrient compounds in the diet have been found to affect the toxicity of AFB_1, although these effects are not necessarily related to modulation of biotransformation. A level of protein in the diet in excess of the minimal level required for normal growth increased the number of preneoplastic foci that developed in rat liver (Dunaif and Campbell, 1987). A similar relationship of increased incidence of hepatocellular carcinoma at higher dietary protein levels was observed in trout treated with AFB_1 (Bailey *et al.*, 1982). In the rat, a low level of dietary protein during AFB_1 treatment was associated with more severe acute toxic effects (bile duct proliferation) but was protective against preneoplastic foci development when given after AFB_1 dosing (Appleton and Campbell, 1983). In monkeys fed either a low-protein (5%˙casein) or high-protein (20% casein) diet, 0.16 ppm AFB_1 induced preneoplastic lesions in liver in only the low protein animals (Mathur and Nayak, 1989). However, at a higher dietary dose of AFB (1 ppm), monkeys on the low-protein diet had large areas of liver cell necrosis but no neoplastic nodules, whereas monkeys on the high-protein diet had neoplastic nodules. Thus, a consistent pattern emerges: low-protein diet enhances liver cell toxicity and necrosis but does not enhance (and, in fact, may protect against) the development of hepatocellular carcinoma.

The effect of dietary protein on tumor development is not likely to be the result of modulation of biotransformation, but such a mechanism may be involved in alterations of acute toxicity. In general, restriction of food intake in rats approximately halved the *in vivo* binding of AFB_1 to hepatic DNA, but a more modest reduction ($\approx 10\%$) was observed *in vitro* (Pegram *et al.*, 1989).

Vitamin deficiencies can alter aflatoxin biotransformation. Liver fractions from rats deficient in vitamin A formed a higher level of DNA adducts with AFB_1 *in vitro* (Bhattacharya *et al.*, 1989). Microsomal activation appeared to be increased and cytosolic conjugation diminished in the vitamin A-deficient animals. Riboflavin deficiency in rats reduced the *in vitro* activation of AFB_1, probably because of diminished levels of NADPH–cytochrome P450 reductase (Prabhu *et al.*, 1989).

Nonnutritive compounds in a number of foods of plant origin modulate AFB_1 biotransformation, binding to DNA, and carcinogenesis. The cruciferous vegetables have been studied the most widely; broccoli (Ramsdell and Eaton, 1988), brussels sprouts (Godlewski *et al.*, 1985; Salbe and Bjeldanes, 1989), and cabbage (Boyd and Stoewsand, 1981; Boyd *et al.*, 1982; Whitty and Bjeldanes, 1987) have been found to be protective in rats *in vivo,* as well as to modulate biotransformation enzyme activities measured *in vitro.* For broccoli, conjugation of AFB_1-8,9-epoxide was increased whereas epoxidation of AFB_1 was not affected significantly, indicating that increased detoxification of the reactive epoxide was the most important effect of the vegetable (Ramsdell and Eaton, 1988). Rats maintained on an AIN76A purified diet developed multiple large preneoplastic foci when given 10 daily doses of 150 μg AFB_1/kg/day and partial hepatectomy, whereas the same treatment regimen in rats maintained on a standard rodent chow diet had no visible preneoplastic nodules and only very few small altered foci were evident (Z. Y. Chen and D. L. Eaton, unpublished observations). These results demonstrate the important role that nonnutritive dietary components can play in the development of hepatotoxicity and hepatocarcinogenicity from aflatoxin B_1.

Plants of the *Allium* genus have been studied extensively for their anticarcinogenic effects. Some work has shown that diallyl sulfide and ajoene, organosulfur compounds found in garlic, reduce binding of AFB_1 to DNA *in vitro* and modulate AFB_1 biotransformation (Tadi *et al.*, 1991). Capsaicin, the active principle of hot chili peppers, also inhibits bioactivation of AFB_1 and DNA binding *in vitro* (Teel, 1991). (For further discussion, see Chapter 10.)

Drug Treatments

Treatment of rats with various compounds has been found to reduce the toxic and carcinogenic effects of aflatoxins. Synthetic antioxidants have been used in numerous studies to modulate the toxicity of aflatoxins. Both butylated hydroxyanisole (BHA; Williams *et al.*, 1986) and ethoxyquin (Cabral and Neal, 1983) were shown to reduce the carcinogenic response to AFB_1 in rats. The covalent binding of AFB_1 to DNA *in vivo* likewise is reduced by pretreatment with these two compounds (Lotlikar *et al.*, 1984a; Kensler *et al.*, 1986; Monroe and Eaton, 1987). Altered biotransformation of AFB_1 is likely to be the mechanism underlying these observations. Pretreatment of rats with BHA increased both the level of GSH conjugation of AFB_1-8,9-epoxide *in vitro* (Monroe and Eaton, 1987) and the amount of conjugate excreted in the bile (Monroe *et al.*, 1986). Similar results have been observed with ethoxyquin (Kensler *et al.*, 1986; Mandel *et al.*, 1987).

Certain dithiolthione compounds are effective in inducing glutathione *S*-transferase and increasing levels of hepatic glutathione (Ansher *et al.*, 1983, 1986; Davies *et al.*, 1987), making them likely candidates for chemoprotection against AFB_1. Indeed, the drug oltipraz has good activity against AFB_1-related hepatotoxicity (Liu *et al.*, 1988) and DNA binding, probably through enhanced detoxification of AFB_1-8,9-epoxide (Kensler *et al.*, 1985, 1987; Roebuck *et al.*, 1991). The presence of dithiolthiones in cabbage frequently has been cited as a potential mechanism for the protective effect of that vegetable against hepatotoxicants including AFB_1 (Ansher *et al.*, 1986; Davies *et al.*, 1987; Kensler *et al.*, 1987), but a re-examination of the constituents of cabbage failed to detect the presence of dithiolthione (Marks *et al.*, 1991). However, relatively low dietary doses (e.g., 0.001%) of 1,2-dithiol-3-thione itself can reduce the size and extent of preneoplastic lesions induced by AFB_1 dramatically (Kensler *et al.*, 1992). The mechanism for this effect is probably induction of glutathione *S*-transferase(s) (Kensler *et al.*, 1992), although other mechanisms such as inhibition of P450-

mediated activation (Putt *et al.,* 1991) also may contribute to this protective effect.

Developing chemoprotective treatments to reduce the incidence of hepatocellular carcinoma in human populations exposed to high levels of dietary aflatoxins would be of interest. BHA is unlikely to be adopted for such use because of its potential toxicity with chronic administration. Oltipraz is a more promising candidate, used therapeutically as an antischistosomal agent. Phase I clinical trials are currently in progress to test its effectiveness against aflatoxin carcinogenesis in humans. (For further discussion, see Chapter 13.)

SUMMARY

The biotransformation of aflatoxins is linked intimately with their toxic and carcinogenic effects. The comparative toxicology that is needed for accurate prediction of the effects of aflatoxin requires a better understanding of the quantitative aspects of its biotransformation. The kinetics of the inactivation as well as activation pathways must be investigated. This information also will facilitate prediction of the best approaches to manipulation of aflatoxin toxicity via modulation of biotransformation.

REFERENCES

Ansher, S. S., Dolan, P., and Bueding, E. (1983). Chemoprotective effects of two dithiolthiones and of butylhydroxyanisole against carbon tetrachloride and acetaminophen toxicity. *Hepatology* **3,** 932–935.

Ansher, S. S., Dolan, P., and Bueding, E. (1986). Biochemical effects of dithiolthiones. *Food Chem. Toxicol.* **24,** 405–415.

Aoyama, T., Yamano, S., Guzelian, P. S., Gelboin, H. V., and Gonzalez, F. J. (1990). Five of 12 forms of vaccinia virus-expressed human hepatic cytochrome P450 metabolically activate aflatoxin B_1. *Proc. Natl. Acad. Sci. U.S.A.* **87,** 4790–4793.

Appleton, B. S., and Campbell, T. C. (1983). Effect of high and low dietary protein on the dosing and postdosing periods of aflatoxin B_1-induced preneoplastic lesion development in the rat. *Cancer Res.* **43,** 2150–2154.

Appleton, B. S., Goetchius, M. P., and Campbell, T. C. (1982). Linear dose–response curve for the hepatic macromolecular binding of aflatoxin B_1 in rats at very low exposures. *Cancer Res.* **42,** 3659–3662.

Baertschi, S. W., Roney, K. D., Stone, M. P., and Harris, T. M. (1988). Preparation of the 8,9-epoxide of the mycotoxin aflatoxin B_1: The ultimate carcinogenic species. *J. Am. Chem. Soc.* **110,** 7929–7931.

Baertschi, S. W., Raney, K. D., Shimada, T., Harris, T., and Guengerich, F. P. (1989). Comparison of rates of enzymatic oxidation of aflatoxin B_1, aflatoxin G_1, and sterigmatocystin and activities of the epoxides in forming guanyl-N^7 adducts and inducing different genetic responses. *Chem. Res. Toxicol.* **2,** 114–122.

Bailey, G., Taylor, M., Selivonchick, D., Eisele, T., Hendricks, J., Nixon, J., Pawlowski, N., and Sinnhuber, R. (1982). Mechanisms of dietary modification of aflatoxin B_1 carcinogenesis. *Basic Life Sci.* **21,** 149–165.

Battista, J. R., and Marnett, L. J. (1985). Prostaglandin H synthase-dependent epoxidation of aflatoxin B₁. *Carcinogenesis,* **6,** 1227–1229.

Bauer, D. H., Lee, D. J., and Sinnhuber, R. O. (1969). Acute toxicity of aflatoxins B₁ and G₁ in the rainbow trout (*Salmo gairdneri*). *Toxicol. Appl. Pharmacol.* **15,** 415–419.

Bhattacharya, R. K., Prabhu, A. L., and Aboobaker, V. S. (1989). *In vivo* effect of dietary factors on the molecular action of aflatoxin B₁: Role of vitamin A on the catalytic activity of liver fractions. *Cancer Lett.* **44,** 83–88.

Boyd, J. N., and Stoewsand, G. S. (1981). Blood α-fetoprotein changes in rats fed aflatoxin B₁ and various levels cabbage. *J. Food Sci.* **46,** 1923–1926.

Boyd, J. N., Babish, J. G., and Stoewsand, G. S. (1982). Modification by beet and cabbage diets of aflatoxin B₁-induced rat plasma α-foetoprotein elevation, hepatic tumorigenesis, and mutagenicity of urine. *Food Chem. Toxicol.* **20,** 47–52.

Buchi, G. H., Muller, P. M., Roebuck, B. D. and Wogan, G. N. (1974). Aflatoxin Q₁: A major metabolite of aflatoxin B₁ produced by human liver. *Res. Commun. Chem. Pathol. Pharmacol.* **8,** 585–592.

Buetler, T. M., and Eaton, D. L. (1992). Complementary DNA cloning, messenger RNA expression, and induction of α-class glutathione *S*-transferases in mouse tissues. *Cancer Res.* **52,** 314–318.

Butler, W. H., Greenblatt, M., and Lijinsky, W. (1969). Carcinogenesis in rats by aflatoxins B₁, G₁, and B₂. *Cancer Res.* **29,** 2206–2211.

Cabral, J. R. P., and Neal, G. E. (1983). The inhibitory effects of ethoxyquin on the carcinogenic action of aflatoxin B₁ in rats. *Cancer Lett.* **19,** 126–132.

Chen, S. C. G., Wei, R. D., and Hsieh, D. P. H. (1981). Purification and some properties of chicken-liver aflatoxin B₁ reductase. *Food Cosmet. Toxicol.* **19,** 19–24.

Ch'ih, J. J., Lin, T., and Devlin, T. M. (1983). Effect of inhibitors of microsomal enzymes on aflatoxin B₁-induced cytotoxicity and inhibition of RNA synthesis in isolated rat hepatocytes. *Biochem. Biophys. Res. Commun.* **115,** 15–21.

Cole, R. J., and Cox, R. H. (1981). "Handbook of Toxic Fungal Metabolites." Academic Press, New York.

Coles, B., Meyer, D. J., Ketterer, B., Stanton, C. A., and Garner, R. C. (1985). Studies on the detoxication of microsomally-activated aflatoxin B₁ by glutathione and glutathione transferases *in vitro. Carcinogenesis* **6,** 693–697.

Coulombe, R. A., Shelton, D. W., Sinnhuber, R. O., and Nixon, J. E. (1982). Comparative mutagenicity of aflatoxins using a *Salmonella*/trout hepatic enzyme activation system. *Carcinogenesis* **3,** 1261–1264.

Crespi, C. L., Penman, B. W., Leakey, J. A. E., Arlotto, M. P., Stark, A., Parkinson, A., Turner, T., Steimel, D. T., Rudo, K., Davies, R. L., and Langenbach, R. (1990). Human cytochrome P450IIA3: cDNA sequence, role of the enzyme in the metabolic activation of promutagens, comparison to nitrosamine activation by human cytochrome P450IIE1. *Carcinogenesis* **11,** 1293–1300.

Crespi, C. L., Penman, B. W., Steimel, D. T., Gelboin, H. V., and Gonzalez, F. J. (1991). The development of a human cell line stably expressing human CYP3A4: Role in the metabolic activation of aflatoxin B₁ and comparison to CYP1A2 and CYP2A3. *Carcinogenesis* **12,** 355–359.

Dalezios, J., Wogan, G. N., and Weinreb, S. M. (1971). Aflatoxin P₁: A new aflatoxin metabolite in monkeys. *Science* **171,** 584–585.

Dann, R. E., Mitscher, L. A., and Couri, D. (1972). The *in vivo* metabolism of ¹⁴C-labeled aflatoxins B₁, B₂, G₁ in rats. *Res. Commun. Chem. Pathol. Pharmacol.* **3,** 667–675.

Davies, M. H., Blacker, A. M., and Schnell, R. C. (1987). Dithiolthione-induced alterations in hepatic glutathione and related enzymes in male mice. *Biochem. Pharmacol.* **36,** 568–570.

Decad, G. M., Dougherty, K. K., Hsieh, D. P. H., and Byard, J. L. (1979). Metabolism of aflatoxin B₁ in cultured mouse hepatocytes: Comparison with rat and effects of cyclohexene oxide and diethylmaleate. *Toxicol. Appl. Pharmacol.* **50,** 429–436.

Degen, G. H., and Neumann, H. G. (1978). The major metabolite of aflatoxin B_1 in the rat is a glutathione conjugate. *Chem. Biol. Interact.* **22,** 239–255.

Degen, G. H., and Neumann, H.-G. (1981). Differences in aflatoxin B_1-susceptibility of rat and mouse are correlated with the capability *in vitro* to inactivate aflatoxin B_1-epoxide. *Carcinogenesis* **2,** 299–306.

Detroy, R. W., and Hesseltine, C. W. (1970). Aflatoxicol: Structure of a new transformation product of aflatoxin B_1. *Can. J. Biochem.* **48,** 830–832.

Dunaif, G. E., and Campbell, T. C. (1987). Dietary protein level and aflatoxin B_1-induced preneoplastic hepatic lesions in the rat. *J. Nutr.* **117,** 1298–1302.

Eaton, D. L., and Ramsdell, H. S. (1992). Species and diet related differences in aflatoxin biotransformation. *In* "Handbook of Applied Mycology: Mycotoxins in Ecological Systems" (D. Bhatnagar, E. B. Lillehoj, and D. K. Arora, eds.), pp. 157–182. Marcel Dekker, New York.

Eaton, D. L., Monroe, D. H., Bellamy, G., and Kalman, D. A. (1988). Identification of a novel dihydroxy metabolite of aflatoxin B_1 produced *in vitro* and *in vivo* in rats and mice. *Chem. Res. Toxicol.* **1,** 108–114.

Essigmann, J. M., Croy, R. G., Nadzan, A. M., Busby, W. F., Reinhold, V. N., Buchi, G., and Wogan, G. N. (1977). Structural identification of the major DNA adduct formed by aflatoxin B_1 *in vitro*. *Proc. Natl. Acad. Sci. U.S.A.* **74,** 1870–1874.

Essigmann, J. M., Green, C. L., Croy, R. G., Fowler, K. W., Buchi, G. H., and Wogan, G. N. (1983). Interactions of aflatoxin B_1 and alkylating agents with DNA: Structural and functional studies. *Cold Spring Harbor Symp. Quant. Biol.* **47,** 327–337.

Faletto, M. B., Koser, P. L., Battula, N., Townsend, G. K., Maccubbin, A. E., Gelboin, H. V., and Gurtoo, H. L. (1988). Cytochrome P_3-450 cDNA encodes aflatoxin B_1-4-hydroxylase. *J. Biol. Chem.* **263,** 12187–12189.

Forrester, L. M., Neal, G. E., Judah, F. J., Glancey, M. J., and Wolf, C. R. (1990). Evidence for involvement of multiple forms of cytochrome P-450 in aflatoxin B_1 metabolism in human liver. *Proc. Natl. Acad. Sci. U.S.A.* **87,** 8306–8310.

Fukuhara, M., Nohmi, T., Mizokami, K., Sunouchi, M., Ishidate, M., and Takanaka, A. (1989). Characterization of three forms of cytochrome P-450 inducible by 3-methylcholanthrene in golden hamster livers with special reference to aflatoxin B_1 activation. *J. Biochem.* **106,** 253–258.

Fukuhara, M., Mizokami, K., Sakaguchi, M., Niimura, Y., Kato, K., Inouye, S. and Takanaka, A. (1990). Aflatoxin B_1-specific cytochrome P-450 isozyme (P-450-AFB) inducible by 3-methylcholanthrene in golden hamsters. *Biochem. Pharmacol.* **39,** 463–469.

Garner, R. C., Miller, E. C., and Miller, J. A. (1972). Liver microsomal metabolism of aflatoxin B_1 to a reactive derivative toxic to *Salmonella typhimurium* TA 1530. *Cancer Res.* **32,** 2058-.

Garner, R. C., Martin, C. N., Smith, J. R. L., Coles, B. F., and Tolson, M. R. (1979). Comparison of aflatoxin B_1 and aflatoxin G_1 binding to cellular macromolecules *in vitro*, *in vivo*, and after peracud oxidation: Characterisation of the major nucleic acid adducts. *Chem. Biol. Interact.* **26,** 57–73.

Godlewski, C. F., Boyd, J. N., Sherman, W. K., Anderson, J. L., and Stoewsand, G. S. (1985). Hepatic glutathione *S*-transferase activity and aflatoxin B_1-induced enzyme altered foci in rats fed fractions of Brussels sprouts. *Cancer Lett.* **28,** 151–157.

Groopman, J. D., Fowler, K. W., Busby, W. F., and Wogan, G. N. (1981). Interaction of aflatoxin B_2 with rat liver DNA and histones *in vivo*. *Carcinogenesis* **2,** 1371–1373.

Gurtoo, H. L., Dahms, R. P., and Paigen, B. (1978). Metabolic activation of aflatoxins related to their mutagenicity. *Biochem. Biophys. Res. Commun.* **81,** 965–972.

Halvorson, M., Safe, S. H., Parkinson, A., and Phillips, T. D. (1988). Aflatoxin B_1 hydroxylation by the pregnenolone-16α-carbonitrile-induced form of rat liver microsomal cytochrome P-450. *Carcinogenesis* **9,** 2103–2108.

Hayes, J. D., Judah, D. J., McLellan, L. I., Kerr, L. A., Peacock, S. D., and Neal, G. E. (1991a). Ethoxyquin-induced resistance to aflatoxin B_1 in the rat is associated with the expression of a novel alpha-class glutathione *S*-transferase subunit, Yc_2, which possesses high catalytic activity for aflatoxin B_1-8,9-epoxide. *Biochem. J.* **279,** 385–398.

Hayes, J. D., Judah, D. J., McLellan, L. I., and Neal, G. E. (1991b). Contribution of the glutathione *S*-transferases to the mechanisms of resistance to aflatoxin B_1. *Pharmacol. Ther.* **50**, 443–472.

Hayes, J. D., Judah, D. J., and Neal, G. (1992). Molecular cloning and heterologous expression of a cDNA encoding a mouse glutathione *S*-transferase Yc subunit possessing high catalytic activity for aflatoxin B_1-8,9-epoxide. *Biochem. J.* **285**, 173–180.

Hendricks, J. D., Sinnhuber, R. O., Nixon, J. E., Wales, J. H., Masri, M. S., and Hsieh, D. P. H. (1980). Carcinogenic response of rainbow trout (*Salmo gairdneri*) to aflatoxin Q_1 and synergistic effect of cyclopropenoid fatty acids. *J. Natl. Cancer Inst.* **64**, 523–527.

Holeski, C. J., Eaton, D. L., Monroe, D. H., and Bellamy, G. M. (1987). Effects of phenobarbital on the biliary excretion of aflatoxin P_1-glucuronide and aflatoxin B_1-*S*-glutathione in the rat. *Xenobiotica* **17**, 139–153.

Holzapfel, C. W., Steyn, P. S., and Purchase, I. F. H. (1966). Isolation and structure of aflatoxins M_1 and M_2. *Tet. Lett.* **25**, 2799–2803.

Hsieh, D. P. H., Salhab, A. S., Wong, J. J., and Yang, S. L. (1974). Toxicity of aflatoxin Q_1 as evaluated with the chicken embryo and bacterial auxotrophs. *Toxicol. Appl. Pharmacol.* **30**, 237–242.

Hsieh, D. P. H., Cullen, J. M., and Ruebner, B. H. (1984). Comparative hepatocarcinogenicity of aflatoxins B_1 and M_1 in the rat. *Food Chem. Toxicol.* **22**, 1027–1028.

Israel-Kalinsky, H., Malca-Mor, L., and Stark, A. A. (1984). Comparative aflatoxin B_1 mutagenesis of *Salmonella typhimurim* TA 100 in metabolic and photoactivation systems. *Cancer Res.* **44**, 1831–1839.

Kensler, T. W., Egner, P. A., Trush, M. A., Bueding, E., and Groopman, J. D. (1985). Modification of aflatoxin B_1 binding to DNA *in vivo* in rats fed phenolic antioxidants, ethoxyquin and a dithiothione. *Carcinogenesis* **6**, 759–763.

Kensler, T. W., Egner, P. A., Davidson, N. E., Roebuck, B. D., Pikul, A., and Groopman, J. D. (1986). Modulation of aflatoxin metabolism, aflatoxin-N^7-guanine formation, and hepatic tumorigenesis in rats fed ethoxyquin: Role of induction of glutathione-*S*-transferases. *Cancer Res.* **46**, 3924–3931.

Kensler, T. W., Egner, P. A., Dolan, P. M., Groopman, J. D., and Roebuck, B. D. (1987). Mechanism of protection against aflatoxin tumorigenicity in rats fed 5-(2-pyrazinyl)-4-methyl-1,2-dithiol-3-thione (oltipraz) and related 1,2-dithiol-3-thiones and 1,2-dithiol-3-ones. *Cancer Res.* **47**, 4271–4277.

Kensler, T. W., Groopman, J. D., Eaton, D. L., Curphey, T. J., and Roebuck, B. D. (1992). Potent inhibition of aflatoxin-induced hepatic tumorigenesis by the monofunctional enzyme inducer 1,2-dithiol-3-thione. *Carcinogenesis* **13**, 95–100.

Koser, P. L., Faletto, M. B., Maccubbin, A. E., and Gurtoo, H. L. (1988). The genetics of aflatoxin B_1 metabolism. *J. Biol. Chem.* **263**, 12584–12595.

Lai, T. S., and Chiang, J. Y. L. (1990). Aflatoxin B_1 metabolism by 3-methylcholanthrene-induced hamster hepatic cytochrome P-450s. *J. Biochem. Toxicol.* **5**, 147–153.

Larsson, P., Larsson, B. S., and Tjalve, H. (1988). Binding of aflatoxin B_1 to melanin. *Food Chem. Toxicol.* **26**, 579–586.

Lin, J.-K., Kennan, K. A., Miller, E. C., and Miller, J. A. (1978). Reduced nicotinamide adenine dinucleotide phosphate-dependent formation of 2,3-dihydro-2,3-dihydroxyaflatoxin B_1 from aflatoxin B_1 by hepatic microsomes. *Cancer Res.* **38**, 2424–2428.

Liu, L., and Massey, T. E. (1992). Bioactivation of aflatoxin B_1 by lipoxygenases, prostaglandin H synthase and cytochrome P450 monooxygenase in guinea pig tissues. *Carcinogenesis* **13**, 533–539.

Liu, L., Daniels, J. M., Stewart, R. K., and Massey, T. E. (1990). *In vitro* prostaglandin H synthase- and monooxygenase-mediated binding of aflatoxin B_1 to DNA in guinea pig tissue microsomes. *Carcinogenesis* **11**, 1915–1919.

Liu, Y., L., Roebuck, B. D., Yager, J. D., Groopman, J. D., and Kensler, T. W. (1988). Protection by 5-(2-pyrazinyl)-4-methyl-1,2-dithiol-3-thione (oltipraz) against the hepatotoxicity of aflatoxin B_1 in the rat. *Toxicol. Appl. Pharmacol.* **93**, 442–451.

Lotlikar, P. D., Clearfield, M. S., and Jhee, E. C. (1984a). Effect of butylated hydroxyanisole on *in vivo* and *in vitro* hepatic aflatoxin B_1-DNA binding in rats. *Cancer Lett.* **24**, 241–250.

Lotlikar, P. D., Jhee, E. C., Insetta, S. M., and Clearfield, M. S. (1984b). Modulation of microsome-mediated aflatoxin B_1 binding to exogenous and endogenous DNA by cytosolic glutathione S-transferases in rat and hamster livers. *Carcinogenesis* **5**, 269–276.

Loveland, P. M., Sinnhuber, R. O., Berggren, K. E., Libbey, L. M., Nixon, J. E., and Pawlowski, N. E. (1977). Formation of aflatoxin B_1 from aflatoxicol by rainbow trout (*Salmo gairdneri*) liver *in vitro*. *Res. Commun. Chem. Pathol. Pharmacol.* **16**, 167–170.

Loveland, P. M., Nixon, J. E., Pawlowski, N. E., Eisele, T. A., Libbey, L. M., and Sinnhuber, R. O. (1979). Aflatoxin B_1 and aflatoxicol metabolism in rainbow trout (*Salmo gairdneri*) and the effects of dietary cyclopropene. *J. Environ. Pathol. Toxicol.* **2**, 707–718.

Loveland, P. M., Nixon, J. E., and Bailey, G. S. (1984). Glucuronides in bile of rainbow trout (*Salmo gairdneri*) injected with [³H]aflatoxin B_1 and the effects of dietary β-naphthoflavone. *Comp. Biochem. Physiol.* **78C**, 13–19.

Loveland, P. M., Wilcox, J. S., Hendricks, J. D., and Bailey, G. S. (1988). Comparative metabolism and DNA binding of aflatoxin B_1, aflatoxin M_1, aflatoxicol and aflatoxicol-M_1 in hepatocytes from rainbow trout (*Salmo gairdneri*). *Carcinogenesis* **9**, 441–446.

Mandel, H. G., Manson, M. M., Judah, D. J., Simpson, J. L., Green, J. A., Forrester, L. M., Wolf, C. R., and Neal, G. E. (1987). Metabolic basis for the protective effect of the antioxidant ethoxyquin on aflatoxin B_1 hepatocarcinogenesis in the rat. *Cancer Res.* **47**, 5218–5223.

Mannervik, B., Alin, P., Guthenberg, C., Jensson, H., Tahir, M. K., Warholm, M., and Jornvall, H. (1985). Identification of three classes of cytosolic glutathione transferase common to several mammalian species: Correlation between structural data and enzymatic properties. *Proc. Natl. Acad. Sci. U.S.A.* **82**, 7202–7206.

Marks, H. S., Leichtweis, H. C., and Stoewsand, G. S. (1991). Analysis of a reported organosulfur carcinogenesis inhibitor 1,2-dithiol-3-thione in cabbage. *J. Agric. Food Chem.* **39**, 893–895.

Masri, M. S., Lundin, R. E., Page, J. R., and Garcia, V. C. (1967). Crystalline aflatoxin M_1 from urine and milk. *Nature (London)* **215**, 753–755.

Masri, M. S., Booth, A. N., and Hsieh, D. P. H. (1974). Comparative metabolic conversion of aflatoxin B_1 to M_1 and Q_1 by monkey, rat, and chicken liver. *Life Sci.* **15**, 203–212.

Mathur, M., and Nayak, N. (1989). Effect of low protein diet on low dose chronic aflatoxin B1 induced hepatic injury in Rhesus monkeys. *J. Toxicol. Toxin Rev.* **8**, 265–274.

Metcalfe, S. A., and Neal, G. E. (1983). The metabolism of aflatoxin B_1 by hepatocytes isolated from rats following the *in vivo* administration of some xenobiotics. *Carcinogenesis* **4**, 1007–1012.

Metcalfe, S. A., Colley, P. J., and Neal, G. E. (1981). A comparison of the effects of pretreatment with phenobarbitone and 3-methylcholanthrene on the metabolism of aflatoxin B_1 by rat liver microsomes and isolated hepatocytes *in vitro*. *Chem. Biol. Interact.* **35**, 145–157.

Monroe, D. H., and Eaton, D. L. (1987). Comparative effects of butylated hydroxyanisole on hepatic *in vivo* DNA binding and *in vitro* biotransformation of aflatoxin B_1 in the rat and mouse. *Toxicol. Appl. Pharmacol.* **90**, 401–409.

Monroe, D. H., Holeski, C. J., and Eaton, D. L. (1986). Effects of single-dose and repeated-dose pretreatment with 2(3)-*tert*-butyl-4-hydroxyanisole (BHA) on the hepatobiliary disposition and covalent binding to DNA of aflatoxin B_1 in the rat. *Food Chem. Toxicol.* **24**, 1273–1281.

Moss, E. J., and Neal, G. E. (1985). The metabolism of aflatoxin B_1 by human liver. *Biochem. Pharmacol.* **34**, 3193–3197.

Neal, G. E., and Colley, P. J. (1978). Some high performance liquid chromatographic studies of the metabolism of aflatoxins by rat liver microsomal preparations. *Biochem. J.* **174**, 839–851.

Neal, G. E., and Colley, P. J. (1979). The formation of 2,3-dihydro-2,3-dihydroxy aflatoxin B_1 by the metabolism of aflatoxin B_1 *in vitro* by rat liver microsomes. *FEBS Lett.* **101**, 382–386.

Neal, G. E., Judah, D. J., Stirpe, F., and Patterson, D. S. P. (1981). The formation of 2,3-dihydroxy-2,3-dihydro-aflatoxin B_1 by the metabolism of aflatoxin B_1 by liver microsomes isolated from certain avian and mammalian species and the possible role of this metabolite in the acute toxicity of aflatoxin B_1. *Toxicol. Appl. Pharmacol.* **58,** 431–437.

Neal, G. E., Nielsch, U., Judah, D. J., and Hulbert, P. B. (1987). Conjugation of model substrates or microsomally-activated aflatoxin B_1 with reduced glutathione, catalysed by cytosolic glutathione-S-transferases in livers of rats, mice, and guinea pigs. *Biochem. Pharmacol.* **36,** 4269–4276.

Newberne, P. M., and Butler, W. H. (1969). Acute and chronic effects of aflatoxin on the liver of domestic and laboratory animals: A review. *Cancer Res.* **29,** 236–250.

O'Brien, K., Moss, E., Judah, D., and Neal, G. (1983). Metabolic basis of the species difference to aflatoxin B_1 induced hepatotoxicity. *Biochem. Biophys. Res. Commun.* **114,** 813–821.

Patterson, D. S. P., and Roberts, B. A. (1970). The formation of aflatoxins B_{2a} and G_{2a} and their degradation products during the *in vitro* detoxification of aflatoxin by livers of certain avian and mammalian species. *Food Cosmet. Toxicol.* **8,** 527–538.

Patterson, D. S. P., and Roberts, B. A. (1971). The *in vitro* reduction of aflatoxins B_1 and B_2 by soluble avian liver enzymes. *Food Cosmet. Toxicol.* **9,** 829–837.

Patterson, D. S. P., and Roberts, B. A. (1972). Aflatoxin metabolism in duck-liver homogenates: The relative importance of reversible cyclopentenone reduction and hemiacetal formation. *Food Cosmet. Toxicol.* **10,** 501–512.

Pegram, R. A., Allaben, W. T., and Chou, M. W. (1989). Effect of caloric restriction on aflatoxin B_1–DNA adduct formation and associated factors in Fischer 334 rats: Preliminary findings. *Mech. Ageing Devel.* **48,** 167–177.

Pohland, A. E., Cushmac, M. E., and Andrellos, P. J. (1968). Aflatoxin B_1 hemiacetal. *J. Assoc. Off. Anal. Chem.* **51,** 907–910.

Pong, R. S., and Wogan, G. N. (1971). Toxicity and biochemical and fine structural effects of synthetic aflatoxins M_1 and B_1 in rat liver. *J. Natl. Cancer Inst.* **47,** 585–592.

Prabhu, A. L., Aboobaker, V. S., and Bhattacharya, R. K. (1989). *In vivo* effect of dietary factors on the molecular action of aflatoxin B_1: Role of riboflavin on the catalytic activity of liver fractions. *Cancer Lett.* **48,** 89–94.

Purchase, I. F. H. (1967). Acute toxicity of aflatoxins M_1 and M_2 in one-day-old ducklings. *Food Cosmet. Toxicol.* **5,** 339–342.

Putt, D. A., Kensler, T., and Hollenberg, P. F. (1991). Effect of three chemoprotective antioxidants, ethoxyquin, oltipraz and 1,2-dithiol-3-thione on cytochrome P-450 levels and aflatoxin B_1 metabolism. *FASEB J.* **5,** A1517.

Quinn, B. A., Crane, T. L., Kocal, T. E., Best, S. J., Cameron, R. G., Rushmore, T. H., Farber, E., and Hayes, M. A. (1990). Protective activity of different hepatic cytosolic glutathione S-transferases against DNA-binding metabolites of aflatoxin B_1. *Toxicol. Appl. Pharmacol.* **105,** 351–363.

Ramsdell, H. S., and Eaton, D. L. (1988). Modification of aflatoxin B_1 biotransformation *in vitro* and DNA binding *in vivo* by dietary broccoli in rats. *J. Toxicol. Env. Health.* **25,** 269–284.

Ramsdell, H. S., and Eaton, D. L. (1990a). Mouse liver glutathione S-transferase isoenzyme activity toward aflatoxin B_1-8,9-epoxide and benzo[a]pyrene-7,8-dihydrodiol-9,10-epoxide. *Toxicol. Appl. Pharmacol.* **105,** 216–225.

Ramsdell, H. S., and Eaton, D. L. (1990b). Species susceptibility to aflatoxin B_1 carcinogenesis: Comparative kinetics of microsomal biotransformation. *Cancer Res.* **50,** 615–620.

Ramsdell, H. S., Parkinson, A., Eddy, A. C., and Eaton, D. L. (1991). Bioactivation of aflatoxin B1 by human liver microsomes: Role of cytochrome P450 IIIA enzymes. *Toxicol. Appl. Pharmacol.* **108,** 436–447.

Raney, K. D., Coles, B., Guengerich, F. P., and Harris, T. P. (1992a). The endo-8,9-epoxide of aflatoxin B_1: A new metabolite. *Chem. Res. Toxicol.* **5,** 333–335.

Raney, K. D., Meyer, D. J., Ketterer, B., Harris, T. M., and Guengerich, F. P. (1992b). Glutathione conjugation of aflatoxin B_1 *exo*- and *endo*-epoxides by rat and human glutathione S-transferases. *Chem. Res. Toxicol.* **5,** 470–478.

Raney, K. D., Shimada, T., Kim, D.-H., Groopman, J. D., Harris, T. M., and Guengerich, F. P. (1992c). Oxidation of aflatoxin and sterigmatocystin by human liver microsomes: Significance of aflatoxin Q_1 as a detoxication product of aflatoxin B_1. *Chem. Res. Toxicol.* **5**, 202–210.

Roebuck, B. D., Siegel, W. G., and Wogan, G. N. (1978). *In vitro* metabolism of aflatoxin B_2 by animal and human liver. *Cancer Res.* **38**, 999–1002.

Roebuck, B. D., Liu, Y.-L., Rogers, A. E., Groopman, J. D., and Kensler, T. W. (1991). Protection against aflatoxin B_1-induced hepatocarcinogenesis in F344 rats by 5-(2-pyrazinyl)-4-methyl-1,2-dithiol-3-thione (oltipraz): Predictive role for short term molecular dosimetry. *Cancer Res.* **51**, 5501–5506.

Ross, R. K., Yuan, J. M., Yu, M. C., Wogan, G. N., Quai, G. S., Tu, J. T., Groopman, J. D., Gao, Y. T., and Henderson, B. E. (1992). Urinary aflatoxin biomarkers and risk of hepatocellular carcinoma. *Lancet* **339**, 943–946.

Sabbioni, G., Skipper, P., Buchi, G., and Tannenbaum, S. R. (1987). Isolation and characterization of the major serum albumin adduct formed by aflatoxin B_1 *in vivo* in rats. *Carcinogenesis* **8**, 819–824.

Salbe, A. D., and Bjeldanes, F. (1989). Effect of diet and route of administration on the DNA binding of aflatoxin B_1 in the rat. *Carcinogenesis* **10**, 629–634.

Salhab, A. S., and Edwards, G. S. (1977). Comparative *in vitro* metabolism of aflatoxicol by liver preparations from animals and humans. *Cancer Res.* **37**, 1016–1021.

Schabort, J. C., and Steyn, M. (1969). Substrate and phenobarbital inducible aflatoxin-4-hydroxylation and aflatoxin metabolism by rat liver microsomes. *Biochem. Pharmacol.* **18**, 2241–2252.

Schoenhard, G. L., Hendricks, J. D., Nixon, J. E., Lee, D. J., Wales, J. H., Sinnhuber, R. O., and Pawlowski, N. E. (1981). Aflatoxicol-induced hepatocellular carcinoma in rainbow trout (*Salmo gairdneri*) and the synergistic effects of cyclopropenoid fatty acids. *Cancer Res.* **41**, 1011–1014.

Shayiq, R. M., and Avadhani, N. G. (1989). Purification and characterization of a hepatic mitochondrial cytochrome P-450 active in aflatoxin B_1 metabolism. *Biochemistry* **28**, 7546–7554.

Shimada, T., and Guengerich, F. P. (1989). Evidence for cytochrome P-450NF, the nifedipine oxidase, being the principal enzyme involved in the bioactivation of aflatoxins in human liver. *Proc. Natl. Acad. Sci. U.S.A.* **86**, 462–465.

Sinnhuber, R. O., Lee, D. J., Wales, J. H., and Landers, M. K. (1974). Hepatic carcinogenesis of aflatoxin M_1 in rainbow trout (*Salmo gairdneri*) and its enhancement by cyclopropene fatty acids. *J. Natl. Cancer Inst.* **53**, 1285–1288.

Sinnhuber, R. O., Hendricks, J. D., Wales, J. H., and Putnam, G. B. (1977). Neoplasms in rainbow trout, a sensitive animal model for environmental carcinogenesis. *Ann. N.Y. Acad. Sci.* **298**, 389–408.

Stark, A. A., Gal, Y., and Shaulsky, G. (1990). Involvement of singlet oxygen in photoactivation of aflatoxins B_1 and B_2 to DNA-binding forms *in vitro*. *Carcinogenesis* **11**, 529–534.

Stoloff, L., Verrett, M. J., Dantzman, J., and Reynaldo, E. F. (1972). Toxicological study of aflatoxin P_1 using the fertile chicken egg. *Toxicol. Appl. Pharmacol.* **23**, 528–531.

Swenson, D. H., Lin, J.-K., Miller, E. C., and Miller, J. A. (1977). Aflatoxin B_1-2,3-oxide as a probable intermediate in the covalent binding of aflatoxins B_1 and B_2 to rat liver DNA and ribosomal RNA *in vivo*. *Cancer Res.* **37**, 172–181.

Tadi, P. P., Teel, R. W., and Lau, B. H. S. (1991). Organosulfur compounds of garlic modulate mutagenesis, metabolism, and DNA binding of aflatoxin B_1. *Nutr. Cancer* **15**, 87–95.

Teel, R. W. (1991). Effects of capsaicin on rat liver S9-mediated metabolism and DNA binding of aflatoxin. *Nutr. Cancer* **15**, 27–32.

Whitty, J. P., and Bjeldanes, L. F. (1987). The effects of dietary cabbage on xenobiotic metabolizing enzymes and the binding of aflatoxin B_1 to hepatic DNA in rats. *Food Chem. Toxicol.* **25**, 581–587.

Williams, G. M., Tanaka, T., and Macura, Y. (1986). Dose-related inhibition of aflatoxin B_1 induced hepatocarcinogenesis by the phenolic antioxidants, butylated hydroxyanisole and butylated hydroxytoluene. *Carcinogenesis* **7,** 1043–1050.

Wogan, G. (1973). Aflatoxin carcinogenesis. *In* "Methods in Cancer Research" (H. Busch, ed.), pp. 309–344. Academic Press, New York.

Wogan, G. N., and Newberne, P. M. (1967). Dose–response characteristics of aflatoxin B_1 carcinogenesis in the rat. *Cancer Res.* **27,** 2370–2376.

Wogan, G. N., Edwards, G. S., and Newberne, P. M. (1971). Structure–activity relationships in toxicity and carcinogenicity of aflatoxins and analogs. *Cancer Res.* **31,** 1936–1942.

Wong, Z. A., and Hsieh, D. P. H. (1978). Aflatoxicol: Major aflatoxin B_1 metabolite in rat plasma. *Science* **200,** 325–327.

Wong, Z. A., and Hsieh, D. P. H. (1980). The comparative metabolism and toxicokinetics of aflatoxin B_1 in the monkey, rat, and mouse. *Toxicol. Appl. Pharmacol.* **55,** 115–125.

Wong, Z. A., Decad, G. M., Byard, J. L., and Hsieh, D. P. H. (1979). Conversion of aflatoxicol to aflatoxin B_1 in rats *in vivo* and in primary hepatocyte culture. *Food Cosmet. Toxicol.* **17,** 481–486.

Yourtee, D. M., and Kirk-Yourtee, C. L. (1986). The mutagenicity of aflatoxin Q_1 to *Salmonella typhimurium* TA100 with or without rat or human liver microsomal preparations. *Res. Commun. Chem. Pathol. Pharmacol.* **54,** 101–113.

Yourtee, D. M., Bean, T. A., and Kirk-Yourtee, C. L. (1987). Human aflatoxin B_1 metabolism: An investigation of the importance of aflatoxin Q_1 as a metabolite of hepatic post-mitochondrial fraction. *Toxicol. Lett.* **38,** 213–224.

4

Pharmacokinetics and Excretion of Aflatoxins

❖

Dennis P. H. Hsieh and Jeffrey J. Wong

INTRODUCTION

Pharmacokinetics, or "toxicokinetics" when toxic rather than therapeutic chemicals are in question, examines the rate of change of concentration with time of a chemical in different compartments of the body of an exposed organism. The relative toxicity of a chemical in an organism is governed by the concentration of the chemical, or its reactive metabolite, in the critical target compartment and by the duration of its presence there. The time-course of this concentration is determined by the amount of toxicant with which the organism comes into contact (dose), the duration of exposure, and the rates of absorption, distribution, metabolism, and excretion.

Studies of the pharmacokinetics and excretion of aflatoxins are useful to (1) determine the body burden of the toxins under different exposure conditions, (2) determine whether an accumulation of the toxins or their reactive metabolites occurs through dietary chronic exposure, and (3) facilitate interspecies comparisons of aflatoxin disposition in different animal models to predict human toxicity. Because most of the toxicological studies on aflatoxins have focused on aflatoxin B_1 (AFB$_1$), the principal and most potent member of the aflatoxin family, this chapter focuses primarily on the toxicokinetics of AFB$_1$ after oral administration of low doses of AFB$_1$, to emulate the comparatively low levels of "real world" dietary exposure to this foodborne mycotoxin.

Sufficient evidence now indicates that AFB$_1$ requires metabolic activation to become toxic, mutagenic, and carcinogenic (Neal, 1987; Chapter 5). Therefore the metabolic disposition is a strong determinant of potency of AFB_1 in an exposed animal. Much has been studied on the hepatic metabolism of AFB_1 in relation to its toxicity; correlations between the potency of AFB_1 and hepatic biotransformation have been attempted (Hsieh *et al.*, 1977; Heathcote and Hibbert, 1978; Wong and Hsieh, 1980). However, these correlations do not fully explain the known species differences in susceptibility to AFB_1 and, in some cases, contradict each other. Thus, other factors also must be considered to fully assess the potency of AFB_1 in an animal. These factors include the fate processes just mentioned and the effects of biotransformation enzyme induction, increased liver size, perfusion rate, bile production and flow, compartmentation within cellular organelles, and extrahepatic biotransformation, especially in the gastrointestinal tract.

Pharmacokinetics involves the determination of the concentrations of a chemical and its metabolites in the blood at different time intervals following administration of the chemical in a single dose or multiple doses. Even within the same species of animal, the concentrations of the parent chemical and its metabolites vary significantly with the dose used, the route of exposure, and other experimental conditions such as the amount and frequency of blood samples collected. Note that information on pharmacokinetics presented is often species and experiment specific; thus, extrapolation to other conditions or species requires careful validation.

ABSORPTION

Since human exposure to AFB_1 generally occurs via ingestion of foodstuffs that are contaminated with parts per billion levels of the toxin, studies of pharmacokinetics associated with chronic administration of such relatively low doses of AFB_1 through the oral route are most relevant. The consideration of dose is of particular significance because fate processes in many cases are readily saturable and experiments using unrealistically high doses may yield information that is of limited relevance. Although the primary route of exposure to aflatoxins in both animals and humans is the gastrointestinal tract, blood time–concentration profiles of AFB_1 in experimental animal models have been obtained through other routes of exposure including gavage, intraperitoneal injection, and intravenous injection. The rate of absorption and biotransformation of AFB_1 may be affected substantially by the route of administration. In the case of oral exposure, the extent of absorption and biotransformation in the entire gastrointestinal tract determines the concentration of AFB_1 and its metabolites in the hepatic portal flow and, subsequently, the degree and rate of hepatic exposure.

Absorption after Oral Administration

Aflatoxin B_1 is a relatively low molecular weight, lipophilic molecule, suggesting efficient absorption after ingestion. Wogan *et al.* (1967) reported no significant difference in the excretion and distribution of radioactivity after either oral or intraperitoneal administration of $[^{14}C]AFB_1$ in male Fischer rats (40–125 g), implying that absorption after oral exposure was complete. Within 24 hr of administration, up to 20 and 60% of a dose of 0.07 mg/kg AFB_1 was eliminated via urinary and fecal routes, respectively. Little radioactivity was excreted within the first 8 hr, suggesting a slow passage through the gastrointestinal tract and extensive biotransformation of the orally administered AFB_1. The liver content of AFB_1-derived radioactivity at 24 hr (7.6% of the dose) accounted for almost the entire residual radioactivity in the carcass, indicating that liver is the principal site of accumulation of AFB_1, metabolites, and/or bound material.

In a similar study of orally administered $[^3H]AFB_1$ (0.6 mg/kg) in Sprague–Dawley rats, Coulombe and Sharma (1985) reported that a total of 15% of initial $[^3H]AFB_1$ was recovered in the urine 23 days after dosing, comparable to the 10–19% urinary excretion of initial AFB_1 collected several days following intraperitoneal (ip) or intravenous (iv) administration of similar doses of AFB_1 (Wong and Hsieh, 1980; Groopman et al., 1988), suggesting that absorption was largely complete. Also, in rhesus monkeys, hepatic retention of radioactivity 4 days after $[^{14}C]AFB_1$ (0.4 mg/kg) dosing was similar following oral (6.8%) and ip (5.6%) administration, again suggesting complete absorption (Dalezios and Wogan, 1972; Dalezios et al., 1973).

In contrast, Degan and Neumann (1978) reported that female Wistar rats excreted 10–30% and 60–65% of total radioactivity into bile within 24 hr of oral and ip administration of $[^{14}C]AFB_1$, respectively. The considerably lower absorption through the oral route might have been caused by a slow passage through the gastrointestinal tract rather than by incomplete absorption. The effect of the slow passage was demonstrated in a study of orally administered $[^{14}C]AFB_1$ (10 mg/kg) by Steyn et al. (1971), who measured the radioactivity remaining in rat and mouse stomach at various time intervals after dosing. The percentage of $[^{14}C]AFB_1$ in the stomach decreased relatively rapidly in the first 30 min in both species. Approximately 22% of $[^{14}C]AFB_1$ was retained in the mouse stomach 30 min after dosing, whereas 39% remained in the rat. After 30 min, the rate of decrease of $[^{14}C]AFB_1$ was much slower; by 8 hr, approximately 8% and 25% of the radioactivity was retained in mouse and rat stomach, respectively. Apparently, the rate of gastric emptying in the rat is rather slow, and is much slower than that in the mouse. In the toxicokinetic studies in the rat conducted by Wong and Hsieh (1978), AFB_1 was given both intravenously and orally. As expected, intravenous administration resulted in significantly higher levels of plasma radioactivity derived from $[^{14}C]AFB_1$ 1 hr after administration, in comparison with oral administration.

Absorption in the Small Intestines

AFB_1 appeared to be absorbed rapidly from the small intestines into the mesenteric venous blood. In an experiment with Wistar rats, Kumagai (1989) injected $[^3H]AFB_1$ directly into the stomach and into various sites of the small intestine of rats and measured radioactivity in the bile 30 min later. The results indicated that the site of absorption was the small intestine; the duodenum was most efficient. The author found that the rate of intestinal absorption of AFB_1 is considerably higher in suckling than in older rats, suggesting a change in the lipid composition of the intestinal epithelium during growth. This result implies that species differences in AFB_1 absorption may be explained in part by differences in the composition of the intestinal epithelium. The author also found that the rate of AFB_1 uptake by intestinal tissue is nearly proportional to AFB_1 concentration, indicating that AFB_1 is absorbed by passive diffusion. When the rate of absorption of AFB_1 was compared with that of AFG_1, a less lipophilic analog, the rate of absorption of the latter was considerably lower, confirming that lipophilicity is a determinant of aflatoxin absorption.

In a similar experiment with Sprague–Dawley rats, we found that over 50% of a dose of 0.05 mg/kg [^{14}C] AFB_1 disappeared from the duodenal region of the small intestine within 1 hr. The absorbed radioactivity was present in the mesenteric venous blood as water-soluble metabolites, protein adducts, and free AFB_1. No chloroform-extractable phase I metabolites of AFB_1 were detected. The proportion of free AFB_1 among the three bloodborne fractions of AFB_1 was less than one-third and decreased with time, indicating that AFB_1 was metabolized actively during the absorption process. The chloroform-nonextractable protein adducts represented over 50% of the total radioactivity in the blood. Water-soluble metabolites were present in only minor quantities, with peak levels representing less than 5% of total radioactivity.

A time-concentration plot of the total radioactivity and the three types of bloodborne AFB_1 metabolites is shown in Figure 1. The decline in the terminal portion of the time–concentration profiles suggests that absorption of AFB_1 from the small intestines into the mesenteric venous blood and conversion of AFB_1 into the various fractions took place as first order processes. The residual radioactivity in the ligated small intestines at 65 min after administration was 45.0% of the administered dose, in which 39% was present as free AFB_1 and the remaining as water-soluble and protein-bound metabolites. No other chloroform-extractable metabolites were found. The AFB_1 residue in the small intestine most likely is derived from the lumen content. The low aqueous-soluble and protein-bound content indicated rapid transport of biotransformation products to the mesenteric venous blood.

Gastrointestinal Metabolism

Although the liver and kidney generally are regarded as the main sites of xenobiotic biotransformation, the gastrointestinal tract also may participate in this function (Hartiala, 1977). The gastrointestinal mucosa, particularly that of the small intestine, appears to possess many of the biotransformation abilities found in the liver, but at a much lower level (Chhabra et al., 1974; Hartiala, 1977; Vainio and Hietanen, 1979). Most absorption of xenobiotics such as AFB_1 is thought to occur at this site, and thus is subjected to biotransformation processes active in the intestinal mucosa.

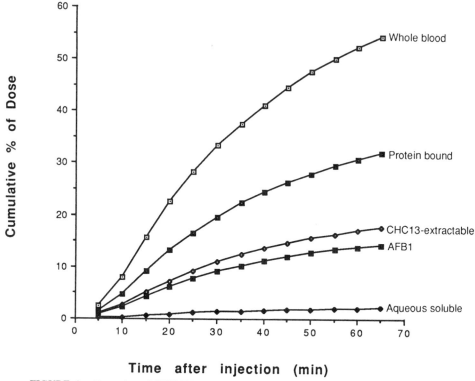

Time after injection (min)

FIGURE 1 Absorption of [^{14}C]AFB$_1$ from a ligated 10- to 15-cm duodenal section of small intestines in male Sprague–Dawley rats. The absorbed radioactivity was shown as the percentage of that in the administered dose (0.05 mg/kg).

The metabolites of AFB$_1$ that are responsible for the active interactions with protein at the gastrointestinal mucosa may be the AFB$_1$-epoxide, the AFB$_1$-dihydrodiol, or AFB$_{2\alpha}$. Each of these AFB$_1$ metabolites is capable of covalent interaction with proteins. The covalent protein adduct of AFB$_1$ in the blood has been identified by other investigators to be the Schiff base formed between the lysine residues of serum albumin and the 8,9-dihydrodiol of AFB$_1$, a hydrolytic product of the active 8,9-epoxide form of AFB$_1$ (Sabbioni *et al.*, 1987). The occurrence of the serum albumin adduct of AFB$_1$ (SAA) in the blood indicates that AFB$_1$ is metabolized to the epoxide either in the gut lumen, the gut wall, or in some blood components. The high levels of protein binding of orally administered, and hence gastrointestinally absorbed, AFB$_1$ in the blood have been observed similarly in experiments with swine (Luthy *et al.*, 1980) and rhesus monkeys (Dalezios *et al.*, 1973). SAA now is being used as a biomarker for individual exposurè to AFB$_1$ because of its high level of occurrence ard its accumulation in the blood (see Chapter 12).

The absence of detectable chloroform-extractable phase I metabolites of

AFB_1 in the blood suggests that these metabolites are conjugated rapidly by phase II enzymes to aqueous-soluble metabolites and/or are protein bound to plasma macromolecules. The presence of AFB_1 as the sole component of the chloroform-extractable fraction in the blood and intestinal residue suggests that intestinal extraction is less than 100% efficient or that biotransformation processes are saturable. The very active phase II metabolism and SAA formation in the intestinal absorption sites indicate that the actual total hepatic exposure to AFB_1 is limited by both incomplete absorption and gastrointestinal metabolism. The total hepatic exposure to AFB_1 present in the mesenteric outflow, and therefore the portal inflow, would represent only 20–25% of the ingested dose. Consequently, the gastrointestinal tract represents a major AFB_1 detoxification site in terms of hepatic toxicity.

Reabsorption of Aflatoxin B_1 Conjugates

To assess the significance of enterohepatic circulation of AFB_1 metabolites, we collected the bile from $[^{14}C]AFB_1$-exposed rats and injected it into the ligated duodenum section to determine the reabsorption of radioactivity in the bile. Reinjection of collected bile (0.15–0.25 μCi) into the small intestines resulted in disappearance of only trace levels of bileborne radioactivity from the small intestines. No radioactivity was detected in the mesenteric outflow and hepatic content, indicating that no detectable amount of the bileborne $[^{14}C]AFB_1$ metabolites was reabsorbed. This result is consistent with those of our previous mutagenicity and analytical studies (D. P. H. Hsieh, J. J. Wong, and C. Wei, *et al.,* unpublished data), which found no detectable mutagenic activity in the bileborne products of AFB_1 metabolism. Thus, no detectable oxidative metabolites of AFB_1 are in the bile when the animals are given the low dose of AFB_1 (0.05 mg/kg) used in our experiment. Our results are in contrast with those of Degen and Neumann (1978), who reported that up to 10% of the bileborne $[^{14}C]AFB_1$ metabolites were in the form of chloroform-extractable and reabsorbable metabolites. Their results might have been caused by the high dose (0.5–2.5 mg/kg) of AFB_1 used in their experiment, which may have saturated *in vivo* detoxification processes.

DISTRIBUTION

Blood Time–Concentration Profiles and Compartment Analyses

To examine the biological fate of free AFB_1 in the blood, Wong and Hsieh (1980) compared the metabolism and toxicokinetics of intravenously administered $[^{14}C]AFB_1$ in the male rhesus monkey (6–8 kg), Sprague–Dawley rat (240–260 g), and Swiss–Webster mouse (40–50 g). Each test animal received a dose that was approximately 10% of the LD_{50} for $[^{14}C]AFB_1$. The LD_{50} values used were 3, 10, and 15 mg/kg for the monkey, rat, and mouse, respectively.

Using the one-compartment pharmacokinetic analysis of the plasma time–concentration profiles for $[^{14}C]AFB_1$, the apparent volume of distribution (V_d) was calculated for the monkey, mouse, and rat after an iv dose of $[^{14}C]AFB_1$. The V_d, an index of the affinity between AFB_1 and animal tissues expressed as a percentage of body weight, was 114%, 47%, and 28% for the monkey, rat, and mouse, respectively. The values of V_d ranked consistently with the relative susceptibility of these species to the acute toxicity of AFB_1. A similar correlation was observed with the first-order rate constant for AFB_1 elimination from plasma (K_E), which were 1.1, 1.4, and 3.2 hr^{-1} for the monkey, rat, and mouse, respectively. The plasma biological half-lives $(t_{\frac{1}{2}},$ in min) were 36.5, 28.9, and 12.9 min for the monkey, rat, and mouse, respectively. Using a two-compartment open model, the calculated toxicokinetic rate constants of $[^{14}C]AFB_1$ were compared for the monkey and the rat. In the monkey, the plasma to tissue transfer rate constant $(K_{1,2},$ in $hr^{-1})$ of 23.1 was approximately three times higher than the tissue to plasma rate constant $(K_{2,1},$ in $hr^{-1})$. The $K_{1,2/2,1}$ value for the rat was only 1.3, compared with 2.9 for the monkey, indicating a slower rate of plasma to tissue transfer. The first order elimination rate constants (K_2) were 5.0 and 3.6 for the monkey and the rat, respectively.

For comparison, similar kinetic studies were conducted for AFB_1 following direct injection into the duodenum of the rat at a lower dose (0.05 mg/kg) to simulate human exposure to AFB_1 as a food contaminant. Blood profile data indicate that the non-chloroform-extractable component that was bound to proteins or bloodborne macromolecules constitutes the major fraction of the total radioactive content in the portal blood. The chloroform-extractable fraction of radioactivity was higher in the portal blood than in blood sampled from the jugular vein. A striking observation was that, at the low dose used, no free AFB_1 was detectable in the blood sampled from the jugular vein. The difference in the chloroform-extractable fraction of radioactivity between the portal blood and the blood from the other sampling site was in part due to the presence and the absence of free AFB_1 in the blood at these two sites. The absence of AFB_1 in the blood from the jugular vein indicates that virtually 100% of the dose was extracted by the liver at the low dose level. Portal blood half-life of the free AFB_1 was approximately 30 min, comparable to that found in the iv injection study just described (28.9 min).

Concentration in the Liver and the Kidney

From the intestine, AFB_1 apparently enters the liver through the hepatic portal blood supply (Wilson *et al.,* 1985). AFB_1 is concentrated heavily in the liver not only after oral administration but after iv and ip dosing as well, because of the high permeability of the hepatocyte membrane for AFB_1, and its active metabolism and subsequent covalent binding with hepatic macromolecules. The kidneys also concentrate AFB_1, although to a much lesser extent.

Wogan *et al.* (1967) reported that, 30 min after an ip dose of $[^{14}C]AFB_1$ (0.07

mg/kg), the liver contained approximately 17% of the radioactivity, the kidneys 5%, and the eviscerated carcass 27%. Adrenal glands, brain, heart, pancreas, spleen, thymus, and testes each contained <0.5% of recovered radioactivity. During the next 90 min, radioactivity in the kidneys and liver decreased rapidly, so the kidneys contained less than 1% and the liver 10% of recovered radioactivity 2 hr after dosing. Similarly, 30 min after administration of [^{3}H]- or [^{14}C]AFB to rats, the amount of radioactivity was 5- to 15-fold greater in the liver than in other tissues. In the male rhesus monkey, Dalezios and Wogan (1972) found that 19% of the administered [^{14}C]AFB$_1$ dose was found in the liver 45 min after ip injection. The kidneys contained the second greatest amount of applied [^{14}C]AFB$_1$ at 0.9%.

This rapid hepatic uptake is consistent with *in vitro* findings. Rapid uptake was observed both in the isolated perfused rat liver and in isolated rat hepatocytes (Unger *et al.*, 1977; Ch'ih and Devlin, 1984). Muller and Petzinger (1988) demonstrated that AFB$_1$ permeation occurs by nonionic diffusion in isolated rat hepatocytes; lipophilicity appears to be a major determinant in the rapid hepatic uptake. These authors suggested that the rate of permeation probably is determined by the composition of the lipid domains in the cell membrane. AFB$_1$ apparently partitions into domains with unsaturated fatty acids. As mentioned earlier, species differences in lipid composition of the membrane would be a factor determining the hepatic uptake of AFB$_1$. For example, the rate of AFB$_1$ uptake was more rapid in rat liver slices than in mouse liver slices (Portman *et al.*, 1970); similarly, [^{3}H]AFB$_1$ uptake by cultured rat hepatocytes was markedly faster than that by mouse hepatocytes (Hanigan and Laishes, 1984).

Within several hours of AFB$_1$ dosing, most of the AFB$_1$ retained in the liver is bound irreversibly to tissue macromolecules. For example, Holeski *et al.* (1987) found that, 2 hr after AFB$_1$ administration (0.25 mg/kg, ip), 15% remained in the liver. Of the radioactivity in the liver, 12% constituted polar metabolites, 3% nonpolar metabolites, and 70% covalently bound adducts. Wong and Hsieh (1980) reported that, 100 hr after [^{14}C]AFB$_1$ dosing (1/10 of LD$_{50}$, iv), 13.6, 6.5, and 1.8% of the administered dose was retained in monkey, rat, and mouse liver, respectively. Dalezios *et al.* (1973) reported that, 4 days after oral dosing of [^{14}C]AFB$_1$ (0.4 mg/kg), rhesus monkeys retained 6.8% of the dose in the liver. For the same length of time after ip dosing, retention of AFB$_1$ was similar (5.6%) (Dalezios and Wogan, 1972). A monkey given an oral dose of 0.015 mg/kg AFB$_1$ retained about 1% of the dose 5 weeks after administration (Dalezios *et al.*, 1973).

Given the high efficiency of the liver to extract free AFB$_1$ from the blood, the binding of AFB$_1$ to serum albumin at the site of intestinal absorption should be considered a major mechanism of detoxification of this carcinogen. At the low doses of AFB$_1$ normally encountered by human consumers, our study on duodenum absorption discussed earlier (Figure 1) found no free AFB$_1$ present in the circulating blood of the dosed rat, primarily because of protein conjugation of the absorbed carcinogen. Intestinal metabolism of AFB$_1$, especially its conjugation reactions, therefore warrants further investigation to elucidate the biochemical basis of differences in susceptibility among animals.

BIOTRANSFORMATION

The aflatoxins, like other xenobiotics, undergo phase I and phase II biotransformation. Detailed description of the metabolism of aflatoxins can be found in Chapter 3. Metabolism or biotransformation of AFB_1 represents a significant fate process by which the parent compound is eliminated from the exposed biological system.

Several pathways of biotransformation have been identified for AFB_1; similar pathways are likely in effect for other aflatoxins. Some pathways in phase I biotransformation activate AFB_1 whereas others reduce its toxicity. The pathways in phase II biotransformation all lead to conjugations of phase I metabolites of AFB_1 that are less toxic than the parent compound. Species differences in the susceptibility to the toxicity and carcinogenicity of AFB_1 are determined largely by how the toxin is biotransformed. Biotransformation in turn is modulated by numerous factors including genetic make-up of species. nutritional conditions, health status, and exposure to metabolic modifiers (see Chapters 10 and 13). The major phase I metabolites mentioned in this chapter are the active form AFB_1-8,9-epoxide, the hydroxy metabolites AFM_1 and AFQ_1, the demethylation product AFP_1, and the reduction product aflatoxicol. The major phase II metabolites are the glutathione, glucuronide, and sulfate conjugates of AFB_1. The DNA, RNA, and serum albumin adducts of AFB_1 have been studied extensively in recent years.

EXCRETION

Excretion of AFB_1 and its metabolites occurs primarily through the biliary pathway, followed by the urinary pathway. In lactating animals, a considerable fraction of the ingested AFB_1 is excreted into the milk in the form of AFM_1 and other metabolites. Aflatoxins in milk and dairy products are ingested by human consumers, in relatively large quantities by the infant populations, constituting a potential food safety problem of worldwide concern (see Chapter 17). In this section, excretion of ingested AFB_1 into feces, urine, and milk is reviewed.

Elimination of AFB_1 appears relatively slow in all species and strains studied. Wong and Hsieh (1980) reported that the total excretion of the administered dose (10% LD_{50}) of $[^{14}C]AFB_1$ 100 hr after iv dosing in the male mouse, rat, and monkey were 80%, 72%, and 73% of the dose, respectively. The excretion of total radioactivity was most extensive during the first 24 hr after dosing. The patterns of urinary metabolites were found to correlate with the relative species susceptibility to the carcinogenic effect of AFB_1. The mouse, a less susceptible species, produced the most water-soluble urinary metabolites. The monkey and the rat, the more susceptible species, produced less. Similar results were observed in studies with different species and different routes of administration.

Biliary Excretion

In Sprague–Dawley rats, Coulombe and Sharma (1985) reported that, after an oral [^{3}H]AFB$_1$ dose (0.6 mg/kg), the apparent plasma half-life for AFB$_1$ radioactivity was 91.8 hr; 55% of [^{3}H]AFB$_1$ radioactivity was excreted cumulatively in the feces, compared with 15% in the urine, 23 days after dosing. Similarly, Wong and Hsieh (1980) reported cumulative excretion of 53% of the administered dose (1 mg/kg) in feces and 19% in urine for the 4 days following iv dosing. Holeski *et al.* (1987) observed rapid biliary excretion of ip-administered [^{3}H]- or [^{14}C]AFB$_1$ (0.25 mg/kg). The rate of biliary excretion peaked at 30 min (6–8% of the administered dose/10 min) and decreased substantially by 2 hr (1% of the administered dose/10 min). The major biliary AFB metabolite was the AFB$_1$–glutathione (AFB$_1$–GSH), which accounted for 49–57% of the total biliary radioactivity. AFP$_1$–glucuronide also was identified in the bile, accounting for 4–15% of total biliary radioactivity in Sprague–Dawley rats.

AFB$_1$–GSH was identified first by Degan and Neumann (1978) as the major AFB$_1$ conjugate in the bile of female Wistar rats orally dosed with AFB$_1$. The conjugate accounted for 10% of a dose of 1 mg/kg and 30% of a dose of 0.8 mg/kg in 24 hr following dosing.

In our study of a low dose (1/800 of LD$_{50}$, or 0.013 mg/kg) of AFB$_1$ administered to Sprague–Dawley rats through intraportal injection, about 58% of the free [^{14}C]AFB$_1$ presented to the liver was excreted in the bile in 4 hr. Analysis of bile by thin layer chromatography (TLC) and high performance liquid chromatography (HPLC) indicated neither AFB$_1$ nor any known oxidative metabolites were present. Mutagen analysis did not indicate the presence of any mutagenic component. Similar patterns of biliary excretion of radioactivity were found with [^{14}C]AFB$_1$ introduced through intraduodenal injection. Biliary elimination of AFB$_1$ was 41% of the dose in 4 hr. Analysis of bile by both TLC and HPLC again revealed neither AFB$_1$ nor any known oxidative metabolites present in the bile. Mutagen analysis did not indicate the presence of any mutagenic component. Our results indicate that conjugation reactions of phase I AFB$_1$ metabolites in the liver are readily saturable in the rat.

Urinary Excretion

Approximately 10–20% of a dose of AFB$_1$ ranging from 0.4 to 1 mg/kg is excreted in urine 20–24 hr after ip administration of the dose to the rat (Raj and Lotlikar, 1984; Groopman *et al.*, 1988). The three major urinary metabolites identified in male Fischer 344 rats dosed with [^{14}C]AFB$_1$ (1 mg/kg, ip) were AFM$_1$, AFP$_1$, and AFB$_1$–N^7-guanine (Groopman *et al.*, 1988). AFM$_1$ was the major recovered metabolite, consisting of 41–50% of recovered radioactivity in urine. AFP_1 constituted less than 10% of the radioactivity. AFB_1-N^7-guanine, the major degradation product of hepatic AFB_1–DNA adducts, represented 16% of recovered radioactivity. A dose-dependent correlation between AFB_1 dose and AFB_1-N^7-guanine excreted in the urine was observed in male Fischer rats (Bennett *et al.*, 1981). Following administration of AFB_1 (1 mg/kg, ip), 80% of the total excretion of the AFB_1-N^7-guanine adduct was determined to occur during a 48-hr period after dosing (Essigmann *et al.*, 1982).

In addition to the major three urinary metabolites just described, thiol metabolites were identified in Sprague–Dawley rat and hamster urine (Raj and Lotlikar, 1984). An unknown $[^3H]AFB_1$ thiol metabolite, possibly the *N*-acetylcysteine (mercapturic acid) conjugate, consisted 1.9% of the dose (0.4 mg/kg) received ip. AFB_1–GSH and AFB_1–Cys–Gly constituted 1.4% and 0.2% of the injected dose, respectively. Excretion of thiol metabolites in the hamster, a species relatively resistant to the carcinogenic effect of AFB_1, was twice as great as that in the rat in 24 hr. Urinary excretion of total AFB_1 metabolites was 15% of the initial dose in the hamster, compared with 10% in the rat. Unlike in rat, monkey, and human, AFM_1 is not a major urinary metabolite in the mouse (Wong and Hsieh, 1980; Wei *et al.*, 1985); instead, AFP_1 is more prevalent. In addition, the mouse is the only species that has been found to excrete AFQ_1 as a urinary metabolite.

In the rhesus monkey, the urinary excretion rate reached a maximum 1 hr after ip dosing with $[^{14}C]AFB_1$ (0.4 mg/kg) (Dalezios and Wogan, 1972). Approximately 35% of the injected dose was excreted in the urine within 96 hr. When $[^{14}C]AFB_1$ was given to the monkey orally (0.4 mg/kg), 40% of the dose was excreted in the urine and 42% in the feces in 7 days after dosing. These results are in agreement with those of Wong and Hsieh (1980) who observed that 37.7% of the ^{14}C label was excreted in the urine and 35.5% in the feces 100 hr after iv administration of $[^{14}C]AFB_1$ (0.3 mg/kg).

AFB_1 was excreted into urine by rhesus monkeys as polar metabolites at a relatively high level (Dalezios and Wogan, 1972; Wong and Hsieh, 1980). Chloroform-extractable metabolites in monkey urine accounted for 15–25% of total metabolites, in contrast to 38–50% in rat urine (Wong and Hsieh, 1980; Wogan *et al.*, 1967). Efficient conjugation of metabolites in the monkey also is indicated by the fact that only 15% of material excreted in the feces was soluble in chloroform (Dalezios *et al.*, 1973).

The major urinary metabolite in monkeys after oral administration of $[^{14}C]AFB_1$ (0.4 mg/kg or 0.015 mg/kg) was identified as AFM_1 (Dalezios *et al.*, 1973). AFM_1 primarily was recovered during the first 24 hr after dosing. The fraction of dose converted to AFM_1 did not change with dose (18.8% at 0.4 mg AFB_1/kg vs. 16.9% at 0.015 mg AFB_1/kg). Conjugated AFP_1 was found to constitute 4.5% of the dose.

AFM_1 was also the major AFB_1 metabolite in rhesus monkeys after iv dosing (0.3 mg/kg), although the proportion formed was much lower (Wong and Hsieh,

1980). Following incubation of the urinary conjugates with sulfatase and β-glucuronidase, 16% of total urinary radioactivity was recovered as AFM_1 (Wei et al., 1985). When urine samples were subjected to acid hydrolysis, 22% of urinary radioactivity was recovered as $AFB_{2\alpha}$.

In humans, AFM_1 is apparently the major AFB_1 metabolite found in the urine of AFB_1-exposed individuals. Campbell et al. (1970) identified AFM_1 in urine of individuals exposed to highly contaminated peanut butter in the Philippines. Approximately 1–4% of initial AFB_1 was recovered in the urine as AFM_1. These results are in close agreement with the findings of Groopman et al. (1985) and Zhu et al. (1987), who analyzed urine of individuals in the Guangxi region of China for AFM_1 excretion. The results indicated that 1.23–2.18% of total AFB_1 in males is excreted as AFM_1; females excrete relatively low levels. The three major AFB_1 human urinary metabolites were AFM_1, AFB_1–N^7-guanine, and AFP_1 (Groopman et al., 1985), the same metabolites found in rat urine.

The AFB_1–N^7-guanine adduct was detected in 122 of 983 human urine samples tested (12.4%) in various parts of Kenya (Autrup et al., 1987). Likewise, Groopman et al. (1985) detected AFB_1–N^7-guanine in the urine of individuals in Guangxi, China, exposed to AFB_1 through their diet. For individuals exposed to the highest dose level of 87.5 μg per person per day, the amount of the adduct in urine was calculated to be 7–10 ng.

Excretion through Milk

Food safety concerns about AFM_1 have prompted numerous studies on the conversion of AFB_1 in dairy cattle feed to AFM_1 in milk (see Chapter 17). When dairy cows with different milk yields (28 and 12 liters/day) were given AFB_1 at 8 mg/animal/day (high dose study) in a daily ration, AFM_1 became detectable in milk 12–24 hr after the first AFB_1 ingestion (Van der Linde et al., 1964). AFM_1 reached a high level a few days thereafter. The amount of AFM_1 in the milk was estimated as less than 1% of ingested AFB_1. After the intake of AFB_1 was stopped, AFM_1 dropped to an undetectable level after 3 days. The detection limit of this study was about 1 μg AFM_1/liter milk. No relationship appeared to exist between AFM_1 content and daily milk yield.

The study by Van der Linde et al. (1964) was followed by various other studies on the carry-over of AFB_1 to milk. The data obtained by several investigators (Van der Linde et al., 1964; Masri et al., 1969; Kiermeier, 1973; McKinney et al., 1973; Polan et al., 1974) suggest that 3–6 days of constant daily ingestion of AFB_1 is required before steady-state excretion of AFM_1 in the milk is achieved, whereas AFM_1 becomes undetectable 2–4 days after withdrawal from contaminated diet. Excretion of AFM_1 in milk appeared to vary from animal to animal, from day to day, and from one milking to the next (Kiermeier et al., 1977).

In a review by Rodricks and Stoloff (1977), the ratios of the concentration of AFB_1 in cattle feed to that of AFM_1 in milk were estimated to be 34–1600 with an average ratio near 300. In another review by Sieber and Blanc (1978), the

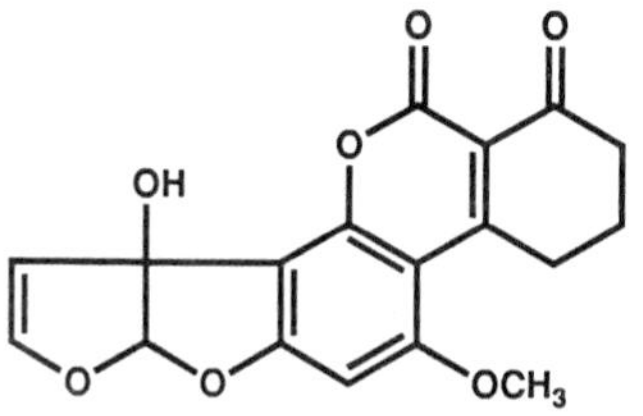

FIGURE 2 Structures of AFM_1 and AFM_4.

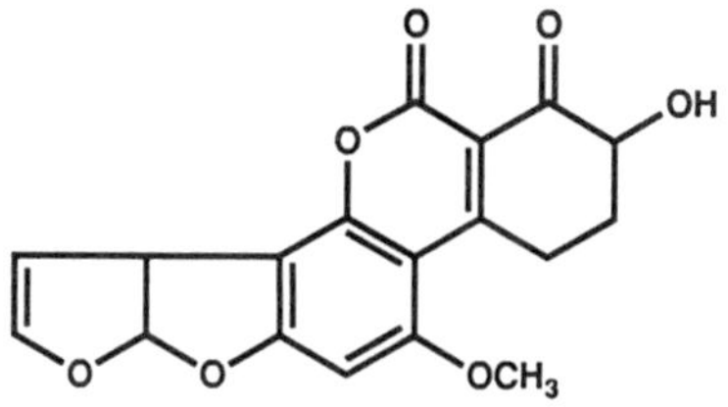

excreted amount of AFM_1 was calculated to be 0–4% of the ingested AFB_1, with an average amount of about 1%. These authors also showed a linear relationship between AFB_1 intake (mg/kg) and AFM_1 in milk (μg/liter). The slope of the linear relationship is approximately 0.7.

Patterson et al. (1980) and Lafont et al. (1980) studied conversion of AFB_1 at or below existing official limits of AFB_1 in the feed, that is, 10–20 μg/kg (Van Egmond, 1989). At 10 μg/kg of AFB_1 (Patterson et al., 1980) in feed, AFM_1 in milk was 0.01–0.33 μg/liter with a mean value of 0.19 μg/liter over a period of 7 days. An average of 2.2% of ingested AFB_1 appeared in the milk as AFM_1. In a comparable study conducted by Lafont et al. (1980), the average conversion ratio of AFB_1 to AFM_1 at low AFB_1 doses was 0.78%, a value considerably smaller than the other estimate. These estimated percentages of 0.78% and 2.2%, however, fall within the range of 0–4% given by Sieber and Blanc (1978), reflecting the wide variation of this ratio.

In addition to AFM_1, AFQ_1 and aflatoxicol were identified as trace metabolites in the milk of goats dosed with very low levels of [^{14}C]AFB_1 (0.23 mg/animal; Helferich et al., 1986). These minor metabolites are most likely of little health significance. However, a new hydroxy derivative of AFB_1 found in milk (Lafont et al., 1986a), known as AFM_4, might be of greater significance (see Chapter 17). This metabolite was found to co-occur with AFM_1 in certain milk at up to 16% of AFM_1 (Lafont et al., 1986b). Tests indicated that AFM_4 might be more toxic and carcinogenic than AFB_1 and AFM_1 (Lafont and Lafont, 1987). Structures of AFM_1 and AFM_4 are shown in Figure 2.

ACKNOWLEDGMENT

The authors thank Paul Kuzumicky for his able assistance in manuscript preparation. This research was supported by the USDA Western Regional Research Project, W-122 and by the NIEHS Center for Environmental Health Sciences at University of California, Davis, Davis grant ES05707.

REFERENCES

Autrup, H., Seremet, T., Wakhisi, J., and Wasunna, A. (1987). Aflatoxin exposure measured by urinary excretion of aflatoxin B_1–guanine adduct and hepatitis B virus infection in areas with different liver cancer incidence in Kenya. *Cancer Res.* **47**, 3430–3433.

Bennett, R. A., Essigmann, J. M., and Wogan, G. N. (1981). Excretion of an aflatoxin–guanine adduct in the urine of aflatoxin B_1-treated rats. *Cancer Res.* **41**, 650–654.

Campbell, T. C., Caedo, J. P., Bulato-Jayme, J., Salamat, L., and Engek, R. W. (1970). Aflatoxin M_1 in human urine. *Nature (London)* **227**, 403–404.

Chhabra, R. S., Pohl, R. J., and Fouts, J. R. (1974). A comparative study of xenobiotic metabolizing enzymes in liver and intestine of various animal species. *Drug Metab. Disp.* **2**, 443–447.

Ch'ih, J. J., and Devlin, T. M. (1984). The distribution and intracellular translocation of aflatoxin B_1 in isolated hepatocytes. *Biochem. Biophys. Res. Commun.* **122**, 1–8.

Coulombe, R. A., and Sharma, R. P. (1985). Clearance and excretion of intratracheally and orally administered aflatoxin B_1 in the rat. *Food Chem. Toxicol.* **23**, 827–830.

Dalezios, J. I., and Wogan, G. N. (1972). Metabolism of aflatoxin B_1 in rhesus monkeys. *Cancer Res.* **32**, 2297–2303.

Dalezios, J. I., Hsieh, D. P. H., and Wogan, G. N. (1973). Excretion and metabolism of orally administered aflatoxin B_1 by rhesus monkeys. *Food. Cosmet. Toxicol.* **11**, 605–616.

Degen, G. H., and Neumann, H. G. (1978). The major metabolite of aflatoxin B_1 in the rat is a glutathione conjugate. *Chem. Biol. Interact.* **22**, 239–255.

Essigmann, J. M., Croy, R. G., Bennett, R. A., and Wogan, G. N. (1982). Metabolic activation of aflatoxin B_1: Patterns of DNA adduct formation, removal and excretion in relation to carcinogenesis. *Drug Metab. Rev.* **13**, 581–602.

Groopman, J. D., Donahue, P. R., Zhu, J., Chen, J., and Wogan, G. N. (1985). Aflatoxin metabolism in humans: detection of metabolites and nucleic acid adducts in urine by affinity chromatography. *Proc. Natl. Acad. Sci. U.S.A.* **82**, 6492–6496.

Groopman, J. D., Cain, L. G., and Kensler, T. W. (1988). Aflatoxin exposure in human populations: Measurements and relationship to cancer. *CRC Clin. Rev. Toxicol.* **19**, 113–145.

Hanigan, H. M., and Laishes, B. A. (1984). Toxicity of aflatoxin B_1 in rat and mouse hepatocytes *in vivo* and *in vitro*. *Toxicology* **30**, 185–193.

Hartiala, K. (1977). Metabolism of foreign substances in the gastrointestinal tract. *In* "Handbook of Physiology Reactions of Environmental Agents" (D. H. K. Lee, H. L. Falk, S. D. Murphy, and S. R. Geiger, eds.), pp. 375–388. American Physiology Society, Bethesda, Maryland.

Heathcote, J. G., and Hibbert, J. R. (1978). Metabolism. "Aflatoxins: Chemical and Biological Aspects," pp. 134–150. Elsevier Scientific, New York.

Helferich, W. G., Baldwin, R. L., and Hsieh, D. P. H. (1986). [^{14}C]-Aflatoxin B_1 metabolism in lactating goats and rats. *J. Anim. Sci.* **62**, 697–705.

Holeski, C. J., Eaton, D. L., Monroe, D. H., and Bellamy, G. M. (1987). Effects of phenobarbitol on the biliary excretion of aflatoxin P_1-glucuronide and aflatoxin B_1-S-glutathione in the rat. *Xenobiotica* **17**, 139–153.

Hsieh, D. P. H., Wong, Z. A., Wong, J. J., Michas, C., and Ruebner, B. H. (1977). Comparative metabolism of aflatoxin. *In* "Mycotoxins in Human and Animal Health" (J. V. Rodricks, C. W. Hesseltine, and M. A. Mehlman, eds.), pp. 37–50. Pathotox, Park Forest South, Illinois.

Kiermeier, F. (1973). Über die Aflatoxin-M-Ausscheidung in Kuhmilch in Abhängigkeit von der aufgenommenen Aflatoxin-B_1-Menge. *Milchwissensch.* **28**, 683–685.

Kiermeier, F., Weiss, G., Behringer, G., Miller, M., and Ranfft, K. (1977). Presence and content of aflatoxin M_1 in milk supplied to a dairy. *Z. Lebensm. Unters. Forsch.* **163**, 71–74.

Kumagai, S. (1989). Intestinal absorption and excretion of aflatoxin in rats. *Toxicol. Appl. Pharmacol.* **97**, 88–97.

Lafont, P., and Lafont, J. (1987). Génotoxicité des hydroxy-aflatoxines du lait. Abstract of lecture at Journées d'Etudes: Moisissures et levures indesirable en industrie agro-alimentaire, November 25–27. Paris, France.

Lafont, P., Lafont, J., Mousset, S., and Frayssinet, C. (1980). Etude de la contamination du lait de vache lors de l'ingestion de faibles quantités d'aflatoxine. *Ann. Nutr. Alim.* **34**, 699–708.

Lafont, P., Platzer, N., Siriwardana, M. G., Sarfati, J., Mercier, J., and Lafont, J. (1986a). Un novel hydroxy-dérivé de l'aflatoxine B_1: l'aflatoxine M_4. I. Production *in vitro*—structure. *Microbiol. Alim. Nutr.* **4**, 65–74.

Lafont, P., Siriwardana, M. G., Sarfati, J., Debeaupuis, J. P., and Lafont, J. (1986b). Un novel hydroxy-dérivé de l'aflatoxine M₄. II. Méthode de dosage mise en évidence de contaminations de laits commerciaux. *Microbiol. Alim. Nutr.* **4**, 141–145.

Luthy, J., Zqeifel, U., and Schlatter, C. (1980). Metabolism and tissue distribution of (¹⁴C)-aflatoxin B₁ in pigs. *Food Cosmet. Toxicol.,* **18**, 253–256.

McKinney, J. D., Cavanaugh, G. C., Bell, J. T., Hoversland, A. S., Nelson, D. M., Pearson, J., and Selkirk, R. J. (1973). Effects of ammoniation on aflatoxins in rations fed lactating cows. *J. Am. Oil Chem. Soc.* **50**, 79–84.

Masri, M. S., Garcia, V. C., and Page, J. R. (1969). The aflatoxin M₁ content of milk from cows fed known amounts of aflatoxin. *Vet. Rec.* **84**, 146–147.

Muller, N., and Petzinger, E. (1988). Hepatocellular uptake of aflatoxin B₁ by nonionic diffusion. Inhibition of bile acid transport by interference with membrane lipids. *Biochim. Biophys. Acta* **938**, 334–344.

Neal, G. E. (1987). Influences of metabolism: Aflatoxin metabolism and its possible role relationships with disease. In "Natural Toxicants in Foods: Progress and Prospects" (D. H. Watson, ed.), pp. 125–168. Ellis Horwood, Chichester.

Patterson, D. S. P., Glancy, E. M., and Roberts, B. A. (1980). The carry-over of aflatoxin M₁ into the milk of cows fed rations containing a low concentration of aflatoxin B₁. *Food Cosmet. Toxicol.* **18**, 35–37.

Polan, C. E., Hayes, J. R., and Campbell, T. C. (1974). Consumption and fate of aflatoxin B₁ by lactating cows. *J. Agric. Food Chem.* **22**, 635–638.

Portman, R. S., Plowman, K. M., and Campbell, T. C. (1970). On mechanisms affecting species susceptibility of aflatoxin. *Biochim. Biophys. Acta* **208**, 487–495.

Raj, H. G., and Lotlikar, P. D. (1984). Urinary excretion of thiol conjugates of aflatoxin B₁ in rats and hamsters. *Cancer Let.* **22**, 125–133.

Rodricks, J. V., and Stoloff, L. (1977). Aflatoxin residues from contaminated feed in edible tissues of foodproducing animals. In "Mycotoxins in Human and Animal Health" (J. V. Rodricks, C. W. Hesseltine, and M. A. Mehlman, eds.), pp. 67–79. Pathotox, Park Forest South, Illinois.

Sabbioni, G., Skipper, P. L., Buchi, G., and Tannenbaum, S. R. (1987). Isolation and characterization of the major serum albumin adduct formed by aflatoxin B₁ in vivo in rats. *Carcinogenesis* **8**, 819–824.

Seiber, R., and Blanc, B. (1978). Zur Ausscheidung von Aflatoxin M₁ in die Milch und dessen Vorkommen in Milch und Milchprodukten—Eine Literaturübersicht. *Mitt. Gebiete Lebensm. Hyg.* **69**, 477–491.

Steyn, M., Pitout, M. J., and Purchase, I. F. H. (1971). A comparative study on aflatoxin B₁ metabolism in mice and rats. *Brit. J. Cancer* **25**, 291–297.

Unger, P. D., Mehendale, H. M., and Hayes, A. W. (1977). Hepatic uptake and disposotion of aflatoxin B₁ in isolated perfused rat liver. *Toxicol. Appl. Pharmacol.* **41**, 523–534.

Vainio, H., and Hietanen, E. (1979). Role of extrahepatic metabolism. In "Concepts in Drug Metabolism" (P. Jenner and B. Testa, eds.), pp. 251–284. Marcel Dekker, New York.

Van der Linde, J. A., Frens, A. M., de Iongh, M., and Vles, R. O. (1964). Inspection of milk from cows fed aflatoxin-containing groundnut meal. *Tijdschr. Diergeneesk,* **89**, 1082–1088.

Van Egmond, H. P. (1989). Current situation on regulations for mycotoxins. Overview of tolerances and status of standard methods of sampling and analysis. *Food Add. Contam.* **6**, 139–188.

Wei, C. I., Marshall, M. R., and Hsieh, D. P. H. (1985). Characterization of water-soluble glucuronide and sulphate conjugates of aflatoxin B₁. 1. Urinary excretion in monkey, rat, and mouse. *Food Chem. Toxicol.* **23**, 809–819.

Wilson, R., Ziprin, R., Ragsdale, S., and Busbee, D. (1985). Uptake and vascular transport of ingested aflatoxin. *Toxicol. Lett.* **29**, 169–176.

Wogan, G. N., Edwards, G. S., and Shank, R. C. (1967). Excretion and tissue distribution of radioactivity from aflatoxin B₁-¹⁴C in rats. *Cancer Res.* **27**, 1729–1736.

Wong, Z. A., and Hsieh, D. P. H. (1978). Aflatoxicol: Major aflatoxin B₁ metabolite in rat plasma. *Science* **200**, 325–327.

Wong, Z. A., and Hsieh, D. P. H. (1980). The comparative metabolism and toxicokinetics of aflatoxin B_1 in the monkey, rat, and mouse. *Toxicol. Appl. Pharmacol.* **55,** 115–125.

Wong, Z. A., Wei, C.-I., Rice, D. W., and Hsieh, D. P. H. (1981). Effects of phenobarbital pretreatment on the metabolism and toxicokinetics of aflatoxin B_1 in the Rhesus monkey. *Toxicol. Appl. Pharmacol.* **60,** 387–397.

Zhu, J., Zhang, L., Hu, X., Chen, J., Xu, Y., Fremy, J., and Chu, F. S. (1987). Correlation of dietary aflatoxin B_1 levels with excretion of aflatoxin M_1 in human urine. *Cancer Res.* **47,** 1848–1852.

5

Nonhepatic Disposition and Effects of Aflatoxin B_1

❖

Roger A. Coulombe, Jr.

INTRODUCTION

Although the vast majority of the work on aflatoxin B_1 (AFB_1) toxicology has focused appropriately on hepatic toxicity, AFB_1 also elicits significant effects in other organ systems. Nonhepatic sequelae may contribute, to varying degrees, to the overall toxicity of AFB_1. In most cases, however, such effects in organs other than the liver are secondary. In any event, AFB_1 distributes into a variety of extrahepatic tissues; subsequent effects, such as neoplasms, are common in several nonhepatic organs. The objective of this chapter is to provide an overview of the disposition and effects of AFB_1 in several organ systems. Particular emphasis is given to the pulmonary, gastrointestinal, renal, nervous, reproductive, and immune systems.

RESPIRATORY SYSTEM

Interest in the possible effects of AFB_1 in the pulmonary tract has been stimulated by survey data indicating that, in addition to the more-important dietary route of exposure, people in several occupational settings are exposed to AFB_1 via the respiratory tract when AFB_1-contaminated dusts are inhaled. Numerous surveys have shown that AFB_1 virtually always is present in grain dusts at levels often exceeding those found in contaminated food (Baxter *et al.*, 1981). For example, aflatoxin concentrations as high as 1814 ppb were found in air-borne dust samples collected from several grain terminals in the midwestern United States (Sorenson *et al.*, 1984). Air sampled from a corn processing plant contained 107 ng/m^3 AFB$_1$. The daily occupational exposure to AFB$_1$ in this plant was between 40 and 856 ng, based on a respiration rate of 1 m^3 per hr (Burg *et al.*, 1981). Sorenson and co-workers (Sorenson *et al.*, 1981) found dust samples containing 130 ppb AFB$_1$ in another survey. Particles 7–11 µm in diameter had an AFB$_1$ content of 695 ppb; one sample under 7 µm in diameter showed 1814 ppb AFB$_1$. Airborne AFB$_1$ concentrations in a grain elevator were much lower in central Illinois, reflecting the relatively low levels of AFB$_1$ generally found in grains grown in the upper Midwest (Zennie, 1984). Dusts generated from peanut shelling and processing operations also were shown to contain AFB$_1$ (Sorenson *et al.*, 1984).

Evidence linking inhaled AFB$_1$ to human lung cancer is tenuous. Two epidemiological studies found that workers in a Dutch peanut oil processing plant exposed to AFB$_1$-contaminated dusts experienced a significantly greater incidence of upper respiratory (trachea and bronchus) as well as liver tumors than unexposed cohorts (Van Nieuwenhuize *et al.*, 1973; Hayes *et al.*, 1984). Isolated case studies also have described the development of tumors in people working with AFB$_1$-contaminated dusts (Deger, 1976; Dvorakova, 1976). A significantly higher amount of aflatoxins was found in the sera of patients with lung cancer (Cusumano, 1991). In that study, no attempt was made to determine history of aflatoxin exposure, but most of the patients lived in agricultural settings where inhalation of aflatoxins in grain dust was possible.

AFB$_1$ induced a 100% incidence of pulmonary adenomas in strain A mice when administered three times weekly for 4 weeks (Wieder *et al.*, 1968). AFB$_1$ was approximately 4 times as active as utrethan and was equally as potent as aniline mustard in inducing lung tumors in this strain (Wieder *et al.*, 1968). In most other studies, however, lung tumors were observed rarely in long-term cancer studies in which AFB$_1$ was given orally or intraperitoneally. In an early study, AFB$_1$ administered via the pulmonary tract caused lung tumors *in vivo*. In that study, tracheal carcinomas were elicited in 3 of 6 rats given 0.3 mg AFB$_1$ administered intratracheally twice weekly for 3 weeks (Dickens *et al.*, 1966). Two pharmacokinetic studies characterized the disposition of AFB$_1$ in rats after pulmonary administration. As might be expected, AFB$_1$ appeared in the blood more rapidly when it was dosed intratracheally instead of orally, although the overall plasma concentration–time plots of AFB$_1$ did not differ significantly between the two routes at later time periods (Coulombe and Sharma, 1985a). Following oral exposure, the amount of AFB$_1$ distributed into lungs and associated tissues was about half that distributed into the liver (Coulombe and Sharma, 1985a). Dust-adsorbed AFB$_1$ administered intratracheally distributed into the liver, kidney, spleen, thymus, and brain as well as into the lung and major airways; at 3 hr postadministration, the level of

AFB$_1$–DNA adducts in the trachea was higher than in the liver (Coulombe *et al.,* 1991).

Tissues derived from various portions of the lung have been used as *in vitro* models to study AFB$_1$ metabolism and activation in the lung. For example, AFB$_1$ is activated metabolically by cultured intact mammalian tracheal epithelium (Coulombe *et al.,* 1984, 1986; Ball *et al.,* 1990; Wilson *et al.,* 1990; Ball and Coulombe, 1991) in cultured human bronchus (Autrup *et al.,* 1979; Stoner *et al.,* 1982), in human epithelioid lung cells (Wang and Cerutti, 1979), and in rabbit lung microsomes (Daniels *et al.,* 1990). These studies have demonstrated that the two major AFB$_1$–DNA adducts, the AFB$_1$–N^7-guanine and the "ring-opened" AFB$_1$–formamidopyrimidine, also are formed in the pulmonary tract in ratios similar to those seen in hepatic tissues of AFB$_1$-treated animals. In addition, the repair of these adducts in cultured lung cells follows similar kinetics, that is, the AFB$_1$–N^7-guanine adduct is repaired preferentially and repair is primarily enzymatic (Ball *et al.,* 1990).

Ultrastructural analysis of upper airway tissues cultured with AFB$_1$ indicates that this mycotoxin is selectively toxic to only one population of airway cells. In cultures of hamster and rabbit tracheal epithelium, AFB$_1$ elicited significant ultrastructural changes in the nonciliated cells but other cell types, such as the ciliated and basal cells, were relatively unaffected (Coulombe *et al.,* 1986; Wilson *et al.,* 1990). This selective toxicity is probably largely the result of the distribution of cytochrome P450 isozymes in the lung and airways, where nonciliated cells contain the majority of the components of the cytochrome P450 system (Coulombe *et al.,* 1986; Plopper *et al.,* 1987). Nonciliated cells in rabbit upper airways contain two forms of rabbit lung pulmonary cytochrome P450 isozymes that are active in converting AFB$_1$ to mutagenic species (Coulombe *et al.,* 1986). Similarly, using immunohistochemical methods, Wild *et al.,* (1991) identified AFB$_1$–DNA adducts only in some individual cells in the main conducting airways of intraperitoneally (ip) dosed rats.

As in the liver, considerable species differences exist with respect to AFB$_1$ activation, detoxification, and action in the lung. Cultured epithelial cells or microsomes from rabbit trachea were more active in converting AFB$_1$ to DNA-binding species than those from hamster, in which AFB$_1$-detoxifying enzymes and reactions predominate (Ball and Coulombe, 1991). Tissues from rat trachea were relatively inactive in AFB$_1$ activation and detoxification (Ball and Coulombe, 1991). Correspondingly, cultured rabbit trachea was more susceptible to the cytotoxic effects of AFB$_1$ than was cultured hamster trachea, whereas cultured rat trachea was relatively refractory (Wilson *et al.,* 1990). Daniels *et al.* (1990) also found the rabbit lung microsomes activated AFB$_1$ to a DNA-binding species, but were relatively inactive in converting the parent compound into stable detoxified metabolites such as AFQ$_1$ and AFM$_1$. Although these *in vivo* and *in vitro* studies indicate that inhaled AFB$_1$ may be a significant health hazard, a firmer determination of risk associated with inhaled AFB$_1$ awaits the results of further study.

RENAL SYSTEM

Various portions of the renal nephron are exposed to AFB_1 or metabolites because a significant portion of ingested mycotoxin is excreted via the urine. For example, in cattle given radiolabeled AFB_1, renal tissue contained the highest concentration of AFB_1 and metabolites (Hayes *et al.*, 1977). Early experiments with laboratory rats demonstrated that oral AFB_1 is a renal carcinogen (Epstein *et al.*, 1969). The ultrastructural characteristics of renal neoplasms later were characterized in Wistar rats receiving 1–3 ppm AFB_1 (Merkow *et al.*, 1973). The tumors, which were present in 50% of the animals receiving 3 ppm AFB_1, were most likely of tubular origin and were characterized by loss of apical orientation of brush borders. Significant strain differences exist in the development of renal tumors in response to AFB_1. For example, the kidneys of Wistar rats are especially prone to AFB_1-induced tumors (Epstein *et al.*, 1969), whereas Fischer rats show no renal neoplasms (Wogan and Newberne, 1967; Wogan *et al.*, 1974; Gurtoo *et al.*, 1985) or neoplasms at a very low incidence (Ward *et al.*, 1975) after long-term AFB_1 given by various routes. Manson *et al.* (1987) also reported that the renal ultrastructure in Fischer rats was unaffected after a single ip dose followed by 23 weeks on a diet containing 1 ppm AFB_1.

Although the majority of AFB_1-induced tumors are of hepatic origin in the rat, this toxin appears to have equal activity in liver and kidneys of Syrian golden hamsters (Herrold, 1969). Oral administration of 0.1 mg of a mixture of aflatoxins B and G twice weekly for up to 11 months induced both renal and hepatic tumors in 80% of the hamsters; ip administration of 0.2 mg weekly for up to 8.5 months induced renal tumors in 90% of the hamsters (Herrold, 1969). Renal lesions in both groups of hamsters were characterized by megalocytosis in the proximal tubules.

Although the mouse is the most resistant animal known to the acute and chronic effects of AFB_1, the kidney appears to be the major target organ of AFB_1 in this species. Shortly after administration, this toxin causes massive hemorrhagic lesions of the kidneys. A single dose of AFB_1 suppressed [^{14}C]orotic acid incorporation into RNA by 50% and inhibited RNA synthesis in mouse kidney, but had none of these effects in the liver (Akao *et al.*, 1971). In agreement with this observation, the level of AFB_1–DNA adducts in the kidney was about 6 times higher than that found in the liver after a single ip injection in the mouse (Croy and Wogan, 1981). Conversely, in the rat, in which the main target organ is the liver, the level of adducts in the kidney was a small fraction of that seen in the liver (Croy and Wogan, 1981).

The susceptibility of poultry to AFB_1 is likely to be caused by the effect of this mycotoxin on the renal system. For example, a fatty and hemorrhagic kidney syndrome that commonly occurs in chickens in Africa has been linked to AFB_1 (Dafalla *et al.*, 1987). Thickening of glomerular basement membrane, abnormal development of glomerular epithelial cells, and degenerative changes in renal tubular cells were observed in male broiler chicks given 2.5–5 ppb AFB_1 (Mollenhauer *et al.*, 1989).

In rats as well as in other species, acute AFB_1 exposure significantly alters renal function. A single dose of AFB_1 (100 μg/kg) significantly decreased glomerular filtration, glucose reabsorption, and tubular transport of electrolytes and organic anions in Wistar rats 24 hr after administration, indicating that AFB_1 might exert its effects on glomerular basement membrane and renal tubules (Grosman *et al.*, 1983). Ikegwuonu *et al.* (1980) focused on AFB_1-induced biochemical changes in the kidney in an effort to explain many of the effects cited in this section. These investigators demonstrated that ip AFB_1 depressed activities of renal glutamate–oxaloacetate and pyruvate transaminases, as well as alkaline phosphatase, in rats. In cultured kidney cell lines, AFB_1 induced aggregation and loss of chromatin, mitochondrial degeneration, and loss of microvilli (Yoneyama *et al.*, 1987).

GASTROINTESTINAL SYSTEM

The gastrointestinal tract is exposed to AFB_1 initially via the diet and subsequently to AFB_1 metabolites from the bile, which is the major route of excretion of this toxin. As mentioned earlier, inhaled AFB_1 has been implicated as a human colon carcinogen (Deger, 1976). Experimentally, AFB_1 is a modest colon carcinogen. A small incidence of colon carcinomas results when AFB_1 is administered to Fischer rats either orally or via the diet (Wogan and Newberne, 1967), but this incidence can be increased by concurrent low dietary vitamin A (Newberne and Rogers, 1973; Newberne and Suphakarn, 1977; Suphakarn *et al.*, 1983). Under these conditions, the increased incidence of AFB_1-induced colon tumors occurs with a concomitant decrease in the incidence of liver tumors. Preneoplastic colonic crypts were observed in female Sprague–Dawley rats given AFB_1, but not in female CF-1 mice receiving a similar dose (Tudek *et al.*, 1989). Small intestinal tumors were observed in 6 of 10 Syrian hamsters receiving 0.2 mg AFB_1 ip weekly for up to 8.5 months (Herrold, 1969).

Cultured colon tissues have been used by Autrup *et al.* (1979, 1980) to study AFB_1 metabolism and activation in this tissue. Colon tissues from rat and human were capable of activating AFB_1 to form DNA adducts. As in hepatic and pulmonary systems, the principal AFB_1 adduct formed in cultured colon was the AFB_1–N^7-guanine adduct (Autrup *et al.*, 1979). Colon tissues formed fewer adducts than liver *in vivo*, although the qualitative patterns of adducts seen in the colon were similar. Rat small intestinal microsomes activated AFB_1 to mutagens and, based on cytochrome P450 content, were as efficient as hepatic microsomes (Walters and Combes, 1985). However, unlike hepatic microsomes, the AFB_1-activating ability of intestinal microsomes was not affected by the *in vivo* administration of phenobarbital or β-naphthoflavone (Walters and Combes, 1985).

The pancreas also is affected by AFB_1. In Fischer rats, decreased numbers of zymogen granules, swollen mitochondria, and dilated rough endoplasmic reticulum were among the ultrastructural effects seen 24 hr or more after a single ip dose of 250 μg AFB_1 (Rao *et al.*, 1974).

In domestic animals, dietary AFB_1 alters many aspects of rumen function. For example, acute AFB_1 exposure decreased rumen motility in cows (Cook *et al.*, 1986). AFB_1 reduced cellulose breakdown and production of volatile fatty acids and ammonia in both *in vivo* and *in vitro* rumen model systems (Mertens, 1977). Such effects on the gastrointestinal tract of domestic animals may contribute to the decreases in weight gain, feed conversion, and milk production observed when animals are fed AFB_1-contaminated feed.

NERVOUS SYSTEM

Investigations into the possible effects of AFB_1 on nervous system function were stimulated in part by finding this toxin in tissue samples from patients with Reye's syndrome, a pediatric disease characterized by cerebral edema and neuronal degeneration (Chaves-Carballo *et al.*, 1976). Ryan *et al.* (1979) reported finding AFB_1 in the blood of six of seven patients during the acute phase of this disease, as well as in the livers of those that subsequently expired. Later case–control studies, however, revealed that serum and urine AFB_1 concentrations in cases of Reye's syndrome and their families and controls were no different (Nelson *et al.*, 1980). Although these and other authors have concluded that AFB_1 probably is not an important cause of Reye's syndrome, experimentally this toxin causes biochemical alterations in the nervous systems of various animal species.

Repeated ip injections of AFB_1 (16 μg/kg daily for 6 weeks) increased central and peripheral nervous system Na^+/K^+-ATPase, β-glucuronidase, and β-galactosidase while inhibiting Mg^{2+}-ATPase in male Wistar rats (Ikegwuonu, 1983). Acute AFB_1 treatment in rats caused decreases in regional brain acetylcholinesterase, whereas chronic treatment resulted in increases in adenohypophyseal acetylcholinesterase (Egbunike and Ikegwuonu, 1984). In chickens, dietary AFB_1 caused a synergistic depression of brain acetylcholinesterase induced by malathion (Ehrich *et al.*, 1985).

Several reports have showed that AFB_1 alters homeostasis of central nervous system biogenic amines. In rats, AFB_1 caused a decrease in brain serotonin and its metabolite 5-hydroxyindoleacetic acid (Weekly *et al.*, 1985) and also inhibited brainstem tryptophan metabolism, as evidenced by decreased tryptophan 2,3-dioxygenase activity (Weekly and Llewellyn, 1984; Weekly *et al.*, 1985). These data suggest that AFB_1 may decrease the free tryptophan pool or alter biosynthesis of serotonin. A similar decrease in whole brain serotonin was observed following 45-day dietary AFB_1 treatment in hamsters (Weekly *et al.*, 1978). In chickens, however, a single dose of 3 mg/kg AFB$_1$ increased brain serotonin concentrations (Ahmed and Singh, 1984).

Concentrations of brain catecholamines are altered by AFB$_1$. In rats, repeated intragastric AFB$_1$ resulted in decreases in dopamine in several brain regions; this effect was accompanied by decreases in the dopamine metabolites homovanillic acid and dihydroxyphenylacetic acid (Coulombe and Sharma, 1985b). In chickens, a single dose of AFB$_1$ caused a depression in the concentration of norepinephrine (Ahmed and Singh, 1984). In ICR mice, a similar decrease in central catecholamines was observed after AFB$_1$ treatment (Weekly, 1991). Another study showed that regional brain catecholamine concentrations were increased by AFB$_1$ treatment in CD-1 mice, although no dose-related changes were noted in activities of enzymes important in biogenic amine synthesis and degradation, for example, DOPA-decarboxylase and monoamine oxidase (Jayasekara et al., 1989). That catecholamine biosynthesis might be depressed by AFB$_1$ was indicated further by the finding that this toxin caused decreases in brain concentrations of tyrosine, the major catecholamine biosynthetic precursor (Weekly et al., 1989). Collectively, these biochemical alterations in the central and peripheral nervous systems have been postulated to contribute to some of the many signs of overt toxicity in domestic animals that are observed after consumption of AFB$_1$-contaminated feeds, for example, changes in feeding patterns and appetite.

In addition to such biochemical effects, exposure to AFB$_1$ also causes tumors of the nervous system. Among tumors in a variety of other sites, ip administration of AFB$_1$ elicited tumors in the central and peripheral nervous systems of pregnant Sprague–Dawley rats (Goerttler et al., 1980). Several nonepithelial neurogenic tumors were observed, for example, schwannomas, gliomas, meningiomas, and granular cell tumors.

REPRODUCTIVE SYSTEM

Aflatoxicosis alters the reproductive efficiency of both male and female domestic animals, particularly poultry. Sexual maturity of young male Japanese quail, as assessed by plasma testosterone concentrations and testicular weight, was delayed when they were fed a diet containing 10 ppm AFB$_1$ soon after hatching (Ottinger and Doerr, 1980). In that study, age differences with respect to AFB$_1$ sensitivity were discovered: quail given the AFB$_1$ diet between 7 and 21 days of age recovered slightly earlier than those treated between 14 and 28 days after hatching. AFB$_1$ also affects the reproductive potential of mature poultry. Reductions in semen volume, testes weight, spermatocrit, and plasma testosterone resulted when mature male leghorn chickens were given a diet containing 20 ppm AFB$_1$ (Sharlin et al., 1981). A depression in egg output as well as pathologic changes in ovaries was observed in laying hens given dietary AFB$_1$ (Hafez et al., 1982).

Some reproductive end points of rats are affected by AFB$_1$. A single ip dose of 150 µg AFB$_1$ significantly depressed Leydig cell function and responsiveness to human chorionic gonadotropin, although neither sperm production nor testosterone secreting ability was affected (Egbunike, 1982). More severe effects were seen in male and female rats chronically treated with AFB$_1$, for example, severe testicular degeneration, impaired spermatocytogenesis, smaller litter sizes, and higher embryo mortality (Egbunike et al., 1980). Biochemical end points such as sperm or testicular enzyme activities have been shown to be altered by in vivo AFB$_1$ treatment (Ikegwuonu et al., 1980; Egbunike, 1981).

IMMUNE SYSTEM

Persistent observations that consumption of AFB$_1$-contaminated feed lowers disease resistance in domestic animals encouraged investigations into the possible effects of this mycotoxin on the immune system. Repeated low-dose administration of AFB$_1$ is now known to modulate several parameters of cell- and antibody-mediated immune function in avian and mammalian species. Importantly, immunotoxic effects often are seen in the absence of gross clinical pathology.

Reductions in cell-mediated immunity, as measured by a delayed hypersensitive skin test, were observed in young broiler chicks receiving 100 ppb or more of a mixture of AFB$_1$ and AFB$_2$ (Giambrone et al., 1985). A depression of phagocytic efficiency was seen in addition to a depression of delayed hypersensitivity in chickens fed 0.3 ppm AFB$_1$ (Kadian et al., 1988). Similar decreases in parameters related to cell-mediated immunity, such as graft vs host and cutaneous basophil hypersensitivity, were observed in chick hatchlings that were treated in ovo with 0.1 µg AFB$_1$ (Dietert et al., 1985). Steers fed a corn ration naturally contaminated with AFB$_1$ showed decreases in cutaneous hypersensitivity (Richard et al., 1983). Long-term AFB$_1$ treatment of weanling rats caused, in addition to a depletion of cell populations of the thymus, reduced bone marrow and red and white cell counts, a depression of macrophage numbers, and reduced phagocytic capacity (Raisuddin et al., 1990).

AFB$_1$ inhibits lymphoblastogenesis in tests that rely on mitogen-stimulated uptake of [^{3}H]thymidine into various lymphocyte populations. For example, in bovine peripheral blood lymphocytes, AFB$_1$ inhibited phytohemagglutinin, pokeweed mitogen, and Mycobacterium bovis antigen-stimulated lymphoblastogenesis (Paul et al., 1977; Bodine et al., 1984). A significant depression was found in T cell-dependent functions of splenic lymphocytes isolated from CD-1 mice that were treated in vivo with up to 0.7 mg/kg AFB$_1$ for 2 weeks (Reddy and Sharma, 1987). In a later study in which the effect of a similar protocol on peripheral blood lymphocytes was investigated, these workers found a similar T cell-specific toxicity (Reddy and Sharma, 1989). In addition, the natural killer cell function of the peripheral blood lymphocytes was affected by AFB_1 treatment.

MISCELLANEOUS

In Syrian hamsters, ip or per os (po) administration of AFB_1 resulted in tumors of the periodontal membrane and Harderian gland, in addition to tumors of the liver, kidney, and small intestine (Herrold, 1969). One report described a variety of extrahepatic tumors in nonhuman primates sampled partway through a long-term study on the effects of AFB_1 (Sieber *et al.*, 1979). In that study, which had been ongoing for 13 years at the time of publication, monkeys (rhesus, cynomolgus, and African green) were given an average total of 363 mg AFB_1, either ip or po, over an average period of 55 months. Of 35 monkeys sampled, 13 developed osteogenic sarcomas of the tibia and radius, gall bladder adenocarcinomas, pancreatic carcinomas, and one papillary carcinoma of the urinary bladder, in addition to hepatocellular carcinomas.

SUMMARY

Although AFB_1 is a classic hepatotoxicant and hepatocarcinogen, clearly tissues other than the liver are affected to varying degrees; extrahepatic effects probably contribute to the sequelae of this natural toxin. Factors such as route of administration, dose, and frequency of dose and species, strain, age, and sex of the animal appear to affect the degree of extrahepatic involvement.

Synergistic factors also appear to increase tissue involvement in response to AFB_1 exposure. For example, when Sprague–Dawley rats were maintained on a diet containing diethylnitrosamine following 15 daily po doses of AFB_1, a high incidence of lung angiosarcomas and stomach squamous carcinomas were observed in addition to hepatocellular carcinomas (Newberne and Connor, 1980). No extrahepatic tumors were observed when animals were given these carcinogens separately. Because real-life exposures to AFB_1 involve simultaneous exposure to other natural and synthetic toxins and carcinogens, AFB_1 also may act synergistically to contribute to a variety of tumor types. The immunosuppressive effects of AFB_1 also may increase susceptibility to a variety of infectious diseases as well as to cancer. Future epidemiological studies may identify extrahepatic cancers associated with dietary AFB_1 exposure.

In addition to the important dietary route of exposure, inhalation of AFB_1-contaminated grain dusts may result in hepatic and extrahepatic tumors. Additional epidemiological and animal studies are needed to determine more accurately and carefully the risk posed by inhalation of AFB_1 in the workplace.

ACKNOWLEDGMENTS

This work was supported in part by Public Health Services Grant NIH ES04813 and by the Utah Agricultural Experiment Station, for which this chapter is designated publication number 4265.

REFERENCES

Ahmed, N., and Singh, U. S. (1984). Effect of aflatoxin B_1 on brain serotonin and catecholamines in chickens. *Toxicol. Lett.* **21,** 365–367.

Akao, M., Kuroda, K., and Wogan, G. N. (1971). Aflatoxin B_1: The kidney as a site of action in the mouse. *Life Sci.* **10,** 495–501.

Autrup, H., Essigman, J. M., Croy, R. G., Trump, B. F., Wogan, G. N., and Harris, C. C. (1979). Metabolism of aflatoxin B_1 and identification of the major aflatoxin B_1–DNA adducts formed in cultured human bronchus and colon. *Cancer Res.* **39,** 694–698.

Autrup, H., Schwartz, R. D., Essigman, J. M., Smith, L., Trump, B. F., and Harris, C. C. (1980). Metabolism of aflatoxin B_1, benzo[*a*]pyrene, and 1,2-dimethylhydrazine by cultured rat and human colon. *Terat. Carc. Mutagen.* **1,** 3–13.

Ball, R. W., and Coulombe, R. A. (1991). Comparative biotransformation of aflatoxin B_1 in mammalian airway epithelium. *Carcinogenesis* **12,** 305–310.

Ball, R. W., Wilson, D. W., and Coulombe, R. A. (1990). Comparative formation and removal of aflatoxin B_1–DNA adducts in cultured mammalian tracheal epithelium. *Cancer Res.* **50,** 4918–4922.

Baxter, C. C., Wey, H. E., and Burg, W. R. (1981). A prospective analysis of the potential risk associated with inhalation of aflatoxin-contaminated grain dusts. *Food Cosmet. Toxicol.* **19,** 765–769.

Bodine, A. B., Fisher, S. F., and Gangjee, S. (1984). Effect of aflatoxin B_1 and major metabolites on phytohemagglutinin-stimulated lymphoblastogenesis of bovine lymphocytes. *J. Dairy Sci.* **67,** 110–114.

Burg, W. R., Shotwell, O. R., and Saltzman, B. E. (1981). Measurements of airborne aflatoxins during the handling of contaminated corn. *Am. Ind. Hyg. Assoc. J.* **42,** 1–11.

Chaves-Carballo, E., Ellefson, R. D., and Gomez, M. R. (1976). An aflatoxin in the liver of a patient with Reye–Johnson syndrome. *Mayo Clin. Proc.* **51,** 48–50.

Cook, W. O., Richard, J. L., Osweiler, G. D., and Trampel, D. W. (1986). Clinical and pathologic changes in acute bovine aflatoxicosis; Rumen motility and tissue and fluid concentrations of aflatoxins B_1 and M_1. *Am. J. Vet. Res.* **47,** 1817–1825.

Coulombe, R. A., and Sharma, R. P. (1985a). Clearance and excretion of intratracheally and orally administered aflatoxin B_1 in the rat. *Food Chem. Toxicol.* **23,** 827–830.

Coulombe, R. A., and Sharma, R. P. (1985b). Effect of repeated dietary exposure of aflatoxin B_1 on brain biogenic amines and metabolites in the rat. *Toxicol. Appl. Pharmacol.* **80,** 496–501.

Coulombe, R. A., Wilson, D. W., and Hsieh, D. P. H. (1984). Metabolism, DNA binding and cytotoxicity of aflatoxin B_1 in tracheal explants from Syrian hamster. *Toxicology* **32,** 117–130.

Coulombe, R. A., Wilson, D. W., Hsieh, D. P., Plopper, and Serabjit-Singh, C. J. (1986). Metabolism of aflatoxin B_1 in the upper airways of the rabbit: The role of the nonciliated tracheal epithelial cell. *Cancer Res.* **46,** 4091–4096.

Coulombe, R. A., Huie, J. M., Ball, R. W., Sharma, R. P., and Wilson, D. W. (1991). Pharmacokinetics of intratracheally administered aflatoxin B_1. *Toxicol. Appl. Pharmacol.* **109,** 196–206.

Croy, R. G., and Wogan, G. N. (1981). Quantitative comparison of covalent aflatoxin-DNA adducts formed in rat and mouse livers and kidneys. *J. Natl. Cancer Inst.* **66,** 761–768.

Cusumano, V. (1991). Aflatoxins in sera from patients with lung cancer. *Oncology* **48,** 194–195.

Dafalla, R., Hassan, Y. M., and Adam, S. E. (1987). Fatty and hemorrhagic liver and kidney syndrome in breeding hens caused by aflatoxin B₁ and heat stress in the Sudan. *Vet. Hum. Toxicol.* **29,** 252–254.

Daniels, J. M., Liu, L., Stewart, R. K., and Massey, T. E. (1990). Biotransformation of aflatoxin B₁ in rabbit lung and liver microsomes. *Carcinogenesis* **11,** 823–827.

Deger, G. E. (1976). Aflatoxin-human colon carcinogenesis? *Ann. Int. Med.* **85,** 204–205.

Dickens, F., Jones, H. E. H., and Waynforth, H. B. (1966). Oral subcutaneous and intratracheal administration of carcinogenic lactones and related substances: The intratracheal administration of cigarette tar in the rat. *Br. J. Cancer* **20,** 134–144.

Dietert, R. R., Qureshi, M. A., Nanna, U. C., and Bloom, S. E. (1985). Embryonic exposure to aflatoxin B₁: Mutagenicity and influence on development and immunity. *Environ. Mutagen.* **7,** 715–725.

Dvorakova, I. (1976). Aflatoxin inhalation and alveolar cell carcinoma. *Br. Med. J.* **1,** 691.

Egbunike, G. N. (1981). Histochemical assessment of 3-β-hydroxysteroid dehydrogenase activity in the testes of rats following acute administration of aflatoxin B₁. *Toxicol. Lett.* **9,** 279–282.

Egbunike, G. N. (1982). Steroidogenic and spermatogenic potentials of the male rat after acute treatment with aflatoxin B₁. *Andrologia* **14,** 440–446.

Egbunike, G. N., and Ikeguonu, F. I. (1984). Effect of aflatoxicosis on acetylcholinesterase activity in the brain and adenohypophysis of the male rat. *Neurosci. Lett.* **52,** 171–174.

Egbunike, G. N., Emerole, G. O., Aire, T. A., and Ikegwuonu, F. I. (1980). Sperm production rates, sperm physiology and fertility in rats chronically treated with sublethal doses of aflatoxin B₁. *Andrologia* **12,** 467–475.

Ehrich, M., Driscoll, C., and Gross, W. B. (1985). Effect of dietary exposure of aflatoxin B₁ on resistance of young chickens to organophosphate pesticide challenge. *Avian Dis.* **29,** 715–720.

Epstein, S. M., Bartus, B., and Farber, E. (1969). Renal epithelial neoplasms induced in male Wistar rats by oral aflatoxin B₁. *Cancer Res.* **29,** 1045–1050.

Giambrone, J. J., Diener, U. L., Davis, N. D., Panangala, V. S., and Hoerr, F. J. (1985). Effects of purified aflatoxin on broiler chickens. *Poultry Sci.* **64,** 852–858.

Goerttler, K. Löhrke, H., Schweizer, H. J., and Hesse, B. (1980). Effects of aflatoxin B₁ on pregnant inbred Sprague–Dawley rats and their F₁ generation. A contribution to transplacental carcinogenesis. *J. Natl. Cancer Inst.* **64,** 1349–1354.

Grosman, M. E., Elias, M. M., Comin, E. J., and Garay, E. A. R. (1983). Alterations in renal function induced by aflatoxin B₁ in the rat. *Toxicol. Appl. Pharmacol.* **69,** 310–325.

Gurtoo, H. L., Koser, P. L., Bansal, S. K., Fox, H. W., Sharma, S. D., Mulhern, A. I., and Pavelic, Z. P. (1985). Inhibition of aflatoxin B₁-hepatocarcinogenesis in rats by β-naphthoflavone. *Carcinogenesis* **6,** 675–678.

Hafez, A. H., Megalla, S. E., Abdel-Fattah, H. M., and Kamel, Y. Y. (1982). Aflatoxin and aflatoxicosis. II. Effects of aflatoxin on ovaries and testicles in mature domestic fowls. *Mycopathologia* **77,** 137–139.

Hayes, J. R., Polan, C. E., and Campbell, T. C. (1977). Bovine liver metabolism and tissue distribution of aflatoxin B₁. *J. Agric. Food Chem.* **25,** 1189–1193.

Hayes, R. B., vanNieuwenhuize, J. P., Raatgever, J. W., and ten Kate, F. J. W. (1984). Aflatoxin exposures in the industrial setting: An epidemiological study of mortality. *Food Chem. Toxicol.* **22,** 39–43.

Herrold, K. M. (1969). Aflatoxin induced lesions in Syrian hamsters. *Br. J. Cancer* **23,** 655–660.

Ikegwuonu, F. I. (1983). The neurotoxicity of aflatoxin B₁ in the rat. *Toxicology* **28,** 247–259.

Ikegwuonu, F. I., Egbunike, G. N., Emerole, G. O., and Aire, T. A. (1980). The effects of aflatoxin B₁ on some testicular and kidney enzyme activity in rat. *Toxicology* **17,** 9–16.

Jayasekara, S., Drown, D. B., Coulombe, R. A., and Sharma, R. P. (1989). Alteration of biogenic amines in mouse brain regions by alklyating agents. I. Effects of aflatoxin B₁ on brain monamines concentrations and activities of metabolizing enzymes. *Arch. Environ. Contam. Toxicol.* **18,** 396–403.

Kadian, S. K., Monga, D. P., and Goel, M. C. (1988). Effect of aflatoxin B_1 on the delayed type hypersensitivity and phagocytic activity of reticuloendothelial system in chickens. *Mycopathologia* **104**, 33–36.

Manson, M. M., Green, J. A., and Driver, H. E. (1987). Ethoxyquin alone induces preneoplastic changes in rat kidney whilst preventing induction of such lesions in liver by aflatoxin B_1. *Carcinogenesis* **8**, 723–728.

Merkow, L. P., Epstein, S. M., Slifkin, M., and Pardo, M. (1973). The ultrastructure of renal neoplasms induced by aflatoxin B_1. *Cancer Res.* **33**, 1608–1614.

Mertens, D. R. (1977). Biological effects of mycotoxins upon rumen function in lactating dairy cows. *In* "Interactions of Mycotoxins in Animal Production," pp. 118–136. National Academy of Sciences, Washington, D.C.

Mollenhauer, H. H., Corrier, D. E., Huff, W. E., Kubena, L. F., Harvey, R. B., and Droleskey, R. E. (1989). Ultrastructure of hepatic and renal lesions in chickens fed aflatoxin. *Am. J. Vet. Res.* **50**, 771–777.

Nelson, D. B., Kimbrough, R., Landrigan, P. S., Hayes, A. W., Yang, G. C., and Benanides, J. (1980). Aflatoxin and Reye's syndrome: a case control study. *Pediatrics* **66**, 865–869.

Newberne, P. M., and Connor, M. (1980). Effects of sequential exposure to aflatoxin B_1 and diethylnitrosamine on vascular and stomach tissue and additional target organs in rats. *Cancer Res.* **40**, 4037–4042.

Newberne, P. M., and Rogers, A. M. (1973). Rat colon carcinomas associated with aflatoxin and marginal vitamin A. *J. Natl. Cancer Inst.* **50**, 439–448.

Newberne, P. M., and Suphakarn, V. (1977). Preventative role of vitamin A in colon carcinogenesis in rats. *Cancer* **40**, 2553–2556.

Ottinger, M. A., and Doerr, J. A. (1980). The early influence of aflatoxin upon sexual maturation in the male Japanese quail. *Poultry Sci.* **59**, 1750–1754.

Paul, P. S., Johnson, D. W., Mirocha, C. J., Soper, F. F., Theon, C. D., Mucosplat, C. C., and Weber, A. F. (1977). *In vitro* stimulation of bovine peripheral blood lymphocytes: Suppression of phytomitogen and specific antigen lymphocyte responses by aflatoxin. *Am. J. Vet. Res.* **38**, 2033–2035.

Plopper, C. G., Cranz, D. L., Kemp, L., Serabjit-Singh, C. J., and Philpot, R. M. (1987). Immunohistochemical demonstration of cytochrome P-450 monoxygenase in Clara cells throughout the tracheobronchiolar airways of the rabbit. *Exp. Lung Res.* **13**, 59–68.

Raisuddin, Singh, K. P., Zaidi, S. I., Saxena, A. K., and Ray, P. K. (1990). Effects of aflatoxin on lymphoid cells of weanling rat. *J. Appl. Toxicol.* **10**, 245–250.

Rao, M. S., Svoboda, D. J., and Reddy, J. K. (1974). The ultrastructural effects of aflatoxin B_1 in the rat pancreas. *Virch. Arch. B. Cell Path.* **17**, 149–157.

Reddy, R. V., and Sharma, R. P. (1987). Studies of immune function of CD-1 mice exposed to aflatoxin B_1. *Toxicology* **43**, 123–132.

Reddy, R. V., and Sharma, R. P. (1989). Effects of aflatoxin B_1 on murine lymphocytic functions. *Toxicology* **54**, 31–44.

Richard, J. L., Pier, A. C., Stubblefield, R. D., Shotwell, O. L., Lyon, R. L., and Cutlip, R. C. (1983). Effect of feeding corn naturally contaminated with aflatoxin on feed efficiency, on physiologic, immunologic, and pathologic changes and on tissue residues in steers. *Am. J. Vet. Res.* **44**, 1294–1299.

Ryan, N. J., Hogan, G. R., Hayes, A. W., Unger, P. D., and Siraj, M. Y. (1979). Aflatoxin B_1: Its role in the etiology of Reye's syndrome. *Pediatrics* **64**, 71–74.

Sharlin, J. S., Howarth, B., Thompson, F. N., and Wyatt, R. D. (1981). Decreased reproductive potential and reduced feed consumption in mature white leghorn males fed aflatoxin. *Poultry Sci.* **60**, 2701–2708.

Sieber, S. M., Correa, P., Dalgard, D. W., and Adamson, R. H. (1979). Induction of osteogenic sarcomas and tumors of the hepatobiliary system in nonhuman primates with aflatoxin B_1. *Cancer Res.* **39**, 4545–4554.

Sorenson, W. G., Simpson, J. P., Peach, M. J., Thedell, T. D., and Olenchock, S. A. (1981). Aflatoxin in respirable corn dust particles. *J. Toxicol. Environ. Health* **7,** 669–672.

Sorenson, W. G., Jones, W., Simpson, J., and Davidson, J. I. (1984). Aflatoxin in respirable airborne peanut dust. *J. Toxicol. Environ. Health* **14,** 525–533.

Stoner, G. D., Daniel, F. B., Schenck, K. M., Schut, H. A. J., Sandwisch, D. W., and Gohara, A. F. (1982). DNA binding and adduct formation of aflatoxin B₁ in cultured human and animal tracheobronchial and bladder tissues. *Carcinogenesis* **3,** 1345–1348.

Supharkarn, V. S., Newberne, P. M., and Goldman, M. (1983). Vitamin A and aflatoxin: Effect on liver and colon cancer. *Nutr. Cancer* **5,** 41–50.

Tudek, B., Bird, R. P., and Bruce, W. R. (1989). Foci of aberrant crypts in the colons of mice and rats exposed to carcinogens associated with foods. *Cancer Res.* **49,** 1236–1240.

VanNieuwenhuize, J. P., Herber, R. F. M., deBruin, A., Meyer, P. B., and Duba, W. C. (1973). Aflatoxinen: Epidemiologisch onderzoek naar carcinogeniteit bij langurige "low level" exposite van een fabriekspopulatie. *T. Soc. Geneesk.* **51,** 754.

Walters, J. M., and Combes, R. D. (1985). Characterization of a microsomal fraction from rat small intestine from metabolic activation of some promutagens. *Carcinogenesis* **6,** 1415–1420.

Wang, T. V., and Cerutti, P. A. (1979). Formation and removal of aflatoxin B₁-induced DNA lesions in epitheliod human lung cells. *Cancer Res.* **39,** 5165–5170.

Ward, J. M., Sontag, J. M., Weisburger, E. K., and Brown, C. A. (1975). Effect of lifetime exposure to aflatoxin B₁ in rats. *J. Natl. Cancer Inst.* **55,** 107–113.

Weekly, L. B. (1991). Aflatoxin B₁ alters central and systemic tryptophan and tyrosine metabolism: influence of immunomodulatory drugs. *Metab. Brain Dis.* **6,** 19–32.

Weekly, L. B., and Llewellyn, G. C. (1984). Activities of tryptophan-metabolizing enzymes in liver and brain of rats treated with aflatoxins. *Food Chem. Toxicol.* **22,** 65–68.

Weekly, L. B., Morgan, J. D., Rea, F. W., Kimbrough, T. D., and Llewellyn, G. C. (1978). Alterations of brain and intestine serotonin levels in hamsters pretreated with dietary aflatoxin. *Cancer Lett.* **2,** 75–80.

Weekly, L. B., Kimbrough, T. D., and Llewellyn, G. C. (1985). Disturbances in tryptophan metabolism in rats following chronic dietary aflatoxin treatment. *Drug Metab. Toxicol.* **8,** 145–154.

Weekly, L. B., O'Rear, C. E., Kimbrough, T. D., and Llewellyn, G. C. (1989). Differential changes in rat brain tryptophan, serotonin and tyrosine levels following acute aflatoxin B₁ treatment. *Toxicol. Lett.* **47,** 173–177.

Wieder, R., Wogan, G. N., and Shimkin, M. B. (1968). Pulmonary tumors in strain A mice given injections of aflatoxin B₁. *J. Natl. Cancer Inst.* **40,** 1195–1197.

Wild, C. P., Montesano, R., Van Benthem, J., Scherer, E., and Den Engelse, L. (1991). Intercellular variation in levels of adducts of aflatoxin B₁ and G₁ in DNA from rat tissues and quantitative immunocytochemical study. *J. Cancer Res. Clin. Oncol.* **116,** 134–140.

Wilson, D. W., Ball, R. W., and Coulombe, R. A., Jr. (1990). Comparative action of aflatoxin B₁ in mammalian airway epithelium. *Cancer Res.* **50,** 2493–2498.

Wogan, G. N., and Newberne, P. M. (1967). Dose–response characteristics of aflatoxin B₁ carcinogenesis in the rat. *Cancer Res.* **27,** 2370–2376.

Wogan, G. N., Paglialunga, S., and Newberne, P. M. (1974). Carcinogenic effects of low dietary levels of aflatoxin B₁ in rats. *Food Chem. Toxicol.* **12,** 681–685.

Yoneyama, M., Sharma, R. P., and Elsner, Y. Y. (1987). Effects of mycotoxins in cultured kidney cells: Cytotoxicity of aflatoxin B₁ in Madin-Darby and primary fetal bovine kidney cells. *Ecotoxicol. Environ. Safety* **13,** 174–184.

Zennie, T. M. (1984). Identification of aflatoxin B₁ in grain elevator dusts in central Illinois. *J. Toxicol. Environ. Health* **13,** 589–593.

6

Carcinogenicity of Aflatoxins in Nonmammalian Organisms

❖

Jerry D. Hendricks

INTRODUCTION

Nonmammalian organisms have played a key role in the history of aflatoxins, potent hepatotoxins and carcinogens that have affected the health of animals and humans. Episodes of mass poisoning of two such animals—turkeys in England, victims of diets containing aflatoxin-contaminated peanut meal (Schoental, 1961), and hatchery-reared rainbow trout in the United States, fed newly formulated pelletized trout rations that contained aflatoxin-contaminated cottonseed meal (Rucker *et al.*, 1961)—first focused attention on the potent lethality and carcinogenicity, respectively, of this previously unknown, naturally occurring group of toxins. Based on these early observations, an avian species, the 1-day-old duckling, was found to be the most sensitive animal for assessing the acute toxicity of aflatoxins, with an LD_{50} of only 0.36 mg/kg body weight (Carnaghan *et al.*, 1963). The rainbow trout has proved to be the most sensitive animal to the carcinogenicity of the aflatoxins; diets containing only few parts per billion of aflatoxin B_1 (AFB_1), fed for as little as 1 day, will cause liver neoplasms in trout 9–12 months later (Lee *et al.*, 1971).

The literature pertaining to the carcinogenicity of aflatoxins in nonmammalian species is disproportionately weighted toward rainbow trout because of their long history as a model for this area of research. In this chapter, I first discuss what is known about aflatoxin carcinogenicity in birds, reptiles, amphibians, and non-salmonid fishes; then I review the work that has been done with rainbow trout. Readers are referred to other chapters in this volume that address the structures of parent compounds and metabolites, as well as the transformation pathways that occur in the various animal models and in humans.

NONMAMMALIAN ANIMALS OTHER THAN SALMONID FISHES

Birds

Information on aflatoxin carcinogenesis is available for three main groups of birds: chickens, turkeys, and ducks.

Chickens

Of these three groups of commercially important birds, chickens are the most resistant to the toxic and carcinogenic effects of aflatoxins (Allcroft, 1969). Exposure to contaminated feeds usually results in reduced growth rates but few mortalities. Sensitivity is greatest in young birds. Carnaghan *et al.* (1966) exposed chickens to a diet containing toxic peanut meal, with AFB_1 present at 1.5 ppm, for 8 weeks. Only one mortality occurred during this period of time; the only external evidence of toxicity was reduced growth. At the tissue level, however, symptoms of liver toxicity were apparent. Grossly, the livers were enlarged, pale, and had petechial hemorrhages whereas, histologically, periportal fatty infiltration, scattered hepatocyte necrosis, bile duct proliferation, and fibrosis were seen. When the experiment was terminated, regenerative foci of hepatocytes, biliary hyperplasia, fibrosis, and lymphocytic hyperplasia were still prominent. Other experiments with chickens have shown recovery over longer periods of time (Asplin and Carnaghan, 1961). After 3 months, livers and kidneys were normal in texture and color; egg production was described as normal when egg laying began.

References to aflatoxin-initiated hepatic neoplasms in chickens are conspicuously missing from the literature, giving the impression that these birds are refractory or at least highly resistant to tumor formation. Through selective breeding, strains of chickens with altered metabolic activities and increased aflatoxin resistance have been produced (Carnaghan *et al.,* 1967; Gumbman *et al.,* 1970; Smith and Hamilton, 1970; Manning *et al.,* 1990). Such efforts may help explain the absence of aflatoxin-initiated hepatic tumors in chickens.

With the development of broiler strains, genetically selected for rapid growth, an absolute as well as a relative increase in the frequency of tumors in young chickens has been seen (Campbell and Appleby, 1966; Hemsley, 1966). Such tumors appear to be predominantly embryonal or genetically determined, however, and diet does not appear to be related to their occurrence.

Turkeys

Although more susceptible to the acute toxicity of aflatoxins than chickens, as evidenced by the outbreak of turkey "X" disease as well as by subsequent experiments (Siller and Ostler, 1961; Wannop, 1961; Allcroft, 1969; Goldblatt, 1969), any reference to hepatic neoplasia in turkeys resulting from aflatoxin exposure is absent from the literature. Siller and Ostler (1961) described the hepatic pathology of aflatoxicosis in turkeys. The condition is similar to that seen in chickens, although more severe. Acute poisoning causes enlarged mottled livers that, histologically, have periportal necrosis, hemorrhaging, and accumulation of fat. Surviving birds have extensive bile duct proliferation, some fibrosis and scarring, nodular regeneration, and lymphoid hyperplasia. These alterations are more persistent than in chickens, but the birds are able to survive, a condition described as "induced tolerance" by Magwood *et al.* (1966).

Ducks

Ducks are more sensitive to both the toxic and the carcinogenic properties of aflatoxin than either turkeys or chickens (Asplin and Carnaghan, 1961; Carnaghan, 1965). The carcinogenicity of aflatoxin to ducks first was demonstrated by Carnaghan (1965) and was confirmed by Newberne (1965). In Carnaghan's experiment, 37 1-week-old Khaki–Campbell ducklings were exposed to a diet containing 0.03 ppm aflatoxin, in Brazilian groundnut meal, for 14 months. During the first 4 weeks, 19 birds died with symptoms of acute aflatoxin poisoning. An additional 7 birds died over the next 7 months, leaving 11 birds for the termination sample. Of these 11 birds, 8 developed hepatic tumors of both hepatocellular and cholangiocellular types, although the relative frequency of each was not stated. The dose administered was high for ducks based on the number of mortalities that occurred. In comparison with rats, ducks appear to be more sensitive to both the acute and the carcinogenic effects of aflatoxin (Caraghan, 1965), but in comparison with rainbow trout, their sensitivity to AFB_1 carcinogenicity is less (Sinnhuber *et al.*, 1977).

Spontaneous liver neoplasms as well as other tumors are extremely rare in the highly inbred white Pekin duck (Rigdon, 1972). This observation is in marked contrast to the increased susceptibility of several strains of inbred mice to spontaneous tumor development and makes the Pekin duck an attractive model for carcinogenesis. In humans, hepatocellular carcinoma is associated with two major risk factors: hepatitis B virus (HBV) infection and aflatoxin exposure (Kew *et al.*, 1979; Beasley *et al.*, 1981; London, 1981). Since the Pekin duck is a natural host for the duck hepatitis B virus (DHBV), a member of the hepadna virus family and closely related to HBV, as well as very sensitive to the carcinogenicity of aflatoxin, this animal has been used as a model to study the role of each risk factor in the occurrence of tumors (Uchida *et al.*, 1988; Cova et al., 1990; Cullen *et al.*, 1990). All three groups of investigators concluded that AFB_1 was a potent carcinogen in ducks, but that DHBV did not contribute to the incidence or severity of the tumors.

Reptiles

The carnivorous nature of reptiles would make it highly unlikely that any reptile would be exposed to aflatoxin in its natural environment. To my knowledge, no documented exposure of a captive reptile to aflatoxin, either experimentally or through contaminated feed, has been made available.

Amphibians

Several species of amphibians have been used in short-term acute toxicity tests with aflatoxins: *Bufo melanostictus* (common garden toad), *Rhacophorus leucomystax maculatus* (tree frog), and *Uperodon* sp. tadpoles (Arseculeratne *et al.*, 1969); *Bombina* and *Rana temporaria* tadpoles (Gabor *et al.*, 1973; Puscaria *et al.*, 1973); and *Triturus alpestris* eggs and larvae (Reiss, 1972). However, none of the surviving animals were examined to determine whether the exposures would have caused tumor development.

Although many attempts have been made to initiate tumors in amphibians with chemical carcinogens, very few tumors have resulted from these exposures (Balls and Rubin, 1964; Balls *et al.*, 1978). In the earliest attempts to produce tumors, polycyclic aromatic hydrocarbons were the carcinogens most often used (Briggs, 1940; Breedis, 1952; Ingram, 1971; Outzen *et al.*, 1976), but nitrosamines and direct-acting nitrosamides have been used with similar results (Ingram, 1972; Montesano *et al.*, 1973; Khudoley, 1977; Balls *et al.*, 1978; Khudoley and Picard, 1980). Balls *et al.* (1968) reported extremely rapid elimination of nitrosamines from *Xenopus*, as well as a slow rate of metabolism. These factors, in addition to a reported inverse relationship between regenerative activity and the capacity to develop neoplasms, may provide a hypothesis for the resistance of amphibians to carcinogenesis. However, no firm conclusions concerning the relationship between regeneration and neoplastic development can be drawn at this time (Balls *et al.*, 1978). None of these early experiments tested the carcinogenicity of aflatoxin in amphibians.

El-Mofty and Sakr (1988) exposed sexually mature male and female Egyptian toads (*Bufo regularis*) to AFB$_1$. The toads were injected subcutaneously, in the dorsal lymph sac, with 0.01 mg AFB$_1$/50 g body weight in 1 ml corn oil, weekly for 15 weeks. Controls received corn oil injections only. Animals were necropsied at 2-week intervals until termination. Cumulative tumor incidence was 19% (12 males and 7 females of 100 total animals). The tumors were diagnosed as hepatocellular carcinomas, some of which metastasized to the kidney. Apparently, this incident is the only reported exposure of an amphibian to AFB$_1$ with the express purpose of testing for carcinogenicity.

In summary, the literature on chemical carcinogenesis in amphibians in general and on aflatoxin carcinogenesis in particular portrays a group of animals that have not been tested as extensively as mammals or fish. However, the tests that have been conducted reveal, in general, a low sensitivity to chemically initiated neoplasia for reasons that currently are not fully understood.

Nonsalmonid Fishes

Elasmobranchs

Investigators have observed that fewer tumors occur in sharks, skates, and rays than in comparable numbers of teleosts. Prior to the establishment of the Registry of Tumors in Lower Animals (Harshbarger, 1969) in 1965, only 7 tumors in elasmobranch species had been reported (Wellings, 1969). Since that time, even with increased awareness and interest in tumor detection, few tumors have been found in these primitive fishes. The reasons for this apparent refractoriness to neoplasia are not clear, although sharks in particular tend to inhabit marine waters that are less impacted by pollution than many of the bottom-dwelling estuarine teleosts in which neoplasia is commonplace.

One would not expect sharks, skates, or rays to be exposed to aflatoxins in their natural environment, but because of their alleged resistance to neoplastic development and the documented potency of AFB_1 as a carcinogen, several attempts to initiate tumors in sharks with AFB_1 have been reported. All attempts to initiate tumors in the nurse shark (*Ginglymostoma cirratum*) and the clearnose skate (*Raja eglanteria*) with AFB_1 have been unsuccessful, but both species proved to be quite sensitive to the acute toxicity of AFB_1 (Luer and Luer, 1981, 1982, 1984). Bodine *et al.* (1985) have shown that liver microsomes from these two species have very low activity for converting AFB_1 to its active epoxide based on the Ames *Salmonella* mutagenesis test (Ames *et al.*, 1975). Compared with calf microsomes, the shark and skate microsomes were only 20% as active in producing mutagenic metabolites. Bodine *et al.* (1989) reported that the major AFB_1 metabolite in the nurse shark and the clearnose skate was aflatoxicol, and that this reaction was reversible, as in trout and other species (Loveland *et al.*, 1977; Salhab and Edwards, 1977). Binding of [^{3}H]AFB_1 was about 3 times greater in calf than in shark liver DNA. Thus, in these two species of elasmobranchs, reduced AFB_1 metabolism and DNA binding provide a basis for the resistance of these animals to AFB_1 carcinogenesis.

Teleosts

For years, most of the experimental chemical (predominantly aflatoxin) carcinogenesis research in fish was performed on the highly sensitive rainbow trout. However, trout require specialized rearing facilities and an abundant supply of high quality cold water, making their use impractical in many locations, especially in urban and tropical or subtropical sites. Thus, small aquarium fish that require smaller holding tanks, generally prefer warmer water temperatures, and require less water volume have emerged as preferred models for cancer research (Matsushima and Sugimura, 1976).

In general, the aflatoxins have not been tested thoroughly in the various species of aquarium fishes, but in those that have been tested, the response or sensitivity to aflatoxin has been much less than for rainbow trout. Guppies (*Lebistes reticulatus*), for instance, required continuous exposure to 6 ppm dietary AFB_1 to produce liver tumors 9–11 months later (Sato *et al.*, 1973). Zebrafish (*Brachydanio rerio*) exposed to AFB_1 at 2 ppm in the water for 3 days failed to develop any tumors within 9 months (Bauer *et al.*, 1972). Matsushima *et al.* (1975) produced cholangiomas in guppies by feeding a diet containing 20 ppm sterigmatocystin for 2 months, but observed no liver tumors after feeding AFB_1 at 30 ppm for 2 months. Sterigmatocystin is a metabolite produced in the biosynthesis of AFB_1 and usually is less potent as a carcinogen than AFB_1 (Hendricks *et al.*, 1980a), but in guppies it appears to be more potent.

In the popular Japanese medaka (*Oryzias latipes*), dietary exposure to 2.5 ppm AFB_1 for 24 weeks produced an 8% incidence of liver cell carcinoma and a 4% incidence of liver cell adenoma. Another 60% of these fish were said to have liver cell nodules. Only a 5% incidence of carcinomas and a 21% incidence of liver cell nodules were produced by 5 ppm AFB_1 for 6 weeks. Aflatoxin G_1 (AFG_1) at 2.5 ppm for 24 weeks resulted in 3% incidences of both liver cell adenomas and carcinomas. Sterigmatocystin, administered in the diet at 5 ppm for 5 weeks, appeared to be equally as potent as or more potent than AFB_1 to medaka, although longer exposure times did not produce greater numbers of tumors (Hatanaka *et al.*, 1982).

Reports of the carcinogenic effects of aflatoxins in other teleost species are rare for several reasons. If we categorize teleosts into two broad groups, the noncultured and the cultured fishes, most of the approximately 25,000 species would fall into the noncultured group. These fishes would never be exposed to the aflatoxins naturally, and no information about their potential sensitivity to this group of natural carcinogens is available. Within the cultured group, some fishes have been adapted as experimental models, others as pets or ornamentals, and still others as commercially important bait or food fishes.

Some of the species used as experimental models have already been discussed. Others, although used for a variety of research purposes including carcinogenesis, have never been exposed to the aflatoxins. For instance, the genus *Poeciliopsis* has been used quite extensively as a cancer model but apparently has never been exposed to AFB_1 (Schultz and Schultz, 1982a,b, 1988). Also the brown bullhead *Ictalurus nebulosis*, although quite sensitive to the hepatocarcinogenic effects of polycyclic aromatic hydrocarbons in both polluted environments and the laboratory (Baumann and Harshbarger, 1985; Baumann *et al.*, 1987), has been exposed to AFB_1 only infrequently. Biba (1983) exposed brown bullhead eggs to a 1-ppm solution of AFB_1 for 1 hr just prior to hatching, but saw no evidence of neoplastic development 8 months later. This experiment is the only AFB_1 exposure identified in this species for which the intent was to initiate tumors.

Aquatic species reared for the pet or ornamental trade usually are fed a formulated flake food that could contain a low level of aflatoxin contamination, depending on the source and quality of the ingredients used in the diet. However, based on the resistance to aflatoxin of the small fish species tested thus far, few or no tumors would be expected in these fish species based on the exposure they would receive through the artificial diet.

Commercially important, intensively cultured species are either insensitive to aflatoxins or a concerted effort is made to limit strictly their exposure to aflatoxins. Channel catfish, *Ictalurus punctatus,* for example, are intensively cultured in the southeastern United States. These fish are fed a diet that contains up to 25% cottonseed meal, an ingredient that nearly always contains some degree of aflatoxin contamination. Catfish growers are not concerned provided the level of aflatoxin does not exceed 50 ppb in the total diet. The fish are fed for 18–24 months on this type of diet before being slaughtered for restaurant or supermarket sale. Millions of catfish are reared annually, yet liver tumors are not observed (H. Dupree, personal communication). This amazing fact emphasizes the extreme resistance of this species to the carcinogenicity of aflatoxin. Experiments determining the acute and subchronic effects of AFB_1 in channel catfish support their resistance to AFB_1 (Jantrarotai and Lovell, 1990; Jantrarotai *et al.,* 1990). The 10-day LD_{50} value for intraperitoneally injected AFB_1 was 11.5 mg/kg body weight. This value compares with 0.81 mg/kg in the rainbow trout (Bauer *et al.,* 1969). Moribund catfish were found to have necrotic hepatocytes and pancreatic acinar cells, sloughed stomach and intestinal mucosa, and greatly reduced blood parameters (hematocrits, hemoglobins, red blood cell (RBC) and white blood cell (WBC) numbers). In fact, Dupree (personal communication) referred to acute AFB_1 poisoning of channel catfish as "no-blood disease." Dietary levels of 100, 464, or 2154 ppb fed for 10 weeks produced no grossly observable effects on fingerling catfish, but 10,000 ppb did reduce growth significantly, produce anemia, and cause liver necrosis. The only liver tumor reported in a channel catfish, for which AFB_1 was suspected as the causative agent, was reported by Ashley (1969). The tumor was found by Sneed in an 11-pound female catfish, known to have been fed a peanut meal-containing diet for the first 2–3 years of life.

SALMONID FISHES

Introduction and Species Sensitivity

That the rainbow trout (formerly named *Salmo gairdneri* but renamed *Oncorhynchus mykiss*) developed epizootic liver cancer in the United States in the early 1960s was no accident. Other salmonid species also were fed aflatoxin-containing diets, but the problem was exposed in the most sensitive species—the rainbow trout. With the aid of hindsight, the explosiveness of the timebomb that, figuratively speaking, was set before hatchery-reared rainbow trout and the certainty that its explosion was only a matter of time is now obvious. First, as previously stated, rainbow trout are the most sensitive animals of any species tested to date to the carcinogenic effects of AFB_1 or any of its metabolites (Lee *et al.*, 1971). Second, this very sensitive animal was exposed unknowingly to relatively high doses of a chemical (i.e., AFB_1) that has been shown to be the most potent naturally occurring carcinogen ever discovered. Third, the aflatoxins were contaminating a feed ingredient, cottonseed meal, that contained unusual cyclopropenoid fatty acids (CPFA), which later were shown to be powerful cocarcinogens with aflatoxin (Lee *et al.*, 1968, 1971; Sinnhuber *et al.*, 1968b, 1974; Hendricks *et al.*, 1980b; Schoenhard *et al.*, 1981) in rainbow trout. These three factors controlled the magnitude of the epizootic that occurred. Fortunately, with the knowledge that contaminated cottonseed meal was the primary carrier of AFB_1 into trout diets, cottonseed meal was eliminated from most diet preparations, especially for starter diets (since the greater sensitivity of younger fish made them particularly vulnerable to AFB_1) and for brood stock rations for which long-term feeding and the promoting effect of female hormones (to be discussed later) in sexually mature female trout increased their sensitivity. Some cottonseed still is used in short-term trout grower rations for trout grown for human consumption, but the duration of feeding is too short, in most cases, for tumor development to be a problem. Nevertheless, an occasional outbreak of liver cancer in cultured rainbow trout still occurs, usually outside the United States (Majeed *et al.*, 1984).

The early experiments that were conducted in the 1960s were primarily discovery experiments, verifying and confirming the fact that naturally occurring AFB_1 and AFG_1 were carcinogenic to rainbow trout. Most of these experiments were conducted at the Western Fish Nutrition Laboratory of the United States Fish and Wildlife Service in Cook, Washington, or at the Department of Food Science and Technology at Oregon State University. The following list includes most, if not all, of the papers that resulted from those early experiments at the two laboratories mentioned, as well as others: Ashley (1967, 1969, 1970, 1973), Ashley and Halver (1963, 1968), Ayres *et al.* (1971), Coates *et al.* (1967), Dollar *et al.* (1967), Engebrecht *et al.* (1965), Halver (1967, 1969), Halver *et al.* (1969), Hueper and Payne (1961), Jackson *et al.* (1968), Lee *et al.* (1968, 1971), Nigrelli and Jakowska (1961), Rucker *et al.* (1961), Scarpelli (1967), Scarpelli *et al.* (1963), Simon *et al.* (1967), Sinnhuber (1967), Sinnhuber *et al.* (1968a,b), Solomon *et al.* (1965), Wales (1967, 1970), Wales and Sinnhuber (1966), Wolf and Jackson (1963, 1967), Wood and Larson (1961), and Yasutake and Rucker (1967).

Species sensitivity within the salmonid group has been discussed rather thoroughly by Wales (1970) and Wolf and Jackson (1967). In general, domesticated strains of rainbow trout, particularly the Shasta strain, are more susceptible to the carcinogenicity of AFB_1 than are wild stocks such as migratory steelhead. Prior to 1989, the salmonid group was conveniently divided into the Pacific salmons (genus *Oncorhynchus*), chars (genus *Salvelinus*), and trouts (genus *Salmo*); AFB_1 sensitivity was assessed qualitatively along generic lines. The salmons were considered very resistant to AFB_1. The chars seemed to be intermediate in sensitivity although only one of the chars, the brook trout (*Salvelinus fontinalis*), had been tested. This fish was less sensitive than rainbow trout but would develop tumors (Wolf and Jackson, 1967). The trouts were the most sensitive but, within the group, rainbows were more sensitive than cutthroats (*Salmo clarki*), which were more sensitive than browns (*Salmo trutta*).

In 1989, Smith and Stearley realigned the western trouts, the rainbows and cutthroats, with the Pacific salmons in the genus *Oncorhynchus* based on greater similarities in electrophoretic patterns with them than with the eastern trouts (the browns) and Atlantic salmon (*Salmo salar*). Now the sensitive rainbow trout is in the same genus as the highly resistant Pacific salmons. Wolf and Jackson (1967) were unable to produce tumors in either chinook (*O. tshawytscha*) or coho (*O. kisutch*) salmon using a diet containing 20% AFB_1-contaminated cottonseed meal, which also would have contained at least some of the synergistic CPFA. Halver *et al.* (1969) also were unable to produce tumors in coho salmon after feeding a diet containing 20 ppb AFB_1 for 20 months. Wales and Sinnhuber (1972) fed sockeye salmon (*O. nerka*) a diet containing 12 ppb AFB_1 plus 50 ppm CPFA for 20 months and produced liver tumors in about 40% of the animals. Bailey *et al.* (1988) exposed coho salmon to AFB_1 in a variety of ways with variable tumor yields: (1) a 30-min static exposure of embryos to a 0.5-ppm solution of AFB_1 resulted in a 9% tumor response 12 months later; (2) 40 ppb AFB_1 in the diet, fed for 4 weeks, failed to initiate any tumors 12 months later; and (3) 5000 ppb dietary AFB_1, fed for 3 weeks, produced an incidence of only 5% hepatic tumors 12 months later. In comparison: (1) a 15-min static exposure of rainbow trout embryos to 0.5-ppm AFB_1 resulted in a 62% incidence of liver tumors 12 months later; (2) 20 ppb dietary AFB_1 for 4 weeks exposure also produced a tumor incidence of 62%; and (3) 5000 ppb dietary AFB_1 for 3 weeks was not attempted with rainbow trout because it would have been lethal. In addition, the tumors occurring in coho salmon in these experiments were diagnosed as benign hepatocellular adenomas rather than the more common mixed or hepatocellular carcinomas observed in rainbow trout. Comparative mutagenicity studies (Coulombe *et al.*, 1984), using liver S20 and isolated hepatocytes from rainbow trout and coho salmon as activating systems, also showed the lower potential for activating AFB_1 in coho salmon liver compared with rainbow trout.

The biochemical basis for this species difference in sensitivity is primarily dependent on a less efficient cytochrome P450 metabolism of AFB_1 to the reactive 8,9-epoxide in the coho salmon. Immunoquantitation of P450 isozymes showed that coho microsomes had much less of the isozyme immunochemically related to the CYP2K1 (formerly LM_2) of rainbow trout, previously shown to be the major isozyme involved in AFB_1 8,9 epoxidation (Bailey *et al.*, 1988). The end result of this difference in AFB_1 metabolism between the two species is that AFB_1 DNA binding was up to 56 times less in coho salmon after intraperitoneal (ip) injection, 20 times less after embryo exposure or a 1-hr incubation of freshly prepared isolated hepatocytes, and 18 times less after a 3-week dietary exposure to 80 ppb. Other parameters, such as DNA adduct persistence, other phase I and phase II metabolism pathways, and AFB_1 elimination were similar in the two species, indicating that AFB_1 epoxidation and the resulting DNA binding probably account for the differences seen in the two species.

Since the early 1970s, research on aflatoxin carcinogenesis has centered on rainbow trout, primarily the highly sensitive Shasta strain; the majority of the effort has taken place at Oregon State University. Research at Oregon State University has followed several themes: (1) metabolism of AFB_1 in rainbow trout, (2) the carcinogenicity of the various metabolites in rainbow trout, (3) the development of alternative routes of exposure to aflatoxins in rainbow trout, (4) the effects of modulating chemicals on the response of rainbow trout to AFB_1, and (5) the pathology of the neoplastic lesions produced in rainbow trout. The rest of this chapter addresses each of these areas of study.

Metabolism of AFB_1 in Rainbow Trout

For several years, fish and other lower vertebrates were thought to lack the necessary mixed function oxidase (MFO) enzymes required to metabolize xenobiotics. Gaudette *et al.* (1958) and Brodie and Maickel (1962) suggested that fish disposed of xenobiotics by direct diffusion through the skin and gills without metabolism. Chan *et al.* (1967) and Dewaide and Henderson (1968), however, proved this theory to be wrong, showing that trout have a well-developed hepatic MFO system located in the microsomal fraction that is capable of metabolizing a variety of xenobiotic compounds.

Schoenhard *et al.* (1976) were the first to show that the major metabolite of AFB_1 in rainbow trout was aflatoxicol (AFL). AFL was formed *in vitro* by the supernatants from both the 20,000 *g* and the 105,000 *g* centrifugations of trout liver homogenates, showing that the compound was produced by a soluble enzyme. These investigators also reported that both AFB_1 and AFL were lethal to *Bacillus subtilis* when incubated with either the trout liver 20,000 *g* supernatant or the top-layer microsomal pellet plus an NADPH-generating system, but that AFB_2 (which lacks the 8,9 position double bond) was not. The results were thus consistent with the report that acid hydrolysis of a rat liver rRNA–AFB_1 adduct had produced 8,9-dihydro-8,9-dihydroxy AFB_1, which was presumptive evidence for an epoxide of AFB_1 as the highly reactive electrophile that would bind to nucleophilic bases of DNA and RNA (Swenson *et al.*, 1973).

Loveland *et al.* (1977) extended the work on AFL and found that the conversion of AFB_1 to AFL by soluble liver enzymes was reversible. When AFL was incubated under the same conditions used for AFB_1, AFB_1 was the major product formed. These researchers hypothesized that the reversibility of this reaction may be a significant factor in the extreme sensitivity of rainbow trout to AFB_1. The investigators also were able to identify one of the minor previously unidentified polar metabolites that was recoverable from AFB_1 incubation, and had been reported by Schoenhard *et al.* (1976) as AFM_1. Subsequently, other investigators

(Salhab and Edwards, 1977) reported the reversible AFB_1–AFL reaction in trout as well as in several mammalian species. The data of this group revealed that human postmitochondrial liver fractions were most active in converting AFL to AFB_1 relative to liver fractions from monkey, dog, rabbit, hamster, guinea pig, mouse, rat, and trout, whereas trout and rabbit fractions were most efficient in converting AFB_1 to AFL. These authors suggested that the enzymatic capability to convert AFB_1 to AFL was related directly to species sensitivity to AFB_1 carcinogenicity, whereas the capability to convert AFL to AFB_1 was related directly to species resistance to AFB_1. These investigators emphasized that the AFB_1–AFL reversible reaction was not a cytochrome P450 (MFO) reaction, but was catalyzed by a soluble reductase/dehydrogenase enzyme system. However, formation of other metabolites such as AFM_1 or the presumptive AFB_1 8,9-epoxide, resulted from microsomal cytochrome P450-mediated MFO reactions.

The variable that was investigated next in the overall study of AFB_1 metabolism in rainbow trout was the effect of MFO inducers. Elcombe and Lech (1978) and Elcombe et al. (1979) showed that polychlorinated biphenyls (PCBs) and β-naphthoflavone (BNF) were potent inducers of MFO enzymes in rainbow trout. Hendricks et al. (1977, 1982) reported that PCBs and BNF inhibited the carcinogenesis of AFB_1 in rainbow trout. Based on these observations, Loveland et al. (1983) exposed 6-month-old rainbow trout to two dietary levels (50 and 500 ppm) of BNF for 4–6 weeks before killing the fish and preparing postmitochondrial fractions (PMF) and microsomal fractions for metabolism studies. Incubation of liver PMF from control diet or 50- or 500-ppm BNF-fed trout with AFB_1 produced several metabolites. As expected, AFL was the major metabolite (45.7 ng/mg protein) in the control group; in this particular experiment, AFM_1 was nondetectable. In the 50-ppm BNF group, AFL was still the major metabolite but less was formed (34.3 ng/mg protein) than in the controls; AFM_1 became detectable at 3.7 ng/mg protein. In the 500-ppm BNF group, AFL formation was approximately the same (33.9 ng/mg protein), but AFM_1 production increased dramatically to 48.9 ng/mg protein to become the major metabolite produced. In addition, another unknown metabolite was found in smaller amounts. By thin layer chromatography (TLC), high performance liquid chromatography (HPLC), nuclear magnetic resonance (NMR), and mass spectrometry (MS) analysis, this metabolite was identified as aflatoxicol M_1 ($AFLM_1$). $AFLM_1$ had been identified previously in dog liver by Salhab et al. (1977), but this report was the first one of $AFLM_1$ in fish. $AFLM_1$ was produced from AFL by both PMF and microsomes; the yield was enhanced greatly by BNF dietary pretreatment. These two features indicate that this product was the result of a cytochrome P450-mediated MFO reaction. Starting with AFM_1, $AFLM_1$ was produced by the PMF and microsomal supernatant, a reaction that was reversible with PMF. Thus, from both structural and formation standpoints, $AFLM_1$ is related to AFM_1 in the same way that AFL is related to AFB_1. In a modified Ames mutagen assay, using Salmonella typhimurium TA98, $AFLM_1$ was 4.1% as mutagenic as AFB_1 (Coulombe et al., 1982). Since the mutagenicity of both $AFLM_1$ and AFM_1 (Coulombe *et al.,* 1982) is much less than that of AFB_1, both are considered detoxification products of AFB_1 metabolism. The fact that both metabolites are produced in much higher quantities after BNF exposure was consistent with the observation that BNF reduced the carcinogenicity of AFB_1.

Statham *et al.* (1976) showed that fish could concentrate certain lipid-soluble xenobiotics in the bile up to several hundred or even several thousand times the concentration found in the water. Loveland *et al.* (1983) suggested that $AFLM_1$ might be an important biliary excretion form of AFB_1, and designed experiments to identify the water-soluble aflatoxin conjugates found in the bile after exposure to AFB_1. These researchers found that the major metabolite in the bile of BNF-exposed trout was $AFLM_1$–glucuronide, whereas in control trout the major product was AFL–glucuronide (Loveland *et al.,* 1984).

Bailey *et al.* (1982a) extended the AFB_1 metabolism studies on liver cell homogenates in rainbow trout to isolated hepatocytes. These studies more closely reflected *in vivo* cellular capabilities for AFB_1 metabolism, since both phase I and II reactions could occur. Hepatocytes also allowed for direct measurement of intracellular DNA adduct formation, which was not possible with cell homogenates. These researchers found a constant rate of dose-responsive AFB_1–DNA adduct formation and unbound metabolite production during the first hour of incubation, when cell viability was highest. Thereafter, the rate of AFL production decreased while the rate of DNA adduct formation increased, for reasons that were not fully determined, although the effect of AFL removal from the system through conjugation was a suggested explanation. Later experiments found isolated hepatocytes to reflect the qualitative changes in AFB_1 metabolism brought about by BNF pretreatment, as well as a corresponding reduction in DNA adduct formation that is consistent with the reduced carcinogenicity of AFB_1 (Bailey *et al.,* 1984). Loveland *et al.* (1988) used freshly isolated trout hepatocytes to study the metabolism and DNA binding of AFB_1 and its phase I metabolites AFL, AFM_1, and $AFLM_1$. All four [^{3}H]-labeled aflatoxins were incubated with isolated hepatocytes for 1 hr, cellular DNA was isolated, and specific activities were determined by scintillation counting. Aflatoxin metabolites in the supernatants were determined by HPLC. DNA binding relative to AFB_1 was 0.53 (pmol aflatoxin bound/μg DNA)/(μmol dose) for AFL, 0.81 for AFM_1, and 0.83 for $AFLM_1$. Based on mutagenicity and carcinogenicity data, the binding levels of AFM_1 and $AFLM_1$ were higher than expected compared with AFB_1 and AFL. HPLC analysis showed that, when hepatocytes were incubated with AFB_1, AFL, AFM_1, or $AFLM_1$ as substrates, the major metabolites were AFL, AFB_1, $AFLM_1$, and AFM_1, respectively.

Shelton *et al.* (1986), using isolated hepatocytes, showed that preexposure to the PCB mixture Aroclor 1254 resulted in a shift in metabolite profile similar to the one seen with BNF, in addition to a corresponding decrease in AFB_1–DNA adduct formation. Therefore, both BNF and Aroclor 1254 have their primary effect on AFB_1 carcinogenicity at the level of cytochrome P450 induction, altered carcinogen metabolism, and reduced DNA adduct formation.

The next development in AFB_1 metabolism studies in rainbow trout was the direct exposure of trout to AFB_1 to determine the metabolite profile and DNA binding *in vivo*. This approach had the added advantage that fish from the same treatment group could be maintained to determine long-term tumor incidences. Whitham *et al.* (1981) first showed that *in vivo* AFB_1–DNA binding was a good estimate of initiation and an accurate predictor of final tumor incidence. Goeger *et al.* (1988) conducted extensive experiments, testing the effects of BNF and butylated hydroxyanisole (BHA) on metabolite formation, DNA binding, and tumor response. BHA had been shown to increase glutathione *S*-transferase and epoxide hydrolase activities in rats and mice (Benson *et al.*, 1979) and to inhibit chemical carcinogenesis in rats (Wattenberg, 1973). A 3-week pretreatment with 500 ppm BNF reduced hepatic AFB_1–DNA adducts to 33–60% of control levels, increased the levels of AFM_1 in the liver 4-fold, reduced the amount of AFL by 50%, and significantly reduced hepatic tumor response. Bile concentrations of AFL–glucuronide were reduced significantly by BNF, but total bile glucuronides were increased 15-fold because of $AFLM_1$–glucuronide. Thus, the effects of BNF on AFB_1 metabolism and DNA adduct formation *in vivo* were comparable to the previous results obtained in hepatocytes (Bailey *et al.*, 1984; Loveland *et al.*, 1988) and liver homogenates (Loveland *et al.*, 1983, 1984). The reduced tumor incidences confirmed the earlier results of Hendricks *et al.* (1982) and Nixon *et al.* (1984). In contrast, BHA at 3000 ppm for a 3-week dietary pretreatment had no effect on AFB_1 metabolism, DNA adduct formation, or final tumor incidence. In rodents, BHA induces the activity of glutathione *S*-transferase activity (Benson *et al.*, 1979), certain isoforms of which have been shown to be highly effective in catalyzing the conjugation of the AFB_1-epoxide to glutathione (GSH), leading to reduced DNA binding and tumor inhibition (Degen and Neumann, 1981). In rainbow trout, Valsta *et al.* (1988) found AFB_1–GSH conjugation to be insignificant in the detoxification of AFB_1. Although both GSH and glutathione *S*-transferase were increased in BNF-fed trout, no relationship appeared to exist between GSH concentration, enzyme activity, and biliary GSH–AFB_1 conjugation. Thus, BHA is ineffective as a tumor inhibitor in trout since its mechanism of action is through GSH conjugation, an ineffective detoxification pathway in trout.

MFO inducers or, in more general terms, chemical modulators of carcinogenesis such as BNF and PCBs have added much to our knowledge of AFB_1 metabolism in rainbow trout. Another compound, indole-3-carbinol (I3C), which occurs naturally as a glucosinolate in cruciferous vegetables, is an effective inhibitor of AFB_1 carcinogenesis (Nixon *et al.*, 1984) but by different mechanisms. I3C also has been extremely valuable in expanding our knowledge about AFB_1 carcinogenesis. Nixon *et al.* (1984) exposed rainbow trout to several flavone and indole compounds, including BNF and I3C, in the diet prior to AFB_1 exposure. At the end of 8 weeks of dietary pretreatment, the activities of several MFO enzymes and cytochrome P450 content were measured. The fish then were exposed to AFB_1 and held for 12 months to develop tumors. The effect of BNF and I3C on *in vivo* DNA binding was determined also. As in previous experiments, BNF induced the MFO system, altered the AFB_1 metabolite profile as described previously, reduced DNA adduct formation *in vivo,* and protected the fish from AFB_1 carcinogenesis. I3C, in contrast, did not induce the MFO system, did not alter the metabolism from that of controls, but did reduce both DNA binding and tumor incidence significantly. Thus, clearly I3C was operating by a different mechanism than BNF.

Goeger *et al.* (1986) investigated the effects of dietary I3C on the *in vitro* and *in vivo* metabolism, distribution, and DNA binding of AFB_1 in rainbow trout. Based on modest differences in AFB_1 metabolism and distribution but a major (70%) reduction in *in vivo* DNA binding, these researchers suggested that I3C was inhibiting DNA binding through a blocking mechanism in one or more of the following ways: (1) reducing the formation of the 8,9-epoxide, (2) enhancing the metabolism and detoxification of AFB_1, (3) reacting directly with the epoxide, or (4) enhancing the enzymatic inactivation of the epoxide. Since the addition of I3C directly to incubation mixtures of control hepatocytes with AFB_1 had no effect on DNA binding, the investigators also suggested that the mechanism might be through a metabolite of I3C rather than through I3C itself. Subsequent studies by Bradfield and Bjeldanes (1987) showed that I3C itself was not active, but that acid condensation products of I3C were. Injection of I3C ip into rats was ineffective in inducing cytochrome P448 (CYP1A1) monooxygenase activity, but oral administration resulted in a 15-fold induction. Treatment of I3C with hydrochloric acid, simulating exposure to stomach acidic conditions, produced condensation products that were effective when administered ip to rats.

Dashwood *et al.* (1988, 1989a) investigated the inhibition of AFB_1 by I3C in rainbow trout in great detail. In an experiment that utilized 10,000 fish, these researchers explored the relationships of dose–response exposures to dietary I3C and AFB_1, with *in vivo* DNA binding and final tumor incidence as the end points. Five doses of I3C and four doses of AFB_1 were used, DNA binding was determined after 7 and 14 days of AFB_1 exposure, and hepatic tumor incidence was determined 12 months after AFB_1 exposure. Linear increases in DNA binding occurred with increasing dose of AFB_1 at each I3C dose level. Increases in I3C dose gave dose-related decreases in AFB_1 DNA binding. At I3C doses of 2000 ppm or less, the inhibitor-altered tumor response was precisely predictable by the DNA adducts formed at each dose level of AFB_1. These results showed that I3C was acting as a pure anti-initiator. Analysis of the AFB_1–DNA adducts supported this conclusion by revealing no qualitative differences in the adducts formed at the various doses of I3C (Dashwood *et al.,* 1988). The adduct formed was the 8,9-dihydro-8-(N^7-guanyl)-9-hydroxyaflatoxin, the same adduct originally reported by Croy *et al.* (1980).

Additional experiments by Fong *et al.* (1990) helped clarify the mechanisms by which I3C exerted its action. These researchers investigated the induction of oxidation and conjugation enzymes, the scavenging of carcinogen electrophiles, and the inhibition of AFB_1 activation as possible mechanisms of anticarcinogenesis by I3C. Liver microsomal 7-ethoxyresorufin O-deethylase activity was not induced significantly by an 8-day exposure to 500–2000 ppm I3C. No changes were seen in the content of cytochrome P450 isozymes CYP2K1 (LM_2) or CYP1A1 (LM_{4b}). Neither microsomal uridine diphosphate–glucuronyl transferase activity nor cytosolic glutathione S-transferase activity was increased. The addition of I3C or its acid reaction products mixture (RXM) did not inhibit the covalent binding of AFB_1-8,9-Cl_2 to calf thymus DNA, but kinetic analysis of microsome-mediated binding of AFB_1 to DNA *in vitro* indicated that the I3C RXM did inhibit the metabolic activation of AFB_1. Thus, acid condensation products formed *in vivo* in the acid conditions of the stomach appeared to be acting as competitive inhibitors for active binding sites on cytochrome P450 in the liver, reducing the activation of AFB_1 to the epoxide and consequently reducing AFB_1 DNA binding. In support of this hypothesis, a specific acid condensation product of I3C, the dimer 3,3′-diindolylmethane (Dashwood *et al.,* 1989b), in addition to I3C and the total RXM, were each co-microinjected with AFB_1 into rainbow trout embryos to determine the effects on DNA binding and tumor incidence. I3C had no significant effect on either DNA binding or final tumor incidence. However, both the dimer and the RXM significantly inhibited DNA binding and the final tumor incidence (Bailey *et al.,* 1992).

Aflatoxin metabolism studies have established clearly the metabolites that are produced by control and induced rainbow trout. Bacterial mutagenicity tests and DNA binding studies in rainbow trout have provided mechanistic bases for predicting the carcinogenicity of AFB_1 and each of the phase I metabolites. This information is in sharp contrast to the state of knowledge in most other fish or other nonmammalian models, for which little is known about carcinogen metabolism.

Development of Alternative Exposure Routes for Rainbow Trout

Dietary
The route of exposure for the fish involved in the original epizootic and all the follow-up experimentation of the 1960s was obviously the dietary route. This route of exposure continues to be valuable experimentally, but has some inherent weaknesses that contraindicate its use for some experiments. (1) It results in uneven exposure since it is voluntary and dependent on the feeding behavior of each individual fish. (2) It requires repeated handling and clean-up of carcinogen-containing diet. (3) A relatively large amount of carcinogen is required to expose a group of animals. (4) It usually results in an extended period of initiation (Hendricks, 1981).

Embryo Immersion
Embryo immersion was developed by Wales (1979) and has proved to be a very valuable route of exposure for AFB_1 as well as other carcinogens. The technique consists of bathing embryos in a solution of a carcinogen for a variable

period of time, usually less than 24 hr and usually 30–60 min for most AFB_1 exposures. A temperature controlled water bath, periodic agitation, subdued lighting, and an air pump and stones to aerate the water for exposures longer than 2 hr are used also. Variables that determine the response include (1) age of the embryo (sensitivity increases with age), (2) concentration of the carcinogenic solution, (3) duration of the treatment, and (4) temperature of the bath (Wales *et al.,* 1978; Wales, 1979; Hendricks *et al.,* 1980c, 1984a; Hendricks, 1981). The technique has a number of advantages, including (1) requiring only a single handling of the carcinogenic solutions, (2) easy containment and detoxification of the carcinogenic solution and rinse water from the eggs by approved methods, (3) more uniform exposure of each embryo to the carcinogen insured by the involuntary exposure and periodic agitation, and (4) requiring less carcinogen to expose a large number of animals than with feeding (Hendricks, 1981). The one primary limitation that this procedure has is the inability to dissolve highly lipophilic compounds in water in sufficient doses to achieve an adequate exposure.

Embryo Microinjection

Metcalfe and Sonstegard (1984) were the first to use embryo microinjection to expose rainbow trout embryos to AFB_1 and other carcinogens. These investigators used drawn glass pipette needles, a 0.1 ml acetone/1.9 ml phosphate buffered saline carrier, a 0.5-μl injection volume, a micromanipulator to facilitate accurate steady insertion of the needle into the egg, rat hepatic S9 mix to augment metabolism of the carcinogen in the embryo, and injection into the perivitelline space of the egg to avoid clogging the needle with yolk. Metcalfe and Sonstegard were able to produce a low incidence of hepatic tumors with AFB_1 at doses of 13 and 25 ng/egg, with or without the S9 mixture and in a non-dose-responsive manner.

Black *et al.* (1985) changed this technique by (1) using dimethylsulfoxide (DMSO) as a carrier, (2) increasing injection volume to 1 μl, (3) using 31 gauge stainless steel needles for injecting, (4) injecting directly into the yolk-sac rather than into the perivitelline space, and (5) using a syringe microburet to deliver the dose to the embryo more accurately. Their technique required a 100 ng/egg dose of AFB_1 to produce a 25% incidence of hepatocellular tumors 8–9 months later.

At Oregon State University, we automated this procedure by interfacing a Hamilton Microlab 900 pump (Hamilton Company, Reno, Nevada) with a computer to permit automatic syringe refilling from a reservoir of injectant and footswitch dispensing of highly accurate 1-μl doses. With two of these injector systems, two people routinely can microinject 8000 embryos in an 8-hr day. We continue to use the 31 gauge needles, but prefer a carrier of 25% acetone/75% vegetable oil for aflatoxin injections. Our first experiment used doses as high as 25 ng AFB_1/μl/egg, based on the results of Metcalfe and Sonstegard (1984) and Black *et al.* (1985), but high mortalities resulted. We reduced the doses to 0.5, 1.0, 2.0, and 4.0 ng/μl/egg and observed hepatic tumor incidences of 26, 34, 45, and 48% 9 months later. This greater response to lower doses may reflect the greater sensitivity of the Shasta strain used at Oregon State University, since the trout used in the two previous studies were of different strains.

The technique of microinjection presents several distinct advantages. (1) Exposure of an individual fish can be accomplished at incredibly low (ng) doses. (2) Highly lipophilic carcinogens can be dissolved in the acetone/oil carrier and administered in this way. (3) Worker safety is very high because of the small, contained amounts of carcinogen used.

A modification of the embryo microinjection technique was developed by Metcalfe *et al.* (1988). Rather than using embryos, they exposed newly hatched rainbow trout sac-fry to carcinogens by microinjection. The procedure was similar to that described previously except the sac-fry were anesthetized by immersion in water saturated with CO_2 before injection. This procedure allowed more accurate placement of the injectant, less trauma to the immobilized sac-fry, lower procedure-related mortalities, and a greater response to a given carcinogen dose, since the fish is about 1 week older than in embryo microinjections. Previous data show increasing sensitivity with increased age of embryos and sac-fry (Hendricks *et al.,* 1980c, 1984a; Hendricks, 1981).

Fry Immersion

Fry immersion has been used for years by researchers using small aquarium fish, but only recently has this technique been utilized for trout fry. The procedure is essentially the same as for embryos, but requires some special adaptations because the animals are free-swimming, have a much greater oxygen demand, and are no longer protected by the embryonic chorion. These modifications are (1) a greater volume of water for a given number of animals, (2) adequate aeration, and (3) a lower dose to achieve comparable tumor incidences. These changes are necessary because the partially protective chorion is gone, the fish gills are in direct contact with the carcinogen-bearing water, and (as do the sac-fry) the fry have an increasingly active hepatic MFO metabolizing system that increases their sensitivity to most carcinogens.

This exposure protocol is particularly useful in experiments that require a dietary pretreatment to a modulating chemical prior to carcinogen initiation, as well as when a short-term point initiation is desirable. The advantages and disadvantages of the procedure are the same as those given for embryo immersion.

Injection of Older Animals

Intraperitoneal injections of AFB_1 or its metabolites have been used frequently for short-term *in vivo* metabolism or DNA binding studies in rainbow trout, but rarely as an exposure technique for carcinogenesis experiments. Scarpelli (1976) successfully exposed a small number of rainbow trout to AFB_1 by ip injections, twice weekly for 25 weeks. Another group of fish received ip injections of the MFO inhibitor SKF-525A 3 hr prior to the AFB_1 injections. Of 20 fish injected, 19 developed tumors in the AFB_1 only group, but only 3 of 17 that received the SKF-525A before AFB_1 had tumors. The AFB_1 dose administered was 0.06 mg/kg body weight.

Other injection routes, such as intramuscular or subcutaneous, are not effective with rainbow trout because of substantial leakage from the injection site. The subcutaneous compartment of fish is not as extensive as that in mammals and does not provide a cavity into which an injected bolus can disperse. Descriptively, fish are "tight-skinned" and the cells of the muscle and the skin do not close on an injection hole, allowing the injectant to leak directly out. Leakage can be a problem even with ip injections if the bolus is not released as far as possible from the injection site.

Gavage

Gavage, force-feeding, or stomach tubing aflatoxins into rainbow trout has, for the most part, been unsuccessful because of their propensity to regurgitate anything irritating to the stomach (Bauer *et al.*, 1969). Dashwood *et al.* (1989b) successfully gavaged rainbow trout with I3C, but attempts to administer large doses of aflatoxins, in capsules or in oil, have resulted in regurgitation in nearly all cases.

Carcinogenicity of Aflatoxin Metabolites in Rainbow Trout

AFM_1

The first published report of the carcinogenicity of a metabolite of AFB_1 in rainbow trout was for AFM_1 by Sinnhuber *et al.* (1974). In addition to being a minor metabolite in control rainbow trout and the major metabolite in BNF-induced trout, AFM_1 is the major metabolite produced by dairy cows and appears in substantial quantities in milk. Therefore, this metabolite has major human health implications in addition to the comparative metabolism interest in the trout. Sinnhuber *et al.* (1974) stated that the AFM_1 used was extremely pure and free from AFB_1 contamination. Their results, using dietary exposure, indicated that the potency of AFM_1 was about 30% that of AFB_1. Canton *et al.* (1975) also produced tumors in rainbow trout with dietary AFM_1, but did not assign a relative potency. Several attempts to initiate tumors in trout embryos with bath exposure to AFM_1 have been unsuccessful (Hendricks *et al.*, 1980c). Wong and Hsieh (1976), using the Ames *Salmonella* mutagen assay system, reported that the mutagenic potency of AFM_1 was about 3% that of AFB_1. Coulombe *et al.* (1982) found AFM_1 to have about 1.6% the potency of AFB_1 as a mutagen in a modified Ames assay in which trout S20 was used rather than rat S9. Both these mutagen assay results show AFM_1 to be considerably less potent than shown by Sinnhuber *et al.* (1974). More definitive dose–response results with AFM_1 will be summarized later in this chapter.

AFL

Schoenhard *et al.* (1981) reported the carcinogenicity of AFL in rainbow trout. In an 8-month sample, the potency of AFL appeared to be about 50% that of AFB_1; however, by 12 months the relative potencies were much closer. Bath exposures of trout embryos to AFL, on the other hand, showed AFL to be equally as carcinogenic as AFB_1 or slightly more so (Hendricks *et al.*, 1980c). Coulombe *et al.* (1982) reported the mutagenicity of AFL to be 66% of that of AFB_1. As for AFM_1, dose–response results with AFL will be presented later in this chapter.

AFQ_1

AFQ_1 is the major *in vitro* microsomal metabolite produced by human and monkey liver. The mutagenic potency of AFQ_1 was only 0.3% compared with AFB_1 (Coulombe *et al.*, 1982); its carcinogenicity in mammals has not been demonstrated. This metabolite has no practical implications in trout but, because of the trout's high sensitivity to aflatoxins, AFQ_1 was tested in the diet and in an embryo bath exposure. Embryo bath exposure at 1.0 ppm for 1 hr failed to initiate any tumors. Dietary exposure to 20 ppb for 12 months also failed to produce any tumors, but 100 ppb in the diet for 12 months did result in a tumor incidence of 10.6% (Hendricks *et al.*, 1980b). This result is the only report of the carcinogenicity of AFQ_1 in any animal, again substantiating the claim that rainbow trout are the most sensitive to aflatoxins of all animals tested.

Sterigmatocystin and Versicolorin A

The mold metabolites sterigmatocystin (ST) and versicolorin A (VA) are biosynthetic precursors of AFB_1 in *Aspergillus flavus* and *Aspergillus parasiticus* (Hsieh *et al.*, 1976). Some other species of *Aspergillus* produce ST as the primary final metabolite (Rabie *et al.*, 1977; Hsieh *et al.*, 1978). Wong *et al.* (1977) tested the mutagenicity of ST and VA and reported values of 10 and 5%, respectively, compared with AFB_1. ST had been shown to be carcinogenic to rats (Purchase and Vander Watt, 1968) and mice (Fugii *et al.*, 1976), whereas VA had not been tested. We exposed trout embryos to ST and VA via the static water bath technique, and found both compounds to be carcinogenic to rainbow trout (Hendricks *et al.*, 1980a). Assigning relative potencies is difficult but, in these experiments, ST would have been about 25% as potent as AFB_1 and VA would have been 2–8% as carcinogenic as AFB_1, depending on the dose given.

AFP_1

AFP_1 is the major *in vivo* urinary metabolite produced by monkey and human liver (Dalezios *et al.*, 1971). The mutagenicity of this metabolite is very low, 0.5% compared with AFB_1 (Coulombe *et al.*, 1982), and its carcinogenicity has not been tested in mammals. Trout embryo bath exposure at two doses, 1 ppm for 24 hr and 5 ppm for 30 min, failed to produce any tumors 12 months later (Hendricks, 1982). Sufficient metabolite has not been available to attempt a dietary exposure with rainbow trout.

AFG_1

Aflatoxin G_1 is a naturally occurring aflatoxin with the 8,9 double bond on the terminal bifuran rings but a dilactone moiety in place of the cyclopentenone ring of AFB_1. Its potency as a mutagen (6.4% of AFB_1) (Coulombe *et al.*, 1982) and a carcinogen (~6.4% of AFB_1) (Ayres *et al.*, 1971) is much less than AFB_1.

Other Metabolites

Several other metabolites, both naturally occurring (AFB_2 and AFG_2) and metabolically derived ($AFB_{2\alpha}$), lack the 8,9 double bond and have not demonstrated any carcinogenic activity in rainbow trout. Aflatoxin H_1 is the reduced form of AFQ_1 and has never been available in sufficient quantities to test for carcinogenicity. The common feature of all the carcinogenic metabolites is the presence of the 8,9 double bond, which is required for epoxide formation leading to DNA adduct formation.

Dose–Response Carcinogenicity of AFB_1, AFL, AFM_1, and $AFLM_1$

Although the carcinogenicity of most of the aflatoxin metabolites has been tested in rainbow trout, simultaneous comparative testing of the four major metabolites still was necessary for several reasons: (1) comparisons of the potency of the four metabolites were unreliable because the various trials had been conducted during different years, with different routes of exposure, slightly different techniques that have improved over time, and different groups of heterozygous outbred fish which can vary from year to year; (2) true dose–response exposures had not been tested, except for AFB_1; and (3) $AFLM_1$ had not been tested for carcinogenicity at all. Thus we designed two large-scale dose–response studies with all four metabolites, using two routes of exposure—embryo microinjection and dietary. The experiments were conducted as follows:

1. Embryo microinjection—Triplicate groups of 150 23-day-old eggs were injected with 4 doses (0.5, 1.0, 2.0, and 4.0 ng/µl) of each metabolite (AFB_1, AFL, AFM_1, and $AFLM_1$) in a carrier of 25:75% acetone:soybean oil. Additional eggs (50) were microinjected with [³H]AFB_1 at the second dose of each metabolite to permit quantitation of DNA adducts/dose. Earlier experiments had shown that DNA binding versus dose followed a linear through-the-origin model over the dose range used in this experiment. Thus, the four aflatoxins could be compared at a single dose. After hatching and swim-up, 100 viable feeding fry in each group were fed the Oregon Test Diet (OTD) (Lee *et al.*, 1991) until necropsy 12 months postexposure. A total of 7400 embryos were microinjected; 4800 fry were started on the OTD. Hepatic tumors were verified and classified by light microscopy to establish final tumor incidences.

2. Dietary exposure—Triplicate groups of 120 feeding fry (1–2 g body weight) were placed on OTD containing aflatoxins. For AFB_1 and AFL, the doses were 4, 8, 16, 32, and 64 ppb; for AFM_1 and $AFLM_1$, the doses were 80, 160, 320, 640, and 1280 ppb. The doses for AFM_1 and $AFLM_1$ were 20× the doses for AFB_1 and AFL in anticipation of the much lower carcinogenicity of these two metabolites. The aflatoxin-containing diets were fed for 14 days. Then the fish were fed the control OTD for the remaining 11.5 months until necropsy.

**TABLE 1   Average Percentage of Hepatic Tumor
Incidences from Embryo Microinjection of Four Doses
of Four Aflatoxins**

Dose (ng/μl)	Average incidence for each metabolite (%)			
	AFB_1	AFL	AFM_1	$AFLM_1$
0.5	18.1	33.3	3.8	0.8
1.0	36.7	50.6	9.8	3.0
2.0	50.2	—[a]	21.0	6.3
4.0	66.8	77.4	34.1	20.1

[a]All fish lost because of plugged nozzle.

In this experiment, 6000 fry were started on the diets. A concurrent DNA binding study using [^{3}H]AFB_1-containing diets was run at one dose level for each metabolite.

Results of the microinjection experiment are presented in Table 1, and reveal the sensitivity of trout to AFB_1; a single exposure of 0.5 ng/egg produced tumors in 18% of the fish. All the metabolites were active at the same range of doses but, based on dose, differences in potency are obvious. AFL appears to be the most potent, followed in order by AFB_1, AFM_1, and $AFLM_1$. However, differences in chemical structure and the route of exposure can affect the pharmacokinetic distribution of carcinogens, especially at the molecular or DNA level, so the effective dose reaching the actual target site, in this case the N7 of guanine, may differ considerably.

Swenberg *et al.* (1987) and Van Zeeland (1988) have proposed the use of target tissue DNA adduction as a better measure of dose received than the dose level administered. This concept, called molecular dosimetry, has been used by Dashwood *et al.* (1989a) to interpret the relationship between DNA binding, final tumor incidence, and the effects of the tumor inhibitor I3C, and by Dashwood *et al.* (1990b) to interpret the relative potencies of AFB_1 and AFL administered to rainbow trout by the embryo bath exposure route. In the latter study, AFL was clearly a more potent carcinogen based on the dose administered. However, when tumor incidence was plotted against DNA adducts formed for both AFB_1 and AFL, the potencies were identical. This result is not without a mechanistic basis, since Loveland *et al.* (1987) showed that the adduct formed when trout are exposed to AFL, either *in vitro* (isolated hepatocytes) or *in vivo,* is not the 8,9-epoxide of AFL but rather the 8,9-epoxide of AFB_1. Therefore, AFL is converted back to AFB_1 before epoxide formation and DNA adduction occur. Similar molecular dosimetry analyses must be completed to address the relative potencies of AFM_1 and $AFLM_1$ by microinjection. This experiment does establish for the first time that $AFLM_1$ is carcinogenic in rainbow trout.

Table 2 contains the results of the dietary dose–response experiment. Obvi-

TABLE 2   Average Percentage of Hepatic Tumor Incidence from Dietary Exposure to Five Doses of Four Aflatoxins

Dose (ppb)	Average incidence (%)		Dose (ppb)	Average incidence (%)	
	AFB_1	AFL		AFM_1	$AFLM_1$
4	12.4	3.1	80	22.5	28.1
8	21.6	23.7	160	38.0	21.1
16	43.2	45.8	320	57.6	30.6
32	68.6	60.8	640	78.1	58.2
64	82.7	84.4	1280	48.9	62.3

ously these data again demonstrate the sensitivity of rainbow trout to AFB_1, since a dose of only 4 ppb, fed for only 2 weeks, produced a significant incidence of tumors. As previously described by Dashwood et al. (1990b), the potencies of AFB_1 and AFL appear very similar, even on a dose administered basis, with dietary exposure. Thus, as Dashwood et al. (1992) stated, the carcinogenic potency of AFB_1 and AFL is influenced by the route of exposure in the classical dose–response approach, but molecular dosimetry effectively integrates the pharmacokinetic differences to reveal similar potencies regardless of exposure route. The data for AFM_1 and $AFLM_1$ are more difficult to explain. The high dose for AFM_1 is unexplained at this time, as is the second dose for $AFLM_1$. Molecular dosimetry also may help explain these unexpected results. Until these analyses are completed, assignment of relative potencies for AFM_1 and $AFLM_1$ are premature.

Effects of Promoters on AFB_1 Carcinogenesis in Rainbow Trout

Cyclopropenoid Fatty Acids

Several compounds have been shown to be effective postinitiation promoters or enhancers of AFB_1 carcinogenesis in rainbow trout, but the mechanisms by which they exert their effect are still unknown. A true promoter is defined as a compound that will not initiate tumors by itself but, when applied once or repeatedly after a subcarcinogenic dose of a true initiator, will cause the development of neoplasms (Berenblum, 1969). One class of compounds, the CPFAs, is better classified as a group of co-carcinogens since these compounds do exert some initiating activity alone (Hendricks et al., 1980d; Hendricks, 1981). However, when administered in the diet after trout embryo bath exposure to AFB_1 in a true promotional protocol, CPFAs did enhance the tumor response in the way a true promoter would. The only difference is that the CPFAs also initiated a low tumor response by themselves in the controls (Hendricks, 1981).

Indole-3-Carbinol

Although I3C has been shown to be a powerful inhibitor of AFB_1 carcinogenesis, Bailey et al. (1987a) showed that it was an equally potent promoter of AFB_1

initiation. Subsequent experiments tested the effects of dose, time lapsed between initiation and I3C exposure, duration, and intermittency of exposure on the promotional properties of I3C. Promotion was related directly to dose, but an immediate start of I3C exposure after AFB_1 initiation had no advantage. In fact, tumor incidence was higher with a 1-month delay between initiation and promotion by I3C, although the delayed group received I3C for 1 month less than the other group. The longer the I3C exposure, the greater the effect, except as in the previous example. However, during a 9-month I3C promotional period, 3 or 6 months of exposure to I3C was more effective at the end of the period than at the beginning, that is, I3C exposure during months 7–9 or 4–9 gave a more effective promotional effect than I3C exposure during months 1–3 or 1–6, respectively. Intermittent exposure was effective in promoting, but less so than continuous exposure (Bailey *et al.*, 1987b; Dashwood *et al.*, 1990a, 1991).

17β-Estradiol

The higher incidence and greater size of neoplasms initiated by AFM_1 (Sinnhuber *et al.*, 1974) or *p,p'*-DDT (Hendricks, 1982) in female rainbow trout that were allowed to mature sexually before necropsy, and the occurrence of "spontaneous" hepatocellular tumors in sexually mature female rainbow trout (Takashima, 1976), caused us to suspect that estrogen hormones may have tumor promoting properties. Nunez *et al.* (1988) exposed rainbow trout embryos to AFB_1 using the static water bath technique; the trout were fed 20 ppm 17β-estradiol for 5 months and were necropsied at 12 months. 17β-Estradiol significantly enhanced the tumor incidence over the AFB_1 positive control. At 20 ppm in the diet, 17β-estradiol was toxic, inhibiting growth and causing mortalities. Use of the hormone was discontinued after 5 months and the fish were fed control diet until necropsy. In more detailed follow-up experiments, Nunez *et al.* (1989) exposed trout embryos to three concentrations of AFB_1, 0.005, 0.025, and 0.125 ppm, in static water baths. Duplicate groups of fry from each dose level were started on diets containing 20 ppm 17β-estradiol 6 weeks later. This diet was fed for 5 weeks, when growth reduction and mortalities began to occur. At that time, the dose level of 17β-estradiol was reduced to 10 ppm and maintained at that level until necropsy at 9 months postexposure. No tumors were found in the 0.005- or 0.025-ppm AFB_1 positive control groups and only 1 fish from the 0.005-ppm group that received 17β-estradiol had a tumor at necropsy. Of the 17β-estradiol-fed 0.025-ppm group, 9% had tumors whereas 17β-estradiol increased the incidence in the 0.125-ppm group from 5 to 60%. 17β-Estradiol caused no tumors by itself in these experiments, so this experiment was the first instance in which true promotion had been demonstrated in rainbow trout. Biochemical assays showed that 17β-estradiol stimulated ornithine decarboxylase activity and DNA synthesis (Nunez *et al.*, 1989). Unpublished observations (J. D. Hendricks) have shown 17β-estradiol to increase the mitotic index in rainbow trout liver greatly. These effects may be mechanisms through which 17β-estradiol exerts its promotional effect on AFB_1 carcinogenesis in trout. 17β-Estradiol promotion was particularly striking since the hormone restricted growth significantly and tumor growth usually is correlated positively with body growth. Despite the growth restriction, tumor incidence, numbers, and size were enhanced in the 17β-estradiol-fed fish. Other experiments have used varying dose levels of 17β-estradiol—5, 10, and 15 ppm—fed every other week with control diet. This exposure regimen was effective in promoting AFB_1 carcinogenesis, while eliminating the growth depression and mortalities associated with continuous 17β-estradiol exposure (Nunez *et al.,* 1989).

Polychlorinated Biphenyls

Although PCBs have been shown to be powerful inducers of MFO activity in rainbow trout and preinitiation inhibitors of AFB_1 carcinogenesis (Hendricks *et al.,* 1977; Shelton *et al.,* 1983, 1984, 1986), Aroclor 1254 has failed to promote AFB_1-initiated carcinogenesis (Hendricks *et al.,* 1980e; J. D. Hendricks, unpublished results). This lack of promotion with AFB_1 is the opposite effect observed with 7,12-dimethylbenz (*a*) anthracene initiation, which is enhanced by postinitiation exposure to PCBs.

β-Naphthoflavone

Bailey *et al.* (1989) reported a single experiment in which BNF enhanced AFB_1 carcinogenesis when fed after initiation.

Dietary Protein

The type and the quantity of protein in the diet of rainbow trout affect tumor incidence. Lee *et al.* (1977) showed that high-protein diets resulted in higher tumor incidence than low-protein diets, and that fish protein concentrate (FPC) produced different results than casein. The low (32%) FPC diet produced a lower incidence than did 32% casein, but the high (49.5%) FPC diet produced more tumors than the high casein. Later experiments used a greater range of protein levels, from 40 to 70%, for both protein types. Dietary and embryo bath exposures to AFB_1 resulted in significant increases in tumor response at each increasing protein level for both protein types. Part of this increased response was thought to be a growth effect, since larger fish usually have a higher incidence of tumors than smaller fish if they are exposed at the same dose of AFB_1 and the fish were larger on the higher protein diets. However, fish were placed into size classes at each protein level and, within a given size class across all four levels of protein, a significantly higher incidence of tumors still occurred in the fish on the higher protein content diets (Bailey *et al.,* 1982b; Hendricks, 1982).

Aflatoxin-Initiated Tumor Pathology

Tumor pathology resulting from aflatoxin initiation has been described periodically since the mid-1970s (Sinnhuber *et al.,* 1977; Hendricks, 1982; Hendricks *et al.,* 1984b), and has been updated and reviewed thoroughly (Hendricks, 1993; see also Chapter 7). Interested readers are referred to these reviews, since only a few brief comments will be made here.

The only documented tumors that have been initiated by aflatoxins in nonmammalian species are of hepatic origin, both hepatocellular and cholangiocellular. In rainbow trout, the predominant neoplasm observed 9–12 months after initiation is a mixed hepato-cholangiocellular carcinoma. These tumors comprise ~60% of all the tumors observed. Another 20–25% of the tumors are pure hepatocellular carcinomas; the remaining 15–20% are a mixture of benign hepatocellular adenomas, small basophilic foci, benign cholangiomas, or malignant cholangiocellular carcinomas.

SUMMARY

As stated at the beginning of this chapter, the available literature on the subject of carcinogenesis in nonmammalian animals is heavily skewed toward the work that has been done with rainbow trout. The sensitivity of rainbow trout to AFB_1 and its metabolites, in addition to the cost advantage of using trout, has made it possible to conduct certain types of experiments with rainbow trout that are not feasible with other animals. Although predicting what the future holds for other nonmammalian species with respect to research on aflatoxins is difficult, I foresee the trout continuing to occupy a strategic role as long as aflatoxin research is considered important to human health.

ACKNOWLEDGMENTS

This work was supported in part by U.S. Public Health Services Grants ES00210, ES03850, and ES04766 from the National Institute of Environmental Health Sciences and Grant DAMD 17-91Z1043 from the Department of the Army.

REFERENCES

Allcroft, R. (1969). Aflatoxicosis in farm animals. *In* "Aflatoxin: Scientific Background, Control, and Implications" (L. A. Goldblatt, ed.), pp. 237–264. Academic Press, New York and London.
Ames, B. N., McCann, J., and Yamasaki, E. (1975). Methods for detecting carcinogens and mutagens with the *Salmonella*/mammalian microsome mutagenicity test. *Mutation Res.* **31**, 347–361.
Arseculeratne, S. N., DeSilva, L. M., Bandunatha, C. H. S. R., Tennekoon, G. E., Wijesundera, S., and Balasubramaniam, K. (1969). The use of tadpoles of *Bufo melanostictus* (Schneider), *Rhacophorus leucomystax maculatus* (Gray), and *Uperodon* sp. in the bioassay of aflatoxins. *Br. J. Exp. Pathol.* **50**, 285–294.
Ashley, L. M. (1967). Histopathology of rainbow trout aflatoxicosis. *In* "Trout Hepatoma Research Conference Papers" (J. E. Halver and I. A. Mitchell, eds.), Vol. 70, pp. 103–120. Bureau of Sport Fisheries and Wildlife, Washington, D.C.
Ashley, L. M. (1969). Experimental fish neoplasia. *In* "Fish in Research" (O. W. Neuhaus and J. E. Halver, eds.), pp. 23–43. Academic Press, New York.
Ashley, L. M. (1970). Pathology of fish fed aflatoxins and other antimetabolites. *In* "A Symposium on Diseases of Fishes and Shellfishes" (S. F. Snieszko, ed.), Vol. 5, pp. 366–379. American Fisheries Society, Washington, D.C.

Ashley, L. M. (1973). Animal model: Liver cell carcinoma in rainbow trout. *Am. J. Pathol.* **72,** 345–348.

Ashley, L. M., and Halver, J. E. (1963). Multiple metastasis of rainbow trout hepatoma. *Trans. Am. Fish. Soc.* **92,** 365–371.

Ashley, L. M., and Halver, J. E. (1968). Hepatoma induction. V. Experiments. *In* "Trout Hepatomagenesis, Supplement to Final Report," pp. 1–10. National Cancer Institute, Bethesda, Maryland.

Asplin, F. D., and Carnaghan, R. B. A. (1961). The toxicity of certain groundnut meals for poultry with special reference to their effect on ducklings and chickens. *Vet. Rec.* **73,** 1215–1219.

Ayres, J. L., Lee, D. J., Wales, J. H., and Sinnhuber, R. O. (1971). Aflatoxin structure and hepatocarcinogenicity in rainbow trout (*Salmo gairdneri*). *J. Natl. Cancer Inst.* **46,** 561–564.

Bailey, G. S., Taylor, M. J., and Selivonchick, D. P. (1982a). Aflatoxin B$_1$ metabolism and DNA binding in isolated hepatocytes from rainbow trout (*Salmo gairdneri*). *Carcinogenesis* **3,** 511–518.

Bailey, G., Taylor, M., Selivonchick, D., Eisele, T., Hendricks, J., Nixon, J., Pawlowski, N., and Sinnhuber, R. (1982b). Mechanisms of dietary modification of aflatoxin B$_1$ carcinogenesis. *In* "Genetic Toxicology" (R. A. Fleck and A. Hollaender, eds.), pp. 149–164. Plenum Publishing, New York.

Bailey, G. S., Taylor, M. J., Loveland, P. M., Wilcox, J. S., Sinnhuber, R. O., and Selivonchick, D. P. (1984). Dietary modification of aflatoxin B1 carcinogenesis: Mechanism studies with isolated hepatocytes from rainbow trout. *Natl. Cancer Inst. Monogr.* **65,** 379–385.

Bailey, G. S., Hendricks, J. D., Shelton, D. W., Nixon, J. E., and Pawlowski, N. E. (1987a). Enhancement of carcinogenesis by the natural anticarcinogen indole-3-carbinol. *J. Natl. Cancer Inst.* **78,** 931–934.

Bailey, G. S., Selivonchick, D. P., and Hendricks, J. D. (1987b). Initiation, promotion, and inhibition of carcinogenesis in rainbow trout. *Environ. Health Perspect.* **71,** 147–153.

Bailey, G. S., Williams, D. E., Wilcox, J. S., Loveland, P. M., Coulombe, R. A., and Hendricks, J. D. (1988). Aflatoxin B$_1$ carcinogenesis and its relation to DNA adduct formation and adduct persistence in sensitive and resistant salmonid fish. *Carcinogenesis* **9,** 1919–1926.

Bailey, G. S., Goeger, D. E., and Hendricks, J. D. (1989). Factors influencing experimental carcinogenesis in laboratory fish models. *In* "Metabolism of Polynuclear Hydrocarbons in the Aquatic Environment" (U. Varanasi, ed.), pp. 253–268. CRC Press, Boca Raton, Florida.

Bailey, G., Hendricks, J., and Dashwood, R. (1992). Anticarcinogenesis in fish. *Mutation Res.* **267,** 243–250.

Balls, M., and Ruben, L. N. (1964). A review of the chemical induction of neoplasms in amphibia. *Experientia* **20,** 241–247.

Balls, M., Clothier, R., and Rubin, L. (1978). Neoplasia of amphibia. *In* "Animal Models of Comparative and Developmental Aspects of Immunity and Disease" (M. E. Gershwin and E. L. Cooper, eds.), pp. 48–62. Pergamon Press, New York.

Bauer, D. H., Lee, D. J., and Sinnhuber, R. O. (1969). Acute toxicity of aflatoxins B$_1$ and G$_1$ in the rainbow trout (*Salmo gairdneri*). *Toxicol. Appl. Pharmacol.* **15,** 415–419.

Bauer, L., Tulusan, A. H., and Müller, E. (1972). Ultrastructural changes produced by the carcinogen, aflatoxin B$_1$, in different tissues. *Virchows Arch. Zellpath.* **10,** 275–285.

Baumann, P. C., and Harshbarger, J. C. (1985). Frequencies of liver neoplasia in a feral fish population and associated carcinogens. *Marine Environ. Res.* **17,** 324–327.

Baumann, P. C., Smith, W. D., and Parland, W. K. (1987). Tumor frequencies and contaminant concentrations in brown bullhead from an industrialized river and a recreational lake. *Trans. Am. Fish Soc.* **116,** 79–86.

Beasley, R. P., Lin, C. C., Hwang, L. Y., and Chien, C. S. (1981). Hepatocellular carcinoma and hepatitis B virus. *Lancet* **2,** 1129–1133.

Benson, A. M., Cha, Y. N., Bueding, E., Heine, H. S., and Talalay, P. (1979). Elevation of extrahepatic glutathione *S*-transferase and epoxide hydratase activities by 2(3)-*tert*-butyl-4-hydroxyanisole. *Cancer Res.* **39,** 2971–2977.

Berenblum, I. A. (1969). Reevaluation of the concept of carcinogenesis. *Prog. Exp. Tumor Res.* **11**, 21–30.

Biba, D. (1983). Effects of aflatoxin on the brown bullhead, *Ictalurus nebulosis.* M.S. Thesis, Auburn University. Auburn, Alabama.

Black, J. J., Maccubbin, A. E., and Schiffert, M. (1985). A reliable, efficient, microinjection apparatus and methodology for the *in vivo* exposure of rainbow trout and salmon embryos to chemical carcinogens. *J. Natl. Cancer Inst.* **75**, 1123–1128.

Bodine, A. B., Luer, C. A., and Gangjee, S. (1985). A comparative study of monooxygenase activity in elasmobranches and mammals: Activation of the model pro-carcinogen aflatoxin B_1 by liver preparations of calf, nurse shark and clearnose skate. *Comp. Biochem. Physiology* **82C**, 255–257.

Bodine, A. B., Luer, C. A., Gangjee, S. A., and Walsh, C. J. (1989). *In vitro* metabolism of the procarcinogen aflatoxin B_1 by liver preparations of the calf, nurse shark and clearnose skate. *Comp. Biochem. Physiol.* **94C**, 447–453.

Bradfield, C. A., and Bjeldanes, L. F. (1987). Structure–activity relationships of dietary indoles: A proposed mechanism of action as modifiers of xenobiotic metabolism. *J. Toxicol. Environ. Health* **21**, 311–323.

Breedis, C. (1952). Induction of accessory limbs and of sarcoma in the newt (*Triturus viridencens*) with carcinogenic substances. *Cancer Res.* **12**, 861–873.

Briggs, R. W. (1940). Tumour induction in *Rana pipiens* tadpoles. *Nature (London)* **146**, 29.

Brodie, B. B., and Maickel, R. P. (1962). Comparative biochemistry of drug metabolism. *Proc. First Int. Pharmacol. Meet.* **1**, 227–324.

Campbell, J. G., and Appleby, E. C. (1966). Tumours in young chickens bred for rapid body growth (broiler chickens): A study of 351 cases. *J. Pathol. Bact.* **92**, 77–90.

Canton, J. H., Kroes, R., Van Logten, M. J., Van Schothorst, M., Stavenuiter, J. F. C., and Verhulsdonk, C. A. H. (1975). The carcinogenicity of aflatoxin M_1 in rainbow trout. *Food Cosmet. Toxicol.* **13**, 441–443.

Carnaghan, R. B. A. (1965). Hepatic tumors in ducks fed a low level of toxic groundnut meal. *Nature (London)* **208**, 308.

Carnaghan, R. B. A., Hartley, R. D., and O'Kelly, J. (1963). Toxicity and fluorescence properties of the aflatoxins. *Nature (London)* **200**, 1101.

Carnaghan, R. B. A., Lewis, G., Patterson, D. S. P., and Allcroft, R. (1966). Biochemical and pathological aspects of groundnut poisoning in chickens. *Pathol. Vet.* **3**, 601–615.

Carnaghan, R. B. A., Herbert, C. N., Patterson, D. S. P., and Sweasy, D. (1967). Comparative biological and biochemical studies in hybrid chicks. 2. Susceptibility to aflatoxin and effects on serum protein constituents. *Br. Poult. Sci.* **8**, 279–284.

Chan, T. M., Gillette, J. W., and Terriere, L. C. (1967). Interaction between microsomal electron transport systems of trout and male rats in cyclodiene epoxidation. *Comp. Biochem. Physiol.* **20**, 731–742.

Coates, J. A., Potts, T. J., and Wilcke, H. L. (1967). Interim hepatoma research report. *In* "Trout Hepatoma Research Conference Papers" (J. E. Halver and I. A. Mitchell, eds.), Vol. 70, pp. 34–38. Bureau of Sport Fisheries and Wildlife, Washington, D.C.

Coulombe, R. A., Shelton, D. W., Sinnhuber, R. O., and Nixon, J. E. (1982). Comparative mutagenicity of aflatoxins using a *Salmonella*/trout hepatic enzyme activation system. *Carcinogenesis* **3**, 1261–1264.

Coulombe, R. A., Jr., Bailey, G. S., and Nixon, J. E. (1984). Comparative activation of aflatoxin B1 to mutagens by isolated hepatocytes from rainbow trout (*Salmo gairdneri*) and coho salmon (*Oncorhynchus kisutch*). *Carcinogenesis* **5**, 29–33.

Cova, L., Wild, C. P., Mehrotra, R., Turusov, V., Shirai, T., Lambert, V., Jacquet, C., Tomatis, L., Trepo, C., and Montesano, R. (1990). Contribution of aflatoxin B1 and hepatitis B virus infection in the induction of liver tumors in ducks. *Cancer Res.* **50**, 2156–2163.

Croy, R. G., Nixon, J. E., Sinnhuber, R. O., and Wogan, G. N. (1980). Investigation of covalent aflatoxin B_1–DNA adducts formed *in vivo* in rainbow trout (*Salmo gairdneri*) embryos and liver. *Carcinogenesis* **1**, 903–909.

Cullen, J. M., Marion, P. L., Sherman, G. J., Hong, X., and Newbold, J. E. (1990). Hepatic neoplasms in aflatoxin B_1-treated, congenital duck hepatitis B virus-infected, and virus-free Pekin ducks. *Cancer Res.* **50,** 4072–4080.

Dalezios, J. I., Wogan, G. N., and Weinreb, S. M. (1971). Aflatoxin P_1: A new aflatoxin metabolite in monkeys. *Science* **171,** 584–585.

Dashwood, R. H., Arbogast, D. N., Fong, A. T., Hendricks, J. D., and Bailey, G. S. (1988). Mechanisms of anti-carcinogenesis by indole-3-carbinol: Detailed *in vivo* DNA binding dose–response studies after dietary administration with aflatoxin B1. *Carcinogenesis* **9,** 427–432.

Dashwood, R. H., Arbogast, D. N., Fong, A. T., Pereira, C., Hendricks, J. D., and Bailey, G. S. (1989a). Quantitative interrelationships between aflatoxin B1 carcinogen dose, indole-3-carbinol anti-carcinogen dose, target organ DNA adduction and final tumor response. *Carcinogenesis* **10,** 175–181.

Dashwood, R. H., Uyetake, L., Fong, A. T., Hendricks, J. D., and Bailey, G. S. (1989b). *In vivo* disposition of the natural anti-carcinogen indole-3-carbinol after PO administration to rainbow trout. *Food Chem. Toxicol.* **27,** 385–392.

Dashwood, R. H., Fong, A. T., Hendricks, J. D., and Bailey, G. S. (1990a). Tumor dose–response studies with aflatoxin B_1 and the ambivalent modulator indole-3-carbinol: Inhibitory versus promotional potency. *In* "Antimutagenesis and Anticarcinogenesis Mechanisms II" (Y. Kuroda, D. M. Shankel, and M. D. Waters, eds.), pp. 361–365. Plenum Publishing, New York.

Dashwood, R. H., Loveland, P. M., Fong, A. T., Hendricks, J. D., and Bailey, G. S. (1990b). Combined *in vivo* DNA binding and tumor-dose response studies to investigate the molecular dosimetry concept. *In* "Mutation and the Environment" (M. Mendelsohn and R. J. Albertini, eds.), pp. 335–344. Wiley-Liss, New York.

Dashwood, R. H., Fong, A. T., Williams, D. E., Hendricks, J. D., and Bailey, G. S. (1991). Promotion of aflatoxin B1 carcinogenesis by the natural tumor modulator indole-3-carbinol: Influence of dose, duration, and intermittent exposure on indole-3-carbinol promotional potency. *Cancer Res.* **51,** 2362–2365.

Dashwood, R. H., Marien, K., Loveland, P. M., Williams, D. E., Hendricks, J. D., and Bailey, G. S. (1992). Formation of aflatoxin–DNA adducts in trout and their use as molecular dosimeters for tumor prediction. *In* "Handbook of Applied Mycology: Mycotoxins in Ecological Systems" (D. Bhatnagar, E. B. Lillehoj, and D. K. Arora, eds.), Vol. 5, pp. 183–211. Marcel Dekker, New York.

Degen, G. H., and Neumann, H-G. (1981). Differences in aflatoxin B_1-susceptibility of rat and mouse are correlated with the capacity *in vitro* to inactivate aflatoxin B1 epoxide. *Carcinogenesis* **2,** 299–306.

Dewaide, J. H., and Henderson, P. T. (1968). Hepatic N-demethylation of aminopyrine in rat and trout. *Biochem. Pharm. Med.* **17,** 1901–1907.

Dollar, A. M., Smuckler, E. A., and Simon, R. C. (1967). Etiology and epidemiology of trout hepatoma. *In* "Trout Hepatoma Research Conference Papers" (J. E. Halver and I. A. Mitchell, eds.), Vol. 70, pp. 1–17. Bureau of Sport Fisheries and Wildlife, Washington, D.C.

Elcombe, C. R., and Lech, J. J. (1978). Induction of monooxygenation in rainbow trout by polybrominated biphenyls: A comparative study. *Environ. Health Perspect.* **23,** 309–314.

Elcombe, C. R., Franklin, R. B., and Lech, J. J. (1979). Induction of hepatic microsomal enzymes in rainbow trout. *In* "Pesticide and Xenobiotic Metabolism in Aquatic Organisms" (M. A. Q. Khan, J. J. Lech, and J. J. Menn, eds.), pp. 319–337. American Chemical Society, Washington, D.C.

El-Mofty, M. M., and Sakr, S. A. (1988). The induction of neoplastic lesions by aflatoxin-B_1 in the Egyptian toad (*Bufo regularis*). *Nutr. Cancer* **11,** 55–59.

Engebrecht, R. H., Ayres, J. L., and Sinnhuber, R. O. (1965). Isolation and determination of aflatoxin B_1 in cottonseed meals. *J. Assoc. Off. Ag. Chem.* **48,** 815–818.

Fong, A. T., Swanson, H. I., Dashwood, R. H., Williams, D. E., Hendricks, J. D., and Bailey, G. S. (1990). Mechanisms of anti-carcinogenesis by indole-3-carbinol. *Biochem. Pharmacol.* **39,** 19–26.

Fugii, K., Kurata, H., Odashima, S., and Hatsuda, Y. (1976). Tumor induction by a single subcutaneous injection of sterigmatocystin in newborn mice. *Cancer Res.* **36**, 1615–1618.

Gabor, M., Puscarin, F., and Deac, C. (1973). Teratogenic action of aflatoxins in frog tadpoles. *Arch. Roum. Pathol. Exp. Microbiol.* **32**, 269–275.

Gaudette, L. E., Maickel, R. P., and Brodie, B. B. (1958). Oxidative metabolism of drugs by vertebrates. *Fed. Proc.* **17**, 370.

Goeger, D. E., Shelton, D. W., Hendricks, J. D., and Bailey, G. S. (1986). Mechanisms of anticarcinogenesis by indole-3-carbinol: Effect on the distribution and metabolism of aflatoxin B1 in rainbow trout. *Carcinogenesis* **7**, 2025–2031.

Goeger, D. E., Shelton, D. W., Hendricks, J. D., Pereira, C., and Bailey, G. S. (1988). Comparative effect of dietary butylated hydroxyanisole and β-naphthoflavone on aflatoxin B1 metabolism, DNA adduct formation, and carcinogenesis in rainbow trout. *Carcinogenesis* **9**, 1793–1800.

Goldblatt, L. A. (1969). Introduction. *In* "Aflatoxin: Scientific Background, Control, and Implications" (L. A. Goldblatt, ed.), pp. 1–11. Academic Press, New York.

Gumbmann, M. R., Williams, S. N., Booth, A. N., Vohra, P., Ernest, R. A., and Bethard, M. (1970). Aflatoxin susceptibility in various breeds of poultry. *Proc. Exp. Biol. Med.* **134**, 683–688.

Halver, J. E. (1967). Crystalline aflatoxin and other vectors for trout hepatoma. *In* "Trout Hepatoma Research Conference Papers" (J. E. Halver and I. A. Mitchell, eds.), Vol. 70, pp. 78–102. Bureau of Sport Fisheries and Wildlife, Washington, D.C.

Halver, J. E. (1969). Aflatoxicosis and trout hepatoma. *In* "Aflatoxin: Scientific Background, Control, and Implications" (L. A. Goldblatt, ed.), pp. 265–306. Academic Press, New York.

Halver, J. E., Ashley, L. M., and Smith, R. R. (1969). Aflatoxicosis in coho salmon. *Natl. Cancer Inst. Monogr.* **31**, 141–155.

Harshbarger, J. C. (1969). The registry of tumors in lower animals. *Natl. Cancer Inst. Monogr.* **31**, 11–16.

Hatanaka, J., Doke, N., Harada, T., Aikawa, T., and Enomoto, M. (1982). Usefulness and rapidity of screening for the toxicity and carcinogenicity of chemicals in medaka, *Oryzias latipes. Japan. J. Exp. Med.* **52**, 243–253.

Hemsley, L. A. (1966). The incidence of tumours in young chickens. *J. Pathol. Bact.* **92**, 91–96.

Hendricks, J. D. (1981). The use of rainbow trout (*Salmo gairdneri*) in carcinogen bioassay, with special emphasis on embryonic exposure. *In* "Phyletic Approaches to Cancer" (C. J. Dawe, J. C. Harshbarger, S. Kondo, T. Sugimura, and S. Takayama, eds.), pp. 227–240. Japan Scientific Societies Press, Tokyo.

Hendricks, J. D. (1982). Chemical carcinogenesis in fish. *In* "Aquatic Toxicology" (L. J. Weber, ed.), pp. 149–211. Raven Press, New York.

Hendricks, J. D. (1993). Histopathology of hepatocellular neoplasms and related lesions in teleost fishes. *In* "An Atlas of Neoplasms and Related Disorders in Fishes" (C. J. Dawe, ed.). Academic Press, New York.

Hendricks, J. D., Putnam, T. P., Bills, D. D., and Sinnhuber, R. O. (1977). Inhibitory effect of a polychlorinated biphenyl (Aroclor 1254) on aflatoxin B_1 carcinogenesis in rainbow trout (*Salmo gairdneri*). *J. Natl. Cancer Inst.* **59**, 1545–1551.

Hendricks, J. D., Sinnhuber, R. O., Wales, J. H., Stack, M. E., and Hsieh, D. P. H. (1980a). Hepatocarcinogenicity of sterigmatocystin and versicolorin A to rainbow trout (*Salmo gairdneri*) embryos. *J. Natl. Cancer Inst.* **64**, 1503–1509.

Hendricks, J. D., Sinnhuber, R. O., Nixon, J. E., Wales, J. H., Masri, M. S., and Hsieh, D. P. H. (1980b). Carcinogenic response of rainbow trout (*Salmo gairdneri*) to aflatoxin Q_1 and synergistic effect of cyclopropenoid fatty acids. *J. Natl. Cancer Inst.* **64**, 523–527.

Hendricks, J. D., Wales, J. H., Sinnhuber, R. O., Nixon, J. E., Loveland, P. M., and Scanlan, R. A. (1980c). Rainbow trout (*Salmo gairdneri*) embryos: A sensitive animal model for experimental carcinogenesis. *Fed. Proc.* **39**, 3222–3229.

Hendricks, J. D., Sinnhuber, R. O., Loveland, P. M., Pawlowski, N. E., and Nixon, J. E. (1980d). Hepatocarcinogenicity of glandless cottonseeds and cottonseed oil to rainbow trout (*Salmo gairdneri*). *Science* **208**, 309–311.

Hendricks, J. D., Putnam, T. P., and Sinnhuber, R. O. (1980e). Null effect of dietary Aroclor 1254 on hepatocellular carcinoma incidence in rainbow trout (*Salmo gairdneri*) exposed to aflatoxin B1 as embryos. *J. Environ. Pathol. Toxicol.* **4,** 9–16.

Hendricks, J. D., Nixon, J. E., Bailey, G. S., and Sinnhuber, R. O. (1982). Inhibition of aflatoxin B_1 carcinogenesis in rainbow trout by dietary β-naphthoflavone and indole-3-carbinol. *Toxicologist* **2,** 102.

Hendricks, J. D., Meyers, T. R., Casteel, J. L., Nixon, J. E., Loveland, P. M., and Bailey, G. S. (1984a). Rainbow trout embryos: Advantages and limitations for carcinogenesis research. *Natl. Cancer Inst. Monogr.* **65,** 129–137.

Hendricks, J. D., Meyers, T. R., and Shelton, D. W. (1984b). Histological progression of hepatic neoplasia in rainbow trout (*Salmo gairdneri*). *Natl. Cancer Inst. Monogr.* **65,** 321–336.

Hsieh, D. P., Lin, M. T., Yao, R. C., and Singh, R. (1976). Biosynthesis of aflatoxin: Conversion of norsolorinic acid and other hypothetical intermediates into aflatoxin B_1. *J. Agr. Food Chem.* **24,** 1170–1174.

Hsieh, D. P., Singh, R., Yao, R. C., and Bennett, J. W. (1978). Anthraquinones in the biosynthesis of sterigmatocystin by *Aspergillus versicolor*. *Appl. Environ. Microbiol.* **35,** 980–982.

Hueper, W. C., and Payne, W. W. (1961). Observations on the occurrence of hepatomas in rainbow trout. *J. Natl. Cancer Inst.* **27,** 1123–1143.

Ingram, A. J. (1971). The reaction to carcinogens in the axolotl (*Ambystoma mexicanum*) in relation to the 'regeneration field control' hypothesis. *J. Embryol. Exp. Morphol.* **26,** 425–441.

Ingram, A. J. (1972). The lethal and carcinogenic effects of dimethylnitrosamine injection in the newt *Triturus helveticus*. *Br. J. Cancer* **26,** 206–215.

Jackson, E. W., Wolf, H., and Sinnhuber, R. O. (1968). The relationship of hepatoma in rainbow trout to aflatoxin contamination and cottonseed meal. *Cancer Res.* **28,** 987–991.

Jantrarotai, W., and Lovell, R. T. (1990). Subchronic toxicity of dietary aflatoxin B_1 to channel catfish. *J. Aquat. Animal Health* **2,** 248–254.

Jantrarotai, W., Lovell, R. T., and Grizzle, J. M. (1990). Acute toxicity of aflatoxin B_1 to channel catfish. *J. Aquat. Animal Health* **2,** 237–247.

Kew, M. C., Desmyter, J., Bradburne, A. F., and Macnab, G. M. (1979). Hepatitis B virus infection in Southern African blacks with hepatocellular cancer. *J. Natl. Cancer Inst.* **62,** 517–520.

Khudoley, V. V. (1977a). The induction of tumours in *Rana temporaria* with nitrosamines. *Neoplasma* **24,** 3.

Khudoley, V. V. (1977b). Tumor induction by carcinogenic agents in anuran amphibian *Rana temporaria*. *Arch. Geschwulsth.* **47,** 385–395.

Khudoley, V. V., and Picard, J. J. (1980). Liver and kidney tumors induced by *N*-nitrosodimethylamine in *Xenopus borealis* (Parker). *Int. J. Cancer* **25,** 679–683.

Lee, B. C., Hendricks, J. D., and Bailey, G. S. (1991). Toxicity of mycotoxins in the feed of fish. *In* "Mycotoxins and Animal Feedingstuff: Natural Occurrence, Toxicity, and Control" (J. E. Smith, ed.), pp. 607–626. CRC Press, Boca Raton, Florida.

Lee, D. J., Wales, J. H., Ayres, J. L., and Sinnhuber, R. O. (1968). Synergism between cyclopropenoid fatty acids and chemical carcinogens in rainbow trout (*Salmo gairdneri*). *Cancer Res.* **28,** 2312–2318.

Lee, D. J., Wales, J. H., and Sinnhuber, R. O. (1971). Promotion of aflatoxin-induced hepatoma growth in trout by methyl malvalate and sterculate. *Cancer Res.* **31,** 960–963.

Lee, D. J., Sinnhuber, R. O., Wales, J. H., and Putnam, G. B. (1977). Effect of dietary protein on the response of rainbow trout (*Salmo gairdneri*) to aflatoxin B_1. *J. Natl. Cancer Inst.* **60,** 317–320.

London, W. T. (1981). Primary hepatocellular carcinoma: Etiology, pathogenesis and prevention. *Human Pathol.* **12,** 1085–1097.

Loveland, P. M., Sinnhuber, R. O., Berggren, K. E., Libbey, L. M., Nixon, J. E., and Pawlowski, N. E. (1977). Formation of aflatoxin B_1 from aflatoxicol by rainbow trout (*Salmo gairdneri*) liver *in vitro*. *Res. Commun. Chem. Path. Pharmacol.* **16,** 167–170.

Loveland, P. M., Coulombe, R. A., Libbey, L. M., Pawlowski, N. E., Sinnhuber, R. O., Nixon, J. E., and Bailey, G. S. (1983). Identification and mutagenicity of aflatoxicol-M_1 produced by metabolism of aflatoxin B_1 and aflatoxicol by liver fractions from rainbow trout (*Salmo gairdneri*) fed β-naphthoflavone. *Food Chem. Toxicol.* **21,** 557–562.

Loveland, P. M., Nixon, J. E., and Bailey, G. S. (1984). Glucuronides in bile of rainbow trout (*Salmo gairdneri*) injected with [^{3}H] aflatoxin B1 and the effects of dietary β-naphthoflavone. *Comp. Biochem. Physiol.* **78C,** 13–19.

Loveland, P. M., Wilcox, J. S., Pawlowski, N. E., and Bailey, G. S. (1987). Metabolism and DNA binding of aflatoxicol and aflatoxin B_1 *in vivo* and in isolated hepatocytes from rainbow trout (*Salmo gairdneri*). *Carcinogenesis* **8,** 1065–1070.

Loveland, P. M., Wilcox, J. S., Hendricks, J. D., and Bailey, G. S. (1988). Comparative metabolism and DNA binding of aflatoxin B_1, aflatoxin M_1, aflatoxicol and aflatoxicol-M_1 in hepatocytes from rainbow trout (*Salmo gairdneri*). *Carcinogenesis* **9,** 441–446.

Luer, C. A., and Luer, W. (1981). Aflatoxicosis in the bonnet-head shark, *Sphyrna tiburo. Fed. Proc.* **40,** 694.

Luer, C. A., and Luer, W. (1982). Acute and chronic exposure of nurse sharks to aflatoxin B_1. *Fed. Proc.* **41,** 925.

Luer, C. A., and Luer, W. (1984). Toxic effects of aflatoxin B1 on reproduction and embryonic development in the clearnose skate (*Raja eglanteria*). *Fed. Proc.* **43,** 1953.

Magwood, S. E., Annau, E., and Corner, A. H. (1966). Induced tolerance in turkeys to aflatoxin poisoning. *Can. J. Comp. Med. Vet. Sci.* **30,** 17–25.

Majeed, S. K., Jolly, D. W., and Gopinath, C. (1984). An outbreak of liver cell carcinoma in rainbow trout, *Salmo gairdneri* Richardson, in the U.K. *J. Fish Dis.* **7,** 165–168.

Manning, R. O., Wyatt, R. D., and Marks, H. L. (1990). Effects of phenobarbital and β-naphthoflavone on the *in vivo* toxicity and *in vitro* metabolism of aflatoxin in an aflatoxin-resistant and control line of chickens. *J. Toxicol. Environ. Health* **31,** 291–311.

Matsushima, T., and Sugimura, T. (1976). Experimental carcinogenesis in small aquarium fishes. *Prog. Exp. Tumor Res.* **20,** 367–379.

Matsushima, T., Sato, S., Hara, K., Sugimura, T., and Takashima, F. (1975). Bioassay of environmental carcinogens with the guppy, *Lebistes reticulatus. Mutation Res.* **31,** 265.

Metcalfe, C. D., and Sonstegard, R. A. (1984). Microinjection of carcinogens into rainbow trout embryos: An *in vivo* carcinogenesis assay. *J. Natl. Cancer Inst.* **73,** 1125–1132.

Metcalfe, C. D., Cairns, V. W., and Fitzsimons, J. D. (1988). Microinjection of rainbow trout at the sac-fry stage: A modified trout carcinogenesis assay. *Aquat. Toxicol.* **13,** 347–356.

Montesano, R., Ingram, A. J., and Magee, P. N. (1973). Metabolism of dimethylnitrosamine by amphibians and fish *in vitro. Experientia* **29,** 599–601.

Newberne, P. M. (1965). Carcinogenicity of aflatoxin contaminated groundnut meal. *In* "Mycotoxins in Food Stuffs" (G. N. Wogan, ed.), pp. 187–208. MIT Press, Cambridge, Massachusetts.

Newberne, P. M., and Butler, W. H. (1969). Acute and chronic effects of aflatoxin on the liver of domestic and laboratory animals: A review. *Cancer Res.* **29,** 236–250.

Nigrelli, R. F., and Jakowska, S. (1961). Fatty degeneration, regenerative hyperplasia, and neoplasia in the livers of rainbow trout, *Salmo gairdneri. Zoologica* **46,** 49–55.

Nixon, J. E., Hendricks, J. D., Pawlowski, N. E., Pereira, C. B., Sinnhuber, R. O., and Bailey, G. S. (1984). Inhibition of aflatoxin B1 carcinogenesis in rainbow trout by flavone and indole compounds. *Carcinogenesis* **5,** 615–619.

Nunez, O., Hendricks, J. D., and Bailey, G. S. (1988). Enhancement of aflatoxin B_1 and *N*-methyl-*N'*-nitro-*N*-nitrosoguanidine hepatocarcinogenesis in rainbow trout *Salmo gairdneri* by 17-β-estradiol and other organic chemicals. *Dis. Aquat. Org.* **5,** 185–196.

Nunez, O., Hendricks, J. D., Arbogast, D. N., Fong, A. T., Lee, B. C., and Bailey, G. S. (1989). Promotion of aflatoxin B_1 hepatocarcinogenesis in rainbow trout by 17β-estradiol. *Aquat. Toxicol.* **15,** 289–302.

Outzen, H. C., Custer, R. P., and Prehn, R. T. (1976). Influence of regenerative capacity and innervation on oncogenesis in the adult frog (*Rana pipiens*). *J. Natl. Cancer Inst.* **57**, 79–84.

Purchase, I. F., and Vander Watt, J. J. (1968). Carcinogenicity of sterigmatocystin. *Food Cosmet. Toxicol.* **6**, 555–556.

Puscaria, F., Papillian, V., Gabor, M., and Deac, C. (1973). Toxicity and morphopathological effects of aflatoxins in frog tadpoles. *Arch. Roum. Pathol. Exp. Microbiol.* **32**, 255–267.

Rabie, C. J., Steyn, M., and Von Schalkwyk, G. C. (1977). New species of *Asperigillus* producing sterigmatocystin. *Appl. Environ. Microbiol.* **33**, 1023–1025.

Reiss, J. (1972). Susceptibility of eggs and larvae of *Triturus alpestris* to aflatoxin B₁. *Toxicon* **10**, 657–658.

Rigdon, R. H. (1972). Tumors in the duck (family Anatidae): A review. *J. Natl. Cancer Inst.* **49**, 467–476.

Rucker, R. R., Yasutake, W. T., and Wolf, H. (1961). Trout hepatoma—A preliminary report. *Prog. Fish-Cult.* **23**, 3–7.

Salhab, A. S., and Edwards, G. S. (1977). Comparative *in vitro* metabolism of aflatoxicol by liver preparations from animals and humans. *Cancer Res.* **37**, 1016–1021.

Salhab, A. S., Abramson, F. P., Geelhoed, G. W., and Edwards, G. S. (1977). Aflatoxicol M₁, a new metabolite of aflatoxicol. *Xenobiotica* **7**, 401–408.

Sato, S., Matsushima, T., Tanaka, N., Sugimura, T., and Takashima, F. (1973). Hepatic tumors in the guppy (*Lebistes reticulatus*) induced by aflatoxin B₁, dimethylnitrosamine, and 2-acetylaminofluorene. *J. Natl. Cancer Inst.* **50**, 765–778.

Scarpelli, D. G. (1967). Ultrastructural and biochemical observations on trout hepatoma. *In* "Trout Hepatoma Research Conference Papers" (J. E. Halver and I. A. Mitchell, eds.), Vol. 70, pp. 60–71. Bureau of Sport Fisheries and Wildlife, Washington, D.C.

Scarpelli, D. G. (1976). Drug metabolism and aflatoxin-induced hepatoma in rainbow trout (*Salmo gairdneri*). *Prog. Exp. Tumor Res.* **20**, 339–350.

Scarpelli, D. G., Greider, M. H., and Frajola, W. J. (1963). Observations on hepatic cell hyperplasia, adenoma, and hepatoma of rainbow trout (*Salmo gairdneri*). *Cancer Res.* **23**, 848–857.

Schoenhard, G. L., Lee, D. J., Howell, S. E., Pawlowski, N. E., Libbey, L. M., and Sinnhuber, R. O. (1976). Aflatoxin B₁ metabolism to aflatoxicol and derivatives lethal to *Bacillus subtilis* GSY 1057 by rainbow trout (*Salmo gairdneri*) liver. *Cancer Res.* **36**, 2040–2045.

Schoenhard, G. L., Hendricks, J. D., Nixon, J. E., Lee, D. J., Wales, J. H., Sinnhuber, R. O., and Pawlowski, N. E. (1981). Aflatoxicol-induced hepatocellular carcinoma in rainbow trout (*Salmo gairdneri*) and the synergistic effects of cyclopropenoid fatty acids. *Cancer Res.* **41**, 1011–1014.

Schoental, R. (1961). Liver changes and primary liver tumours in rats given toxic guinea pig diet (M. R. C. Diet 18). *Br. J. Cancer* **15**, 812–815.

Schultz, M. E., and Schultz, R. J. (1982a). Induction of hepatic tumors with 7,12-dimethylbenzo[*a*]anthracene in two species of viviparous fishes (Genus *Poeciliopsis*). *Environ. Res.* **27**, 337–351.

Schultz, M. E., and Schultz, R. J. (1982b). Diethylnitrosamine-induced hepatic tumors in wild vs. inbred strains of a viviparous fish. *J. Heredity* **73**, 43–48.

Schultz, M. E., and Schultz, R. J. (1988). Differences in response to a chemical carcinogen within species and clones of the livebearing fish, *Poeciliopsis*. *Carcinogenesis* **9**, 1029–1032.

Shelton, D. W., Coulombe, R. A., Pereira, C. B., Casteel, J. L., and Hendricks, J. D. (1983). Inhibitory effect of Aroclor 1254 on aflatoxin initiated carcinogenesis in rainbow trout and mutagenesis using a *Salmonella*/trout hepatic activation system. *Aquat. Toxicol.* **3**, 229–238.

Shelton, D. W., Hendricks, J. D., Coulombe, R. A., and Bailey, G. S. (1984). Effect of dose on the inhibition of carcinogenesis/mutagenesis by Aroclor 1254 in rainbow trout fed aflatoxin B₁. *J. Toxicol. Environ. Health* **13**, 649–657.

Shelton, D. W., Goeger, D. E., Hendricks, J. D., and Bailey, G. S. (1986). Mechanisms of anticarcinogenesis: The distribution and metabolism of aflatoxin B₁ in rainbow trout fed Aroclor 1254. *Carcinogenesis* **7**, 1065–1071.

Siller, W. G., and Ostler, D. C. (1961). The histopathology of an interohepatic syndrome of turkey poults. *Vet. Rec.* **73**, 134–138.

Simon, R. C., Dollar, A. M., and Smuckler, E. A. (1967). Descriptive classification on normal and altered histology of trout livers. *In* "Trout Hepatoma Research Conference Papers" (J. E. Halver and I. A. Mitchell, eds.), Vol. 70, pp. 18–28. Bureau of Sport Fisheries and Wildlife, Washington, D.C.

Sinnhuber, R. O. (1967). Aflatoxin in cottonseed meal and liver cancer in rainbow trout. *In* "Trout Hepatoma Research Conference Papers" (J. E. Halver and I. A. Mitchell, eds.), Vol. 70, pp. 48–55. Bureau of Sport Fisheries and Wildlife, Washington, D.C.

Sinnhuber, R. O., Lee, D. J., Wales, J. H., Ayres, J. L., and Amend, D. L. (1968a). Dietary factors and hepatoma in rainbow trout (*Salmo gairdneri*). 1. Aflatoxins in vegetable protein feedstuffs. *J. Natl. Cancer Inst.* **41**, 711–718.

Sinnhuber, R. O., Lee, D. J., Wales, J. H., and Ayres, J. L. (1968b). Dietary factors and hepatoma in rainbow trout (*Salmo gairdneri*). II. Cocarcinogenesis by cyclopropenoid fatty acids and the effect of gossypol and altered lipids on aflatoxin-induced liver cancer. *J. Natl. Cancer Inst.* **41**, 1293–1301.

Sinnhuber, R. O., Lee, D. J., Wales, J. H., Landers, M. K., and Keyl, A. C. (1974). Hepatic carcinogenesis of aflatoxin M1 in rainbow trout (*Salmo gairdneri*) and its enhancement by cyclopropene fatty acids. *J. Natl. Cancer Inst.* **53**, 1285–1288.

Sinnhuber, R. O., Hendricks, J. D., Wales, J. H., and Putnam, G. B. (1977). Neoplasms in rainbow trout, a sensitive animal model for environmental carcinogenesis. *Ann. N.Y. Acad. Sci.* **298**, 389–408.

Smith, G. R., and Stearley, R. F. (1989). The classification and scientific names of rainbow and cutthroat trouts. *Fisheries* **14**, 4–10.

Smith, J. W., and Hamilton, P. B. (1970). Aflatoxicosis in the broiler chicken. *Poultry Sci.* **49**, 207–215.

Solomon, G. C., Jensen, R., and Tanner, H. (1965). Hepatic changes in rainbow trout (*Salmo gairdneri*) fed diets containing peanut, cottonseed, and soybean meals. *Am. J. Vet. Res.* **26**, 764–770.

Statham, C. N., Melancon, M. J., Jr., and Lech, J. J. (1976). Bioconcentration of xenobiotics in trout bile: A proposed monitoring aid for some waterborne chemicals. *Science* **193**, 680–681.

Swenberg, J. A., Richardson, F. C., Tyeryar, L., Deal, F., and Boucheron, J. (1987). The molecular dosimetry of DNA adducts formed by continuous exposure of rats to alkylating hepatocarcinogens. *Prog. Exp. Tumor Res.* **31**, 42–51.

Swenson, D. H., Miller, J. A., and Miller, E. C. (1973). 2,3-Dihydro-2,3-dihydroxy-aflatoxin B_1: An acid hydrolysis product of an RNA–aflatoxin B_1 adduct formed by hamster and rat liver microsomes *in vitro*. *Biochem. Biophys. Res. Commun.* **53**, 1260–1267.

Takashima, F. (1976). Hepatoma and cutaneous fibrosarcoma in hatchery-reared trout and salmon related to gonadal maturation. *Prog. Exp. Tumor Res.* **20**, 351–366.

Uchida, T., Suzuki, K., Esumi, M., Arii, M., and Shikata, T. (1988). Influence of aflatoxin B_1 intoxication on duck livers with duck hepatitis B virus infection. *Cancer Res.* **48**, 1559–1565.

Valsta, L. M., Hendricks, J. D., and Bailey, G. S. (1988). The significance of glutathione conjugation for aflatoxin B_1 metabolism in rainbow trout and coho salmon. *Food Chem. Toxicol.* **26**, 129–135.

Van Zeeland, A. A. (1988). Molecular dosimetry of alkylating agents: Quantitative comparisons of genetic effects on the basis of DNA adduct formation. *Mutagenesis* **3**, 179–191.

Wales, J. H. (1967). Degeneration and regeneration of liver parenchyma accompanying hepatomagenesis. *In* "Trout Hepatoma Research Conference Papers" (J. E. Halver and I. A. Mitchell, eds.), Vol. 70, pp. 56–59. Bureau of Sport Fisheries and Wildlife, Washington, D.C.

Wales, J. H. (1970). Hepatoma in trout. *In* "A Symposium on Diseases of Fishes and Shellfishes" (S. F. Snieszka, ed.), Vol. 5, pp. 351–365. American Fisheries Society, Washington, D.C.

Wales, J. H. (1979). Induction of hepatoma in rainbow trout *Salmo gairdneri* Richardson by the egg bath technique. *J. Fish Dis.* **2**, 563–566.

Wales, J. H., and Sinnhuber, R. O. (1966). An early hepatoma epizootic in rainbow trout, *Salmo gairdneri. Calif. Fish Game* **52,** 85–91.

Wales, J. H., and Sinnhuber, R. O. (1972). Hepatomas induced by aflatoxin in the Sockeye salmon (*Oncorhynchus nerka*). *J. Natl. Cancer Inst.* **48,** 1529–1530.

Wales, J. H., Sinnhuber, R. O., Hendricks, J. D., Nixon, J. E., and Eisele, T. A. (1978). Aflatoxin B_1 induction of hepatocellular carcinoma in the embryos of rainbow trout (*Salmo gairdneri*). *J. Natl. Cancer Inst.* **60,** 1133–1139.

Wannop, C. C. (1961). The histopathology of turkey "X" disease in Great Britain. *Avian Dis.* **5,** 371–381.

Wattenberg, L. W. (1973). Inhibition of chemical carcinogen-induced pulmonary neoplasia by butylated hydroxyanisole. *J. Natl. Cancer Inst.* **50,** 1541–1544.

Wellings, S. R. (1969). Neoplasia and primitive vertebrate phylogeny: Echinoderms, prevertebrates, and fishes—A review. *Natl. Cancer Inst. Monogr.* **31,** 59–128.

Whitham, M., Nixon, J. E., and Sinnhuber, R. O. (1981). Liver DNA bound *in vivo* with aflatoxin B_1 as a measure of hepatocarcinoma initiation in rainbow trout. *J. Natl. Cancer Inst.* **68,** 623–628.

Wolf, H., and Jackson, E. W. (1963). Hepatomas in rainbow trout: Descriptive and experimental epidemiology. *Science* **142,** 676–678.

Wolf, H., and Jackson, E. W. (1967). Hepatoma in salmonids: The role of cottonseed products and species differences. *In* "Trout Hepatoma Research Conference Papers" (J. E. Halver and I. A. Mitchell, eds.), Vol. 70, pp. 29–33. Bureau of Sport Fisheries and Wildlife, Washington D.C.

Wong, J. J., and Hsieh, D. P. H. (1976). Mutagenicity of aflatoxins related to their metabolism and carcinogenic potential. *Proc. Natl. Acad. Sci.* **73,** 2241–2244.

Wong, J. J., Singh, R., and Hsieh, D. P. (1977). Mutagenicity of fungal metabolites related to aflatoxin biosynthesis. *Mutation Res.* **44,** 447–450.

Wood, E. M., and Larson, C. P. (1961). Hepatic carcinoma in rainbow trout. *Arch. Pathol.* **71,** 471–479.

Yasutake, W. T., and Rucker, R. R. (1967). Nutritionally induced hepatomagenesis of rainbow trout. *In* "Trout Hepatoma Research Conference Papers" (J. E. Halver and I. A. Mitchell, eds.), Vol. 70, pp. 39–47. Bureau of Sport Fisheries and Wildlife, Washington, D.C.

Yokosuka, O., Omata, M., Zhou, Y. Z., Imazaki, F., and Okuda, K. (1985). Duck hepatitis B virus DNA in liver and serum of Chinese ducks: Integration of viral DNA in a hepatocellular carcinoma. *Proc. Natl. Acad. Sci. U.S.A.* **82,** 5180–5184.

7

Role of Aflatoxin–DNA Adducts in the Cancer Process

❖

George S. Bailey

AFLATOXIN B_1 COVALENT BINDING TO DNA *IN VIVO*

DNA Adducts Formed

Aflatoxin B_1 (AFB$_1$) reacts *in vivo* with the DNA in target cells to give primarily *trans*-8,9-dihydro-8-(N^7-guanyl)-9-hydroxyaflatoxin B_1 (Essigman *et al.*, 1977; Lin *et al.*, 1977; Martin and Garner, 1977). Studies on the reaction of synthesized AFB$_1$-8,9-epoxide (Baertschi *et al.*, 1988) with DNA *in vitro* strongly suggest that adduction *in vivo* proceeds via a precovalent intercalation complex between double-stranded DNA and the highly electrophilic, unstable AFB$_1$-*exo*-8,9-epoxide isomer (Gopalakrishnan *et al.*, 1990). The presence of a positive charge on the imidazole portion of the initial N^7-guanyl adduct (Figure 1) gives rise to a ring-opened formamidopyrimidine (FAPY) derivative with distinct chromatographic behavior (Croy and Wogan, 1981a). Accumulation of this derivative is time dependent, nonenzymatic, and of some potential biological importance because of its apparent persistence in DNA.

Other minor N^7-guanyl adducts can arise through enzymatic oxidation of aflatoxin P_1, M_1, and other AFB$_1$ phase I metabolites unsaturated in the 8,9 position (Croy and Wogan, 1979; J. D. Groopman, personal communication). Subsequent conversion of these initial N^7-guanyl adducts to FAPY derivatives can add considerable complexity to aflatoxin–DNA adduct chromatographic profiles. Production of such adducts would vary among species according to the

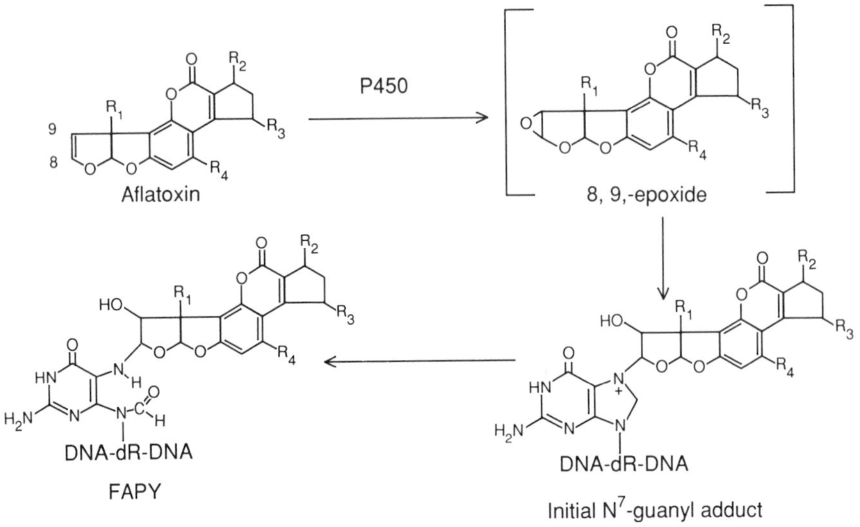

FIGURE 1 Metabolic pathway leading to DNA adduction and formation of FAPY adducts for aflatoxins. AFB_1, AFM_1, and AFP_1 are known to produce DNA adducts directly; AFL is known not to do so. AFB_1: R_1 = H, R_2 = O, R_3 = H, R_4 = OCH_3; AFM_1: R_1 = OH, R_2 = O, R_3 = H, R_4 = OCH_3; AFL: R_1 = H, R_2 = OH, R_3 = H, R_4 = OCH_3; AFP_1: R_1 = H, R_2 = O, R_3 = H, R_4 = OH; $AFLM_1$: R_1 = OH, R_2 = OH, R_3 = H, R_4 = OCH_3; AFQ_1: R_1 = H, R_2 = O, R_3 = OH, R_4 = OCH_3.

balance of specific P450 isozymes. Not all aflatoxins with intact 8,9 double bonds form adducts directly. For example, aflatoxicol (AFL) is a highly carcinogenic phase I metabolite that appears to be a reservoir for extending AFB_1 carcinogenicity in certain species, especially trout (Schoenhard *et al.*, 1981). Metabolic studies with AFL tritiated at the cyclopentenol carbon indicated at least 95% back-conversion to AFB_1 prior to DNA adduction (Loveland *et al.*, 1987). Subsequent studies have failed to identify any AFL-derived adduct other than 8,9-dihydro-8-(N^7-guanyl)-9-hydroxyaflatoxin B_1 (G. S. Bailey and J. D. Groopman, unpublished observations). Two testable hypotheses are that the cyclopentenol hydroxyl renders AFL a poor P450 substrate for oxidation and that the 8,9-epoxide, if formed, is unable to form an effective precovalent intercalation intermediate with DNA.

Minor adduction of AFB_1 to adenosine (D'Andrea and Haseltine, 1978) and cytosine (Yu *et al.*, 1991) in DNA *in vitro* has been reported, but evidence is lacking that these adducts form *in vivo* or have any importance in aflatoxin carcinogenesis.

Target Organ and Species Specificity

AFB_1 is a suspect human hepatocarcinogen; its principal target organ in susceptible mammalian, avian, and fish species is also the liver. AFB_1–DNA adduction is greater in liver than in other organs; the level of liver DNA adduction per unit AFB_1 dosage generally correlates with species susceptibility (Lutz *et al.*, 1980; Cole *et al.*, 1988). For example, F344 rats exposed to a single dose of AFB_1 display 10-fold greater adduction in liver than in kidney, a minor target organ in rats (Croy and Wogan, 1981b). In the same study, CD-1 Swiss mice, which are relatively resistant to AFB_1, had only 2% of the rat AFB_1–DNA adducts in liver at a 12-fold greater dosage and 3-fold *lower* adduction in liver than in kidney. A strikingly similar result has been reported in salmonid fish. The sensitive rainbow trout displayed 25-fold greater liver AFB_1–DNA adduction than the highly resistant coho salmon, but both species showed comparable low-level adduction in the kidney, a non-target organ (Bailey *et al.*, 1988). Interestingly, prolonged AFB_1 exposure can induce renal tumors in rats (Epstein *et al.*, 1969) but not in trout (G. S. Bailey and J. D. Hendricks, unpublished data), despite nonnegligible trout kidney AFB_1–DNA adduction and the ability of several other genotoxins to induce nephroblastoma in trout.

The lung has received attention as an alternative target because it constitutes a potentially important route of AFB_1 exposure for agricultural workers in the grain and nut industries. Coulombe and co-workers have shown that nonciliated tracheal cells from rodents and primates have varying metabolic capacities for activating AFB_1 to form N^7-guanyl adducts (see Ball *et al.*, 1990; Wilson *et al.*, 1990; Chapter 5). The ability of primates to form these adducts, albeit at low levels, is consistent with epidemiologic data indicating AFB_1-contaminated grain dust as a risk factor for upper airway carcinogenesis in humans.

Dose–Response

An understanding of AFB_1 dose–response relationships is critical in evaluating human risk, especially the possibility of a no-adduct threshold at low AFB_1 dose. Single-dose exposures of 10–1000 ng/kg in male F344 rats produced liver AFB_1 DNA binding curves that were linear at low dose and less than linear at high dose (Appleton *et al.*, 1982). Chronic exposure over the same dose range in the same model produced dose-linear steady-state levels of adducts (Buss *et al.*, 1990). Rainbow trout treated with carcinogenic doses of AFB_1 for 2–4 weeks also showed dose-linear accumulation of liver AFB_1–DNA adducts (Dashwood *et al.*, 1988). These studies involve dosages that extend as low as human exposure levels, and provide no indication of a threshold dose below which AFB_1 exposure might impose no genotoxic risk.

REPAIR AND PERSISTENCE OF AFLATOXIN B_1–DNA ADDUCTS: MECHANISMS OF ADDUCT LOSS

The initial *trans*-8,9-dihydro-8-(N^7-guanyl)-9-hydroxyaflatoxin B_1 DNA adduct may decrease in concentration through several processes: active DNA repair, spontaneous reactions, death of heavily damaged cells, and growth dilution. Wang and Cerrutti (1980) established approximate rate constants and pH dependence for three spontaneous reactions with AFB_1-modified DNA at 37°C *in vitro:* release of the AFB_1–guanyl adduct to form apurinic sites, release of AFB_1-8,9-dihydrodiol with the presumed restoration of guanylic sites in DNA, and conversion of the initial adduct to the FAPY derivative. Formation of FAPY was appreciable only at neutral or alkaline pH, whereas dihydrodiol release was the dominant reaction at all pH values. At neutral pH, depurination occurred with a half-life estimated at 12 hr. Loss of initial adduct was much slower, and accumulation of the FAPY derivative more extensive, in xeroderma pigmentosum A (XPA) cells than in normal human fibroblasts, suggesting that this excision repair pathway is a major mechanism for enzymatic repair of AFB_1 adducts (Leadon *et al.*, 1981). An important observation in this study was that AFB_1 loss in repair-deficient XPA cells was much slower than in purified AFB_1-adducted DNA *in vitro,* indicating that chromatin configuration per se confers stability to AFB_1–DNA adducts and that *in vitro* stability controls are not adequate in assessing the repair capacity of cells, organs, or organisms. In general, these and other studies indicate the relative resistance of the FAPY derivative to loss by spontaneous or enzymatic processes. Evidence has been presented for an FAPY–DNA glycosylase activity in bacteria and rat liver (Chetsanga and Frenette, 1983), but the importance of this repair pathway remains unclear.

In rat liver, the pseudo-half-life for loss of initial AFB_1 adduct as a result of these processes has been reported at 7.5 hr, with approximately 20% conversion of initial adduct to FAPY in the first 24 hr after exposure (Croy and Wogan, 1981a). By comparison, AFB_1–DNA adducts show far greater persistence in the rainbow trout liver, with an overall approximate half-life of 21 days in fish held at 12°C (Bailey *et al.*, 1988). This operationally defined low capacity for removal of bulky DNA adducts presumably contributes to the high sensitivity of rapidly growing juvenile trout to AFB_1 and other bulky carcinogens.

Since the persistent FAPY derivative may be present during more than one round of DNA replication, it often has been suggested to be especially important in tumor initiation. However, several reasons to question this view exist. Transcriptionally active DNA has been shown consistently to be repaired more readily than total genomic DNA, suggesting that persistent adducts may accumulate predominantly in noninformative DNA that is not important in the cancer process. Further, initial differences in carcinogen–DNA adduct levels, rather than differences in adducts that persist over time, have been found to correlate with differential tumor response. Rainbow trout exposed to AFB_1 have significantly greater initial liver AFB_1–DNA adduction than do coho salmon (Bailey *et al.*, 1988), as do ethoxyquin-treated rats compared with controls (Kensler *et al.*, 1986). In both cases, the initial difference in adducts tends toward zero with time; the difference in initial adduct levels, not the levels of persistent adducts, correlates with the observed difference in tumorigenic response. The fact that certain adducts for aflatoxins (and other genotoxins) may persist in total genomic DNA does not constitute convincing evidence for their participation in the carcinogenesis process.

NONRANDOM GENOMIC DAMAGE BY AFLATOXINS

Influence of Local Chromatin Structure

Studies relating tumor response to carcinogen–DNA adduction routinely measure adducts in total genomic DNA. This approach will not, however, provide a direct measure of adduction at specific genes or sequences of interest because eukaryotic DNA is not uniformly susceptible to adduction by AFB_1 or other bulky carcinogens. At the most fundamental level of chromatin organization, guanyl sites residing within nucleosomal core DNA have been shown to be protected 5-fold against AFB_1 adduction *in vivo* compared with internucleosomal DNA (Bailey *et al.*, 1980). Studies with sequence-phased, uniform nucleosomal preparations indicate that protection within the nucleosomal core structure is distributed uniformly, suggesting a nonlocalized protection mechanism such as torsional suppression of precovalent AFB_1 epoxide intercalation (Moyer *et al.*, 1989). Transcriptionally active ribosomal RNA genes in rat liver are 4- to 5-fold more susceptible to AFB_1 adduction than is total nuclear DNA (Irvin and Wogan, 1984), presumably as a result of the less constricted chromatin structure of active gene sequences. This phenomenon is not unique to multicopy genes, since active β-globin and histone 5 genes in chick erythrocytes are similarly more susceptible to AFB_1 damage than is the transcriptionally inactive vitellogenin gene (Delcuve *et al.*, 1988). From these results, clearly the processes (e.g., cell cycle progression, development, transcriptional induction, chromosomal deletions and rearrangements, excision DNA repair) that might temporarily or permanently alter the chromatin configuration of *ras, p53,* and other genes that are important in carcinogenesis would alter the potential of these genes to be damaged by AFB_1 and other bulky procarcinogens.

Local DNA Sequence Effects

AFB_1 shows sequence selectivity as do most other DNA-interactive carcinogens and drugs (for a review, see Warpehoski and Hurley, 1988). The potential of any particular guanine to react with the AFB_1-epoxide is influenced by its neighboring sequences, such that a 10-fold or greater variation in adduction frequency exists among guanines within defined sequences of DNA (Misra *et al.*, 1983; Muench *et al.*, 1983). The molecular basis for adduction preference remains to be defined, but has been suggested to reflect sequence effects on precovalent association rate constants as well as lesser sequence context effects (e.g., alterations in target guanine nucleophilicity) on the rate constant for the covalent reaction. Although empirical rules have been offered that generally predict AFB_1–guanine adduction preferences within certain trinucleotide sequence contexts (e.g., Misra *et al.*, 1983), these rules have not been found to predict adduction preferences reliably within gene sequences of particular interest (Marien *et al.*, 1987). Currently, modification preference at any guanine site of interest must be determined experimentally.

Comparison among Aflatoxins

One basis for the differing carcinogenicity among aflatoxins might be a difference in preferential adduction at guanine targets critical to tumor initiation per unit total genomic DNA adduction. This hypothesis was explored for the four aflatoxins AFB_1, AFM_1, AFL, and $AFLM_1$. Reaction of two cloned DNA sequences with the four aflatoxin epoxides revealed largely similar sequence selectivity profiles (Marien *et al.*, 1987), although some differences were observed. The sequence selectivity for AFB_1 and AFM_1 was compared further across the human Ha-*ras* proto-oncogene exon 1 region *in vitro* (Marien *et al.*, 1989). The results indicated that guanines in codons 12 and 13 were above average in adduction susceptibility, and that the middle guanine of codon 13 was most susceptible to adduction within this region of the gene. However, no significant difference was seen in the selectivity profiles for AFB_1 and AFM_1 across this region. This finding does not support the hypothesis that the differences in AFB_1 and AFM_1 carcinogenicity reflect a difference in selective adduction of target *ras* gene guanines. In addition, since codon 13 mutant alleles occur much less frequently than codon 12 alleles in aflatoxin-induced hepatic tumors (see subsequent discussion), differences in allele biological dysfunction or some parameter other than initial adduction profile must determine mutant *ras* allele frequencies among tumors.

MOLECULAR DOSIMETRY AND PROTO-ONCOGENE ACTIVATION IN EXPERIMENTAL ANIMAL TUMORIGENESIS

AFB_1 Dosimetry in Higher and Lower Vertebrates

Typical dose–response analyses examine tumor risk (on linear, logit, or probit scales) against carcinogen dose applied (on linear or log scales). An increasingly used alternative examines risk against target organ DNA adducts formed, that is, dose effectively received. This procedure integrates all variables that intervene between exposure and DNA damage, including recent exposure to enzyme inducers or inhibitors, genetic variation in metabolic capacity for carcinogen activation or detoxification, route of exposure, repair capacity, and exposure dose itself. This approach holds particular promise for assessing human risk when exposure amounts and other factors are unknown or immeasurable. In the rat, steady-state or maximal level of liver DNA adducts (dose effectively received) has been shown to vary linearly with AFB_1 dose applied, whether given as 1 dose orally, 10 doses orally, or chronically in drinking water (Buss *et al.*, 1990). By molecular dosimetry analysis, tumor risk also shows an apparent linear increase with rat liver AFB_1–DNA adducts by chronic exposure; 50% tumor incidence is reached at a steady-state level of approximately 1 adduct per 10^6 nucleotides (Bechtel, 1989). Interestingly, molecular dosimetry data for tumor risk in rainbow trout after 2 weeks of AFB_1 exposure were found to be collinear with the rat data. One interpretation is that AFB_1–DNA adducts in trout and rats induce tumors with approximately equal efficiency. This relationship can be regarded only as approximate because trout tumors were measured at 1 year whereas rat tumors were measured after 2 years chronic exposure. However, the results do indicate that the overall probability for progression of initiated cells to frank hepatocellular carcinoma does not differ greatly between higher and lower vertebrates. One study establishing collinearity between AFB_1 dose applied, liver DNA adducts formed, and 24-hr urinary AFB_1–N^7-guanine adduct released in rats indicates that measurement of these adducts in human urine will provide a useful individual surrogate dosimeter for hepatic risk following human AFB_1 exposure (Groopman *et al.*, 1992).

Tumor-Initiating Potency of Adducts from Various Aflatoxins

Differing carcinogenic potencies among aflatoxins can arise from several sources including differential uptake, metabolic activation, and detoxification; sequence selectivity for modification of critical guanines; adduct persistence; and the overall mutagenic potential of each adduct at each modified site. Little information has been available comparing these factors for the primary dietary aflatoxins AFB_1 and AFM_1 or for their phase I metabolites such as AFL and $AFLM_1$. We have taken the molecular dosimetry approach to compare the tumorigenic potencies and patterns of *ras* proto-oncogene activation in the rainbow trout model for these four aflatoxins. By dietary exposure, AFB_1 and AFL were found to be equally carcinogenic, as shown by their equivalent TD_{50} values in a full dose–response experiment (Figure 2A). In comparison, AFM_1 and $AFLM_1$ were approximately 10-fold and 23-fold less carcinogenic, respectively, based on dose given. However, when plotted as tumor risk vs. liver DNA adducts formed at each applied dose (Figure 2B), the DNA adducts from AFM_1 were equivalent to the AFB_1–N^7-guanyl adduct derived from AFB_1 and AFL in initiating tumors. Despite some imprecision in the data, the results give an indication that adducts derived from dietary $AFLM_1$ are approximately 1/3 less efficient in inducing tumors in this model (Figure 2B). The types of DNA adducts formed by $AFLM_1$, and the molecular basis of an apparently different initiating potency in trout, have yet to be defined. These results are consistent with the finding that AFL provides only AFB_1 adducts and, more pertinently, that AFB_1 and AFM_1 epox-

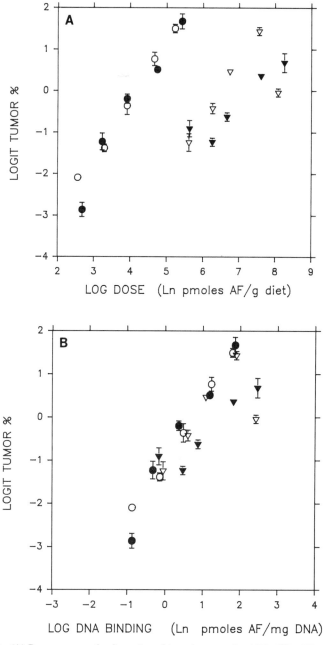

FIGURE 2 (A) Dose–response for formation of hepatic tumors by AFB$_1$ (○), AFL (●), AFM$_1$ (▽), and AFLM$_1$ (▼) in rainbow trout. Each data point represents three groups of ~100 individuals each, fed the specified level of aflatoxin for 4 weeks and held for tumor analysis 9 months later. Details will be published elsewhere. (B) Molecular dosimetry plot for the experimental groups shown in A. Relative covalent binding indices for each aflatoxin were determined from the slopes of liver DNA binding vs. dietary dose curves using tritiated aflatoxins, and were used to calculate log dose received based on doses given. Details will be published elsewhere.

ides do not differ in selectivity for guanine adduction across the *ras* gene exon 1 region.

Activated c-*ras* Proto-Oncogenes in Aflatoxin-Initiated Hepatic Tumors

Previous studies have indicated that many of the AFB_1-initiated hepatic tumors in trout carry cells with activating point mutations in the Ki-*ras* proto-oncogene (Chang *et al.*, 1991). We have begun to examine tumors from the molecular dosimetry tumor study just described to compare the types and incidences of activated Ki-*ras* alleles elicited by the four aflatoxins. Preliminary data from these comparisons are shown in Table 1. A similar proportion of tumors from all four aflatoxins was found to contain activated Ki-*ras* alleles; the dominant species was a codon 12 GGA → GTA transversion for all aflatoxins. Only one G → A transition mutation has been detected to date for any aflatoxin. This bias toward G → T transversions for AFB_1 (and, by extension, AFL) is compatible with its known mutagenic properties (Foster *et al.*, 1983). That AFM_1 and $AFLM_1$ show a similar bias suggests a mutagenic mechanism *in vivo* similar to that for AFB_1. For comparison, *N*-methyl-*N′*-nitro-*N′*-nitrosoguanidine (MNNG)-induced trout hepatic tumors also carry a similarly high proportion of activated Ki-*ras* alleles (Table 1). The predominance of G → A transition mutations is consistent with the well-known mutagenesis mechanism involving O^6-methylguanine mispairing for this genotoxin, and demonstrates the carcinogen-specific basis for the Ki-*ras* mutations observed in the trout model. Hepatic tumors from AFB_1-treated rats also have been reported to carry analogous Ki-*ras* mutant alleles (McMahon *et al.*, 1990; see also Chapter 9), although at lower incidence and with less bias toward G → T transversions than in the trout study.

The trout molecular dosimetry and Ki-*ras* activation studies indicate that all four aflatoxins have a similar molecular basis for tumor initiation and, with the possible exception of $AFLM_1$, have different carcinogenicities that reflect solely

TABLE 1   Aflatoxin-Activated Ki-*ras* Alleles in Trout Hepatic Tumors

Carcinogen		Codon 12 (GGA)			Codon 13 (GGT)			
		AGA	GTA	GAA	AGA	TGT	GAT	GTT
AFB_1[a]	(of 14)	1	7	0	0	0	0	2
AFL	(of 12)	0	8	ND	ND	ND	ND	0
AFM_1	(of 12)	0	9	ND	ND	ND	ND	0
$AFLM_1$	(of 12)	0	5	ND	ND	ND	ND	1
MNNG[a]	(of 15)	3	0	10	0	0	0	0
MNNG	(of 28)	ND[b]	ND	24	ND	ND	ND	ND

[a]Results by oligonucleotide hybridization and sequencing. Others by 3′-mismatch polymerase chain reaction (PCR) amplification only.
[b]ND, Not yet determined.

the differences in pharmacokinetic and metabolic processes leading to initial target organ DNA damage.

SUMMARY

The predominant DNA adduct for all aflatoxins studied derives from covalent bond formation between C8 of aflatoxin 8,9-epoxides and N^7 of guanine bases in nucleic acids. The reaction appears to proceed through a precovalent intercalation transition state complex, the formation of which is influenced by local sequence context. After a single exposure, concentration of the initial AFB_1–DNA adduct decreases at varying rates in different organisms or cell types, based on conversion to persistent FAPY ring-opened derivatives, enzymatic DNA repair, spontaneous depurination, and growth dilution. Formation of initial adduct is linear over the low-dose range in all species examined, and liver is the primary target organ. In trout and rats, hepatic tumor formation is also dose linear, with no indication of a substantial low-dose threshold for tumor formation. Molecular dosimetry studies in trout indicate that, with the possible exception of $AFLM_1$, aflatoxins studied to date generate DNA adducts that are equally potent at eliciting hepatic tumor formation. In the trout model, many aflatoxin-initiated hepatic tumors carry Ki-*ras* alleles activated through codon 12 or 13 G $\rightarrow$ T transversions.

ACKNOWLEDGMENTS

The original studies in this article were supported in part by Grant CA44317 from the National Cancer Institute and Grants ES00210, ES03850, and ES04766 from the National Institute of Environmental Health Sciences. This is technical paper number 10023 from the Oregon State University Agricultural Experiment Station.

REFERENCES

Appleton, B. S., Goetchius, M. P., and Campbell, T. C. (1982). Linear dose–response curve for the hepatic macromolecular binding of aflatoxin B_1 in rats at very low exposure. *Cancer Res.* **42**, 3659–3662.

Baertschi, S. W., Raney, K. D., Stone, M. P., and Harris, T. M. (1988). Preparation of the 8,9-epoxide of the mycotoxin aflatoxin B_1: The ultimate carcinogenic species. *J. Am. Chem. Soc.* **110**, 7929–7931.

Bailey, G. S., Nixon, J. E., Hendricks, J. D., Sinnhuber, R. O., and van Holde, K. E. (1980). Carcinogen aflatoxin B_1 is located preferentially in internucleosomal deoxyribonucleic acid following exposure *in vivo* in rainbow trout. *Biochemistry* **19**, 5836–5842.

Bailey, G. S., Williams, D. E., Wilcox, J. S., Loveland, P. M., Coulombe, R. A., and Hendricks, J. D. (1988). Aflatoxin B_1 carcinogenesis and its relation to DNA adduct formation and adduct persistence in sensitive and resistant salmonid fish. *Carcinogenesis* **9**, 1919–1926.

Ball, R. W., Wilson, D. W., and Coulombe, R. A., Jr. (1990). Comparative formation and removal of aflatoxin B_1–DNA adducts in cultured mammalian tracheal epithelium. *Cancer Res.* **50,** 4918–4922.

Bechtel, D. H. (1989). Molecular dosimetry of hepatic aflatoxin B_1–DNA adducts: Linear correlation with hepatic cancer risk. *Req. Toxicol. Pharmacol.* **10,** 74–81.

Buss, P., Caviezel, M., and Lutz, W. K. (1990). Linear dose–response relationship for DNA adducts in rat liver from chronic exposure to aflatoxin B_1. *Carcinogenesis* **11,** 2133–2135.

Chang, Y. J., Mathews, C., Mangold, K., Marien, K., Hendricks, J., and Bailey, G. (1991). Analysis of *ras* gene mutations in rainbow trout liver tumors initiated by aflatoxin B_1. *Mol. Carcinogen.* **4,** 112–119.

Chetsanga, C. J., and Frenette, G. P. (1983). Excision of aflatoxin B_1-imidazole ring opened guanine adducts from DNA by formamidopyrimidine-DNA glycosylase. *Carcinogenesis* **4,** 997–1000.

Cole, K. E., Jones, T. W., Lipsky, M. M., Trump, B. F., and Hsu, I-C. (1988). *In vitro* binding of aflatoxin B_1 and 2-acetylaminofluorene to rat, mouse, and human hepatocyte DNA: The relationship of DNA binding to carcinogenicity. *Carcinogenesis* **9,** 711–716.

Croy, R. G., and Wogan, G. N. (1979). Identification of an aflatoxin P_1–DNA adduct formed *in vivo* in rat liver. *Proc. Am. Assoc. Cancer Res.* **20,** 182.

Croy, R. G., and Wogan, G. N. (1981a). Temporal patterns of covalent DNA adducts in rat liver after single and multiple doses of aflatoxin B_1. *Cancer Res.* **41,** 197–203.

Croy, R. G., and Wogan, G. N. (1981b). Quantitative comparison of covalent aflatoxin–DNA adducts formed in rat and mouse livers and kidneys. *J. Natl. Cancer Inst.* **66,** 761–768.

D'Andrea, A. D., and Haseltine, W. A. (1978). Modification of DNA by aflatoxin B_1 creates alkali-labile lesions in DNA at positions of guanine and adenine. *Proc. Natl. Acad. Sci. U.S.A.* **75,** 4120–4124.

Dashwood, R. H., Arbogast, D. N., Fong, A. T., Hendricks, J. D., and Bailey, G. S. (1988). Mechanisms of anticarcinogenesis by indole-3-carbinol: Detailed *in vivo* DNA binding dose–response studies, after dietary administration with aflatoxin B_1. *Carcinogenesis* **9,** 427–432.

Delcuve, G. P., Moyer, R., Bailey, G., and Davie, J. R. (1988). Gene-specific differences in the aflatoxin B_1 adduction of chicken erythrocyte chromatin. *Cancer Res.* **48,** 7146–7149.

Epstein, S. M., Bartus, B., and Farber, E. (1969). Renal epithelial neoplasms induced in male Wistar rats by oral aflatoxin B_1. *Cancer Res.* **29,** 1045.

Essigman, J. M., Croy, R. G., Nadzan, A. M., Busby, W. F., Reinhold, V. N., Buchi, G., and Wogan, G. N. (1977). Structural identification of the major DNA adducts formed by aflatoxin B_1 *in vitro*. *Proc. Natl. Acad. Sci. U.S.A.* **74,** 1870–1874.

Foster, P. L., Eisenstadt, E., and Miller, J. H. (1983). Base substitution mutations induced by metabolically actived aflatoxins B1. *Proc. Natl. Acad. Sci. U.S.A.* **80,** 2695–2698.

Gopalakrishnan, S., Harris, T. M., and Stone, M. P. (1990). Intercalation of aflatoxin B_1 in two oligodeoxynucleotide adducts: comparative ^{1}H NMR analysis of d(ATC^{AFB}GAT): d(ATCGAT) and d(AT^{AFB}GCAT)$_2$. *Biochemistry* **29,** 10438–10448.

Groopman, J. D., Hasler, J. A., Trudel, L. J., Pikul, A., Donahue, P. R., and Wogan, G. N. (1992). Molecular dosimetry in rat urine of aflatoxin-N^7-guanine and other aflatoxin metabolites by multiple monoclonal antibody affinity chromatography and immunoaffinity/high performance liquid chromatography. *Cancer Res.* **52,** 267–274.

Irvin, T. R., and Wogan, G. N. (1984). Quantitation of aflatoxin B_1 adduction within the ribosomal RNA gene sequences of rat liver DNA. *Proc. Natl. Acad. Sci. U.S.A.* **81,** 664–668.

Kensler, T. W., Egner, P. A., Davidson, N. E., Roebuck, B. D., Pikul, A., and Groopman, J. D. (1986). Modulation of aflatoxin metabolism, aflatoxin-N-7-guanine formation, and hepatic tumorigenesis in rats fed ethoxyquin: Role of induction of glutathione S-transferases. *Cancer Res.* **46,** 3924–3931.

Leadon, S. A., Tyrrell, R. M., and Cerutti, P. A. (1981). Excision repair of aflatoxin B_1–DNA adducts in human fibroblasts. *Cancer Res.* **41,** 5125–5129.

Lin, J. K., Miller, J. A., and Miller, E. C. (1977). 2,3-Dihydro-2-(guan-7-yl)-3-hydroxy-aflatoxin B_1, a major acid hydrolysis product of aflatoxin B_1–DNA or –ribosomal RNA adducts formed in hepatic microsome-mediated reactions and in rat liver *in vivo*. *Cancer Res.* **37**, 4430–4438.

Loveland, P. M., Wilcox, J. S., Pawlowski, N. E., and Bailey, G. S. (1987). Metabolism and DNA binding of aflatoxicol and aflatoxin B_1 *in vivo* and in isolated hepatocytes from rainbow trout (*Salmo gairdneri*). *Carcinogenesis* **8**, 1065–1070.

Lutz, W. K., Jaggi, W., Luthy, J., Sagelsdorff, P., and Schlatter, C. (1980). *In vivo* covalent binding of aflatoxin B_1 and aflatoxin M_1 to liver DNA of rat, mouse, and pig. *Chem. Biol. Interact.* **32**, 249–256.

McMahon, G., Davis, E. F., Huber, L. J., Kim, Y., and Wogan, G. N. (1990). Characterization of c-Ki-*ras* and N-*ras* oncogenes in aflatoxin B_1-induced rat liver tumors. *Proc. Natl. Acad. Sci. U.S.A.* **87**, 1104–1108.

Marien, K., Moyer, R., Loveland, P., van Holde, K., and Bailey, G. (1987). Comparative binding and sequence interaction specificities of aflatoxin B_1, aflatoxicol, aflatoxin M_1, and aflatoxicol M_1 with purified DNA. *J. Biol. Chem.* **262**, 7455–7462.

Marien, K., Mathews, K., van Holde, K., and Bailey, G. (1989). Replication blocks and sequence interaction specificities in the codon 12 region of the c-Ha-*ras* proto-oncogene induced by four carcinogens *in vitro*. *J. Biol. Chem.* **264**, 13226–13232.

Martin, C. N., and Garner, R. C. (1977). Aflatoxin B_1-oxide generated by chemical or enzymic oxidation of aflatoxin B_1 causes guanine substitution in nucleic acids. *Nature (London)* **267**, 863–865.

Misra, R. P., Muench, K. F., and Humayun, Z. (1983). Covalent and noncovalent interactions of aflatoxin with defined deoxyribonucleic acid sequences. *Biochemistry* **22**, 3351–3359.

Moyer, R., Marien, K., van Holde, K., and Bailey, G. (1989). Site-specific aflatoxin B_1 adduction of sequence-positioned nucleosome core particles. *J. Biol. Chem.* **264**, 12226–12231.

Muench, K. F., Misra, R. P., and Humayun, Z. (1983). Sequence specificity in aflatoxin B_1–DNA interactions. *Proc. Natl. Acad. Sci. U.S.A.* **80**, 6–10.

Schoenhard, G. L., Hendricks, J. D., Nixon, J. E., Lee, D. J., Wales, J. H., Sinnhuber, R. O., and Pawlowski, N. E. (1981). Aflatoxicol-induced hepatocellular carcinoma in rainbow trout (*Salmo gairdneri*) and the synergistic effects of cyclopropenoid fatty acids. *Cancer Res.* **41**, 1011–1014.

Wang, T. V., and Cerutti, P. (1980). Spontaneous reactions of aflatoxin B_1 modified deoxyribonucleic acid *in vitro*. *Biochemistry* **19**, 1692–1698.

Warpehoski, M. A., and Hurley, L. H. (1988). Sequence selectivity of covalent DNA modification. *Chem. Res. Toxicol.* **1**, 315–333.

Wilson, D. W., Ball, R. W., and Coulombe, R. A., Jr. (1990). Comparative action of aflatoxin B_1 in mammalian airway epithelium. *Cancer Res.* **50**, 2493–2498.

Yu, F-L., Huang, J-X., Bender, W., Wu, Z., and Chang, J. C. S. (1991). Evidence for the covalent binding of aflatoxin B_1-dichloride to cytosine in DNA. *Carcinogenesis* **12**, 997–1002.

8

Mechanisms by Which Aflatoxins and Other Bulky Carcinogens Induce Mutations

❖

Edward L. Loechler

INTRODUCTION

The aflatoxins are among the most potently mutagenic and carcinogenic substances known (McCann *et al.*, 1975; Foster *et al.*, 1983; Busby and Wogan, 1985; Sambamurti *et al.*, 1988; Levy *et al.*, 1992; Trottier *et al.*, 1992). In general, mutagenic substances are structurally diverse and range from simple compounds such as methylating agents to the so-called bulky agents (Singer and Grunberger, 1983; Loechler *et al.*, 1990). The category of bulky mutagens/carcinogens (Figure 1) is ill-defined, but is regarded generally to include the aflatoxins, especially aflatoxin B_1 (AFB$_1$) which is considered as the most important of this family of toxins; the polycyclic aromatic hydrocarbons, including benzo[*a*]pyrene (B[*a*]P) and benzo[*c*]phenanthrene (B[*c*]Ph); the aromatic amines, including 2-aminofluorene (2-AF) and 2-acetylaminofluorene (2-AAF); and the anticancer drug *cis*-diamminedichloroplatinum(II) (*cis*-DDP). Aspects of the mutagenesis of each of these compounds are considered in this chapter.

Mutagenesis underlies the process of carcinogenesis based on the observation that oncogenes are derived from their normal cellular counterparts, proto-oncogenes, via mutation (Balmain and Brown, 1988; Bishop, 1991; Hollstein *et al.*, 1991; Vogelstein and Kinzler, 1992). Spontaneous and mutagen-induced processes are likely to contribute to carcinogenesis (Hollstein *et al.*, 1991; Vogelstein and Kinzler, 1992). Different mutagens appear to induce different

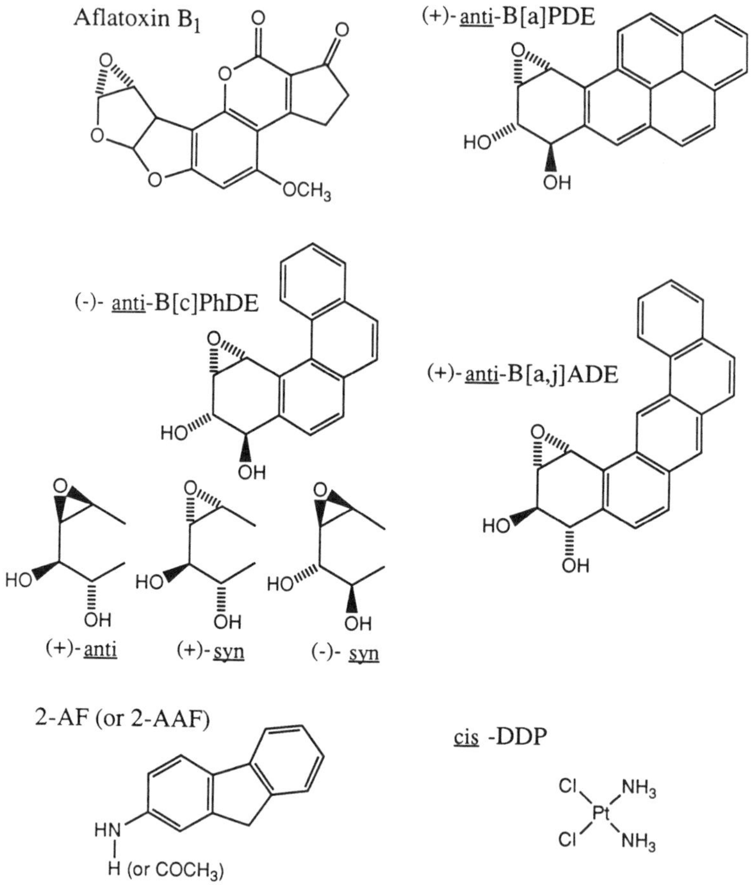

FIGURE 1   Structures for the mutagens/carcinogens discussed in this chapter.

kinds of mutations preferentially, that is, they show a different mutagenic specificity (Miller, 1980, 1983; and references cited subsequently).

For a chemical (or radiation) to cause a cell to become tumorigenic, a large
body of literature suggests that (1) it should react readily with DNA to give DNA
adducts and (2) these adducts (or their breakdown products) must be efficient in
causing mutations (Singer and Grunberger, 1983; Loechler *et al.*, 1990). Thus,
the study of the principles of adduct formation and the principles of adduct-
induced mutagenesis is essential to understanding the carcinogenic process as
initiated by mutagens/carcinogens. Much has been learned about the adducts
formed in DNA by mutagens/carcinogens (Singer and Grunberger, 1983) and
about the mutational specificity of these mutagens and carcinogens (Eisenstadt *et
al.*, 1982; Foster *et al.*, 1983; Bichara and Fuchs, 1985; Levin and Ames, 1986;
Burnouf *et al.*, 1987; Yang *et al.*, 1987, 1990; Mazur and Glickman, 1988;
Sambamurti *et al.*, 1988; Bigger *et al.*, 1989, 1992; Sahasrabudhe *et al.*, 1989,
1990; Bernelot-Moens *et al.*, 1990; Carothers and Grunberger, 1990; Chen *et al.*,

1990, 1991; Schaaper *et al.*, 1990; Melchiore and Beland, 1991; Wei *et al.*, 1991; Anderson *et al.*, 1992; Levy *et al.*, 1992; Trottier *et al.*, 1992; Rodriguez and Loechler, 1993a,b). However, little is known about the relationship between these two, that is, which adduct induces which mutation and why (Basu and Essigmann, 1988; Loechler, 1989, 1991a; Loechler *et al.*, 1990; Singer and Essigmann, 1991). In most cases, even less is known about the structures that adducts adopt in relation to biological end points such as mutation. For example, more is known about O^6-methylguanine (O^6MeGua) than about any other pre-mutagenic lesion (Loechler *et al.*, 1984; Chambers *et al.*, 1985; Bhanot and Ray, 1986; Hill-Perkins *et al.*, 1986; Ellison *et al.*, 1989; Mitra *et al.*, 1989; Singer *et al.*, 1989; Dosanjh *et al.*, 1991), yet the structural basis for its induction of guanine to adenine mutations is controversial (Patel *et al.*, 1986; Basu and Essigmann, 1988; Basu and Essigmann, 1990); although some work suggests that the original proposal of Loveless (1969) based on a simple O^6MeGua:T base pair may not be unreasonable (Leonard *et al.*, 1990; Loechler, 1991b).

This chapter discusses issues that are germane to the question of how afla-toxins induce mutations. In this regard, the data from all bulky muta-gens/carcinogens are likely to be interrelated, so results for a number of these compounds are included. More information is available regarding mutagenesis in bacteria, so this topic is reviewed more comprehensively, but examples from eukaryotes are included also.

STUDIES ON INDIVIDUAL ADDUCTS OF BULKY
MUTAGENS/CARCINOGENS

No site-directed study has been performed on any aflatoxin adducts to date, largely because of their chemical lability. However, since (e.g.) AFB_1 shows an extraordinary preference for reaction at N^7-guanine, all mutations induced by activated AFB_1 are likely to arise via AFB_1–N^7-guanine adducts (Essigmann *et al.*, 1977, 1983; Gopalakrishnan *et al.*, 1989, 1990) or their corresponding im-idazole ring-opened derivative AFB_1–N^7-formamidopyrimidine (FAPY) (Lin *et al.*, 1977). Assuming this to be true, a mutation frequency of ~3% per adduct was estimated for AFB_1–N^7-guanine (Stark *et al.*, 1979). This study was done in a uvr^- strain, so this value will not be higher unless a repair pathway other than excision repair can lead to a wild-type outcome. Recombinational repair is one possibility, as discussed in Section II,F.

Five individual bulky adducts have been studied in *E. coli* using site-directed methods, and the results are summarized below.

Aminofluorene–C8-Guanine

Using a double-stranded phage genome, principally G → T mutations were induced at a frequency of 1–3% in SOS-induced uvr^+ cells. The mutation frequency in non-SOS-induced cells was considerably (~15-fold) lower (Reid *et al.,* 1990). No estimate was made of what this frequency would be in *uvr⁻* cells, but a value no higher than 5–15% seems reasonable based on work with other adducts (see Section II,F).

(+)-Anti-benzo[*a*]pyrene–N²-Guanine

Using a double-stranded plasmid genome, principally G → T mutations were obtained at a frequency of 0.18% in SOS-induced *uvr⁺* cells. The mutation frequency was ∼4-fold lower in cells not SOS induced. The frequency in *uvr⁻* cells was estimated to be ∼0.2% and 1–2% in SOS uninduced and induced cells, respectively (Mackay *et al.,* 1992).

cis-Diamminedichloroplatinum–{d(ApG)}

Using a double-stranded phage genome, principally A → T mutations were induced in *uvr⁺* cells at a frequency of 1–3%. The mutations appeared to be extremely (> 30-fold) dependent on SOS induction (Burnouf *et al.,* 1989). No estimate was made for this frequency in *uvr⁻* cells, but a value no higher than 5–15% seems reasonable based on work with other adducts.

cis-Diamminedichloroplatinum–{d(GpG)}

Using a single-stranded phage genome, principally G → T mutations were induced in *uvr⁺* cells at a frequency of 1–3%. The mutations appeared to be extremely (> 30-fold) dependent on SOS induction. DNA repair is not likely to be relevant because a single-stranded vector was used (Bradley *et al.,* 1993, unpublished observations).

2-Acetylaminofluorene–C8-Guanine

Using a double-stranded phage genome, deletions of a 5′-GC-3′ sequence were observed when this adduct was situated in the third—but not the first or second guanine—in a *Nar*I sequence (5′-G₁G₂CG₃CC-3′) in *uvr⁺* cells (Burnouf *et al.,* 1987). SOS induction was essential. The authors do not favor a DNA polymerase-based explanation for the mechanism of mutation; rather, a mechanism based on a topoisomerase or nuclease acting in this sequence, which appears to be capable of adopting a Z-DNA-like conformation, seems more likely. Using a double stranded plasmid genome, the same adduct was studied when placed in runs of G:C base pairs in *E. coli* (Lambert *et al.,* 1992). Deletion of a single G:C base pair was observed, and MF increased in SOS-induced cells, in excision repair deficient cells, and as the adduct was moved from the 5′- to the 3′-guanine in the run. In another study (Moriya *et al.,* 1988), principally G → T and G → C were obtained at ∼7% in non-SOS-induced *uvr⁺* cells.

Additional Comments

Several of these site-directed studies were conducted in double-stranded vectors. Investigators have observed that if DNA damage occurs in only one of the two strands, progeny plasmids are derived preferentially from the strand containing no adducts, a phenomenon referred to as "strand bias" (Koffel-Schwartz *et al.*, 1987). To minimize this problem, lesions such as UV damage were incorporated into the strand not containing the specific adduct under investigation (Burnouf *et al.*, 1987, 1989; Reid *et al.*, 1990; Mackay *et al.*, 1992). The mechanism for strand bias has never been investigated, but it is likely to result from recombinational repair (i.e., postreplication repair) as originally proposed by Witkin (1976). Thus, the inclusion of UV (or other) damage in the strand not containing the specific adduct is likely to minimize the bypass of the adduct via recombinational repair to give a wild-type outcome.

It would be tempting to conclude from these results that we know the mutagenic specificity for each of these adducts and now can assess the contribution of each adduct to the mutagenic specificity of the chemical from which it is derived. However, the discussion in the next section suggests that the kind of mutation induced by a particular adduct may be defined by its sequence context, which may complicate the relationship between adduct structure and mutagenic outcome. However, this information also may provide important clues that will help us understand the mechanism by which bulky adducts induce mutations.

CAN DNA SEQUENCE CONTEXT INFLUENCE MUTATIONS INDUCED BY BULKY ADDUCTS?

For conceptual purposes, I have chosen to distinguish between sequence-specific effects in mutation that are quantitative vs. qualitative. The concept that sequence context affects mutagenesis quantitatively is well accepted and forms the basis for the concepts of mutational "hot spots" and "mutagenic spectra" (Miller, 1980, 1983). In principle, these effects may occur via sequence-dependent mutagen reactivity with DNA, DNA repair, or efficiency of mutational bypass during replication. Qualitative effects in mutagenesis refers to the actual kinds of mutations (e.g., GC → TA vs. GC → AT vs. GC → CG), which are usually associated with the notion of "mutagenic specificity" and is frequently compiled for the collected base substitution mutations for a particular mutagen/carcinogen (Miller, 1980, 1983). Principles that govern both quantitative and qualitative effects may be revealed by searching for rules. (The term "rules" is used, although "trends" might be better because the effects are not likely to be absolute.) Note that rules governing qualitative and quantitative sequence-dependent changes may not be the same.

A major thesis of this chapter is that the mutagenic specificity of a particular bulky mutagen/carcinogen can be influenced by its DNA sequence context.

More specifically, the kind of base substitution mutation induced by a particular bulky DNA adduct appears to be affected by DNA sequence context. Results from four different studies are discussed in this section. Although not an exhaustive survey, these data sets illustrate the important concepts.

As cited above, the notion that DNA sequence context influences the qualitative pattern of mutation has been clearly demonstrated for deletion mutations, where Fuchs and colleagues have shown that 2-acetylaminofluorene-C8-guanine causes either -1 or -2 frameshift mutations depending upon sequence context (Burnouf, *et al.*, 1987; Lambert, *et al.*, 1992). The likely basis for this is that this adduct can adopt different conformations depending upon sequence context, as discussed below.

Sequence Context and AFB_1 Mutagenesis

That sequence context actually might define mutation first came to light in the work of Humayan and colleagues in their studies of AFB_1 mutagenesis (Sambamurti *et al.*, 1988; Sahasrabudhe *et al.*, 1989). Early work on aflatoxin mutagenesis, using the classic *lacI* nonsense assay developed by Miller for *Escherichia coli*, showed that GC → TA transversion mutations were induced almost exclusively (>90%) by microsomally activated AFB_1 (Foster *et al.*, 1983). However, using a plasmid-based system with a *lacZ'* mutational target, Humayan observed that GC → TA and GC → AT contributed approximately equally to the base-substitution mutations induced by AFB_1, either when activated to its epoxide by microsomes or as the 7,8-dichloride. Some of the data from Humayan's group suggested that the GC → AT mutations were unlikely to result from the corresponding imidazole ring-opened AFB_1–N^7-FAPY adducts derived inadvertently from AFB_1–N^7-guanine (Sambamurti *et al.*, 1988). The prevalence of GC → AT mutations in the *lacZ'* system but not the *lacI* nonsense system could be the result of differences between chromosomal and plasmid mutational mechanisms in *E. coli*. However, this explanation seems unlikely since replication involves the same enzymes. A more likely explanation is the one originally proposed by Humayan: significant sequence contexts differences exist in the two studies.

Hot spots for AFB_1 mutagenesis in the plasmid-based *lacZ'* assay were found predominantly in relatively GC-rich regions of DNA, which also are known to be hot spots for AFB_1 reaction with N^7-guanine (Benasutti *et al.*, 1988, and references therein). In contrast, studies involving the *lacI* nonsense system principally involved relatively AT-rich regions of DNA because of the nature of the system, which relies on codons that can be converted to premature stop (i.e., nonsense) codons (Coulondre and Miller, 1977).

We have evaluated these data sets more systematically to consider whether the *lacI* nonsense system has any of the sequences that were found to be hot spots for GC → AT mutagenesis in the *lacZ'* system. In *lacZ'*, 39/50 of the GC → AT mutations (~80%) were found in 5'-CG̲C-3', 5'-G̲GC-3', 5'-CG̲T-3' or 5'-G̲GA-3' sequences. (The underlined base throughout this text indicates the site of the mutation.) The basis of the *lac*I nonsense system precludes the detection of GC → AT mutations in sequences except with a 5'-flanking thymidine (i.e., in 5'-TG̲-3' sequences), which does not correspond to the hot spots for GC → AT mutations detected in the *lac*Z' system. It is worth noting that, in both the *lac*I nonsense and the *lac*Z' data sets, GC → AT mutations in 5'-TG̲-3' sequences were observed with AFB₁, although they contributed in a minor way to both data sets. (One of the drawbacks in the *lac*Z' system is that a list of the known mutations that are phenotypically detectable has not been compiled, which precludes normalization for target size.)

This analysis makes it reasonable to propose that the *lac*I nonsense system does not have sequence contexts available to it that can detect most GC → AT mutations with AFB₁. However, these data do not allow us to conclude definitively that a single adduct is involved, because plasmid DNA was adducted randomly with AFB₁ in all cases. This topic is discussed at greater length in a subsequent section.

The notion that bulky N⁷-guanine adducts induce GC → AT mutations is bolstered further by results with ICR-191 (a half mustard) and *cis*-DDP. Both chemicals principally react at N⁷-guanine and both have been shown in random mutagenesis studies to induce GC → AT as well as GC → TA mutations (Burnouf *et al.*, 1987; Sahasrabudhe *et al.*, 1990).

Aflatoxin B₁ Mutagenesis in Human Cells and in Tumors

Of considerable significance is the role of AFB₁ mutagenesis in human cells and in tumors, an idea that is considered in this section. Despite its significance, this topic represents a digression because the data are not readily subject to the kind of analysis presented being discussed.

Using a plasmid-based system (pS189) with a *supF* mutational target, mutagenesis by AFB₁, either as the oxide (Levy *et al.*, 1992) or when activated by P450 IA2 (Trottier *et al.*, 1992), has been studied in human cells in culture. Overall mutagenic specificity is similar in both studies, in which GC → TA mutations predominate (~65%), but GC → CG mutations (~20%) and GC → AT mutations (~11%) are obtained also. These results are different than in bacteria, in which GC → AT and GC → CG mutations are more and less prevalent, respectively. Preliminary analysis does not reveal any striking patterns that are comparable to those observed with (+)-*anti*-B[*a*]PDE (B[*a*]P diol epoxide) or with the B[*c*]PhDEs p. 157 except that mutations are ~3-fold more likely to occur at a G:C base pair flanked on either side by another G:C base pair than at a G:C base pair flanked on both sides by an A:T base pair. This result is consistent with the pattern of AFB₁ oxide reactivity with guanine residues *in vitro* in duplex DNA (Benasutti *et al.*, 1988). One striking result was that tandem double mutations were induced by AFB$_1$ oxide at a remarkably high frequency (~15%), but only in repair deficient [i.e., XP12BE (SV40)] human cells (Levy *et al.*, 1992).

Mutations in codon 12 of c-Ki-*ras* and N-*ras* were identified in liver tumors from rats treated with AFB$_1$ (McMahon *et al.*, 1986, 1990). GC → TA mutations in a 5'-CTG̱GT-3' and GC → AT mutations in a 5'-TGG̱TG-3' were detected. (Note that a tandem mutation also was obtained in codon 13.) These sequence contexts do not permit a simple comparison with any of the mutagenesis data obtained in bacterial or human cells.

A strong case can be made for AFB$_1$ involvement in the etiology of some human liver cancer, in which activation of the *p53* oncogene appears to be important. In fact, an extraordinary preference is seen for the formation of a GC → TA mutation at codon 249 in *p53*, where the sequence context is 5'-ĀG̱-ḠCC-3' (Bressac *et al.*, 1991; Hsu *et al.*, 1991). (The overbars indicate the crucial Arg 249 codon; the underlining indicates the mutated guanine.) Five sites studied by Humayan in *lacZ'* have the 5'-GG̱C-3' sequence motif. The ratio of GC → TA:GC → AT:GC → CG mutations for these sites is 12:17:3 or 38%:53%:9%. [5'-GG̱C-3' sequence contexts in *supF* are problematic (Rodriguez and Loechler, 1993a), so a comparison to the mutational results in human cells is difficult.] As noted earlier (and indicated by this ratio), this sequence is one in which GC → AT mutations can be formed, raising the issue of why they were not found. The obvious explanation is that a GC → AT mutation at the underlined base in the Arg codon, 5'-AGG̱-3', changes it to 5'-AGA̱-3', which is still an Arg codon. Thus, GC → AT mutations at codon 249 would be phenotypically silent. The exclusivity of GC → TA mutations therefore may be more apparent than real. Note that a 5'-GG̱C-3' sequence context is one of the few sequence contexts in which GC → CG̱ was observed by Humayan, and one GC → CG mutation was obtained at codon 249 in human liver tumors.

Sequence Context and O⁶-Benzylguanine Mutagenesis

Moschel, Barbacid, and colleagues studied the mutational consequences of O⁶-alkylguanine (O⁶AlkGua) adducts using site-specific means in the first and the second guanines in codon 12 of the Ha-*ras* oncogene (codon sequence 11–13: 5'-GCC GGC GGT-3') using a shuttle plasmid (Mitra *et al.*, 1989). After transfection of NIH 3T3 cells with the adduct-containing plasmids, mutations in Ha-*ras* were isolated because they lead to cellular transformation and foci are formed. Previously, researchers noted that tumors caused by methylating carcinogens invariably had Ha-*ras* activated via a GC → AT mutation in the second, but not the first, guanine in codon 12 although virtually any mutation at either guanine is known to generate an oncogenic Ha-*ras* gene. This study showed that O⁶MeGua, specifically located by site-directed means at either the first or the second guanine, caused GC → AT mutations at approximately the same frequency. These data argued against differential DNA repair or mutational bypass as the explanation for preferential mutagenesis in the second guanine in codon 12 of Ha-*ras* in the tumors. The more likely explanations were either differential methylation of the second guanine, for which some theoretical and experimental evidence exists (Buckley, 1987; Dolan *et al.*, 1988; Richardson *et al.*, 1989), or preferential growth of a tumor cell containing a mutation in the second guanine.

In another facet of this work, O^6benzylguanine, a bulky lesion, also was situated at either the first or the second guanine of codon 12 of Ha-*ras*. GC → AT transition mutations were observed with the adduct at the second guanine. However, GC → AT, GC → TA, and GC → CG at approximately the same frequency were obtained with the adduct at the first guanine. This result is the most definitive demonstration that the same bulky adduct may induce different mutations depending on its sequence context. That this result is in error is unlikely, since all the work was done in parallel and O^6MeGua induced only GC → AT mutations in the first guanine of codon 12 of Ha-*ras*.

Sequence Context and (+)-Anti-benzo[*a*]pyrene Diol Epoxide Mutagenesis

From site-directed work (Mackay *et al.*, 1992), we determined that the adduct (+)-*anti*-B[*a*]P-N^2-guanine induces G → T mutations, at least in the sequence context of our study, which was 5-CCTGCAG-3′. We realized that we could only understand the relevance of this observation if we knew the kinds of mutations induced by (+)-*anti*-B[*a*]PDE itself. This experiment was performed in *E. coli* using a plasmid-based system we developed using a *supF* mutational target gene (Rodriguez *et al.*, 1992). The plasmid (pUB3) was adducted with (+)-*anti*-B[*a*]PDE *in vitro*. Immediately thereafter, it was transformed into *E. coli* (ES87) cells that were SOS induced (Rodriguez and Loechler, 1993a).

Note that one of the major advantages of using *supF* as a mutational target is that ~69% of all base-pairing mutations have been detected phenotypically (Kraemer and Seidman, 1989; Rodriguez and Loechler, 1993a). Thus, if (e.g.) only a GC → CG mutation is obtained at a particular site, then frequently we can conclude that only that mutation is induced by the chemical of interest at that site, not that the other mutations are phenotypically silent.

Most mutations were found at G:C base pairs, a result that is consistent with the preference (+)-*anti*-B[*a*]PDE shows for adduction at G:C base pairs (Singer and Grunberger, 1983) and with mutagenic studies involving the mixture of (+)-*anti*-B[*a*]PDE and (−)-*anti*-B[*a*]PDE enantiomers (Eisenstadt *et al.*, 1982; Levin and Ames, 1986; Yang *et al.*, 1987, 1990; Mazur and Glickman, 1988; Bernelot-Moens *et al.*, 1990; Carothers and Grunberger, 1990; Chen *et al.*, 1990, 1991). GC → TA mutations were enhanced preferentially by SOS induction.

With respect to quantitative aspects of mutagenesis, hot spots for mutation were found in 5′-GG-3′ sequences either when located in 5′-GGCC-3′ contexts

TABLE 1   Selected Base Substitition Mutations
Induced by (+)-*anti*-B[*a*]PDE in a *supF* Gene
Using a Plasmid-Based Assay in SOS-Induced *E. coli*[a]

	T	A	C	Heat[b]
5'-T<u>G</u>-3' (all)[c]	13[d]	0	0	−
	14	1	1	+
5'-G<u>G</u>-3' (all)[e]	16	5	2	−
	15	6	8	+
5'-C<u>G</u>-3' (position 115)[f]	13	1	1	−
	15	7	11	+

[a]Mutations are presented for a guanine (underlined) undergoing mutation. The columns represent the transition made: G→T (GC→TA), G→A(GC→AT), or G→C(GC→CG).

[b]Heating was done on (+)-*anti*-B[*a*]PDE adducted pUB3 at 80°C for 10 min prior to transformation.

[c]All sites in *supF* with a thymidine on the 5' side of a guanine undergoing mutation were considered collectively.

[d]The number of mutations in each case: 83 base substitution mutations (of 187 total mutants), were isolated without heat (Rodriguez and Loechler, 1993a); 115 base substitution mutations (of 270 total mutants) were isolated with heat (Rodriguez and Loechler, 1993b).

[e]All sites in *supF* with a guanine on the 5' side of a guanine undergoing mutation were considered collectively.

[f]Data on the mutation obtained at the base pairing mutation hotspot at G_{115}, which has the sequence context 5'-C<u>G</u>CC-3'.

or when located in runs of three or more G:C base pairs. It has been noted that hot spots for adduction appear to be 5'-GG-3' sequence contexts (Boles and Hogan, 1986).

Some of our data for base-pairing mutagenesis after SOS induction with (+)-*anti*-B[*a*]PDE are given in Table 1. The second row in each set of data shows results when the plasmid (pUB3), after its adduction with (+)-*anti*-B[*a*]PDE, was heated at 80°C for 10 min prior to transformation (Rodriguez and Loechler, 1993b). The significance of this effect is discussed later.

During our analysis, we noted that the base on the 5' side of the guanine undergoing mutation was critical in defining qualitative patterns of mutagenesis. Table 1 shows that heating had little effect on the pattern of mutation when all 5'-TG-3' sequence contexts are considered together; G → T mutations predominated. As discussed on the previous page, when (+)-*anti*-B[*a*]P–N²-guanine was studied in a site-directed manner, a 5'-TG-3' sequence context was used and G → T mutations also predominated. Thus, a correlation exists, strongly suggesting that (+)-*anti*-B[*a*]P–N²-guanine is responsible for G → T mutations, at least in 5'-TG-3' sequence contexts. [We also have conducted preliminary studies on the mutations induced by (+)-*anti*-dibenz[*a, j*]anthracene diol epoxide (Gill *et al.*, in press) and, although G → A mutations predominated (51%), 17/19 mutations in 5'-TG-3' sequence contexts were G → T.]

In contrast, although G → T mutations are important in 5′-GG-3′ sequence contexts, G → A and G → C mutations become more prevalent (Table 1). [This result is also observed in 5′-AG-3′ and 5′-CG-3′ sequence contexts and is statistically significant (Rodriguez and Loechler, 1993b).] Several questions are raised by these observations. Does our adduct-specific study (Mackay *et al.,* 1992) imply that (+)-*anti*-B[*a*]P–N²-Gua induces only G → T mutations; in which case, what adduct is responsible for G → A and G → C mutations? Alternatively, if (+)-*anti*-B[*a*]P–N²-Gua can induce G → A and G → C mutations, then why are these mutations not observed to a significant extent in 5′-TG-3′ sequence contexts?

In this regard, the results at the major base substitution hot spot, G_{115} may be revealing. G_{115} → T mutations predominate prior to heating, whereas G_{115} → A and G_{115} → C mutations become statistically significantly more prevalent after heating. (Heating had at most a marginal effect on the results in 5′-TG-3′ and 5′-GG-3′ sequence contexts.) One explanation (Model 1) is that the adduct responsible for mutations at G_{115} is in one conformation prior to heating (possibly a conformation similar to 5′-TG-3′ sequences, based on Table 1) and is in another conformation after heating (possibly a conformation similar to 5′-TG-3′ sequences, based on Table 1).

At least one alternative explanation exists for the results at G_{115} (Model 2). An N⁷-guanine adduct of (+)-*anti*-B[*a*]PDE could cause the mutations before heating. This adduct is labile (King *et al.,* 1979; Osborne and Merrifield, 1985), so heating should generate an apurinic site, which might give the results seen after heating. The data are not inconsistent with one pattern of apurinic (AP) site mutagenesis, as noted by Lawrence *et al.* (1990).

Preliminary data (E. Drouin and E. L. Loechler, 1993) are more consistent with a heat-induced conformational change in a heat-stable adduct (i.e., Model 1). [Note that Drinkwater *et al.* (1980) estimated that AP sites are responsible for <1% of mutations induced by (+/−)-*anti*-B[*a*]PDE.] However, either model is potentially significant. Model 1 implies that the same adduct could induce different mutations in the *identical* sequence context, depending on adduct conformation, whereas Model 2 implies that (+)-*anti*-B[*a*]P–N⁷-guanine adducts might be important for (+)-*anti*-B[*a*]PDE mutagenesis, which is not generally considered.

Sequence Context and Mutagenesis with Benzo[*c*]phenanthrene Diol Epoxides

Bigger, Dipple, and their colleagues have generated an extensive data set on the mutations induced by the four enantiomers of B[*c*]PhDE (Figure 1) in the *supF* gene of the pTZ189 plasmid in human Ad293 cells (Bigger *et al.,* 1989, 1992). Tables 2 and 3 show the mutations obtained for each of the enantiomers at many of the prominent hot spots. No doubt numerous quantitative differences exist. For example, relatively fewer mutations occur at G:C base pairs with (+)-*syn*-B[*c*]PhDE, a result that is likely to reflect the fact that this compound

**TABLE 2   Base Substitution Mutations at G:C Base Pairs
Induced by the Four Stereoisomers of B[c]PhDE at Specific Sequences
in the supF Gene Using a Shuttle Plasmid (pZ189) in Human Ad293 Cells[a,b]**

Base number	Sequence	T	A	C	Base number	Sequence	T	A	C
122	AAGGG	5	0	0	155	TCGAA	5	1	1
		3	0	0			3	1	1
		1	0	0			5	1	0
		1	0	0			1	0	1
123	AGGGA	1	1	1	156	TCGAA	3	0	2
		1	1	0			4	2	4
		2	1	2			4	1	3
		0	1	0			1	1	1
127	CTGCT	3	0[c]	0	159	AAGGT	2	1	0
		1	0[c]	0			2	1	0
		8	0[c]	3			2	1	0
		0	0[c]	1			1	0	0
133	TAGAG	7	0	5	164	TCGAA	5	0[c]	3
		4	0	2			5	0[c]	2
		6	0	2			3	0[c]	4
		1	0	3			4	0[c]	1
139	CAGAT	5	0	4	168	AGGAT	0	0	1
		2	0	0			1	3	1
		2	0	1			2	1	1
		2	1	0			1	3	1

[a]Columns are designated as in Table 1.

[b]The four entries in each set of mutations are (top to bottom) from the stereoisomers (+)-*anti*, (−)-*anti*, (−)-*syn*, and (+)-*syn*, respectively (Dipple *et al.*, 1987; Bigger *et al.*, 1992).

[c]This mutation has not been detected phenotypically to date in *supF*.

reacts less extensively at guanines (Dipple *et al.*, 1987; Bigger *et al.*, 1992). However, all four stereoisomers are remarkably similar in terms of qualitative results. For example, all four stereoisomers cause GC → TA and GC → CG but not GC → AT mutations at G_{133}, whereas all three types of mutations are detected at G_{156}; in both cases, all three mutations are phenotypically detectable. The most obvious exceptions to this generalization involve (+)-*syn*-B[c]PhDE and include an unusually high fraction of AT → CG mutations at G_{134} (Table 3) as well as a near absence of GC → TA mutations at G_{133}, G_{155}, and G_{127}, although the latter two may reflect quantitative effects, given the small number of mutations.

The qualitative similarities have led us to merge the data for all the stereo-isomers for the purposes of analysis. To a first approximation, this decision is not unreasonable based on the preceding discussion. Although merging may obscure some subtleties, if sequence context rather than the exact chemical component of the adduct dominates the mutational specificity at any particular base pair, then

**TABLE 3   Base Substitution Mutations at A:T Base Pairs
Induced by the Four Stereoisomers of B[c]PhDE at Specific Sequences
in the supF Gene Using a Shuttle Plasmid (pZ189) in Human Ad293 Cells**[a,b]

Base number	Sequence	T	G	C	Base number	Sequence	T	G	C
106	GAAAC	3	2	2	136	TAAAT	1	0	1
		1	0	0			1	0	0
		2	0	1			1	0	0
		5	0	3			7	0	0
112	CGAGC	12	3	1	137	AAATC	0	2	0[c]
		2	0	0			0	0	0[c]
		8	1	0			0	1	0[c]
		8	0	1			0	1	0[c]
120	CAAAG	2	0	0	140	GCAGA	1	0	3
		1	0	0			0	0	0
		2	0	0			1	0	5
		5	1	0			0	0	3
134	TTAGA	6	0	1	161	GAACC	2	0	2
		6	0	1			1	0	2
		5	1	0			3	0	1
		6	1	6			6	0	1
135	CTAAA	2	0	0	162	CGAAC	8	0	0[c]
		1	0	0			1	0	0[c]
		2	0	1			6	0	0[c]
		4	0	1			4	1	0[c]

[a]Mutations are presented for an adenine (underlined) undergoing mutation. The columns represent the transition mode: A→T (AT→TA), A→G (AT→GC), A→C (AT→CG).
[b]The four entries in each set of mutations are (top to bottom) from the stereoisomers (+)-anti, (−)-anti, (−)-syn, and (+)-syn, respectively (Dipple et al., 1987; Bigger et al., 1992).
[c]This mutation has not been detected phenotypically to date in supF.

the exact stereochemistry of the chemical moiety of the adduct may be relatively less important.

GC → TA and AT → TA mutations are observed at virtually every site at which any mutations are obtained which, for guanine and adenine adducts, implies dATP insertion during replication. This result suggests that dATP incorporation is ubiquitous and may hinge on quantitative, not qualitative, factors. Some comments in this direction are presented beginning on p. 165. From the perspective of qualitative changes in mutational specificity, one question arises. Do sequence context effects exist that appear to enhance the formation of mutations other than GC → TA and AT → TA?

Table 4 shows all 5'-AG-3' sequences in supF and the kinds of mutations induced. Adenine residues at positions 112, 134, and 140 are hot spots, whereas the other adenine residues listed in Table 4 show fewer mutations. All the sites with fewer mutations have a longer run of purines on the 5' side of the 5'-AG-3'

TABLE 4　Base Substitution Mutations at A:T Base Pairs in 5′-A̲G-3′ Sequences Induced by B[c]PhDE at Specific Sequences in the *supF* Gene when Studied in Shuttle Plasmid (pZ189) in Human Ad293 Cells[a]

Base number	Sequence[b]	T	G	C	Total purine[c]	5′-Purine[d]
112	tccCGA̲GCggc	30	4	2	3	1
134	gatTTA̲GAgtc	23	8	2	4	0
140	acgGCA̲GAttt	2	0	11	3	0
128	ggaGCA̲GActc	1	0[e]	0[e]	3	0
158	ttcGAA̲GGttc	3	0	2	5	2
132	tttAGA̲GTtta	1	1	0	4	2
121	gccAAA̲GGgag	1	0[e]	0[e]	8	2
125	aagGGA̲GCaga	1	0[e]	0[e]	7	5
170	gggGAA̲GGatt	0	0	0[e]	9	5
153	ttcGAA̲GTtcc	1	0[e]	0[e]	4	2

[a]Columns T, G, and C as in Table 3.
[b]An expanded sequence context is provided to show the number of contiguous purines.
[c]The total number of contiguous purines that include the adenine undergoing mutation.
[d]The number of purines on the 5′ side of the 5′-A̲G-3′ sequence.
[e]This mutation has never been detected in *supF,* which may indicate that it is phenotypically silent.

sequence than do the sites associated with the hot spots. Thus, one possibility is that fewer adducts are formed at adenine residues located in runs of purines. (Of course other possibilities exist.) This effect is quantitative.

A second observation is that AT → TA mutations decrease and AT → CG mutations increase at A_{140}, which is in a 5′-CA̲G-3′ sequence (Table 4). This result suggests that having a cytosine on the 5′ side of a 5′-A̲G-3′ sequence changes the mutational specificity from predominantly AT → TA to predominantly AT → CG. At one other 5′-CA̲G-3′ sequence in *supF,* at A_{128}, no AT → CG mutations were obtained. However, A_{128} → T mutations decreased, although this sequence does not conform to the rule regarding decrease in mutation frequency in runs of purines. Thus, AT → TA mutations may decrease and AT → CG mutations may increase at A_{128} (as they do at A_{140}) but the A_{128} → C mutation may be phenotypically silent. This notion is consistent with the fact that A_{128} → C mutations have never been isolated in *supF* (Kraemer and Seidman, 1989).

Table 5 shows mutations in 5′-GA̲-3′ sequences in *supF.* Two trends are apparent. First, GC → CG mutations are seen consistently in 5′-GA̲-3′ sequence contexts. When one considers the whole data set, 51/67 (~76%) of the GC → CG mutations are observed in 5′-GA̲-3′ sequences. When normalized for target size, GC → CG mutations are ~5 times more likely to occur in 5′-GA̲-3′ than in 5′-G̲(C/G/T)-3′ sequences. No other sequence context provided as obvious a pattern for GC → CG mutations. Note that more GC → CG mutations were obtained at G_{133} with B[c]PhDE than at any other sequence; this site was also a hot spot for GC → CG mutations observed with (+)-*anti*-B[a]PDE (Rodriguez and Loechler, 1993a) and (+)-*anti*-DB[a, j]ADE (Gill *et al.,* in press).

TABLE 5 Base Substitution Mutations at G:C Base Pairs in 5′-G̲A-3′ Sequences Induced by B[c]PhDE at Specific Sequences in the *supF* Gene when Studied in a Shuttle Plasmid (pZ189) in Human Ad293 Cells[a]

Base number	Sequence[b]	T	A	C	Total purine[c]	5′-Purine[d]
164	ggtTCG̲AAtcc	18	0[e]	10	3	0
156	actTCG̲AAggt	12	4	10	5	0
133	attTAG̲AGtct	8	0	12	4	1
155	cctTCG̲AAgtc	14	3	3	4	0
139	cggCAG̲ATtta	11	1	5	3	1
168	ggaAGG̲ATtcg	4	7	5	7	5
124	caaGGG̲AGcag	2	0	3	7	4
172	tggGGG̲AAgga	2	1	2	10	4
129	gagCAG̲ACtct	1	0	4	3	1
163	gatTCG̲AAtcc	1	1	1	3	0
111	ttcCCG̲AGcgg	1	0	0[e]	3	0

[a]Columns T, A, and C as in Table 1.
[b]An expanded sequence context is provided to show the number of contiguous purines.
[c]The total number of contiguous purines that include the guanine undergoing mutation.
[d]The number of purines on the 5′ side of the 5′-G̲A-3′ sequence.
[e]This mutation has never been detected in *supF*, which may indicate that it is phenotypically silent.

Runs of purines on the 5′ side of the 5′-G̲A-3′ target sequence appear to decrease the frequency of GC → CG as well as GC → TA mutations (Table 5). However, this result does not explain all the sequences in which mutation frequency is low. Mutations at G:C base pairs in 5′-G̲A-3′ sequences and mutations at A:T base pairs in 5′-A̲G-3′ sequences appear to be inhibited by runs of purines on the 5′ side. This correlation is intriguing and suggests that some thread might link these two observations.

The most striking pattern for GC → AT mutations is that 5′-AA̲G-3′ and 5′-AG̲G-3′ sequences appear to be hot spots. For example, 6/7 mutations were in 5′-A̲AG-3′ rather than in 5′-(C/G/T)AG-3′ sequences; 17/19 were in 5′-AGG-3′ rather than in 5′-(C/G/T)GG-3′ sequences. In addition, 5′-G̲AA-3′, which is the inversion of 5′-AAG-3′, also may be of significance because 7/8 mutations in 5′-CG̲-3′ sequences were in 5′-CG̲AA-3′ sequences. Finally, only 2 GC → AT mutations were found in 5′-TG̲-3′ sequences.

HOW MUTAGENIC SPECIFICITY MIGHT BE CONTROLLED BY SEQUENCE CONTEXT

That sequence context may have an influence on the probability of mutagenesis (i.e., create hot spots) is well established and well accepted, although exact rules do not always exist. That sequence context may define the kind of mutation induced at a particular base pair is not well established and certainly not generally accepted. However, if such selectivity exists, determining the molecular basis for it will be important. Three possibilities are considered.

1. Different adducts (e.g., at N^7-guanine rather than N^2-guanine) may be formed preferentially in different sequence contexts. Each adduct may cause a different kind of mutation. Although this model is less likely to be correct, as outlined later, it frequently has been proposed (e.g., Loechler, 1989). Clearly, imagining that a single bulky adduct can induce different mutations in different sequence contexts is not unreasonable, given the site-directed study results discussed on p. 157 involving O^6-benzylguanine (Mitra *et al.,* 1989).

AFB_1, no matter how it is activated, reacts virtually exclusively at the N^7 position of guanine (Essigmann *et al.,* 1977, 1983; Lin *et al.,* 1977; Gopalakrishnan *et al.,* 1989, 1990), making it unlikely that the changes in mutational specificity with changing sequence context (Foster *et al.,* 1983; Sambamurti *et al.,* 1988; Sahasrabudhe *et al.,* 1989) for AFB_1 can be attributed to the presence of adducts bound to DNA bases at different atoms that differ in their mutagenic specificities.

Although N^2-guanine and N^6-adenine adducts dominate for the B[c]PhDEs, each of the four different stereoisomers gives significant amounts of both *trans* and *cis* addition adducts in different ratios (Dipple *et al.,* 1987). Of interest is the fact that the ratio of *cis/trans* adducts for a particular stereoisomer does not seem to influence mutagenic outcome. The *cis/trans* ratios ($R_{c/t}$) for N^2-guanine adducts with $(+)$-*syn* ($R_{c/t} = 0.9$) and with $(-)$-*syn* ($R_{c/t} = 0.4$) are much higher than with $(-)$-*anti* ($R_{c/t} = 0.1$) and with $(+)$-*anti* ($R_{c/t} = 0.04$). If *cis*-N^2-guanine adducts caused a particular kind of mutation compared with *trans*-N^2-guanine adducts, then this mutation should be more prevalent with $(+)$-*syn*- and $(-)$-*syn*-B[c]PhDE. No such trend is apparent (Bigger *et al.,* 1992). In fact the fraction of GC → TA vs GC → AT vs GC → CG mutations and AT → TA vs AT → GC vs AT → CG mutations is remarkably similar for all four stereoisomers, with the possible exception of an increase in the fraction of GC → AT at the expense of GC → TA mutations with $(+)$-*syn*- and $(+)$-*anti*-B[c]PhDE (Bigger *et al.,* 1992). Thus, although formally possible, the effects of sequence context on mutagenesis are unlikely to be explained by the differential formation of *cis* and *trans* addition adducts with B[c]PhDEs in different sequence contexts. Additionally, N^7-guanine adducts do not appear to form with any of the four stereoisomers of B[c]PhDE.

In our work with $(+)$-*anti*-B[a]PDE (Rodriguez and Loechler, 1993a,b), the results are less certain. Different adducts (e.g., B[a]P–N^2-guanine vs B[a]P–N^7-guanine) may, in fact, be playing a role in some cases in changes in the mutational specificity. However, as noted on p. 159, preliminary results argue against this explanation.

2. Sequence context in the vicinity of a DNA adduct formed from a bulky mutagen or carcinogen may affect adduct conformation, which in turn may affect the mutagenic specificity induced by that particular bulky adduct. The notion that

DNA sequence context might influence adduct conformation has been shown for $(+)$-*anti*-B[*a*]P–N²-guanine in linear dichroism studies with alternating DNA copolymers (Roche *et al.,* 1991). Evidence also suggests that 2-AAF–C8-guanine adopts different conformations depending on its sequence context (Veaute and Fuchs, 1991; Belguise and Fuchs, 1991).

Other data can be interpreted in this light as well. For example, tandem double mutations induced by AFB$_1$ oxide were lost preferentially when comparing repair deficient to repair proficient human cells, as cited earlier (Levy *et al.,* 1992). A particular conformation of the adduct AFB$_1$–N⁷-guanine may have been responsible for inducing these tandem double mutations, and this conformation of the adduct is repaired preferentially compared with other conformations in repair proficient cells.

3. Rules governing what bases insert opposite a particular adduct may be defined by a DNA polymerase. The choice of the enzyme may be influenced by the sequence context surrounding the adducted base. If this were the case, the rules governing mutational specificity for different mutagens or carcinogens might have common threads. Note that the rules might be similar for one species, such as *E. coli,* but might vary among species. In addition, as stressed above, this explanation does not suggest that sequence context effects will be quantitatively similar for different mutagens or carcinogens, that is, hot spots may vary.

POSSIBLE MECHANISMS OF MUTAGENESIS BY BULKY ADDUCTS

Introduction

Mechanisms of mutagenesis involving DNA adducts can be placed in one of three categories (Table 6; Loechler *et al.,* 1990).

1. A misinformational mechanism implies that a DNA polymerase attempts to "read" the base moiety of a DNA adduct but misinterprets it.

TABLE 6   Potential Mutagenic Mechanisms

Misinformational
Chemical perturbations
Adduct-induced base tautomerization
Adduct-induced base ionization
Structural perturbations
Adduct-induced base rotation
Adduct-induced base wobble
Noninformational
Other

2. A noninformational mechanism implies that a DNA polymerase encounters an adduct that is uninterpretable and chooses to incorporate a particular deoxynucleoside triphosphate (dNTP) for reasons other than an attempt to "read" the adduct (Strauss *et al.,* 1982; Strauss, 1989). Reasons for choosing a particular dNTP could range from it being dictated by an inherent property of the DNA polymerase to the formation of a DNA structure that best can be accommodated by the active site of a DNA polymerase.

3. Other mechanisms constitute a general category that includes all mechanisms that do not involve DNA polymerase bypass of primary adducts and include (1) the possibility that many lesions are processed to a common intermediate that is then the mutagenic species, for example, AP sites (see subsequent discussion; Loeb, 1984), and (2) mechanisms in which DNA polymerase is not involved in the fixation of the mutation (e.g., mutagenic mechanisms involving topoisomerases have been suggested in several cases; Ripley *et al.,* 1988; Burnouf *et al.,* 1989).

Mutations involving misinformational lesions can, in principle, be divided into several categories (Table 6). First, the carcinogen moiety of an adduct may induce a chemical perturbation in the base moiety of the adduct, which may improve the probability of misreplication. Second, the carcinogen moiety of an adduct also could affect the position of the base moiety of the adduct, resulting in a structural perturbation that might improve the probability of misreplication. These possibilities are discussed here.

Adduct-Induced Base Tautomerization

The alkyl group in O^6AlkGua adducts chemically locks the guanine moiety of the adduct in its enol (i.e., imidate ester) tautomeric form, which is the classic example of adduct-induced base tautomerization (Topal and Fresco, 1976). O^6AlkGua, at least when the methyl group is *anti* with respect to the N^1 position, can base pair with thymine, which simply explains why O^6AlkGua adducts might induce G $\rightarrow$ A transition mutations (Loveless, 1969; Leonard *et al.,* 1990; Loechler, 1991b).

Adduct-Induced Base Ionization

Adduction at certain atoms can perturb the pK_a of the base moiety of the adduct inductively (Topal and Fresco, 1976). For example, N^7-guanine adducts are formally positively charged, which lowers the pK_a of the N^1 proton. This effect in turn increases the probability that N^1 is deprotonated, which might increase the probability that base pairing will occur between the zwitterionic adduct and lead to base pairing with thymine during replication. This mechanism has been proposed for AFB_1 mutagenesis (Sambamurti *et al.,* 1988; Sahasrabudhe *et al.,* 1989), and is evaluated on p. 168.

Adduct-Induced Base Rotation

Adduct-induced base rotation involves *anti* to *syn* base rotation. The major 2-AF adduct (AF–C8-guanine) was shown to form a base pair with adenine after an *anti* to *syn* base rotation (Norman *et al.*, 1989), and may be related to the mechanism by which 2-AF induces GC → TA transversion mutations (Bichara and Fuchs, 1985; Reid *et al.*, 1990). This idea is discussed at greater length in the section beginning on p. 169 in another context.

Adduct-Induced Base Wobble

A second example of an adduct-induced structural perturbation is adduct-induced base wobble. By this mechanism, the carcinogen residue perturbs the position of the base moiety of the adduct with respect to the helix axis. This concept has been discussed at some length based on molecular modeling (Loechler, 1989). For example, if the B[*a*]P moiety of B[*a*]P–N^2-guanine, which is in the minor groove, could pull the guanine moiety of the adduct toward the minor groove, then a simple G:T mispair might be more likely to form since G:T mispairs are known to have the guanine base moved toward the minor groove and the thymine base moved toward the major groove. No firmly established example of this mechanism exists, although it was offered as one possible mechanism by which *cis*-thymine glycol might induce T → C mutations (Basu *et al.*, 1989).

Noninformational Lesions

The exact definition of a noninformational lesion is elusive, but the term implies (1) that such a lesion is uninterpretable when encountered by DNA polymerase and (2) that rules other than simple base-pairing schemes govern which base is incorporated opposite the lesion. Noninformational lesions have been suggested to share several characteristics. (1) Noninformational lesions block the progress of DNA polymerase in most primer extension studies *in vitro* (Strauss *et al.*, 1982; Strauss, 1989) (2) Noninformational lesions are not mutagenic in bacteria in the absence of the induction of the SOS response (McCann *et al.*, 1975; Walker, 1984; Loeb and Preston, 1986). (3) Adenine (i.e., dATP) appears to be incorporated preferentially opposite noninformational lesions (Strauss *et al.*, 1982; Loeb and Preston, 1986; Strauss, 1989), although other bases are inserted based on results with AP sites (Kunkel, 1984; Lawrence *et al.*, 1990). One unifying, but indirect, hypothesis has been offered to account for these observations: all mutations involving noninformational lesions share an AP site as a requisite intermediate (Loeb, 1984). No satisfying direct mechanism(s) has been proposed to account for these apparent characteristics of noninformational lesions.

DOES AFB₁ INDUCE G → A MUTATIONS VIA ADDUCT-INDUCED BASE IONIZATION?

Adduct-induced base ionization has been proposed by Humayan to explain the induction of GC → AT mutations with AFB$_1$ (Sambamurti et al., 1988; Sahasrabudhe et al., 1989). The argument is that simple alkyl N[7]-guanine adducts such as Me–N[7]-guanine are not mutagenic because they are bypassed sufficiently rapidly by DNA polymerases, so N[1] deprotonation does not occur after single-strand formation at the replication fork. In contrast, AFB$_1$–N[7]-guanine adducts block DNA replication and, thereby, permit time for the N[1] position to be deprotonated. Several arguments can be raised against this mechanism.

After conversion of AFB$_1$–N[7]-guanine to AFB$_1$–N[7]-FAPY, the fraction of GC → AT mutations decreases in SOS-induced cells, a result that was cited, as evidence by Humayan for the N[1] deprotonation mechanism because AFB$_1$–N[7]-FAPY adducts have a normal pK_a for N[1] deprotonation. However, GC → AT mutations do not disappear entirely when AFB$_1$–N[7]-guanine is converted to AFB$_1$–N[7]-FAPY, as Table 7 attests. In fact, the fraction of GC → AT mutations is virtually unaffected in SOS-uninduced cells and decreases only slightly in SOS-induced cells. Thus, AFB$_1$–N[7]-guanine to AFB$_1$–N[7]-FAPY conversion does not eliminate GC → AT mutations completely, as predicted by the N[1]-deprotonation mechanism, which argues against this explanation for GC → AT mutations.

Second, cis-DDP induces GC → AT as well as GC → TA mutations (Burnouf et al., 1987). Although cis-DDP forms adducts at N[7]-guanine, the adduct bond is a ligand bond that apparently does not increase the positive charge in the imidazole ring as extensively, based on the fact that the pK_a of the N[1] proton is not perturbed, at least in the dinucleotide (Chu et al., 1978). [Note that trans-DDP adducts in oligonucleotides appear to decrease the pK_a at N[1]-guanine between 1.0 and 1.5 units (van der Veer et al., 1986; Gibson and Lippard, 1987; Lepre et al., 1987; Lepre and Lippard, 1990; Lepre et al., 1990), which may undercut to some extent the argument made here.] This result also argues that N[7]-guanine adducts induce GC → AT mutations independent of whether they enhance N[1] deprotonation.

TABLE 7 Fraction of GC→TA and GC→AT Mutations Induced by AFB₁ Activated as the 7,8-Dichloride[a]

	SOS uninduced E. coli			SOS induced E. coli		
Adduct	GC→TA (%)	GC→AT (%)	Other (%)	GC→TA (%)	GC→AT (%)	Other (%)
AFB₁-N[7]-Gua	22	26	52	28	22	50
AFB₁-N[7]-FAPY	21	29	50	16	28	56

[a]These data are extracted from Table 2 in Sambamurti et al. (1988).

Adduct-induced base ionization via N^1 deprotonation is an extremely attractive mechanism for AFB_1 induction of GC $\rightarrow$ AT mutations. However, as currently formulated, it does not seem likely. N^1 deprotonation may be involved, but models (and experiments) based on a lower pK_a of an N^1 proton in an adduct would seem to be irrelevant.

MISINFORMATIONAL COMPARED WITH NONINFORMATIONAL MECHANISMS OF MUTAGENESIS

Most of the bulky mutagens/carcinogens including B[*a*]PDE, 2-AF, AFB_1 oxide, and *cis*-DDP, appear to meet the requirements of compounds that yield noninformational lesions. They block the progress of DNA polymerases *in vitro* (Pinto and Lippard, 1985; Refolo *et al.*, 1985; Jacobsen *et al.*, 1987; Strauss, 1989); their mutations are either dependent on or enhanced by SOS induction (pp. 151–152); and they appear to follow the A-insertion rule preferentially (Eisenstadt *et al.*, 1982; Foster *et al.*, 1983; Kunkel, 1984; Bichara and Fuchs, 1985; Levin and Ames, 1986; Burnouf *et al.*, 1987, 1990; Yang *et al.*, 1987, 1990; Mazur and Glickman, 1988; Sambamurti *et al.*, 1988; Bigger *et al.*, 1989, 1992; Sahasrabudhe *et al.*, 1989; Bernelot-Moens *et al.*, 1990; Carothers and Grunberger, 1990; Chen *et al.*, 1990, 1991; Lawrence *et al.*, 1990; Reid *et al.*, 1990; Schaaper *et al.*, 1990; Melchiore and Beland, 1991; Wei *et al.*, 1991; Andersson *et al.*, 1992; Mackay *et al.*, 1992; Naser *et al.*, 1993). Does this mean that noninformational pathways of mutagenesis really exist?

Site-directed studies have indicated that the primary consequence of the replication of bulky adducts (i.e., >90%) is no mutation, that is, the base moiety of the adduct is read correctly (pp. 151–152). Further, this value is even higher in cells that are not SOS induced. (Note that all these adducts do not disrupt the hydrogen bonding face of the base moiety of the adduct, which means that, for example, a guanine adduct would be expected to pair with cytosine based simply on hydrogen bonding potential.) These results lead to another question: How can one entertain the notion of a "noninformational lesion" when the lesion is correctly informational >90% of the time?

This apparent contradiction raises the possibility that these lesions block DNA replication and may have their mutations enhanced by SOS induction, but are essentially misinformational. Perhaps dATP incorporation opposite a lesion always occurs via a misinformational mechanism, that is, DNA polymerase does attempt to "read" the adduct. (For the moment, we focus on the incorporation of dATP opposite an adduct because it seems to dominate the mutational mechanism for most bulky mutagens/carcinogens.) This thinking is bolstered by the report of a structure for a 2-AF–C8-guanine:A base pair by NMR (Norman *et al.*, 1989), in which the guanine moiety of the adduct adopts the *syn* conformation and traps a proton. This structure, and the induction of G $\rightarrow$ T mutations, suggest that this mechanism of 2-AF–C8-guanine mutagenesis exists, that is, a polymerase reads the *syn*-AF–C8-guanine adduct as thymine during replication and pairs it with dATP—a misinformational mechanism (Loechler, 1989).

In fact, three distinctly different structures for G:A mispairs have been reported experimentally (Figure 2; Kan *et al.,* 1983; Patel *et al.,* 1984; Kennard, 1985; Hunter *et al.,* 1986; Prive *et al.,* 1987; Brown *et al.,* 1989) suggesting multiple ways in which the carcinogen moiety of an adduct might contribute to the misreplication of the base moiety of that adduct, such that dATP would be incorporated.

One argument against this line of reasoning follows. *cis*-DDP{d(ApG)} directs dATP incorporation opposite the adenine moiety of the adduct, presumably requiring base pairing between the adducted adenine and dATP. Further, as discussed beginning on p. 159, E, B[*c*]PheDEs as well as other polycyclic aromatic hydrocarbons cause AT → TA mutations and form N^6-adenine adducts, so imagining that N^6-adenine adducts cause AT → TA mutations seems reasonable.

To try to understand the basis for AT → TA mutations, the structure(s) for an A:A mispair could be considered; none have been reported, but Figure 2 shows the best possible base-pairing schemes for A:A mispair. The problem with the structure in Figure 2F is that the adenine moiety on the left would be more likely to pair with cytosine, which would eliminate the purine:purine pairing. The problem with the structure in Figure 2E is that it is intuitively unreasonable because it requires *anti* → *syn* base rotation, a trapped proton, and an imino tautomer. In fact, both structures in Figure 2E and F require that the adduct enhance the formation of the imino tautomer of the adenine moiety, which is not likely to occur via inductive or resonance effects. An N^6-adenine adduct might enhance rotation about the C6–N^6 bond; however, this rotation would decrease rather than increase the probability of an amine to imine tautomerization. Thus, no really satisfying model for an A:A mispair is available. In fact, this result is in keeping with the observation that A:A mispairs destabilize the DNA helix and are not likely to be hydrogen bonded (Ikuta *et al.,* 1987).

Based on the preceding discussion, bulky adenine adducts seem very unlikely to induce A → T transversion mutations via structures involving A:A mispairs with reasonable hydrogen bonding (i.e., base pairing) interactions, suggesting that the basis for A → T mutagenesis with adenine adducts cannot be that DNA polymerase misreads an adenine adduct as thymine, and would support the notion of a noninformational lesion and a noninformational mechanism of mutagenesis.

For the adducts of bulky carcinogens, *in vitro* work also suggests that certain DNA polymerases can insert dATP opposite guanine *and* adenine adducts (Reardon *et al.,* 1990). Preferential enhancement of GC → TA mutations after SOS induction in the case of (+)-*anti*-B[*a*]PDE (Rodriguez and Loechler, 1993a) and the complete dependence on SOS induction for the formation of GC → TA mutations with 2-AF also suggest that SOS induction may enhance GC → TA mutations preferentially (Bichara and Fuchs, 1985). These observations are con-

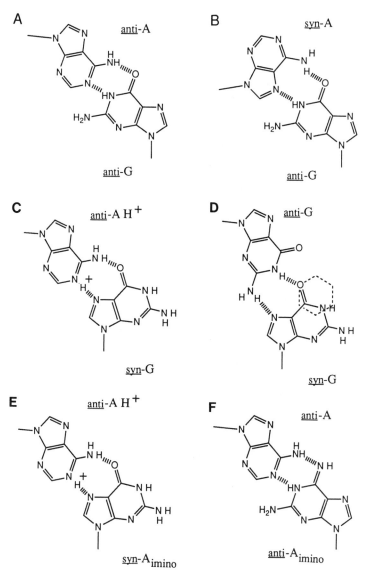

FIGURE 2 Mispairing schemes. (A–C) Three different structures for G:A mispairs that have been reported (Kan *et al.,* 1983; Patel *et al.,* 1984; Kennard 1985; Hunter *et al.,* 1986; Prive *et al.,* 1987; Brown, *et al.,* 1989). (D) Likely structure for a G:G mispair. (E, F) Two possible structures for an A:A mispair.

sistent with the observations that spontaneous mutagenesis is enhanced in SOS-induced cells and that GC → TA and AT → TA transversion mutations are enhanced specifically, which may result from a spontaneous version of the same process (Miller and Low, 1984; Yatagai *et al.,* 1991).

On the other hand, when considering the mutational spectra of bulky mutagens/carcinogens, the numerous examples cited earlier make it clear that the A-insertion rule is not absolute. In addition, our preliminary work with (+)-*anti*-DB[*a, j*]ADE (Gill *et al.*, in press) has shown that SOS induction preferentially enhances GC → AT mutations, which is a novel result. Thus, some sequence contexts are likely to exist in which bulky agents are misreplicated, perhaps in an SOS-dependent fashion, by one of the misinsertional mechanisms discussed earlier. This result also makes it possible that GC → TA mutations may be induced by different mechanisms.

Finally, none of these pathways of mutagenesis is enforced, since results suggest that the adducts principally do not cause mutations when replicated. Thus, either some sort of polymerase "glitch" occurs during insertion or different adduct conformations are responsible for correct as opposed to incorrect base insertion during replication.

CONCLUSIONS

Clearly the patterns of mutation induced by bulky mutagens/carcinogen, including the aflatoxins, are complex. This complexity can be separated into two categories: quantitative effects and qualitative effects. Both clearly are influenced by sequence context. The influence of sequence context on quantitative aspects of mutagenesis is well recognized (i.e., as mutagenic "hot spots"). With respect to qualitative effects, it seems reasonable to imagine that sequence context in the vicinity of a DNA adduct derived from a bulky mutagen/carcinogen influences adduct conformation, and this effect in turn affects the mutagenic specificity of that particular bulky adduct. Evidence suggests that several bulky adducts can adopt different conformations depending on their sequence contexts (Roche *et al.*, 1991; Veaute and Fuchs, 1991; Belguise and Fuchs, 1991) and that a particular bulky adduct can induce different mutations depending on its sequence context (Burnouf *et al.*, 1987; Mitra *et al.*, 1989; Lambert, *et al.*, 1992). Finally, a noninformational pathway of mutagenesis seems likely to exist for the adducts of bulky mutagens/carcinogens, but misinformational pathways may exist in parallel.

ACKNOWLEDGMENTS

I thank Audry Halpern for helping in some of the preliminary analysis of sequence specificity of mutagenesis, and John Essigmann and Anthony Dipple for reading portions of this chapter. I also thank John Essigmann, Anita Bigger, Anthony Dipple, Alan Anderson, Daniel Levy, Michael Seidman, and Kevin Kraemer for making unpublished data available to me. Work in my laboratory on mutagenesis by (+)-*anti*-B[*a*]PDE was supported by American Cancer Society Grant CN-54 and NIH Grant ES03775.

REFERENCES

Andersson, B., Felt, S., and Lambert, B. (1992). Strand specificity for mutations induced by (+)-*anti* BPDE in the *hprt* gene in human T-lymphocytes. *Mutation Res.* **269**, 129–140.

Balmain, A., and Brown, K. (1988). Oncogene activation in chemical carcinogenesis. *Adv. Cancer Res.* **51**, 147–182.

Basu, A. K., and Essigmann, J. M. (1988). Site-specifically modified oligonucleotides as probes for the structural and biological effects of DNA-damaging agents. *Chem. Res. Toxicol.* **1**, 1–18.

Basu, A. K., and Essigmann, J. M. (1990). Site-specifically alkylated oligodeoxynucleotides: Probes for mutagenesis, DNA repair and the structural effects of DNA damage. *Mutation Res.* **233**, 189–201.

Basu, A. K., Loechler, E. L., Leadon, S. A., and Essigmann, J. M. (1989). Genetic effects of *cis*-thymine glycol: Site-specific mutagenesis and molecular modeling studies. *Proc. Natl. Acad. Sci. U.S.A.* **86**, 7677–7681.

Belguise-Valladier, P., and Fuchs, R. P. P. (1991). Strong sequence-dependent polymorphism in adduct-induced DNA structure: Analysis of single N-2-acetylaminofluorene residues bound within the *NarI* mutation hot spot. *Biochemistry,* **30**, 10091–10100.

Benasutti, M., Ejadi, S., Whitlow, M. D., and Loechler, E. L. (1988). Mapping the binding site of aflatoxin B_1 in DNA: Systematic analysis of the reactivity of aflatoxin B_1 with guanines in different DNA sequences. *Biochemistry* **27**, 472–481.

Bernelot-Moens, C., Glickman, B. W., and Gordon, A. J. E. (1990). Induction of specific frameshift and base substitution events by benzo[*a*]pyrene diol epoxide in excision-repair-deficient *Escherichia coli. Carcinogenesis* **11**, 781–785.

Bhanot, O. S., and Ray, A. (1986). The *in vivo* mutagenic frequency and specificity of O^6-methylguanine in ϕX174 replicative form DNA. *Proc. Natl. Acad. Sci. U.S.A.* **83**, 7348–7352.

Bichara, M., and Fuchs, R. P. P. (1985). DNA binding and mutation spectra of the carcinogen N-2-aminofluorene in *Escherichia coli:* A correlation between the conformation and the pre-mutagenic lesion and the mutation specificity. *J. Mol. Biol.* **183**, 341–351.

Bigger, C. A. H., Strandberg, J., Yagi, H., Jerina, D. M., and Dipple, A. (1989). Mutagenic specificity of a potent carcinogen, benzo[*c*]phenanthrene (4R, 3S)-dihydrodiol, (2S, 1R)-epoxide, which reacts with adenine and guanine in DNA. *Proc. Natl. Acad. Sci. U.S.A.* **86**, 2291–2295.

Bigger, C. A. H., St. John, J., Yagi, H., Jerina, D. M., and Dipple, A. (1992). Mutagenic specificities of four stereoisomeric benzo[*c*]phenanthrene dihydrodiol epoxides. *Proc. Natl. Acad. Sci. U.S.A.* **89**, 368–372.

Bishop, J. M. (1991). Molecular themes in oncogenesis. *Cell* **64**, 235–248.

Boles, T. C., and Hogan, M. E. (1986). High-resolution mapping of carcinogen binding sites on DNA. *Biochem.* **25**, 3039–3043.

Bradley, L. J. N., Yarema, K. J., Lippard, S. J., and Essigmann, J. M. (1993). Mutagenicity and genotoxicity of the major DNA adduct of the antitumor drug *cis*-diamminedichloroplatinum (II). *Biochemistry* **32**, 982–988.

Bressac, B., Kew, M., Wands, J., and Ozturk, M. (1991). Selective G to T mutations of p53 gene in hepatocellular carcinoma from southern Africa. *Nature (London)* **350**, 429–431.

Brown, T., Leonard, G. A., Booth, E. D., and Chambers, J. (1989). Crystal structure and stability of a DNA duplex containing A(*anti*):G(*syn*) base-pairs. *J. Mol. Biol.* **207**, 455–457.

Buckley, N. (1987). A regioselective mechanism for mutagenesis and oncogenesis caused by al-kylnitrosourea sequence specific DNA alkylation. *J. Am. Chem. Soc.* **109**, 7918–7920.

Burnouf, D., Daune, M., and Fuchs, R. P. P. (1987). Spectrum of cisplatinum-induced mutations in *Esherichia coli. Proc. Natl. Acad. Sci. U.S.A.* **84**, 3758–3762.

Burnouf, D., Koehl, P., and Fuchs, R. P. P. (1989). Single adduct mutagenesis: Strong effect of the position of a single acetylaminofluorene adduct within a mutation hot spot. *Proc. Natl. Acad. Sci. U.S.A.* **86**, 4147–4151.

Burnouf, D., Gauthier, C., Chottard, J. C., and Fuchs, R. P. P. (1990). Single d(ApG)*cis*-

diamminedichloroplatinum (II) adduct-induced mutagenesis in *Escherichia coli. Proc. Natl. Acad. Sci. U.S.A.* **87,** 6087–6091.

Busby, W. F., Jr., and Wogan, G. N. (1985). Aflatoxins. *In* "Chemical Carcinogenesis" (C. E. Searle, ed.), Vol. 2, pp. 945–1136. American Chemical Society, Washington, D.C.

Carothers, A. M., and Grunberger, D. (1990). DNA base changes in benzo[*a*]pyrene diol epoxide-induced dihydrofolate reductase mutants of Chinese hamster ovary cells. *Carcinogenesis,* **11,** 189–192.

Chambers, R. W., Sledziewska-Gojska, E., Hirani-Hojatti, S., Borowy-Borowski, H. (1985). *uvrA* and *recA* mutations inhibit a site-specific transition produced by a single O^6-methylguanine in gene G of bacteriophage φX174. *Proc. Natl. Acad. Sci. U.S.A.* **82,** 7173–7177.

Chen, R.-H., Maher, V. M., and McCormick, J. J. (1990). Effects of excision repair by diploid human fibroblasts on the kinds and locations of mutations induced by (+/−)-7beta,8alpha-dihydroxy-9alpha,10alpha-epoxy-7,8,9,10-dihydrobenzo[*a*]pyrene in the coding region of the *HPRT* gene. *Proc. Natl. Acad. Sci. U.S.A.* **87,** 8680–8684.

Chen, R.-H., Maher, V. M., and McCormick, J. J. (1991). Lack of a cell cycle-dependent strand bias for mutation induced in the *HPRT* gene by (+/−)-7beta,8alpha-dihydroxy-9alpha,10alpha-epoxy-7,8,9,10-dihydrobenzo[*a*]pyrene in excision repair-deficient human cells. *Cancer Res.* **51,** 2587–2592.

Chu, G. Y. H., Mansy, S., Duncan, R. E., and Tobias, R. S. (1978). Heavy metal nucleotide interactions. 11. Stereochemical and electronic effects in the electrophilic attack of *cis-* and *trans*-diammineplatinum(II) on 5′-guanosine monophosphate and polyguanylate in aqueous solution. *J. Am. Chem. Soc.* **100,** 595–601.

Coulondre, C., and Miller, J. H. (1977). Genetic studies of the *lac* repressor IV. Mutagenic specificity in the *lacI* gene of *Escherichia coli. J. Mol. Biol.* **117,** 577–606.

Dipple, A., Pigott, M. A., Agarwal, S. K., Yagi, H., Sayer, J. M., and Jerina, D. M. (1987). Optically active benzo[*c*]phenanthrene diol epoxide bind extensively to adenine in DNA *Nature (London)* **327,** 535–536.

Dolan, M. E., Oplinger, M., and Pegg, A. E. (1988). Sequence specificity of guanine alkylation and repair. *Carcinogenesis* **9,** 2139–2143.

Dosanjh, M. K., Galeros, G., Goodman, M. F., and Singer, B. (1991). Kinetics of extension of O^6-methylguanine paired with cytosine or thymine in defined oligonucleotide sequences. *Biochemistry* **30,** 11595–11599.

Drinkwater, N. R., Miller, E. C., and Miller, J. A. (1980). Estimation of apurinic/apyrimidinic sites and phosphotriesters in deoxyribonucleic acid treated with electrophilic carcinogens and mutagens *Biochemistry,* **19,** 5087–5092.

Drouin, E., and Loechler, E. (1993). AP-sites are not significantly involved in mutagenesis by the (+)-*anti*-diol epoxide of benzo[*a*]pyrene: Evidence that the complexity of its mutagenic specificity arises from adduct conformational polymorphism. *Biochemistry* **32,** 6555–6562.

Eisenstadt, E., Warren, A. J., Porter, J., Atkins, D., and Miller, J. H. (1982). Carcinogenic epoxides of benzo[*a*]pyrene and cyclopenta[cd]pyrene induce base substitutions via specific transversions. *Proc. Natl. Acad. Sci. U.S.A.* **82,** 1945–1949.

Ellison, K. S., Dogliotti, E., Connors, T. D., Basu, A. K., and Essigmann, J. M. (1989). Site-specific mutagenesis by O^6-alkylguanines located in the chromosomes of mammalian cells: Influence of the mammalian O^6-methylguanine–DNA alkyltransferase. *Proc. Natl. Acad. Sci. U.S.A.* **86,** 8620–8624.

Essigmann, J. M., Croy, R. G., Nadzan, A. M., Busby, W. F., Jr., Reinhold, V. N., Buchi, G., and Wogan, G. N. (1977). Structural identification of the major DNA adduct formed by aflatoxin B_1 *in vitro. Proc. Natl. Acad. Sci. U.S.A.* **74,** 1870–1874.

Essigmann, J. M., Green, C. L., Croy, R. G., Fowler, K. W., Buchi, G., and Wogan, G. N. (1983). Interactions of aflatoxin B_1 and alkylating agents with DNA: Structural and functional studies. *Cold·Spring Harbor Symp. Quant. Biol.* **47,** 327–337.

Foster, P. L., Eisenstadt, E., and Miller, J. H. (1983). Base substitution mutations induced by metabolically activated aflatoxin B_1 *Proc. Natl. Acad. Sci. U.S.A.* **80,** 2695–2698.

Gibson, D., and Lippard, S. J. (1987). Synthesis and NMR structural studies of the adduct of *trans*-diamminedichloroplatinum(II) with the DNA fragment d(GpCpG). *Inorg. Chem.* **26,** 2277–2279.

Gill, R. D., Rodriguez, H., Cortez, C., Harvey, R. G., Loechler, E. L., and DiGiovanni, J. (1993) Mutagenic specificity of (+)-*anti*-diol epoxide of dibenz[a,j]anthracene in the *supF* gene of an *Escherichia coli* plasmid. *Molec. Carcinogenesis*, in press.

Gopalakrishnan, S., Stone, M. P., and Harris, T. M. (1989). Preparation and characterization of an aflatoxin B_1 adduct with the oligonucleotide d(ATCGAT)$_2$. *J. Am. Chem. Soc.* **111,** 7232–7239.

Gopalakrishnan, S., Harris, T. M., and Stone, M. P. (1990). Intercalation of aflatoxin B_1 in two oligonucleotide adducts: Comparative ^{1}H NMR analysis of d(ATC^{AFB}GAT)·d(ATCGAT) and d(AT^{AFB}GCAT). *Biochemistry* **29,** 10438–10448.

Hill-Perkins, M., Jones, M. D., and Karran, P. (1986). Site-specific mutagenesis *in vivo* by a single methylated or deaminated purine base. *Mutation Res.* **162,** 153–163.

Hollstein, M., Sidransky, D., Vogelstein, B., and Harris, C. C. (1991). p53 mutations in human cancers. *Science* **253,** 49–53.

Hsu, I. C., Metcalf, R. A., Sun, T., Welsh, J. A., Wang, N. J., and Harris, C. C. (1991). Mutational hotspot in the p53 gene in human hepatocellular carcinomas. *Nature (London)* **350,** 427–428.

Hunter, W. N., Brown, T., and Kennard, O. J. (1986). Structural features and hydration of d(CGCGAATTAGCG); a double helix containing two G:A mispairs. *Biomol. Struct. Dynam.* **4,** 173–191.

Ikuta, S., Wallace, R. B., and Itakura, K. (1987). Dissociation kinetics of 19 base paired oligonucleotide-DNA duplexes containing different single mismatched base pairs. *Nucleic Acids Res.* **15,** 797–811.

Jacobsen, J. S., Refolo, L. M., Sambamburti, K., and Humayan, M. Z. (1987). DNA replication-blocking properties of adducts formed by aflatoxin B_1-2,3-dichloride and aflatoxin B_1-2,3-oxide. *Mutation Res.* **179,** 89–101.

Kan, L.-S., Chandrasegaran, S., Pulford, S. M., and Miller, P. S. (1983). Detection of a guanine:adenine base pair in a decadeoxyribonucleotide by proton magnetic resonance spectroscopy. *Proc. Natl. Acad. Sci. U.S.A.* **80,** 4263–4265.

Kennard, O. J. (1985). Structural studies of DNA fragments: The G:T wobble base pair in A, B and Z DNA: The G:A base pair in B-DNA. *Biomol. Struct. Dynam.* **3,** 205–226.

King, H. W. S., Osborne, M. E., and Brookes, P. (1979). The *in vitro* and *in vivo* reaction at the N7-position of guanine of the ultimate carcinogen derived from benzo[*a*]pyrene. *Chem. Biol. Interact.* **24,** 345–353.

Koffel-Schwartz, N., Verdier, J.-M., Bichara, M., Freund, A.-M., Daune, M. P., and Fuchs, R. P. P. (1987a). Carcinogen-induced mutation spectrum in wild-type, *uvrA* and *umuC* strains of *Escherichia coli*. *J. Mol. Biol.* **177,** 33–51.

Koffel-Schwartz, N., Maenhaut-Michel, G., and Fuchs, R. P. P. (1987b). Specific strand loss in N-2-acetylaminofluorene-modified DNA. *J. Mol. Biol.* **193,** 651–659.

Kraemer, K. H., and Seidman, M. M. (1989). Use of *supF*, an *Escherichia coli* tyrosine suppressor tRNA gene, as a mutagenic target in shuttle-vector plasmids. *Mutation Res.* **220,** 61–72.

Kunkel, T. A. (1984). Mutational specificity of depurination. *Proc. Natl. Acad. Sci. U.S.A.* **81,** 1494–1498.

Lambert, I. B., Napolitano, R. L., Fuchs, R. P. P. (1992). Carcinogen-induced frameshift mutagenesis in repetitive sequences. *Proc. Natl. Acad. Sci. U.S.A.* **89,** 1310–1314.

Lawrence, C. W., Borden, A., Banerjee, S. K., and LeClerc, J. E. (1990). Mutation frequency and spectrum resulting from a single abasic site in a single-stranded vector. *Nucleic Acids Res.* **18,** 2153–2157.

Lepre, C. A., and Lippard, S. J. (1990). Interaction of platinum antitumor compounds with DNA. *Nucleic Acids Mol. Biol.* **4,** 9–38.

Lepre, C. A., Strothkamp, K. G., and Lippard, S. J. (1987). Synthesis and ^{1}H NMR spectroscopic characterization of *trans*-[Pt(NH$_3$) {d(ApGpGpCpCpT)-N7-A(1),N7-G(3)}]. *Biochemistry* **26,** 5651–5657.

Lepre, C. A., Chassot, L., Cortello, C. E., and Lippard, S. J. (1990). Synthesis and characterization of *trans*-[Pt(NH$_3$)$_2$Cl$_2$] adducts of d(CCTCGAGTCTCC)·d(GGAGACTCGAGG). *Biochemistry* **29,** 811–823.

Leonard, G. A., Thompson, J., Watson, W. P., and Brown, T. (1990). High resolution structure of a mutagenic lesion in DNA. *Proc. Natl. Acad. Sci. U.S.A.* **87,** 9573–9576.

Levin, D. E., and Ames, B. A. (1986). Classifying mutagens as to their specificity in causing the six possible transition and transversions: A simple analysis using the *Salmonella* mutagenicity assay. *Environ. Mut.* **8,** 9–28.

Levy, D. D., Groopman, J. D., Lim, S. E., Seidman, M. M., and Kraemer, K. H. (1992). Sequence specificity of aflatoxin B$_1$-induced mutations in a plasmid replicated in xeroderma pigmentosum and DNA repair proficient human fibroblasts. *Cancer Res.* **52,** 5668–5673.

Lin, J.-K., Miller, J. A., and Miller, E. C. (1977). 2,3-Dihydro-2-(guan-7-yl)-3-hydroxy-aflatoxin B$_1$, a major acid hydrolysis product of aflatoxin B$_1$–DNA or –ribosomal RNA adducts formed in hepatic microsome-mediated reactions and in rat liver *in vitro*. *Cancer Res.* **37,** 4430–4438.

Loeb, L. A. (1984). Apurinic sites as mutagenic intermediates. *Cell* **40,** 483–485.

Loeb, L. A., and Preston, B. D. (1986). Mutagenesis by apurinic/apyrimidinic sites. *Ann. Rev. Genet.* **20,** 201–230.

Loechler, E. L. (1989). Adduct-induced base-shifts: A mechanism by which the adducts of bulky carcinogens might induce mutations. *Biopolymers* **28,** 909–927.

Loechler, E. L. (1991a). Molecular modeling in mutagenesis and carcinogenesis. *Meth. Enzymol.* **203,** 458–476.

Loechler, E. L. (1991b). Rotation about the C6–O^6 bond in O^6-methylguanine: The *syn* and *anti* conformers can be of similar energies in duplex DNA as estimated by molecular modeling techniques. *Carcinogenesis* **12,** 1693–1699.

Loechler, E. L., Green, C. L., and Essigmann, J. M. (1984). *In vivo* mutagenesis by O^6-methylguanine built into a unique site in a viral genome. *Proc. Natl. Acad. Sci. U.S.A.* **81,** 6271–6275.

Loechler, E. L., Benasutti, M., Basu, A. K., Green, C. L., and Essigmann, J. M. (1990). The role of carcinogen DNA adduct structure in the induction of mutations. *In* "Progress in Clinical and Biological Research" (M. L. Mendelsohn and R. J. Albertini, eds.), Vol. 340A, pp. 51–60. Wiley-Liss, New York.

Loveless, A. (1969). Possible relevance of O-6 alkylation of deoxyguanosine to the mutagenicity and carcinogenicity of nitrosamine and nitrosamide. *Nature (London)* **223,** 206–207.

McCann, J., Choi, E., Yamasaki, E., and Ames, B. N. (1975). Detection of carcinogens in the *Salmonella*/microsome test: Assay of 300 chemicals. *Proc. Natl. Acad. Sci. U.S.A.* **72,** 5135–5139.

Mackay, W., Benasutti, M., Drouin, E., and Loechler, E. L. (1992). Mutagenesis by (+)-*anti*-BP-N^2-Gua, the major adduct of activated benzo[*a*]pyrene, when studied in an *Escherichia coli* plasmid using site-directed methods. *Carcinogenesis* **13,** 1415–1425.

McMahon, G., Hanson, L., Lee, J.-J., and Wogan, G. N. (1986). Identification of an activated c-Ki-*ras* oncogene in rat liver tumors induced by aflatoxin B$_1$. *Proc. Natl. Acad. Sci. U.S.A.* **83,** 9418–9422.

McMahon, G., Davis, E. F., Huber, L. J., Kim, Y., and Wogan, G. N. (1990). Characterization of c-Ki-*ras* and N-*ras* oncogenes in aflatoxin B$_1$-induced rat liver tumors. *Proc. Natl. Acad. Sci. U.S.A.* **87,** 1104–1108.

Mazur, M., and Glickman, B. W. (1988). Sequence specificity of mutations induced by benzo[*a*]pyrene-7,8-diol-9,10-epoxide at endogenous *aprt* gene in CHO cells. *Som. Cell Mol. Genet.* **14,** 393–400.

Miller, J. H. (1980). The *lacI* gene. *In* "The Operon" (J. H. Miller and W. S. Reznikoff eds.), 2d Ed., pp. 31–88. Cold Spring Harbor Press, Cold Spring Harbor, New York.

Miller, J. H. (1983). Mutational specificity in bacteria. *Ann. Rev. Genet.* **17,** 215–238.

Miller, J. H., and Low, K. B. (1984). Specificity of mutagenesis resulting from the induction of the SOS system in the absence of mutagenic treatment. *Cell* **37,** 675–682.

Mitra, G., Pauly, G. T., Kumar, R., Pei, G. K., Hughes, S. H., Moschel, R. C., and Barbacid, M. (1989). Molecular analysis of O⁶-substituted guanine-induced mutagenesis of *ras* oncogenes. *Proc. Natl. Acad. Sci. U.S.A.* **86**, 8650–8654.

Moriya, M., Takeshita, M., Johnson, F., Peden, K., Will, S., and Grollman, A. P. (1988). Targeted mutations induced by a single acetylaminofluorene DNA adduct in mammalian cells and bacteria. *Proc. Natl. Acad. Sci. U.S.A.* **85**, 1586–1589.

Norman, D., Abuaf, P., Hingerty, B. E., Live, D., Grunberger, D., Broyde, S., and Patel, D. (1989). NMR and computational characterization of the *N*-(deoxyguanosin-8-yl) aminofluorene adduct [(AF)G] opposite adenosine in DNA: (AF)G[*syn*]:A[*anti*] pair formation and its pH dependence. *Biochemistry* **28**, 7462–7476.

Osborne, M., and Merrifield, K. (1985). Depurination of benzo[*a*]pyrene-diol epoxide treated DNA. *Chem. Biol. Interact.* **53**, 183–195.

Patel, D. J., Koslowski, S. A., Ikuta, S., and Itakura, K. (1984). Deoxyguanosine-deoxyadenosine pairing in the d(CGAGAATTCGCG) duplex: Conformation and dynamics at and adjacent to the dG:dA mismatch site. *Biochemistry* **23**, 3207–3217.

Patel, D. J., Shapiro, L., Kozlowski, S. A., Gaffney, B. L., and Jones, R. A. (1986). Structural studies of the O⁶-meG-T interaction in the d(C-G-T-G-A-A-T-T-C-O⁶meG-C-G) duplex. *Biochemistry* **25**, 1036–1042.

Pinto, A. L., and Lippard, S. J. (1985). Sequence-dependent termination of *in vitro* DNA synthesis by *cis*- and *trans*-diamminedichloroplatinum(II). *Proc. Natl. Acad. Sci. U.S.A.* **82**, 4616–4619.

Prive, G. G., Heinemann, U., Kan, L. S., Chandrasegaran, S., and Dickerson, R. E. (1987). Helix geometry, hydration, and G:A mismatch in a B-DNA decamer. *Science* **238**, 498–504.

Reardon, D. B., Bigger, C. A. H., and Dipple, A. (1990). DNA polymerase action on bulky deoxyguanosine and deoxyadenosine adducts. *Carcinogenesis* **11**, 165–168.

Refolo, L. M., Conley, M. P., Sambamurti, K., Jacobsen, J. S., and Humayan, M. Z. (1985). Sequence context effects in DNA replication blocks induced by aflatoxin B₁. *Proc. Natl. Acad. Sci. U.S.A.* **82**, 3096–3100.

Reid, T. M., Lee, M.-S., and King, C. M. (1990). Mutagenesis by site-specific arylamine adducts in plasmid DNA: Enhancing replication of the adducted strand alters mutation frequency. *Biochemistry* **29**, 6153–6161.

Richardson, F. C., Boucheron, J. A., Skopek, T. R., and Swenberg, J. A. (1989). Formation of O⁶-methyldeoxyguanosine at specific sites in a synthetic oligonucleotide designed to resemble a known mutagenic hotspot. *J. Biol. Chem.* **264**, 838–841.

Ripley, L. S., Dubins, J. S., deBoer, J. G., DeMarini, D. M., Bogerd, A. M., and Kreuezer, K. N. (1988). Hotspot sites for acridine-induced frameshift mutations in bacteriophage T4 correspond to sites of action of the T4 type II topoisomerase. *J. Mol. Biol.* **200**, 665–680.

Roche, C. J., Jeffrey, A. M., Mao, B., Alfano, A., Kim, S. K., Ibanez, V., and Geacintov, N. E. (1991). Dependence of conformations of benzo[*a*]pyrene diol epoxide–DNA adducts derived from stereoisomers of different tumorigenicities on base sequence. *Chem. Res. Toxicol.* **4**, 311–317.

Rodriguez, H., and Loechler, E. L. (1993a). Mutational specificity of the (+)-*anti*-diol epoxide of benzo[*a*]pyrene in a *supF* gene of an *Escherichia coli* plasmid: DNA sequence context influences hot-spots, mutagenic specificity, and the extent of SOS enhancement of mutagenesis. *Carcinogenesis* **14**, 373–383.

Rodriguez, H., and Loechler, E. L. (1993b). Mutagenesis by the (+)-*anti*-diol epoxide of benzo[*a*]pyrene: What controls mutagenic specificity? *Biochemistry* **32**, 1759–1769.

Rodriguez, H., Bhat, U., Snow, E. T., and Loechler, E. L. (1992). An *Escherichia coli* plasmid-based mutational system in which *supF* mutants are selectable: Insertion elements dominate the spontaneous spectra. *Mutation Res.* **270**, 219–231.

Sahasrabudhe, S. R., Sambamurti, K., and Humayun, M. Z. (1989). Base-substitution mechanisms and the origin of strand bias. *Mol. Gen. Genet.* **217**, 20–25.

Sahasrabudhe, S. R., Luo, X., and Humayun, M. A. (1990). Induction of G:C to A:T transition by the acridine half-mustard ICR-191 supports a mispairing mechanism for mutagenesis by some bulky mutagens. *Biochemistry* **29**, 10899–10905.

Sambamurti, K., Callahan, J., Perkins, C. P., Jacobson, J. S., and Humayan, M. Z. (1988). Mechanisms of mutagenesis by bulky DNA lesions at the guanine N7 position. *Genetics* **120**, 863–873.

Schaaper, R. M., Koffel-Schwartz, N., and Fuchs, R. P. P. (1990). *N*-Acetoxy-*N*-acetyl-2-aminofluorene-induced mutagenesis in the *lacI* gene of *Escherichia coli*. *Carcinogenesis* **11**, 1087–1095.

Singer, B., and Essigmann, J. M. (1991). Site-specific mutagenesis: Retrospective and prospective. *Carcinogenesis* **12**, 949–955.

Singer, B., and Grunberger, D. (1983). "Molecular Biology of Mutagens and Carcinogens." Plenum Press, New York.

Singer, B., Chavez, F., Goodman, M. F., Essigmann, J. M., and Dosanjh, M. K. (1989). Effect of 3′-flanking neighbors on kinetics of pairing of dCTP or dTTP opposite O^6-methylguanine in a defined primed oligonucleotide when *Escherichia coli* DNA polymerase I is used. *Proc. Natl. Acad. Sci. U.S.A.* **86**, 8271–8274.

Stark, A. A., Essigmann, J. M., Demain, A. L., Skopek, T. R., and Wogan, G. N. (1979). Aflatoxin B$_1$ mutagenesis, DNA binding, and adduct formation in *Salmonella typhimurium*. *Proc. Natl. Acad. Sci. U.S.A.* **76**, 1343–1347.

Strauss, B. S. (1989). *In vitro* mutagenesis and SOS repair. *Ann. Inst. Super. Sanit.* **25**, 177–190.

Strauss, B., Rabkin, S., Sagher, D., and Moore, P. (1982). The role of DNA polymerase in base substitution mutagenesis on non-instructional templates. *Biochemie* **64**, 829–838.

Topal, M. D., and Fresco, J. R. (1976). Complementary base pairing and the origin of substitution mutations. *Nature (London)* **263**, 285–289.

Trottier, Y., Waithe, W. I., and Anderson, A. (1992). Kinds of mutations induced by aflatoxin B$_1$ in a shuttle vector replicating in human cells transiently expressing cytochrome P450IA2 cDNA. *Mol. Carcinogen.* **6**, 140–147.

van der Veer, J. L., Peters, A. R., and Reedijk, J. (1986). Reaction products from platinum(IV) amine compounds and 5′-GMP are mainly bis(5′-GMP) platinum(II) amine adducts. *J. Inorg. Chem.* **26**, 137–142.

Veaute, X., and Fuchs, R. P. P. (1991). Polymorphism in *N*-2-acetylaminofluorene induced DNA structure as revealed by DNaseI footprinting. *Nucleic Acids Res.* **19**, 5603–5606.

Vogelstein, B., and Kinzler, K. W. (1992). Carcinogens leave fingerprints. *Nature (London)* **355**, 209–210.

Walker, G. C. (1984). Mutagenesis and inducible response to deoxyribonucleic acid damage in *Escherichia coli*. *Microbiol. Rev.* **48**, 60–93.

Wei, S.-J. C., Chang, R. L., Wong, C.-Q., Bhachech, N., Cui, X. X., Hennig, E., Yagi, H., Sayer, J. M., Jerina, D. M., Preston, B. D., and Conney, A. H. (1991). Dose-dependent differences in the profile of mutations induced by the ultimate carcinogen from benzo[*a*]pyrene. *Proc. Natl. Acad. Sci. U.S.A.* **88**, 11227–11230.

Witkin, E. M. (1976). Ultraviolet mutagenesis and inducible DNA repair in *Escherichia coli*. *Bacteriol. Rev.* **40**, 869–907.

Yang, J. L., Maher, V. M., and McCormick, J. J. (1987). Kinds of mutations formed when a human shuttle vector containing adducts of (+/−)-7beta,8alpha-dihydroxy-9alpha,10alpha-epoxy-7,8,9,10-dihydrobenzo[*a*]pyrene replicates in human cells. *Proc. Natl. Acad. Sci. U.S.A.* **84**, 3787–3791.

Yang, J.-L., Chen, R.-H., Maher, V. M., and McCormick, J. J. (1990). Kinds and locations of mutations induced by (+/−)-7beta,8alpha-dihydroxy-9alpha,10alpha-epoxy-7,8,9,10-dihydrobenzo[*a*]pyrene in the coding region of the hypoxanthine (guanine) phosphoribosyltransferase gene in diploid human fibroblasts. *Carcinogenesis* **12**, 71–75.

Yatagai, F., Horsfall, M. J., and Glickman, B. W. (1991). Specificity of SOS mutagenesis in native M13*lacI* phage. *J. Bacteriol.* **173**, 7996–7999.

9

Aflatoxin Carcinogenesis in the Context of the Multistage Nature of Cancer

❖

Yvonne P. Dragan and Henry C. Pitot

HISTORICAL PERSPECTIVE ON DEVELOPMENT OF MODELS OF EXPERIMENTAL CARCINOGENESIS

Continual Carcinogen Administration

Studies of chemically induced carcinogenesis performed by Yamagiwa and Ichikawa (1915) in the early 1900s first demonstrated that the application of a chemical (coal tar) to the skin of rabbits could result in tumors at the site of application. In the mid-1930s, Sasaki and Yoshida (1935) showed that the feeding of the azo dye *o*-aminoazotoluene resulted in the development of liver tumors, thus extending the field of chemical carcinogenesis to the induction of neoplasms at a site distant from their site of application. Next, Kinosita (1936) demonstrated that chronic exposure to 4-dimethylaminoazobenzene in the diet also resulted in liver cancer. These observations that chronic exposure to some azo dyes could result in liver tumors later were extended to other chemical classes, including the aromatic amines and the nitrosamines (Rabes *et al.*, 1972). Thus, the early studies in chemical carcinogenesis were typified by the chronic administration of the suspected carcinogen for 1–2 years until death or until a large percentage of the animals presented with tumors.

Less Than Lifetime Carcinogen Administration

The study of the development of carcinomas by analysis of successive sacrifices during chronic exposure to a variety of agents indicated the presence of precursor lesions. Subsequently, numerous investigators suggested that the short-term administration of a suspected carcinogen could result in carcinoma formation. Short-term exposure to 2-acetylaminofluorene, with repeated but less than lifetime exposure, was demonstrated to result in carcinoma formation by Firminger and Reubner (1961) and later by Teebor and Becker (1971). Later, Craddock (1978) found that the duration of exposure (and thus the cumulative dose) of diethylnitrosamine related directly to whether carcinomas would develop during the lifetime of the animal. For determination of the minimal exposure (dose and duration) to a complete carcinogen such as aflatoxin, which would result in carcinoma formation, exposure protocols of a shorter duration were employed (Wogan, 1966; Wogan and Newberne, 1967; Butler and Barnes, 1968; Butler *et al.*, 1969; Epstein *et al.*, 1969; Kalengayi *et al.*, 1975; Godoy *et al.*, 1976).

Multistage Nature of Carcinogenesis

During the last 40 years, studies of the epidemiology of human cancers and the experimental examination of carcinogenesis have demonstrated that many, if not all, cancers result from a multistage process of development (Pitot *et al.*, 1989). In experimental systems, cancer development can be divided functionally into the stages of initiation, promotion, and progression (Slaga, 1983; Pitot, 1989a). Human and experimentally induced animal tumors share a similarity in the natural history of their development and with respect to their gross and microscopic pathology. In addition, preneoplastic and malignant human lesions express enzymatic changes analogous to those observed in animal models for tumors of the colon, gastrointestinal tract, esophagus, oral cavity, liver, lung, and mammary tissue (Pitot, 1989a). These areas of preneoplastic change often have been the site of origin of malignant growth (Pitot, 1989a).

In experimental systems, a latent period occurs between exposure to a carcinogen and the development of cancers. This latency period is analogous to the time and/or age dependency for the development of several common human cancers (Moolgavkar and Knudson, 1981). The latency period comprises a series of stages that can be characterized both in experimental protocols and in humans (Table 1). Initiation has been described as the irreversible genetic damage that results from exposure to a carcinogen (Pitot, 1986). Several studies have suggested that initiation is induced most effectively during G_1-S and early S phase (Rabes *et al.*, 1986; Kaufmann *et al.*, 1987). Therefore, when coupled with a proliferative stimulus, carcinogen exposure could result in the formation of an initiated cell population with an increased susceptibility to further neoplastic development. The clonal proliferative growth of such initiated cells initially constitutes the succeeding stage of promotion and accounts for part of the time

TABLE 1   Characteristics of the Stages of Carcinogenesis Common
to the Mouse Epidermis and Rat Liver Models

Initiation	Promotion	Progression
Irreversible	Reversible	Irreversible
Additive	Maximal response	Aneuploidy
No threshold	Threshold	Karyotypic instability
Environmentally modifiable	Environmentally modifiable	Initially response to environmental stimuli; later, relative autonomy

necessary for the development of human and experimental neoplasms. Thus, in the stage of promotion, an acceleration of the growth of previously initiated cells occurs. During progression, additional genetic damage is accompanied by the development of the characteristics of aggressive malignant growth. Thus, in experimental models, carcinogenesis has been divided into the stages of initiation, promotion, and progression. Agents with a potential carcinogenic risk can be classified as acting at any one or combination of these stages (Pitot *et al.*, 1989).

The demonstration by Mottram (1944) that a single dose of a carcinogen was sufficient to initiate carcinogenesis suggested that a limited exposure to a complete carcinogen could result in the development of preneoplastic lesions. Coupling a brief exposure to a carcinogen with the subsequent administration of a promoting stimulus then resulted in the clonal growth of preneoplastic skin papillomas (Rous and Kidd, 1941; Berenblum and Shubik, 1949). The first demonstration of the stages of initiation and promotion in the rat liver was reported by Peraino *et al.* (1971), who showed that a brief exposure to 2-acetylaminofluorene followed by phenobarbital administration resulted in both preneoplastic and neoplastic changes. Concurrent administration of the initiating and promoting agents resulted in the alteration of the effectiveness of tumor induction, because many promoting agents are enzyme inducers (Peraino *et al.*, 1975) and, thus, may alter the ratio of detoxification to activation pathways (Schulte-Hermann, 1985). In addition, reversal of the order of administration of the initiating and promoting agent often resulted in an inhibition of development of preneoplastic lesions when the dose of the initiating agent was subcarcinogenic (Scherer and Emmelot, 1976; Emmelot and Scherer, 1980).

Characteristics of the Stages of Initiation and Promotion

Because initiated cells cannot be distinguished readily from normal cells morphologically, the growth of clonally derived preneoplastic lesions after promotion typically is used as an endpoint for the detection of initiation (Pitot, 1990). In fact, the putative progeny of initiated hepatocytes have been detected after only four rounds of DNA synthesis (Scherer *et al.*, 1972). The mechanisms underlying tumor promotion are as yet unclear, but may involve alterations of gene expression (Boutwell, 1964), increased DNA synthesis (Schulte-Hermann *et al.,* 1981), and/or decreased apoptosis (Bursch *et al.,* 1984). The primary characteristic of promotion that distinguishes it from initiation and progression is the phenotypic instability of the altered genetic expression. The irreversibility of initiation by a carcinogenic insult has been demonstrated in both mouse skin (Boutwell, 1964) and rat liver (Xu *et al.,* 1990) by the administration of the promoting stimulus after a lengthy delay following exposure to the initiating carcinogen. In addition, suspension of administration of the promoting stimulus from previously initiated cells, followed by its readministration, results in the rapid reinstatement of the clonal cell population (Boutwell and Baldwin, 1960; Hendrich *et al.,* 1986), further suggesting the irreversible nature of initiation. The property of additivity of initiating agents has been determined from studies on the continuous administration of complete carcinogens (e.g., Miller *et al.,* 1948), the chronic but intermittent treatment with initiating agents (Teebor and Becker, 1971), and the linearity of dose–response relationships for initiating agents (Scherer and Emmelot, 1976; Zerban *et al.,* 1988). The reversible nature of promotion is based on several types of studies including the dose–response characteristics of promoting agents (Verma and Boutwell, 1980; Pitot *et al.,* 1987). The dependence of preneoplastic lesions, including transplanted focal hepatocytes, on the continued presence of the promoting stimulus (Hanigan and Pitot, 1985; Hendrich *et al.,* 1986) is compatible with the reversible nature of the promotion process. Further, the dependence of the growth of preneoplastic lesions on the frequency of administration of the promoting agent (Boutwell, 1964; Xu *et al.,* 1991) indicates the presence of a threshold for the dose response of tumor promoter action.

Progression

Foulds (1954) suggested that the entire process in the development of cancer after initiation be called progression. However, since the irreversible changes that occur late in the carcinogenic process are distinct from the reversible characteristics of the period of promotion, the stage of progression is a distinct third stage of carcinogenesis (Scherer, 1984; Pitot, 1989b,1991). The process of progression involves additional irreversible genetic damage that is coupled with increasing karyotypic instability associated with increased growth rate, metastasis, invasiveness, and autonomy (Pitot, 1989b). Experimental studies have attempted to assess this stage by determining the number and multiplicity of malignant lesions that occur after a variety of manipulations. More recently, the initiation–promotion–initiation model (Potter, 1981) has been employed to determine processes involved in progression and to identify agents that will enhance the conversion of benign to malignant lesions (Hennings *et al.,* 1983; Scherer, 1984; O'Connell *et al.,* 1986a,b; Pitot, 1991). Since epidemiologic studies have suggested that cancer is the result of a minimum of two genetic insults (Moolgavkar and Knudson, 1981), the development of many human cancers can be mimicked by the initiation–promotion–progression model of carcinogenesis (Dragan *et al.,* 1992). Just as the study of carcinogenesis has progressed from the continuous administration or short-term feeding of carcinogens to the assessment of the influence of diet, other carcinogens, and growth-stimulating factors and their role in the stages of cancer development, so too have studies of aflatoxin-induced carcinogenicity progressed. Thus, within the context of the natural history of cancer development, the mechanism of induction of, and ultimately the prevention of cancers induced by the complete carcinogen aflatoxin can be assessed.

AFLATOXINS AS CARCINOGENS

Early Studies of Aflatoxin Carcinogenicity

The aflatoxins are mycotoxins elaborated by *Aspergillus flavus* and *Aspergillus parasiticus*. These compounds result in toxicity to humans and to a variety of other animal species when contaminated food supplies are ingested. Since the discovery of these molecules in 1960, as the result of acute toxicity of poultry from contaminated food supplies, and their subsequent evaluation as carcinogens to rodents (Lancaster *et al.*, 1961; Schoental, 1961) and trout (Ashley *et al.*, 1964,1965), aflatoxins have been recognized as potent environmental toxicants and carcinogens to many species (Newberne and Butler, 1969; IARC, 1976,1987; WHO, 1979; Busby and Wogan, 1984). Originally the term aflatoxin generically included any aflatoxin contaminant found in the food, but now is used for aflatoxin B_1 unless stated otherwise. These widely disseminated environmental carcinogens cause liver parenchymal tumors as well as tumors at several other organ sites including the colon (Wogan and Newberne, 1967), glandular stomach (Butler and Barnes, 1966), and kidney (Epstein *et al.*, 1969; see also Chapter 7). In nonhuman primates, a low incidence of several other neoplasms has been observed (Sieber *et al.*, 1979). Although several aflatoxin metabolites and congeners have been tested for their carcinogenicity, aflatoxin B_1 is the most potent (Hsieh *et al.*, 1984; Cullen *et al.*, 1987). Examples of human aflatoxicosis have been reported. The International Agency for Research on Cancer (IARC) now considers the data for human liver cancer after aflatoxin exposure to be sufficient to warrant its classification as a human carcinogen (IARC, 1987).

The mutagenicity of aflatoxin also has been well characterized (see review by Busby and Wogan, 1984). Aflatoxin B_1 is a procarcinogen that must be activated metabolically to the 8,9-epoxide, the putative ultimate carcinogen (Lin *et al.*, 1977). The primary nucleic acid adduct resulting from aflatoxin B_1 administration is the N^7-guanine derivative 2,3-dihydro-2-(N^7-guanyl)-3-hydroxyaflatoxin B_1 (Muench *et al.*, 1983). These adducts have been found primarily in GC-rich regions of DNA (Muench *et al.*, 1983). The resultant primary DNA mutation (>90%) in at least one *in vitro* model is a GC → TA transversion (Foster *et al.*, 1983). Croy *et al.* (1978) have shown that the N^7-guanine adduct occurs in a dose-dependent manner in animals administered aflatoxin and have suggested that the N^7-guanine adduct of aflatoxin is unstable, leading to depurination and, hence, to a G → A transversion. Another study found more adducts in liver mitochondrial DNA than in nuclear DNA (Niranjan *et al.*, 1982). Yang *et al.* (1985) have suggested that aflatoxin-derived adducts bind preferentially to GC-rich *Alu* sequences in high-molecular-weight human DNA. Perhaps most importantly, one study has shown half the human hepatocellular carcinomas examined (8/16) from areas of high aflatoxin exposure to contain a G → A transversion at base pair 3 of codon 249 in the *p53* gene whereas, of 6 mutations in this gene in 22 tumors from a low aflatoxin exposure area, none was at the codon 249 hotspot (Hollstein *et al.*, 1991). These observations in the human suggest that aflatoxin exposure can enhance carcinogenicity by increasing activation of dominant oncogenes or, as is the case with p53, through the structural alteration of a tumor suppressor gene.

Chronic Administration Studies of Aflatoxin Carcinogenicity

The acute liver toxicity of aflatoxin-containing groundnut meal to poultry and fish led to an examination of its carcinogenicity on chronic administration to several species. The earliest studies of aflatoxin-induced carcinogenicity in rats employed an unspecified type, amount, and mixture of aflatoxins as components of a natural feedlot-based diet (Lancaster *et al.*, 1961; Schoental, 1961). Salmon and Newberne (1963) determined by successive extraction that a contaminant in the peanut-meal component of the diet probably caused the observed neoplasms. Soon thereafter, Butler and Barnes (1964) used known amounts of aflatoxin to demonstrate its causal dose-dependent role in the induction of rat liver neoplasms. Similar studies were performed by Carnaghan (1965) in ducks. Several nonhuman primates also have been assessed for their sensitivity to aflatoxin-induced liver neoplasms (Lin *et al.*, 1974; Reddy *et al.*, 1976; Sieber *et al.*, 1979). Newberne and Butler (1969) reviewed studies on the carcinogenicity of aflatoxins in a variety of species and suggested that the trout and the rat were the most sensitive, whereas the adult mouse was relatively resistant to aflatoxin-induced carcinogenicity. Newberne (1965) decreased in a step-wise manner the amount of aflatoxin administered to the rat until malignant tumors were not observed after a 1-year exposure. This study demonstrated that continuous exposure of rats to 0.005 ppm aflatoxin in the diet did not result in malignant tumors in a 1-year time frame, although preneoplastic liver changes were observed.

Dickens *et al.* (1966) added a purified mixture of aflatoxins that had been found as contaminants in meal to the drinking water of rats to demonstrate their carcinogenicity. Later studies with purified aflatoxin B_1 demonstrated a dose response for liver tumors after addition to the diet (Wogan and Newberne, 1967; Newberne and Wogan, 1968) or to drinking water (Butler *et al.*, 1969). These dose–response studies suggested that increasing the dose decreased the latency for tumor development (Wogan *et al.*, 1974). In addition, several studies have examined different purified aflatoxins and assessed their carcinogenicity relative to aflatoxin B_1. Butler *et al.* (1969) compared aflatoxins B_1, G_1, and B_2, whereas

Hsieh *et al.* (1984) and Cullen *et al.* (1987) compared aflatoxin M_1 with B_1. These studies found aflatoxin B_1 to be more potent than the congeners and metabolites tested. Studies on the chronic administration of aflatoxin suggested that it is one of the most potent complete carcinogens known for rodent and trout liver.

Several multigeneration studies with known quantities of aflatoxin have indicated that preneoplastic altered hepatic foci or hepatomas result even from transplacental exposure to aflatoxin (Alfin-Slater *et al.* 1969; Grice *et al.*, 1973; Ward *et al.*, 1975). Specifically, the administration of aflatoxin B_1 to pregnant dams and to their offspring resulted in a 75% hepatoma incidence in the offspring (Ward *et al.*, 1975). In addition, aflatoxin given only to the pregnant dam could result in hepatocarcinogenesis in the offspring (Goerttler *et al.*, 1980). Thus, aflatoxin is a potent animal carcinogen in both adult and neonatal animals. Further, exposure to aflatoxin in utero is sufficient to result in development of liver cancer in later life.

Start/Stop or Discontinuous Administration of Aflatoxin

Although complete carcinogens including aflatoxins result in malignancies on chronic exposure, the processes that occur during human carcinogenesis are not as readily distinguishable and may result without chronic exposure. Thus, other model systems that limit hepatotoxicity and in which the stages of carcinogenesis can be separated were developed for the study of aflatoxin carcinogenicity. Since continuous lifetime exposure to aflatoxin was unnecessary for the development of tumors (Wogan, 1966), shorter-term exposure to aflatoxin has been assessed for its carcinogenic activity. The early studies of aflatoxin-induced liver carcinogenesis in rats indicated the presence and sequential development of precursor lesions (Lancaster *et al.*, 1961), also suggesting the potential role of a limited exposure to aflatoxin in the development of cancer. Specifically, sequential analyses during chronic exposure to aflatoxin demonstrated first basophilic cell hyperplasia and later emergence of large pale eosinophilic cells. This change was followed by nodular and later by hepatocellular carcinoma development (Newberne and Wogan, 1968). The cells of origin of the liver neoplasms were, however, species specific; in some cases, cholangiocarcinomas or mixed-cell carcinomas resulted from aflatoxin administration. In additional studies, Newberne and Wogan (1968) compared continuous exposure to a known amount of aflatoxin in contaminated feed with exposure for 16 weeks and subsequent withdrawal of aflatoxin. In this study, the incidence of neoplasia was the same for short-term administration as for chronic exposure, but the latency was longer and more variable. Sequential analysis of lesion development in these animals led the authors to conclude that an irreversible change occurred in hepatocytes of treated rats after 12 weeks of aflatoxin administration.

Purified aflatoxins then were administered for less than a lifetime, and tumor yields were assessed. Administration of purified aflatoxin B_1 in drinking water for 5 months resulted in a dose-related increase in the incidence of hepatocellular carcinomas in the lifetime of the animal, despite the cessation of carcinogen administration (Epstein *et al.*, 1969). Butler *et al.* (1969) compared the hepato-carcinogenicity at 1 year of aflatoxins B_1, G_1, and B_2 following a 10- to 20-week exposure and demonstrated that their relative carcinogenic potency was aflatoxin $B_1 > G_1 >> B_2$. Godoy *et al.* (1976) demonstrated that 3 weeks of administration of 5 mg aflatoxin B_1/kg in the diet was insufficient to induce neoplasms 1 year later, whereas exposure to the same diet for 6 weeks resulted in a 100% incidence. The hepatocarcinogenicity of this dietary concentration of aflatoxin was confirmed by studies by Butler and Hempsall (1981). With exposure to the same dietary aflatoxin concentration for 1–6 weeks, Butler *et al.* (1981) also observed liver neoplasms at 1 year. Their studies described basophilic foci at 3 weeks and γ-glutamyltranspeptidase-positive (GGT^+) foci by 6 weeks. Additional studies by Wogan (1966) found that a 30-day exposure to graded doses of aflatoxin in the diet wrought a 100% incidence of liver neoplasms at 10 months when the daily intake was 100 μg in a 1-month period, whereas only precursor lesions were observed with a 30-day exposure to daily aflatoxin levels of 37.5 or 15 μg. Thus, several studies suggest that a limited exposure to aflatoxin can result in later liver neoplasms and that subcarcinogenic doses can result in detectable precursor lesions.

Carnaghan (1967) demonstrated that a single exposure to 0.5 mg aflatoxin (a mixture of aflatoxins G_1 and B_1) per rat was sufficient to result in a 100% incidence in lifetime studies. This study, in combination with that of Wogan (1966), suggests that the effective carcinogenic dose of aflatoxin is less than 0.5 mg per rat. Similar studies by Kalengayi *et al.* (1975), employing the LD_{50}, found that a single administration resulted in GGT^+ foci in 15 weeks and carcinomas by 44 weeks. Thus, precursor lesions could be observed after a single exposure to aflatoxin. Further, Wogan *et al.* (1971) determined that administration of 0.4 mg/rat resulted in precursor lesions but no malignancies after 1 year. Dividing this subcarcinogenic dose into 5 administrations also failed to produce malignancies in 82 weeks, whereas division into 10 treatments resulted in a low incidence of neoplasms at 82 weeks. This incidence could be enhanced if the time between administrations was increased. In addition, an increase in the dose of each administration increased the acute toxicity but decreased the latency until the first detectable tumor. In similar studies, Wogan and Newberne (1967) found a 100% incidence of liver neoplasms in rats administered 5 mg purified aflatoxin B_1/kg, whereas the application of 400 μg per rat ($\sim$1.2 mg/kg) in 10 equal daily doses resulted in a low incidence of liver neoplasms. Further, although 0.015–1 ppm aflatoxin B_1 in the diet for 40 weeks resulted in carcinomas, 1 ppm administered for 14 days was insufficient to induce carcinomas by 82 weeks. Thus, the incidence of liver neoplasms was higher for repeated administration of smaller doses (0.015 ppm) than for a brief exposure to a higher total dose (1 ppm). In addition, the potent rat hepatocarcinogen aflatoxin, administered subchronically, results in precursor lesions at doses that do not result in hepatocellular carcinomas in the lifetime of the animal.

Multistage Models of Aflatoxin-Induced Liver Cancer

Numerous studies (for review, see Busby and Wogan, 1984) have examined the effectiveness in trout and rats of the coadministration of aflatoxin and a number of suspected promoting regimens. Pretreatment or concurrent administration of aflatoxin with Aroclor 1254 (Hendricks *et al.,* 1977a), flavones (Nixon *et al.,* 1984; Gurtoo *et al.,* 1985; Goeger *et al.,* 1988), butylated hydroxytoluene (Goeger *et al.,* 1988), butylated hydroxyanisole (Tanaka *et al.,* 1985; Williams *et al.,* 1986), α benzene hexachloride (Angsubhakorn *et al.,* 1981a), or phenobarbital (McLean and Marshall, 1971) resulted in a decrease in the number of animals with neoplasms compared with animals receiving only aflatoxin. Other enzyme inducers such as ethoxyquin (Manson *et al.,* 1987) and oltipraz (Kensler *et al.,* 1987) also have been shown to diminish aflatoxin-induced carcinomas and GGT^+foci. The high incidence of neoplasms observed with aflatoxin administration alone can preclude the detection of cocarcinogens, as may have been true for studies of aflatoxin cocarcinogenicity with dieldrin in trout (Hendricks *et al.,* 1977b) and cyclopropenoid fatty acids in rats (Nixon *et al.,* 1974) or trout (Lee *et al.,* 1971; Sinnhuber *et al.,* 1974). Thus, suspected promoting agents that are antioxidants or have enzyme-inducing action can cause a decrease in neoplasms or preneoplastic lesions in studies with concurrent carcinogen administration. This ploy of coadministration of an enzyme inducer (e.g., oltipraz) with aflatoxin exposure has been proposed as a chemopreventive to decrease hepatocarcinogenesis in humans (see Chapter 13). Lotlikar (1989) has suggested that chemopreventive measures should be by induction of phase II enzymes and not by manipulation of phase I enzymes, since phase I enzymes may result in increased activation of aflatoxin to its ultimate form. Antioxidants, which have this latter characteristic, are tumor promoters for several organs and, hence, have not been tested for this purpose in humans.

Ethanol and oral contraceptives are believed to be promoting agents in the rat liver. Evidence suggests that they may act in this manner in the human as well (Pitot, 1989a). Ethanol administration prior to and concurrent with aflatoxin administration resulted in an enhanced carcinogenicity in rat liver when iron-excluding lesions and other morphological end points were measured (Tanaka *et al.,* 1989), despite the lack of an observed cocarcinogenic effect in an earlier study (Newberne *et al.,* 1966). The earlier study employed a high dose of aflatoxin, which may have precluded recognition of a cocarcinogenic effect. The demonstration by Tanaka *et al.* (1989) of a cocarcinogenic effect of ethanol and aflatoxin suggests that, in animal models in which the alcohol is not contaminated with aflatoxin, ethanol alters the toxification/detoxification ratio in favor of an increased carcinogenicity of aflatoxin. This result is supported further by studies demonstrating a lack of promotion with ethanol when ethanol is administered after aflatoxin (Misslbeck *et al.*, 1984). The studies of Kamdem *et al.* (1983) with concurrent aflatoxin and oral contraceptives indicated that oral contraceptive administration stimulates the hepatocarcinogenicity of aflatoxin.

For initiation, carcinogen administration must be coupled with a proliferative stimulus. Two prevalent models of multistage carcinogenesis in the rat liver use either the neonatal liver (Peraino *et al.*, 1981) or a partial hepatectomy (Pitot *et al.*, 1978) as the proliferative stimulus. As for many other carcinogens, aflatoxin exposure has been coupled with increased cellular proliferation to enhance its carcinogenicity. In the mouse, which is relatively resistant to the carcinogenic action of aflatoxin, the physiological proliferation associated with normal growth in the neonatal state has been exploited by Vessilinovitch *et al.* (1972) to demonstrate that a single or brief neonatal exposure to aflatoxin results in a high incidence of hepatic neoplasms later in life. In this neonatal model, male mice are more susceptible than female mice to the hepatocarcinogenic effect of aflatoxin. Interestingly, 7-day-old mice had a lower incidence of aflatoxin-induced carcinogenesis than did 4-day-old mice. Similar results have been observed in rats with other carcinogens (Peraino *et al.*, 1981). Additionally, transplacental initiation by aflatoxin increased the incidence of liver neoplasms later in life (Goerttler *et al.*, 1980). Further, Metcalfe and Sonstegard (1984) demonstrated that 25 ng aflatoxin B_1 injected into a trout embryo was sufficient to result in liver neoplasms in the adult. Thus, brief exposure to aflatoxin during a period of physiological growth, such as early in life, is sufficient to result in liver neoplasms.

Necrosis in the adult also induces proliferation. The hepatocarcinogenicity of aflatoxin is enhanced when cell proliferation caused by a partial hepatectomy is coupled with aflatoxin exposure in mice (Dix, 1984) and rats (Scherer and Emmelot, 1976). Although Newberne *et al.* (1966) did not observe an increase in neoplasms with aflatoxin exposure after a partial hepatectomy, aflatoxin alone induced 100% incidence of neoplasms at the dose employed. This group did, however, observe an increase in the multiplicity of neoplasms when a cirrhotic stimulus such as iron supplementation or a choline-deficient diet was followed by an adequate diet. In addition, necrosis has been induced with carbon tetrachloride administration and ethanol- or hepatitis B-induced cirrhosis. Administration of carbon tetrachloride to mice followed by aflatoxin exposure 48 hr later resulted in hepatomas (Arora, 1981). An acceleration of aflatoxin-induced hepatoma formation by carbon tetrachloride also has been observed in rats (Scotto *et al.*, 1975). Sun *et al.* (1971) have compared the effectiveness of carbon tetrachloride and ethanol as proliferative stimuli for both aflatoxin B_1 and G_1. Although neoplasms resulted with carbon tetrachloride administration, such changes were not observed when ethanol was administered. Several groups have discussed the potential contribution of virus-induced cellular proliferation in promoting previously initiated cells (Rous and Kidd, 1941; Harris and Tsung-Tang, 1986; Weinstein, 1989). Lin *et al.* (1974) have shown virus-induced cirrhosis to promote aflatoxin carcinogenicity in nonhuman primates. As noted earlier, initiation is not effected unless mutation is coupled with a proliferative stimulus (Scherer *et al.*, 1972; Columbano *et al.*, 1987), which can occur if the dose of the carcinogen employed is acutely toxic, since the ensuing necrosis can increase the number of initiated cells (Solt and Farber, 1976; Solt *et al.*, 1977; Ying *et al.*, 1981). Thus, initiation can result from high-dose exposure to a complete carcinogen (Solt and Farber, 1976), repeated administration of a complete carcinogen (Watanabe and Williams, 1978; Peraino *et al.*, 1980), or administration in conjunction with a proliferative stimulus (Sun *et al.*, 1971; Scotto *et al.*, 1975; Scherer and Emmelot, 1976; Arora, 1981).

Two-Stage Models of Aflatoxin Carcinogenesis

Many studies have examined the evolution of precursor lesions in the liver after subacute administration of aflatoxin or at sequential sacrifices during its chronic administration. For example, Newberne and Wogan (1968), Butler and Hempsall (1981), and Butler *et al.* (1981) have described preneoplastic changes during early aflatoxin administration. Pritchard and Butler (1988) have noted the progression from altered hepatic foci (AHF) to nodules to hepatocellular carcinoma. In addition, Kalengayi and Desmet (1975) described periportal necrosis after administration of an LD_{50} dose of aflatoxin as well as the regenerating lesions, which lacked glycogen, that appeared later. Manson (1983) detailed the appearance of GGT^+ foci 4 weeks after aflatoxin administration. This group noted that a single administration of aflatoxin resulted in bile duct proliferation, whereas repeated administration resulted in AHF. GGT^+ foci also were noted by Baldwin and Parker (1987) after aflatoxin administration. In addition, both eosinophilic and basophilic AHF were observed; the eosinophilic lesions were apparently the precursor lesions for the hepatocellular carcinomas. This observation agreed with earlier work that had shown the early appearance of basophilic lesions and later eosinophilic ones (Newberne and Wogan, 1968). In similar studies, Bannasch *et al.* (1985) showed that a single dose of 5 mg aflatoxin/kg resulted in a distinctive AHF population devoid of GGT, but with normal levels of canalicular adenosine triphosphatase and glucose 6-phosphatase. These AHF also had decreased glycogen synthase and increased glucose 6-phosphate dehydrogenase activity. Morphologically, these lesions were described as tigroid in appearance and had a 100-fold higher labeling incidence. Similarly, Gebhardt *et al.* (1989) did not observe glutamine synthase-positive lesions following administration of aflatoxin. Several common phenotypic markers of liver preneoplasia have been examined for the detection of aflatoxin-induced liver lesions that are precursors of neoplasia.

The observation of precursor lesions following subchronic exposure to aflatoxin and during the sequential analysis of aflatoxin-induced carcinogenesis suggested that classic initiation–promotion protocols that employed aflatoxin as the initiator should be examined further. The demonstration that single and cumulative doses of less than 0.5 mg aflatoxin/rat did not result in neoplasms, in concert with the observation of AHF, suggested that this dose could serve as an initiating dose. As noted in an early section, ethanol and oral contraceptives appear to stimulate the growth of aflatoxin-initiated hepatocytes, although this relationship has not been tested rigorously. Studies of initiation–promotion protocols with aflatoxin as the initiating agent examined dietary conditions such as low protein, high fat, and selenium deficiency as potential promoting stimuli. Early studies by Rogers and Newberne (1969) suggested that high dietary fat content increased both initiation and promotion of aflatoxin-induced hepatic neoplastic lesions. These studies also suggested that the type of fat was important in these considerations, since corn oil was more effective than beef fat. Misslbeck *et al.* (1984) also reported evidence that high fat increased the size and number of GGT⁺ foci observed after aflatoxin exposure. Baldwin and Parker (1987), however, suggested that a high dietary fat content was important for initiation but not for promotion. In addition, low dietary selenium levels during promotion enhanced aflatoxin-initiated foci (Milks *et al.,* 1985; Baldwin and Parker, 1987).

The protein level and quality during the initiation and promotion phase affect the later development of aflatoxin-induced precancerous lesions. Studies by Madhaven and Gopalan (1968a,b) and Temcharoen *et al.* (1978) suggest that protein deficiency concurrent with aflatoxin exposure can result in aflatoxicosis, with ensuing cirrhosis, but not in precursor lesions. Dietary protein content does, however, affect aflatoxin-induced carcinogenesis since a high-protein diet resulted in a higher incidence of carcinomas in rats than did a low-protein diet (Madhaven and Gopalan, 1968a,b; Wells *et al.,* 1976). Appleton and Campbell (1983) found an increased number of GGT⁺ foci when high protein levels were present during promotion. Schulsinger *et al.* (1989) demonstrated that low-quality protein (wheat gluten) supported the growth of smaller and fewer GGT⁺ foci than did high-quality protein (casein or low-quality protein + lysine).

The role of vitamin insufficiency also has been examined in the development of aflatoxin-induced carcinomas. A fourfold increase above the normal vitamin content was not protective against aflatoxin-induced carcinogenicity (Hamilton *et al.,* 1974). A decrease in riboflavin and vitamin D_3, however, resulted in an increase in the sensitivity to aflatoxin-induced carcinogenesis in that study. Vitamin A deficiency increased the incidence of liver or colon cancer after aflatoxin administration to rats (Newberne and Suphakarn, 1977; Suphakarn *et al.,* 1983). In addition, supplementation of the rat diet with a number of vegetables, including green beans, beets, and squash, increased the incidence of GGT⁺ foci when administered concurrently with aflatoxin, whereas cabbage decreased their incidence (Boyd *et al.,* 1983). Studies in fish with the active ingredient in cabbage, indole-3-carbinol, using neoplasms as the end point, suggested that this compound is an effective promoting agent when administered after but not before aflatoxin. The effectiveness of promotion increased with the duration of exposure and persisted or increased as the interval between initiation and promotion was prolonged. Further, intermittent exposure to indole-3-carbinol resulted in a reduced but detectable promoting effect (Dashwood *et al.,* 1991).

Hormonal regulation of aflatoxin-induced carcinogenesis has been examined extensively, because males are more sensitive than females. Castration (Cardeilhac and Nair, 1973) and chemical feminization with diethylstilbestrol (Newberne and Williams, 1969) resulted in a lower level of aflatoxin-induced carcinomas than was observed in intact males. In addition, hypophysectomy decreased the necrosis induced by aflatoxin (Neal and Judah, 1978), whereas adrenal corticotropic hormone (ACTH) administration enhanced aflatoxin-induced carcinogenicity through adrenal stimulation (Chedid et al., 1980). Despite the lower incidence of aflatoxin-induced neoplasms in intact female than male rats, administration of oral contraceptives to female rats enhances the carcinogenicity of aflatoxin (Kamdem et al., 1983).

A dose-dependent increase of canalicular adenosine triphosphatase-deficient AHF was observed when aflatoxin was administered after a partial hepatectomy (Scherer and Emmelot, 1976). Milks et al. (1985) found a dose-dependent increase in GGT+ foci, another common marker of preneoplasia, with aflatoxin exposure. These two studies suggest that aflatoxin should be an effective initiating agent in a two-stage model of liver cancer. However, Moore et al. (1982) observed a lack of promotion by phenobarbital of aflatoxin-induced preneoplastic liver lesions, a result that may have been the result of the effectiveness of aflatoxin as a complete carcinogen at the dose employed. In a similar study, Shirai et al. (1985) also were unable to demonstrate the promotion of aflatoxin-initiated foci with phenobarbital. In addition, Moore et al. (1988) used a single injection of aflatoxin as the initiating agent followed by 1% dietary butylated hydroxytoluene as the potential promoter and found a diminution in the number of glutathione S-transferase-positive (PGST) foci compared with the number resulting from aflatoxin administration alone. The use of an excessive dose of the initiating agent may have prevented observation of a statistical increase in AHF with further manipulation. In addition, the use of inappropriate markers of precursor lesions may have been a factor in these experiments. A study in neonatal rats demonstrated that initiation within a few days of birth with 1 μg aflatoxin/g followed by phenobarbital administration after weaning resulted in an effective development of preneoplastic foci and hyperplastic nodules in the group receiving both the initiating and the promoting agents (Reubner et al., 1986). This study suggests that neonatal rats are more sensitive than adult rats. Studies by Kraupp-Grasl et al. (1990) supported the contention that, in adult rats, initiation with a single dose of aflatoxin and promotion with nafenopin resulted in the growth of a subpopulation of initiated hepatocytes that were diffusely basophilic and weakly eosinophilic but lacked GGT expression. In these studies, male rats responded more strongly than female rats to the promoting effect of nafenopin. The studies of these investigators also showed phenobarbital to increase the size of some AHF. Earlier studies of aflatoxin-induced carcinogenicity had suggested that the effectiveness of aflatoxin-induced initiation was the same in male and female rats, indicating that the males may have been more responsive to the promotion stage than the females. Several investigators have suggested that promoting agents affect only a subpopulation of initiated cells and that this subpopulation is specific for each promoting agent (Columbano *et al.*, 1982; Dragan *et al.*, 1991).

Studies of the effects of aflatoxin on initiation are also of interest. These investigations include, but are not limited to, an examination of the dose-dependent induction by aflatoxin of initiated foci and the dose-dependent promotion by nafenopin and other peroxisome-proliferating agents of these initiated cells. Since hepatocytes expressing PGST have been proposed to be initiated cells (Moore *et al.*, 1987), the induction of single hepatocytes expressing this marker should be examined after aflatoxin administration since AHF and hepatic nodules overexpressing PGST have been observed after its administration (Chen and Eaton, 1991). Other promoting agents, including dioxin, polychlorinated biphenyls (PCBs), oral contraceptives, and ethanol, should be examined for their ability to enhance preneoplastic foci initiated by aflatoxin. In addition, properties of promotion such as the reversibility of the preneoplastic lesions, the influence of intermittent exposure to the promoting agent, and threshold levels for promoting agents of environmental or therapeutic concern should be assessed. These initiation–promotion protocols should allow for the rapid initial screening of agents that might act as anticarcinogens by inhibiting either the initiation or the promotion stage of cancer development.

Aflatoxin in Cytotoxic Models of Chemical Carcinogenesis

Several hepatotoxic regimens have been coupled with the administration of a carcinogen to accelerate the growth of the initiated cells, including the concurrent or subsequent administration of ethionine, orotic acid (Columbano *et al.*, 1982), or a lipotrope-deficient diet, specifically one depleted of choline (Yokoyama *et al.*, 1985) or methionine (Mikol *et al.*, 1983). These experimental formats are called selection protocols, because the AHF that develop appear to have a resistance to their cytotoxicity. Typically, these selection agents induce tumors without an added carcinogen, although orotic acid is apparently an exception. Lipotrope deficiency often has been coupled with aflatoxin administration. Rogers and Newberne (1969) demonstrated the appearance of nodules 6 months after administration of a nonnecrogenic dose (375 μg) of aflatoxin and a severely lipotrope-deficient diet. A marginally lipotrope-deficient high-fat diet increased the incidence of aflatoxin B_1-induced tumors (Rogers *et al.*, 1980). Too large a dose of aflatoxin G_1 was used in a different study to permit observation of an increased incidence as a result of a marginally lipotrope-deficient diet (Rogers, 1975). Ethionine administration with aflatoxin resulted in an increase in the carcinogenicity observed with either agent alone (Newberne *et al.*, 1966). Thus, cytotoxicity induced by a lipotrope-deficient diet enhances the carcinogenicity of aflatoxin.

Selection procedures couple the administration of two carcinogenic agents. Aflatoxin administered with lasiocarpine did not result in an increased incidence of carcinomas (Reddy and Svoboda, 1972; Elastoff *et al.*, 1987) although lasiocarpine administration resulted in necrosis, but this result was not observed with aflatoxin. Administration of aflatoxin with cycasin, dipentylnitrosamine, or urethane was not synergistic (Newberne *et al.*, 1967; Elastoff *et al.*, 1987), perhaps because of enhanced toxicity. Cotreatment with aflatoxin B_1 and ochratoxin A resulted in a synergistic increased incidence of liver and kidney tumors (Kanisawa and Suzuki, 1978). Studies by Angsubhakorn *et al.* (1981b) showed that the combination of dimethylnitrosamine and aflatoxin administration resulted in a greater than additive incidence of liver tumors, but did not alter the latency of their appearance. Newberne and Connor (1980) coupled the administration of aflatoxin with diethylnitrosamine and observed an increased incidence of liver neoplasms that was dependent on the dose of diethylnitrosamine when the dose of aflatoxin was held constant. Neoplasms were observed at additional sites with this combination.

Perhaps the best studied selection protocol is that of Solt and Farber (1976), which couples administration of a necrogenic dose of the complete carcinogen diethylnitrosamine with a brief administration of a noncytotoxic dose of the complete carcinogen 2-acetylaminofluorene. A proliferative stimulus is provided at the midpoint of the 2-acetylaminofluorene administration by either carbon tetrachloride administration or by partial hepatectomy. Ito *et al.* (1980) have demonstrated that aflatoxin B_1 at 2 ppm in the drinking water for 6 weeks is sufficient for initiation when the 2-acetylaminofluorene/partial hepatectomy selection is employed. In fact, with hyperplastic nodules as the end point, aflatoxin B_1 was the most potent initiating agent tested for increasing the number and the size of these lesions (Ito *et al.*, 1980). When a necrogenic dose of diethylnitrosamine was used as the initiating agent and aflatoxin was used as the selecting agent, aflatoxin was an efficient promoting stimulus, as are many other complete carcinogens that have been tested in this model (Ito *et al.*, 1980). A dose response for aflatoxin as the initiating agent (substitution for diethylnitrosamine) in the Solt–Farber procedure, in which 2-acetylaminofluorene was used as the promoting agent, was performed by Imaida *et al.* (1981). This study confirmed that aflatoxin was the most potent of the five carcinogens tested for initiating action in this bioassay. In fact, the dose of 0.25 mg/kg body weight, equivalent to administration of 1 ppm aflatoxin B_1 in the diet for 3 days, was effective in inducing initiation. This study also suggested that different initiating agents induced distinct types of initiated cells that are differentially responsive to selection or promotion. In addition, this study supports the findings of Scherer and Emmelot (1976) and of Misslbeck *et al.* (1984) on the dose-dependent induction by aflatoxin of canalicular adenosine triphosphatase-deficient and GGT$^+$ foci, respectively. Moore *et al.* (1988) extended these studies by comparing initiation with a single injection of aflatoxin to a single dose of two dialkylnitrosamines. At 3 days and at 3 weeks after administration of aflatoxin, fewer PGST$^+$ foci were located in the periportal area, in congruence with the site of aflatoxin-induced necrosis, than were observed for the dialkylnitrosamines. The PGST$^+$ lesions that did result were large and approached nodule size within the 3-week observation period. Initiation with aflatoxin also can be detected by promotion with the peroxisome-proliferating agent nafenopin (Kraupp-Grasl *et al.*, 1990). In addition, ethanol (Tanaka *et al.*, 1989) and oral contraceptives (Kamdem *et al.*, 1983) may promote aflatoxin-initiated foci, although this proposal has yet to be tested rigorously. Aflatoxin also can provide the initiation event in a selection procedure such as that of Solt and Farber (1976). Although aflatoxin-initiated nodules responded to phenobarbital by P_{450} induction, the nodules did not increase in size in response to phenobarbital (Chen and Eaton, 1991). Cellular proliferation can be enhanced by aflatoxin if a dose that results in necrosis is employed or if its administration is coupled with a cirrhosis-inducing event such as carbon tetrachloride, partial hepatectomy, or ethanol- or hepatitis-induced cirrhosis. Further, aflatoxin also can provide the selection stimulus for a modified Solt-Farber procedure. Thus, despite the fact that acute administration of aflatoxin can result in an inhibition of DNA synthesis, aflatoxin can increase cellular proliferation under some circumstances (Neal and Cabral, 1980).

Aflatoxin Effects during Progression in Carcinogenesis Models

During the stage of progression, characteristics designated as the hallmarks of malignant growth such as aneuploidy, karyotypic instability, and more autonomous growth have been observed (Pitot, 1989b). Studies in the mouse epidermis (Hennings *et al.*, 1983; O'Connell *et al.*, 1986a,b; Rotstein and Slaga, 1988) and later in the rat liver (Scherer *et al.*, 1984; Pitot, 1991) demonstrated that progression is a distinct stage in the cancer development process and that experimental carcinogenesis can be divided into the three stages of initiation, promotion, and progression. In both the mouse skin and the rat liver, the action of agents as putative progressors has been tested by imposing administration of a second "initiating agent" onto the initiation–promotion model, as was suggested by Potter (1981). Although aflatoxin—in fact, any agent that requires metabolic activation—has not been examined as a putative progressor agent in these models, several observations indicate that complete carcinogens serve as progressor agents, in addition to their action as initiating agents. In fact, studies in which aflatoxin was administered as a single large dose or administered chronically or intermittently resulted in progression when the dose and interval of observation were sufficient. Specifically, the divided dose experiment by Wogan and Newberne (1967), in which an increased interval between administrations of aflatoxin resulted in a greater incidence of carcinoma formation than with a single administration or with repeated doses administered closer in time, also suggests that aflatoxin could result in progression as well as initiation. In addition, the studies on the coadministration of other carcinogens with aflatoxin exemplified the additivity of the effects of two complete carcinogens.

Alterations in the experience of oncogenes have been found to occur during the progression stage of cancer development (Pitot, 1989b). Several studies have examined the activation of oncogenes by aflatoxin exposure. Sinha *et al.*, (1988) examined four hepatic neoplasms induced in male Fischer rats by exposure to aflatoxin B_1 and G_1 and found three with N-*ras* activation and one with a G → A transversion at codon 12 in Ki-*ras*. Tashiro *et al.* (1986) found an increased expression of c-*myc* and c-*ras* in all the hepatomas resulting from administration of 40 injections totaling 1.5 mg aflatoxin B_1. In one tumor, amplification and rearrangement of c-*ras* was observed. In a different study, which examined 12 tumors resulting from exposure to the same amount and format of administration of aflatoxin B_1, McMahon *et al.* (1986) found that the genomic DNA of 10 tumors was transforming. Of these transformation-inducing tumor DNAs, two of eight had an activated Ki-*ras,* in contrast to studies in the mouse in which Bauer-Hofmann *et al.* (1990) found that 40% of CF1 mice exposed to 6 μg aflatoxin B_1 at 7 days of age had neoplasms containing a *ras* mutation in codon 61, whereas Wiseman *et al.* (1986) found a 70% incidence of such mutations in B6C3F1 mice. Thus, aflatoxin exposure can activate dominant oncogenes, although the relevance of these observations is diminished by the relative resistance of mice to the carcinogenic potential of AFB_1.

Interaction of aflatoxin exposure with viral infection has been implicated in the progression of human hepatocellular carcinoma. In addition to activity in promotion, the integration of virus into the host genome could increase the expression of a dominant oncogene or decrease the expression of a tumor suppressor gene. Integration of hepatitis B virus has been reported adjacent to a glucocorticoid response element in a sequence homologous to v-*erb*A in at least one human tumor (Dejean *et al.,* 1986), in support of the idea that virus integration can activate dominant oncogenes. In addition, several studies have demonstrated an enhanced carcinogenic effect of aflatoxin exposure in hepatitis-infected ducks (Cova *et al.,* 1990; Cullen *et al.,* 1990). A promising new direction of study is the treatment of a transgenic mouse expressing hepatitis B virus with aflatoxin (Sell *et al.,* 1991). The relative resistance of mice to aflatoxin-induced tumorigenesis reinforces the additive effect of hepatitis B virus and aflatoxin exposure through the observation of a rapid development of hepatocellular carcinomas in these animals with congenital hepatitis and later aflatoxin exposure (Sell *et al.,* 1991). These studies support the contention that aflatoxin exposure can act as a progressor agent in combination with chemical or viral factors.

SUMMARY AND CONCLUSIONS

Experimental study of aflatoxin-induced carcinogenesis supports the description of cancer as a multistage process. The complete carcinogen aflatoxin demonstrates the properties common to initiating agents (Newberne and Butler, 1969; Busby and Wogan, 1984). Studies on the sequential analysis of hepatocellular carcinoma formation have suggested that AHF and hyperplastic nodules may be precursor lesions for such carcinoma development (Watanabe and Williams, 1978; Kaufmann *et al.,* 1986; Harada *et al.,* 1989). Precursor lesions have been observed after short-term aflatoxin exposure (Butler and Barnes, 1968; Epstein *et al.,* 1969; Godoy *et al.,* 1976; Pritchard and Butler, 1988), after low-dose exposure (Wogan, 1966), and with sequential analysis during chronic exposure (Lancaster *et al.,* 1961; Newberne and Wogan, 1968). Newberne and Wogan (1968) described first basophilic and later eosinophilic lesions. Bannasch *et al.* (1985) described tigroid cells, whereas Kraupp-Grasl *et al.* (1990) detected a small population of AHF with diffuse basophilia and some eosinophilia after aflatoxin administration. In addition to the morphological evidence provided by the tinctorial hematoxylin and eosin stain, enzyme-altered hepatic foci have been found after aflatoxin administration. Specifically, GGT^+ (Kalengayi *et al.,* 1975; Butler *et al.,* 1981; Manson, 1983; Milks *et al.,* 1985; Baldwin and Parker, 1987) and ATPase-deficient AHF (Scherer and Emmelot, 1976) have been observed. Aflatoxin results in altered carbohydrate metabolism in some AHF after aflatoxin administration (Bannasch *et al.,* 1985). Thus, several markers of liver tumor precursor lesions have been described in the rodent after aflatoxin administration.

Aflatoxin carcinogenesis can be promoted by a variety of factors, including an enhanced proliferative rate and alterations in dietary or hormonal factors. Administration of aflatoxin during natural periods of growth and prior to compensatory regeneration can enhance its carcinogenicity. Necrosis also can increase the incidence of hepatic tumors in aflatoxin-treated animals. In addition, several dietary factors have been found to modulate the growth of tumors in aflatoxin-treated rodents. For example, high levels of dietary fat enhance initiation and promote the growth of preneoplastic liver lesions induced by aflatoxin. High dietary protein levels of high-quality protein in the diet of aflatoxin-treated rats also can increase hepatic neoplasm incidence. Vitamin and mineral balance additionally can promote the formation of liver tumors in rats administered aflatoxin. Factors affecting hormonal balance can influence the severity of aflatoxin-induced liver tumors in the rat further. Several studies on the carcinogenicity of the divided doses of aflatoxin suggest that endogenous promotion also contributes to the carcinogenicity of aflatoxin. Hepatocarcinogenicity induced by aflatoxin can be modulated by various factors during the stage of promotion, suggesting the possible role for chemopreventative strategies after human exposure.

Although precursor lesions for hepatocellular development have been observed in numerous studies, AHF rarely have been used as an end point in studies of aflatoxin-induced carcinogenicity. Aflatoxin is a very potent rat hepatocarcinogen, and many studies assessing the effect of agents on aflatoxin carcinogenicity have employed such a high dose of aflatoxin that the large incidence of carcinomas resulting from aflatoxin administration precludes the detection of changes wrought by the second agent. Additionally, concurrent administration of a promoting agent that results in the induction of the P_{450} mixed function oxidase system with a complete carcinogen such as aflatoxin often results in a decrease in initiation while stimulating promotion. The differential action of many inducers of biotransformation enzymes on the relative proportion of activation and detoxification pathways often results in an inhibition of carcinogenic action by complete carcinogens such as aflatoxin, which must be metabolized to an active form or for which the carcinogenic intermediate is detoxified efficiently by inducible enzymes (McLean and Marshall, 1971; Angsubhakorn *et al.*, 1981a,b; Gurtoo *et al.*, 1985; Tanaka *et al.*, 1985; Williams *et al.*, 1986; Kensler, *et al.*, 1987; Manson *et al.*, 1987). However, pretreatment or co-treatment with ethanol (Tanaka *et al.*, 1989) and oral contraceptives (Kamdem *et al.*, 1983) increases the carcinogenicity of aflatoxin exposure, suggesting that these agents to which humans are exposed may be effective promoting agents of aflatoxin carcinogenicity.

Aflatoxin carcinogenicity has not been examined thoroughly in classic initiation–promotion models, primarily because of the potency of initiation by aflatoxin and the reluctance to use precursor lesions such as AHF as an end point in such studies. In addition, the prototypical liver tumor promoter phenobarbital may not enhance the growth of some aflatoxin-initiated hepatocytes (Shirai *et al.*, 1985; Moore *et al.*, 1988; Chen and Eaton, 1991), although Kraupp-Grasl *et al.* (1990) have found that chronic phenobarbital administration will promote some aflatoxin-induced nodules and that nafenopin will enhance the growth of a separate subpopulation of aflatoxin-initiated hepatocytes. These studies suggest that initiation–promotion studies with low-dose aflatoxin as the initiating agent may be fruitful in determining promoting agents with relevance to human exposure to aflatoxin. Initiation with aflatoxin and selection with a partial hepatectomy and acetylaminofluorene administration, as in the Solt–Farber protocol (1976), has proved aflatoxin to be an effective initiating agent (Ito *et al.*, 1980; Imaida *et al.*, 1981). Thus, aflatoxin is a potent initiating agent. Further studies with promoting agents relevant to human exposure to aflatoxin should be performed.

Aflatoxicosis can result in cytotoxicity, including necrosis, and thus stimulate cell division. Neal and Cabral (1980) found that aflatoxin increased cell proliferation, and Ito *et al.* (1980) demonstrated aflatoxin to be an effective selection agent in the Solt–Farber protocol (1976). Thus, the complete carcinogen aflatoxin can display promoting action under certain conditions.

Combination of aflatoxin with cytotoxic agents or with a second complete carcinogen could lead to progression if both carcinogens are used at a sufficiently low dose to see the enhancement. When aflatoxin is given as one of the two carcinogens during the Solt–Farber procedure (1976), growth of precursor lesions is observed. Thus, a lipotrope-deficient diet (Rogers and Newberne, 1969; Rogers, 1975; Rogers *et al.*, 1980), addition of ethionine (Newberne *et al.*, 1966), or addition of a nitrosamine (Newberne and Connor, 1980; Angsubhakorn *et al.*, 1981a) enhanced aflatoxin-induced carcinogenicity. Thus, aflatoxin has not been examined specifically for action as a progressor agent. Its examination in the initiation–promotion–progression model should provide information on the complete risk of aflatoxin exposure to the development of human liver cancer.

The complete carcinogen aflatoxin can act during all three stages of the carcinogenesis process to result in experimental or human cancer development. Specifically, low-level exposure to aflatoxin can result in initiation when this carcinogen is activated to its ultimate carcinogenic form during a period of cell proliferation. This enhanced proliferation may result from necrosis subsequent to aflatoxicosis or from cirrhosis due to alcohol ingestion, viral infection, or other sources. Exposure of spontaneously initiated cells, those with an altered expression of a dominant oncogene or a tumor suppressor gene, and previously initiated cells to aflatoxin may result in the progression of those cells into malignant growth. Thus, within the context of the natural history of the progression of cancer development, human aflatoxin exposure and other risk factors for the etiology of human liver cancer can best be assessed.

REFERENCES

Alfin-Slater, R. B., Aftergood, L., Hernandez, H. J., Stern, E., and Melnick, D. (1969). Studies of long-term administration of aflatoxin to rats as a natural food contaminant. *J. Am. Oil Chem. Soc.* **46**, 493–497.

Angsubhakorn, S., Bhamarapravati, N., Ramruen, K., Sahaphong, S., Thamavit, W., and Miyamoto, M. (1981a). Further study of α benzene hexachloride inhibition of aflatoxin B_1. *Br. J. Cancer* **43**, 881–883.

Angsubhakorn, S., Bhamarapravati, N., Romruen, K., and Sahaphong, S. (1981b). Enhancing effect of dimethylnitrosamine on aflatoxin B_1 hepatocarcinogenesis in rats. *Int. J. Cancer* **28**, 621–626.

Appleton, B. S., and Campbell, T. C. (1983). Effect of high and low dietary protein on the dosing and postdosing periods of aflatoxin B_1-induced hepatic preneoplastic lesion development in the rat. *Cancer Res.* **43**, 2150–2154.

Arora, R. G. (1981). Enhanced susceptibility to aflatoxin B_1 toxicity in weanling mice pretreated with carbon tetrachloride. *Acta Pathol. Microbiol. Scand. Sect. A* **89**, 303–308.

Ashley, L. M., Halver, J. E., and Wogan, G. N. (1964). Hepatoma and aflatoxicosis in trout. *Fed. Proc.* **23**, 105.

Ashley, L. M., Halver, J. E., Gardner, W. K., and Wogan, G. N. (1965). Crystalline aflatoxins cause trout hepatomas. *Fed. Proc.* **24**, 627.

Baldwin, S., and Parker, R. S. (1987). Influence of dietary fat and selenium in initiation and promotion of aflatoxin B_1-induced preneoplastic foci in rat liver. *Carcinogenesis* **8**, 101–107.

Bannasch, P., Benner, U., Enzmann, H., and Hacker, H. J. (1985). Tigroid cell foci and neoplastic nodules in the liver of rats treated with a single dose of aflatoxin B_1. *Carcinogenesis* **6**, 1641–1648.

Bauer-Hofmann, R., Buchmann, A., Wright, A. S., and Schwarz, M. (1990). Mutations in the Ha-*ras* proto-oncogene in spontaneous and chemically induced liver tumors of the CF1 mouse. *Carcinogenesis* **11**, 1875–1877.

Berenblum, I., and Shubik, P. (1949). The persistence of latent tumor cells induced in the mouse's skin by a single application of 9,10-dimethyl-1,2-benzanthracene. *Br. J. Cancer* **3**, 109–118.

Boutwell, R. K. (1964). Some biological aspects of skin carcinogenesis. *Prog. Exp. Tumor Res.* **4**, 207–250.

Boutwell, R. K., and Baldwin, H. H. (1960). A demonstration of threshold and reversibility in tumor formation. *Proc. Am. Assoc. Cancer Res.* **3**, 97.

Boyd, J. N., Misslbeck, N., and Stoewsand, G. S. (1983). Changes in preneoplastic response to aflatoxin B_1 in rats fed green beans, beets, or squash. *Food Chem. Toxicol.* **21**, 37–40.

Bursch, W., Lauer, B., Timmermann-Trosiener, I., Barthel, G., Schuppler, J., and Schulte-Hermann, R. (1984). Controlled death of normal and putative preneoplastic cells in rat liver following withdrawal of tumor promoters. *Carcinogenesis* **5**, 453–458.

Busby, W. E., and Wogan, G. N. (1984). Aflatoxins. *In* "Chemical Carcinogenesis" (C. E. Searle, ed.), pp. 945–1136. American Chemical Society, New York.

Butler, W. H., and Barnes, J. M. (1964). Toxic effects of groundnut meal containing aflatoxin to rats and guinea-pigs. *Br. J. Cancer* **17**, 699–710.

Butler, W. H., and Barnes, J. M. (1966). Carcinoma of the glandular stomach in rats given diets containing aflatoxin. *Nature (London)* **206**, 90.

Butler, W. H., and Barnes, J. M. (1968). Toxic effects of groundnut meal containing aflatoxin to rats and guinea pigs. *Food Cosmet. Toxicol.* **9**, 135–141.

Butler, W. H., and Hempsall, V. (1981). Histochemical studies of hepatocellular carcinomas in the rat induced by aflatoxin. *Hepatology* **134**, 157–170.

Butler, W. H., Greenblatt, M., and Lijinsky, W. (1969). Carcinogenesis in rats by aflatoxins B_1, G_1, and B_2. *Cancer Res.* **29**, 2206–2211.

Butler, W. H., Hempsall, V., and Stewart, M. G. (1981). Histochemical studies on the early proliferative lesion in the rat liver by aflatoxin. *J. Pathol.* **133**, 325–340.

Cardeilhac, P. T., and Nair, K. P. C. (1973). Inhibition by castration of aflatoxin-induced hepatoma in carbon tetrachloride-treated rats. *Toxicol. Appl. Pharmacol.* **26**, 393–397.

Carnaghan, R. B. A. (1965). Hepatic tumors in ducks fed a low level of toxic groundnut meal. *Nature (London)* **208**, 308.

Carnaghan, R. B. A. (1967). Hepatic tumors and other chronic liver changes in rats following a single oral administration of aflatoxin. *Br. J. Cancer* **21**, 811–814.

Chedid, A., Halfman, C. J., and Greenberg, S. R. (1980). Hormonal influences on chemical carcinogenesis: Studies with the aflatoxin B_1 hepatocarcinoma model in the rat. *Digest Dis. Sci.* **25**, 869–874. ·

Chen, Z. Y., and Eaton, D. L. (1991). Differential regulation of cytochrome(s) P450 B1/2 by phenobarbital in hepatic hyperplastic nodules induced by aflatoxin B_1 or diethylnitrosamine plus 2-acetylaminofluorene in male F344 rats. *Toxicol. Appl. Pharmacol.* **111**, 132–144.

Columbano, A., Ledda, G. M., Rao, P. M., Rajalakshmi, S., and Sarma, D. S. R. (1982). Initiation of experimental liver carcinogenesis by chemicals: Are the carcinogenic altered hepatocytes stimulated to grow into foci by different selection procedures identical? *In* "Chemical Carcinogenesis" (C. Nicolini, ed.), pp. 167–178. Plenum Publishing, New York.

Columbano, A., Ledda-Columbano, G. M., Rao, P. M., Rajalakshmi, S., and Sarma, D. S. R. (1987). Inability of mitogen-induced liver hyperplasia to support the induction of enzyme-altered islands induced by liver carcinogens. *Cancer Res.* **47**, 5557–5559.

Cova, L., Wild, C. P., Mehrota, R., Turusov, V., Shirai, T., Lambert, V., Jaquet, C., Tomatis, L., Trepo, C., and Montesano, R. (1990). Contribution of aflatoxin B_1 and hepatitis B virus infection in the induction of liver tumors in ducks. *Cancer Res.* **50**, 2156–2163.

Craddock, V. M. (1978). Cell proliferation and induction of liver cancer. *In* "Primary Liver Tumors" (H. Remmer, P. Bannasch, and H. Potter, eds.), pp. 377–383. MTP Press, Lancaster, England.

Croy, R. G., Essigmann, J. M., Reinhold, V. N., and Wogan, G. N. (1978). Identification of the principal aflatoxin B_1–DNA adduct formed *in vivo* in rat liver. *Proc. Natl. Acad. Sci. U.S.A.* **75**, 1745–1749.

Cullen, J. M., Reubner, B. H., Hsieh, L. S., Hyde, D. M., and Hsieh, L. S. (1987). Carcinogenicity of dietary aflatoxin M_1 in male Fischer rats compared to aflatoxin B_1. *Cancer Res.* **47**, 1913–1917.

Cullen, J. M., Marion, P. L., Sherman, G. J., Hong, X., and Newbold, J. E. (1990). Hepatic neoplasms in aflatoxin B_1-treated, congenital duck hepatitis B virus-infected and virus-free Pekin ducks. *Cancer Res.* **50**, 4072–4080.

Dashwood, R. H., Fong, A. T., Williams, D. E., Hendricks, J. D., and Bailey, G. S. (1991). Promotion of aflatoxin B_1 carcinogenesis by the natural tumor modulator indole-3-carbinol: Influence of dose, duration, and intermittent exposure on indole-3-carbinol promotional potency. *Cancer Res.* **51**, 2362–2365.

Dejean, A., Bougueleret, L., Grzeschik, K.-H., and Tiollais, P. (1986). HBV integration in a sequence homologous to v-*erb*A and steroid receptor genes in a hepatocellular carcinoma. *Nature (London)* **322**, 70–71.

Dickens, F., Jones, H. E. H., and Waynforth, H. B. (1966). Oral, subcutaneous, and intratracheal administration of carcinogenic lactones and related substances: The intratracheal administration of cigarette tar in the rat. *Br. J. Cancer* **20**, 134–144.

Dix, K. M. (1984). The development of hepatocellular tumours following aflatoxin B_1 exposure of the partially hepatectomized mouse. *Carcinogenesis* **5**, 385–390.

Dragan, Y. P., Rizvi, T. A., Xu, Y. H., Hully, J. R., Maronpot, R. R., Campbell, H. A., and Pitot, H. C. (1991). An initiation-promotion assay in the rat liver as a potential complement to the two-year carcinogenesis bioassay. *Fund. Appl. Toxicol.* **16**, 525–547.

Dragan, Y. P., Xu, Y.-H., Xu, Y. I., Sargent, L. M., and Pitot, H. C. (1992). Multistage hepatocarcinogenesis in the rat: Insights into risk estimation. *In* "Relevance of Animal Studies to the Evaluation of Human Cancer Risk" (R. D'Amato, T. J. Slaga, W. H. Farland, and C. Henry, eds.), pp. 261–279. Wiley–Liss, New York.

Elastoff, R. M., Fears, T. R., and Schneiderman, M. A. (1987). Statistical analysis of a carcinogen mixture experiment. I. Liver carcinogens. *J. Natl. Cancer Inst.* **79**, 509–526.

Emmelot, P., and Scherer, E. (1980). The first relevant cell stage in rat liver carcinogenesis a quantitative approach. *Biochim. Biophys. Acta* **605**, 247–304.

Epstein, M., Bartus, B., and Farber, E. (1969). Renal epithelial neoplasms induced in male Wistar rats by oral aflatoxin B_1. *Cancer Res.* **29**, 1045–1050.

Firminger, H. I., and Reubner, M. D. (1961). Influence of adrenocortical, androgenic, and anabolic hormones on the development of carcinoma and cirrhosis of the liver in A × C rats fed *N*-2-fluorenyldiacetamide. *J. Natl. Cancer Inst.* **27**, 559–572.

Foster, P. L., Eisenstadt, E., and Miller, J. H. (1983). Base substitution mutations induced by metabolically activated aflatoxin B_1. *Proc. Natl. Acad. Sci. U.S.A.* **80**, 2695–2698.

Foulds, L. (1954). The experimental study of tumor progression: A review. *Cancer Res.* **14**, 327–339.

Gebhardt, R., Tanaka, T., and Williams, G. M. (1989). Glutamine synthetase heterogeneous expression as a marker for the cellular lineage of preneoplastic and neoplastic liver populations. *Carcinogenesis* **10**, 1917–1923.

Godoy, H. M., Judah, D. J., Arona, H. L., Neal, G. E., and Jones, G. (1976). The effects of prolonged feeding with aflatoxin B_1 on adult rat liver. *Cancer Res.* **36**, 2399–2407.

Goeger, D. E., Shelton, D. W., Hendricks, J. D., Pereira, C., and Bailey, G. S. (1988). Comparative effect of dietary butylated hydroxyanisole and β-naphthoflavone on aflatoxin B_1 metabolism, DNA adduct formation, and carcinogenesis in rainbow trout. *Carcinogenesis* **9**, 1793–1800.

Goerttler, K., Lohrke, H., Schweizer, H. J., and Hesse, B. (1980). Effects of aflatoxin B_1 on pregnant inbred Sprague–Dawley rats and their F_1 generation. *J. Natl. Cancer Inst.* **64**, 1349–1354.

Grice, H. C., Moodie, C. A., and Smith, D. C. (1973). The carcinogenic potential of aflatoxin B_1 or its metabolism in rats from dams fed aflatoxin pre- and post partum. *Cancer Res.* **33**, 262–268.

Gurtoo, H. L., Koser, P. L., Bansal, S. K., Fox, H. W., Sharma, S. D., Mulhern, A. I., and Pavelic, Z. P. (1985). Inhibition of aflatoxin B_1 hepatocarcinogenesis in rats with β-naphthoflavone. *Carcinogenesis* **6**, 675–678.

Hamilton, P. B., Tung, H. T., Wyatt, R. D., and Donaldson, W. E. (1974). Interaction of dietary aflatoxin with some vitamin deficiencies. *Poultry Sci.* **53**, 871–877.

Hanigan, M. H., and Pitot, H. C. (1985). Growth of carcinogen-altered rat hepatocytes in the liver of syngeneic recipients promoted with phenobarbital. *Cancer Res.* **45**, 6063–6070.

Harada, T., Maronpot, R. R., Morris, R. W., and Boorman, G. A. (1989). Observations on altered hepatic foci in National Toxicology Program two-year carcinogenicity studies in rats. *Toxicol. Pathol.* **17**, 690–708.

Harris, C. C., and Tsung-Tang, S. (1986). Interactive effects of chemical carcinogens and hepatitis B virus in the pathogenesis of hepatocellular carcinoma. *Cancer Surveys,* **5**, 765–780.

Hendrich, S., Glauert, H. P., and Pitot, H. C. (1986). The phenotypic stability of altered hepatic foci: Effects of withdrawal and subsequent readministration of phenobarbital. *Carcinogenesis* **7**, 2041–2045.

Hendricks, J. D., Putnam, T. P., Bills, D. B., and Sinnhuber, R. O. (1977a). Inhibitory effect of polychlorinated biphenyls (Aroclor 1254) on aflatoxin B_1 carcinogenesis in rainbow trout (*Salmo gairdneri*). *J. Natl. Cancer Inst.* **53**, 1285–1288.

Hendricks, J. D., Putnam, T. P., and Sinnhuber, R. O. (1977b). Effect of dietary dieldrin on aflatoxin B_1 carcinogenesis in rainbow trout (*Salmo gairdneri*). *J. Environ. Pathol. Toxicol.* **2**, 719–728.

Hennings, H., Shores, R., Wenk, M. L., Spangler, E. F., Tarone, R., and Yupsa, S. H. (1983). Malignant conversion of mouse skin tumours is increased by tumour initiators and unaffected by tumour promoters. *Nature (London)* **304**, 67–69.

Hollstein, M., Sidransky, D., Vogelstein, B., and Harris, C. C. (1991). p53 Mutations in human cancers. *Science* **253**, 49–53.

Hsieh, D. P. H., Cullen, J. M., and Reubner, B. H. (1984). Comparative hepatocarcinogenicity of aflatoxins B_1 and M_1 in the rat. *Food Chem. Toxicol.* **22**, 1027–1028.

International Agency for Research on Cancer (1976). Aflatoxins. In: IARC Monograph on the Evaluation of Carcinogenic Risks to Humans, Vol. 10, pp. 55–64. IARC, Lyon.

International Agency for Research on Cancer (1987). Aflatoxins. In: IARC Monograph on the Evaluation of Carcinogenic Risks to Humans, Supplement 7, pp. 83–87. IARC, Lyon.

Imaida, K., Shirai, T., Tatematsu, M., Takano, T., and Ito, N. (1981). Dose response of five hepatocarcinogens for the initiation of rat hepatocarcinogenesis. *Cancer Lett.* **14**, 279–283.

Ito, N., Tatematsu, M., Nakanishi, K., Hasegawa, R., Takano, T., Imaida, K., and Ogiso, T. (1980). The effects of various chemicals on the development of hyperplastic liver nodules in hepatectomized rats treated with *N*-nitrosodiethylamine or *N*-2-fluorenylacetamide. *Gann* **71**, 832–842.

Kalengayi, M. M. R., and Desmet, V. J. (1975). Sequential histological and histochemical study of the rat liver after single dose of aflatoxin B_1 intoxication. *Cancer Res.* **35**, 2836–2844.

Kalengayi, M. M. R., Ronchi, G., and Desmet, V. J. (1975). Histochemistry of γ-glutamyltranspeptidase in rat liver during aflatoxin B_1-induced carcinogenesis. *J. Natl. Cancer Inst.* **55**, 579–588.

Kamdem, L., Magdalou, J., Siest, G., Ban, M., and Zissu, D. (1983). Induced hepatotoxicity in female rats by aflatoxin B_1 and ethynylestradiol interaction. *Toxicol. Appl. Pharmacol.* **67**, 26–40.

Kanisawa, M., and Suzuki, S. (1978). Induction of renal and hepatic tumors in mice by ochratoxin A, a mycotoxin. *Gann* **69**, 599–600.

Kaufmann, W. K., MacKenzie, S. A., Rahija, R. J., and Kaufman, D. G. (1986). Quantitative relationship between initiation of hepatocarcinogenesis and induction of altered islands. *J. Cell Biol.* **30**, 1–9.

Kaufmann, W. K., Rice, J. M., Wenk, M. L., Devor, D., and Kaufman, D. G. (1987). Cell-cycle dependent initiation of hepatocarcinogenesis in rats by methyl(acetoxymethyl)nitrosamine. *Cancer Res.* **47**, 1263–1266.

Kensler, T. W., Egner, P. A., Dolan, P. M., Groopman, J. D., and Roebuck, B. D. (1987). Mechanism of protection against aflatoxin tumorigenicity in rats fed 5-(2-pyrazinyl)-4-methyl-1,2-dithiol-3-thione (oltipraz) and related 1,2-dithiol-3-thiones and 1,2-dithiol-3-ones. *Cancer Res.* **47**, 4271–4277.

Kinosita, R. (1936). Researches on the cancerogenesis of the various chemical substances. *Gann* **30**, 423–426.

Kraupp-Grasl, B., Huber, W., Putz, B., Gerbracht, U., and Schulte-Hermann, R. (1990). Tumor promotion by the peroxisome proliferator nafenopin involving a specific subtype of altered foci in rat liver. *Cancer Res.* **50**, 3701–3708.

Lancaster, M. C., Jenkins, F. P., and Philip, J. M. (1961). Toxicity associated with certain samples of groundnuts. *Nature (London)* **192**, 1095–1096.

Lee, D. J., Wales, J. H., and Sinnhuber, R. O. (1971). Promotion of aflatoxin-induced hepatoma growth in trout by methyl melvalate and sterculate. *Cancer Res.* **31**, 960–963.

Lin, J. J., Liu, C., and Svoboda, D. J. (1974). Long-term effects of aflatoxin B_1 and viral hepatitis on marmoset liver. *Lab. Invest.* **30**, 267–278.

Lin, J. K., Miller, J. A., and Miller, E. C. (1977). 2,3-Dihydro-2-(guan-7-yl)-3-hydroxy-aflatoxin B₁, a major acid hydrolysis product of aflatoxin B₁–DNA or –ribosomal RNA adducts formed in hepatic microsome mediated reactions and in rat liver *in vivo*. *Cancer Res.* **37**, 4430–4438.

Lotlikar, P. D. (1989). Metabolic basis for susceptibility and resistance to aflatoxin B₁ hepatocarcinogenesis in rodents. *J. Toxicol. Toxin Revs.* **8**, 97–109.

McLean, A. E. M., and Marshall, A. (1971). Reduced carcinogenic effects of aflatoxin in rats given phenobarbitone. *Br. J. Exp. Pathol.* **52**, 322–329.

McMahon, G., Hanson, L., Lee, J. J., and Wogan, G. N. (1986). Identification of an activated c-Ki-*ras* oncogene in rat liver tumors identified by aflatoxin B1. *Proc. Natl. Acad. Sci. U.S.A.* **83**, 9418–9422.

Madhaven, T. V., and Gopalan, C. (1968a). The effect of dietary protein on carcinogenesis of aflatoxin. *Arch. Pathol.* **80**, 123–126.

Madhaven, T. V., and Gopalan, C. (1968b). Effect of dietary protein on carcinogenesis of aflatoxin. *Arch. Pathol.* **85**, 133–141.

Manson, M. M. (1983). Biphasic early changes in rat liver γ-glutamyltranspeptidase in response to aflatoxin B₁. *Carcinogenesis* **4**, 467–472.

Manson, M. M., Green, J. A., and Driver, H. E. (1987). Ethoxyquin alone induces preneoplastic changes in rat kidney whilst preventing induction of such lesions in liver by aflatoxin B₁. *Carcinogenesis* **8**, 723–728.

Metcalfe, C. D., and Sonstegard, R. A. (1984). Microinjection of carcinogens into rainbow trout embryos: An *in vivo* carcinogenesis assay. *J. Natl. Cancer Inst.* **73**, 1125–1132.

Mikol, Y. B., Hoover, K. I., Creasia, D., and Poirier, L. A. (1983). Hepatocarcinogenesis in rats fed methyl-deficient, amino-acid defined diets. *Carcinogenesis* **4**, 1619–1629.

Milks, M. M., Wilt, S. R., Ali, I. I., and Couri, D. (1985). The effects of selenium on the emergence of aflatoxin B₁-induced enzyme altered foci in rat liver. *Toxicol. Appl. Toxicol.* **5**, 320–326.

Miller, J. A., Miller, E. C., Kline, D. E., and Rusch, H. P. (1948). Correlation of the level of hepatic riboflavin with the appearance of liver tumors in rats fed amino azo dyes. *J. Exp. Med.* **88**, 89–102.

Misslbeck, N. G., Campbell, T. C., and Roe, D. A. (1984). Effect of ethanol consumed in combination with high or low fat diets on the postinitiation phase of hepatocarcinogenesis in the rat. *J. Nutr.* **114**, 2311–2323.

Moolgavkar, S. H., and Knudson, A. G. (1981). Mutation and cancer: A model for carcinogenesis. *J. Natl. Cancer Inst.* **66**, 1037–1052.

Moore, M. A., Pitot, H. C., Miller, E. C., and Miller, J. A. (1982). Cholangiocellular carcinomas induced in syrian golden hamsters administered aflatoxin B₁ in large doses. *J. Natl. Cancer Inst.* **68**, 271–278.

Moore, M. A., Nakagawa, K., Satoh, K., Ishikawa, T., and Sato, K. (1987). Single GST-positive liver cells—Putative initiated hepatocytes. *Carcinogenesis* **8**, 483–486.

Moore, M. A., Nakagawa, K., and Ishikawa, T. (1988). Selection pressure and altered hepatocellular islands after a single injection of aflatoxin B₁. *Jpn. J. Cancer Res.* **79**, 187–194.

Mottram, J. C. (1944). A developing factor in experimental blastogenesis. *J. Pathol. Bacteriol.* **56**, 181–187.

Muench, K. F., Misra, R. P., and Humayun, M. Z. (1983). Sequence specificity in aflatoxin B₁–DNA interactions. *Proc. Natl. Acad. Sci. U.S.A.* **80**, 6–10.

Neal, G. E., and Cabral, J. R. P. (1980). Effect of partial-hepatectomy on the response of rat liver to aflatoxin B₁. *Cancer Res.* **40**, 4739–4743.

Neal, G. E., and Judah, D. J. (1978). Effect of hypophysectomy and aflatoxin B₁ on rat liver. *Cancer Res.* **38**, 3460–3467.

Newberne, P. M. (1965). Carcinogenicity of aflatoxin-contaminated peanut meals. *In* "Mycotoxins in Foodstuffs" (G. N. Wogen, ed.), pp. 187–208. MIT Press, Cambridge, Massachusetts.

Newberne, P. M., and Butler, W. H. (1969). Acute and chronic effects of aflatoxin on the liver of domestic and laboratory animals: A review. *Cancer Res.* **29**, 236–250.

Newberne, P. M., and Connor, M. (1980). Effects of sequential exposure to aflatoxin B₁ and diethylnitrosamine on vascular and stomach tissue and additional target organs in rats. *Cancer Res.* **40**, 4037–4042.

Newberne, P. M., and Suphakarn, V. S. (1977). Preventative role of Vitamin A in colon carcinogenesis in the rat. *Cancer* **40**, 2553–2556.

Newberne, P. M., and Williams, G. (1969). Inhibition of aflatoxin carcinogenesis by diethylstilbesterol in male rats. *Arch. Environ. Health* **19**, 489–498.

Newberne, P. M., and Wogan, G. N. (1968). Sequential morphological changes in aflatoxin B_1 carcinogenesis in the rat. *Cancer Res.* **28**, 770–781.

Newberne, P. M., Harrington, D. H., and Wogan, G. N. (1966). Effects of cirrhosis and other liver insults on induction of liver tumors by aflatoxin in rats. *Lab. Invest.* **13**, 1962–1969.

Newberne, P. M., Hunt, C. E., and Wogan, G. N. (1967). Neoplasms in the rat associated with administration of urethan and aflatoxin. *Exp. Mol. Pathol.* **6**, 285–299.

Niranjan, B. G., Bhat, N. K., and Avadhanti, N. G. (1982). Preferential attack of mitochondrial DNA by aflatoxin B_1 during hepatocarcinogenesis. *Science* **215**, 73–74.

Nixon, J. E., Sinnhuber, R. O., Lee, D. J., Landers, M. K., and Harr, J. R. (1974). Effect of cyclopropenoid compounds on the carcinogenic activity of diethylnitrosamine and aflatoxin B_1 in rats. *J. Natl. Cancer Inst.* **53**, 453–458.

Nixon, J. E., Hendricks, J. D., Pawlowski, N. E., Pereira, C. B., Sinnhuber, R. O., and Bailey, G. S. (1984). Inhibition of aflatoxin B_1 carcinogenesis in rainbow trout by flavone and indole compounds. *Carcinogenesis* **5**, 615–619.

O'Connell, J. K., Klein-Szanto, A. J. P., DiGiovanni, D., Fries, J. W., and Slaga, T. J. (1986a). Malignant progression of mouse skin papillomas treated with ethylnitrosourea, *N*-methyl-*N'*-nitro-*N*-nitrosoguanidine, or 12-*O*-tetradecanoylphorbol-13-acetate. *Cancer Lett.* **30**, 269–274.

O'Connell, J. K., Klein-Szanto, A. J. P., DiGiovanni, D., Fries, J. W., and Slaga, T. J. (1986b). Enhanced malignant progression of mouse skin tumors by the free radical generator benzoyl peroxide. *Cancer Res.* **46**, 2863–2865.

Peraino, C., Fry, R., and Staffeld, E. F. (1971). Reduction and enhancement by phenobarbital of hepatocarcinogenesis in the rat by 2-acetylaminofluorene. *Cancer Res.* **31**, 1506–1512.

Peraino, C., Fry, R. J. M., and Staffeldt, E. (1975). Effects of phenobarbital, amobarbital, diphenylhydantoin, and dichlorodiphenyltrichloroethane on 2-acetylaminofluorene-induced hepatic tumorigenesis in the rat. *Cancer Res.* **35**, 2884–2890.

Peraino, C., Staffeldt, E. F., Haugen, D. A., Lombard, L. S., Stevens, F. J., and Fry, R. J. M. (1980). Effects of varying the dietary concentration of phenobarbital on its enhancement of 2-acetylaminofluorene-induced tumorigenesis. *Cancer Res.* **40**, 3268–3273.

Peraino, C., Staffeldt, E. F., and Ludeman, V. A. (1981). Early appearance of histochemically altered hepatocyte foci and liver tumors in female rats treated with carcinogens one day after birth. *Carcinogenesis* **2**, 463–465.

Pitot, H. C. (1986). "Fundamentals of Oncology," 3d Ed. Marcel Dekker, New York.

Pitot, H. C. (1989a). Principles of carcinogenesis: Chemical. *In* "Cancer. Principles and Practice of Oncology" (V. DeVita, S. Hellman, and S. A. Rosenberg, ed.), 3d Ed., pp. 116–135. J. B. Lippincott Co., Philadelphia.

Pitot, H. C. (1989b). Progression: The terminal stage in carcinogenesis. *Jpn. J. Cancer Res.* **80**, 599–607.

Pitot, H. C. (1990). Altered hepatic foci: Their role in murine hepatocarcinogenesis. *Annu. Rev. Pharmacol. Toxicol.* **30**, 465–500.

Pitot, H. C. (1991). Characterization of the stage of progression in hepatocarcinogenesis in the rat. *In* "Boundaries between Promotion and Progression" (O. Sudilovsky, L. Liotta, and H. C. Pitot, ed.), pp. 3–18. Plenum Press, New York.

Pitot, H. C., Barsness, L., Goldsworthy, T., and Kitagawa, T. (1978). Biochemical characterization of the stages of hepatocarcinogenesis after a single dose of diethylnitrosamine. *Nature (London)* **271**, 456–458.

Pitot, H. C., Goldsworthy, T. L., Moran, S., Keenan, W., Glauert, H. P., Maronpot, R. R., and

Campbell, H. A. (1987). A method to quantitate the relative initiating and promoting poten-cies of hepatocarcinogenic agents in their dose-response relationships to altered hepatic foci. *Carcinogenesis* **8,** 1491–1499.

Pitot, H. C., Campbell, H. A., Maronpot, R., Bawa, N., Rizvi, T. A., Xu, Y.-H., Sargent, L., Dra-gan, Y., and Pyron, M. (1989). Critical parameters in the quantitation of the stages of initia-tion, promotion, and progression in one model of hepatocarcinogenesis in the rat. *Toxicol. Pathol.* **17,** 594–612.

Potter, V. R. (1981). A new protocol and its rationale for the study of initiation and promotion of carcinogenesis in rat liver. *Carcinogenesis* **2,** 1375–1379.

Pritchard, D. J., and Butler, W. H. (1988). The ultrastructural features of aflatoxin B_1-induced lesions in rat liver. *Br. J. Exp. Pathol.* **69,** 793–804.

Rabes, H. M., Scholze, P., and Jantsch, B. (1972). Growth kinetics of diethylnitrosamine-induced, enzyme-deficient "preneoplastic" liver cell population *in vivo* and *in vitro. Cancer Res.* **32,** 2577–2586.

Rabes, H. M., Muller, L., Hartmann, A., Kerler, R., and Schuster, C. (1986). Cell cycle dependent initiation of adenosine triphosphatase-deficient populations in adult rat liver by a single dose of *N*-methyl-*N*-nitrosourea. *Cancer Res.* **46,** 645–650.

Reddy, J. K., and Svoboda, D. (1972). Effects of lasisocarpine on aflatoxin B_1 carcinogenicity in rat liver. *Arch. Pathol.* **93,** 55–60.

Reddy, J. K., Svoboda, D. J., and Ray, M. S. (1976). Induction of liver tumors by aflatoxin B_1 in the tree shrew (*Tupia glis*), a nonhuman primate. *Cancer Res.* **36,** 151–160.

Reubner, B. H., Cullen, I., Hsieh, L., and Hsieh, D. H. P. (1986). Carcinogenicity of aflatoxin B_1 in infant rats. Effect of phenobarbital. Abstract 4663. *Fed. Proc.* **45,** 956.

Rogers, A. E. (1975). Variable effects of a lipotrope-deficient high-fat diet on chemical carcinogen-esis in rats. *Cancer Res.* **35,** 2469–2474.

Rogers, A. E., and Newberne, P. M. (1969). Aflatoxin B_1 carcinogenesis in lipotrope-deficient rats. *Cancer Res.* **29,** 1965–1972.

Rogers, A. E., Lenhardt, G., and Morrison, G. (1980). Influence of dietary lipotrope and lipid content on aflatoxin B_1, *N*-2-acetylaminofluorene, 1-2-dimethylhydrazine carcinogenesis in rats. *Can-cer Res.* **40,** 2802–2807.

Rotstein, J. B., and Slaga, T. J. (1988). Acetic acid, a potent agent of tumor progression in the multistage mouse skin model for chemical carcinogenesis. *Cancer Lett.* **42,** 87–90.

Rous, P., and Kidd, J. G. (1941). Conditional neoplasms and subthreshold neoplastic states. A study of tar tumors in rabbits. *J. Exp. Med.* **73,** 365–384.

Salmon, W. D., and Newberne, P. M. (1963). Occurrence of hepatomas in rats fed diets containing peanut meal as a major source of protein. *Cancer Res.* **23,** 571–575.

Sasaki, T., and Yoshida, T. (1935). Experimentelle Erzeugung des Lebercarcinomas durch Fütterung mit *o*-Amidoazotoluol. *Virchows Arch. Pathol. Anat.* **295,** 175–200.

Scherer, E. (1984). Neoplastic progression in experimental hepatocarcinogenesis. *Biochim. Biophys. Acta* **738,** 219–249.

Scherer, E., and Emmelot, P. (1976). Kinetics of induction and growth of enzyme-deficient islands involved in hepatocarcinogenesis. *Cancer Res.* **36,** 2544–2554.

Scherer, E., Hoffman, M., Emmelot, P., and Friedrich-Freksa, M. (1972). Quantitative study on foci of altered liver cells induced in the rat by a single dose of diethylnitrosamine and partial hepatectomy. *J. Natl. Cancer Inst.* **49,** 93–106.

Scherer, E., Feringa, A. W., and Emmelot, P. (1984). Initiation–promotion–initiation. Induction of neoplastic foci within islands of precancerous liver cells in the rat. *In* "Models, Mechanisms, and Etiology of Tumor Promotion" (M. Bozsornyi, K. Lapis, N. E. Day, and H. Yamasaki, ed.), pp. 57–66. IARC Science Publishers, Lyon.

Schoental, R. (1961). Liver changes in primary liver tumours in rats given toxic guinea pig diet (MRC Diet 18). *Br. J. Cancer* **15,** 812–815.

Schulsinger, D. A., Root, M. M., and Campbell, T. C. (1989). Effect of dietary protein quality on development of aflatoxin B$_1$-induced hepatic preneoplastic lesions. *J. Natl. Cancer Inst.* **81,** 1241–1245.

Schulte-Hermann, R. (1985). Tumor promotion in the liver. *Arch. Toxicol.* **57,** 147–158.

Schulte-Hermann R., Ohde, G., Schuppler, J., and Timmermann-Trosiener, I. (1981). Enhanced proliferation of putative preneoplastic cells in rat liver following treatment with the tumor promoters phenobarbital, hexachlorohexane, steroid compounds, and nafenopin. *Cancer Res.* **41,** 2556–2562.

Scotto, J. M., Stralin, H. G., Lageron, A., and Lemmonier, F. J. (1975). Influence of carbon tetrachloride or riboflavin on liver carcinogenesis with a single dose of aflatoxin B$_1$. *Br. J. Pathol.* **56,** 133–138.

Sell, S., Hunt, J. M., Dunsford, H. A., and Chiseri, F. V. (1991). Synergy between HBV expression and chemical hepatocarcinogenesis in transgenic mice. *Cancer Res.* **51,** 1278–1285.

Shirai, T., Imaida, K., Ohshima, M., Fukushima, S., Lee, M. S., King, C. M., and Ito, N. (1985). Different responses to phenobarbital promotion in the development of γ-glutamyltranspeptidase-positive foci in the liver of rats initiated with diethylnitrosamine, *N*-hydroxy 2-acetylaminofluorene, and aflatoxin B$_1$. *Jpn. J. Cancer Res.* **76,** 16–19.

Sieber, S. M., Correa, P., Dalgard, D. W., and Adamson, R. H. (1979). Induction of osteogenic sarcomas and tumors of the hepatobiliary system in nonhuman primates with aflatoxin B$_1$. *Cancer Res.* **39,** 4545–4554.

Sinha, S., Webber, C., Marshall, C. J., Knowles, M. A., Proctor, A., and Barras, N. C. (1988). Activation of ras oncogene in aflatoxin-induced rat liver carcinogenesis. *Proc. Natl. Acad. Sci. U.S.A.* **85,** 3673–3677.

Sinnhuber, R. O., Lee, D. J., Wales, J. H., Landers, M. K., and Keyl, A. C. (1974). Hepatic carcinogenicity of aflatoxin M$_1$ in rainbow trout and its enhancement by cyclopropene fatty acids. *J. Natl. Cancer Inst.* **53,** 1285–1288.

Slaga, T. J. (1983). Overview of tumor promotion in animals. *Environ. Health Perspect.* **50,** 3–14.

Solt, D., and Farber, E. (1976). New principle for the analysis of chemical carcinogenesis. *Nature (London)* **263,** 701–703.

Solt, D. B., Medline, A., and Farber, E. (1977). Rapid emergence of carcinogen-induced hyperplastic lesions in a new model for the sequential analysis of liver carcinogenesis. *Am. J. Pathol.* **88,** 595–618.

Sun, S.-C., Wei, R.-D., and Schaeffer, B. T. (1971). The influence of postnecrotic cirrhosis on aflatoxin carcinogenesis in rats. *Lab. Invest.* **24,** 368–372.

Suphakarn, V. S., Newberne, P. M., and Goldman, M. (1983). Vitamin A and aflatoxin: Effect on liver and colon cancer. *Nutr. Cancer* **5,** 41–50.

Tanaka, T., Maeura, Y., and Weisberger, J. H. (1985). Inhibition of aflatoxin B$_1$ induced hepatocarcinogenesis by butylated hydroxytoluene and butylated hydroxyaniline (Abstract #472). *Proc. Am. Assoc. Cancer Res.* **26,** 119.

Tanaka, T., Nishikawa, A., Iwata, H., Mori, Y., Hara, A., Hiromo, I., and Mori, H. (1989). Enhancing effect of ethanol on aflatoxin B$_1$-induced hepatocarcinogenesis in male ACI/N rats. *Jpn. J. Cancer Res.* **80,** 526–530.

Tashiro, F., Morimura, S., Hayashi, K., Makino, R., Kawamura, H., Horikoshi, N., Nemoto, K., Ohtsubo, K., Sugimura, T., and Ueno, Y. (1986). Expression of the c-Ha-*ras* and c-*myc* genes in aflatoxin B$_1$-induced hepatocellular carcinomas. *Biochem. Biophys. Res. Commun.* **138,** 858–864.

Teebor, G. W., and Becker, F. F. (1971). Regression and persistence of hyperplastic nodules induced by *N*-2 fluoroacetylacetamide and their relationship to hepatocarcinogenesis. *Cancer Res.* **31,** 1–3.

Temcharoen, P., Anukarahanonta, T., and Bhamarapravati, N. (1978). Influence of dietary protein and vitamin B$_{12}$ on the toxicity and carcinogenicity of aflatoxin in rat liver. *Cancer Res.* **38,** 2185–2192.

Verma, A. K., and Boutwell, R. K. (1980). Effects of dose and duration of treatment with the tumor promoting agent, 12-*O*-tetradecanoylphorbol-13-acetate, on mouse skin carcinogenesis. *Carcinogenesis* **1**, 271–276.

Vessilinovitch, S. D., Mikhailovich, N., Wogan, G. N., Lombardi, L. S., and Rao, K. V. N. (1972). Aflatoxin B$_1$, a hepatocarcinogen in the infant mouse. *Cancer Res.* **32**, 2289–2291.

Ward, J. M., Sontag, J. M., Weisburger, E. K., and Brown, C. A. (1975). Effect of lifetime exposure to aflatoxin B$_1$ in rats. *J. Natl. Cancer Inst.* **55**, 107–113.

Watanabe, K., and Williams, G. M. (1978). Enhancement of rat hepatocellular-altered foci by the liver tumor promoter phenobarbital: Evidence that foci are precursors of neoplasms and that the promoter acts on carcinogen-induced lesions. *J. Natl. Cancer Inst.* **61**, 1311–1314.

Weinstein, I. B. (1989). Synergistic interactions between chemical carcinogens, tumor promoters, and viruses and their relevance to human liver cancer. *Cancer Detect. Prevent.* **14**, 253–260.

Wells, P., Alftergood, L., and Alfin-Slater, R. B. (1976). Effects of varying levels of dietary protein on tumor development and lipid metabolism in rats exposed to aflatoxin. *J. Am. Oil Chem. Soc.* **53**, 559–562.

Williams, G. M., Tanaka, T., and Maeura, Y. (1986). Dose related inhibition of aflatoxin B$_1$-induced hepatocarcinogenesis by the phenolic antioxidants, butylated hydroxytoluene and butylated hydroxyanisole. *Carcinogenesis* **7**, 1043–1050.

Wiseman, R., Stowers, S., Miller, E., Anderson, M., and Miller, J. (1986). Activating mutations of the c-Ha-*ras* protooncogene in chemically induced hepatomas of the male B6C3F1 mouse. *Proc. Natl. Acad. Sci. U.S.A.* **83**, 5825–5829.

Wogan, G. N. (1966). Chemical nature and biological effects of aflatoxins. *Bacteriol. Rev.* **30**, 460–470.

Wogan, G. N., and Newberne, P. M. (1967). Dose–response characteristics of aflatoxin B$_1$ carcinogenesis in the rat. *Cancer Res.* **27**, 2370–2376.

Wogan, G. N., Edwards, G. S., and Newberne, P. M. (1971). Structure–activity relationships in toxicity and carcinogenicity of aflatoxin and analogs. *Cancer Res.* **31**, 1936–1942.

Wogan, G. N., Paglialunga, S., and Newberne, P. M. (1974). Carcinogenic effects of low dietary levels of aflatoxin B$_1$ in rats. *Food Cosmet Toxicol.* **12**, 681–685.

World Health Organization. (1979). "Environmental Health Criteria II. Mycotoxins." World Health Organization, Geneva.

Xu, Y. H., Maronpot, R., and Pitot, H. C. (1990). Quantitative stereologic study of the effects of varying the time between initiation and promotion on four histochemical markers in rat liver during hepatocarcinogenesis. *Carcinogenesis* **11**, 267–272.

Xu, Y. D., Dragan, Y. P., Young, T., and Pitot, H. C. (1991). The effect of the format of administration and the total dose of phenobarbital on altered hepatic foci following initiation in female rats with diethylnitrosamine. *Carcinogenesis* **12**, 1009–1016.

Yamagiwa, K., and Ichikawa, K. (1915). Experimentelle Studie über die Pathogenese der Epithelialgeschwulste. *Mitt. Med. Facultat. Kaiserl. Univ. Tokyo* **15**, 295.

Yang, S. S., Taub, J., Modali, R., Viera, W., Yasei, P., and Yang, G. C. (1985). Dose dependency of aflatoxin B$_1$ binding on human high molecular weight DNA in the activation of proto-oncogenes. *Environ. Health Perspect.* **62**, 231–238.

Ying, T. S., Sarma, D. S. R., and Farber, E. (1981). Role of acute hepatic necrosis in the induction of early steps in liver carcinogenesis. *Cancer Res.* **41**, 2096–2102.

Yokoyama, S., Sells, M. A., Reddy, T. V., and Lombardi, B. (1985). Hepatocarcinogenic and promoting action of a choline-devoid diet in the rat. *Cancer Res.* **45**, 2834–2842.

Zerban, H., Preussmann, R., and Bannasch, P. (1988). Dose–time relationship of the development of preneoplastic liver lesions induced in rats by *N*-nitrosodiethylamine. *Carcinogenesis* **9**, 607–610.

10

Nutritional Modulation of Aflatoxin Carcinogenesis

❖

Adrianne E. Rogers

INTRODUCTION

Even before aflatoxins were isolated and identified as toxins and carcinogens, a dietary effect on their carcinogenicity had been reported. Investigators studying dietary deficiency of methyl group donors or lipotropes (the combined deficiencies of choline and methionine, sometimes with additional deficiency of folate) reported the appearance of hepatocellular carcinomas in deficient rats and, with much lower incidence, in methyl-supplemented controls (Copeland and Salmon, 1946; Newberne *et al.,* 1966). Both control and deficient diets contained peanut meal, later shown to be contaminated with aflatoxins, to which the deficient rats were highly susceptible (Newberne and Rogers, 1986,1990). Subsequent studies confirmed the initial observations (Table 1) and extended them to examination of mechanisms by which the deficiency influences aflatoxin B_1 (AFB_1) hepatocarcinogenesis and of effects of the deficiency on carcinogenesis by other chemicals in liver (Table 2) and other organs. In addition to its effects on liver, methyl group deficiency increases carcinogenesis, preneoplastic changes, or both induced by chemical carcinogens in the colon, esophagus, mammary gland, and pancreas under some experimental conditions (Rogers *et al.,* 1991).

Other dietary and nutritional effects on AFB_1 carcinogenesis in laboratory animals are known. AFB_1 hepatocarcinogenesis in rats is increased by a high

TABLE 1   Aflatoxin B₁ Carcinogenesis in Methyl-Deficient Male Rats

Rat strain	AFB₁ (μg total, route)	Hepatocellular carcinomas (%)			Reference
		Control diet	Borderline deficient diet[a]	Severely deficient diet[b]	
AES	unknown, diet	0	–	30	Copeland and Salmon (1946)
Sprague–Dawley	0.07 ppm, diet	14	–	33	Newberne et al. (1966)
Sprague–Dawley	240, ig[c]	0	–	29	Newberne et al. (1968)
F344	375, ig	–	83	38	Rogers and Newberne (1969)
F344	350, ig	6	22	–	Rogers and Newberne (1971)
Sprague–Dawley	375, ig	70	–	100	Newberne and Rogers (1973b)
F344	375, ig	11	87	–	Rogers (1975)

[a]Borderline deficient rats have mildly fatty livers but grow at normal rates and do not develop hepatic fibrosis or cirrhosis.
[b]Severely deficient rats grow at a reduced rate and develop fibrosis and cirrhosis over a period of 2–6 months.
[c]ig, Intragastric.

dietary level (20% or more by weight) of corn oil (Newberne et al., 1979) and is reduced by dietary deficiency of protein (casein) (Dunaif and Campbell, 1987), by a high level (20–28% by weight) of beef fat (Newberne et al., 1979; Rogers and Newberne, 1981), and, variably, by increased selenium ingestion (Lei et al., 1990). AFB₁ colon carcinogenesis is increased by dietary deficiency of vitamin A, but liver carcinogenesis is not affected (Newberne and Rogers, 1973b; Newberne and Suphakarn, 1977,1983). In the mouse, AFB₁ hepatocarcinogenesis is reduced by dietary methyl donors fed in excess of the nutrient requirement

TABLE 2   Effects of Methyl Deficiency on Hepatocarcinogenesis by Chemicals Other than Aflatoxin B₁ in Male Sprague–Dawley Rats

Carcinogenic regimen[a]	Hepatocellular carcinoma (%)		Reference
	Control diet	Methyl-deficient diet	
DENA, 40 ppm, diet, 12 wk	24	60	Rogers (1975)
DBN, 3.7 mg/kg, sc	24	64	Rogers et al. (1974)
DDCP, 195 mg/rat, ig	40	67	Rogers (1975)
AAF, 0.02, 0.01%, diet	19	41	Rogers (1975)

[a]DENA, N-Nitrosodiethylamine; DBN, N-nitrosodibutylamine; DDCP, 3,3-diphenyl 3-dimethyl-carbamoyl-1-propyne; AAF, N-2-fluorenylacetamide. sc, Subcutaneous; ig, intragastric.

(Newberne *et al.*, 1990). In rainbow trout, indole-3-carbinol (I3C) and β-naphthoflavone, both nonnutrient dietary components, reduce AFB_1 hepato-carcinogenesis (Nixon *et al.*, 1984; Dashwood *et al.*, 1989,1990), although I3C also can increase AFB_1 hepatocarcinogenesis under certain conditions (Bailey *et al.*, 1987).

DIETARY DEFICIENCY OF METHYL DONORS

Methyl-deficient diets have many effects on the livers of laboratory animals, including induction of hepatocellular tumors in rats in the absence of known carcinogenic exposure. The deficiency has marked effects on the morphological, biochemical, and molecular responses of the liver to chemical carcinogens. Many of the effects have been summarized and reviewed (Table 3; Zeisel, 1990; Rogers *et al.*, 1991).

Mechanisms of Action

The deficiency affects AFB_1 metabolism, AFB_1–DNA adduct formation, AFB_1-induced hyperplasia, and the formation of hyperplastic or enzyme-altered foci (EAF). Several methyl donor-deficient diets have been studied in different laboratories and shown to have similar effects on chemical hepatocarcinogenesis (Rogers and Newberne, 1969,1971,1981; Newberne and Rogers, 1973a; Rogers 1975; Poirier, 1986; Shinozuka and Katyal, 1986; Chandar *et al.*, 1987; Chandar and Lombardi, 1988). The mechanism by which the diet alone is carcinogenic is not known. Endogenous DNA adduct content is not increased in livers of deficient rats (Gupta *et al.*, 1987). Increased DNA strand breaks and reduced capacity for DNA repair were reported in lymphocytes cultured from methyl-deficient rats (James and Yin, 1989); no known increase in hematopoietic tumors could be

TABLE 3   Abnormalities in Methyl-Deficient Rodent Liver[a]

Triglyceride accumulation and decreased VLDL secretion
Decreased content of choline, S-adenosylmethionine, folate
Hyperplasia of hepatocytes
Increased lipid peroxidation
Reduced xenobiotic metabolism
Reduced nucleic acid and protein methylases
Undermethylation of DNA
Oncogene activation and hypomethylation (c-H-*ras*; c-*myc*)
Increased plasma membrane 1,2-*sn*-diacylglycerol
Increased protein kinase C
Increased activity of phospholipases A_2 and C
Increased prostaglandin E_2

[a]Reprinted with permission from Rogers *et al.* (1991).
Copyright CRS Press, Inc., Boca Raton, Florida.

related to that finding. The "I-compounds," DNA modifications that are related inversely to tumor risk in laboratory rodents, are decreased in livers of methyl-deficient rats (Li *et al.,* 1990). Their function and mechanisms of action in protecting against carcinogenesis are unknown.

In addition to the abnormalities summarized in Table 3, two reports may be useful in investigating dietary modulation of AFB_1 carcinogenesis. Feeding rats a methyl-deficient diet with or without phenobarbital after carcinogen (*N*-nitrosodiethylamine, DENA) exposure induced early development of aneuploid cells (hyperdiploid, hypertetraploid) in the EAF and increased the ratio of tetraploid and heterogeneous EAF to diploid EAF, compared with results obtained in rats fed complete diet plus phenobarbital to promote development of EAF. (Rats fed control diet without phenobarbital did not develop EAF; Wang *et al.,* 1990.) In the second study, deficient rats given DENA expressed mutant p53 whereas control-fed rats did not (Smith *et al.,* 1991).

Xenobiotic Metabolism

A major contribution to carcinogenesis enhancement by methyl donor deficiency is likely to be made by alterations of xenobiotic metabolism and conjugation. Alterations have been demonstrated in biochemical and mutagenesis studies, many of them focused on AFB_1. Marker enzymes for microsomal oxidation reactions, as well as total microsomal protein and cytochromes P450, are reduced in deficient rats (Rogers and Newberne, 1971; Murray *et al.,* 1986). Evidence of alterations among the different P450 enzymes exists (Murray *et al.,* 1986), but specific changes have not been reported. Because of a reduction in activity of marker enzymes for mixed function oxygenases in methyl-deficient rats (Rogers and Newberne, 1971), bacterial mutagenesis studies were performed using S9 fractions from deficient or control livers to activate AFB_1. The results showed significantly less activation by the liver fractions or liver homogenates from deficient rats than by preparations from control rats, whether or not the rats had been given multiple doses of AFB_1; the results were confirmed using the lambda prophage induction test (Table 4; Suit *et al.,* 1977). Therefore, *in vitro* evidence favors a reduction rather than an increase in AFB_1 activation in livers of deficient rats and indicates no difference between single and multiple doses in AFB_1 activation to a microbial mutagen. In a related observation, Bhattacharya *et al.* (1987) reported that addition of folic acid to a microbial assay reduced mutagenesis by AFB_1. The other major methyl nutrients were not tested (Table 4).

Alteration of AFB_1–DNA Adducts

Carcinogenesis by AFB_1 in rats is virtually dependent on multiple exposures, although partial hepatectomy has been reported to permit carcinogenesis by a single dose (Dix, 1984; Rizvi *et al.,* 1989); single highly toxic doses induce adenomas and, perhaps, carcinomas in Wistar rats (Angsubhakorn *et al.,* 1990). In rats, the number of exposures required may be as few as 5; the total dose may be as low as 50 μg (Rogers *et al.,* 1971). Correlations have been demonstrated in

TABLE 4   Inhibition of AFB_1 Mutagenicity by Selected Diet Components in Microbial Test Systems

Diet component	Salmonella typhimurium strain	Inhibition (%)	Reference
Methyl deficiency[a]	TA98	41–73	Suit et al. (1977)
Folic acid	TA100	55	Bhattacharya et al. (1987)
Retinal	TA100	100	
Retinol	TA100	98	
Ellagic acid	TA100	50–>90[b]	Mandal et al. (1987)
Ellagic acid	TA100	None	Francis et al. (1989b)
Cu^{2+}	TA98	84–100	Francis et al. (1988)
	TA100	62–100	
$Fe^{2+,3+}$	TA98	35–64	
	TA100	—	
Mn^{2+}	TA98	11–45	
	TA100	55–73	
SeO_3^-	TA98	38–69	
	TA100	29–47	
Zn^{2+}	TA98	31–61	
	TA100	46–78	
Olive oil	TA98	None 23[b,c]	Brennan-Craddock et al. (1990)
Beef fat	TA98	None 14[b,c]	

[a]S9 prepared from liver of deficient rats compared with S9 from control rats.
[b]Calculated using data read from graphs in reference.
[c]In vivo test system: BALB/c mouse.

trout between adducts and tumors in response to AFB_1 given with or without nonnutrient dietary components that block AFB_1 carcinogenesis (Nixon et al., 1984; Dashwood et al., 1989; Table 5). Bechtel (1989) calculated from published studies the steady-state AFB_1–DNA adducts in Fischer rats fed AFB_1 in the diet for 2 years and in trout exposed in tank water for 1–20 days, and showed that tumor risk correlated with adduct content. Using a multiple dose exposure, Schrager et al. (1990) showed that total AFB_1–DNA adduct burden over the 10-day exposure period was increased 41% in methyl-deficient compared with control Fischer rats, and that adducts persisting 11 days after completion of exposure were significantly greater in deficient than in control rats (Table 5). The adduct content of the DNA varied considerably over the exposure period, as would be expected from interactions of AFB_1 metabolism, adduction, and repair of adducts. Clearly the sum of the interactions was a greater level of adduction in deficient than in control rats. The increased rate of hepatocyte proliferation and turnover in deficient rats (Rogers and Newberne, 1969,1971; Chandar et al., 1987) presumably reduces the number of detectable adducts while contributing to the fixation of DNA damage caused by them. Studies are needed to determine whether the DNA adduct:tumor risk ratio is the same for control and deficient

TABLE 5   Influence of Selected Diet Components on AFB$_1$ Carcinogenesis and Adduction to Hepatic DNA

Diet component	Animal	AFB$_1$ (dose, route)	AFB$_1$–DNA adducts	Hepatocarcinoma incidence	Reference
			Inhibition of (%)		
Indole-3-carbinol	Rainbow trout	20 ppb, diet, 2 wk	45	90	Nixon et al. (1984)
β-Naphthoflavone	Rainbow trout	20 ppb, diet, 2 wk	55	84	
Indole-3-carbinol	Rainbow trout	10–320 ppb, diet, 2 wk	45–80[a]	30–80[a]	Dashwood et al. (1989)
Adequate methyl	F344 rat	25 μg × 10, ip[b]	70	—[c]	Schrager et al. (1990)

[a]Estimated from graphs in publication for nontoxic doses indole-3-carbinol (I3C) given prior to AFB$_1$ exposure, I3C given after exposure may increase tumorigenesis (Bailey et al., 1987).

[b]ip, Intraperitoneal.

[c]See Table 1 for carcinoma incidences in methyl-deficient and methyl-adequate rats.

rats and what the relationships are between the adduct peaks, cell division, and DNA repair.

Hepatocyte Hyperplasia

The induction of hepatocyte hyperplasia by methyl deficiency appears to be one likely mechanism by which AFB$_1$ and other chemical hepatocarcinogens act (Table 6). This hypothesis led to studies of the influence of partial hepatectomy on AFB$_1$ carcinogenesis in which conflicting results have been obtained. In rats given carcinogenic doses of AFB$_1$ by gastric gavage in a multiple dose regimen, partial hepatectomy, either before or after AFB$_1$ administration, had no significant effect on tumorigenesis, even when very small doses that resulted in low tumor incidences were given. AFB$_1$ is initially an inhibitor of cell division in the rat liver. A single dose of 100 μg given to 9- to 10-week-old F344 rats immediately or 14 hr after two-thirds hepatectomy completely suppressed the 24-hr peak in DNA synthesis. The same dose given 30 hr after hepatectomy reduced DNA synthesis and mitosis 48 hr after surgery to one-third the level in rats given no AFB$_1$. Administration of AFB$_1$ in divided doses totaling 50, 100, or 375 μg to rats beginning 6, 24, or 48 hr or 6 days after two-thirds hepatectomy induced the expected incidences of hepatocellular carcinoma, with no detectable effect of the time after hepatectomy at which AFB$_1$ was given. The times were chosen to look for effects of exposure before, or at the time of, the major peaks in DNA synthesis and mitosis following hepatectomy, or after cell division was essentially completed and the cell number, if not the full cell mass, had returned to normal. In a second experiment using the full dose of AFB$_1$ (375 μg) given 6 hr or 6 days after one-third rather than two-thirds hepatectomy, again no significant

TABLE 6   [³H]Thymidine-Labeled Hepatocytes in Control or Marginally or Severely Methyl-Deficient F344 Rats Given Aflatoxin B₁[a]

| | [³H]Thymidine-labeled hepatocytes[b] (%) | | |
| | Vehicle control[c] | Aflatoxin B₁ | |
Diet		25 μ × 3	25 μg × 14–15
Experiment₁[d]			
Control	0.05 ± 0.02[e]	0.82 ± 0.42	0.30 ± 0.08
Marginally deficient	0.40 ± 0.17	2.5 ± 1.3	1.4 ± 0.2
Experiment₂[f]			
Marginally deficient	0.06 ± 0.01	—	0.16 ± 0.04[g]
Severely deficient	2.4 ± 0.4–3.0 ± 0.5[h]	—	3.4 ± 0.9–6.9 ± 1.4[h]

[a]Data from Rogers and Newberne (1969, 1971).
[b]24 hr after last AFB₁ dose; not including hyperplastic enzyme-altered foci (EAF).
[c]Dimethylsulfoxide.
[d]Diets fed 3 wks (from age 4 wks) before AFB₁ administration.
[e]All values are mean ± S.E.
[f]Diets fed 2 months (from age 6 wks) before AFB₁ administration.
[g]Hyperplastic EAF had 15.6 ± 2.4% labeled hepatocytes.
[h]Fatty and nonfatty nodules, respectively; EAF were not distinguishable in these fatty cirrhotic livers.

effect was seen on tumorigenesis of the time after hepatectomy at which AFB₁ was given or of hepatectomy compared with sham surgery. These results could argue that hepatectomy had a negative effect on tumorigenesis because the dose of AFB₁ to the liver remnant was greater than the dose to the whole liver of the sham-operated controls. In Sprague–Dawley rats given a two-thirds hepatectomy 3 months after exposure to subcarcinogenic doses of AFB₁, a low incidence of tumors was found in the resected groups, but the incidence was not significantly greater than the zero incidence in nonresected groups. In all studies, the liver appeared fully restored despite AFB₁ exposure (Table 7; Rogers *et al.,* 1971).

In contrast, Rizvi *et al.,* (1989) reported a marked increase in hepatocellular carcinomas in rats that survived a single, large, intraperitoneal dose of AFB₁ (0.25 mg/kg) given 24 hr after partial hepatectomy compared with rats given the AFB₁ and no surgery (Table 7). Mortality from AFB₁ toxicity was 55% in the hepatectomy group and 20% in the controls. Similarly, male mice given a single, large, intraperitoneal dose of AFB₁ 46 hr after partial hepatectomy developed hepatocellular carcinomas whereas nonoperated controls did not (Table 7; Dix, 1984). Under these experiment conditions, the effective dose of AFB₁ to the liver remnant was highly toxic; the hepatectomy effect may be the result of a more marked and sustained increase in DNA synthesis than was induced by hepatectomy alone in the earlier studies or of different metabolism and kinetics of AFB₁ given in one large intraperitoneal dose rather than in multiple small intragastric doses. Neal and Cabral (1980) found toxicity of a single dose of AFB₁ adminis-

TABLE 7 Partial Hepatectomy and Aflatoxin B_1 Carcinogenesis in Male Rats and Mice

Strain	Species	Age[a] (wks)	AFB_1 regimen	Partial hepatectomy		Hepatocellular carcinoma (% incidence)	Reference
				%	Time[b]		
F344	Rat	8	$5\ \mu g/d \times 10,20^c$	67	6 hr	14	Rogers *et al.* (1971)
					24 hr	0	
					48 hr	17	
					6 d	10	
			$5\ \mu g/d \times 20 + 25\ \mu g/d \times 11^c$	67	6 hr	50	
					24 hr	80	
					48 hr	60	
					6 d	86	
F344	Rat	6	$5\ \mu g/d \times 5 + 10\ \mu g/d \times 10 + 25\ \mu g/d \times 10^c$	33	6 hr	$21^d,56^e$	Rivzi *et al.* (1989)
					6 d	$23^f,72^g$	
AS_2	Rat	—	$0.25\ mg/kg \times 1^l$	67	24 hr	91^i	
$STCF_1$	Mouse	9	$6\ \mu g/g \times 1^l$	67	46 hr	56^i	Dix (1984)
Sprague–Dawley	Rat	6	$25\ \mu g/d \times 5^c$	67	3 mo[k]	0^j	Rogers *et al.* (1971)
			$25\ \mu g/d \times 10^c$			8^j	
			$25\ \mu g/d \times 15^c$			10^j	

[a] At first AFB_1 dose.
[b] Before 1st AFB_1 dose, except k below.
[c] By gastric gavage.
[d] Incidence up to 10 mo; 13% of sham-operated rats had tumors at that time.
[e] Incidence at 14 mo; 43% of sham-operated rats had tumors at that time.
[f] Incidence up to 10 mo; no sham-operated rats had tumors at that time.
[g] Incidence at 14 mo; 81% of sham-operated rats had tumors at that time.
[h] By intraperitoneal injection.
[i] In rats given the same AFB_1 dose but no surgery, the incidence was 8%.
[j] No hepatocellular carcinomas were found in animals given the same AFB_1 dose but no surgery.
[k] Hepatectomy after AFB_1 exposure.

tered to rats before or shortly after partial hepatectomy increased compared with toxicity in sham-operated rats. These researchers could demonstrate no effect of the surgery on overall AFB_1 metabolism but did not examine specific metabolites or adducts. In that experiment, toxicity and early mortality precluded determination of carcinogenesis, but early histologic indicators of preneoplasia were increased in the hepatectomy group compared with the sham-operated group. Therefore, the partial hepatectomy studies indicate that carcinogenicity of a single dose of AFB_1 is enhanced by prior hepatectomy but that a multiple dose regimen is not enhanced. This conclusion does not support the idea that methyl deficiency-induced hyperplasia explains the effect of the deficiency on AFB_1 carcinogenesis, although the persistent hyperplasia in deficient rats may have an effect different from that of the short-term hyperplasia after partial hepatectomy.

In an early study (Rogers and Newberne, 1969), rats fed a severely methyl-deficient diet or the same diet supplemented with choline to make it borderline in methyl supply were given, in divided doses, a carcinogenic dose (375 μg) of AFB_1. The borderline-deficient rats had mildly fatty livers and showed marked development of hyperplastic, histologically abnormal EAF by the end of the 3-week period of AFB_1 administration, and rapid development of a high incidence of hepatocarcinomas. The severely deficient rats had cirrhosis, and many succumbed to the liver, kidney, and myocardial damage that is the result of the severe deficiency. The animals did not, however, show either the early hepatic abnormalities or the high incidence of hepatocarcinoma seen in the borderline-deficient animals. This result also weakens the hypothesis that hyperplasia determines the susceptibility of deficient hepatocytes to AFB_1 carcinogenicity, since the severely deficient liver was markedly more hyperplastic than the marginally deficient liver prior to and immediately after AFB_1 exposure (Experiment 2; Table 6). In rats fed severely methyl-deficient diets, cell division and cell turnover remain high throughout prolonged periods of feeding (Rogers and MacDonald, 1965; Chandar and Lombardi, 1988).

An interesting feature of hepatocarcinogenesis in either severely or borderline methyl-deficient rats is that, once the initiating events have taken place [whether induced by AFB_1 (Newberne et al., 1977), by a different carcinogen l-azaserine (Rogers et al., 1987), or by the deficient diet alone (Chandar and Lombardi, 1988)], return of the animals to a fully adequate diet appears to enhance tumorigenesis compared with continuation on the deficient diet. Therefore, the deficiency is effective in the early stages of carcinogenesis but may reduce the promotion and progression of tumors. This conclusion is consistent with the observation that phenobarbital reduces the diffuse hyperplasia induced by the deficiency but increases development of EAF in deficient rats (Shinozuka and Katyal, 1985). Phenobarbital may inhibit mitosis by blocking translocation of protein kinase C (PKC), an effect demonstrated in hepatocytes following exposure to phorbol ester (Brockenbrough et al., 1991). If initiated hepatocytes have a different response than uninitiated hepatocytes to growth factors acting through PKC, and if phenobarbital represses certain responses preferentially, then conditions may arise that favor the development of EAF from the initiated cells. Diacylglycerol (DAG) and activated PKC are increased early in liver of methyl-deficient rats (Blusztajn and Zeisel, 1989; Zeisel, 1990); their long-term responses and relationship to carcinogenesis are not known.

In summary, AFB_1 hepatocarcinogenesis induced by a multiple dose regimen is highly responsive to dietary methyl deficiency. The deficiency alone induces hepatocyte hyperplasia and exaggerates the hyperplasia induced by AFB_1, but the increased carcinogenic response is not correlated entirely with the hepatocyte hyperplasia induced by the deficiency. Multiple dose AFB_1 carcinogenesis is not responsive to hyperplasia induced immediately before carcinogen exposure by partial hepatectomy, but single high-dose carcinogenesis is responsive to partial hepatectomy.

AFB_1 Toxicity

Differences in response to single and multiple doses of AFB_1 are readily apparent in evaluation of AFB_1 toxicity in methyl-deficient and normal male rats. Deficiency renders the rat resistant to toxicity of a single large dose of AFB_1 (Table 8; Rogers and Newberne, 1971), but highly susceptible to toxicity of repeated small doses of AFB_1 (Rogers and Newberne, 1971; Rogers, 1975,1978). For example, mortality in deficient rats given AFB_1 (25 µg/day, intragastric, 14 doses) was 50% whereas in control rats mortality was 4% (Rogers and Newberne, 1971). Similar increases in toxicity of multiple doses of other hepatocarcinogens were observed in deficient rats. Mortality was 82% in deficient rats given aflatoxin G_1 (AFG_1, 50 µg for 9 doses over 3 weeks) whereas all controls given the same dose survived; 2-acetylaminofluorene (AAF) fed at 0.02% in the diet caused marked reduction in weight gain in deficient but not in control rats; ethionine fed at 0.1% in the diet caused a loss of 7 g body weight in deficient rats, whereas controls fed the same level gained 50 g (Rogers, 1975). The explanation for the differences in response is not known.

Factors that determine AFB_1 toxicity and carcinogenicity include xenobiotic

TABLE 8   Toxicity of a Single Dose of Aflatoxin B_1 in Male Rats Fed Methyl-Deficient Diets[a,b]

Diet	Strain	Route of administration[c]	Mortality (%)
Control	Sprague–Dawley	ig	60
		ip	100
Methyl-deficient	Sprague–Dawley	ig, ip	0[d]
Control	F344	ig	100
Methyl-deficient	F344	ig	0

[a]Data from Rogers and Newberne (1971).
[b]Dose was 7 mg AFB_1/kg.
[c]ig, Intragastric; ip, intraperitoneal.
[d]Also no deaths at 9 mg/kg.

metabolic and conjugating pathways that interact with tissue and cell factors. A lack of correlation between AFB_1 toxicity and carcinogenicity, perhaps related to the differences between single and multiple dose toxicity in methyl-deficient rats, can be found in studies of nutrient or dietary effects on AFB_1 mutagenicity, teratogenicity, and other evidence of genetic damage. Failure of microbial mutagenesis assays to mirror carcinogenesis in dietary studies was discussed earlier. Investigation of AFB_1 effects on rat embryos *in vitro* showed that activation of AFB_1 by male rat liver S9 induced by phenobarbital or 3-methylcholanthrene increased embryolethality but did not change, either quantitatively or qualitatively, the induction of dysmorphogenesis compared with effects of exposure to AFB_1 without the exogenous metabolizing system. The authors concluded that the embryolethal and the dysmorphogenic effects of AFB_1 may be induced by different AFB_1 metabolites, and that exogenous biotransformation altered only the level of embryolethal mediator(s) (Geissler and Faustman, 1988). The same authors reviewed other studies of *in vitro* bacterial mutagenesis, AFB_1–DNA adduction, and AFB_1 inhibition of RNA synthesis that demonstrated greater effects using phenobarbital-induced rat hepatic metabolizing systems than using uninduced systems, and concluded that the genotoxic effects of AFB_1 might be induced by the embryolethal metabolite(s) whereas the dysmorphogenesis was induced by different metabolite(s).

In another test system, when whole hepatocytes from butylated hydroxyanisole (BHA)-treated rats were used to metabolize AFB_1 in cultures of V79 (hamster lung) cells, mutation frequencies were reduced but indicators of chromosome damage (sister chromatid exchange, micronuclei) were not altered compared with cultures in which uninduced hepatocytes were used. Biochemical assays showed that glutathione *S*-transferases and glutathione (GSH) both were increased in the BHA-induced hepatocytes whereas cytochrome P450 was not (Rogers *et al.*, 1990). The authors reviewed earlier studies demonstrating BHA induction of glutathione *S*-transferase and reduction *in vivo* and *in vitro* of formation of AFB_1–DNA adducts, presumably the result of increased conjugation of AFB_1-epoxide with GSH. Sulfhydryl binding and detoxification of AFB_1 metabolites has been exploited in chemoprevention studies using synthetic food antioxidants such as BHA and ethoxyquin and the drug oltipraz, all compounds that increase tissue GSH, glutathione *S*-transferase, or both (Cabral and Neal, 1983; Ansher *et al.*, 1986; Kensler *et al.*, 1986, 1987; Williams *et al.*, 1986; Liu *et al.*, 1988; Roebuck *et al.*, 1991). GSH is reduced in methyl-deficient rats only if the cysteine supplementation of the diet is reduced (Rogers *et al.*, 1987) and is presumably not responsible for the changes in AFB_1 toxicity in the studies cited earlier.

Other Considerations

The existence and importance of interactions between hepatocarcinogens, nutrients responsible for methyl supply and metabolism, and tumor development are suggested by several additional observations. Evidence from several sources suggests that hepatocarcinogenic chemicals induce methyl deficiency, possibly by interfering with methylation and folate pathways (Rogers, 1975; Poirier, 1986; Rogers et al., 1991). The cancer chemotherapeutic agent procarbazine (PCZ), which is carcinogenic for mammary gland and hematopoietic tissue in rats, seriously disrupts hepatic choline metabolism. Mammary gland carcinogenesis by PCZ is enhanced in methyl-deficient rats (Rogers et al., 1990).

Elucidating interactions of carcinogens and methyl group metabolism in vitro may be possible. In vitro resistance of hepatocytes to toxicity of AFB_1 is increased by prior exposure of animals to AFB_1 or to methotrexate (MTX), which blocks methyl metabolism in the folate pathways; feeding a methyl-deficient diet for 3 months before taking the hepatocytes for culture induced a low level of resistance to toxicity of MTX but not to AFB_1 or any other drug tested (Carr, 1987).

The methyl donor nutrients discussed are important also in phospholipid and protein synthesis, synthesis and maintenance of cell membranes, sulfhydryl metabolism, and other cellular and metabolic reactions. The complexity of their interactions makes it difficult to study any one nutrient in isolation. Mechanisms not yet studied may exist by which these compounds influence carcinogenesis.

DIETARY PROTEIN DEFICIENCY

Protein deficiency reduces AFB_1 induction of EAF and hepatocarcinogenesis in rats (Table 9). In toxicity and tumorigenesis studies in rats that were severely protein deficient (5% casein) throughout the study, multiple dose AFB_1 toxicity was increased but carcinogenicity and weight gain were decreased compared with the control rats (Madhaven and Gopalan, 1965, 1968). Preston et al. (1976) reported reduced AFB_1 binding to hepatic DNA in similarly protein-deficient (5% casein) Sprague–Dawley rats given a single dose of labeled AFB_1. Effects of dietary protein on binding of multiple doses used in the carcinogenic regimens have not been reported. In the same laboratory, development of EAF after exposure to multiple doses of AFB_1 was found to be reduced in Fischer 344 rats fed 5% casein compared with rats fed 20% casein. However, the reduction appeared to be attributable largely to postinitiation protein deficiency rather than to deficiency during AFB_1 exposure; its relationship to AFB_1 DNA binding is not clear (Appleton and Campbell, 1983). Mirmomeni et al. (1979) fed AFB_1 in 25 or 12% casein diet to Sprague–Dawley rats for 6 months and reported significantly less AFB_1 toxicity, as measured by serum enzymes, as well as reduced development of hyperplastic nodules in the rats fed 12% casein. Similarly, Dunaif and Campbell (1987), studying closely graded dietary levels of casein fed only after exposure to AFB_1, showed a marked reduction in the number of EAF and the volume of liver they occupied in rats fed 8–10% casein, diets that did not result in significantly reduced weight gain although they were low to borderline in protein, compared with foci in rats fed 20% casein. Foci similarly were reduced in

TABLE 9   Aflatoxin B_1 Carcinogenesis in Rats Fed Protein-Deficient Diets

| Rats | | Total AFB$_1$ dose (mg/kg), route[a] | Dietary casein (%) | Liver lesions | | Reference |
Strain	Sex			EAF (No. per cm^3)	Hepatocellular carcinoma (%)	
—	M,F	0.44–2.35,ig	20	—	37	Madhaven and
			5		0[b]	Gopalan (1968)
Sprague–	M	2,diet,6mo	25	many	—	Mirmomeni et al.
Dawley			12	none		(1979)
F344	M	2.5,ig	20	96	—	Appleton and
			5[b,c]	9[b]		Campbell
						(1983)
F344	M	2.5,ig	20	120	—	Dunaif and Camp-
			12[c]	66[d]		bell (1987)
			10[c]	21[d]		
			4[c]	31[b,d]		

[a]ig, Intragastric.
[b]Body weight was significantly lower than in controls fed 20% casein.
[c]Fed after AFB$_1$ administration completed; 20% casein fed before and during AFB$_1$ administration in all groups.
[d]Significantly lower than 20% casein group.

more severely deficient rats fed 4–6% casein. The authors postulated that reduced hepatocyte proliferation in deficient rats might account for the reduction in development of foci (Dunaif and Campbell, 1987). The reduction in EAF occurs at about the same dietary protein level as the reduction in feed efficiency ratio (Horio et al., 1991). The influence of dietary protein on AFB$_1$ carcinogenesis itself, rather than on induction of EAF, and the timing and mechanisms of the protein influence are promising areas for further investigation.

DIETARY FAT

The effect of dietary fat on AFB$_1$ carcinogenesis has been studied. Interesting and opposite effects on toxicity and carcinogenicity have been reported. In Sprague–Dawley rats fed diets either moderate (15% by wt) or high (30% by wt) in corn oil or beef fat content, lethality of a single doses of AFB$_1$ was increased markedly by beef fat fed at either level compared with corn oil but was not influenced by the amount of fat (Newberne et al., 1979; Table 10). Similarly, in male Wistar rats, mortality after a single AFB$_1$ dose was doubled by feeding 28% beef fat and 2% corn oil compared with 30% corn oil (Nyandieka, 1987; Table 10). In contrast, liver tumor incidence after multiple dose of AFB$_1$ was increased in Sprague–Dawley rats fed high corn oil compared with rats fed high beef fat, and EAF were increased in rats fed high corn oil compared with rats fed lower

**TABLE 10   Toxicity of a Single Dose of Aflatoxin B₁ in Male Rats
Fed High Fat Diets**[a]

Diet	Strain	Route of administration[b]	Mortality (%)	Reference
15% corn oil	Sprague–Dawley	ig	25	Newberne et al.
13% beef fat + 2% corn oil	Sprague–Dawley	ig	84	(1979)
30% corn oil	Sprague–Dawley	ig	21	
28% beef fat + 2% corn oil	Sprague–Dawley	ig	63	
30% corn oil	Wistar	ig	20	Nyandieka (1987)
28% beef fat + 2% corn oil	Wistar	ig	49	

[a]Dose was 7 mg AFB₁/kg.
[b]ig, Intragastric.

concentrations of corn oil (Table 11, Newberne *et al.*, 1979; Misslbeck *et al.*, 1984; Baldwin and Parker, 1987). An apparent suppression of AFB₁ hepatocarcinogenesis by beef fat was reported also in studies investigating methyldeficient diets (Rogers and Newberne, 1981).

In vivo bacterial mutagenesis by AFB₁ in mice was reduced by feeding 1% safflower oil plus 25% (by wt) olive oil or beef fat compared with mice fed 1% safflower oil as the only fat. Ability of an hepatic S9 fraction from the mice to alter AFB₁ mutagenesis *in vitro* was not altered by either high fat diet (Table 4; Brennan-Craddock *et al.*, 1990).

VITAMINS AND MINERALS IN THE DIET

Vitamin A

Vitamin A, presented as retinal or retinol, inhibits hepatic microsome-mediated AFB₁ bacterial mutagenesis and *in vitro* adduction to DNA; folic acid inhibits mutagenesis, and β-carotene inhibits adduction (Bhattacharya *et al.*, 1984,1987; Tables 4, 12). In an early study of AFB₁ carcinogenesis in rats, deficiency of vitamin A increased AFB₁ induction of colon tumors but had no effect on liver tumors; hypersupplementation with vitamin A had no effect on AFB₁ carcinogenesis at either site (Newberne and Rogers, 1973b). Subsequently the increase in colon carcinogenesis in deficient rats was confirmed, as was the failure of increased dietary vitamin A to protect against colon or hepatocarcinogenesis (Newberne and Suphakarn, 1977,1983). The influence of folate alone on AFB₁ carcinogenesis is not known, but it is an important component in consideration of effects of methyl deficiency.

Trace Minerals

Selenium (Se) in the diet influences AFB₁ carcinogenesis. Lei *et al.* (1990) found that 3 ppm Se in drinking water reduced AFB₁-induced EAF and carcino-

TABLE 11 Dietary Fat and Aflatoxin B$_1$ Carcinogenesis in Male Sprague–Dawley Rats

Total intragastric dose of AFB$_1$	Dietary fat (%, Type)	Liver lesions (% incidence)		Reference
		Enzyme-altered foci	Hepatocellular carcinoma	
375 μg	28, Beef fat + 2, corn oil	—	53 (51)[a]	Newberne et al. (1979)
	30, Corn oil		100 (60)[a]	
3.8 mg/kg	20, Corn oil[a]	Increased	—	Misslbeck et al. (1984)
	6, Corn oil			
4 mg/kg	20, Corn oil[a,b]	100		Baldwin and Parker
	2, Corn oil	77		(1987)

[a]Diets fed only after AFB$_1$ administration.
[b]High corn oil diet fed during initiation followed by phenobarbital during promotion significantly increased the volume of EAF in the liver.

genesis in rats without inducing evidence of Se toxicity; Se at 6 ppm also was effective but was toxic (Table 13). The absence or reduction of an anticarcinogenic effect of SE at higher toxic doses was reported earlier by Newberne et al. (1987), who found protection by levels of 1–2 ppm Se in drinking water compared with either lower levels (0.05, 0.10 ppm) or higher levels (3.5, 5 ppm; Table 13). Using EAF as the end point, Milks et al. (1985) reported that either 2 or 5 ppm Se in drinking water at the time of administration of a single dose of AFB$_1$ reduced the number of EAF in Sprague–Dawley rats given a promoting regimen of phenobarbital and partial hepatectomy. A Se-deficient diet, fed during AFB$_1$ administration in conjunction with phenobarbital fed after AFB$_1$ exposure,

TABLE 12 Selected Diet Components that Influence AFB$_1$ Adduction to DNA in Vitro

Diet component	Test system	Inihibition of AFB$_1$–DNA adducts (%)	Reference
Ellagic acid	Rat trachea explant	9–57	Mandal et al. (1987)
	Human tracheobronchial explant	24–79	
BHA, BHT	DNA and rat microsomes	80–98	Bhattacharya et al. (1984)
Benzoflavones	DNA and rat microsomes	65–74	
Caffeic acid	DNA and rat microsomes	63	
β-Carotene	DNA and rat microsomes	51	
Copper^{2+}	DNA and rat microsomes	92	
Retinal, retinol	DNA and rat microsomes	90	
Selenite	DNA and rat microsomes	58	
Capsaicin	DNA and rat microsomes	19–71	Teel (1991)
Garlic extracts	DNA and rat microsomes	45–76	Tadi et al. (1991)

TABLE 13   AFB$_1$ Carcinogenesis and Selenium Intake in Male Rats

Se (ppm in water)	Rat strain	Total AFB$_1$ dose (mg)	Reduction by Se of number of enzyme-altered foci (%)	Hepatocellular carcinoma (% incidence)	Reference
0.05, 0.10	Sprague–Dawley	0.375	—	42, 55	Newberne et al.
1.00, 2.00			54	16, 22	(1987)
3.5, 5.00[a]			None	46, 50	
0	Wistar	2.3–3.4	—	61	Lei et al.
3			79	0	(1990)
6[a]			61	0	

[a]Toxic as judged by weight gain, clinical appearance, or liver histology.

was associated with evidence of increased EAF compared with a Se-sufficient diet. If the Se-deficient diet was fed after AFB$_1$ administration, no significant increase in foci was seen (Baldwin and Parker, 1987).

Less extensive investigation has been done of effects of trace minerals other than Se on aflatoxin toxicity and carcinogenicity. Syrian hamsters fed mixed aflatoxins (19–37 ppm), of which 13% was AFB$_1$ and 84% was AFG$_1$, and manganese (Mn; 1149 ppm) in a natural product diet for 10 weeks had less histological evidence of aflatoxin-induced changes in the liver than hamsters fed aflatoxin and Mn (51 ppm) in the diet (Hastings and Llewellyn, 1987). Many metal salts were tested in vitro against AFB$_1$ in the Ames assay and showed significant inhibition of mutagenesis. Copper (Cu^{2+}) was the most active, but several others, including Mn and Se, had significant activity (Francis et al., 1988; Table 4). Copper and selenium as well as nonnutrient food components discussed subsequently also reduce AFB$_1$ adduction to DNA in vitro (Bhattacharya et al., 1984; Mandal et al., 1987; Teel, 1991; Table 12).

ETHANOL

Ethanol as a carcinogen or cocarcinogen for the liver is, and has been, a subject of great interest because risk for hepatocellular carcinoma in the United States is associated with alcohol intake. Many studies have been performed using several carcinogens, but no clear experimental demonstration has been made that ethanol increased hepatocarcinogenesis (reviewed by Rogers and Conner, 1986). Several studies have been done to evaluate AFB$_1$–ethanol interactions. Misslbeck et al. (1984) fed 35% of calories as ethanol in a fully nutritious diets to rats after they had been given a carcinogenic dose of AFB$_1$ and found no effect of ethanol on numbers of EAF. Toskulkao, Glinsukon, and co-workers (1986, 1988,1990) reported increased hepatotoxicity of single doses of AFB$_1$ of 0.75–2 mg/kg in Wistar male rats pretreated with 4 doses of ethanol of 4 g/kg between 48 and 21 hr earlier. No long-term studies were performed. Rats given ethanol in drinking water for 10 days, or given a single ethanol dose, and then given a single dose of labeled AFB_1 were studied for formation of hepatic AFB_1–DNA adducts. Alcohol by either regimen had no effect on the level of adduction (Marinovich and Lutz, 1985).

NONNUTRIENT FOOD COMPONENTS

Nonnutrient food components, present naturally in or added to the diet, influence AFB_1 mutagenicity, DNA adduction, and carcinogenicity. Inhibition of microbial mutagenesis by microsomes from and of hepatocarcinogenesis in, animals fed large amounts of whole plant products have been reported (Stoewsand *et al.*, 1978; Boyd *et al.*, 1982) and have led to studies in which purified food components have been studied (Tables 4, 5, 12). I3C, a component of cruciferous vegetables that inhibits DNA adduction and carcinogenesis by polycyclic aromatic hydrocarbons (PAH) in laboratory rodents, inhibited AFB_1–DNA adduction and AFB_1-induced hepatocarcinogenesis in rainbow trout. When I3C was given in a nontoxic dose before and during AFB_1 exposure, the level of hepatic AFB_1–DNA adducts correlated well with subsequent tumor incidence; the effect was characterized as "pure anti-initiating activity" (Nixon *et al.*, 1984; Dashwood *et al.*, 1989). If I3C is given after AFB_1 exposure, however, an opposite, possibly slightly weaker, effect is seen and tumorigenesis is increased (Bailey *et al.*, 1987; Dashwood *et al.*, 1990,1991).

The quantitative relationships between I3C dose and inhibition and promotion of AFB_1 carcinogenesis in trout have been investigated intensively. Enhancement of tumorigenesis in trout by feeding I3C after AFB_1 exposure is dose responsive, can be demonstrated by feeding I3C after delays of as long as 12 and possibly 24 weeks after AFB_1 exposure, and does not require continuous exposure to I3C. It does require at least 12 weeks of exposure at the dose tested (2000 ppm). The final outcome of exposure to I3C in the diet would be expected to depend on other dietary components as well as on dose and timing of I3C ingestion (Dashwood *et al.*, 1991).

Ellagic acid, a phenolic component of fruits (strawberries, grapes) and nuts, inhibits bacterial and mammalian cell mutagenesis and DNA adduct formation by activated benz[*a*]pyrene (B[*a*]P) *in vitro* and, in some but not all studies, inhibits B[*a*]P-induced tumorigenesis *in vivo*. Ellagic acid is effective against the mutagenicity of other PAHs as well. Ellagic acid inhibits B[*a*]P metabolism in epidermal cells *in vitro* and reduces hepatic and pulmonary xenobiotic metabolism and conjugation in the mouse (Mandal *et al.*, 1987). In investigations of interactions of ellagic acid with AFB_1 bacterial mutagenesis, both inhibition (Mandal *et al.*, 1987) and no evidence of interaction (Francis *et al.*, 1989b) have been reported using similar test systems (Table 4). However, bone marrow cell chromosome aberrations induced in rats by AFB_1 were reduced if the rats were given prior gastric gavage with ellagic acid or with green tea extract that contained ellagic acid. Caffeine, also present in green tea extract, reduced the aberrations as well (Ito *et al.,* 1989). Plant flavonoids, present in fruits and vegetables, have among their effects both anticarcinogenic and antimutagenic properties. In bacterial mutagenesis, they are active against AFB_1 as well as PAHs and AAF, probably by interfering with microsomal enzyme activation (Francis *et al.,* 1989a).

Garlic extracts, which have proven to be strongly anticarcinogenic in the gastrointestinal tract exposed to nitrosamine or hydrazine carcinogenic compounds (Wargovich *et al.,* 1988), reduced AFB_1 microbial mutagenesis, altered *in vitro* metabolism and conjugation of AFB_1, and reduced AFB_1–DNA adducts formed *in vitro* (Tadi *et al.,* 1991; Table 12). Protection by the extracts against AFB_1 carcinogenesis has not been reported.

Capsaicin, a component of peppers, reduced *in vitro* formation of AFB_1–DNA adducts; its effects on carcinogenesis are unknown (Teel, 1991; Table 12).

FEED RESTRICTION

Feed restriction reduces AFB_1 hepatocarcinogenesis in rats (Newberne and Rogers, 1986) and prevents the induction of glutathione *S*-transferase by AFB_1 that is found in full-fed control rats (Rajpurohit and Krishnasnamy, 1988).

METHYL-SUPPLEMENTED DIET

Methyl supplementation of a nutritionally complete diet has been reported to reduce AFB_1 carcinogenesis in male B6C3F1 mice (Newberne *et al.,* 1990) and long has been known to suppress ethionine (Farber and Ichinose, 1958) and AAF (Miller and Miller, 1972) hepatocarcinogenesis in rats. In contrast, methyl supplementation of a complete diet did not reduce hepatocellular carcinoma induction in rats by DENA plus phenobarbital or DDT, although the supplementation did prevent the reduction of hepatic *S*-adenosylmethionine induced by the two promoters (Shivapurkar *et al.,* 1986). In mice, methyl supplementation of a complete diet had no effect on spontaneous liver tumor incidence or on tumor induction by DENA. However, the supplementation did reduce malignant tumors in mice given phenobarbital with or without DENA. Liver adenomas were increased in the same groups, so total liver tumors were not reduced significantly (Fullerton *et al.,* 1990).

Increasing the supply of *S*-adenosylmethionine (Adomet) by parenteral injection of Adomet itself after carcinogen exposure can reduce hepatocarcinogenesis by DENA and phenobarbital (Feo *et al.,* 1988; Garcea *et al.,* 1989a,b); Adomet has not been tested against AFB_1.

CONCLUSION

Aflatoxin carcinogenesis in laboratory animals is susceptible to dietary and nutritional modulation. The earliest and probably the most extensively studied modulator is the group of nutrients central to methyl metabolism (choline, methionine, folate). Deficiency induces a marked increase in the potency of AFB_1 as a hepatocarcinogen in rats; hypersupplementation with the nutrients reduces AFB_1 carcinogenicity in mice. The mechanism by which the deficiency acts is not known, but several potential mechanisms have been demonstrated.

The other major nutritional effects known are inhibition of AFB_1 hepatocarcinogenesis by moderate selenium supplementation and increased AFB_1 colon carcinogenicity (but not hepatocarcinogenicity) in vitamin A-deficient rats. In addition, nonnutrient diet components have potentially important effects, particularly indole-3-carbinol which can either inhibit or enhance AFB_1 carcinogenicity depending on the time at which it is fed. *In vitro* studies using measures of genetic toxicity of AFB_1 suggest that other nutrient or nonnutrient diet components may be effective against AFB_1 carcinogenesis, but such effects have not yet been demonstrated. Clearly, many opportunities for research exist in this area that can lead to important mechanistic explanations and ultimately to public health measures in aflatoxin-exposed populations.

REFERENCES

Angsubhakorn, S., Get-Ngern, P., Miyamoto, M., and Bhamarapravati, N. (1990). A single dose–response effect of aflatoxin B_1 on rapid liver cancer induction in two strains of rats. *Int. J. Cancer* **46,** 664–668.

Ansher, S. S., Bueding, E., and Dolan, P. (1986). Biochemical effects of dithiolthiones. *Food Chem. Toxicol.,* **24,** 405–415.

Appleton, B. S., and Campbell, T. C. (1983). Effect of high and low dietary protein on the dosing and postdosing periods of aflatoxin B_1-induced hepatic preneoplastic lesion development in the rat. *Cancer Res.* **43,** 2150–2154.

Bailey, G. S., Hendricks, J. D., Shelton, D. W., Nixon, J. E., and Pawlowski, M. E. (1987). Enhancement of carcinogenesis by the natural anticarcinogen indole-3-carbinol. *J. Natl. Cancer Inst.* **78,** 931–934.

Baldwin, S., and Parker, R. S. (1987). Influence of dietary fat and selenium in initiation and promotion of aflatoxin B_1-induced preneoplastic foci in rat liver. *Carcinogenesis* **8,** 101–107.

Bechtel, D. H. (1989). Molecular dosimetry of hepatic aflatoxin B_1–DNA adducts: Linear correlation with hepatic cancer risk. *Reg. Toxicol. Pharmacol.* **10,** 74–81.

Bhattacharya, R. K., Aboobaker, V. S., and Firozi, P. F. (1984). Factors modulating the formation of DNA adduct by aflatoxin B_1 *in vitro. Carcinogenesis* **5,** 1359–1362.

Bhattacharya, R. K., Francis, A. R., and Shetty, T. K. (1987). Modifying role of dietary factors on the mutagenicity of aflatoxin B_1: *In vitro* effect of vitamins. *Mutation Res.* **188,** 121–128.

Blusztajn, J. K., and Zeisel, S. H. (1989). 1,2-*sn*-Diacylglycerol accumulates in choline-deficient liver. A possible mechanism of hepatic carcinogenesis via alteration in protein kinase C activity? *FEBS Lett.* **243,** 267–270.

Boyd, J. N., Babish, J. G., and Stoewsand, G. S. (1982). Modification by beet and cabbage diets of aflatoxin B₁-induced rat plasma beta-foetoprotein elevation, hepatic tumorigenesis, and mutagenicity of urine. *Food Chem. Toxicol.* **20**, 47.

Brennan-Craddock, W. E., Coutts, T. M., Rowland, I. R., and Alldrick, A. J. (1990). Dietary fat modifies the *in vivo* mutagenicity of some food-borne carcinogens. *Mutation Res.* **230**, 49–54.

Brockenbrough, J. S., Meyer, S. A., Li, C., and Jirtle, R. (1991). Reversible and phorbol ester-specific defect of protein kinase C translocation in hepatocytes isolated from phenobarbital-treated rats. *Cancer Res.* **51**, 130–136.

Cabral, J. R. P., and Neal, G. E. (1983). The inhibitory effects of ethoxyquin on the carcinogenic action of aflatoxin B₁ in rats. *Cancer Lett.* **19**, 125–132.

Carr, B. I. (1987). Pleiotropic drug resistance in hepatocytes induced by carcinogens administered to rats. *Cancer Res.* **47**, 5577–5583.

Chandar, N., and Lombardi, B. (1988). Liver cell proliferation and incidence of hepatocellular carcinomas in rats fed consecutively a choline-devoid and a choline-supplemented diet. *Carcinogenesis* **9**, 259–263.

Chandar, N., Amenta, J., Kandala, J. C., and Lombardi, B. (1987). Liver cell turnover in rats fed a choline-devoid diet. *Carcinogenesis* **8**, 669–673.

Copeland, D. H., and Salmon, W. D. (1946). The occurrence of neoplasms in the liver, lungs and other tissues of rats as a result of prolonged choline deficiency. *Am. J. Pathol.* **22**, 1059–1067.

Dashwood, R. H., Arbogast, D. N., Fong, A. T., Pereira, C., Hendricks, J. D., and Bailey, G. S. (1989). Quantitative inter-relationships between aflatoxin B₁ carcinogen dose, indole-3-carbinol anti-carcinogen dose, target organ DNA adduction, and final tumor response. *Carcinogenesis* **10**, 175–181.

Dashwood, R. H., Fong, A. T., Hendricks, J. D., and Bailey, G. S. (1990). Tumor dose–response studies with aflatoxin B₁ and the ambivalent modulator indole-3-carbinol: Inhibitory versus promotional potency. *Basic Life Sci.* **52**, 361–365.

Dashwood, R. H., Fong, A. T., Williams, D. E., Hendricks, J. D., and Bailey, G. S. (1991). Promotion of aflatoxin B₁ carcinogenesis by the natural tumor modulator indole-3-carbinol: Influence of dose, duration, and intermittent exposure on indole-3-carbinol promotional potency. *Cancer Res.* **51**, 2362–2365.

Dix, K. M. (1984). The development of hepatocellular tumours following aflatoxin B₁ exposure of the partially hepatectomized mouse. *Carcinogenesis* **5**, 385–390.

Dunaif, G. E., and Campbell, T. C., (1987). Dietary protein level and aflatoxin B₁-induced preneoplastic hepatic lesions in the rat. *J. Nutr.* **117**, 1298–1302.

Farber, E., and Ichinose, H. (1958). The prevention of ethionine-induced carcinoma of the liver in rats by methionine. *Cancer Res.* **18**, 1209–1213.

Feo, F., Garcea, R., Daino, L., Pascale, R., Frassetto, S., Cozzolino, P., Vannini, M. G., Ruggiu, M. E., Simile, M. M., and Puddu, M. (1988). S-Adenosylmethionine antipromotion and antiprogression effect in hepatocarcinogenesis. Its association with DNA methylation and inhibition of gene expression. *In* "Chemical Carcinogenesis. Models and Mechanisms" (F. Feo, P. Pani, A. Columbano, and R. Garcea, eds.), pp. 407–423. Plenum Publishing, New York.

Francis, A. R., Shetty, T. K., and Bhattacharya, R. K. (1988). Modifying role of dietary factors on the mutagenicity of aflatoxin B₁: *In vitro* effect of trace elements. *Mutation Res.* **199**, 85–93.

Francis, A. R., Shetty, T. K., and Bhattacharya, R. K. (1989a). Modifying role of dietary factors on the mutagenicity of aflatoxin B₁: *In vitro* effect of plant flavonoids. *Mutation Res.* **222**, 393–401.

Francis, A. R., Shetty, T. K., and Bhattacharya, R. K. (1989b). Modification of the mutagenicity of aflatoxin B₁ and N-methyl-N′-nitro-N-nitrosoguanidine by certain phenolic compounds. *Cancer Lett.* **45**, 177–182.

Fullerton, F. R., Hoover, K., Mikol, Y. B., Creasia, D. A., and Poirier, L. A. (1990). The inhibition by methionine and choline of liver carcinoma formation in male C3H mice dosed with diethylnitrosamine and fed phenobarbital. *Carcinogenesis* **11**, 1301–1305.

Garcea, R., Daino, L., Pascale, R., Simile, M. M., Puddu, M., Frassetto, S., Cozzolino, P., Seddaiu,

M. A., Gaspa, L., and Feo, F. (1989a). Inhibition of promotion and persistent nodule growth by *S*-adenosyl-L-methionine in rat liver carcinogenesis: Role of remodeling and apoptosis. *Cancer Res.* **49**, 1850–1856.

Garcea, R., Daino, L., Pascale, R., Simile, M. M., Puddu, M., Ruggiu, M. E., Seddaiu, M. A., Satta, G., Sequenza, M. J., and Feo, F. (1989b). Protooncogene methylation and expression in regenerating liver and preneoplastic liver nodules induced in the rat by diethylnitrosamine: Effect of variations of *S*-adenosylmethionine:*S*-adenosylhomocysteine ratio. *Carcinogenesis* **10**, 1183–1192.

Geissler, F., and Faustman, E. M. (1988). Developmental toxicity of aflatoxin B_1 in the rodent embryo *in vitro:* Contribution of exogenous biotransformation systems to toxicity. *Teratology* **37**, 101–111.

Gupta, R. C., Barley, K., Locker, J., and Lombardi, B. (1987). ^{32}P-Postlabeling analysis of liver DNA adducts in rats chronically fed a choline-devoid diet. *Carcinogenesis* **8**, 187–9.

Hastings, C. E., Jr., and Llewellyn, G. C. (1987). Reduced aflatoxicosis in livers of hamsters fed a manganese sulfate supplement. *Nutr. Cancer* **10**, 67–77.

Horio, F., Youngman, L. D., Bell, R. C., and Campbell, T. C. (1991). Thermogenesis, low-protein diets, and decreased development of AFB_1-induced preneoplastic foci in rat liver. *Nutr. Cancer* **16**, 31–41.

Ito, Y., Ohnishi, S., and Fujie, K. (1989). Chromosome aberrations induced by aflatoxin B_1 in rat bone marrow cells *in vivo* and their suppression by green tea. *Mutation Res.* **222**, 253–261.

James, S. J., and Yin, L. (1989). Diet-induced DNA damage and altered nucleotide metabolism in lymphocytes from methyl-donor-deficient rats. *Carcinogenesis* **10**, 1209–1214.

Kensler, T. W., Egner, P. A., Davidson, N. E., Roebuck, B. D., Pikul, A., and Groopman, J. D. (1986). Modulation of aflatoxin metabolism, aflatoxin-N^7-guanine formation, and hepatic tumorigenesis in rats fed ethoxyquin: Role of induction of glutathione *S*-transferases. *Cancer Res.* **46**, 3924–3931.

Kensler, T. W., Egner, P. A., Dolan, P. M., Groopman, J. D., and Roebuck, B. D. (1987). Mechanisms of protection against aflatoxin tumorigenicity in rats fed 5-(2-pyrazinyl)-4-methyl-1,2-dithiol-3-thione (oltipraz) and related 1,2-dithiol-3-thiones and 1,2-dithiol-3-ones. *Cancer Res.* **47**, 4271–4277.

Lei, D. N., Wang, L. Q., Ruebner, B. H., Hsieh, D. P. H., Wu, B. F., Zhu, C. R., and Du, M. J. (1990). Effect of selenium on aflatoxin hepatocarcinogenesis in the rat. *Biomed. Environ. Sci.* **3**, 65–80.

Li, D., Xu, D., Chandar, N., Lombardi, B., and Randerath, K. (1990). Persistent reduction of indigenous DNA modification (I-compound) levels in liver DNA from male Fischer rats fed choline devoid diet and in DNA of resulting neoplasms. *Cancer Res.* **50**, 7577–7580.

Liu, Y. L., Roebuck, B. D., Yager, J. D., Groopman, J. D., and Kensler, T. W. (1988). Protection by 5-(2-pyrazinyl)-4-methyl-1,2-dithiol-3-thione (oltipraz) against the hepatotoxicity of aflatoxin B_1 in the rat. *Toxicol. Appl. Pharmacol.* **93**, 442–451.

Madhavan, T. V., and Gopalan, C. (1965). Effect of dietary protein aflatoxin liver injury in weanling rats. *Arch. Pathol.* **80**, 123–126.

Madhavan, T. V., and Gopalan, C. (1968). The effect of dietary protein on carcinogenesis of aflatoxin. *Arch. Pathol.* **85**, 133–137.

Mandal, S., Ahuja, A., Shivapurkar, N. M., Cheng, S. J., Groopman, J. D., and Stoner, G. D. (1987). Inhibition of aflatoxin B_1 mutagenesis in *Salmonella typhimurium* and DNA damager in cultured rat and human tracheobronchial tissues by ellagic acid. *Carcinogenesis* **8**, 1651–1656.

Marinovich, M., and Lutz, W. K. (1985). Covalent binding of aflatoxin B_1 to liver DNA in rats pretreated with ethanol. *Experientia* **41**, 1338–1340.

Milks, M. M., Wilt, S. R., Ali, I. I., and Couri, D. (1985). The effects of selenium on the emergence of aflatoxin B_1-induced enzyme-altered foci in rat liver. *Fund. Appl. Toxicol.* **5**, 320–326.

Miller, E. C., and Miller, J. A. (1972). Approaches to the mechanisms and control of chemical carcinogenesis. Bertner Foundation Award Lecture in Environment and Cancer, 24th Annual Symposium. pp. 5–39. Williams and Wilkins, Baltimore.

Mirmomeni, M. H., Suzangar, M., Wise, A., Messripour, M., and Emami, H. (1979). Biochemical studies during aflatoxin B_1-induced liver damage in rats fed different levels of dietary protein. *Int. J. Cancer* **24**, 471–476.

Misslbeck, N. G., Campbell, T. C., and Roe, D. A. (1984). Effect of ethanol consumed in combination with high or low fat diets on the postinitiation phase of hepatocarcinogenesis in the rat. *J. Nutr.* **114**, 2311–2323.

Murray, M., Zaluzny, L., and Farrell, G. C. (1986). Drug metabolism in cirrhosis. Selective changes in cytochrome P-450 isozymes in the choline-deficient rat model. *Biochem. Pharmacol.* **35**, 1817–1824.

Neal, G. E., and Cabral, J. R. P. (1980). Effect of partial hepatectomy on the response of rat liver to aflatoxin B_1. *Cancer Res.* **40**, 4739–4743.

Newberne, P. M., and Rogers, A. E. (1973a). Nutrition, monocrotaline, and aflatoxin B_1 in liver carcinogenesis. *Plant Foods Man* **1**, 23–31.

Newberne, P. M., and Rogers, A. E. (1973b). Rat colon carcinomas associated with aflatoxin and marginal vitamin A. *J. Natl. Cancer Inst.* **50**, 439–448.

Newberne, P. M., and Rogers, A. E. (1986). Labile methyl groups and the promotion of cancer. *Ann. Rev. Nutr.* **6**, 407–32.

Newberne, P. M., and Rogers, A. E. (1990). Lipotropic factors and carcinogenesis. *In* "Cancer and Nutrition" (Alfin-Slater and Kritchevsky, eds.), pp. 159–185. Plenum Publishing, New York.

Newberne, P. M., and Suphakarn, V. (1977). Preventive role of vitamin A in colon carcinogenesis in rats. *Cancer* **40**, 2553–2556.

Newberne, P. M., and Suphakarn, V. (1983). Nutrition and cancer. A review, with emphasis on the role of vitamins C and E and selenium. *Nutr. Cancer* **5**, 107–119.

Newberne, P. M., Harrington, D. H., and Wogan, G. N. (1966). Effects of cirrhosis and other liver insults on induction of liver tumors by aflatoxin in rats. *Lab. Invest.* **15**, 962–969.

Newberne, P. M., Rogers, A. E., and Wogan, G. N. (1968). Hepatorenal lesions in rats fed a low lipotrope diet and exposed to aflatoxin. *J. Nutr.* **94**, 331–343.

Newberne, P. M., Weigert, J., and Kula, N. (1979). Effects of dietary fat on hepatic mixed-function oxidases and hepatocellular carcinoma induced by aflatoxin B_1 in rats. *Cancer Res.* **39**, 3986–3991.

Newberne, P. M., Punyarit, P., DeCamargo, J., and Suphakarn, V. (1987). The role of necrosis in hepatocellular proliferation and liver tumors. *Arch. Toxicol. Supple.* **10**, 54–67.

Newberne, P. M., Suphiphat, V., Locniskar, M., and de Camargo, J. L. V. (1990). Inhibition of hepatocarcinogenesis in mice by dietary methyl donors methionine and choline. *Nutr. Cancer* **14**, 175–181.

Nixon, J. E., Hendricks, J. D., Pawlowski, N. E., Percera, C. B., Sinnhuber, R. O., and Baily, G. S. (1984). Inhibition of aflatoxin B_1 carcinogenesis in rainbow trout by flavone and indole compounds. *Carcinogenesis* **5**, 615–619.

Nyandieka, H. S. (1987). Effects of dietary fat and aflatoxin B_1 on microsomal monooxygenase activity. *Arch. Toxicol.* **60**, 59–60.

Poirier, L. A. (1986). The role of methionine in carcinogenesis in vivo. *In* "Essential Nutrients in Carcinogenesis" (L. Poirier, M. Pariza, and P. M. Newberne, eds.), pp. 269–282. Plenum Press, New York.

Preston, R. S., Hayes, J. R., and Campbell, T. C. (1976). The effect of protein deficiency in the *in vivo* binding of aflatoxin B_1 to rat liver macromolecules. *Life Sci.* **19**, 1191–1198.

Rajpurohit, R., and Krishnaswamy, K. (1988). Differences in response of glucuronide and glutathione conjugating enzymes to aflatoxin B_1 and *N*-acetylaminofluorene in underfed rats. *J. Toxicol. Environ. Health* **24**, 103–109.

Rivzi, T. A., Mathur, M., and Nayak, N. C. (1989). Enhancement of aflatoxin B_1-induced hepatocarcinogenesis in rats by partial hepatectomy. *Virchows Arch. B Cell Pathol.* **56**, 345–350.

Roebuck, B. D., Liu, Y. L., Rogers, A. E., Groopman, J. D., and Kensler, T. W. (1991). Protection against aflatoxin B_1-induced hepatocarcinogenesis in F344 rats by 5-(2-Pyrazinyl)-4-methyl-1,2-dithiol-3-thione (oltipraz): Predictive role for short-term molecular dosimetry. *Cancer Res.* **51**, 5501–5506.

Rogers, A. E. (1975). Variable effects of a lipotrope-deficient, high-fat diet on chemical carcinogenesis in rats. *Cancer Res.* **35**, 2469–2474.

Rogers, A. E. (1978). Dietary effects chemical carcinogenesis in the livers of rats. *In* "Rat Hepatic Neoplasia" (P. Newberne and W. H. Butler, eds.), pp. 242–262. MIT Press, Cambridge, Massachusetts.

Rogers, A. E., and Conner, M. W. (1986). Alcohol and cancer. *In* "Essential Nutrients in Carcinogenesis" (L. Poirier, P. M. Newberne, and M. W. Pariza, eds.), pp 473–496. Plenum Press, New York.

Rogers, A. E., and MacDonald, R. A. (1965). Hepatic vasculature and cell proliferation in experimental cirrhosis. *Lab. Invest.* **14**, 1710–1726.

Rogers, A. E., and Newberne, P. M. (1969). Aflatoxin B_1 carcinogenesis in lipotrope-deficient rats. *Cancer Res.* **29**, 1965–1972.

Rogers, A. E., and Newberne, P. M. (1971). Diet and aflatoxin B_1 toxicity in rats. *Toxicol. Appl. Pharmacol.* **20**, 113–121.

Rogers, A. E., and Newberne, P. M. (1981). Lipotrope deficiency in experimental carcinogenesis. *Nutr. Cancer* **2**, 104.

Rogers, A. E., Kula, N. S., and Newberne, P. M. (1971). Absence of an effect of partial hepatectomy on aflatoxin B_1 carcinogenesis. *Cancer Res.* **31**, 491–495.

Rogers, A. E., Sanchez, O., Feinsod, F. M., and Newberne, P. M. (1974). Dietary enhancement of nitrosamine carcinogenesis. *Cancer Res.* **34**, 96–99.

Rogers, A. E., Nields, H. M., and Newberne, P. M. (1987). Nutritional and dietary influences on liver tumorigenesis in mice and rats. *Arch. Toxicol. Suppl.* **10**, 231–243.

Rogers, A. E., Akhtar, R., and Zeisel, S. H. (1990). Procarbazine carcinogenicity in methotrexate-treated or lipotrope-deficient male rats. *Carcinogenesis* **11**, 1491–1495.

Rogers, A. E., Zeisel, S. H., and Akhtar, R. (1991). Choline, methionine, folate and chemical carcinogenesis. *In* "Vitamins and Minerals in the Prevention and Treatment of Cancer" (M. M. Jacobs, ed.), pp. 123–143. CRC Press, Boca Raton, Florida.

Rogers, C. G., Boyes, B. G., and Lok, E. (1990). Comparative genotoxicity of 3 procarcinogens in V79 cells as related to glutathione *S*-transferase activity of hepatocytes from untreated rats and those fed 2% butylated hydroxyanisole. *Mutation Res.* **244**, 163–171.

Schrager, T. F., Newberne, P. M., Pikul, A. H., and Groopman, J. D. (1990). Aflatoxin–DNA adduct formation in chronically dosed rats fed a choline-deficient diet. *Carcinogenesis* **11**, 177–180.

Shinozuka, H., and Katyal, S. L. (1986). Choline deficiency and chemical carcinogenesis. *In* "Essential Nutrients in Carcinogenesis" (L. A. Poirier, P. M. Newberne, and M. W. Pariza, eds.), pp. 253–267. Plenum Press, New York.

Shivapurkar, N., Hoover, K. L., and Poirier, L. A. (1986). Effect of methionine and choline on liver tumor promotion by phenobarbital and DDT in diethylnitrosamine-initiated rats. *Carcinogenesis* **7**, 548–550.

Smith, M. L., Yeleswarapu, L., Locker, J., and Lombardi, B. (1991). Expression of p53 mutant protein(s) in diethylnitrosamine-induced foci of enzyme-altered hepatocytes in male Fischer-344 rats. *Carcinogenesis* **12**, 1137–1141.

Stoewsand, G. S., Babish, J. B., and Wimberly, H. C. (1978). Inhibition of hepatic toxicities from polybrominated biphenyls and aflatoxin B_1 in rats fed cauliflower. *J. Environ. Pathol. Toxicol.* **2**, 399.

Suit, J. L., Rogers, A. E., Jetten, M. E. R., and Luria, S. E. (1977). Effects of diet on conversion of aflatoxin B_1 to bacterial mutagen(s) by rats *in vivo* and by rat hepatic microsomes *in vitro*. *Mutation Res.* **46**, 313–323.

Tadi, P. P., Teel, R. W., and Lau, B. H. S. (1991). Organosulfur compounds of garlic modulate mutagenesis, metabolism, and DNA binding of aflatoxin B_1. *Nutr. Cancer* **15**, 87–95.

Teel, R. W. (1991). Effects of capsaicin on rat liver S9-mediated metabolism and DNA binding of aflatoxin. *Nutr. Cancer* **15**, 27–32.

Toskulkao, C., and Glinsukon, T. (1988). Hepatic lipid peroxidation and intracellular calcium accumulation in ethanol potentiated aflatoxin B_1 toxicity. *J. Pharmacobio.-Dyn.* **11**, 191–197.

Toskulkao, C., and Glinsukon, T. (1990). Hepatic mitochondrial function and lysosomal enzyme activity in ethanol-potentiated aflatoxin B_1 hepatotoxicity. *Toxicol. Lett.* **52**, 179–190.

Toskulkao, C., Yoshida, T., Glinsukon, T., and Kuroiwa, Y. (1986). Potentiation of aflatoxin B_1 induced hepatotoxicity in male Wistar rats with ethanol pretreatment. *J. Toxicol. Sci.* **11**, 41–51.

Wang, J. H., Hinrichsen, L. I., Whitacre, C. M., Cechner, R. L., and Sudilovsky, O. (1990). Nuclear DNA content of altered hepatic foci in a rat liver carcinogenesis model. *Cancer Res.* **50**, 7571–7576.

Wargovich, M. J., Woods, C., Eng, V. W. S., Stephens, C., and Gray, K. (1988). Chemoprevention of *N*-nitrosomethylbenzylamine-induced esophageal cancer in rats by the naturally occurring thioether, dially sulfide. *Cancer Res.* **48**, 6872–6875.

Williams, G. M., Tanaka, T., and Maeura, Y. (1986). Dose-related inhibition of aflatoxin B_1 induced hepatocarcinogenesis by the phenolic antioxidants, butylated hydroxyanisole, and butylated hydroxytoluene. *Carcinogenesis* **7**, 1043–1050.

Zeisel, S. H. (1990). Choline deficiency. *J. Nutr. Biochem.* **1**, 332–349.

Part II

❖

Human Carcinogenicity and Toxicity

11

Epidemiology of Aflatoxin-Related Disease

❖

Andrew J. Hall and Christopher P. Wild

INTRODUCTION

The potent biological effects of aflatoxins in experimental animals and the exposure of large human populations to aflatoxin in the diet have led to a number of hypotheses regarding the importance of this exposure in the etiology of human disease. For some diseases, for example, hepatocellular carcinoma, the evidence for involvement of aflatoxin is relatively strong, whereas for other diseases considered in this chapter the evidence is weaker. However, the examination of each of the hypotheses has been hindered by the difficulties of measuring exposure in the populations concerned and by the epidemiological difficulties of controlling for confounding factors. In addition, the interpretation of causal direction has proved controversial in retrospective studies. The intent of this chapter is to examine these difficulties and to suggest how they may be resolved using new biochemical and molecular techniques. The latter include exposure measurements, for example, aflatoxin bound to albumin and aflatoxin-specific mutations in the *p53* tumor suppressor gene, and measures of susceptibility to aflatoxin-induced disease, for example, polymorphisms in drug metabolizing enzymes. Finally, some of the continuing difficulties of epidemiological design will be explored.

The Toxicology of Aflatoxins:
Human Health, Veterinary, and Agricultural Significance   **233**

EXPOSURE MEASUREMENT

The major source of exposure to aflatoxin is ingestion of contaminated food (Stoloff, 1983a). A much smaller population, an occupational group involved in the processing of contaminated cereals or groundnuts, is exposed through inhalation of dust containing the toxin. Although some information is available on the outcome of this exposure, it is highly atypical and the extent to which the effects of inhalation of aflatoxin can be extrapolated to the more general situation in which exposure is of dietary origin is unclear (Van Neuwenhuize *et al.*, 1973; see Bosch and Peers, 1991).

Human exposure to aflatoxin has been assessed in the context of epidemiological studies in two ways: by questionnaire, aimed at estimating quantity and frequency of intake of dietary items assumed to be commonly contaminated with aflatoxin, or by a combination of questionnaire and laboratory measurements made on, hopefully, representative food samples collected from the population concerned. Exposure measurement is one area in which biochemical and molecular techniques, if appropriately applied, promise to strengthen epidemiological studies by reducing misclassification of individuals with respect to exposure. The applications envisioned are presented briefly here. For more detailed information on the techniques themselves, reference should be made to Chapter 12.

Food

Laboratory methods of food analysis to determine the aflatoxin content have been available for many years. However, the precision of these techniques is mitigated by the difficulties of measuring actual ingestion of each food item in human populations. The difficulties of measuring dietary exposures are well known [Medical Research Council (MRC), 1984]. However, aflatoxin presents particular problems that arise from the societies in which major exposure occurs. These populations are rural agricultural communities in the economically poor world. Classical methods of dietary measurement at the individual level, for example, short-term recall food frequency questionnaires, daily diary recording of food items, and actual measurement of food intake by weighing of food before consumption combined with laboratory analysis of matched food samples, are particularly difficult to apply in these communities. Eating is ritualized in some of these societies and any intrusion that disturbs the social context may be resented. Measurement may lead to abnormal eating behaviors when recording is being done. For example, in many West African societies, specific groups of individuals will eat from a common food bowl, either by hand or using a spoon. The introduction of plates to measure individual portions is unlikely to be acceptable. Even observers recording handfuls or spoonfuls per person may be unacceptably intrusive. If this form of recording is accepted and is thought to be reliable, an additional problem is that "snack foods" may represent a significant exposure. Thus, in groundnut-producing societies, these nuts often are eaten raw, boiled, or roasted in the course of the day's work. Measurement of these exposures is particularly difficult. Thus, assessment of exposure based on food intake is problematic, even given newer more specific methods of aflatoxin measurement (Hudson *et al.,* 1992).

Another problem of measuring food intake as an exposure assessment is that, for some of the disease outcomes, the relevant exposure was some months or years in the past. This problem is compounded by the fact that the disease itself may affect diet. Thus, measurement of intake once disease is established is highly misleading. The measurement of exposure, with follow-up to ascertain disease, requires large numbers of individual measurements, which presents major logistic problems.

An alternative to measurements at the individual level is the evaluation of population exposure, which can be done through appropriate selection of representative individuals in the population to reflect the mean and range of intake values. Such values can be compared to disease incidence in the population. This approach does not require such precision if the differences in exposure between the populations are large, as they are for aflatoxin exposure at an international level (Table 1), and even among some population groups within countries (Van Rensburg *èt al.,* 1985). If, however, exposure and disease incidence are measured in the same time period, this approach does not overcome the possibility that exposure has changed during the long latent period between exposure and disease development.

Some estimates of population exposure have been made based on food analysis; examples for different countries in Africa and southeast Asia and for the United States are given in Table 1. The estimates range up to 200 ng/kg body wt/day, except in one report in southern China which showed 10 times this level. The estimated exposure levels by Stoloff (1983b) for the southern United States are, unexpectedly, as high as the African and southeast Asian levels. Reexamining this latter observation using the biological markers discussed subsequently would be valuable. In countries with a relatively uniform diet (e.g., in west Africa), aflatoxin contamination of staple foods would lead to a chronic exposure, something that would not occur in countries in which the diet is more heterogeneous (e.g., United States).

Another problem in comparing exposures, whether in individuals or in populations, is the marked seasonal variation seen in food levels (Goldblatt, 1969). This seasonality is a reflection of both climatic change and the food harvesting and storage practices of the societies concerned. Increased temperature and humidity increase the likelihood that the responsible fungi, *Aspergillus flavus* and *Aspergillus parasiticus,* will grow on the crop (Smith and Moss, 1985). Thus, contamination of crops is increased in the wet hot seasons. In addition to seasonal variations, annual variations in exposure also could increase misclassification when the time period between exposure and disease is long.

In view of these difficulties, scientific efforts have addressed biological measurements of aflatoxin exposure.

TABLE 1   Estimated Population Exposures Based on Analysis
of Aflatoxins in Food

Country	Period of sampling	Food source[a]	Range of estimated aflatoxin exposure[b] (ng/kg/day)	Reference
Kenya	1969–1970	H	3.5–14.8	Peers and Linsell (1973)
Swaziland	1972–1973	H	5.1–43.1	Peers et al. (1976)
	1982–1983	H	11.4–158.6	Peers et al. (1987)
Mozambique	1969–1974	P	38.6–183.7	Van Rensburg et al. (1985)
Transkei	1976–1977	P	16.5	Van Rensburg et al. (1985)
The Gambia	1988	P	4–115	Wild et al. (1992b)
China (southern Guangxi)	1978–1984	M	11.7–2027	Yeh et al. (1989)
Thailand	1969–1970	P	6.5–53.0	Shank et al. (1972)
United States				
(Southeast)	1910–1934	M	197[c]	Stoloff (1983a,b)
	1935–1959		108[c]	
(North and	1910–1934		0.22	
West)	1935–1959		0.34	
United States	1960–1979	M	2.7	Bruce (1990)

[a]H, Uncooked food samples from the home; P, samples of cooked food from the plate; M, samples from the market.

[b]AFB_1 except Kenya (B_1 and B_2) and The Gambia (B_1, B_2, G_1, G_2).

[c]Aflatoxin measurements in corn and peanuts 1964–1981 extrapolated to intakes based on estiamted consumption over different time periods.

Biological Samples

After ingestion, aflatoxin is metabolized by cytochrome P450 enzymes, predominantly in the hepatocyte. The epoxide generated binds covalently to nucleic acids and proteins (Busby and Wogan, 1984). The major nucleic acid adduct is aflatoxin–N^7-guanine, a proportion of which subsequently is excreted in the urine (Bennett et al., 1981; Groopman et al., 1993). As well as binding to nucleic acids, aflatoxin epoxide also binds covalently to proteins, for example, to form an albumin adduct that is found in peripheral blood (Wild et al., 1986,1990a; Sabbioni et al., 1987; Gan et al., 1988), and to glutathione, the products of which can be excreted in the feces or urine (Degan and Neumann, 1978). In addition to epoxidation, aflatoxin can be hydroxylated to more polar, less carcinogenic metabolites such as AFM_1, which also can be excreted in the urine or further metabolized to a reactive epoxide. The elucidation of the metabolism of aflatoxin in humans has provided a rationale for using some of the products of aflatoxin metabolism found in urine, blood, or tissues as markers of exposure. One of the premises for using aflatoxin macromolecular adducts as biomarkers is that the measurement made reflects not only the dietary exposure of the individual but also the characteristics of that individual for the uptake, distribution, metabolism, detoxification, and removal of adducts. This dual contribution to the end point measured is a key point to consider when interpreting the data obtained in epidemiological studies.

The available biomarkers can be considered to give information at three levels. First, the analysis of free aflatoxin in the body gives a quantitative measure of internal dose. Second, the analysis of aflatoxin covalently bound to cellular macromolecules (DNA, RNA, proteins) after metabolism gives a quantitative measure of biologically effective dose. Third, the analysis of mutation spectra in specific genes may give a qualitative indication of past exposure and demonstrate a biological effect of that exposure.

Urine

Unmetabolized aflatoxins (AFB_1, AFG_1), hydroxylated metabolites (AFM_1, AFQ_1), and demethylated metabolites (AFP_1) have been measured in human urine. More emphasis has been focused on the nucleic acid adduct aflatoxin–N^7-guanine, because this metabolite reflects DNA damage in the presumed target cell for aflatoxin carcinogenesis (Groopman et al., 1993), the hepatocyte, and may be more relevant to exposure-related risk (Bechtel, 1989). Unfortunately, the levels of the urinary metabolites and adduct reflect intake on the previous day. Although this measure is excellent for studies of acute exposure, it may not reflect chronic intake by individuals or by populations. The misclassification of chronic exposure introduced by single urinary measurements will tend to disguise true associations between exposure and disease, a problem that may be overcome by multiple measurements on the same individual, as has been done for the similarly fluctuating urinary sodium excretion to determine its relationship to hypertension (Liu et al., 1979). However, such multiple measurements present logistic problems.

Peripheral Blood

In a manner similar to the urinary studies, initial work focused on the measurement of free aflatoxins in sera or plasma (Hendrickse et al., 1982; Tsuboi et al., 1984) but the subsequent identification of a covalent adduct of aflatoxin, in this case an albumin adduct (Gan et al., 1988; Wild et al., 1990b), has offered considerable advantages. As does the aflatoxin–N^7-guanine adduct described earlier, the albumin adduct appears to reflect DNA damage in the hepatocytes in experimental systems (Wild et al., 1986). Collection of fingerprick blood samples is also logistically easier than repeated urine collection. An additional advantage is that the adduct, at least in experimental animals, is as long-lived as albumin (Wild et al., 1986; Sabbioni et al., 1987). Thus, adduct quantification provides a measure over a considerable period of time (2–3 months) based on the half-life of albumin in humans. This assay provides some measurement of chronic exposure that is not available from other markers such as the aflatoxin–N^7-guanine adduct in urine. A single measurement in time should provide a much better representation of average exposure, as illustrated in Figure 1, in which the

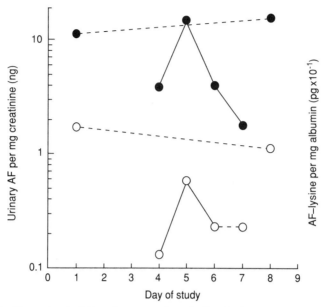

FIGURE 1 Fluctuation in daily urinary excretion of aflatoxin in comparison with aflatoxin–albumin adducts. Urinary aflatoxin levels (———) were determined in four consecutive 24-hr collections from day 4 to 7 inclusive. Aflatoxin–albumin adduct levels in peripheral blood (– – –) were measured in the same individuals on days 1 and 8. Dietary measurements of aflatoxin intake were made on the same subjects. The solid circles and open circles represent two different individuals. These studies were performed in The Gambia and are reported fully elsewhere (Groopman *et al.*, 1992; Wild *et al.*, 1992).

daily fluctuations in urinary excretion of aflatoxin metabolites are shown to vary up to 10-fold over a 4-day period but the aflatoxin–albumin adduct is shown to be relatively stable (less than 2-fold variation). The method has provided a considerable database for aflatoxin exposure in different countries (Wild *et al.*, 1993). However, even this measurement cannot overcome the problem of disease influencing intake, nor the long latency period between the relevant exposure and disease onset for some conditions. Investigators hoped that, by examining the distribution of levels in a population, individuals could be stratified into levels of consumption so a single measurement would characterize an individual's exposure over time and throughout the year. This relationship would be analogous to the situation with growth (Sinclair, 1978) or hypertension (Fixler, 1985), for which the level at any one point in time determines the centile in which a person's growth or blood pressure continues to increase. This phenomenon is known as tracking. Unfortunately, in the one study that has examined tracking for aflatoxin exposure, this hope has not been fulfilled. Children were characterized in terms of albumin adduct levels at the beginning and end of the rainy season (6 months apart) and no evidence of tracking was found (Figure 2). This result probably reflects a number of factors, including the half-life of albu-

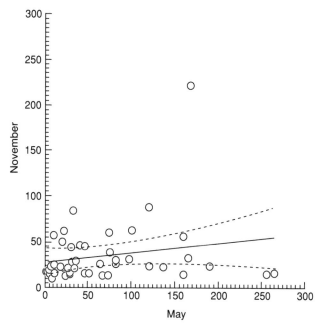

FIGURE 2   Intraindividual correlation between serum aflatoxin–albumin adduct levels (pg AFB_1-lysine eq. per mg albumin) in samples taken 6 months apart. Sera were collected from a series of 46 children in The Gambia in May and again in November 1988. The aflatoxin–albumin adduct levels were determined as described previously (Wild *et al.,* 1990b). Linear regression is plotted for the intraindividual correlation ($r^2 = 0.0367$; $p = 0.213$; two-tailed t test not significant).

min and the large variations in daily intake of aflatoxin, suggesting that the contribution to albumin adduct levels of variation in dietary intake may outweigh the contribution of interindividual differences in aflatoxin adduct formation and loss. Nutritional factors and the presence of other diseases also may have influenced the aflatoxin–albumin adduct levels. The establishment of tracking requires more detailed study over several half-lives of albumin and preferably, given the seasonal variations mentioned earlier, requires study annually at times when the dietary exposure contributes a large fraction of total annual exposure. The data at present suggest that measurements will have to be made at several points in the year and over time to characterize some form of individual cumulative exposure. Nevertheless, provided comparisons are made for the same season, this biomarker should provide a better means of comparison between average population exposures than either food intake or urinary excretion.

Tissue

Levels of free aflatoxin in tissues, most notably the liver, have been available for some years. Specific DNA adducts of aflatoxin have been detected as well (Hsieh *et al.,* 1988). These adducts are more satisfactory from the biological perspective because they represent the postulated pathogenic pathway for a number of disease outcomes, but both approaches suffer from a lack of controls. The measures usually are carried out on biopsy or postmortem specimens. Obtaining similar samples from appropriate controls usually is not possible, leading to great difficulty in interpretation. In addition, the tissue levels may be affected by the disease under study, whether through disease-induced changes in diet or through alterations in carcinogen metabolism.

One indication of this latter effect has been observed with the aflatoxin–albumin adduct. A case control study in the People's Republic of China examined aflatoxin exposure by two methods: aflatoxin in household food items and aflatoxin–albumin adducts. The measures showed a high degree of correlation in controls but not in the cases, suggesting that changes had occurred in diet, metabolism of aflatoxin, or both (Figure 3; Yu, Y.-Z., Wild, C. P., unpublished data).

Two new approaches have been developed to examine the potential effects of aflatoxin without direct measurement of exposure: measures of mutation spectra and measures of genetic variation in metabolism.

Mutation Spectra

The *p53* tumor suppressor gene is mutated in more than 50% of all human tumors (Hollstein *et al.*, 1991). In this gene, functional mutant proteins can result from mutation at a large number of base sites. One consequence is that a study of the frequency of the different types and sites of mutations may give indications about the exposures that induced those mutations. In experimental systems using other genes, namely hypoxanthine phosphoribosyltransferase (HPRT) and adenine phosphoribosyltransferase (ART) genes in mammalian cells, a high degree of specificity has been observed between the mutation spectra and a given mutagen (Meuth, 1990; Skandalis and Glickman, 1990). The most striking specificity in such mutation spectra reported to date in human tumors is in the *p53* gene with respect to hepatocellular carcinoma. In tumors from areas with a supposedly high aflatoxin exposure, occurrence of a G → T transversion at codon 249 of the *p53* gene is frequent. In some sets of tumors, for example from Qidong, People's Republic of China, and from Mozambique, this frequency reaches 30–50% (Bressac *et al.*, 1991; Hsu *et al.*, 1991). The same mutation has, to date, been observed rarely in North America, Europe, or Japan (Ozturk *et al.*, 1991). The mutation does not appear to be explained by hepatitis B virus infection. Aflatoxin induces predominantly G → T transversions in experimental studies (Foster *et al.*, 1983). Using the human *p53* gene in an *in vitro* assay, codon 249 has been shown to be a preferential site for formation of aflatoxin–N^7-guanine adducts (Pusieux *et al.*, 1991), evidence consistent with a role for aflatoxin in the mutations seen in human tumors.

This finding is potentially important in evaluating the significance of aflatoxin exposure in addition to exposure to other carcinogens. However, to date work is

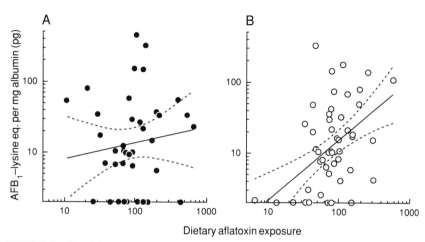

FIGURE 3 Correlation of estimated dietary intake of aflatoxin and measurement of aflatoxin–albumin adducts in hepatocellular carcinoma cases (A) and matched controls (B) in The People's Republic of China. A case-control study of hepatocellular carcinoma in males and females was conducted in Fusui County, Guangxi, China, in collaboration with Shun-Zhang Yu at the School of Public Health, Shanghai Medical University, Shanghai. Aflatoxin exposure was measured (1) by the analysis of foods (rice, maize, and peanut oil) in the home of cases and controls and calculating the intake (μg per person per day) and (2) by measuring the aflatoxin–albumin adduct level in peripheral blood as described previously (Wild *et al.*, 1992). Linear regressions are presented for the cases (A) and controls (B). In the cases, no correlation is seen between dietary levels and aflatoxin–albumin adduct levels ($r = 0.0086$; $p = 0.959$; two-tailed t test) whereas in controls a significant correlation was seen ($r = 0.317$; $p = 0.021$).

limited because the cases studied are small in number. Whether the cases selected are representative of the population is not clear. No controls are available to assess how often changes in *p53* might be related to exposure independent of disease. Finally, the changes still must be related to population variations in exposure. This last point introduces all the problems of measurement presented earlier. Additional possible advances using this methodology are discussed in subsequent sections.

Genetic Variation in Metabolism

Considerable progress has been made in identifying the cytochrome P450 (Kitada *et al.*, 1989; Shimada and Guengerich, 1989; Aoyama *et al.*, 1990; Forrester *et al.*, 1990; see Chapter 3) and, to a lesser extent, the glutathione *S*-transferase (Liu *et al.*, 1991) enzymes involved in the metabolic activation and detoxification of aflatoxin in human liver. A major cytochrome P450 involved in aflatoxin epoxidation appears to be CYP3A4 (Shimada and Guengerich, 1989; Ramsdell *et al.*, 1991), although others clearly have the potential to metabolize this toxin (Aoyama *et al.*, 1990; Forrester *et al.*, 1990). Important kinetic differences may exist in the ability of specific human P450 enzymes to activate aflatoxin at low versus high concentrations (Ramsdell and Eaton, 1990). Previously an excess of extensive metabolizers of the marker drug debrisoquine were seen among cancer cases, including liver cancer, in Nigeria (Idle *et al.*, 1981). A suggestion was made that the enzyme responsible for metabolism of this drug, CYP2D6, could be involved in aflatoxin metabolism, thereby conferring an increased risk on extensive metabolizers in this population. In fact, from *in vitro* assays, CYP2D6 does not seem to play a major role in aflatoxin metabolism, at least in Caucasian individuals (e.g., Forrester *et al.*, 1990). With respect to detoxification by enzymatic conjugation of aflatoxin epoxide to glutathione, two studies assigned a significant role to glutathione *S*-transferase mu (Liu *et al.*, 1991; Raney *et al.*, 1992). This result is of particular interest, since a varying fraction of individuals in the populations examined does not have expression of this enzyme because of a gene deletion in one allele inherited in an autosomal dominant manner (Broad *et al.*, 1991). In parallel to the situation with cytochrome P450s, isoenzymes other than glutathione *S*-transferase mu are likely to be involved in conjugation of the aflatoxin epoxide.

The knowledge concerning enzyme specificity has been complemented by the sequencing of the genes involved and by identification of specific substrates (endogenous or exogenous) for certain of the enzymes, thus providing a basis to develop pheno- and genotyping assays. Of specific interest concerning aflatoxin are the assays for CYP3A4, CYP1A2, CYP2D6, and glutathione *S*-transferase mu. Cortisol is a substrate for CYP3A4 and is metabolized to 6β-hydroxycortisol; the ratio of these two steroids in the urine, measured by immunoassay or high performance liquid chromatography (HPLC), may provide a measure of hepatic CYP3A4 activity (Ged *et al.*, 1989). Alternatively, the calcium antagonist nifedipine can be used as a test drug for this P450. Caffeine and debrisoquine are substrates for CYP1A2 and CYP2D6, respectively, and can be used for *in vivo* phenotyping. In addition, identification of the most frequent mutations responsible for CYP2D6 polymorphism has allowed the development of rapid genotyping assays based on the polymerase chain reaction (PCR; Gough *et al.*, 1990; Wolf *et al.*, 1990). Similarly, for glutathione *S*-transferase mu polymorphism, a specific substrate for an *in vitro* assay—*trans*-stilbene oxide—and a PCR-based genotyping assay are available (Zhong *et al.*, 1991).

This knowledge of aflatoxin metabolism and the development of appropriate assays allows individuals to be characterized in terms of their likely response to the harmful effects of this toxin, that is, their susceptibility. This characterization may be complicated, however, by the rather broad substrate specificity of the enzymes and a present lack of information with respect to the relative roles of the different enzymes *in vivo*. Also, most studies have been performed on samples from Caucasians. This database must be extended and validated in populations likely to be studied for aflatoxin-related disease. A valuable and necessary step will be to examine the expression of the enzymes with respect to the level of aflatoxin DNA or protein adducts in exposed people. In this way, establishing the relative importance of the different enzymes involved may be possible.

One important consideration for application of these methods to epidemiological studies is that the genotype will be unaffected by disease. Genotyping assays, therefore, can be applied to cases of disease and to controls in an unbiased way, whereas the phenotype may be influenced by disease or other concomitant exposures, for example, by altering the expression of specific cytochrome P450s. This relationship must be taken into account. Note, with respect to aflatoxin and hepatitis B, that carriers of the virus have increased CYP3A4 activity as measured using the 6β-hydroxycortisol assay mentioned (Geubel *et al.*, 1987).

The interpretation and significance of the data on individual susceptibility would, of course, be influenced by the degree of exposure to aflatoxin. In a rural tropical region, this latter parameter may be relatively uniform among individuals if integrated over an extended time period.

Despite the limitations, the general approach of characterizing the susceptibility of the individual is likely to be a powerful one for examining the contribution of aflatoxin to disease because potential confounders can be controlled simultaneously, as discussed further in a subsequent section. For example, a case–control study of liver cancer can control for past exposure to hepatitis B virus through serological markers. The strength of the geno- and phenotyping approaches will be dependent on the relative importance of genetic variation and the variation in chronic exposure found within the population. Neither of these variables is known with precision for any population to date.

BIOLOGICAL EFFECTS OF EXPOSURE

A considerable amount of information about the mechanism of action of aflatoxin has been obtained experimentally. Cross-reference should be made to Chapters 1–10 in this book. For the purpose of this chapter, note that aflatoxin can induce DNA damage (and mutation), toxicity (resulting in cell death and a consequent increase in cell replication), and immunosuppression (see Busby and Wogan, 1984). The mechanism of action in any one aflatoxin-related disease may differ.

Whereas the mutagenic and toxic effects of aflatoxin have been the focus of much research attention, the immunosuppressive effects have been studied less extensively. In fact, virtually no data are available on the immunosuppressive effects of aflatoxin in human populations. In contrast, extensive work has been done in animals that has been reviewed by Denning (1987) and Pier and McLoughlin (1985). These data can be considered in two parts: studies that examined changes in susceptibility or severity of disease because of specific infectious agents and studies that examined specific immunological functions in exposed animals. Original references are found in the two review articles mentioned unless otherwise cited.

In chickens, a clear effect of aflatoxin on susceptibility to infection with *Candida albicans, Salmonella,* and Marek's disease virus is seen. In turkeys, increased infection with *Pasteurella multocida* has been observed. In pigs, susceptibility to *treponemes,* and in calves, increased pulmonary infection with *Fasciola hepatica* is seen. Of particular interest is the fact that the protective effects of vaccines against Marek's disease virus in chickens and *P. multocida* in turkeys are reduced by aflatoxin exposure. Chicken erythrocytes from aflatoxin-exposed birds were more sensitive to β-hemolysin (a virulence factor for many infectious agents) produced by staphylococci (Doerr *et al.,* 1987).

In terms of measurements of immunity other than infectious challenge, aflatoxin has been shown to affect immunoglobin synthesis in chickens and mice; to reduce complement levels in guinea pigs (Pier *et al.,* 1989), chickens, and pigs; to impair phagocytosis in chickens; to reduce locomotion and bactericidal activity of neutrophils in chickens; to reduce monocyte locomotion and phagocytosis in rats, chickens, and rabbits; and to reduce several measures of cell-mediated immunity in chickens (Kadian *et al.,* 1988), guinea pigs (Pier *et al.,* 1989), turkeys, cows, and mice (Reddy *et al.,* 1987). The data on lymphocyte transformation by aflatoxin are less clear in cows, but this measure is suggested to be reduced in mouse and human.

Thus, good reason exists to examine in more detail the immunosuppressive effects of aflatoxins in humans. This effect may be of relevance, in conjunction with other factors, to the high rates of infection seen in children in developing countries that have high levels of exposure to aflatoxin. This effect also may be relevant to the marked geographical variations seen in responses to some vaccines.

PUTATIVE DISEASE OUTCOMES

The potential disease outcomes in relation to exposure largely have been derived from animal models (Purchase, 1974; WHO, 1979). Very few studies have been done in which exposure has been the starting point and a range of outcomes has been studied, largely because exposure occurs particularly in societies in which routine health statistics are underdeveloped. Thus, the starting point for most studies has been the identification of people suffering from diseases potentially associated with aflatoxin. In general, very little epidemiological work has been done, except in relation to hepatocellular carcinoma. Four major diseases have been studied.

Acute Hepatitis

Aflatoxin has been blamed for outbreaks of acute hepatitis in a few situations. The most extensively studied, in northwest India in 1974, has been the subject of two investigations and three papers (Krishnamachari *et al.,* 1975; Tandon *et al.,* 1977,1978). Clearly ample opportunity existed for very high levels of exposure to aflatoxin in moldy maize. Aflatoxin was found in some of the individuals studied, particularly in their tissues; the hepatic disease had some characteristic features. A viral cause was excluded. A subsequent outbreak of fatal hepatitis in Kenya, which also was attributed to aflatoxin-contaminated maize, lends support to this finding (Ngindu *et al.*, 1982). Also several other case reports have been reviewed in the literature (Campbell and Stoloff, 1974). Establishing causality in these situations is very difficult, but the circumstantial evidence, the measurably high levels of aflatoxins in food and human tissues, and the histology are all persuasive. The availability of the biological exposure markers mentioned in Section II,B should be of great value in investigating any future outbreaks. As has been pointed out by others (WHO, 1979), these populations with documented acute high exposure (although probably on a chronic low level background) could provide valuable information on the long-term sequelae of such exposure. Unfortunately, no long-term follow-up has been published.

Kwashiorkor

The etiology of protein–energy malnutrition and recommendations for its control have been discussed (Latham, 1990). With respect to kwashiorkor, Hendrickse (1982,1991) has been the champion of aflatoxin as the cause of this form of malnutrition. He has pointed out that this disease is limited to the tropics and has an increased incidence in the wet season, both features of aflatoxin exposure (Hendrickse, 1991). Studies in Kenya (De Vries *et al.*, 1987) and Sudan (Hendrickse *et al.*, 1982) have demonstrated high levels of urinary excretion of aflatoxin in affected children compared with marasmic or normal children. However, these studies were small and did not always control for age, and whether the controls were matched appropriately was not clear from the information given. Nevertheless, one study showed a striking pattern of excretion of aflatoxins by children with kwashiorkor after hospitalization that was not seen in other hospitalized children. In South Africa, a case–control study of 66 hospitalized children with kwashiorkor found that only one control and one case were positive for any serum level of aflatoxin B_1, B_2, G_1, G_2, M_1, P_1, Q_1, or aflatoxicol using highly sensitive chromatographic methods (P. G. Thiel, personal communication). This finding suggests that if aflatoxin does contribute to kwashiorkor, other causes predominate in some geographical areas. Note that significant amounts of aflatoxin occur in breast milk (Zarba *et al.*, 1991) and that aflatoxin crosses the placenta, resulting in *in utero* exposure (Denning *et al.*, 1990; Wild *et al.*, 1991). The worldwide magnitude of the problem of childhood protein–energy malnutrition, with millions of children affected, and the observations of childhood exposure to aflatoxin give ample reason to continue to explore the hypothesis (see subsequent discussion), but appropriate epidemiological methods are needed.

Reye's Syndrome

Reye's syndrome occurs throughout the world. A few hundred cases of the disease occur per year in the United States (Hurwitz, 1989) without an apparent geographical association with aflatoxin exposure. Four countries—Czechoslovakia (Dvorockova *et al.,* 1977), New Zealand (Becroft, 1966), Thailand (Shank *et al.,* 1971), and the United States (Ryan *et al.,* 1979; Nelson *et al.,* 1980)—have been the sites of studies of the association between Reye's syndrome and aflatoxin. Only Thailand has a reputed high exposure to aflatoxin, although other studies using the aflatoxin–albumin adduct test suggests that the exposure is relatively low compared with east and west Africa and southern China (Wild *et al.,* 1992). The evidence for association rests on the finding of aflatoxin in tissues from cases. In the United States, even this finding has not been consistent (Rogan *et al.,* 1985). The accumulation of aflatoxin in cases could be a result of the disease affecting metabolism rather than an etiological factor. The finding that aspirin consumption is a major risk factor for Reye's syndrome requires further studies to account for this additional factor (Hurwitz, 1989).

Hepatocellular Carcinoma

The association between liver cancer and aflatoxin has been the subject of many studies and many reviews (e.g., Edmonson and Craig, 1987; Bosch and Munoz, 1989; Stoloff, 1989; Bruce, 1990). Reviewing these is not the intent of this chapter. However, the studies that have been done can be divided into two groups—those that are ecological, that is, they compare populations in terms of average aflatoxin exposure and liver cancer rate, and those that use the case–control approach. In this approach, cases of liver cancer are compared with suitable controls in terms of aflatoxin exposure and other risk factors. The ecological approach has used various measures of exposure, including food measurements with or without assessment of individual food intake, urinary excretion of metabolites, and *p53* gene mutations. Case–control studies have used dietary intake by recall as the major method, although some studies have used aflatoxin–albumin adducts (Srivatanakul *et al.,* 1991); other uncontrolled studies have examined tissue levels of aflatoxin.

The ecological approach has produced the most convincing evidence of an association. Liver cancer rates increase with increasing levels of exposure in all the populations studied. A study in China by Campbell *et al.* (1990) is in contrast to all other ecological studies because neither urinary aflatoxin metabolites nor recalled dietary intake of foods potentially contaminated with aflatoxin was associated with primary liver cancer rates. Hepatitis B carrier prevalences were, however, strongly correlated. One of the reasons for this result is likely to be the reliability of the method for characterizing hepatitis carrier status, namely, blood markers, compared with the imprecise urinary measurement and dietary recall methods for aflatoxin (Wild and Montesano, 1991), highlighting the fact that the major difficulty in interpretation of all these studies is the possibility of confounding. Since hepatitis B is now recognized to be a major cause of liver cancer in the countries with high rates of carrier status, the extent to which the association between liver cancer and aflatoxin can be explained by an association between hepatitis B and aflatoxin is not clear. This relationship has been addressed in several studies (Peers *et al.*, 1987; Yeh *et al.*, 1989; Campbell *et al.*, 1990) that, in general, conclude that an effect of aflatoxin in addition to hepatitis B is functioning. However, ecological studies cannot determine whether this association is an interaction between hepatitis B and aflatoxin or the result of independent effects on risk. Does aflatoxin modify the risk of liver cancer only in individuals who are hepatitis B carriers or is it a risk factor in noncarriers on its own? Studies specifically designed to address this question have not been published, although they are on-going and some are in an advanced stage of planning (see Chapter 12 for further discussion of this point).

Others

Indian childhood cirrhosis was thought to be related to aflatoxin on the basis of chemical analysis of urine specimens (Robinson, 1967). However, subsequently the chemical detected has been found not to be aflatoxin (WHO, 1979). The epidemiology of the disease did not concur with exposure patterns in other studies (Yadgiri *et al.*, 1970) and the two are unlikely to be associated.

Other cancers, notably colon cancer and renal tumors (see Busby and Wogan, 1984), occur in animals exposed to aflatoxin. Of note in the studies of carcinogenicity in primates is the common occurrence of osteosarcomas (Sieber *et al.*, 1979). For human colon tumors and osteosarcomas, the geographical pattern of occurrence does not suggest that aflatoxin is a major carcinogen. An association with renal tumors has not been explored in populations with high toxin exposure.

CAUSALITY

Causality in epidemiology is impossible to demonstrate conclusively without intervention studies. Unsuspected confounding may always be an explanation of observed associations. For example, other mycotoxins such as the *Fusaria* toxins (fumonisins) are known to be carcinogenic (Gelderblom *et al.*, 1991), yet these have not been examined in the same samples as aflatoxin and could confound the relationships described for it. Interventions overcome this confounding by randomizing groups to balance all confounders, both known and unknown. This point is considered further in the section under Possible Future Research Studies.

Less well recognized is the fact that epidemiology identified two different classes of causal factors (Rose, 1985). Rose has pointed this out for cardiovascular disease, but the principle applies to all potential causal relationships. The point Rose makes is that causal factors exist that determine risks within populations as well as between populations. Within populations, what determines why one person develops the disease and the other does not ("My grandfather smoked 40 cigarettes a day all his life and lived to be a hundred")? The epidemiological studies that address these issues are case–control and cohort studies within populations. Between populations, what factors determine why one country has a different rate of disease than another? These factors are examined by ecological studies, as described for hepatocellular carcinoma. A risk factor may, of course, act in both ways. Thus, hepatitis B virus persistence appears to determine the risk of one individual within a society of developing liver cancer; in addition, the prevalence of such persistent infection in a population determines the population's rate of liver cancer. When factors do act in both ways, their detection is much simpler since all epidemiological types of study produce consistently positive findings, although the magnitude of the effect may differ. Rose also pointed out that genetic susceptibility, whatever the mechanism of its action, often will be a major source of individual variation in risk within populations, but rarely between populations. This variation is illustrated by aflatoxin, for which the very large variations in exposure between different countries, for example, the United Stated compared with Mozambique, might be associated with the variation in liver cancer rates. In contrast, within a country, whether with a uniformly high or low exposure, the variation in exposure may be so small that it is not the major determinant of disease risk, which may be determined by the rate and type of metabolism of the aflatoxin, that is, by the genotype of the individual.

If the epidemiological studies associating aflatoxin and disease are re-examined in this light, great contrasts between the diseases can be seen. In kwashiorkor, the evidence is based primarily on case–control studies. Virtually no studies have been done of ecological variation in aflatoxin exposure and incidence of kwashiorkor, despite the demonstration of marked variations in incidence of the disease even within countries. In contrast, the association with liver cancer has been determined almost exclusively through such studies, and strongly suggests that aflatoxin is a population determinant of incidence rate. Thus, for these disease outcomes, examining the other alternatives becomes crucial.

Once one has identified a risk factor, by whatever form of study, one must examine whether an explanation for the association exists other than the causal one. These other explanations fall into four areas: chance, bias, confounding, and reverse causation. In the first of these, chance or random variation has produced the finding; the likelihood of this effect is examined by tests of statistical significance. In some of the reports on kwashiorkor, statistical tests have not been applied; the probability that they were due to chance is unclear. Bias is introduced by study or sampling design and measurement. For example, measurement is a particular problem for studies in which people are asked to recall their diet. People suffering from disease may have a different intensity of recall than those who are well. In the case–control studies of liver cancer, for example, knowing how much the results are influenced by this bias is difficult. In ecological studies, sampling design is much more important than measurement. Errors in laboratory analyses of food are relatively small compared with the errors introduced by a biased sample of individuals. Thus, bias can be reduced both by appropriate design and by choice of measurement method. Confounding, in contrast, is something that is present in the population. Those who are carriers of hepatitis B also may be those who have the highest aflatoxin exposures (Allen *et al.*, 1992). This confounding must be controlled for in the study design, that is, by matching control selection, or in the data analysis, by examining the influence of one variable with another. In most of the ecological studies of liver cancer, this controlling has not been done. Even when controlling has been carried out, that the confounding was measured appropriately and controlled for in the analysis is not clear. Finally, reverse causation is a situation in which a causal association exists but the direction is misinterpreted, that is, cause is mistaken for effect. This situation may exist in the studies in which tissue levels of aflatoxin are found to be high in those with liver disease, whether kwashiorkor, Reye's syndrome, or liver cancer. The altered metabolism associated with the disease may lead to the increased level of aflatoxin.

These problems of observational epidemiology must be answered by consistency of results in different populations, using different methodologies and using different study designs. In terms of the latter, each of the three basic epidemiological designs offers different advantages. The next section considers how these designs might be utilized best in the future in pursuing the epidemiology of aflatoxin-related disease.

POSSIBLE FUTURE RESEARCH STRATEGIES

In epidemiology, three basic observational approaches to explore risk factors are available. Interventions are then necessary to establish causality beyond doubt. These techniques are considered here.

Geographical Studies

Geographical or so-called ecological studies have provided reasonably strong evidence of association between aflatoxin and liver cancer. Further studies using the albumin-bound marker of aflatoxin exposure may give a better estimate of the importance of this factor in liver cancer than did the dietary and urinary methods used to date. In addition, some of the measures of intermediate effects may illuminate the pathogenesis of the tumor. One study compared the *p53* mutation in different geographical areas and related this change to measures of aflatoxin intake (Ozturk *et al.*, 1991). Unfortunately, the numbers of tumors studied in any one place were very limited and whether they were representative of tumors occurring in the relevant population was not clear. Better designed studies of this type should prove rewarding.

In contrast, no ecological studies have been carried out in relation to kwashiorkor, Reye's syndrome, or cirrhosis. A major reason for this absence of studies is that few studies are available on the incidence of these conditions in different populations, whereas cancer registries clearly have demonstrated the variability of liver cancer rates. However, in Ethiopia, good evidence is available of geographical variation in the incidence of kwashiorkor (Lindtjorn, 1987). Thus, ecological studies in this condition and perhaps in Reye's syndrome in Thailand, where apparent heterogeneity of incidence exists, should be done. This approach is particularly attractive since it identifies population risk factors that may be key to public health intervention. The objective in these studies would be to select geographical areas in which either exposure or disease was known to vary in incidence and then to determine the other. Ideally, one would want minimal variation in terms of potential confounding factors, such as poverty, among the areas. The first step for such studies might be to carry out measures of aflatoxin exposure in populations using, for example, the aflatoxin–albumin marker, or to establish surveillance for the disease in question. The data so generated would be of great value, even if no great variation were detected.

Case Control Studies

As discussed earlier, case–control studies have the particular difficulty of interpretation, since most of the diseases under consideration potentially affect hepatic metabolism, so any biological measure of aflatoxin in individuals may be a result of the disease. However, the genetic markers of metabolism now being identified are not altered by the disease state. These markers can be utilized in such studies. However, the underlying assumption must be that the genetically determined metabolism is dominant over variability of exposure within the population as the cause of individual risk of disease. In rural farming communities, this is likely to be true because all individuals are exposed to high levels of contaminated food. The choice of controls in this situation is unlikely to be a problem, provided that matching includes ethnicity or tribe. Confounding and interaction will need to be addressed, in particular with respect to hepatitis B and C persistent infections, since these could be associated with certain genotypes as well as being causes of disease. Such studies will require reasonably large sample sizes. Estimates of the prevalence of debrisoquine poor metabolizers in west Africa are about 10% of the population (Idle *et al.*, 1981). Most populations studied appear to have a population with approximately 50% with the glutathione *S*-transferase mu polymorphism, although this varies significantly worldwide (Broad *et al.*, 1991). These figures give estimates of sample size of 200–300 cases to detect a relative risk of 2 for both polymorphisms at conventional levels of statistical significance and power (95%, 90%) in non-virus-carriers. Since these noncarrier cases make up approximately 20% of all cases in endemic regions of Asia and Africa, one would need case–control studies with approximately 1200 cases. Such studies also would provide sufficient power to examine interactions between virus-carrier and genetic status.

Studies of mutations would be applicable to case–control studies if the analyses could be performed on suitable control populations. Assays of the frequency and type of somatic cell mutations using peripheral blood cells have been developed (Albertini *et al.*, 1990) for marker genes. Mutation analysis in genes such as *p53*, which appears to be of importance in the carcinogenic process, would be of value also. An assay for circulating antibodies to the *p53* mutant protein would offer the possibility of determining the frequency of mutation in control populations, something that is not currently possible with the tissue-based assay for *p53*. However, such an assay would require validation and determination of its sensitivity and specificity in the populations to be studied.

Cohort Studies

Cohort studies are the most promising method for examining the relationship between individual exposure and subsequent disease. In particular, the newer methods of determining exposure, which will give a measure of chronic exposure, should allow more precise characterization of individual risk. However, these studies are demanding in terms of resources and time. Large numbers of individuals must be entered into the study since most of the expected outcomes are relatively rare. Multiple measurements over time on each individual are likely to be necessary since tracking of exposure levels appears to be nonexistent. In addition, groups of individuals who are both carriers and noncarriers of the hepatitis B virus and potentially the hepatitis C virus must be examined to determine the risk in the presence and absence of these factors. Such studies of interaction multiply the required sample size by a factor of 4. Sample size estimates are not straightforward, since no threshold of exposure below which there is no risk of neoplasia has been determined. Approximately 80 incidence cases would be required in the lowest exposure category to determine interactions within reasonable confidence limits. In this context, exposure data show 5% of Gambians to be negative for albumin-bound aflatoxin on a single measurement, compared with 90% of Thais (Wild *et al.*, 1992). A cohort would have to insure that sufficient numbers of this fraction of the population were recruited so 80 cancer cases subsequently developed within it. Despite these disadvantages of cohort studies, major benefits do exist. The problem of reverse causation is resolved, provided the measurement of exposure preceded disease by a sufficient period. Multiple outcomes can be examined if a suitable surveillance system is incorporated into the study. Parallel cohort studies can be carried out in different geographical areas with varying levels of exposure to obtain a measure of the population importance of the factor. This methodology is needed for all the possible associations outlined earlier. For liver cancer and kwashiorkor, the sample sizes required are large but feasible. Efficiency can be improved further by carrying out case–control studies within the cohort. This method then only requires analysis of stored samples from the cases and a suitable sample of controls. Ross *et al.* (1992) have demonstrated the successful application of this technique in their publication from a Shanghai cohort. The aflatoxin assay used in this setting was the urinary aflatoxin–N^7-guanine adduct. The interaction with hepatitis B was large, despite the problems of day-to-day variation in this exposure method. For Reye's syndrome, it is unlikely that a sufficient cohort size could be assembled, unless this cohort were part of a study with additional objectives. These objectives might be to study the importance of immunosuppressive effects on risk of infections and the accompanying morbidity and mortality. As Denning (1987) has pointed out, this effect may constitute a major factor in the high disease incidence in children in developing countries and deserves urgent attention.

Interventions

The key issue in causality is to show that the disease incidence is altered by a change in the level of exposure. Since increasing a population's exposure to aflatoxin is clearly unethical, the appropriate intervention is some form of reduction, which would be difficult to attain at the individual level and would need to be done in communities. This method has the added advantage that the intervention could be applied in a pragmatic way, allowing the additional evaluation of the public health strategy (Hall and Aaby, 1990). However, valid results require the randomization of a suitable number of communities to the intervention. The design of these studies is likely to be highly complex, the studies are expensive to carry out, and they require a multidisciplinary team (Smith and Morrow, 1991). Aflatoxin–albumin adduct measurements now provide a means of measuring the efficacy of this form of intervention. The number of communities required will be determined by the between-community variation in exposure. As with cohort studies, multiple outcomes could be measured and the causal relationship of aflatoxin to kwashiorkor, cirrhosis, and liver cancer could be examined simultaneously, although liver cancer would require considerable follow-up unless the exposure were acting primarily as a late promoter in carcinogenesis. The major advantage of this randomized intervention is that it removes confounding by other factors such as hepatitis B virus.

Intervention against aflatoxin can be envisioned in several ways: (1) preharvest, by controlling irrigation and pest infestation and by producing genetic crop variants resistant to fungal infection; (2) postharvest, by improved storage and sorting techniques; and (3) postingestion, by modifying the biological effects of aflatoxins. Oltipraz has been proposed as one method of reducing aflatoxin-induced DNA damage and has been shown to be effective in preventing aflatoxin-induced hepatocellular carcinoma in rats (Roebuck *et al.,* 1991). This antischistosomal drug (Bueding *et al.,* 1982) induces glutathione *S*-transferase levels, which leads to increased inactivation of aflatoxin in rats (Kensler *et al.,* 1992). However, the glutathione *S*-transferase isoenzymes induced in humans, compared with those in rodents, still require further investigation. In addition, repeated regular doses of oltipraz probably are needed to have a lasting effect. More likely interventions are effects on the harvesting, storage, and sorting techniques employed. These changes will require close collaborative efforts between agricultural and medical scientists for their implementation and evaluation. Since the gains from this collaboration are likely to be large in terms of the public health, they must begin now.

ACKNOWLEDGMENTS

The authors acknowledge the helpful comments of R. Montesano and M. Parkin and the contribution of Elspeth Perez to the preparation of the manuscript.

REFERENCES

Albertini, R. J., Nicklas, J. A., O'Neill, J. P., and Robinson, S. H. (1990). *In vivo* somatic mutations in humans: Measurement and analysis. *Ann. Rev. Genet.* **24,** 305–326.

Allen, S. J., Wild, C. P., Wheeler, J. G., Riley, E. M., Montesano, R., Bennett, S., Whittle, H. C., Hall, A. J., and Greenwood, B. M. (1992). Aflatoxin exposure, malaria, and hepatitis B infection in rural Gambian children. *Transact. R. Soc. Trop. Med.* **86,** 426–430.

Aoyama, T., Yamano, S., Guzelian, P. S., Gelboin, H. V., and Gonzalez, F. J. (1990). Five of 12 forms of vaccinia virus-expressed human hepatic cytochrome P450 metabolically activate aflatoxin B_1. *Proc. Natl. Acad. Sci. U.S.A.* **87,** 4790–4793.

Bechtel, D. H. (1989). Molecular dosimetry of hepatic aflatoxin B_1–DNA adducts; Linear correlation with hepatic cancer risk. *Reg. Toxicol. Pharmacol.* **10,** 74–81.

Becroft, D. M. O. (1966). Syndrome of encephalopathy and fatty degeneration of the viscera in New Zealand children. *Br. Med J.* **2,** 135–140.

Bennett, R. A., Essigmann, J. M., and Wogan, G. N. (1981). Excretion of an aflatoxin–guanine adduct in the urine of aflatoxin B_1-treated rats. *Cancer Res.* **41,** 650–654.

Board, P., Coggan, M., Johnstone, P., Ross, V., Suzuki, T., and Webb, G. (1990). Genetic heterogeneity of the human glutathione transferases: a complex of gene families. *Pharmacol. Ther.* **48,** 357–369.

Bosch, F. X., and Munoz, N. (1989). Epidemiology of hepatocellular carcinoma. *In* "Liver Cell Carcinoma" (P. Bannasch, D. Keppler, and G. Weber, eds.), pp. 3–14. Kluwer Academic Publishers, Dordrecht, The Netherlands.

Bosch, F. X., and Peers, F. (1991). Aflatoxins: Data on human carcinogenic risk. *In* "Relevance to Human Cancer of *N*-Nitroso Compounds, Tobacco Smoke, and Mycotoxins" (I. K. O'Neill, J. Chen, and H. Bartsch, eds.), pp. 48–53. International Agency for Research on Cancer, Lyon.

Bressac, B., Kew, M., Wands, J., and Ozturk, M. (1991). Selective G to T mutations of *p53* gene in hepatocellular carcinoma from southern Africa. *Nature (London)* **350,** 429–431.

Bruce, R. D. (1990). Risk assessment for aflatoxin: II. Implications of human epidemiology data. *Risk Anal.* **10,** 561–569.

Bueding, E., Dolan, P., and Leroy, J. P. (1982). The antischistosomal activity of oltipraz. *Res. Commun. Chem. Pathol. Pharmacol.* **37,** 293–303.

Busby, W. F., and Wogan, G. N. (1984). Aflatoxins. *In* "Chemical Carcinogens" (C. D. Searle, ed.), pp. 945–1136. American Chemical Society, Washington, D.C.

Campbell, T. C., and Stoloff, L. (1974). Implication of mycotoxins for human health. *J. Agr. Food Chem.* **22,** 1006–1015.

Campbell, T. C., Chen, J., Liu, C., Li, J., and Parpia, B. (1990). Nonassociation of aflatoxin with primary liver cancer in a cross-sectional ecological survey in the Peoples' Republic of China. *Cancer Res.* **50,** 6882–6893.

Degan, G. H., and Neumann, H-G. (1978). The major metabolite of aflatoxin B$_1$ in the rat is a glutathione conjugate. *Chem. Biol. Interact.* **22,** 239–255.

Denning, D. W. (1987). Aflatoxin and human disease. *Adv. Drug React. Ac. Pois. Rev.* **4,** 175–209.

Denning, D. W., Allen, R., Wilkinson, A. P., and Morgan, M. R. A. (1990). Transplacental transfer of aflatoxin in humans. *Carcinogenesis* **11,** 1033–1035.

De Vries, H. R., Lamplugh, S. M., and Hendrickse, R. G. (1987). Aflatoxins and kwashiorkor in Kenya: A hospital based study in a rural area of Kenya. *Ann. Trop. Paediatr.* **7,** 249–257.

Doerr, J. A., Huff, W. E., and Hamilton, P. B. (1987). Increased sensitivity to Staphylococcal beta hemolysins of erythrocytes from chickens during aflatoxicosis. *Poultry Sci.* **66,** 1929–1933.

Dvorockova, I., Kusak, V., Vesely, D., Vesela, J., and Nesmidal, P. (1977). Aflatoxin and encephalopathy with fatty degeneration of viscera (Reye). *Ann. Nutr. Alim.* **31,** 977–990.

Edmondson, H. A., and Craig, J. R. (1987). Neoplasms of the liver. *In* "Diseases of the Liver" (L. Schiff and E. R. Schiff, eds.), pp. 1109–1158. Lippincott, Philadelphia.

Fixler, D. E. (1985). Blood pressure in children and adolescents. *In* "Epidemiology of Hypertension" (C. J. Bulpitt, ed.), pp. 35–50. Elsevier, Amsterdam.

Forrester, L. M., Neal, G. E., Judah, D. J., Glancey, M. J., and Wolf, C. R. (1990). Evidence for involvement of multiple forms of cytochrome P-450 in aflatoxin B$_1$ metabolism in human liver. *Proc. Natl. Acad. Sci. U.S.A.* **87,** 8306–8310.

Foster, P. L., Eisenstadt, E., and Miller, J. H. (1983). Base substitution mutations induced by metabolically activated aflatoxin B$_1$. *Proc. Natl. Acad. Sci. U.S.A.* **80,** 2695–2698.

Gan, L.-S., Skipper, P. L., Peng, X., Groopman, J. D., Chen, J., Wogan, G. N., and Tannenbaum, S. R. (1988). Serum albumin adducts in the molecular epidemiology of aflatoxin carcinogenesis: Correlation with aflatoxin B$_1$ intake and urinary excretion of aflatoxin M$_1$. *Carcinogenesis* **9,** 1323–1325.

Ged, C., Rouillon, J. M., Pichard, L., Combalbert, J., Bressot, N., Bories, P., Michel, H., Beaune, P., and Maurel, P. (1989). The increase in urinary excretion of 6β-hydroxycortisol as a marker of human hepatic cytochrome P450IIIA induction. *Br. J. Clin. Pharmacol.* **28,** 373–387.

Gelderblom, W. C. A., Kriek, N. P. J., Marasas, W. F. O., and Thiel, P. G. (1991). Toxicity and carcinogenicity of the *Fusarium moniliforme* metabolite, fumonisin B$_1$, in rats. *Carcinogenesis* **12,** 1247–1251.

Geubel, A. P., Pauwels, S., Buchet, J. P., Dumont, E., and Dive, C. (1987). Increased cyt P-450 dependent function in healthy HBsAg carriers. *Pharmacol. Ther.* **33,** 193–196.

Goldblatt, L. A. (1969). "Aflatoxin." Academic Press, New York.

Gough, A. C., Miles, J. S., Spurr, N. K., Moss, J. E., Gaedigk, A., Eichelbaum, M., and Wolf, C. R. (1990). Identification of the primary gene defect at the cytochrome P450 *CYP2D* locus. *Nature (London)* **347,** 773–776.

Groopman, J. D., Wild, C. P., Hasler, J., Chen, J., Wogan, G. N., and Kensler, T. W. (1993). Molecular epidemiology of aflatoxin exposures: Validation of aflatoxin-N^7-guanine levels in urine as a biomarker in experimental rat models and humans. *Environ. Health Perspect.* **99,** 107–113.

Groopman, J. D., Hall, A. J., Whittle, H., Hudson, G. J., Wogan, G. N., Montesano, R., and Wild, C. P. (1992). Molecular dosimetry of aflatoxin-N^7-guanine in human urine obtained in The Gambia, West Africa. *Cancer Epidemiol. Biomarkers Prev.* **1,** 221–227.

Hall, A. J., and Aaby, P. (1990). Tropical trials and tribulations. *Int. J. Epidemiol.* **19,** 777–781.

Hendrickse, R. G. (1991). Clinical implications of food contaminated by aflatoxins. *Ann. Acad. Med.* **20,** 84–90.

Hendrickse, R. G., Coulter, J., Lamplugh, S., MacFarlane, S., Williams, T., Omer, M., and Suliman, G. (1982). Aflatoxins and kwashiorkor: A study in Sudanese children. *Br. Med. J.* **285,** 843–846.

Hollstein, M. C., Sidransky, D., Vogelstein, B., and Harris, C. C. (1991). p53 mutations in human cancers. *Science* **253,** 49–53.

Hsieh, L. L., Hsu, S. W., Chen, D. S., and Santella, R. M. (1988). Immunological detection of aflatoxin B$_1$–DNA adducts formed *in vivo. Cancer Res.* **48,** 6328–6331.

Hsu, I. C., Metcalf, R. A., Sun, T., Welsh, J. A., Wang, N. J., and Harris, C. C. (1991). Mutational hotspot in the *p53* gene in human hepatocellular carcinomas. *Nature (London)* **350,** 427–428.

Hudson, G. J., Wild, C. P., Zarba, A., and Groopman, J. D. (1992). Aflatoxins isolated by preparative immunoaffinity chromatography from foods consumed in The Gambia, West Africa. *Natural Toxins* **1,** 100–105.

Hurwitz, E. S. (1989). Reye's syndrome. *Epidemiol. Rev.* **11,** 249–253.

Idle, J. R., Mahgoub, A., Sloan, T. P., Smith, R. L., Mbanefo, C. O., and Bababunmi, E. A. (1981). Some observations on the oxidation phenotype status of Nigerian patients presenting with cancer. *Cancer Lett.* **11,** 331–338.

Kadian, S. K., Monga, D. P., and Goel, M. C. (1988). Effect of aflatoxin B_1 on the delayed type hypersensitivity and phagocytic activity of reticuloendothelial system in chickens. *Mycopathologia* **104,** 33–36.

Kensler, T. W., Groopman, J. D., Eaton, D. L., Curphey, T. J., and Roebuck, B. D. (1992). Potent inhibition of aflatoxin-induced hepatic tumorigenesis by the monofunctional enzyme inducer 1,2-dithiole-3-thione. *Carcinogenesis* **13,** 95–100.

Kitada, M., Taneda, M., Ohi, H., Komori, M., Itahashi, K., Nagao, M., and Kamataki, T. (1989). Mutagenic activation of aflatoxin B_1 by P-450 HFLa in human fetal livers. *Mutat. Res.* **227,** 53–58.

Krishnamachari, K. A. V. R., Bhat, R. V., Nagarajan, V., and Tilak, T. B. G. (1975). Investigations into an outbreak of hepatitis in parts of western India. *Indian J. Med. Res.* **63,** 1036–1048.

Latham, M. C. (1990). Protein–energy malnutrition—Its epidemiology and control. *J. Environ. Pathol. Toxicol. Oncol.* **10,** 168–180.

Lindtjorn, B. (1987). Famine in Ethiopia 1983–1985: Kwashiorkor and marasmus in four regions. *Ann. Trop. Paediatr.* **7,** 1–5.

Liu, K., Cooper, R., McKeever, J., McKeever, P., Byington, R., Salteri, I., Stamler, R., Gosch, F., Stevens, E., and Stamler, J. (1979). Assessment of the association between habitual salt intake and high blood pressure: Methodological problems. *Am. J. Epidemiol.* **110,** 219–224.

Liu, Y. H., Taylor, J., Linko, P., Lucier, G. W., and Thompson, C. L. (1991). Glutathione *S*-transferase mu in human lymphocyte and liver: Role in modulating carcinogen-derived DNA adducts. *Carcinogenesis* **12,** 2269–2275.

Medical Research Council Environmental Epidemiology Unit (1984). "The Dietary Assessment of Populations." MRC, Southampton.

Meuth, M. (1990). The structure of mutation in mammalian cells. *Biochim. Biophys. Acta* **1032,** 1–17.

Nelson, D. B., Kimborough, R., Landrigan, P. S., Hayes, A. W., Yang, G. C., and Benanides, J. (1980). Aflatoxin and Reye's syndrome: A case control study. *Pediatrics* **66,** 965–969.

Ngindu, A., Johnson, B. K., Kenya, P. R., Ngira, J. A., Ocheng, D. M., Nandwa, H., Omondi, T. N., Jansen, A. J., Ngare, W., Kaviti, J. N., Gatei, D., and Siongok, T. A. (1982). Outbreak of acute hepatitis caused by aflatoxin poisoning in Kenya. *Lancet* **1,** 1346–1348.

Ozturk, M., *et al.* (1991). p53 mutation in hepatocellular carcinoma after aflatoxin exposure. *Lancet* **338,** 1356–1359.

Peers, F. G., and Linsell, C. A. (1973). Dietary aflatoxins and liver cancer. *Br. J. Cancer* **27,** 473–484.

Peers, F. G., Gilman, G. A., and Linsell, C. A. (1976). Dietary aflatoxins and human liver cancer. A study in Swaziland. *Int. J. Cancer* **17,** 167–176.

Peers, F., Bosch, X., Kaldor, J., Linsell, A., and Pluijmen, M. (1987). Aflatoxin exposure, hepatitis B virus infection and liver cancer in Swaziland. *Int. J. Cancer* **39,** 545–553.

Pier, A. C., Belden, E. L., Ellis, J. A., Nelson, E. W., and Maki, L. R. (1989). Effects of cyclopiazonic acid and aflatoxin singly and in combination on selected clinical, pathological and immunological responses of guinea pigs. *Mycopathologia* **105,** 135–142.

Pier, A. C., and McLoughlin, M. E. (1985). Mycotoxic suppression of immunity. *In* "Trichothecenes and Other Mycotoxins" (J. Lacey, ed.), pp. 507–519. John Wiley & Sons, Chichester.

Purchase, I. F. H. (1974). "Mycotoxins." Elsevier, New York.

Pusieux, A., Lim, S., Groopman, J. D., and Ozturk, M. (1991). Selective targeting of p53 mutational hotspots in human cancers by etiologically defined carcinogens. *Cancer Res.* **51,** 6185–6189.

Ramsdell, H. S., and Eaton, D. L. (1990). Species susceptibility to aflatoxin B_1 carcinogenesis: Comparative kinetics of microsomal biotransformation. *Cancer Res.* **50,** 615–620.

Ramsdell, H. S., Parkinson, A., Eddy, A. C., and Eaton, D. L. (1991). Bioactivation of aflatoxin B1 by human liver microsomes: Role of cytochrome P450 IIIA enzymes. *Toxicol. Appl. Pharmacol.* **108,** 436–447.

Raney, K. D., Meyer, D. J., Ketterer, B., Harris, T. M., and Guengerich, F. P. (1992). Glutathione conjugation of aflatoxin B_1 *exo-* and *endo*-epoxides by rat and human glutathione *S*-transferases. *Chem. Res. Toxicol.* **5,** 470–478.

Reddy, R. V., Taylor, M. J., and Sharma, R. P. (1987). Studies of immune function of CD-1 mice exposed to aflatoxin B_1. *Toxicology* **43,** 123–132.

Robinson, P. (1967). Infantile cirrhosis of the liver in India with special reference to probable aflatoxin aetiology. *Clin. Pediatr.* **6,** 57–62.

Roebuck, B. D., Liu, Y-L., Rogers, A. E., Groopman, J. D., and Kensler, T. W. (1991). Protection against aflatoxin B_1-induced hepatocarcinogenesis in F344 rats by 5-(2-prazinyl)-4-methyl-1,2-dithiole-3-thione (oltipraz): Predictive role for short-term molecular dosimetry. *Cancer Res.* **51,** 5501–5506.

Rogan, W. J., Yang, G. C., and Kimbrough, R. D. (1985). Aflatoxin and Reye's syndrome: A study of livers from deceased cases. *Arch. Environ. Hlth.* **40,** 91–95.

Rose, G. (1985). Sick individuals and sick populations. *PAHO Epidemiol. Bull.* **6,** 1–8.

Ross, R. K., Yuan, J., Yu, M. C., Wogan, G. N., Qian, G., Tu, J., Groopman, J. D., Gao, Y., and Henderson, B. E. (1992). Urinary aflatoxin biomarkers and risk of hepatocellular carcinoma. *Lancet* **339,** 943–946.

Ryan, N. J., Hogan, G. R., Hayes, A. W., Unger, P. D., and Siraj, M. Y. (1979). Aflatoxin B_1: Its role in the etiology of Reye's syndrome. *Pediatrics* **64,** 71–75.

Sabbioni, G., Skipper, P. L., Buchi, G., and Tannenbaum, S. R. (1987). Isolation and characterisation of the major serum albumin adduct formed by aflatoxin B_1 *in vivo* in rats. *Carcinogenesis* **8,** 819–824.

Shank, R. C., Bourgeois, C. H., Keschamras, N., and Chandavimol, P. (1971). Aflatoxins in autopsy specimens from Thai children with an acute disease of unknown aetiology. *Fd. Cosmet. Toxicol.* **9,** 501–507.

Shank, R. C., Gordon, J. E., Wogan, G. N., Nondasuta, A., and Subhamani, B. (1972). Dietary aflatoxins and human liver cancer. III. Field survey of rural Thai families for ingested anflatoxins. *Fd. Cosmet. Toxicol.* **10,** 71–84.

Shimada, T., and Guengerich, F. P. (1989). Evidence for cytochrome P450NF, the nifidipine oxidase, being the principal enzyme involved in the bioactivation of aflatoxins in human liver. *Proc. Natl. Acad. Sci. U.S.A.* **86,** 462–465.

Sieber, S. M., Correa, P., Dalgard, D. W., and Adamson, R. H. (1979). Induction of osteogenic sarcoma and tumours of the hepatobiliary system in nonhuman primates with aflatoxin B_1. *Cancer Res.* **39,** 4545–4554.

Sinclair, D. (1978). "Human Growth after Birth." Oxford University Press, Oxford.

Skandalis, A., and Glickman, B. W. (1990). Endogenous gene systems for the study of mutational specificity in mammalian cells. *Cancer Cells* **2,** 79–83.

Smith, J. E., and Moss, M. O. (1985). "Mycotoxins: Formation, Analysis and Significance." John Wiley and Sons, Chichester.

Smith, P. G., and Morrow, R. H. (eds.) (1991). "Methods for Field Trials of Interventions against Tropical Diseases. A Toolbox." Oxford Medical Publications, Oxford.

Srivatanakul, P., Parkin, D. M., Khlat, M., Chenvidhya, D., Chotiwan, P., Insiripong, S. L., Abbé, K. A., and Wild, C. P. (1991). Liver cancer in Thailand. II. A case-control study of hepatocellular carcinoma. *Int. J. Cancer* **48,** 329–332.

Stoloff, L. (1983a). Mycotoxins as potential environmental carcinogens. *In* "Carcinogens and Muta-gens in the Environment" (H. R. Stich, ed.), Vol. 1, pp. 97–120. CRC Press, Boca Raton, Florida.

Stoloff, L. (1983b). Aflatoxin as a cause of primary liver-cell cancer in the United States: A proba-bility study. *Nutr. Cancer* **5,** 165–186.

Stoloff, L. (1989). Aflatoxin is not a probable human carcinogen: The published evidence is suffi-cient. *Reg. Toxicol. Pharmacol.* **10,** 272–283.

Tandon, B. N., Krischnamurthy, L., Koshy, A., Tandon, H. D., Ramalinsasami, V., Bhandari, J. R., Mathur, M. M., and Mathur, P. D. (1977). Study of an epidemic of jaundice, presumably due to toxic hepatitis, in northwest India. *Gastroenterology* **72,** 488–494.

Tandon, H. D., Tandon, B. N., and Ramalingaswami, V. (1978). Epidemic of toxic hepatitis in India of possible mycotoxic origin. *Arch. Pathol. Lab. Med.* **102,** 372–376.

Tsuboi, S., Nakagawa, T., Tomita, M., Soo, T., Ono, H., Kawamura, K., and Iwamura, N. (1984). Detection of aflatoxin B_1 in serum samples of male Japanese subjects by radioimmunoassay and high performance liquid chromatography. *Cancer Res.* **44,** 1231–1234.

Van Neuwenhuize, J. P., Herbern, F. M., De Bruin, A., Meyern, P. B., and Duba, W. C. (1973). Cancers in men following a long-term low-level exposure to aflatoxins in indoor air. *T. Soc. Geneeskd.* **51,** 754–760.

Van Rensburg, S. J., Cook-Mozaffari, P., Van Schalkwyk, D. J., Van Der Watt, J. J., Vincent, T. J., and Purchase, I. F. (1985). Hepatocellular carcinoma and dietary aflatoxin in Mozambique and Transkei. *Br. J. Cancer* **51,** 713–726.

Wild, C. P., and Montesano, R. (1991). Correspondence re: T. Colin Campbell *et al.,* Nonassociation of aflatoxin with primary liver cancer in a cross-sectional ecological survey in the People's Republic of China. *Cancer Res.,* 50, 6882–6893, 1990. *Cancer Res.* **51,** 3825–3827.

Wild, C. P., Garner, R. C., Montesano, R., and Tursi, F. (1986). Aflatoxin B_1 binding to plasma albumin and liver DNA upon chronic administration to rats. *Carcinogenesis* **7,** 853–858.

Wild, C. P., Jiang, Y-Z., Sabbioni, G., Chapot, B., and Montesano, R. (1990a). Evaluation of methods for quantitation of aflatoxin–albumin adducts and their application to human exposure assess-ment. *Cancer Res.* **50,** 245–251.

Wild, C. P., Jiang, Y-Z., Allen, S. J., Jansen, L. A. M., Hall, A. J., and Montesano, R. (1990b). Aflatoxin-albumin adducts in human sera from different regions of the world. *Carcinogenesis* **11,** 2271–2274.

Wild, C. P., Rasheed, F. N., Jawla, M. F. B., Hall, A. J., Jansen, L. A. M., and Montesano, R. (1991). *In utero* exposure to aflatoxin in West Africa. *Lancet* **337,** 1602.

Wild, C. P., Jansen, L. M., Cova, L., and Montesano, R. (1993). Molecular dosimetry of aflatoxin exposure: contribution to understanding the multifactorial aetiopathogenesis of primary hepa-tocellular carcinoma (PHC) with particular reference to hepatitis B virus. *Environ. Health Perspect.* **99,** 115–122.

Wild, C. P., Hudson, G. J., Sabbioni, G., Chapot, B., Hall, A. J., Wogan, G. N., Whittle, H., Montesano, R., and Groopman, J. D. (1992). Dietary intake of aflatoxins and the level of albumin-bound aflatoxin in peripheral blood in The Gambia, West Africa. *Cancer Epidemiol. Biomarkers Prev.* **1,** 229–234.

Wolf, C. R., Moss, J. E., Miles, J. S., Gough, A. C., and Spurr, N. K. (1990). Detection of debriso-quine hydroxylation phenotypes. *Lancet* **336,** 1452–1453.

World Health Organization (1979). "Mycotoxins: Environmental Health Criteria," Vol. 11. WHO, Geneva.

Yadgiri, B., Reddy, V., Tulpule, P. G., Srikantia, S. G., and Gopalan, C. (1970). Aflatoxin and Indian childhood cirrhosis. *Am. J. Clin. Nutr.* **23,** 94–98.

Yeh, F-S, Yu, M. C., Mo, C. C., Luo, S., Tong, M. J., and Henderson, B. (1989). Hepatitis B virus, aflatoxins, and hepatocellular carcinoma in Southern Guangxi, China. *Cancer Res.* **49,** 2506–2509.

Zarba, A., Wild, C. P., Hall, A. J., Montesano, R., and Groopman, J. D. (1992). Aflatoxin M_1 in human breast milk from The Gambia, West Africa quantified by combined monoclonal antibody immunoaffinity chromatography and HPLC. *Carcinogenesis* **13,** 891–894.

Zhong, S., Howie, A. F., Ketterer, B., Taylor, J., Hayes, J. D., Beckett, G. J., Wathem, G., Wolf, C. R., and Spurr, N. K. (1991). Glutathione-*S*-transferase mu locus: Use of genotyping and phenotyping assays to assess association with lung cancer susceptibility. *Carcinogenesis* **12,** 1533–1537.

12

Molecular Dosimetry Methods for Assessing Human Aflatoxin Exposures

❖

John D. Groopman

INTRODUCTION

The development of methods to biomonitor human exposure to aflatoxins has been justified by the association of aflatoxin exposure with human liver cancer. Although in some reports the associations between dietary exposure and liver cancer have been strong, definitive investigations have been hindered by the lack of precise dosimetry data on aflatoxin exposure and metabolism, and the general poor quality of worldwide cancer incidence and mortality data. Since epidemiological studies cannot reveal information about an individual's risk to develop disease, the use of molecular biomarkers of aflatoxin exposures in individuals may provide insights for the prediction of risk for liver cancer and other disease outcomes. In addition, with these validated tools, appropriate prevention strategies can be devised and assessed.

The discrete steps in the progressive nature of many chronic diseases suggest that molecular biological markers can be identified for events falling anywhere along the continuum from an exposure to disease outcome. A joint committee of the National Academy of Sciences and the National Research Council (NRC, 1987) has developed a useful paradigm for this process (Figure 1). This model groups molecular biomarkers into those reflective of susceptibility, internal dose, biologically effective dose (dose to critical macromolecules or at site of toxic action), early biological effect, altered structure/function, and clinical disease. In

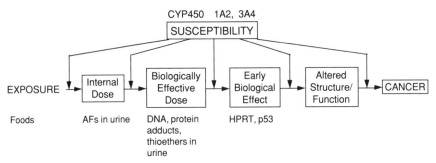

FIGURE 1   Paradigm for biomarker analysis.

more general terms, molecular biological markers can be considered to fall into the broad categories of markers of exposure, effect, and susceptibility. Markers of exposure indicate exposure to a parent compound, markers of effect signal biological responses to an exposure, and markers of susceptibility provide information about inherent sensitivity of the person to an environmental agent. By definition, some of these markers are chemical-agent specific, for example, a carcinogen–DNA adduct, whereas others are biological-process specific, for example, the expression of an oncogene protein.

The development of molecular biomarkers for aflatoxins is based on knowledge of the metabolism and critical genetic macromolecular adduct formation of these compounds and of possible target sites. These molecular biological markers also may help identify interindividual differences in susceptibility, so they can be accounted for as confounders or effect modifiers in data interpretation. The basic mechanisms of action of the aflatoxins occur after they are metabolized by the microsomal mixed function oxygenase system (reviewed by Busby and Wogan, 1985; Groopman *et al.,* 1991). These enzymes catalyze the oxidative metabolism of aflatoxin B_1 (AFB_1), resulting in the formation of various hydroxylated derivatives as well as an unstable, highly reactive epoxide metabolite. Synthesis of the 8,9-epoxide has been accomplished, confirming this structure (Baertschi *et al.,* 1988). Microsomes from human and rat liver have been demonstrated to catalyze the production of both the *exo* and *endo* forms of the 8,9-epoxide (Raney *et al.,* 1992). The toxicological significance of these two forms of the epoxide is under active investigation. Detoxification of AFB_1 is accomplished by enzymatic conjugation of the hydroxylated metabolites with sulfate or glucuronic acid to form water-soluble sulfate or glucuronide esters that are excreted in urine or bile in conjunction with the unconjugated compounds such as AFB_1 and aflatoxin M_1 (AFM_1). In addition, during lactation AFM_1 can be excreted in the milk. An alternative route for removal of AFB_1 from the organism involves the enzyme-catalyzed reaction of the epoxide metabolite with glutathione and its subsequent excretion in the bile. Some of the known detoxification pathways of AFB_1 metabolism have been summarized in Figure 2.

For a biomarker to be valid for any of the uses described earlier, careful

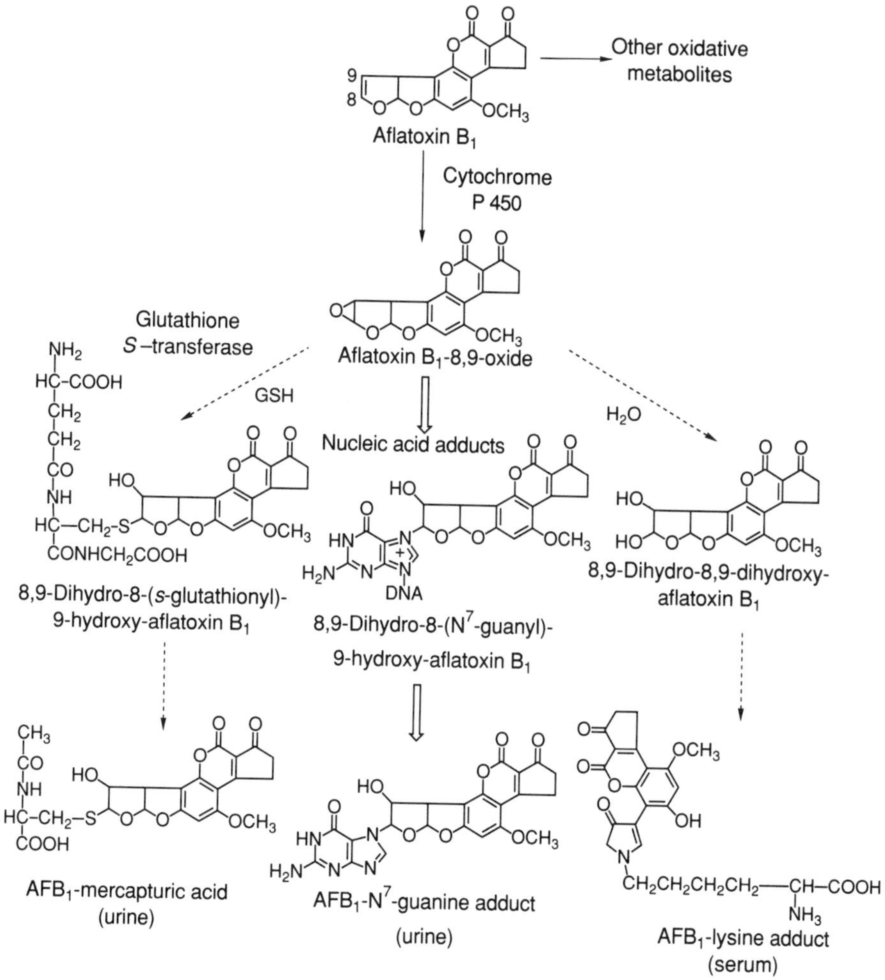

FIGURE 2   Metabolic pathways for aflatoxin metabolism and biomarker development.

laboratory development must be followed by application in field tests to ascertain its rigor. Ideally, an appropriate animal model is used to determine the role of the marker in the disease pathway and to establish relationships between dose and response. The developed marker then can be used in pilot human studies in which sensitivity, specificity, accuracy, and reliability parameters are established. The critical next step is the use of the marker in transitional epidemiological studies (Hulka, 1991) to help validate the marker. These studies can assess intra- or interindividual variability, background levels, and relationship of the marker to external dose or to disease status as well as issues of feasibility for use in larger population-based studies. Establishing a connection, in humans, between the biological marker and the exposure or the outcome of interest is important. To fully interpret the information the marker can provide, prospective epidemiological studies may be necessary to demonstrate the role the marker plays in the overall pathogenesis of the disease.

This chapter reviews the current status of the molecular dosimetry field for aflatoxins and concentrates on studies impacting on the identification of internal dose, biologically effective dose, and early biological effects using both experimental and human investigations. A vast literature exists on exposure assessment through the diet that is comprehensively described in other chapters in this volume. In addition, a developing literature is available on the identification of aflatoxin adducts in human tissues by immunohistocytochemical methods (Garner *et al.*, 1988; Hsieh *et al.*, 1989; Wild *et al.*, 1990a). However, these techniques are not yet adequately sensitive or specific for routine analyses in human samples, so they are not discussed in detail in this chapter. The literature on the toxicology of aflatoxins has been covered extensively by Busby and Wogan (1985). Reviews focused on biological monitoring and epidemiological considerations have been published by Groopman *et al.* (1991) and Bosch and Munoz (1988) (see also Chapter 11).

INTERNAL DOSE MEASUREMENTS IN RELATION TO EXPOSURE MEASUREMENTS

Several methods have been developed for the measurement of AFB_1 and the oxidative metabolite AFM_1 in biological samples. Some of these studies are reviewed in the context of urinary markers, metabolites in milk, and detection of the parent compound in blood. All these studies have been included as internal doses since none of the cited investigations involve the measurement of DNA or protein adducts.

Urinary Markers

Studies reflective of internal dose in humans can be separated into two categories: those that measure the presence of an aflatoxin biomarker and those that compare the internal dose measurement with an independent measure of exposure. Numerous observational reports are available of the presence of aflatoxins in urine. These studies used thin layer chromatography to detect the presence of AFM_1 in human urine (Campbell *et al.*, 1970). Many other studies in human urine are summarized by Garner *et al.* (1985). Sun *et al.* (1986) used monoclonal antibody immunoaffinity columns and high performance liquid chromatography (HPLC) for the analysis of AFM_1 in human urine samples.

The largest cross-sectional survey of AFM_1 excretion into urine was reported by Nyathi *et al.* (1987). Over 1200 urine samples were collected from different areas of Zimbabwe and were analyzed for aflatoxin contamination. The urine samples were extracted with chloroform and the resultant aflatoxins quantified by thin layer chromatography and HPLC. The most commonly observed contaminant was AFM_1, at an average concentration of 4.2 ng/ml of urine. Although the national average of urine samples contaminated was 4.3%, in some areas up to 10% of the urine samples were contaminated. These data indicate that AFM_1 is one of the most predominant aflatoxin metabolites in human urine.

Several studies also have described the measurement of aflatoxin metabolites in addition to AFM_1 in urine. In some cases, these measures have been good internal dosimeters reflective of exposure. For example, Wild *et al.* (1986b,1992) have used immunopurification to extract aflatoxins from urine and polyclonal antibody to quantitate in a highly sensitive enzyme-linked immunosorbent assay (ELISA). In a study reported from The Gambia (Wild *et al.*, 1992), the ELISA analysis reflected aflatoxin intake in a dose-dependent manner with a correlation coefficient of 0.65, in contrast to reports from a study in China by Groopman *et al.* (1992a) that found no relation between exposure and urinary aflatoxin metabolites. However, the differences in the specificities of the antibodies used in these different analyses may account for these results.

One major study has been reported that compares exposure to aflatoxins with the urinary excretion of AFM_1 (Zhu *et al.*, 1987). This research group analyzed a total of 252 urine samples from 32 households in Fushui county of the Guangxi Autonomous Region of the People's Republic of China. A good correlation ($r = 0.65$) between total dietary AFB_1 intake and total AFM_1 excretion in human urine was observed during a 3-day period. Between 1.2 and 2.2% of dietary AFB_1 was estimated to be present as AFM_1 in these samples. These findings have been confirmed using a different method for AFM_1 quantitation (Groopman *et al.*, 1992a). In addition, the percentage of AFM_1 excreted into the urine of these people living in Guangxi Province was similar to data collected by Campbell *et al.* (1970) from the urine of Filipinos ingesting peanut butter that was heavily contaminated with approximately 500 µg AFB_1/kg. An estimated 1–4% of the ingested aflatoxin was excreted as this metabolite.

Internal Dose Markers in Milk

Aflatoxin metabolites are excreted not only in urine but also in lactating animals and, thus, can be found in breast milk. To date several literature reports have described the presence of aflatoxins in human milk (Coulter *et al.*, 1984; Lamplugh *et al.*, 1988; Maxwell *et al.*, 1989). Recently, an ELISA and a complementary fluorescence-based HPLC method have been developed to quantify AFM_1 in human milk (Wild *et al.*, 1987). In one study, 54 samples were obtained from women living in Zimbabwe; of those samples, 11% were found to contain up to 50 pg AFM_1 per ml milk. No positive samples were found in 42 milk samples obtained from women living in France. Thus, this technique can be applied to human samples as part of an overall integrated effort to measure aflatoxin metabolites in human samples.

Zarba *et al.* (1992) also reported mother-to-child exposure of AFM_1 in breast milk from a molecular dosimetry study in The Gambia, West Africa, that was initiated to explore the relationships between dietary intake of aflatoxins and a number of aflatoxin biomarkers. In the breast milk study, five lactating women were identified and milk samples were collected by hand expression once a day during days 3–7 for three women and during days 3–6 for the two other women. AFM_1 in human milk was measured in all five subjects by a preparative mono-clonal antibody immunoaffinity column/HPLC method. In three of the five women, aflatoxin G_1 (AFG_1) was found. Estimates of the percentage of aflatoxin in the diet excreted as AFM_1 in milk ranged from 0.09 to 0.43%. Collectively, these data indicate that methods exist to assess the levels of AFM_1 excretion in human milk and to use these approaches in the assessment of child exposure to this carcinogen.

Free Aflatoxins in Blood

Several reports in the literature suggest that AFB_1 can be quantitated in the blood of humans after the consumption of meals containing this mycotoxin. For example, Tsuboi *et al.* (1984) in Japan and Denning *et al.* (1988) in Nigeria have used an ELISA to detect AFB_1 at levels up to 3 ng per ml serum. Similar levels (up to 1.5 ng/ml) have been reported by HPLC–fluorescence techniques in sera from Sudanese children (Hendrickse *et al.*, 1982). These approaches may be useful for the rapid screening of samples for recent exposures. However, measurement of AFB_1 in blood probably will have limited use in studies that obtained biosamples amenable to metabolite analysis.

MARKERS OF BIOLOGICALLY EFFECTIVE DOSE

Experimental Animal Studies

Among the various markers of biologically effective dose that have been developed, the measurement of carcinogen–DNA adducts is of primary interest because these adducts represent damage to a critical macromolecular target. The primary AFB_1–DNA adduct was identified by Essigmann *et al.* (1977) as 2,3-dihydro-2-(N^7-guanyl)-3-hydroxy-AFB_1 (AFB–N^7-guanine), a major product liberated from DNA modified *in vitro* with AFB_1. The presence of this molecule subsequently was confirmed *in vivo* (Croy *et al.*, 1978). The binding of AFB_1 residues to DNA *in vivo* was found to be essentially linear after single-dose exposures (Appleton *et al.*, 1982). The AFB–N^7-guanine adduct is removed from DNA rapidly with a half-life in rats of 8–10 hr (Groopman *et al.*, 1980), and is excreted exclusively in urine of exposed rats (Bennett *et al.*, 1981).

Carcinogen–protein adducts are also very useful because they are surrogate measures of covalent binding to DNA. Because of their longer half-lives than some DNA adducts, these markers can reveal exposures integrated over longer time periods. Sabbioni *et al.* (1987) have identified the structure of the major AFB$_1$–albumin adduct found *in vivo*. The adduct forms with serum albumin by the binding of the epoxide, with subsequent formation of the dihydrodiol and sequential oxidation to the dialdehyde, and condensation with the ε amino group of lysine. This adduct is a Schiff base that undergoes Amadori rearrangement to an α-amino ketone. This protein adduct is a completely modified AFB$_1$ structure that retains only the coumarin and cyclopentenone rings of the parent compound. Sabbioni and Wild (1991) found a major AFG$_1$–albumin adduct in rats after exposure to AFG$_1$. The product isolated from a Pronase® digest of *in vivo*-modified albumin was identical by chromatographic retention time to the synthetic product obtained by the acylase-catalyzed deacetylation product of *N*-α-acetyl-L-lysine with 8,9-dihydro-8,9-dibromo-AFG$_1$. A competitive ELISA for this adduct was established using polyclonal antibodies to AFB$_1$. This assay was used in conjunction with an HPLC–fluorescence technique to quantitate the *in vivo* formation of AFG$_1$–albumin adducts in comparison to AFB$_1$. A linear dose–response relationship was observed in rats after single exposures to 0.1–3 mg AFG$_1$/kg body weight. The levels of AFG$_1$–albumin adducts were determined to be 5.7- and 2.8-fold lower than with equivalent doses of AFB$_1$, as determined by immunoassay and HPLC–fluorescence, respectively. The authors suggest that the adduct could be used as an internal standard for methods that use the measurement of aflatoxin–albumin adducts in humans.

Additional interest in the analysis of aflatoxin–albumin adducts has been kindled by the finding that albumin is the only protein in serum that binds AFB$_1$ to any significant extent in monkeys and rats (Dalezios *et al.*, 1971; Wild *et al.*, 1986a). Of a single dose, 1–3% is bound to serum albumin after 24 hr in rats. A constant relationship between AFB$_1$ bound to plasma albumin and liver DNA has been observed in Wistar rats after single dose (3.5–200 μg/kg) and multiple doses (3.5 μg/kg) of AFB$_1$ (Wild *et al.*, 1986a). This result suggested that albumin-bound AFB$_1$ might be a particularly valuable indicator of DNA damage in the liver.

Numerous animal studies in many different species have been conducted over the years to examine the relationship between AFB$_1$ dose and the formation of DNA and protein adducts. A comprehensive review of these investigations is beyond the scope of this chapter. However, relatively few studies have been done examining the relationship between chronic exposure to AFB$_1$ and macromolecular adduct formation. These studies form an important basis for future human analyses, and the reader is urged to examine these efforts (Croy *et al.*, 1981; Kensler *et al.*, 1986; Wild *et al.*, 1986a; Goeger *et al.*, 1988). The following discussion describes some research to develop molecular dosimetry methods for AFB–N^7-guanine excretion in urine and serum albumin adduct formation.

Studies were done in rats after a single exposure to AFB$_1$ to characterize excretion kinetics of specific metabolites. Urine was collected over 24 hr from 12 male F344 rats dosed orally at levels ranging from 0.030 to 1.00 mg AFB$_1$/kg body weight (Groopman *et al.*, 1992b). Aliquots of each urine were purified preparatively by immunoaffinity chromatography and individual metabolites were quantified by HPLC analysis. $AFB-N^7$-guanine, AFQ_1, AFP_1, AFM_1, and AFB_1 accounted for 7.5, 3.0, 31.5, 2.2, and 0.3% of the total aflatoxins injected on the HPLC, respectively. The relationship between AFB_1 dose and the excretion of the $AFB-N^7$-guanine adduct over the 24 hr after exposure was determined to have a correlation coefficient of 0.99. This analysis demonstrates an excellent linear correspondence between oral dose and excretion of a biologically relevant metabolite in urine. In contrast, other oxidative metabolites such as AFP_1 revealed no linear excretion characteristics. Finally, at 24 hr postdosing, the residual level of AFB_1 liver DNA adducts was determined and compared with $AFB-N^7$-guanine excretion in urine; the correlation coefficient was 0.98, which supports the concept that measurement of the $AFB-N^7$-guanine adduct in urine reflects DNA damage in the primary target organ.

The risk for AFB_1 hepatocarcinogenesis can be modified in animals by using a number of chemoprotective agents including phenolic antioxidants, ethoxyquin, and dithiolethiones. For example, Cabral and Neal (1983) found that ethoxyquin (EQ) can inhibit hepatocarcinogenesis in the rat completely. Therefore, determining whether DNA adduct formation by AFB_1 in liver corresponded to this protective effect was of interest. The effects of dietary administration of EQ on AFB_1 metabolism, DNA adduct formation and removal, and hepatic tumorigenesis were examined in male Fischer rats (Kensler et al., 1986). Rats were fed a semipurified diet containing 0.4% EQ for 1 week, gavaged with 250 μg AFB_1 per kg 5 times a week during the next 2 weeks, and finally restored to the control diet 1 week after cessation of dosing. At 4 months, focal areas of hepatocellular alteration were identified and quantitated by staining sections of liver for gamma-glutamyl transpeptidase (GGT). Treatment with EQ reduced both area and volume of liver occupied by GGT^+ foci by more than 95%.

Using the same multiple dosing protocol, patterns of covalent modifications of DNA by AFB_1 were determined. EQ produced a dramatic reduction in the binding of AFB_1 to hepatic DNA: 18-fold initially and 3-fold at the end of the dosing period. Although binding was detectable at 3 and 4 months postdosing, no effect of EQ was observed, suggesting that these persistent adducts are not of primary relevance to AFB_1 carcinogenesis. The inhibitory effect of EQ on AFB_1 binding to DNA and tumorigenesis appeared to be related to induction of detoxification enzymes, since rats fed 0.4% EQ for 7 days showed a 5-fold increase in hepatic cytosolic glutathione S-transferase specific activities. Correspondingly, biliary elimination of AFB_1–glutathione conjugates was increased 4.5-fold in animals on the EQ diet during the first 2 hr after administration of AFB_1. Thus, induction by EQ of enzymes important to AFB_1 detoxification, for example, glutathione S-transferase, can lead to enhanced carcinogen elimination as well as to reductions of AFB_1–DNA adduct formation and subsequent expression of preneoplastic lesions and, ultimately, neoplasia.

These studies have been extended to the cancer chemoprotective agent oltipraz, a substituted dithiolethione (Roebuck et al., 1991). Male F344 rats were fed a purified diet supplemented with 0.075% w/w oltipraz for 4 weeks. In this study, rats received 10 intragastric doses of AFB_1 (25 μg/rat/day; days 8–12 and 15–19). This 10-dose exposure to AFB_1 produced an 11% incidence of hepatocellular carcinoma at 23 months; an additional 9% of the rats exhibited hyperplastic nodules in their livers. In contrast, feeding rats a diet supplemented with 0.075% oltipraz for a 4-week period surrounding the time of AFB_1 exposure afforded complete protection against AFB_1-induced hepatocellular carcinomas and hyperplastic nodules. None of these lesions was observed in the oltipraz-fed AFB_1-treated animals. Further, no tumors were found at secondary extrahepatic sites for AFB_1 carcinogenesis such as the colon and the kidney. At the same time, animals exposed to oltipraz exhibited a dramatic (75–80%) reduction in liver DNA damage, which was mirrored by the reduction in the amount of AFB_1–N^7-guanine excreted in their urine. These data strongly support a conclusion that the AFB_1–N^7-guanine adduct is a relevant measure for risk of tumor development. Thus, the chemoprotection model affords us an experimental system in which to ask basic questions about the relationships between levels of AFB–N^7-guanine in urine in high and low risk animals.

Structure–activity studies with dithiolethione indicated that the cancer chemoprotective activity of oltipraz is embodied exclusively in the 1,2-dithiole-3-thione nucleus of the molecule. Several experiments indicated that the unsubstituted congener of oltipraz, 1,2-dithiole-3-thione, is also an effective inhibitor of AFB_1-induced tumorigenesis, as determined by analyses for preneoplastic foci expressing either GGT^+ or glutathione S-transferase π (Kensler *et al.*, 1992). These data justified an examination of the impact of chemoprotection by 1,2-dithiole-3-thione on the molecular dosimetry of AFB–N^7-guanine in urine and a comparison of the modulation of this biomarker with levels of AFB_1–DNA adduct in the liver. The effects of 1,2-dithiole-3-thione on the kinetics of hepatic AFB_1–DNA adduct formation and removal in rats receiving 250 μg AFB_1/kg by gavage on days 0–4 and 7–11 are shown in Figure 3A. Maximal levels of carcinogen binding were achieved following the third dose in the control group, and declined thereafter despite continued exposure to AFB_1. This diminution of binding, particularly during the second dosing cycle, has been observed previously (Croy *et al.*, 1981; Kensler *et al.*, 1986; Wild *et al.*, 1986a) and may be a consequence of the induction of glutathione S-transferases or other enzymes involved in AFB_1 detoxification after chronic exposure to AFB_1 (Kensler *et al.*, 1992). Inclusion of 0.03% 1,2-dithiole-3-thione in the diet, beginning 1 week prior to dosing with AFB_1, resulted in substantially lower levels of hepatic AFB_1–DNA adducts throughout the exposure period. Binding was reduced by 76% over the initial 18-day period.

The levels of total AFB_1 equivalents in 24-hr urine samples collected over the 2-week exposure period showed no remarkable differences between rats fed the control AIN-76A diet and those fed the 1,2-dithiole-3-thione-supplemented diet (Groopman *et al.*, 1992c). Urinary aflatoxin levels rise rapidly after dosing with AFB_1 and drop equally quickly after cessation of dosing, reflecting the overall

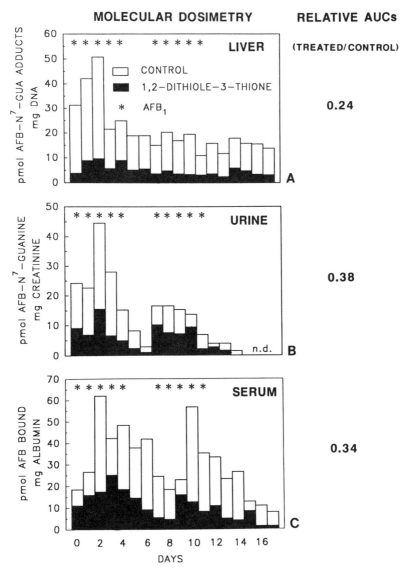

FIGURE 3 Total urinary aflatoxins in rats exposed to AFB_1.

short *in vivo* half-life of this carcinogen. The lack of an effect by 1,2-dithiole-3-thione is not surprising, since exposures to AFB_1 were identical in both dietary groups. However, a distinctly different pattern emerges when the urines are subjected to sequential monoclonal antibody immunoaffinity chromatography and HPLC. Shown in Figure 3B are the levels of $AFB-N^7$-guanine in serial 24-hr urine samples collected from rats undergoing a chemoprotective intervention with 0.03% 1,2-dithiole-3-thione. The highest level of $AFB-N^7$-guanine excretion occurred on day 2 in both groups after the third dosage of AFB_1. This

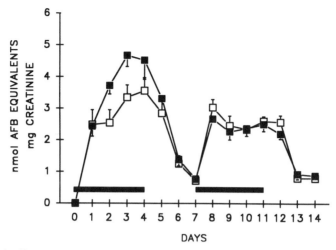

FIGURE 4 Chemoprotection model data in rats exposed to AFB_1. ■ ABB_1; □ AFB_1 + 1,2-dithiole-3-thione.

outcome is identical to that observed with hepatic levels of AFB_1–DNA adducts and with the serum albumin adduct formation (Figure 3C). Over the 15-day collection period in which AFB–N^7-guanine adducts were detectable in the urine, feeding of 1,2-dithiole-3-thione produced an overall reduction of 62% in the elimination of this AFB_1–DNA adduct excision product, mirroring the data on the overall levels of hepatic AFB_1–DNA adducts. The amount of AFB–N^7-guanine in urine represents only 1% of the total aflatoxin metabolites in urine and explains why the dramatic differences seen between treatment groups in AFB–N^7-guanine levels are not reflected in the levels of total urinary aflatoxin metabolites shown in Figure 4. Thus, these data indicate that the excreted DNA adduct in urine and the formation of the serum albumin adduct accurately reflect the amount of genotoxic damage at the target organ site in the liver. In addition, these data indicate that the measurement of these adducts may reflect risk for disease development in the animal.

Human Studies

Despite its development and validation in the susceptible rat model, a systematic evaluation of the AFB–N^7-guanine adduct in several human populations has supported its validity as a measure of relevant exposure. The first studies by Autrup et al. (1983,1987) used synchronous fluorescence spectroscopy (SFS) for the analysis of AFB–DNA adducts in urine. Data were collected data on AFB–N^7-guanine in urine samples from in Murang'a district, Kenya. Over 1000 urine samples were analyzed, 12.6% of which were positive for excretion of AFB–N^7-guanine. These researchers found a regional variation in the excretion levels; the highest rate of AFB_1–DNA adducts was found in the Western Highlands and Central Province.

Additional studies have been carried out in the Guanxgi Autonomous Region, which is among the areas of highest liver cancer incidence in China. This study had the advantage of having both dietary intake data and the measure of urinary aflatoxin biomarkers. The dietary intake of aflatoxins was monitored for 1 week in a study group consisting of 30 males and 12 females ranging in age from 25 to 64 years (Groopman *et al.,* 1992a). The vast majority of AFB_1 exposure was from contaminated corn. The average male intake of AFB_1 was 48.4 μg per day, giving a total mean exposure over the study period of 276.8 μg. The average female intake was 92.4 μg per day resulting in a total average exposure over the 7-day period of 542.6 μg AFB_1. The maximum intake for the male and female subjects over the 1-week collection period was 963.9 and 1035 μg, respectively; the minimum exposure for male and female subjects was 56.7 and 90 μg AFB_1, respectively. Total 24-hr urine samples were collected starting on the fourth day as consecutive 12-hr fractions.

Total aflatoxin metabolites in the urine samples were measured by competitive radioimmunoassay using a monoclonal antibody that recognizes AFB_1, AFM_1, and aflatoxin P_1 (AFP_1) with equal affinity and cross-reacts with $AFB-N^7$-guanine with 5- to 10-fold less affinity. The relationship between AFB_1 intake per day and total aflatoxin metabolite excretion per day revealed a correlation coefficient of 0.26, with a statistical significance level of 0.10. Thus, total metabolites in urine as measured by this method did not provide data to indicate that total metabolites were an appropriate dosimeter measurement for exposure status.

These results prompted analysis of the urine samples by combined preparative monoclonal antibody affinity chromatography and analytical HPLC to determine levels of individual aflatoxin metabolites; nearly 550 individual analytical HPLCs were performed. $AFB-N^7$-guanine, AFM_1, AFP_1, and AFB_1 were the aflatoxins most commonly detected and quantified in the urine samples. The linear regression analyses for the urinary levels of each of these individual aflatoxins was compared with aflatoxin intake to yield correlation coefficients for $AFB-N^7$-guanine, AFM_1, AFP_1, and AFB_1 of 0.65, 0.55, 0.02, and -0.10, respectively. Also, investigators noted that AFP_1 contributes substantially to the overall levels of aflatoxin in the urine samples as detected by competitive radioimmunoassay, thereby masking the association between exposure and minor metabolites. However, the resolution of the total aflatoxin metabolite content in the urine reveals the association of $AFB-N^7$-guanine and AFM_1 excretion as biomarkers of exposure.

One objective of this study was determining the number of samples required from an individual and the time-frame for sample collection necessary to validate a biomarker as reflecting a biologically effective dose of AFB_1 in humans. Particular interest arose to characterize the molecular dosimetry of $AFB-N^7$-guanine because of the putative relationship of this metabolite with the cancer initiation process. Figure 5 shows total $AFB-N^7$-guanine excretion in the urine of the male and female subjects over the complete urine collection period, plotted

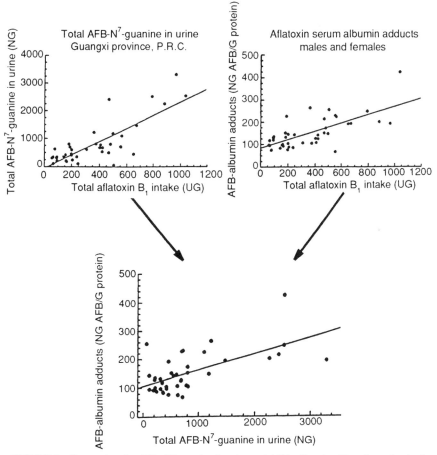

FIGURE 5 Human data for AFB–N[7]-guanine in urine and AFB–albumin adduct formation in the same individuals.

against the total AFB_1 exposure in the diet for each of the subjects. This analysis smooths the day-to-day variations in both intake and excretion of AFB–N[7]-guanine and reveals a correlation coefficient of 0.80 and a P value of <0.0000001. This analysis clearly demonstrates that a summation of excretion and exposure status provides a stronger association between exposure and the molecular dosimetry marker than was seen in prior statistical analyses, and supports the concept that quantitation of the AFB–N[7]-guanine adduct in urine is a reliable biomarker for AFB_1 exposures.

In the subjects just described, Gan *et al.* (1988) examined serum albumin adduct formation as well. Immunoreactive products were purified by immunoaffinity chromatography and quantified by competitive radioimmunoassay. A highly significant correlation of adduct level with intake ($r = 0.69$, P

<0.000001) was observed. From the slope of the regression line for adduct level as a function of intake, 1.4–2.3% of ingested AFB_1 was calculated to become covalently bound to serum albumin, a value very similar to that observed when rats are administered AFB_1. When the data for DNA adduct excretion in urine and serum albumin are compared (Figure 5), a statistically significant relationship is seen that has a correlation coefficient of 0.73. Thus, both these markers appear to be valid in human monitoring studies.

Similar studies of DNA adduct excretion and serum albumin adduct formation have been done in The Gambia. These studies reveal similar relationships between dietary exposure and adduct formation (Groopman *et al.*, 1992d; Wild *et al.*, 1992). The correlation coefficients for DNA adduct excretion and serum albumin adduct formation were 0.82 and 0.55, respectively. Thus, in two populations at high risk for liver cancer, living in very different geographic regions and having different dietary sources of aflatoxins, these dosimetry markers appear to be rigorous for exposure determination.

The investigations described here revealed that both the DNA and the albumin adducts of aflatoxin are useful molecular dosimetry markers. However, from a practical perspective, the measurement and quantification of the serum albumin adduct offers an approach that can be used to screen very large numbers of people. These methods have been validated extensively in experimental and human sample analysis. The technique is described in detail in Wild *et al.* (1990c). Using this method, an extensive geographic study was carried out on hundreds of samples (Wild *et al.*, 1990b). Measurements of aflatoxin bound to serum albumin in children and adults from various African countries show that 12–100% contain aflatoxin–albumin adducts, with levels up to 350 pg AFB_1–lysine equivalent/mg albumin. In Thailand, lower levels and prevalence of this adduct were observed, while no positive sera were detected in France or Poland. Data are presented showing that exposure to this carcinogen can occur throughout life. Significant seasonal variation, consistent with high and low aflatoxin levels in foods, was observed. Collectively, the data from the ELISA approach validate it as an important screening tool for epidemiological studies.

MOLECULAR DOSIMETRY FOR MARKERS OF SUSCEPTIBILITY

Metabolic susceptibility factors can come into play at almost any time between an exposure and a disease outcome. Certainly differences in individual absorption and distribution, metabolism, and repair all have the potential to influence individual sensitivity to a particular agent. In turn, these specific differences may account for the wide variations in incidence of cancer between and within groups of people. Thus, the availability of biomarkers to assess the individual's intrinsic status to metabolize aflatoxins may be important for risk analysis.

The mechanistic basis for metabolic phenotyping for aflatoxin exposure is derived from the work reported by Shimada and Guengerich (1989). Their data with human liver microsomes indicated that the major cytochrome P450 involved in the bioactivation of AFB_1 to its genotoxic epoxide derivative is cytochrome P450 3A4. This previously characterized protein also catalyzes the oxidation of nifedipine and other dihydropyridines, quinidine, macrolide antibiotics, various steroids, and other compounds. Evidence was obtained using activation of AFB_1 (as monitored by *umuC* gene expression response in *Salmonella typhimurium* TA1535/pSK1002), enzyme reconstitution, immunochemical inhibition, and correlation of response with levels of P450 3A4 (nifedipine oxidase) activity in different liver samples. Liver samples with increasing levels of this enzyme produced higher amounts of $AFB–N^7$-guanine in DNA *in vitro*.

Biomarkers for the phenotypic determination of cytochrome P450 3A4 have been reported. In particular, the ratio of endogenous urinary 6β-hydroxycortisol to cortisol is a marker used to assess the activity of cytochrome P450 3A4 (Joellenbeck *et al.*, 1992). Cytochrome P450 3A4 is a major liver enzyme involved in the activation of benz[*a*]pyrene diol epoxide and several other carcinogens. Thus, the development of biomarkers for this enzyme will be applicable to many other epidemiological studies (Guengerich *et al.*, 1992).

MARKERS OF EARLY BIOLOGICAL EFFECT

Some of the most exciting prospects for markers of biological effect or pre-clinical disease arise from tools developed in the field of molecular biology. The increasing mechanistic understanding of the genetic alterations that underlie the progression from initiation to tumor formation has permitted the beginning development of sensitive tests for diagnosis of tumors. One example of this emerging field is presented by the studies of the tumor suppressor gene *p53*, which is the most common mutated gene detected in human cancers. The number and type of mutations in this gene are not equally distributed, but occur in specific hot spots that vary with tumor type (Hollstein *et al.*, 1991). Different patterns in mutations among tumors are consistent with different etiologies for the specific tumor types. For example, two independent studies of *p53* mutations in hepatocellular carcinomas occurring in populations exposed to aflatoxin revealed high frequencies of G → T transversions, with clustering at codon 249 (Bressac *et al.*, 1991; Hsu *et al.*, 1991). On the other hand, studies of *p53* mutations in hepatocellular carcinomas from Japan and other areas in which little exposure to aflatoxin occurs revealed no mutations at codon 249 (Ozturk *et al.*, 1991). Previous data in bacteria had shown that AFB_1 exposure causes almost exclusively G → T transversions (Foster *et al.*, 1983). Also, some work has shown that aflatoxin-epoxide can bind to this particular codon 249 of *p53* in a plasmid *in vitro*, providing additional indirect evidence for a putative role of aflatoxin exposure in *p53* mutagenesis (Puisieux *et al.*, 1991). This information provides the circumstantial, but as yet unproven, link between this signature mutation of aflatoxin exposure and the events detected in *p53* in liver tumors from China and southern Africa.

The knowledge of mutations in the tumor suppressor gene *p53* has been used as a possible diagnostic tool for early stage cancers, such as bladder cancer. In these studies, mutations detected in the *p53* gene from bladder epithelial cells shed into the urine of bladder cancer patients (Sidransky *et al.,* 1991) suggest future development of tests for genetic alterations by examining proteins shed in urine. Additional studies are required to determine the temporal relationship between these mutations and disease status. Perhaps in the future, a mutational spectrum in a gene such as *p53* can serve as a marker of exposure to and damage from other environmental agents such as AFB_1.

EPIDEMIOLOGICAL STUDIES

The molecular dosimetry markers developed for aflatoxin exposure can be used in a variety of epidemiological investigations, including cross-sectional, case–control, and cohort or prospective investigations. A report of a cross-sectional study by Campbell *et al.,* (1990) revealed a nonassociation between composite aflatoxin metabolites in human urine and liver cancer disease rates. These findings are consistent with the animal and human data described previously in this chapter, and provide further proof that the composite measurement in urine is not a good marker for exposure and risk from aflatoxins.

A case–control study of potential risk factors for hepatocellular carcinoma was reported for northeast Thailand (Srivatanakul *et al.,* 1991). In this study, 65 cases from three hospitals and matched controls were examined. Infection with hepatitis B virus was the major risk factor identified; chronic carriers of hepatitis surface antigen had an estimated relative risk of 15.2. No increase in risk was found with recent aflatoxin intake, as estimated by consumption of possibly contaminated foods or by measuring aflatoxin–albumin adducts in serum. Regular use of alcohol (2 or more glasses of spirits per week) was associated with a nonsignificant elevation in risk (*o.r.* = 3.4, 95% c.i. 0.8–14.6), but the number of regular drinkers in the population was small. Thus, in this case–control study, the contribution of recent exposure to aflatoxin to liver cancer was not significant.

In most instances, the most rigorous test of an association between an agent and a disease outcome is found in prospective epidemiological studies, in which healthy people are followed up until the disease is diagnosed. Data analyzed from a prospective nested case–control study, begun in Shanghai in 1986 to examine the relationship between markers for aflatoxin and hepatitis B virus and the development of liver cancer, have been published (Ross *et al.,* 1992). Over a 3.5-year period, 18,244 urine samples were collected from healthy males between the ages of 45 and 64. Of these individuals, 22 subsequently developed liver cancer. Their urine samples were age-matched with 5–10 controls and analyzed for aflatoxin biomarkers and hepatitis B virus surface antigen status. The data revealed a highly significant increase in the relative risk (RR = 4.9) for those liver cancer cases in which AFB–N[7]-guanine was detected. Also, elevated risks for other aflatoxin urinary markers were seen. The relative risk for people who tested positive for the hepatitis B virus surface antigen was also ~5, but individuals with both urinary aflatoxins and positive hepatitis B virus surface antigen status had a relative risk for developing liver cancer of over 60. These results show, for the first time, the relationship of a chemical carcinogen-specific biomarker with cancer risk and also demonstrate for the first time a synergistic interaction between two major risk factors for liver cancer—hepatitis B virus and AFB_1.

FUTURE PROSPECTS

Primary hepatocellular carcinoma is one of the most lethal and common cancers in the world. Several epidemiological studies have associated the exposure status of people to AFB_1 with the etiology of liver cancer. However, until the report by Ross *et al.* (1992), these studies relied on the criteria of presumptive intake data rather than on quantitative analyses of aflatoxin–DNA adduct and metabolite contents obtained by monitoring biological fluids from exposed people. Information obtained by monitoring exposed individuals for specific DNA adducts and metabolites appears to be a good predictor of risk to develop liver cancer. Although additional work must be done, both experimental animal studies and human analyses support the concept that measurement of the major rapidly excised AFB–N[7]-guanine adduct in tissues and fluids and of the more persistent aflatoxin–albumin adduct is appropriate dosimetry for estimating exposure status and risk in individuals consuming this mycotoxin.

The individual molecular biomarkers discussed in this chapter, whether already being applied in population studies or still in the earliest phases of exploration, reflect only one aspect of the exposure-to-disease continuum depicted in Figure 1. These markers indicate exposure, such as internal dose (i.e., urinary aflatoxins) or biologically effective dose (i.e., AFB–N[7]-guanine in urine and aflatoxin–albumin adducts), or may indicate the development of altered structure/function or disease, or possible susceptibility to disease (possibly mutated *p53*). The major developmental challenge facing the field is the knowledge of the connection between biologically effective dose and biological effect. To date, only through supposition is this linkage made. Even in cases in which all the data are consistent with a specific etiological agent and mutagenic event, for example, aflatoxin and *p53* mutations, the specific data remain to be elucidated. Such information eventually will be gleaned from experimental studies and molecular epidemiological investigations in human populations. In general, with continued rapid development of the technologies for biomarker studies, these goals should be attained in the near future. However, even without information connecting a marker of exposure with a genetic effect, the molecular biomarkers of exposure, susceptibility, and effect provide improved tools for epidemiological investigations compared with previously available measures such as the survey questionnaire or ambient air measurement of an agent. Thus, these tools should evolve quickly in the policy and risk assessment setting, since they should confirm precisely whether a level of exposure set by regulations is safe.

ACKNOWLEDGMENTS

This work was supported by grants from the U.S. Public Health Service (CA54114, CA39416, CA48409, ES06052, and K04 CA 01517).

REFERENCES

Appleton, S., Coetschius, M. P., and Campbell, T. C. (1982). Linear dose–response curve for hepatic macromolecular binding of aflatoxin B_1 in rats at very low exposures. *Cancer Res.* **42,** 3659–3662.

Autrup, H., Bradley, K. A., Shamsuddin, A. K. M., Wakhisi, J., and Wasunna, A. (1983). Detection of putative adduct with fluorescence characteristics identical to 2,3-dihydro-2-(7'-guanyl)-3-hydroxyaflatoxin B_1 in human urine collected in Murang'a district, Kenya. *Carcinogenesis* **4,** 1193–1195.

Autrup, H., Seremet, T., Wakhisi, J., and Wasunna, A. (1987). Aflatoxin exposure measured by urinary excretion of aflatoxin B_1–guanine adduct and hepatitis B virus infection in areas with different liver cancer incidence in Kenya. *Cancer Res.* **47,** 3430–3433.

Baertschi, S. W., Raney, K. D., Stone, M. P., and Harris, T. M. (1988). Preparation of the 8,9-epoxide of the mycotoxin aflatoxin B_1: The ultimate carcinogenic species. *J. Am. Chem. Soc.* **110,** 7929–7931.

Bennett, R. A., Essigmann, J. M., and Wogan, G. N. (1981). Excretion of an aflatoxin–guanine adduct in the urine of aflatoxin B_1-treated rats. *Cancer Res.* **41,** 650–654.

Bosch, F. X., and Munoz, N. (1988). Prospects for epidemiological studies on hepatocellular cancer as a model for assessing viral and chemical interactions. *IARC. Sci. Publ.* **89,** 427–438.

Bressac, B., Kew, M., Wands, J., and Ozturk, M. (1991). Selective G to T mutations of *p53* gene in hepatocellular carcinoma from Southern Africa. *Nature (London)* **350,** 429–431.

Busby, W. F., and Wogan, G. N. (1985). Aflatoxins. *In* "Chemical Carcinogens" (C. E. Searle, ed.), 2d Ed., pp. 945–1136. American Chemical Society, Washington, D.C.

Cabral, J. R., and Neal, G. E. (1983). The inhibitory effects of ethoxyquin on the carcinogenic action of aflatoxin B_1 in rats. *Cancer Lett.* **19,** 125–132.

Campbell, T. C., Caedo, J. P., Bulatao-Jayme, J., Salamat, L., and Engel, R. W. (1970). Aflatoxin M_1 in human urine. *Nature (London)* **227,** 403–404.

Campbell, T. C., Chen, J., Liu, C., Li, J., and Parpia, B. (1990). Nonassociation of aflatoxin with primary liver cancer in a cross-sectional ecological survey in the People's Republic of China. *Cancer Res.* **50,** 6881–6893.

Coulter, J. B., Lamplugh, S. M., Suliman, G. I., Omer, M. I., and Hendrickse, R. G. (1984). Aflatoxins in human breast milk. *Ann. Trop. Paediatr.* **4,** 61–66.

Croy, R. G., and Wogan, G. N. (1981). Temporal patterns of covalent DNA adducts in rat liver after single and multiple doses of aflatoxin B_1. *Cancer Res.* **41,** 197–203.

Croy, R. G., Essigmann, J. M., Reinhold, V. N., and Wogan, G. N. (1978). Identification of the principal aflatoxin B_1–DNA adduct formed *in vivo* in rat liver. *Proc. Natl. Acad. Sci. U.S.A.* **75,** 1745–1749.

Dalezios, J. I. (1971). Aflatoxin P_1: A new metabolite of aflatoxin B_1. Its isolation and identification. Ph.D. Thesis. Massachusetts Institute of Technology, Cambridge.

Denning, D. W., Onwubalili, J. K., Wilkinson, A. P., and Morgan, M. R. (1988). Measurement of aflatoxin in Nigerian sera by enzyme-linked immunosorbent assay. *Trans. R. Soc. Trop. Med. Hyg.* **82,** 169–71.

Essigmann, J. M., Croy, R. G., Nadzan, A. M., Busby, W. F., Jr., Reinhold, V. N., Buchi, G., and Wogan, G. N. (1977). Structural identification of the major DNA adduct formed by aflatoxin B_1 *in vitro. Proc. Natl. Acad. Sci. U.S.A.* **74,** 1870–1874.

Foster, P. L., Eisenstadt, E., and Miller, J. H. (1983). Base substitution mutations induced by metabolically activated aflatoxin B_1. *Proc. Natl. Acad. Sci. U.S.A.* **80,** 2695–2698.

Gan, L.-S., Skipper, P.-L., Peng, X.-C., Groopman, J. D., Chen, J.-S., Wogan, G. N., and Tannenbaum, S. R. (1988). Serum albumin adducts in the molecular epidemiology of aflatoxin carcinogenesis: Correlation with aflatoxin B_1 intake and urinary excretion of aflatoxin M_1. *Carcinogenesis* **9,** 1323–1325.

Garner, R. C., Ryder, R., and Montesano, R. (1985). Monitoring of aflatoxins in human body fluids and application to filed studies. *Cancer Res.* **45,** 922–928.

Garner, R. C., Dvorackova, I., and Tursi, F. (1988). Immunoassay procedures to detect exposure to aflatoxin B_1 and benzo(a)pyrene in animals and man at the DNA level. *Int. Arch. Occup. Environ. Health.* **60,** 145–150.

Goeger, D. E., Shelton, D., Hendricks, J. D., Pereira, C., and Bailey, G. S. (1988). Comparative effect of dietary butylated hydroxyanisole and β-naphthoflavone on aflatoxin B_1 metabolism, DNA adduct formation and carcinogenesis in rainbow trout. *Carcinogenesis* **9,** 1793–1800.

Groopman, J. D., Busby, W. F., and Wogan, G. N. (1980). The nuclear distribution of aflatoxin B_1 and its interaction with histones *in vivo. Cancer Res.* **40,** 4343–4351.

Groopman, J. D., Sabbioni, G., and Wild, C. P. (1991). Molecular dosimetry of aflatoxin exposures. *In* "Molecular Dosimetry of Human Cancer: Epidemiological, Analytical, and Social Considerations" (J. D. Groopman and P. Skipper, eds.), pp. 302–324. CRC Press, Boca Raton, Florida.

Groopman, J. D., Zhu, J., Donahue, P. R., Pikul, A., Zhang, L.-S., Chen, J-S., and Wogan, G. N. (1992a). Molecular dosimetry of urinary aflatoxin DNA adducts in people living in Guangxi Autonomous Region, People's Republic of China. *Cancer Res.* **52,** 45–51.

Groopman, J. D., Hasler, J., Trudel, L. J., Pikul, A., Donahue, P. R., and Wogan, G. N. (1992b). Molecular dosimetry in rat urine of aflatoxin-N^7-guanine and other aflatoxin metabolites by multiple monoclonal antibody affinity chromatography and HPLC. *Cancer Res.* **52,** 267–274.

Groopman, J. D., Egner, P., Love-Hunt, A., DeMatos, P., and Kensler, T. W. (1992c). Molecular dosimetry of aflatoxin DNA and serum albumin adducts in chemoprotection studies using 1,2-dithiole-3-thione in rats. *Carcinogenesis* **13,** 101–106.

Groopman, J. D., Hall, A., Whittle, H., Hudson, G., Wogan, G. N., Montesano, R., and Wild, C. P. (1992d). Molecular dosimetry of aflatoxin-N^7-guanine in human urine obtained in The Gambia, West Africa. *Cancer Epidemiol. Biomarkers Prev.* **1,** 221–228.

Guengerich, F. P., Shimada, T., Raney, K. D., Yun, C-H., Meyer, D. J., Ketterer, B., Harris, T. M., Groopman, J. D., and Kadlubar, F. F. (1992). Elucidation of catalytic specificities of human cytochrome P-450 and glutathione S-transferase enzymes and relevance to molecular epidemiology studies. *Environ. Health Perspec.* **98,** 75–80.

Hendrickse, R. G., Coulter, J. B., Lamplugh, S. M., Macfarlane, S. B., Williams, T. E., Omer, M. I., and Suliman, G. I. (1982). Aflatoxins and kwashiorkor: A study in Sudanese children. *Br. Med. J.* **285,** 843–846.

Hollstein, M., Sidransky, D., Vogelstein, B., and Harris, C. C. (1991). p53 Mutations in human cancers. *Science* **253,** 49–53.

Hsieh, L. L., Hsu, S. W., Chen, D. S., and Santella, R. M. (1989). Immunological detection of aflatoxin B_1–DNA adducts formed *in vivo. Cancer Res.* **48,** 6328–6331.

Hsu, I. C., Metcalf, R. A., Sun, T., Wesh, J. A., Wang, N. J., and Harris, C. C. (1991). Mutational hotspot in the *p53* gene in human hepatocellular carcinomas. *Nature (London)* **350,** 427–428.

Hulka, B. (1991). Epidemiological studies using biological markers: Issues for epidemiologists. *Cancer Epidemiol. Biomarkers Prev.* **1,** 13–19.

Joellenbeck, L., Qian, Z., Zarba, A., and Groopman, J. D. (1992). Urinary 6β-hydroxycortisol/cortisol ratios measured by HPLC for use as a biomarker for the human cytochrome P-450 3A family. *Cancer Epidemiol. Biomarkers Prev.* **1,** 567–572.

Kensler, T. W., Egner, P. A., Davidson, N. E., Roebuck, B. D., Pikul, A., and Groopman, J. D. (1986). Modulation of aflatoxin metabolism, aflatoxin-N^7-guanine formation, and hepatic tumorigenesis in rats fed ethoxyquin: Role of induction of glutathione *S*-transferase. *Cancer Res.* **46,** 3924–3931.

Kensler, T. W., Groopman, J. D., Eaton, D. L., Curphey, T. J., and Roebuck, B. D. (1992). Potent inhibition of aflatoxin-induced hepatic tumorigenesis by the monofunctional enzyme inducer 1,2-dithiole-3-thione. *Carcinogenesis* **12,** 95–100.

Lamplugh, S. M., Hendrickse, R. G., Apeagyei, F., and Mwanmut, D. D. (1988). Aflatoxins in breast milk, neonatal cord blood, and serum of pregnant women. *Br. Med. J.* **296,** 968.

Maxwell, S. M., Apaegyei, F., deVries, H. R., Mwanmut, D. D., and Hendrickse, R. G. (1989). Aflatoxins in breast milk, neonatal cord blood, and sera of pregnant women. *J. Toxicol. Toxin Rev.* **8,** 19–29.

National Research Council Committee on Biological Markers (1987). Biological markers in environmental health research. *Environ. Health Perspect.* **74,** 3–9.

Nyathi, C. B., Mutiro, C. F., Hasler, J. A., and Chetsanga, C. J. (1987). A survey of urinary aflatoxin in Zimbabwe. *Int. J. Epidemiol.* **16,** 516–519.

Ozturk, M., and collaborators (1991). p53 Mutation in hepatocellular carcinoma after aflatoxin exposure. *Lancet* **338,** 1356–1359.

Puisieux, A., Lim, S., Groopman, J. D., and Ozturk, M. (1991). Selective targeting of p53 gene mutational hotspots in human cancers by etiologically defined carcinogens. *Cancer Res.* **51,** 6185–6189.

Raney, K. D., Meyer, D. J., Ketterer, B., Harris, T. M., and Guengerich, F. P. (1992). Glutathione conjugation of aflatoxin B_1 exo- and endo-epoxides by rat and human glutathione *S*-transferases. *Chem. Res. Toxicol.* **5,** 470–478.

Roebuck, B. D., Liu, Y-L., Rogers, A. E., Groopman, J. D., and Kensler, T. W. (1991). Protection against aflatoxin B_1-induced hepatocarcinogenesis in F344 rats by 5-(2-pyrazinyl)-4-methyl-1,2-dithiol-3-thione (oltipraz): Predictive role for short-term molecular dosimetry. *Cancer Res.* **51,** 5501–5506.

Ross, R., Yuan, J-M., Yu, M., Wogan, G. N., Qian, G-S., Tu, J-T., Groopman, J. D., Gao, Y.-T., and Henderson, B. E. (1992). Urinary aflatoxin biomarkers and risk of hepatocellular carcinoma. *Lancet* **339,** 943–946.

Sabbioni, G., and Wild, C. P. (1991). Identification of an aflatoxin G_1–serum albumin adduct and its relevance to the measurement of human exposure to aflatoxins. *Carcinogenesis* **12,** 97–103.

Sabbioni, G., Skipper, P., Buchi, G., and Tannenbaum, S. R. (1987). Isolation and characterization of the major serum albumin adduct formed by aflatoxin B_1 *in vivo* in rats. *Carcinogenesis* **8,** 819–824.

Shimada, T., and Guengerich, F. P. (1989). Evidence for cytochrome P-450NF, the nifedipine oxidase, being the principal enzyme involved in the bioactivation of aflatoxins in human liver. *Proc. Natl. Acad. Sci. U.S.A.* **86,** 462–465.

Sidransky, D., Von Eschenback, A., Tsai, Y. C., Jones, P. *et al.* (1991). Identification of *p53* gene mutations in bladder cancers and urine samples. *Science* **252,** 706.

Srivatanakul, P., Parkin, D. M., Khlat, M., Chenvidhya, D., Chotiwan, P., Insiripong, S., L'Abbe, K. A., and Wild, C. P. (1991). Liver cancer in Thailand. II. A case-control study of hepatocellular carcinoma. *Int. J. Cancer* **48,** 329–332.

Sun, T. T., Chu, Y.-R., Hsia, C.-C., Wei, Y.-P., and Wu, S.-M., (1986). Strategies and current trends of etiologic prevention of liver cancer. *In* "Biochemical and Molecular Epidemiology" (C. C. Harris, ed.), pp. 283–292. Liss, New York.

Tsuboi, S., Nakagawa, T., Tomita, M., Seo, T., Ono, H., Kawamura, K., and Iwamura, N. (1984). Detection of aflatoxin B_1 in serum samples of male Japanese subjects by radioimmunoassay and high-performance liquid chromatography. *Cancer Res.* **44,** 1231–1234.

Wild, C. P., Garner, R. G., Montesano, R., and Tursi, F. (1986a). Aflatoxin B_1 binding to plasma albumin and liver DNA upon chronic administration to rats. *Carcinogenesis* **7,** 853–858.

Wild, C. P., Umbenhauer, D., Chapot, B., and Montesano, R. (1986b). Monitoring of individual human exposure to aflatoxins (AF) and *N*-nitrosamines (NNO) by immunoassays. *J. Cell. Biochem.* **30,** 171–179.

Wild, C. P., Pionneau, F., Montesano, R., Mutiro, C. F., and Chetsanga, C. J. (1987). Aflatoxin detected in human breast milk by immunoassay. *Int. J. Cancer* **40,** 328–333.

Wild, C. P., Montesano, R., Van Benthem, J., Scherer, E., and Den Engelse, L. (1990a). Intercellular variation in levels of adducts of aflatoxin B_1 and G_1 in DNA from rat tissues: A quantitative immunocytochemical study. *J. Cancer Res. Clin. Oncol.* **116,** 134–140.

Wild, C. P., Jiang, Y. Z., Allen, S. J., Jansen, L. A., Hall, A. J., and Montesano, R. (1990b). Aflatoxin–albumin adducts in human sera from different regions of the world. *Carcinogenesis* **11,** 2271–2274.

Wild, C. P., Jiang, Y. Z., Sabbioni, G., Chapot, B., and Montesano, R. (1990c). Evaluation of methods for quantitation of aflatoxin–albumin adducts and their application to human exposure assessment. *Cancer Res.* **50,** 245–251.

Wild, C. P., Hudson, G., Sabbioni, G., Wogan, G. N., Whittle, H., Montesano, R., and Groopman, J. D. (1992). Correlation of dietary intake of aflatoxins with the level of albumin bound aflatoxin in peripheral blood in The Gambia, West Africa. *Cancer Epidemiol. Biomarkers Prev.* **1,** 229–234.

Zarba, A., Wild, C. P., Hall, A., Montesano, R., and Groopman, J. D. (1992). Aflatoxin M_1 in human breast milk from The Gambia, West Africa isolated by combined monoclonal antibody immunoaffinity chromatography and HPLC. *Carcinogenesis* **13,** 891–894.

Zhu, J.-Q., Zhang, L.-S., Hu, X., Xiao, Y., Chen, J.-S., Xu, Y.-C., Fremy, J., and Chu, F. S. (1987). Correlation of dietary aflatoxin B_1 levels with excretion of aflatoxin M_1 in human urine. *Cancer Res.* **47,** 1848–1852.

13

Strategies for Chemoprotection against Aflatoxin-Induced Liver Cancer

❖

Thomas W. Kensler, Elaine F. Davis, and Mary G. Bolton

INTRODUCTION

Hepatocellular carcinoma is one of the most common cancers in parts of China and sub-Saharan Africa and results in an estimated 250,000 deaths annually. Epidemiological studies have established hepatitis B virus infection and aflatoxin exposure as major risk factors for liver cancer (Beasley, 1982; Yeh *et al.*, 1989; Ross *et al.*, 1992). Primary prevention strategies offer the best long-term approach for reducing liver cancer incidence. Primary prevention targets healthy individuals and may be achieved by avoiding exposure to risk factors or by stimulating the defense mechanisms of the host to interfere with the carcinogenic process. Toward this end, hepatitis B virus vaccination programs have been established in several countries. However, to break the cycle that begins with infection at birth, universal vaccination must be carried out for at least two generations. The need for a lengthy intervention coupled with high cost greatly restricts the use of hepatitis B vaccines in most areas with high incidences of liver cancer. Food surveys indicate a lower risk of exposure to aflatoxins in technologically developed countries than in developing ones, primarily because of the prevention of contamination of foods in the field or during processing. Approaches such as rapid postharvest drying of crops and controlled storage conditions are effective but costly means for reducing aflatoxin content of foods. Thus, the control of aflatoxin contamination of food supplies, although commonplace in developed nations, is not economically feasible in much of the world. In practice then, additional affordable primary prevention strategies are needed to have significant worldwide impact on mortality rates from liver cancer.

The use of chemical or dietary interventions to alter the susceptibility of humans to the actions of carcinogens and to block, retard, or reverse carcinogenesis is an emerging strategy for disease prevention and has been termed alternatively chemoprophylaxis, chemoprevention, or chemoprotection. Experimental observations in the mid-1970s that seemingly innocuous preservatives used in the Western diet could protect dramatically against diverse carcinogens at distal sites in animals sparked the development of chemoprotection as a viable approach for the reduction of human cancers (see Wattenberg, 1985). In fact, chemoprotection is now known to be an effective approach to cancer prevention in humans (Kraemer *et al.*, 1988; Hong *et al.*, 1990). Many indications exist that this strategy for disease prevention is extremely effective in aflatoxin-exposed laboratory animals. Consequently, the prospect that chemoprotection against aflatoxin-induced liver cancer can be achieved in high-risk human populations appears promising. From this perspective, the objectives of this chapter are twofold: to review the status of experimental chemoprotection against aflatoxin-induced liver disease and to consider the prospects and problems associated with possible chemoprotective interventions in humans at high risk for aflatoxin exposure.

FACTORS AFFECTING EXPERIMENTAL AFB$_1$ HEPATOCARCINOGENESIS

Several factors have been reported to modify the carcinogenic response of animals to AFB$_1$, including modified nutritional and endocrine status as well as numerous pharmacological interventions. In this section, we summarize many of the prescriptive and proscriptive interventions that have served to reduce AFB$_1$ acute hepatotoxicity and carcinogenesis in several experimental models. Although the influence of diet on aflatoxin hepatocarcinogenesis has been covered in Chapter 10, special note will be made of the experimental results outlining the effects of cruciferous vegetables on AFB$_1$ tumorigenesis.

Chemoprotection in the Rat

The susceptibility of the rat to the hepatocarcinogenic effects of AFB$_1$ has made this species a valuable model for chemoprotection studies. The incidence of hepatic cancers induced by aflatoxin is reduced significantly when rats are treated concurrently with a variety of synthetic agents such as β-naphthoflavone (Swenson *et al.*, 1977), α-hexachlorocyclohexane (Angsubhakorn *et al.*, 1978, 1989), γ-hexachlorocyclohexane (Angsubhakorn *et al.*, 1989), ethoxyquin (Cabral and Neal, 1983), phenobarbital (McLean and Marshall, 1971; Gurtoo *et al.*, 1985; butylated hydroxyanisole (Williams *et al.*, 1986), butylated hydroxytoluene (Williams *et al.*, 1986), and oltipraz (Roebuck *et al.*, 1991). The experimental designs and results from these studies are summarized in Table 1 and the structures of the chemoprotectors are shown in Figure 1. Although all these agents engender substantial protection against cancer, most noteworthy are the observations that hexachlorocyclohexanes (α-HCH and γ-HCH) and oltipraz produced complete protection against hepatocellular carcinomas at reasonably low dietary concentrations. Thus, the prospect of realizing chemoprotection in humans with pharmacologically achievable doses of appropriate agents seems reasonable.

Of particular interest are the mechanisms of chemoprotection by these agents against AFB_1 hepatocarcinogenesis. All the chemoprotection protocols tested to date have involved concomitant exposure to AFB_1 and chemoprotector. Therefore these agents are likely to affect the metabolism or disposition of AFB_1. Thus, possible mechanisms to explain the observed protective effects might include (1) induction of phase I enzymes (i.e., cytochromes P450) to enhance carcinogen detoxication; (2) inhibition of phase I enzymes to retard metabolic activation; (3) induction of phase II xenobiotic-metabolizing enzymes (e.g., UDP-glucuronosyl transferases, glutathione *S*-transferases), to enhance carcinogen detoxication and elimination; and (4) nucleophilic trapping of reactive intermediates. Postinitiation effects of these agents also may be important, but have not been examined in anticarcinogenesis bioassays to date.

The known pathways involved in the metabolic activation of AFB_1 are summarized in Figure 2. The ultimate carcinogenic species of aflatoxin, aflatoxin-8,9-epoxide, is formed by cytochrome P450 monooxygenase(s). The aflatoxin-8,9-epoxide reacts avidly with the N^7 atom of guanine in DNA and RNA so 10-fold higher levels of AFB_1 binding to nucleic acids than to protein are observed in the liver of rats after oral administration (Wild *et al.*, 1986). Aflatoxin-8,9-epoxide can undergo nonenzymatic hydrolysis to form 8,9-dihydro-8,9-dihydroxyaflatoxin B_1. This dihydrodiol can, in turn, form an adduct with the ϵ-amino group of lysine, principally in albumin, through a series of condensation and rearrangement reactions (Sabbioni *et al.*, 1987). In addition to binding to macromolecules, the aflatoxin-8,9-epoxide can undergo conjugation with glutathione to form 8,9-dihydro-8-(*S*-glutathionyl)-9-hydroxyaflatoxin B_1. This aflatoxin–glutathione conjugate is a major biliary metabolite formed through the catalytic actions of a family of isozymes, the glutathione *S*-transferases, and is considered a detoxication product (Degan *et al.*, 1978). Subsequent metabolism of the glutathione conjugate in the liver and kidney yields aflatoxin–mercapturic acid, a urinary metabolite.

A common effect of butylated hydroxyanisole (BHA), butylated hydroxytoluene (BHT), ethoxyquin, and oltipraz is the substantial reduction in the levels of aflatoxin–DNA adducts formed in liver (Kensler *et al.*, 1985,1987). Feeding studies with these compounds at levels used in anticarcinogenesis bioassays produced 60–90% reductions in the amount of AFB_1 bound to hepatic DNA.

TABLE 1 Chemoprotection of Aflatoxin Hepatocarcinogenesis in the Rat

Chemoprotector	Strain, sex, and age or weight	Conditions	Response	Reference
Diethylstilbestrol (DES)	Charles River CD, M, 3 weeks	0.2 ppm AFB_1 in diet and 0.0004% DES in diet for lifetime	71% incidence of tumors in AFB_1 treatment group at 16 months; 20% incidence of tumors in AFB_1 + DES group	Newberne and Williams (1969)
α-Hexachlorocyclohexane (α-HCH)	Fischer, M, 150–200 g	1 ppm AFB_1 in diet and 0.05% α-HCH in diet for 20 weeks	100% tumor incidence with AFB_1 alone; no tumors in AFB_1 + α-HCH group	Angsubhakorn et al. (1978)
γ-Hexachlorocyclohexane (γ-HCH; lindane)	Buffalo, F, 100 g	1ppm AFB_1 in diet and 0.01% γ-HCH in diet for 15 weeks	32% incidence of tumors in AFB_1 treatment group at 82 weeks; no tumors in AFB_1 + γ-HCH group	Angsubhakorn et al. (1989)
Phenobarbital (PB)	Fischer, M, 120 g	0.3 ppm AFB_1 in diet and 0.1% sodium PB in drinking water for 15 months	61% incidence of tumors in AFB_1 treatment group at 15 months; 11% incidence of tumors in AFB_1 + PB group	Swenson et al. (1977)
β-Napthoflavone (β-NF)	Fischer, M, 60–80 g	25 μg AFB_1/rat/day per os (p.o.), 5 days/week for 8 weeks and 0.015% β-NF in diet for 9 weeks	100% incidence of tumors in AFB_1 treatment group at 42 weeks; 25% incidence of tumors in AFB_1 + β-NF group	Gurtoo et al. (1985)
Ethoxyquin (EQ)	Fischer, M, 8 weeks	4 ppm AFB_1 and 0.5% EQ in diet for 6 weeks	20% incidence of tumors in AFB_1 treatment group at 48 weeks; no tumors in AFB_1 + EQ group	Cabral and Neal (1983)
Butylated hydroxyanisole (BHA)	Fischer, M, 150 g	25 μg AFB_1/kg p.o., 3 days/week for 20 weeks and 0.1 or 0.6% BHA in diet	63% incidence of tumors in AFB_1 treatment group at 44 weeks; 13% incidence of tumors in AFB_1 + 0.1% BHA group; 8% incidence of tumors in AFB_1 + 0.6% BHA group	Williams et al. (1986)
Butylated hydroxytoluene (BHT)	Fischer, M, 150 g	25 μg AFB_1/kg p.o., 3 days/weeks for 20 weeks and 0.1 or 0.6% BHT in diet	63% incidence of tumors in AFB_1 treatment group at 44 weeks; 12% incidence of tumors in AFB_1 + 0.1% BHT group; no tumors in AFB_1 + 0.6% BHT group	Williams et al. (1986)
Oltipraz	Fischer, M, 100 g	25 μg AFB_1 rat p.o., 5 days/week for 2 weeks and 0.075% oltipraz in diet	20% incidence of tumors in AFB_1 treatment group at 23 months; no tumors in AFB_1 + 0.075% oltipraz group	Roebuck et al. (1991)

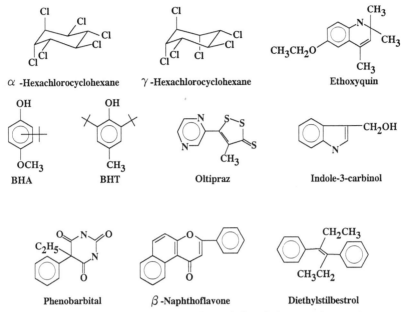

α-Hexachlorocyclohexane γ-Hexachlorocyclohexane Ethoxyquin

BHA BHT Oltipraz Indole-3-carbinol

Phenobarbital β-Naphthoflavone Diethylstilbestrol

FIGURE 1 Structures of some inhibitors of aflatoxin hepatocarcinogenesis.

Alteration in the balance of competing pathways of aflatoxin-8,9-epoxide reactions could modulate the availability of the 8,9-epoxide for binding to DNA directly. Correspondingly, another common feature of these chemoprotectors is their ability to induce hepatic glutathione S-transferases. Feeding ethoxyquin has been shown to increase the activity of glutathione S-transferase 3 to 5-fold, as measured by glutathione conjugation of chlorodinitrobenzene, in the cytosol of rat livers (Kensler *et al.*, 1986; Mandel *et al.*, 1987). However, although changes in the specific activity of an enzyme do not always presage changes in catalytic rate *in vivo,* concomitant with the increased glutathione S-transferase specific activity and decreased levels of aflatoxin–DNA adducts was a 3-fold increase in the biliary excretion of aflatoxin–glutathione conjugates in ethoxyquin-treated rats (Kensler *et al.*, 1986). Thus, ethoxyquin treatment clearly affects the metabolism and disposition of AFB_1 *in vivo*. Feeding BHA induced a 2 to 3-fold increase in glutathione S-transferase activity toward chlorodinitrobenzene as well as a correlative increase in thiol conjugates of aflatoxin (Chang and Bjeldanes, 1987). In this study, a 6-fold reduction of aflatoxin–DNA adducts in the liver was observed. An inverse relationship also has been shown between decreased binding of aflatoxin to DNA and induction of glutathione S-transferase activity in the livers of rats pretreated with oltipraz (Kensler *et al.*, 1987).

In vitro studies have shown that purified rat glutathione S-transferases of the alpha class will conjugate aflatoxin-8,9-epoxide to glutathione (Coles *et al.*, 1985). The mu class also shows some activity toward aflatoxin (Guengerich *et*

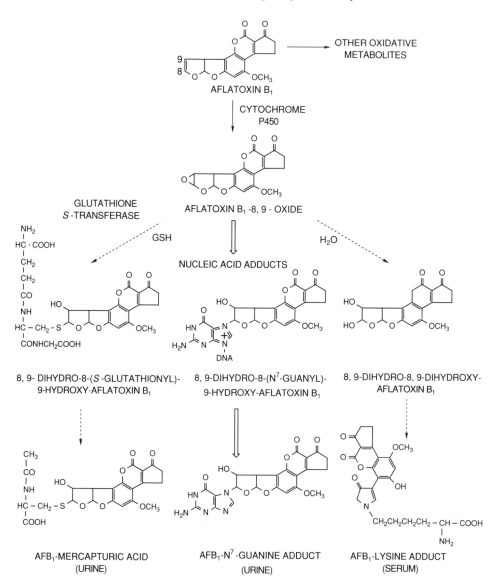

FIGURE 2 Pathways involved in the metabolic activation of AFB_1 and the subsequent formation of exposure/risk biomarkers.

al., 1992). Moreover, *in vitro* studies involving hepatic cytosols from rats pretreated with agents known to induce glutathione *S*-transferase protect against DNA damage produced by the metabolites of aflatoxin (Quinn *et al.,* 1990). Although the molecular mechanisms involved in glutathione *S*-transferase induction by anticarcinogens are not fully resolved, oltipraz has been shown to increase the expression of the rat glutathione *S*-transferase Ya subunit at the transcriptional level (Davidson *et al.*, 1990). Increases in the steady-state levels of mRNA correlate with the increased activity of glutathione *S*-transferase toward chlorodinitrobenzene in the cytosols of treated rats. A novel inducible subunit of the alpha-class glutathione *S*-transferases designated Yc_2, has been identified in rats pretreated with ethoxyquin and exhibits a 25-fold greater activity toward the aflatoxin-epoxide than do other isoforms characterized to date (Hayes *et al.*, 1991). Characterization of this subunit shows a 95% sequence identity over 60% of the primary structure with the isoform believed to be responsible for the high activity of mouse cytosols toward the aflatoxin epoxide (Hayes *et al.*, 1991; McLellan *et al.*, 1991). Since mice are naturally resistant to the hepatocarcinogenic effects of AFB_1 (Busby and Wogan, 1984), an understanding of the factors, such as glutathione *S*-transferases, that control natural sensitivity and resistance to AFB_1 should provide insights into developing more effective and selective protective strategies.

Some chemoprotectors also affect the oxidative metabolism of aflatoxins. For example, incubation of microsomes prepared from livers of ethoxyquin-treated rats with AFB_1 produce larger amounts of AFQ_1 and AFM_1 than do control microsomes (Mandel *et al.*, 1987). An extensive evaluation by Putt *et al.* (1991) of the effects of feeding BHA, BHT, ethoxyquin, and oltipraz on the induction of different classes of hepatic cytochrome P450s and on the metabolism of AFB_1 indicate that the food antioxidants, but not oltipraz, significantly elevate the levels of several forms of P450, as determined by Western blotting analyses. Moreover, incubation of AFB_1 with microsomes prepared from BHA-, BHT-, or ethoxyquin-fed animals enhanced the rates of both the oxidative activation and the inactivation of AFB_1. Direct addition of any of the four anticarcinogens to control microsomes significantly inhibited the oxidative metabolism of AFB_1. Ch'ih *et al.* (1989) also have reported that direct addition of BHA to isolated rat hepatocytes blocks the oxidative metabolism of AFB_1. Guengerich and Kim (1990) have suggested that naringenin, a flavone found in grapefruit juice that inhibits the activation of AFB_1 by human liver microsomes, might be a useful candidate chemoprotective agent.

Proscriptive measures, such as changes in dietary preferences, can have dramatic effects on the health of populations. Epidemiological evidence has indicated that consumption of vegetables including cabbage, Brussels sprouts, broccoli, and turnips reduces the risk of cancer in humans (Graham, 1983). Diets containing lyophilized cabbage (Boyd *et al.*, 1982; Whitty and Bjeldanes, 1987), beets (Boyd *et al.*, 1982), cauliflower (Stoewsand *et al.*, 1978), and broccoli (Ramsdell and Eaton, 1988), as well as an extract from these vegetables, R-goitrin (Chang and Bjeldanes, 1987), have shown protective effects against aflatoxin–DNA binding and tumorigenesis in the rat. As with the prescriptive chemical interventions, consumption of these vegetables leads to increases in hepatic glutathione *S*-transferase activity and aflatoxin–thiol conjugation (Chang and Bjeldanes, 1987; Whitty and Bjeldanes, 1987; Ramsdell and Eaton, 1988). However, the nature of the enzyme inducing or chemoprotective principles in most of these vegetables remains unknown. Zhang *et al.* (1992) have identified the isothiocyanate sulforaphane as a major and very potent inducer of electrophile detoxication enzymes contained in Saga broccoli. Because sulforaphane induces glutathione *S*-transferases in many rodent tissues, including liver, this isothiocyanate may prove to be an effective inhibitor of AFB_1 carcinogenesis. Induction of enzymes in the intestine, as seen with the administration of cabbage in the diet (Whitty and Bjeldanes, 1987), leads to a decrease in the amount of parent compound available to the liver (Bradfield and Bjeldanes, 1984).

Most mechanistic and bioassay studies have focused on the role of decreasing initiation in protection. Nonetheless, the inhibition of the hepatotoxic auto-promoting effects of aflatoxin by chemoprotective agents should not be ignored as a possible mechanism of modulation of carcinogenesis. Hepatocellular damage clearly has been implicated in the development of primary hepatocellular carcinoma in transgenic mice expressing hepatitis B surface antigen (Chisari *et al.*, 1989). Recurrent hepatotoxicity is likely to play an important role in AFB_1 carcinogenesis, as evidenced in part by the observations that AFB_1 is a poor single-dose carcinogen. Rats pretreated with phenobarbital, BHT, and oltipraz are less susceptible to the toxic effects of aflatoxin (MgBodile *et al.*, 1975; Salocks *et al.*, 1981; Liu *et al.*, 1988). For example, hepatocytes isolated from rats treated with BHT exhibited a significantly lower level of lactose dehydrogenase activity leaking into the medium after exposure to AFB_1 than did control hepatocytes. Pretreatment of rats with oltipraz reduced the mortality produced by a single per os (po) dose of 10 mg AFB_1/kg from 83 to 36%. In a subchronic dosing study, treatment of rats with 500 μg AFB_1/kg/day for 10 days produced 100% mortality within 1 week after cessation of dosing. However, animals fed a diet supplemented with 300 ppm 1,2-dithiole-3-thione were protected completely from the lethal effects of AFB_1 (Kensler *et al.*, 1992a). The feeding of crocin dyes to rats inhibits the acute hepatic damage induced by AFB_1 (Lin and Wang, 1986; Wang *et al.*, 1991). Crocin dyes are isolated from the fruits of *Gardenia* and are used in traditional Chinese medicines as well as in color additives in foods and beverages. The protective effects of crocin dyes appear to be the result of elevations in hepatic glutathione content and glutathione *S*-transferase activity (Wang *et al.*, 1991). Overall, the inhibition of cytotoxicity and the subsequent compensatory cell proliferation elicited by AFB_1 may be a major component of the protective actions of these compounds and certainly requires further study.

Chemoprotection in the Rainbow Trout

The rainbow trout is also highly susceptible to AFB_1 hepatocarcinogenesis. Interest in this species originated in the early 1960s with epizootic liver cancer in hatchery trout that was traced to aflatoxin-contaminated cottonseed meal in the feed. Since that time, the rainbow trout has been studied extensively as a model of aflatoxin carcinogenesis and anticarcinogenesis (Bailey *et al.*, 1984). As summarized in Table 2, several compounds that alter the metabolism and disposition

TABLE 2 Chemoprotection of Aflatoxin Hepatocarcinogenesis in Rainbow Trout

Chemoprotector	Conditions	Response	Reference
Aroclor 1254	6 ppb AFB_1 plus 100 ppm Aroclor 1254 in the diet for 1 year	70% tumor incidence with AFB_1 alone; 30% incidence of tumors in AFB_1 + Aroclor 1254	Hendricks *et al.* (1977)
Aroclor 1254	1, 4, or 8 ppb AFB_1 plus 50 ppm Aroclor 1254 in the diet for 1 year	23, 54, or 83% tumor incidence with 1, 4, or 8 ppb AFB_1 alone; 12, 31, or 75% incidence with 50 ppm Aroclor + 1, 4, or 8 ppb AFB_1, respectively	Shelton *et al.* (1984)
β-Naphthoflavone (β-NF)	20 ppb AFB_1 in the diet for 2 weeks plus 50 or 500 ppm β-NF in diet for 16 weeks	38% tumor incidence with AFB_1 alone; 18% or 6% incidence in AFB_1 + 50 or 500 ppm β-NF, respectively	Nixon *et al.* (1984)
β-Naphthoflavone	10 ppb AFB_1 in the diet for 4 weeks plus 0.05% or 0.005% β-NF in the diet for 8 weeks	23% tumor incidence with AFB_1 alone; 4% or 0.6% incidence with AFB_1 + 0.05 or 0.005% β-NF	Goeger *et al.* (1988)
Indole-3-carbinol (I3C)	20 ppb AFB_1 in the diet for 2 weeks plus 100 ppm I3C in the diet for 16 weeks	38% tumor incidence with AFB_1 alone; 4% incidence with AFB_1 + I3C	Nixon *et al.* (1984)

of aflatoxin are effective inhibitors of AFB_1 hepatocarcinogenesis in the rainbow trout. Compounds such as β-naphthoflavone and polychlorinated biphenyls (Aroclor 1254), which strongly induce cytochrome P450 activities in the trout, are potent anticarcinogens in this model (Hendricks *et al.*, 1977; Nixon *et al.*, 1984; Shelton *et al.*, 1984; Goeger *et al.*, 1988). Pretreatment with these agents leads to altered oxidative metabolism of AFB_1 (particularly enhanced AFM_1 formation) and a decrease in the formation of aflatoxin–DNA adducts in the liver. Interestingly, the feeding of Aroclor 1254 after AFB_1 exposure has no protective effect, further supporting the importance of altered xenobiotic metabolism in its action (Hendricks *et al.*, 1980).

Indole-3-carbinol, which is found as a glucosinolate in cruciferous vegetables, is a potent inhibitor of AFB_1 hepatocarcinogenesis in the rainbow trout (Nixon *et al.*, 1984). When fed at 0.2%, indole-3-carbinol produced a 70% reduction in average *in vivo* hepatic DNA binding of AFB_1 administered over a 21-day period compared with controls (Goeger *et al.*, 1986). Feeding indole-3-carbinol also increases the accumulation of aflatoxicol M_1–glucuronide 7-fold in the bile of trout 24 hr after injection of aflatoxin, suggesting that induction of conjugating enzymes may be an important component of the protective mechanism. The major conjugation product in the bile of trout pretreated with Aroclor 1254 and β-naphthoflavone is also the aflatoxicol M_1–glucuronide (Loveland *et al.*, 1984). In contrast to observations in rodents, conjugation of aflatoxin to glutathione does not appear to be an important detoxification pathway in the trout (Vlasta *et al.*, 1988). Surprisingly, however, the role of glucuronosyl transferases in chemoprotection against AFB_1 hepatocarcinogenesis in the rat has not been pursued vigorously although activities of these enzymes are elevated to degrees comparable to the glutathione *S*-transferases.

One of the advantages of the rainbow trout model is that large numbers of fish can be used in experiments. Dashwood *et al.* (1988, 1989) have conducted large-scale studies, not economically feasible in rodents, detailing the *in vivo* inhibitory potency of indole-3-carbinol as a function of dose of indole-3-carbinol and dose of AFB_1. Reduction of covalent binding of AFB_1 to hepatic DNA and final tumor response were employed as study end points. This approach of using inhibitor and carcinogen as covariables administered repeatedly in the diet demonstrated that increases in dose of indole-3-carbinol resulted in corresponding dose-related decreases in aflatoxin–DNA binding and tumor response at each of four dietary concentrations of AFB_1. Moreover, a comparison of tumor incidence versus log concentration of aflatoxin–DNA adducts revealed a single dose–response curve for all data sets between 0 and 2000 ppm indole-3-carbinol. These results indicate that the level of aflatoxin–DNA binding in any treatment group was related quantitatively to, and predictive of, tumor response. Thus, the mechanism of chemoprotection by indole-3-carbinol was independent of AFB_1 dose and was effective over a range of tumor incidences from 80% to less than 1%.

Chemoprotection against Aflatoxin Toxicities in Domestic Animals

Mycotoxin contamination of feeds has an enormous impact on the health of domestic animals. For example, economic losses in the poultry industry arising from poor growth and efficiency of feed conversion have been associated with aflatoxins in feed (Jones *et al.*, 1982; see also Chapter 23). Much attention has been directed toward the prevention of aflatoxin contamination of feeds through agricultural practices and the development of effective economical methods for the detoxification of aflatoxins in feeds (Hagler, 1990; see also Chapter 18). Although perhaps not adaptable for commercial use, chemoprotective interventions have been examined in the experimental setting with poultry and livestock. Larsen *et al.* (1985) examined the effect of feeding BHT to chickens on the ability of these birds to resist the adverse effects of dietary AFB_1 on weight gain and feed conversion efficiency. Chicks on aflatoxin-contaminated feed did not gain weight as rapidly or as efficiently as birds on uncontaminated feed; the depression of growth caused by AFB_1 was substantially less in birds given BHT than in those fed the diet lacking the antioxidant. Manning *et al.* (1990) have investigated the effect of pretreatment with either phenobarbital or β-naphthoflavone on the toxicity of AFB_1 to chickens. Phenobarbital offered dramatic protection from the lethal effects of AFB_1 whereas pretreatment with β-naphthoflavone exacerbated AFB_1 toxicity. This differential effect appears to relate to alterations in the oxidative metabolism of AFB_1 by these two inducers of cytochromes P450. These studies further highlight the difficulties associated with modulation of phase I enzyme activity as a reasonable strategy for chemoprotection.

Note that an entirely different approach to chemoprotection might be exploitable. Although most work has focused on altering the activation/detoxication balance of AFB_1 in the liver, the internal dose of AFB_1 could be reduced through the use of inorganic adsorbent materials in the diet that act to sequester and immobilize aflatoxins in the gastrointestinal tract of livestock and poultry. Hydrated sodium calcium aluminosilicate diminishes aflatoxin uptake and distribution to target organs; prevents aflatoxicosis in chickens, turkeys, and swine; and decreases levels of AFM_1 residues in contaminated milk from lactating dairy cattle (Colvin *et al.,* 1989; Harvey *et al.,* 1989; Beaver *et al.,* 1990; Phillips *et al.,* 1990). The selective chemisorption appears to involve the formation of a complex between the β-carbonyl system of the aflatoxins and uncoordinated "edge site" aluminum ions in the aluminosilicate (Sarr *et al.,* 1990).

CLINICAL CONSIDERATIONS FOR CHEMOPROTECTION IN HUMANS

Although insights into basic mechanisms of aflatoxin carcinogenesis and the development of new chemoprotective agents have flourished in experimental settings, the challenge now lies in applying chemoprotection to the human population. That chemoprotection against cancer can work in humans is without doubt. Several investigations have established significant site- and chemoprotector-specific effects in the prevention of invasive skin and upper aerodigestive tract cancers in extremely high risk groups (Kraemer *et al.,* 1988; Hong *et al.,* 1990). However, a variety of issues must be addressed to permit the optimal application of chemoprotective strategies in humans. The most important of these issues include identification of target populations, continued development of new potential chemoprotective agents, and identification and evaluation of biomarkers as intermediate end points for assessing the clinical efficacy of interventions.

Identifying Target Populations for Interventions

Conceptually, chemoprotective agents can be used in several different settings. At one extreme are the majority of the present clinical trials that use chemoprotective agents in individuals at high risk for cancer development. To date, identification of such high risk individuals has been based on known exposures to carcinogenic agents (e.g., tobacco, asbestos), familial syndromes with a genetic predisposition to cancer (e.g., familial polyposis in the colon, familial breast cancer), and most notably prior pathology increasing the probability of a second primary cancer. The continuing identification and development of biomarkers to define those individuals at highest risk for cancer is crucial to the expanded application of chemoprotection against cancer. Presently, over a score of clinical trials testing β-carotene, synthetic retinoids, folic acid, selenium, tamoxifen, and vitamins A, C, and E have been initiated or completed, principally in groups at high risk for upper aerodigestive, lung, skin, and breast cancers (Lippman *et al.*, 1991; Malone, 1991). However, with the maturity of chemoprotection as an approach to cancer reduction, designing and conducting interventions for chemoprotection in the general population is becoming possible. Such large-scale public health interventions may have significant impact on reducing overall rates of morbidity and mortality from cancer. Although these interventions are likely to be prescriptive initially, to provide scientific credibility through controlled clinical trials, proscriptive changes in diet are likely to have the greatest impact on the general population in the long run.

Individuals exposed to aflatoxins represent a group at high-risk in which the prospects for effective prescriptive interventions appear promising. Several epidemiological studies have examined the relationship of the estimated dietary intake of aflatoxin to the incidence of primary human liver cancer in various parts of the world (see Chapter 11). Overall, a positive association between the two parameters appears to exist because high intakes of aflatoxin were associated consistently with high incidence rates. More compelling evidence is from a prospective study conducted in Shanghai that demonstrated a significantly increased risk for subsequent development of liver cancer in those individuals excreting measurable levels of aflatoxins in their urine (Ross *et al.*, 1992). Moreover, a 60-fold higher risk for liver cancer was observed in individuals positive for both aflatoxin exposure and hepatitis B virus status than in those individuals expressing neither risk factor. Thus, identification of people with aflatoxins in their biofluids (urine, blood) who are hepatitis B virus-positive might define those individuals likely to benefit most from a chemoprotective intervention. Given the dramatic seasonal variation in aflatoxin contamination of foodstuffs, targeting interventions to the postharvest seasons when the major share of the annual intake of aflatoxin occurs may be sufficient. Such an approach would reduce the costs of interventions dramatically and would likely enhance compliance.

Considering metabolic susceptibility factors when identifying high-risk individuals for interventions also may be prudent. Enzymatic differences among individuals have been shown to influence risk for several types of cancers (Guengerich and Shimada, 1991). Humans are known to vary widely ($>$100-fold) in rates of metabolism of xenobiotics and in the levels of individual cytochrome P450s. AFB_1 requires metabolic activation to exert its toxicological effects. The metabolism of AFB_1 to its epoxide in human liver appears to involve cytochrome P450 3A4 principally, although other P450s undoubtedly are involved as well (Shimada and Guengerich, 1989; see also Chapter 3). Cytochrome P450 3A4 activity can be modulated by drugs or dietary components and individuals can be phenotyped readily by noninvasive means with both endogenous and exogenous substrates (Guengerich and Shimada, 1991). The 6β-hydroxycortisol/cortisol ratios in the urine have been found to be correlated to the level of aflatoxin–N^7-guanine adducts excreted into the urine by people exposed to roughly equivalent levels of aflatoxin (J. D. Groopman, personal communication). Although these findings suggest a potentially important relationship, the striking differences in species susceptibility to AFB_1 is attributed, in large part, to altered capacity to conjugate aflatoxin-8,9-epoxide to glutathione (Quinn *et al.*, 1990). Several isozymes of human glutathione *S*-transferases appear to conjugate glutathione to the aflatoxin epoxide, although at lower rates than observed in rodents (Guengerich *et al.*, 1992). Although the overall contributions of the different classes of glutathione *S*-transferase to aflatoxin detoxication remain to be resolved, the mu form of human glutathione *S*-transferase, which is active in AFB_1–glutathione conjugation, is polymorphic in human liver. The polymorphism can be characterized by high or low conjugation rates of *trans*-stilbene oxide (*t*SBO). Aflatoxin–DNA binding was inhibited to a greater extent when human liver cytosols with a high rate of glutathione conjugation were incubated *in vitro* with AFB_1, microsomes and DNA than when cytosols from people with low glutathione *S*-transferase mu activity were used (Liu *et al.*, 1991). Inasmuch as the level of glutathione *S*-transferase mu activity in lymphocytes may mirror activity in liver, measurement of this enzyme may serve as a marker for susceptibility to aflatoxin-induced DNA damage. Given the overall importance of the metabolic balance between activating and detoxifying reactions in aflatoxin–DNA adduct formation, those individuals or populations that exhibit high expression of cytochrome P450 coupled with low expression of glutathione *S*-transferases may be at significantly higher risk than those individuals exhibiting other phenotypes for these enzymes. Human monitoring studies examining the interrelationships between individual metabolic phenotype and excretion patterns of aflatoxin-epoxide-derived products in biofluids should clarify the importance of these potential determinants of susceptibility.

Criteria for Selecting Chemoprotective Agents

Irrespective of the risk factors used for selecting individuals for prescriptive interventions, the effective utilization of chemoprotective agents in humans requires that a number of pharmacologic and toxicologic criteria be met. Ideally, these agents will have a targeted defined molecular activity that provides efficacy at low concentrations. Chemoprotectors that utilize receptor-mediated pathways, for example, enzyme inducers, retinoids, and tamoxifen, serve as current examples. Chemoprotectors preferably should exhibit oral bioavailability, achieve effective tissue levels at target sites at low doses, and have a relatively long biological half-life. Means to assess the pharmacodynamic actions of the agent in target tissues—for example, enzyme induction, decreased rates of cell proliferation, or altered differentiation—are also extremely useful. These short-term end points permit optimization of dose and schedule parameters in clinical trials. Organ specificity is desirable with chemoprotection in specific high-risk groups, whereas involvement of multiple organ sites is generally preferable for chemoprotection in the general population. Nonetheless, the agents should not exhibit acute or subacute pharmacology in nontarget tissues and should have a generous therapeutic index if they are not devoid of acute or chronic toxicities. Side effects that alter the quality of life will preclude use of interventions for all but very high risk individuals. The spectrum and severity of toxicities that can be tolerated in chemoprotective agents is very different from those of drugs used for therapeutic purposes. In addition, issues of developmental and reproductive toxicity must be addressed carefully if the target population includes fertile women. Finally, economic considerations require that these agents be potent, stable, and easy to synthesize and formulate. Examples of compounds that are currently under development and that meet some, but not all, of these criteria include retinoids [N-4-(hydroxyphenyl)-retinamide], micronutrients (selenium), minerals (sodium molybdate), purified dietary constituents (ellagic acid), dithiolethiones (oltipraz), endogenous steroids (dehydroepiandrosterone and analogs), estrogen antagonists (tamoxifen), protease inhibitors (Bowman–Birk inhibitor from soybeans), and prostaglandin (piroxicam) and polyamine (difluoromethylornithine) synthesis inhibitors. As indicated in Tables 1 and 2, few of these promising compounds have been examined in models of aflatoxin hepatocarcinogenesis.

Conversely, most of the chemoprotective agents discussed in Section II as inhibitors of aflatoxin-induced toxicities in experimental models are most likely unsuitable for use in humans. For example, issues of toxicity or lack of potency may preclude the use of phenolic antioxidants, ethoxyquin, indole-3-carbinol, and hexachlorocyclohexanes. Many of these compounds are carcinogenic in their own right and can act as tumor promoters at several organ sites (Ito and Hirose, 1987; Dashwood *et al.*, 1991). Moreover, extremely high concentrations—often 50-fold greater than those approved for use in foodstuffs—are required for optimal efficacy with the food antioxidants BHA or BHT. A significant concern with agents such as β-naphthoflavone and phenobarbital lies in their actions as bifunctional enzyme inducers. Bifunctional inducers elevate the activities of both phase I enzymes such as cytochrome P450 and phase II enzymes such as glutathione S-transferases. In contrast, the use of monofunctional inducers that only elevate the activities of phase II conjugating enzymes is likely to minimize the potential complications of enhanced procarcinogen activation through the ancillary induction of certain P450s. Although suitable for use against aflatoxins, monofunctional inducers might exacerbate toxicities of compounds such as alkyl halides that undergo metabolic activation through conjugation reactions.

A candidate drug for use in interventions in humans ingesting aflatoxins is the substituted 1,2-dithiole-3-thione oltipraz. Dithiolethiones have antioxidant, radioprotective, antiviral, chemotherapeutic, and chemoprotective properties (Kensler *et al.*, 1992a; Prochaska *et al.*, 1993). In addition to protecting against AFB_1-induced hepatocarcinogenesis in rodents, oltipraz exhibits remarkable chemoprotective activity in carcinogenesis models targeting the breast, colon, pancreas, lung, forestomach, skin, and bladder. Oltipraz also has been used in single-dose chemotherapy against schistosomiasis in people. As a consequence, oltipraz currently is undergoing evaluation in phase I clinical trials to determine pharmacokinetic parameters during chronic administration and potential dose-limiting side effects or toxicities. Pending the favorable outcome of these studies, as well as continuing animal bioassays for chronic toxicities, this drug is poised for study in small-scale interventions in the near future. As part of an ongoing collaboration with the Shanghai Cancer Institute, investigators at Johns Hopkins University are developing protocols for phase II interventions with oltipraz in Qidong City, People's Republic of China. Several features highlight this population as an appropriate group for initial studies. First, this population is at high risk for hepatocellular carcinoma (60 deaths/100,000 per year) (Zhu *et al.*, 1989). Second, aflatoxins are a contaminant in the food supply of this region (Zhu *et al.*, 1989). Third, aflatoxin–albumin adducts are detected in the blood of many of these people (G-S. Qian and J. D. Groopman, personal communication). Fourth, mutations typical of those produced by AFB_1 (G → T transversions) are found with high frequency ($>50\%$) in codon 249 of the *p53* gene isolated from hepatocellular carcinomas obtained from this population (Hsu *et al.*, 1991; Scorsone *et al.*, 1992). Other populations at risk for aflatoxin exposure in Africa and southeast Asia represent additional groups for potential interventions. However, the use of oltipraz or other prescriptive interventions against low-level aflatoxin exposures in the general population is unwarranted because of cost and other risk versus benefit considerations.

Development of Intermediate Biomarkers

The development of biomarkers as intermediate end points in chemoprotection trials is another important and necessary goal. Cancer incidence or mortality as the end point of a chemoprotective intervention insures slow progress, be it in experimental models or in humans. Biomarkers can reflect a spectrum of intermediary stages in the neoplastic process from exposure to disease. Moreover, these intermediate end points can include genetic markers (carcinogen–DNA adducts, mutated genes, sister chromatid exchanges, micronuclei), differentiation markers (cytokeratins, involucrin, blood group antigens), growth regulatory factors (EGF, TGF-β), or proliferation markers (polyamines, proliferating cell nuclear antigen, thymidine-labeling index) as well as morphologic indices of precancerous lesions (Hulka *et al.*, 1990; Lippman *et al.*, 1990). Intriguing preliminary results have been achieved in chemoprotection trials using markers of clastogenic damage and dysregulated proliferation in accessible tissues such as buccal mucosal cells and colonic mucosa (Lipkin and Newmark, 1985; Stich *et al.*, 1991).

Development of intermediate biomarkers for internal largely inaccessible tissues such as the liver is considerably more problematic. As a consequence, the

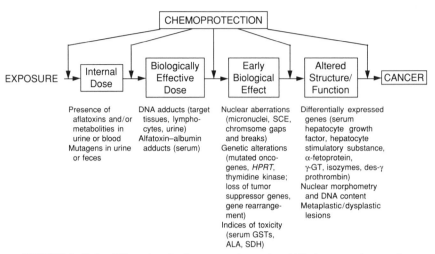

FIGURE 3   Role of biomarkers in chemoprotection against AFB_1 hepatocarcinogenesis.

diagnosis of hepatocellular carcinoma usually is made late in the course of the disease, contributing to the poor prognosis. This outcome reflects the fact that hepatocellular carcinomas produce no early pathognomic symptoms or signs and are not associated with distinctive patterns of altered hepatic function. However, as shown in Figure 3, some possible intermediate biomarkers are worth considering. In particular, the availability of specific serum markers of liver cancer would allow the detection of presymptomatic tumors, thereby improving an otherwise poor prognosis, and would facilitate long-term surveillance as well as intervention studies in high-risk individuals. Although no ideal tumor marker has been identified, several serum factors appear to be indicative of either regenerating liver or transformation (Tew, 1989; Michalopoulos, 1990). Several gene products are expressed differentially in transformed hepatocytes. These molecules may be secreted or leaked from the cell into the circulating plasma. Re-expression of normally quiescent embryonal genes in hepatocellular carcinoma leads to elevated serum levels of α-fetoprotein and carcinoembryonic antigen. Tumor cell-specific isoforms of γ-glutamyl transpeptidase (γ-GT) also have been identified in serum and differ from normal γ-GT isozymes by their carbohydrate composition. Altered intermediary metabolism in malignant cells also leads to the accumulation of atypical products. For example, the prothrombin precursor des-γ-prothrombin, is thought to accumulate in malignant hepatocytes because of failure of the transformed cell to express the carboxylase gene (Shah *et al.,* 1987). Several liver-specific growth factors are elaborated during the regeneration induced by partial hepatectomy and chemical insult. Several hepatic growth factors or hematopoietins have been identified and cloned that stimulate mitogenesis and hypertrophy in hepatocytes (Michalopoulos, 1990). These hematopoietins can be measured in human serum by enzyme-linked immunosorbent assay (ELISA). Another substance that functions as a complete hepatocyte mitogen is hepatic stimulatory substance (LaBrecque *et al.*, 1987). Alterations in the circulating levels of these growth factors might be indicative of ongoing regenerative hyperplasia consequent to aflatoxin hepatotoxicity.

The emerging field of molecular dosimetry provides some particularly exciting opportunities for the development of carcinogen-specific biomarkers. Work in molecular dosimetry of aflatoxins has focused on developing methods to quantify aflatoxin residues in serum and urine (see Chapter 12). Aflatoxin is known to bind to DNA and proteins, so measurement of the aflatoxin-8,9-epoxide-derived products may provide a reasonable estimate of the biologically effective dose of this carcinogen in target tissues. Although routinely assaying aflatoxin–DNA adduct levels in human tissues is not possible, the major DNA adduct species formed *in vivo* in mammals, aflatoxin–N^7-guanine, is excised rapidly and eliminated as a modified guanine base in urine. Several studies have established that the excretion of aflatoxin–N^7-guanine adduct in the urine occurs in a dose-dependent manner after single-dose exposures of rats to AFB_1 and that the amount of adduct excreted is proportional to the amount of aflatoxin–DNA adduct formed in the liver (Bennett *et al.*, 1981; Groopman *et al.*, 1992a). Groopman *et al.* (1992b) also have observed a strong correlation between levels of aflatoxin–N^7-guanine in the urine and exposure levels in humans consuming aflatoxin-contaminated foods. Because the ability to cause genetic damage indicates potential carcinogenicity, these exposure markers should assume greater utility in estimating future cancer risk. Figure 4 depicts the results of an initial attempt to extend the monitoring of levels of aflatoxin–N^7-guanine in urine and aflatoxin–albumin adducts in serum from exposure indices to quantitative short-term end points for estimating risk of aflatoxin-induced disease. The cancer chemoprotective agent 1,2-dithiole-3-thione was added to the diet of rats to attenuate the risk for development of tumors following exposure to AFB_1. Rats were monitored for levels of urinary and serum biomarkers over the course of a 2-week exposure to AFB_1. This exposure protocol for AFB_1 induces a 100% incidence of preneoplastic foci (glutathione *S*-transferase P- and γ-glutamyl transpeptidase-positive) within 3 months, and a 20% incidence of hepatocellular carcinoma in lifetime bioassays (Roebuck *et al.*, 1991). Half the rats were maintained on a diet supplemented with 0.03% 1,2-dithiole-3-thione, an intervention that has been shown to reduce the hepatic burden of aflatoxin-induced preneoplastic lesions by >99% (Kensler *et al.*, 1992b). A striking similarity is observed in the effects of 1,2-dithiole-3-thione on the levels of (1) aflatoxin–DNA adducts in liver, (2) aflatoxin–N^7-guanine in urine, and (3) aflatoxin–albumin adducts in serum. Levels of all three markers are reduced 65–75% by the intervention with 1,2-dithiole-3-thione. Thus, dramatic reductions in cancer risk within a small population of animals are reflected by marked reductions in the levels of DNA adducts in liver as well as in levels of epoxide-derived biomarkers obtained from biofluids. More extensive dose–response comparisons with graded concentrations of chemoprotectors producing a series of intermedi-

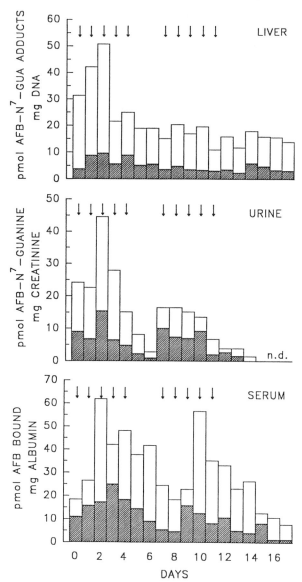

FIGURE 4   Effect of 1,2-dithiole-3-thione on levels of aflatoxin–DNA adducts in liver, aflatoxin–N^7-guanine in urine, and aflatoxin–albumin adducts in serum of rats fed either control or 1,2-dithiole-3-thione-supplemented diets after multiple administrations of AFB_1 (arrows). Modified with permission from Groopman *et al.* (1992c).

ate reductions in tumorigenesis will be required to provide a more quantitative estimate of the relationship between diminution of biomarker levels and altered cancer outcome. Because of the short half-life ($t_{1/2} = 10$ hr) of the aflatoxin–N^7-guanine adduct in hepatic DNA, levels of the excised adduct in urine are likely to reflect DNA damage mediated by recent AFB$_1$ exposures. In contrast, the longer 18 to 21-day half-life of serum albumin in humans indicates that this protein-based dosimeter is likely to reflect weeks of exposure history and may be particularly useful in trying to correlate macromolecular damage to risk of cancer.

Clearly, production of genetic damage by aflatoxins is not a sine qua non for cancer. Many other factors including recurrent cytotoxicity, cell proliferation, and nutritional status can exert strong postinitiation effects to either enhance or retard tumorigenesis. Probably all the chemoprotective agents used to date as inhibitors of aflatoxin carcinogenesis act through multiple mechanisms. Thus, strategies to monitor solely genetic damage in individuals are likely to underestimate the degree of protection afforded by these agents. In the case of AFB$_1$-exposed animals fed dithiolethiones, hepatic aflatoxin–DNA adducts are detectable in animals that are nonetheless at very low risk for tumorigenesis. Similarly, the degree of diminution of the levels of the biomarkers over the entire AFB$_1$ exposure period underestimates subsequent risk reduction for neoplasia. Strategies to account for other modifying factors must be developed.

Although these initial findings with aflatoxin biomarkers are encouraging, note that the ultimate validation of any marker is its ability to predict disease occurrence. To date, no intermediate biomarker has been validated in the context of a chemoprotection trial with modulation of cancer as the end point. Validation procedures for promising biomarkers include assessments of the sensitivity and selectivity of the biological markers (Hulka *et al.,* 1990; Hulka, 1991). Sensitivity evaluates the proportion of the study population that is identified correctly (true positives) by the biomarker, whereas selectivity examines the occurrence of many or few false-positive results among the unexposed or disease-free individuals. Preliminary validation work in animals is necessary to develop assays, generate dose–response curves, and gain an understanding of whether the marker is involved directly in carcinogenesis or merely as a secondary adaptive response of the host. However, epidemiological studies in humans are required to confirm the biomarker–cancer link. In this respect, both observational (case–control, prospective cohort) and intervention studies can be used to determine whether the marker is a mediator of the pathogenic process (Schatzkin *et al.,* 1990). Case–control and cohort studies can suggest a relationship between a biomarker and cancer whereas intervention studies can assess the extent to which a change in the biomarker is predictive of an observed intervention effect. The selection of prospective intermediate biomarkers for chemoprotective interventions requires that several issues be considered. Is there a causal relationship (i.e., issues of strength, consistency, specificity, temporality, and coherence) between the intermediate biomarker and cancer? Does the biomarker appear at a defined stage of carcinogenesis? Can the biomarker be modulated by chemoprotective agents? Do the biomarker and its assay provide acceptable sensitivity, specificity, and accuracy? Is the biomarker stable and technically easy to measure? Can the biomarker be obtained by noninvasive techniques? Clearly, the identification and validation of intermediate biomarkers is in itself a difficult process but, as noted earlier, one that is critical to the timely advancement of chemoprotection in human populations. The ongoing strategies used for the validation of the aflatoxin biomarkers aflatoxin–N[7]-guanine in urine and aflatoxin–albumin adducts in serum are anticipated to serve as templates for the identification and development of biomarkers for other environmental chemical carcinogens.

CONCLUSIONS

The history of medicine clearly has established that preventive strategies offer the best long-term means for eliminating morbidity and mortality from infectious diseases. Chemically induced diseases such as some cancers are likely to respond similarly. Opportunities for prevention of aflatoxin-induced diseases include reducing exposure and interfering with the toxicological processes through pharmacologic or nutritional interventions. In practice, multiple strategies will be required to reduce mortality from liver cancer substantially throughout the world. No single approach—be it universal hepatitis B vaccination, improved storage of foodstuffs, chemoprotective interventions, earlier diagnosis, or more effective therapy—is destined to eliminate the problem. However, each of these approaches, singly and in combination, can have significant impact on mortality and morbidity resulting from aflatoxin exposure. With several decades of chemoprotection research in animal models as a support, the effectiveness of chemoprotection in humans is now ready for testing. This chapter highlighted the results of experimental chemoprotection against aflatoxin-induced liver disease in animals and considered some of the prospects and problems associated with conducting chemoprotective interventions in humans at high risk for aflatoxin exposure. Although the prospects for effective interventions in these individuals appear bright, more importantly the window of opportunity to conduct chemoprotection interventions has been opened.

ACKNOWLEDGMENTS

We gratefully acknowledge financial support for our studies on cancer chemoprotection from the National Institutes of Health (CA 39416, CA 44530, and ES 07141). T. W. Kensler is recipient of NIH Research Career Development Award CA 01230. We also are indebted to our co-workers and collaborators: K. Baumgartner, T. Curphey, N. Davidson, P. Dolan, P. Egner, P. Hollenberg, D. MacMillan, D. Putt, A. Rogers, and P. Talalay.

REFERENCES

Angsubhakorn, S., Bhamarapravati, N., Romruen, K., Sahaphong, S., and Thamavit, W. (1978). Alpha benzene hexachloride inhibition of aflatoxin B_1-induced hepatocellular carcinoma. A preliminary report. *Experientia* **34**, 1069–1070.

Angsubhakorn, S., Bhamarapravati, N., Pradermwong, A., Im-Emgamol, N., and Sahaphong, S. (1989). Minimal dose and time protection by lindane (γ-isomer of 1,2,3,4,5,6-hexachlorohexane) against liver tumors induced by aflatoxin B_1. *Int. J. Cancer* **43**, 531–534.

Bailey, G. S., Hendricks, J. D., Nixon, J. E., and Pawlowski, N. E. (1984). The sensitivity of rainbow trout and other fish to carcinogens. *Drug Metab. Rev.* **15**, 725–750.

Beasley, R. P. (1982). Hepatitis B virus as the etiologic agent in hepatocellular carcinoma—Epidemiologic considerations. *Hepatology* **2**, 21S–26S.

Beaver, R. W., Wilson, D. M., James, M. A., Haydon, K. D., Colvin, B. M., Sangster, L. T., Pikul, A. H., and Groopman, J. D. (1990). Distribution of aflatoxins in tissues of growing pigs fed an aflatoxin-contaminated diet amended with a high affinity aluminosilicate sorbent. *Vet. Hum. Toxicol.* **32**, 16–18.

Bennett, R. A., Essigmann, J. M., and Wogan, G. N. (1981). Excretion of an aflatoxin-guanine adduct in the urine of aflatoxin B_1-treated rats. *Cancer Res.* **41**, 650–654.

Boyd, J. N., Babish, J. G., and Stoewsand, G. S. (1982). Modification by beet and cabbage diets of aflatoxin B_1-induced rat plasma α-foetoprotein elevation, hepatic tumorigenesis, and mutagenicity of urine. *Fd. Chem. Toxicol.* **20**, 47–52.

Bradfield, C. A., and Bjeldanes, L. F. (1984). Effect of dietary indole-3-carbinol on intestinal and hepatic monooxygenase, glutathione S-transferase and epoxide hydrolase activities in the rat. *Fd. Chem. Toxicol.* **22**, 977–982.

Busby, W. F., Jr., and Wogan, G. N. (1984). Aflatoxins. *In* "Chemical Carcinogens" (C. E. Searle, ed.), 2d Ed., Vol. 2, pp. 945–1136. American Chemical Society, Washington, D.C.

Cabral, J. R., and Neal, G. E. (1983). The inhibitory effects of ethoxyquin on the carcinogenic action of aflatoxin B_1 in rats. *Cancer Lett.* **19**, 125–132.

Chang, Y., and Bjeldanes, L. F. (1987). *R*-Goitrin- and BHA induced modulation of aflatoxin B_1 binding to DNA and biliary excretion of thiol conjugates in rats. *Carcinogenesis* **8**, 585–590.

Ch'ih, J. J., Biedrzycka, D. W., Lin, T., Khoo, M. O., and Devlin, T. M. (1989). 2(3)-*tert*-Butyl-4-hydroxyanisole inhibits oxidative metabolism of aflatoxin B_1 in isolated rat hepatocytes. *Proc. Soc. Exp. Biol. Med.* **192**, 35–42.

Chisari, F. V., Klopchin, K., Moriyama, T., Pasquinelli, C., Dunsford, H. A., Sell, S., Pinkert, C. A., Brinster, R. L., and Palmiter, R. D. (1989). Molecular pathogenesis of hepatocellular carcinoma in hepatitis B virus transgenic mice. *Cell* **59**, 1145–1156.

Coles, B., Meyer, D. J., Ketterer, B., Stanton, C. A., and Garner, R. C. (1985). Studies on the detoxification of microsomally-activated aflatoxin B_1 by glutathione and glutathione transverses *in vitro*. *Carcinogenesis* **6**, 693–697.

Colvin, B. M., Sangster, L. T., Haydon, K. D., Beaver, R. W., and Wilson, D. M. (1989). Effect of a high affinity aluminosilicate sorbent on prevention of aflatoxicosis in growing pigs. *Vet. Hum. Toxicol.* **31**, 46–48.

Dashwood, R. H., Arbogast, D. N., Fong, A. T., Hendricks, J. D., and Bailey, G. S. (1988). Mechanisms of anti-carcinogenesis by indole-3-carbinol: Detailed *in vivo* DNA binding dose–response studies after dietary administration with aflatoxin B_1. *Carcinogenesis* **9**, 427–432.

Dashwood, R. H., Arbogast, D. N., Fong, A. T., Hendricks, J. D., and Bailey, G. S. (1989). Quantitative interrelationships between aflatoxin B_1 carcinogen dose, indole-3-carbinol anti-carcinogen dose, target organ DNA adduction and final tumor incidence. *Carcinogenesis* **10**, 175–181.

Dashwood, R. H., Fong, A. T., Williams, D. E., Hendricks, J. D., and Bailey, G. S. (1991). Promotion of aflatoxin B_1 carcinogenesis by the natural tumor modulator indole-3-carbinol: Influence of dose, duration, and intermittent exposure on indole-3-carbinol promotional potency. *Cancer Res.* **51**, 2362–2365.

Davidson, N. E., Egner, P. A., and Kensler, T. W. (1990). Transcriptional control of glutathione S-transferase gene expression by the chemoprotective agent 5-(2-pyrazinyl)-4-methyl-1,2-dithiole-3-thione (oltipraz) in rat liver. *Cancer Res.* **50**, 2251–2255.

Degen, G. H., and Neumann, H. G. (1978). The major metabolite of aflatoxin B_1 in the rat is a glutathione conjugate. *Chem. Biol. Interact.* **22**, 239–255.

Goeger, D. E., Shelton, D. W., Hendricks, J. D., and Bailey, G. S. (1986). Mechanisms of anticarcinogenesis by indole-3-carbinol: Effects on the distribution and metabolism of aflatoxin B_1 in rainbow trout. *Carcinogenesis* **7**, 2025–2031.

Goeger, D. E., Shelton, D. W., Hendricks, J. D., Pereira, C., and Bailey, G. S. (1988). Comparative effect of dietary butylated hydroxyanisole and β-naphthoflavone on aflatoxin B_1 metabolism, DNA adduct formation, and carcinogenesis in rainbow trout. *Carcinogenesis* **9**, 1793–1800.

Graham, S. (1983). Results of case-control studies of diet and cancer in Buffalo, New York. *Cancer Res.* **43**, 2409s–2413s.

Groopman, J. D., Hasler, J. A., Trudel, L. J., Pikul, A., Donahue, P. R., and Wogan, G. N. (1992a). Molecular dosimetry in rat urine of aflatoxin-N^7-guanine and other aflatoxin metabolites by multiple monoclonal antibody affinity chromatography and immunoaffinity/high performance liquid chromatography. *Cancer Res.* **52**, 267–274.

Groopman, J. D., Zhu, J., Donahue, P. R., Pikul, A., Zhang, L., Chen, J-S., and Wogan, G. N. (1992b). Molecular dosimetry of urinary aflatoxin-DNA adducts in people living in Guangxi autonomous region, People's Republic of China. *Cancer Res.* **52**, 45–54.

Groopman, J. D., DeMatos, P., Egner, P. A., Love-Hunt, A., and Kensler, T. W. (1992c). Molecular dosimetry of urinary aflatoxin-N^7-guanine and serum aflatoxin-albumin adducts predicts chemoprotection by 1,2-dithiole-3-thione in rats. *Carcinogenesis* **13**, 101–106.

Guengerich, F. P., and Kim, D-H. (1990). *In vitro* inhibition of dihydropyridine oxidation and aflatoxin B_1 activation in human liver microsomes by naringenin and other flavoids. *Carcinogenesis* **11**, 2275–2279.

Guengerich, F. P., and Shimada, T. (1991). Oxidation of toxic and carcinogenic chemicals by human cytochrome P-450 enzymes. *Chem. Res. Toxicol.* **4**, 391–407.

Guengerich, F. P., Raney, K. D., Kin, D-H., Shimada, T., Meyer, D. J., Ketterer, B., and Harris, T. M. (1992). Oxidation and conjugation of aflatoxins by humans and experimental animals. *In* "Relevance of Animal Studies to the Evaluation of Human Cancer Risk" (R. D'Amato, T. J. Slaga, W. H. Farland, and C. Henry, eds.), pp. 157–165. Wiley-Liss, New York.

Gurtoo, H. L., Koser, P. L., Bansal, S. K., Fox, H. W., Sharma, S. D., Mulhern, A. I., and Pavelic, Z. P. (1985). Inhibition of aflatoxin B_1-hepatocarcinogenesis in rats by β-naphthoflavone. *Carcinogenesis* **6**, 675–678.

Hagler, W. M., Jr. (1990). Potential for detoxification of mycotoxin-contaminated commodities. *In* "Mycotoxins, Cancer, and Health" (G. A. Bray and D. H. Ryan, eds.), pp. 253–269. Louisiana State University Press, Baton Rouge.

Harvey, R. B., Kubena, L. F., Phillips, T. D., Huff, W. E., and Corrier, D. E. (1989). Prevention of aflatoxicosis by addition of hydrated sodium calcium aluminosilicate to the diets of growing barrows. *Am. J. Vet. Res.* **50**, 416–420.

Hayes, J. D., Judah, D. J., McLellan, L. I., Kerr, L. A., Peacock, S. D., and Neal, G. E. (1991). Ethoxyquin-induced resistance to aflatoxin B_1 in the rat is associated with the expression of a novel Alpha-class glutathione S-transferase subunit, Yc_2, which possesses high catalytic activity for aflatoxin B_1-8,9-epoxide. *Biochem. J.* **279**, 385–398.

Hendricks, J. D., Putnam, T. P., Bills, D. D., and Sinnhuber, R. O. (1977). Inhibitory effect of a polychlorinated biphenyl (Aroclor 1254) on aflatoxin B_1 carcinogenesis in rainbow trout (*Salmo gairdneri*). *J. Natl. Cancer Inst.* **59**, 1545–1550.

Hendricks, J. D., Putnam, T. P., and Sinnhuber, R. O. (1980). Null effect of dietary Aroclor 1254 on hepatocellular carcinoma incidence in rainbow trout (*Salmo gairdneri*) exposed to aflatoxin B₁ as embryos. *J. Environ. Pathol. Toxicol.* **4**, 9–16.

Hong, W. K., Lippman, S. M., Itri, L. M., Karp, D. D., Lee, J. S., Byers, R. M., Schantz, S. P., Kramer, A. M., Lotan, R., Peters, L. J., Dimery, I. W., Brown, B. W., and Goepfert, H. (1990). Prevention of secondary primary tumors with isotretinoin in squamous-cell carcinoma of the head and neck. *N. Engl. J. Med.* **323**, 795–801.

Hsu, I. C., Metcalf, R. A., Sun, T., Welsh, J. A., Wang, N. J., and Harris, C. C. (1991). Mutational hotspot in the p53 gene in human hepatocellular carcinomas. *Nature (London)* **350**, 427–428.

Hulka, B. S. (1991). Epidemiological studies using biological markers: Issues for epidemiologists. *Cancer Epidemiol. Biomarkers Prev.* **1**, 13–19.

Hulka, B. S., Wilcosky, T. C., and Griffith, J. D. (eds.) (1990). "Biological Markers in Epidemiology." Oxford University Press, New York.

Ito, N., and Hirose, M. (1987). The role of antioxidants in chemical carcinogenesis. *Jpn. J. Cancer. Res.* **78**, 1011–1026.

Jones, F. T., Hagler, W. M., and Hamilton, P. B. (1982). Association of low levels of aflatoxin in feed with productivity losses in commercial broiler operations. *Poultry Sci.* **61**, 861–868.

Kensler, T. W., Egner, P. A., Trush, M. A., Bueding, E., and Groopman, J. D. (1985). Modification of aflatoxin B₁ binding to DNA *in vivo* in rats fed phenolic antioxidants, ethoxyquin and a dithiolthione. *Carcinogenesis* **6**, 759–763.

Kensler, T. W., Egner, P. A., Davidson, N. E., Roebuck, B. D., Pikul, A., and Groopman, J. D. (1986). Modulation of aflatoxin metabolism, aflatoxin-N^7-guanine formation, and hepatic tumorigenesis in rats fed ethoxyquin: Role of induction of glutathione *S*-transferases. *Cancer Res.* **46**, 3924–3931.

Kensler, T. W., Egner, P. A., Dolan, P. M., Groopman, J. D., and Roebuck, B. D. (1987). Mechanism of protection against aflatoxin tumorigenicity in rats fed 5-(2-pyrazinyl)-4-methyl-1,2-dithiol-3-thione (oltipraz) and related 1,2-dithiol-3-thione and 1,2-dithiol-3-ones. *Cancer Res.* **47**, 4271–4277.

Kensler, T. W., Groopman, J. D., and Roebuck, B. D. (1992a). Chemoprotection by oltipraz and other dithiolethiones. *In* "Cancer Chemoprevention" (L. W. Wattenberg, M. Lipkin, C. Boone, and G. Kelloff, eds.), pp. 205–226. CRC Press, Boca Raton, Florida.

Kensler, T. W., Groopman, J. D., Eaton, D. L., Curphey, T. J., and Roebuck, B. D. (1992b). Potent inhibition of aflatoxin-induced hepatic tumorigenesis by the monofunctional enzyme inducer 1,2-dithiole-3-thione. *Carcinogenesis* **13**, 95–100.

Kraemer, K. H., DiGiovanna, J. J., Moshell, A. N., Tarone, R. E., and Peck, G. L. (1988). Prevention of skin cancer in xeroderma pigmentosum with the use of oral isotretinoin. *N. Engl. J. Med.* **318**, 1633–1637.

LaBrecque, D. R., Steele, G., Fogerty, S., Wilson, M., and Barton, J. (1987). Purification and physical–chemical characteristics of hepatic stimulator substance. *Hepatology* **7**, 100–106.

Larsen, C., Ehrich, M., Driscoll, C., and Gross, W. B. (1985). Aflatoxin-antioxidant effects on growth of young chicks. *Poultry Sci.* **64**, 2287–2291.

Lin, J-K., and Wang, C-J. (1986). Protection of crocin dyes on the acute hepatic damage induced by aflatoxin B₁ and dimethylnitrosamine in rats. *Carcinogenesis* **7**, 595–599.

Lippman, S. M., Lee, J. S., Lotan, R., Hittelman, W., Wargovich, M. J., and Hong, W. K. (1990). Biomarkers as intermediate end points in chemoprevention trials. *J. Natl. Cancer Inst.* **82**, 555–560.

Lippman, S. M., Hittelman, W. N., Lotan, R., Pastorino, U., and Hong, W. K. (1991). Recent advances in cancer chemoprevention. *Cancer Cells* **3**, 59–65.

Lipkin, M., and Newmark, H. (1985). Effect of added dietary calcium on colonic epithelial cell proliferation in subjects at high risk for familial colonic cancer. *New Engl. J. Med.* **313**, 1381–1384.

Liu, L-Y., Roebuck, B. D., Yager, J. D., Groopman, J. D., and Kensler, T. W. (1988). Protection by

5-(2-pyrazinyl)-4-methyl-1,2-dithiol-3-thione (oltipraz) against the hepatotoxicity of aflatoxin B$_1$ in the rat, *Toxicol. Appl. Pharmacol.* **93**, 442–451.

Liu, Y. H., Taylor, J., Linko, P., Lucier, G. W., and Thompson, C. L. (1991). Glutathione *S*-transferase μ in human lymphocyte and liver: Role in modulating formation of carcinogen-derived DNA adducts. *Carcinogenesis* **12**, 2269–2275.

Loveland, P. M., Nixon, J. E., and Bailey, G. S. (1984). Glucuronides in bile of rainbow trout (*Salmo gairdneri*) injected with [^{3}H]aflatoxin B$_1$ and the effects of dietary β-naphthoflavone. *Comp. Biochem. Physiol.* **78C**, 13–19.

McLean, A. E., and Marshall, A. (1971). Reduced carcinogenic effects of aflatoxin in rats given phenobarbitone. *Br. J. Exp. Pathol.* **52**, 323–329.

McLellan, L. I., Kerr, L. A., Cronshaw, A. D., and Hayes, J. D. (1991). Regulation of mouse glutathione *S*-transferases by chemoprotectors. Molecular evidence for the existence of three distinct alpha-class glutathione *S*-transferase subunits, Ya1, Ya2, and Ya3, in mouse liver. *Biochem. J.* **276**, 461–469.

Malone, W. F. (1991). Studies evaluating antioxidants and β-carotene as chemopreventives. *Am. J. Clin. Nutr.* **53**, 305S–313S.

Mandel, H. G., Manson, M. M., Judah, D. J., Simpson, J. L., Green, J. A., Forrester, L. M., Wolf, C. R., and Neal, G. E. (1987). Metabolic basis for the protective effect of the antioxidant ethoxyquin on aflatoxin B$_1$ hepatocarcinogenesis in the rat. *Cancer Res.* **47**, 5218–5223.

Manning, R. O., Wyatt, R. D., and Marks, H. L. (1990). Effects of phenobarbital and β-napthoflavone on the *in vivo* toxicity and *in vitro* metabolism of aflatoxin in an aflatoxin-resistant and control line of chickens. *J. Toxicol. Environ. Health* **31**, 292–311.

Mgbodile, M. U., Holscher, M., and Neal, R. A. (1975). A possible protective role for reduced glutathione in aflatoxin B$_1$ toxicity: Effect of pretreatment of rats with phenobarbital and 3-methylcholanthrene on aflatoxin toxicity. *Toxicol. Appl. Pharmacol.* **34**, 128–142.

Michalopoulos, G. K. (1990). Liver regeneration: Molecular mechanisms of growth control. *FASEB J.* **4**, 176–187.

Newberne, P. M., and Williams, G. (1969). Inhibition of aflatoxin carcinogenesis by diethylstilbestrol in male rats. *Arch. Environ. Health* **19**, 489–498.

Nixon, J. E., Hendricks, J. D., Pawloski, N. E., Pereira, C. B., Sinnhuber, R. O., and Bailey, G. S. (1984). Inhibition of aflatoxin B$_1$ carcinogenesis in rainbow trout by flavone and indole compounds. *Carcinogenesis* **5**, 615–619.

Phillips, T. D., Sarr, B. A., Clement, B. A., Kubena, L. F., and Harvey, R. B. (1990). Prevention of aflatoxicosis in farm animals via selective chemisorption of aflatoxin. *In* "Mycotoxins, Cancer, and Health" (G. A. Bray and D. H. Ryan, eds.), pp. 223–237. Louisiana State University Press, Baton Rouge.

Prochaska, H. J., Yeh, Y., Baron, P., and Polsky, B. (1993). Oltipraz, an inhibitor of human immunodeficiency virus type 1 replication. *Proc. Natl. Acad. Sci. U.S.A.* **90**, 3953–3957.

Putt, D. A., Kensler, T. W., and Hollenberg, P. F. (1991). Effect of three chemoprotective antioxidants, ethoxyquin, oltipraz and 1,2-dithiole-3-thione on cytochrome P-450 levels and aflatoxin B$_1$ metabolism. *FASEB J.* **5**, A1517.

Quinn, B. A., Crane, T. L., Kocal, T. E., Best, S. J., Cameron, R. G., Rushmore, T. H., Farber, E., and Hayes, M. A. (1990). Protective activity of different hepatic cytosolic glutathione *S*-transferases against DNA-binding metabolites of aflatoxin B$_1$. *Toxicol. Appl. Pharmacol.* **105**, 351–363.

Ramsdell, H. S., and Eaton, D. L. (1988). Modification of aflatoxin B$_1$ biotransformation *in vitro* and DNA binding *in vivo* by dietary broccoli in rats. *J. Toxicol. Environ. Health* **25**, 269–284.

Roebuck, B. D., Liu, Y.-L., Rogers, A. E., Groopman, J. D., and Kensler, T. W. (1991). Protection against aflatoxin B$_1$-induced hepatocarcinogenesis in F344 rats by 5-(2-pyrazinyl)-4-methyl-1,2-dithiole-3-thione (oltipraz), predictive role for short-term molecular dosimetry. *Cancer Res.* **51**, 5501–5506.

Ross, R. K., Yuan, J-M., Yu, M. C., Wogan, G. N., Qian, G-S., Tu, J-T., Groopman, J. D., Gao, Y-T., and Henderson, B. H. (1992). Urinary aflatoxin biomarkers and risk of hepatocellular carcinoma. *The Lancet* **339**, 943–946.

Sabbioni, G., Skipper, P. L., Buchi, G., and Tannenbaum, S. R. (1987). Isolation and characterization of the major serum albumin adduct formed by aflatoxin B_1 *in vivo* in rats. *Carcinogenesis* **8**, 819–834.

Salocks, C. B., Hsieh, D. P. H., and Byard, J. L. (1981). Butylated hydroxytoluene pretreatment protects against cytotoxicity and reduced covalent binding of aflatoxin B_1 in primary hepatocyte cultures. *Toxicol. Appl. Pharmacol.* **59**, 331–345.

Sarr, A. B., Clement, B. A., and Phillips, T. D. (1990). Effects of molecular structure on the chemisorption of aflatoxin B_1 and related compounds by hydrated sodium calcium aluminosilicate. *Toxicologist* **10**, 163.

Schatzkin, A., Freedman, L. S., Schiffman, M. H., and Dawsey, S. M. (1990). Validation of intermediate end points in cancer research. *J. Natl. Cancer Inst.* **82**, 1746–1752.

Scorcone, K. A., Zhou, Y-Z., Butel, J., and Slagle, B. (1992). *p53* Mutations cluster at codon 249 in hepatitis B virus-positive hepatocellular carcinomas from China. *Cancer Res.* **52**, 1635–1638.

Shah, D. V., Engelke, J. A., and Suttie, J. W. (1987). Abnormal prothrombin in the plasma of rats carrying hepatic tumour. *Blood* **69**, 850–854.

Shelton, D. W., Hendricks, J. D., Coulombe, R. A., and Bailey, G. S. (1984). Effect of dose on the inhibition of carcinogenesis/mutagenesis by Aroclor 1254 in rainbow trout fed aflatoxin B_1. *J. Toxicol. Environ. Health* **13**, 649–657.

Shimada, T., and Guengerich, F. P. (1989). Evidence for cytochrome P-450$_{NF}$, the nifedipine oxidase, being the principal enzyme involved in the bioactivation of aflatoxins in human liver. *Proc. Natl. Acad. Sci. U.S.A.* **86**, 462–465.

Stich, H. F., Mathew, B., Sankaranarayanan, R., and Nair, M. K. (1991). Remission of oral precancerous lesions of tobacco/areca nut chewers following administration of β-carotene or vitamin A, and maintenance of the protective effect. *Cancer Detect. Prevent.* **15**, 93–98.

Stoewsand, G. S., Babish, J. B., and Wimberly, H. C. (1978). Inhibition of hepatic toxicities from polybrominated biphenyls and aflatoxin B_1 in rats fed cauliflower. *J. Environ. Pathol. Toxicol.* **2**, 399–406.

Swenson, D. H., Lin, J. K., Miller, E. C., and Miller, J. A. (1977). Aflatoxin B_1-2,3-oxide as a possible intermediate in the covalent binding of aflatoxin B_1 and B_2 to rat liver DNA and ribosomal RNA *in vivo. Cancer Res.* **37**, 172–181.

Tew, M. C. (1989). Tumour markers of hepatocellular carcinoma. *J. Gastroenter. Hepatology* **4**, 373–384.

Vlasta, L. M., Hendricks, J. D., and Bailey, G. S. (1988). The significance of glutathione conjugation for aflatoxin B_1 metabolism in rainbow trout and coho salmon. *Fd. Chem. Toxicol.* **26**, 129–135.

Wang, C-J., Shiow, S-J., and Lin, J-K. (1991). Effects of crocetin on the hepatotoxicity and hepatic DNA binding of aflatoxin B_1 in rats. *Carcinogenesis* **12**, 459–462.

Wattenberg, L. W. (1985). Chemoprevention of cancer. *Cancer Res.* **45**, 1–8.

Whitty, J. P., and Bjeldanes, L. F. (1987). The effects of dietary cabbage on xenobiotic-metabolizing enzymes and the binding of aflatoxin B_1 to hepatic DNA in rats. *Fd. Chem. Toxicol.* **25**, 581–587.

Wild, C. P., Garner, R. C., Montesano, R., and Tursi, F. (1986). Aflatoxin B_1 binding to plasma albumin and liver DNA upon chronic administration to rats. *Carcinogenesis* **7**, 853–858.

Williams, G. M., Tanaka, T., and Maeura, Y. (1986). Dose-related inhibition of aflatoxin B_1 induced-hepatocarcinogenesis by the phenolic antioxidants, butylated hydroxytoluene and butylated hydroxyanisole. *Carcinogenesis* **7**, 1043–1050.

Yeh, F-S., Yu, M. C., Mo, C-C., Tong, M-J., and Henderson, B. E. (1989). Hepatitis B virus, aflatoxins, and hepatocellular carcinoma in southern Guangxi, China. *Cancer Res.* **49**, 2506–2509.

Zhang, Y., Talalay, P., Cho, G-C., and Posner, G. (1992). A major inducer of anticarcinogenic protective enzymes from broccoli: Isolation and elucidation of structure. *Proc. Natl. Acad. Sci. U.S.A.* **89,** 2399–2403.

Zhu, Y.-R., Chen, J.-G., and Huang, X.-Y. (1989). Hepatocellular carcinoma in Qidong County. *In* "Primary Liver Cancer" (Z. Y. Tang, ed.), pp. 204–222. China Academic Publisher, Springer-Verlag, Berlin.

Part III

❖

Agricultural and Veterinary Problems

14

Factors Affecting
Aspergillus flavus Group
Infection and Aflatoxin
Contamination of Crops

❖

David M. Wilson and Gary A. Payne

INTRODUCTION

Aflatoxin contamination is of interest because of possible human and animal health problems caused by ingestion of contaminated foods and feeds. This chapter considers the fungi that produce the aflatoxins, the major aflatoxins that are found in foods and feeds, and the current state of knowledge about aflatoxin contamination of corn, peanuts, and cottonseed. No definitive studies are available on the mechanisms of infection of crops by the toxigenic species of the *Aspergillus flavus* group other than infection of corn silks and seeds. Several other possible avenues of infection may contribute to colonization and aflatoxin contamination. Therefore, we have attempted to develop the most likely possibilities while realizing that some are not supported by adequate experimental evidence.

The species of the *A. flavus* group that produce the aflatoxins include *A. flavus, Aspergillus parasiticus,* and *Aspergillus nomius. Aspergillus flavus* and/or *A. parasiticus* may colonize a food or feed and subsequently contaminate the product with one or more of the aflatoxins. The relative importance of *A. nomius* in aflatoxin contamination of crops is unknown at present. Specific strains of *A. flavus* and *A. parasiticus* are capable of producing aflatoxins B_1, B_2, G_1, G_2, and M_1, and several other related compounds. Strains of *A. flavus* and *A. parasiticus* also have been identified that produce no aflatoxins.

The Toxicology of Aflatoxins:
Human Health, Veterinary, and Agricultural Significance

309

Selected research studies on aflatoxin contamination of corn, peanuts, and cottonseed are evaluated critically in this chapter because more information is available on these crops than on other commodities also known to be subject to aflatoxin contamination. Aflatoxin contamination is an additive process; it may begin in the preharvest crop and accumulate further during harvesting, drying, storage, and processing. Contamination also may occur at any given stage after harvest.

AFLATOXIGENIC FUNGI

Taxonomy

Aflatoxin contamination of foods and feeds occurs when aflatoxigenic species of the *A. flavus* group successfully colonize and grow in a commodity, and subsequently produce the aflatoxin secondary metabolites. The three species in the *A. flavus* group that produce aflatoxins are *A. flavus* Link ex. Fries, *A. parasiticus* Speare, and *A. nomius* Kurtzman, Horn, and Hesseltine. The toxigenic *Aspergillus toxicarius* Murakami was considered synonymous with *A. parasiticus* by Christensen (1981) and Wicklow (1983). The other species in the *A. flavus* group, including the domesticated Koji molds *Aspergillus oryzae* (Ahlburg) Cahn and *Aspergillus sojae* Sakaguichi and Yamada, do not produce aflatoxins (Wang and Hesseltine, 1982).

The taxonomy of the *A. flavus* group is unsettled at the present time. These fungi are difficult to identify because differences are often inconclusive. *Aspergillus flavus* and *A. parasiticus* are taxonomically distinct species according to Wicklow (1983) and Klich and Pitt (1988). However, Hesseltine *et al.* (1970) described several isolates that were taxonomically intermediate between *A. flavus* and *A. parasiticus*. How these isolates would be identified by Wicklow or Klich and Pitt is not clear.

Kurtzman *et al.* (1986) have complicated the taxonomic issue further by proposing that both *A. flavus* and *A. parasiticus* should be considered varieties of *A. flavus* based on DNA relatedness. The toxic species, as proposed by Kurtzman *et al.* (1986), are *A. flavus* var. *flavus, A. flavus* var. *parasiticus,* and *A. nomius*.

The species designation is dependent on the taxonomist, and currently differences of opinion exist about the nomenclature of species of the *A. flavus* group. These differences are probably only important when one is concerned with the taxonomy of specific isolates because the preponderance of literature on *A. flavus* and aflatoxin contamination does not distinguish the species but refers to the *A. flavus* group.

Inoculum Sources

The primary source of *A. flavus* group inoculum is not readily apparent. These aspergilli have no sexual stage; thus, the initial inoculum generally is thought to be conidial. Indeed, perhaps no one primary source of inoculum as described by standard plant pathological terminology exists because describing the etiology and pathology of *A. flavus* and aflatoxin contamination using traditional plant pathology definitions is impossible. In nature, *A. flavus* group inoculum may be airborne, soilborne, and insect-vectored propagules. The propagules may be conidia, mycelia, sclerotia, or combinations of various propagule types. Each source may be important individually, depending on the circumstance, and one or more may constitute the initial inoculum.

Abdalla (1988) monitored *A. flavus* in the air for 12 months in Sudan and reported that frequency was related to total fungal counts and dusty weather. The aspergilli composed 68% of the total fungal isolates and *A. flavus* constituted 31% of the aspergilli during the hot, dry, dusty summer. Airborne *A. flavus* propagules were detected infrequently during the rainy and winter seasons. Bothast *et al.* (1978) and Ilag (1975) found low populations of airborne *A. flavus* group conidia in corn fields whereas Holtmeyer and Wallin (1981) trapped airborne aflatoxin-producing isolates at eight widely separated sites in or near corn fields in Missouri. Jones *et al.* (1980) demonstrated a positive correlation of airborne conidia with irrigation in corn plots in North Carolina. These researchers also found that corn hybrids pollinated during weeks with high airborne populations were likely to contain a high percentage of ears with visible *A. flavus* at harvest. Ilag (1975) reported that infection of corn was related directly to the amount of *A. flavus* in the air, and that airborne populations were least in corn fields, higher in the drying area, and highest in Phillipine warehouses. Hill *et al.* (1985) sampled airborne dusts associated with farm and elevator operations in Georgia and found more *A. flavus* in 1980, a dry year, than in 1979 and 1982. In geographic areas where aflatoxin contamination is a recurring problem, airborne propagules may be the most important source of inoculum. Airborne inoculum may be less important in temperate regions, such as the United States corn belt, where aflatoxin contamination is infrequent and generally only associated with drought conditions.

The soil is also an important potential source of preharvest *A. flavus* group inoculum. *Aspergillus flavus* propagules may survive in the soil as conidia, mycelia associated with organic matter, or sclerotia. Lillehoj *et al.* (1980) observed viable *A. flavus* and *A. parasiticus* in soil or in washes of soil insects collected in corn plots in Iowa, Illinois, Missouri, and Georgia. The *A. flavus* group populations were not associated with any particular location, corn maturity group, or environmental factor. Wicklow (1983) proposed that sclerotia serve as the primary inoculum and conidia and mycelia as the secondary inoculum. However, many *A. flavus* and *A. parasiticus* isolates do not produce sclerotia in culture. Angle *et al.* (1989) did not find any sclerotia in soil amended with conidia of a non-sclerotia-producing *A. flavus,* but found that the total colony forming units (cfu) per gram of soil remained essentially unchanged, at about 1.3 $\times$ 10^4 cfu/g after an initial decline from 5.5 $\times$ 10^4 cfu/g, over a 1-year period. These experiments demonstrated that sclerotia are not necessary for long-term survival of *A. flavus* in soil. M. E. Will and D. M. Wilson (unpublished data) monitored soil throughout the growing season in two continuously irrigated and two continuously dryland peanut fields and in two continuously irrigated and two continuously dryland corn fields in Georgia in 1991 and recovered no sclerotia from soil. Viable *A. flavus* propagules often were associated with particulate organic matter in the soil. The soil populations of the *A. flavus* group were higher in the corn and peanut dryland fields than in the irrigated fields. Wilson *et al.* (1989a) reported that irrigation lowered the percentage of *A. flavus* recovered from peanut shells and kernels relative to peanuts grown under drought conditions. In Missouri, soil populations of *A. flavus* were related to the cropping history (Angle *et al.*, 1982).

The evidence that insects can vector *A. flavus* group inoculum is convincing. Insects that visit the susceptible plant part need not be pests that cause damage to transport inoculum to infection sites. Widstrom (1979) reviewed the broad inter-regional role of insect vectors in aflatoxin contamination of corn, cotton, and peanuts. McMillian (1983) reviewed the role of arthropods in aflatoxin contamination of corn, and listed bees, *Heliothis zea, Sitophilus oryzae, Ostrinia nubilalis, Spodoptera frugiperda, Zeadiatraea grandiosella, Tribolium confusum, Sitotroga cerealella,* and species of *Nitidulidae* as well as other unidentified insects as being capable of serving as vectors because of internal or external *A. flavus* infestation. McMillian (1983) concluded that the dissemination of *A. flavus* group inoculum is enhanced significantly by insects, but the activity of any specific insect is not necessarily related to its ability to act as a vector. Also, the insect–*A. flavus* association is influenced strongly by location and environmental conditions. *Aspergillus flavus* was found more often on insects inhabiting corn ears and *A. parasiticus* was found more often on soil insects (Lillehoj *et al.*, 1980). The incidence of insect infestation depends on the geographic location, the season of the year, and the insect species. However, *A. flavus* infestation of insects is widespread whereas aflatoxin contamination is localized (Lillehoj *et al.*, 1978).

Aflatoxin Production: Aflatoxins B_1, B_2, G_1, and G_2

The B and G aflatoxins are produced by all toxigenic *A. parasiticus* isolates whereas most, but not all, *A. flavus* isolates produce only the B aflatoxins (Pitt, 1989). Also, many morphologically atypical *A. flavus* isolates and isolates taxonomically intermediate between *A. flavus* and *A. parasiticus* produce the B and G aflatoxins (Hesseltine *et al.*, 1970). Wicklow and Shotwell (1983) demonstrated that the sclerotia of five *A. flavus* strains contained the B and G aflatoxins whereas the conidia of four of the five strains contained only the B aflatoxins. Production of the G aflatoxins may occur in *A. flavus* primarily in sclerotial cells.

Diener *et al.* (1987) and Pitt (1989) stated that *A. flavus* typically produces aflatoxins B_1 and B_2 whereas *A. parasiticus* produces B_1, B_2, G_1, and G_2, suggesting that when only the B aflatoxins are seen *A. flavus* is responsible and that when the B and G aflatoxins are seen *A. parasiticus* or a mixture of both species is responsible for the contamination. Generally, commodities that contain only the B aflatoxins may be contaminated with *A. flavus,* but the literature does not support the contention that *A. parasiticus* is the only source of the G aflatoxins. Therefore, all products with the B and G aflatoxins should not be assumed to be a result of *A. parasiticus* contamination.

Aspergillus flavus and *A. parasiticus* both may be present as mixed inoculum. These fungi may interact to produce unexpected aflatoxin contamination patterns. Calvert *et al.* (1978) demonstrated that, with mixed *A. flavus* and *A. parasiticus* inoculum, the production of B_1 in preharvest corn was not affected by different inoculum ratios whereas G_1 production was dependent on the percentage of *A. parasiticus* in the inoculum. These investigators found that little G_1 was produced when *A. flavus* constituted 75% of the inoculum. D. M. Wilson and J. K. King (unpublished data) found that small amounts of aflatoxins G_1 and G_2 were produced when *A. flavus* constituted greater than 25% of the inoculum in mixed liquid cultures seeded with mixtures of *A. flavus* and *A. parasiticus* conidia. Apparently, *A. flavus* may be able to inhibit the production of G_1 and G_2 by *A. parasiticus.* The percentage of *A. parasiticus* isolated from Georgia corn rarely is over 10% and varies in Georgia peanuts from 10 to 30%; *A. flavus* constitutes 70–90% of the *A. flavus* group isolates (Hill *et al.,* 1985). Thus, that little G_1 is seen in naturally contaminated samples is not surprising. However, that *A. flavus* alone is responsible for contamination of samples with only aflatoxin B_1 and B_2 is not necessarily true.

Over 4000 aflatoxin-contaminated corn and peanut samples were analyzed at the mycotoxin laboratory in Tifton, Georgia; 76.9% of corn and 61.8% of peanut samples contained only aflatoxin Bs whereas 21.9% of corn and 35.7% of peanut samples contained Bs and Gs (Hill *et al.,* 1985). Inspection of data on aflatoxin-contaminated samples collected by the Food and Drug Administration (FDA) over 6 years revealed that less than 1% of cottonseed samples contained both the B and G aflatoxins whereas over 50% of the brazil nut samples contained both the B and G aflatoxins (Wilson and Abramson, 1992). Much more research on the incidence and importance of the different aflatoxins and the prevalence of *A. flavus* and *A. parasiticus* in different crops is needed before we can begin to understand the biological factors leading to aflatoxin contamination.

FACTORS AFFECTING *ASPERGILLUS FLAVUS* GROUP INOCULUM AND INITIAL COLONIZATION

The plant pathological concepts of primary and secondary inoculum have little meaning with *A. flavus* group colonization and subsequent aflatoxin production, primarily because describing the disease based on visible symptoms is difficult. Susceptible crops probably are exposed to a continual source of inoculum in areas where preharvest aflatoxin contamination is endemic. In climates in which aflatoxin contamination is rare or only seen occasionally, different sources of airborne, soilborne, or insect-carried inoculum may be present, and each may be more or less important. However, inoculum density may increase rapidly in years favorable for aflatoxin contamination.

Corn

The first comprehensive study of *Aspergillus* ear molds was published in 1920 by Taubenhaus. The black and yellow ear molds were identified as being caused by *Aspergillus niger* and *A. flavus* in insect-damaged ears. The fungi often occurred together in the insect-damaged ears. Taubenhaus (1920) demonstrated that inoculations of wounded tissue were successful during the milky kernel stage and not successful when the kernels were fully mature, as measured by visible sporulation. However, because *A. flavus* was not an economically important ear-rot pathogen in temperate climates, where most corn is grown, this work was not widely recognized and *A. flavus* became known as a storage mold.

Christensen (1974) separated the fungi that occur in stored grains into two groups—the field fungi and the storage fungi. The field and storage fungi concept was based on extensive work on grain produced and stored in temperate climates. The aspergilli were classified as storage fungi because they could grow at moisture contents in equilibrium with relative humidities of 70–90% with no free moisture present (Christensen, 1974). The field and storage fungi concept works well in general in temperate climates, but in warm humid, subtropical, and tropical climates the storage fungi often become the field as well as the storage fungi (Wilson and Abramson, 1992).

Occurrence of *A. flavus* in United States corn is much more common in the Southeast than in the Midwest. For example, an extensive study in central Illinois revealed no *A. flavus* in 1974 (Hesseltine and Bothast, 1977), whereas *A. flavus* averaged 0.6 and 1.6% in physically damaged corn grown in Indiana in 1971 and 1972 (Rambo *et al.*, 1974). In a survey of South Carolina corn in 1973, Hesseltine *et al.* (1976) found *A. flavus* in 60% of 305 corn samples. Hill *et al.* (1985) found that up to 56% of preharvest ears in Georgia contained *A. flavus*. Apparently climatic and regional differences determine the incidence of the *A. flavus* group in the field as well as in storage. Aflatoxin contamination of stored corn was recognized early but aflatoxin contamination in the field was not considered to be serious until Anderson *et al.* (1975) published their landmark paper. Since then, aflatoxin contamination of preharvest corn has been reported in at least 21 states (Wilson *et al.*, 1989b).

Payne and co-workers at North Carolina State University have emphasized studies on direct infection of corn silks, one of the possible mechanisms of infection. Marsh and Payne (1984a) documented the colonization of pollinated corn silks using a color mutant of *A. flavus. Aspergillus flavus* spores placed on pollinated yellow-brown silks germinated within 4–8 hr and hyphae penetrated the silks directly or through cracks and intercellular gaps. Hyphal growth within the silk tissue was parallel to the silk axis and was restricted to parenchymatous tissue. The fungus began to sporulate within 24 hr. Once the external silks were colonized, hyphae reached the base of the ear in 4–8 days (Marsh and Payne, 1984a,b; Payne *et al.*, 1988a).

Silk inoculation could result from airborne inoculum, or the inoculum could be introduced by insects carrying *A. flavus* that visit the ear during silking. The insects that could be *A. flavus* vectors do not necessarily have to be pests that cause damage (McMillian, 1983). Depending on the stage of corn development, successful inoculation by insect-vectored *A. flavus* could occur easily.

Payne (1992) suggested that insects may facilitate infection of preharvest ears by transporting inoculum to the ears, transporting inoculum already on the silk into the ears, disseminating inoculum within the ears, and creating a favorable *A. flavus* habitat through injury associated with feeding. Insects may act as vectors by carrying *A. flavus* internally and depositing inoculum in the frass. (McMillian, 1983). McMillian (1987) reported that corn earworm moths collected in Tifton, Georgia, over several years were contaminated with *A. flavus* group species an average of 48, 53, and 50% for the months of June, July, and August, respectively. Lussenhop and Wicklow (1990) demonstrated that nitidulid beetles could transport *A. flavus* from the soil to corn ears in Georgia. In a 2-year study, Wilson *et al.* (1986) found that maize weevils infested with *A. flavus* and placed on ears 30 days after full silk increased the percentage of *A. flavus* recovered from kernels from 52 to 78%.

Hill *et al.* (1985) monitored the incidence of *A. flavus* in corn kernels at the tip, middle, and base of the ear every 15 days from full silk until harvest and found that, in Georgia, the incidence increased during growing season in both 1978 and 1979. In 1978, but not in 1979, insecticide treatments on two of three planting dates reduced the incidence of *A. flavus* in the tip and middle but not in the base of ears at harvest.

Inoculum of the *A. flavus* group in corn may originate as airborne propagules, be introduced by insect activity, or be introduced by a combination of these sources continuously over the growing season. After the inoculum is introduced into the ear, the prevailing environmental conditions determine whether *A. flavus* can colonize the corn ear successfully.

Peanut

The peanut flower is fertilized above ground; then a peg develops that grows into the soil, and the fruit develops underground. The initial *A. flavus* group inoculum could be airborne during flowering, as suggested by Wells and Kreutzer (1972), or it could be in the soil. In Virginia, *A. flavus* was recovered from 1.5–3% of aerial pegs (Griffin and Garren, 1976); in Georgia, peanuts grown in plots to induce extreme drought stress had an infection of aerial and subterrain pegs of 75 and 80%, respectively (Sanders *et al.*, 1981). Cole *et al.* (1989) reported that they found no airborne propagules around the environmental plots in which peanuts were contaminated heavily with aflatoxins.

Initial inoculum in peanuts most likely originates in the soil. Successful colonizations are probably dependent on soil *A. flavus* populations, as well as on the water content and temperature of the soil. Wilson and Stansell (1983) found that plots with drought during the last 40–75 days of the growing season had higher *A. flavus* group populations in the soil than irrigated plots. Because the developing peanut fruit is in contact with the soil, direct invasion is likely (Griffin and Garren, 1976).

Aucamp (1969) demonstrated that mites could be vectors of *A. flavus*. Lynch and Wilson (1991) verified that lesser cornstalk borer larvae were vectors of *A. flavus* and *A. parasiticus,* as suggested by Dickens *et al.* (1973). Internal and external pod damage by lesser cornstalk borer increased the percentage of kernels from which *A. flavus* could be recovered (Lynch and Wilson, 1991).

Inoculation and colonization of the developing peanut fruit could happen at flowering but inoculation with the *A. flavus* group is more likely to occur repeatedly in the soil. Soil inoculations most likely are influenced by peanut maturity, pod and kernel damage, soil organic matter, soil moisture, and soil temperature, as well as by insect activity. Invasion can occur at anytime until the peanuts are too dry to support *A. flavus* growth.

Cottonseed

Aspergillus flavus inoculum in cotton bolls probably comes from soilborne and wind-driven airborne propagules (Ashworth *et al.,* 1969a; Lee *et al.,* 1986). The soil in the desert areas of California contains *A. flavus*-infested organic particles that may be reservoirs for insect-vectored *A. flavus* inoculum (Kiyomoto and Ashworth, 1974). *Aspergillus flavus* has been isolated from up to 79% of stink and lygus bugs tested (Stephenson and Russell, 1974). Simpson and Batra (1984) isolated *A. flavus* from flowers, floral nectaries, bracts, and bractioles, and observed the fungus growing on leaves with insect honeydew stimulating growth.

Irrigated cotton grown in the desert in California and Arizona probably is exposed continuously to *A. flavus* inoculum from flowering through maturation. The most likely sources of inoculum are wind-driven and insect-vectored propagules.

CONDITIONS AFFECTING *ASPERGILLUS FLAVUS* INVASION, GROWTH, AND AFLATOXIN ACCUMULATION

Susceptible crop plants that are inoculated successfully and colonized by members of the *A. flavus* group may become internally colonized by the fungus. Such colonization may lead to aflatoxin production. Alternatively, the colonization may remain quiescent or extensive fungal growth may occur without aflatoxin formation. These possibilities depend on the *A. flavus* isolate, the crop, and the environmental conditions.

Corn

Aflatoxin contamination of corn occurs worldwide. Aflatoxins can be produced in preharvest as well as in stored corn. Payne (1992), McMillian *et al.* (1991), and Wilson *et al.* (1989b) have reviewed the literature on aflatoxins in corn. *Aspergillus flavus* colonization was not observed at the full silk stage of uninoculated Georgia corn but colonization was noted 15 days later (Hill *et al.,* 1985). Marsh and Payne (1984a,b) showed that, in silk-inoculated corn, the surface of kernels and surrounding tissue was colonized readily without invasion of the kernels. Smart *et al.* (1990) characterized the route of infection after colonization. *Aspergillus flavus* may enter the undamaged seed in one of two ways. The fungus may grow superficially and invade at the junction of the bracts with their rachillas or the fungus may grow up through the cob into the spikelet. The hyphae seem to be restricted to parenchymatous tissue. Smart *et al.* (1990) and Payne *et al.* (1988a) found that invasion of undamaged kernels occurred late in the season when the plants were at or near physiological maturity. These studies clearly demonstrated that *A. flavus* can be a parasite as well as a saprophyte in corn. Payne *et al.* (1988b) examined the effect of temperature on infection of silk-inoculated corn and found that, with a day/night regime of 34/30°C, 28% of the kernels were infected; at 34/22°C, up to 7% of the kernels were infected. The inoculum, however, must be present to have silk- or insect-vectored infection. Inoculum must build rapidly in the Midwest in the dry hot years that result in aflatoxin contamination.

Because the *A. flavus* group can colonize the surface of corn kernels and surrounding tissue readily, a relationship between insect injury and aflatoxin contamination is not surprising. The relationship between insect damage and aflatoxin contamination is related to enhanced aflatoxin production in damaged areas and not in undamaged areas of the ear (Lee *et al.,* 1980). Insect control reduces but does not eliminate aflatoxin contamination (Widstrom *et al.,* 1976), because at least two infection mechanisms are operative. One is related to the direct pathogenic *A. flavus* invasion discussed earlier that does not involve insect activity and the other is related to *A. flavus* growth and toxin accumulation in damaged tissues.

Widstrom *et al.* (1990) assessed the effect of planting date on aflatoxin concentration in wound-inoculated corn during a 5-year study in Georgia. Aflatoxin concentrations were not related to insect damage or to the percentage of ears with visible sporulation. However, the wound inoculation was so severe that it may have masked any insect correlations. Aflatoxin concentrations were correlated highly with maximum and minimum daily temperatures and net daily evaporation during the 20- to 40-day and 40- to 60-day periods after full silk. Regressions of aflatoxin concentrations on planting date revealed a linear decrease in aflatoxin from early to late planting. Early plantings in Georgia are believed to be at high risk because the critical grain filling period beginning 20 days after full silk falls when temperatures are highest and when net evaporation is at its peak.

Aflatoxin accumulation can be very rapid in wounded and inoculated corn. Thompson *et al.* (1983) found 238 ng/g 2 days after pinboard wound inoculation, which peaked at 2500 ng/g 11 days after inoculation. Payne *et al.* (1988a) compared aflatoxin accumulation in silk- and wound-inoculated corn and reported that aflatoxin content increased linearly for 7–9 weeks in wound-inoculated corn; in silk-inoculated corn, the accumulation occurred later but also increased linearly. Widstrom *et al.* (1986) stated that aflatoxin accumulation was linear between 30 and 50 days after full silk in two different hybrids. Each hybrid had different accumulation rates and the maximal difference in aflatoxin content between hybrids occurred near physiological maturity.

Color mutants were used in a Georgia study comparing the effectiveness of knife-wound inoculation 20 days after full silk, silk inoculation 3–5 days after full silk, and silk inoculation followed by placement of 10 *A. flavus*-infested maize weevils per ear of corn 25 days later (Wilson *et al.*, 1986). Three *A. flavus* color mutants, one *A. parasiticus* color mutant, and one wild-type *A. flavus* isolate were used for the 2-year study. One white *A. flavus* mutant did not produce aflatoxins in pure culture; the other three *A. flavus* strains produced B_1 and B_2. The *A. parasiticus* mutant produced B_1, B_2, G_1, and G_2 in pure culture. Knife-wound inoculation resulted in a higher percentage recovery of each color mutant from kernels at harvest than silk inoculation. The mean percentage of mutants recovered after silk inoculation ranged from 0.7 to 14.1%, whereas the range was 12.6–54.3% after knife-wound inoculation. Large differences were seen in the ability of these mutants to infect corn, especially with the wound inoculation. With silk inoculation followed by infestation with wild-type *A. flavus*-contaminated weevils, the percentage of kernels infected with the color mutants did not increase; however, the percentage of wild-type *A. flavus* recovered from the kernels was higher. Wound inoculation resulted in the highest concentration of total aflatoxins with three of the five strains. Silk inoculation followed by treatment with the infested weevils resulted in the highest concentrations in the remaining two strains. Wound inoculation with the atoxigenic strain resulted in less aflatoxin than in the uninoculated check.

To our knowledge, this combination silk–insect inoculation study is the only one published. The data can be used to illustrate several important points. First, silk inoculation may be *A. flavus* group isolate dependent; second, wound inoculation usually, but not always, results in higher aflatoxin contents than silk inoculation; third, the fungi that colonize tissue after silk inoculation do not necessarily restrict or limit the growth of other *A. flavus* isolates introduced later in the season (Wilson *et al.*, 1986). Inoculation, colonization, and aflatoxin synthesis are most likely continual processes in developing corn (Thompson *et al.*, 1983; Widstrom *et al.*, 1986; Payne *et al.*, 1988a).

The incidence and severity of aflatoxin contamination is highly dependent on inoculum and environmental conditions. The most serious outbreaks of contamination in corn have occurred in years with above average temperatures and drought conditions. However, aflatoxin contamination in the southeastern United

States is a continuing problem (McMillian *et al.*, 1991; Payne, 1992). A significant incidence of aflatoxin contamination was seen in Indiana in 1983 and in Iowa and other corn belt states in 1988. Average July–August, 1988, temperatures were 2–3°C higher than normal in Iowa with the southern half of Iowa receiving about 6 inches less rainfall than normal. Aflatoxin contamination in 1988 Iowa corn was localized in areas with high temperature and rainfall deficits (Hurbergh, 1991).

Temperature and moisture are important but many other factors also influence aflatoxin accumulation. Payne *et al.* (1986) studied the effects of irrigation and tillage and demonstrated that irrigation or subsoiling to break the root impenetrable zone both resulted in less aflatoxin contamination. In this study, drought stress led to an increased number of infected kernels in silk-inoculated ears. Other stresses including too little or too much nitrogen, excessive plant populations, insect damage, and poor irrigation practices influence aflatoxin accumulation (Anderson *et al.*, 1975; Fortnum and Manwiller, 1985; Payne *et al.*, 1989; Wilson *et al.*, 1989a).

Peanut

Environmental conditions are very important in aflatoxin contamination of peanuts. In 1963 and 1964, the extent of shell and kernel invasion by *A. flavus* and other fungi at digging in Alabama, Georgia, and North Carolina was determined by Diener *et al.* (1965). Both years had well-distributed rainfall. In 1963, the percentage of kernels and shells invaded with *A. flavus* increased with time and maturity; in 1964, low levels of *A. flavus* were found in peanuts grown in all three states. Cole *et al.* (1989) summarized 6 years of work at the U.S. Department of Agriculture (USDA) National Peanut Research Laboratory at Dawson, Georgia, on the environmental conditions necessary for preharvest contamination of peanuts. The dual roles of temperature and moisture were established by Cole and co-workers, who determined that no aflatoxin contamination occurred in visually sound kernels when the mean soil temperature was below 25°C or above 32°C. Aflatoxin concentration in edible grade peanuts grown under drought stress was maximal when the mean soil temperature was 30.5°C. Kernel colonization by the *A. flavus* group was also maximal at 30.5°C.

The relationship between kernel colonization and temperature was more variable than that between soil temperature and aflatoxin content. Edible grade kernels from irrigated peanuts contained no aflatoxins (Wilson and Stansell, 1983; Cole *et al.*, 1989). Any kind of damage to kernels or pods increased the incidence of the *A. flavus* group and occasionally aflatoxin, regardless of soil moisture content (Hill *et al.*, 1985; Lynch and Wilson, 1991). Larger, more mature kernels required more drought stress to become contaminated than did immature kernels (Cole *et al.*, 1989). Pitt (1989) reported that the correlations between preharvest *A. flavus* invasion and aflatoxin were not effective predictors of aflatoxin in harvested peanuts.

The role that phytoalexins play in protecting peanuts from invasion by the *A. flavus* group was studied by Wooton and Strange (1987). These researchers reported that fungal colonization of kernels grown with no drought stress virtually ceased after 3 days when the phytoalexin concentration exceeded 50 μg/g. Colonization and aflatoxin production were related inversely to water supply in kernels grown under drought stress, indicating that drought-stressed kernels had an impaired phytoalexin response. Thus, drought conditions in peanuts decrease the phytoalexin response, increase the probability of high soil *A. flavus* group populations, increase the probability of lesser cornstalk borer damage, and increase the probability of kernel invasion by the *A. flavus* group with subsequent aflatoxin contamination.

Other factors that may increase aflatoxin content of preharvest peanuts include calcium deficiencies (Wilson *et al.,* 1989a), microscopic shell damage (Porter *et al.,* 1986), and external feeding damage (scarification) of the shell by lesser cornstalk borer larvae (Lynch and Wilson, 1991).

Cottonseed

Aspergillus flavus boll rot was associated with lint discoloration and problems with spinning and dyeing of fabrics before aflatoxin contamination was recognized (Bollenbacher and Marsh, 1954). The *A. flavus* inoculum appears to be insect transported or airborne because of wind-driven soilborne propagules (Ashworth *et al.,* 1969a,b; Stephenson and Russell, 1974; Lee *et al.,* 1986). *Aspergillus flavus* has been isolated from flowers, floral nectaries, bracts, and bractioles (Simpson and Barta, 1984). Klich (1986) recovered *A. flavus* as early as 25 days after anthesis from green undamaged bolls that arose from flowers inoculated with *A. flavus.* Cotty (1989, 1990) reported that differences in pathogenic aggressiveness of *A. flavus* contribute to the ability of an isolate to infect cotton and produce aflatoxin, and that atoxigenic isolates may reduce the overall aflatoxin content.

Bolls collected from the lower third of plants contained most of the *A. flavus*-infected seed and almost all the seed with aflatoxin (Ashworth *et al.,* 1969b). The incidence of aflatoxin in Arizona was least from fully fluffed bolls (193 ng/g), intermediate from tight loculed bolls (934 ng/g), and most from bollworm damaged bolls (3521 ng/g) (Russell, 1980). For cottonseed as well as corn and peanuts, aflatoxin contamination is localized. Insect injury enhances but does not cause aflatoxin contamination.

Temperature appears to be extremely important in preharvest aflatoxin contamination in irrigated western-grown cotton. Temperatures during maturation are probably too low for maximal aflatoxin contamination in the southeastern states. High night temperatures from 25 to 34°C and high day temperatures favor aflatoxin contamination in Arizona and California, especially in bolls opening in August (Ashworth *et al.,* 1969a,b; Russell, 1980). In a 3-year study, Russell *et al.* (1976) found that aflatoxin accumulation in Arizona cottonseed was influenced significantly by the timing of irrigation termination and by the degree of pink bollworm infestation. Termination of irrigation in early August resulted in less toxin than when two additional irrigations were applied.

AFLATOXIN CONTAMINATION OF STORED PRODUCTS

The fungi that produce the aflatoxins are capable of growing in and contaminating almost any stored product with aflatoxins if *A. flavus* group inoculum is present and the equilibrium relative humidity of the stored product is above 85% for an extended period of time. Wilson and Abramson (1992) reviewed the factors leading to and the commodities susceptible to aflatoxin contamination in storage. Water activity, a_w, roughly corresponds to the equilibrium relative humidity in stored products. Fungi in general will not grow at water activities below 0.70. At water activities slightly above 0.70, fungi will grow slowly and increase the water content, leading to a wetter product and allowing more rapid growth. The rapid growth also may raise the temperature of the stored product.

Water activity, grain temperature, and aeration are all important determinants in safe grain and oilseed storage (Christensen, 1974). The fungal successions of stored barley grain illustrate the processes involved in grain deterioration (Hill, 1979). At water activities below 0.70, grains and oilseeds can be stored safely indefinitely without danger of fungal deterioration. However, growth of *Aspergillus restrictus* begins at a water activity of 0.70 or slightly higher. At an a_w of 0.80–0.85, growth by the *Aspergillus glaucus* group begins; at water activities above 0.85, many *Penicillium* and *Aspergillus* species, including species of the *A. flavus* group, are capable of growing in products stored in ambient conditions.

Aflatoxin production in storage may follow; contamination is affected by the commodity, the temperature, oxygen availability, and the initial inoculum density. Therefore, aflatoxin contamination of susceptible crops in storage is primarily a result of storage of the commodity at an a_w of 0.85 or above. Insufficient drying, insect and rodent activity, moisture migration, roof leaks, wind-driven rain, and other warehousing problems all may contribute to *A. flavus* group growth and localized areas of heavy aflatoxin contamination.

SUMMARY

1. The aflatoxigenic species of the *A. flavus* group include *A. flavus, A. parasiticus,* and *A. nomius.*
2. The *A. flavus* group inoculum sources include airborne, soilborne, and insect-vectored propagules. The propagules may be conidia, sclerotia, or mycelia.
3. *Aspergillus flavus* group colonization of corn, peanuts, and cottonseed may occur any time after flowering. The inoculum may come from the air, the

soil, or insects. Inoculation and colonization may occur without infection of the seed and aflatoxin production.

4. Infection and aflatoxin contamination of susceptible preharvest crops depend on the environmental conditions, especially temperature and moisture. Insect damage often results in increased aflatoxin contamination. Aflatoxins B_1, B_2, G_1, and G_2 all occur in preharvest crops; B_1 and B_2 are the most common.

5. Aflatoxin contamination of stored commodities is related primarily to the moisture content and temperature of the product.

6. Additional studies on the biology and plant pathology of the A. *flavus* group are critical to developing effective control strategies, especially for possible biological control approaches using A. *flavus* or A. *parasiticus* strains that do not produce aflatoxins.

ACKNOWLEDGMENTS

We thank Darlene Morrison for her help in typing the manuscript and Susan Wilson for her help in proofreading the manuscript. This work was supported by State and Hatch funds allocated to the Georgia and North Carolina Agriculture Experiment Stations.

REFERENCES

Abdalla, M. H. (1988). Prevalence of airborne *Aspergillus flavus* in Khartoum (Sudan) airspora with reference to dusty weather and inoculum survival in simulated summer conditions. *Mycopathologia* **104**, 137–141.

Anderson, H. W., Nehring, E. W., and Wichser, W. R. (1975). Aflatoxin contamination of corn in the field. *J. Agric. Food Chem.* **23**, 775–782.

Angle, J. S., Dunn, K. A., and Wagner, G. H. (1982). Effect of cultural practices on the soil population of *Aspergillus flavus* and *Aspergillus parasiticus*. *Soil Sci. Soc. Am. J.* **46**, 301–304.

Angle, J. S., Lindgren, R. L., and Gilbert-Effiong, D. (1989). Survival of *Aspergillus flavus* in soil. *Biodeterioration Res.* **2**, 245–250.

Ashworth, L. J., McMeans, J. L., and Brown, C. M. (1969a). Infection of cotton by *Aspergillus flavus*: Epidemiology of the disease. *J. Stored Prod. Res.* **5**, 193–202.

Ashworth, L. J., McMeans, J. L., and Brown, C. M. (1969b). Infection of cotton by *Aspergillus flavus*: Time of infection and the influence of fiber moisture. *Phytopathology* **59**, 383–385.

Aucamp, J. L. (1969). The role of mite vectors in the development of aflatoxin in groundnuts. *J. Stored Prod. Res.* **5**, 245–249.

Bollenbacher, K., and Marsh, P. B. (1954). A preliminary note on a fluorescent fiber condition in row cotton. *Plant Dis. Rep.* **38**, 375–379.

Bothast, R. J., Beuchat, L. R., Emswiller, B. S., Johnson, M. G., and Pierson, M. D. (1978). Incidence of airborne *Aspergillus flavus* spores in corn fields of five states. *Appl. Environ. Microbiol.* **35**, 627–628.

Calvert, O. H., Lillehoj, E. B., Kwolek, W. F., and Zuber, M. S. (1978). Aflatoxin B_1 and G_1 production in developing *zea mays* kernels from mixed inocula of *Aspergillus flavus* and A. *parasiticus*. *Phytopathology* **68**, 501–506.

Christensen, C. M. (1974). "Storage of Cereal Grains and Their Products," 2nd Ed. American Association of Cereal Chemists, St. Paul, Minnesota.

Christensen, M. (1981). A synoptic key and evaluation of species in the *Aspergillus flavus* group. *Mycologia* **73**, 1056–1084.

Cole, R. J., Sanders, T. H., Dorner, J. W., and Blankenship, P. D. (1989). Environmental conditions required to induce preharvest aflatoxin contamination of groundnuts: Summary of six years' research. *In* "Aflatoxin Contamination of Groundnuts" (D. McDonald and V. K. Mehan, eds.), pp. 279–287. ICRISAT, Patancheru, India.

Cotty, P. J. (1989). Virulence and cultural characteristics of two *Aspergillus flavus* strains pathogenic on cotton. *Phytopathology* **79**, 808–814.

Cotty, P. J. (1990). Effect of atoxigenic strains of *Aspergillus flavus* on aflatoxin contamination of developing cottonseed. *Plant Dis.* **74**, 233–235.

Dickens, J. W., Satterwhite, J. B., and Sneed, R. E. (1973). Aflatoxin contaminated peanuts produced on North Carolina farms in 1968. *Proc. Am. Peanut Res. Ed. Assoc.* **5**, 48–58.

Diener, U. L., Jackson, C. R., Cooper, W. E., Stipes, R. J., and Davis, N. D. (1965). Invasion of peanut pods in the soil by *Aspergillus flavus. Plant Dis. Reptr.* **49**, 931–935.

Diener, U. L., Cole, R. J., Sanders, T. H., Payne, G. A., Lee, L. S., and Klich, M. A. (1987). Epidemiology of aflatoxin formation by *Aspergillus flavus. Ann. Rev. Phytopath.* **25**, 249–270.

Fortnum, B. A., and Manwiller, A. (1985). Effects of irrigation and kernel injury on aflatoxin B_1 production in selected maize hybrids. *Plant Dis.* **69**, 262–265.

Griffin, G. J., and Garren, K. H. (1976). Colonization of aerial peanut pegs by *Aspergillus flavus* and *A. niger* group fungi under field conditions. *Phytopathology* **66**, 1161–1162.

Hesseltine, C. W., and Bothast, R. J. (1977). Mold development in ears of corn from tasseling to harvest. *Mycologia* **69**, 328–340.

Hesseltine, C. W., Shotwell, O. L., Smith, M., Ellis, J. J., Van Dergraft, E., and Shannon, G. (1970). Production of various aflatoxins by strains of the *Aspergillus flavus* series. *In* "Toxic Micro-Organisms" (M. Herzberg, ed.), pp. 202–210. United States–Japan Cooperative Program in Natural Resources (UJNR) and Department of the Interior, U.S. Government Printing Office, Washington, D.C.

Hesseltine, C. W., Shotwell, O. L., Kwolek, W. F., Lillehoj, E. B., Jackson, W. K., and Bothast, R. J. (1976). Aflatoxin occurrence in 1973 corn at harvest. II. Mycological studies. *Mycologia* **68**, 341–353.

Hill, R. A. (1979). The microflora of barley grain with special reference to *Penicillium* species. Ph.D. Thesis. University of Reading, England.

Hill, R. A., Wilson, D. M., McMillian, W. W., Widstrom, N. W., Cole, R. J., Sanders, T. H., and Blankenship, P. D. (1985). Ecology of the *Aspergillus flavus* group and aflatoxin formation in corn and peanuts. *In* "Trichothenes and Other Mycotoxins" (J. Lacey, ed.), pp. 79–95. John Wiley and Sons, Chichester, England.

Holtmeyer, M. G., and Wallin, J. R. (1981). Incidence and distribution of airborne spores of *Aspergillus flavus* in Missouri. *Plant Dis.* **65**, 58–60.

Hurburgh, C. R. (1991). Aflatoxin in midwestern corn. *In* "Aflatoxin in Corn—New Perspectives" (O. L. Shotwell and C. R. Hurburgh, eds.), pp. 343–350. Research Bulletin 599. Iowa Agriculture and Home Economics Experiment Station, Iowa State University, Ames.

Ilag, L. L. (1975). *Aspergillus flavus* infection of pre-harvest corn, drying corn and stored corn in the Philippines. *Philippine Phytopath.* **74**, 37–41.

Jones, R. K., Duncan, H. E., Payne, G. A., and Leonard, K. J. (1980). Factors influencing infection by *Aspergillus flavus* in silk inoculated corn. *Plant Dis.* **64**, 859–863.

Kiyomoto, R. K., and Ashworth, L. J. (1974). Status of cotton boll rot in San Joaquin Valley of California following simulated pink bollworm injury. *Phytopathology* **64**, 259–260.

Klich, M. A. (1986). Presence of *Aspergillus flavus* in developing cotton bolls and its relation to contamination of mature seeds. *Appl. Environ. Microbiol.* **52**, 963–965.

Klich, M. A., and Pitt, J. I. (1988). Differentiation of *Aspergillus flavus* from *A. parasiticus* and other closely related species. *Trans. Br. Mycol. Soc.* **91**, 99–108.

Kurtzman, C. P., Smiley, M. J., Robnett, C. J., and Wicklow, D. T. (1986). DNA relatedness among wild and domesticated species in the *Aspergillus flavus* group. *Mycologia* **78**, 955–959.

Lee, L. S., Lillehoj, E. B., and Kwolek, W. F. (1980). Aflatoxin distribution in individual corn kernels from intact ears. *Cereal Chem.* **57**, 340–343.

Lee, L. S., Lee, L. V., and Russell, T. E. (1986). Aflatoxin in Arizona cottonseed: Field inoculation of bolls by *Aspergillus flavus* spores in wind-driven soil. *J. Am. Oil Chem. Soc.* **63**, 530–532.

Lillehoj, E. B., Fennell, D. I., Kwolek, W. F., Adams, G. L., Zuber, M. S., Horner, E. S., Widstrom, N. W., Warren, H., Guthrie, W. D., Saner, D. B., Findley, W. R., Manwiller, A., Josephson, L. M., and Bockholt, A. J. (1978). Aflatoxin contamination of corn before harvest: *Aspergillus flavus* association with insects collected from developing ears. *Crop. Sci.* **18**, 921–924.

Lillehoj, E. B., McMillian, W. W., Guthrie, W. D., and Barry, D. (1980). Aflatoxin producing fungi in preharvest corn: Inoculum source in insects and soils. *J. Environ. Qual.* **9**, 691–694.

Lussenhop, J., and Wicklow, D. T. (1990). Nitidulid beetles (Nitidulidae:Coleoptera) as vectors of *Aspergillus flavus* in preharvest maize. *Trans. Mycol. Soc. Jpn.* **31**, 63–74.

Lynch, R. E., and Wilson, D. M. (1991). Enhanced infection of peanut, *Arachis hypogaea* L., seeds with *Aspergillus flavus* group due to external scarification of peanut pods by the lesser cornstalk borer, *Elasmopalpus lignosellus* (Zeller). *Peanut Sci.* **18**, 110–116.

McMillian, W. W. (1983). Role of arthropods in field contamination. *In* "Aflatoxin and *Aspergillus flavus* in Corn" (U. L. Diener, R. L. Asquith, and J. W. Dickens, eds.), pp. 20–22. Alabama Agricultural Experimental Station, Auburn University, Auburn.

McMillian, W. W. (1987). Relation of insects to aflatoxin contamination in maize grown in the southeastern U.S.A. *In* "Aflatoxin in Maize" (M. S. Zuber, E. B. Lillehoj, and B. L. Renfro, eds.), pp. 194–220. CIMMYT, Mexico City, Mexico.

McMillian, W. W., Widstrom, N. W., Beaver, R. W., and Wilson, D. M. (1991). Aflatoxin in Georgia: Factors associated with its formation in corn. *In* "Aflatoxin in Corn—New Perspectives" (O. L. Shotwell and C. R. Hurburgh, eds.), pp. 329–334. Research Bulletin 599. Iowa Agriculture and Home Economics Experiment Station, Iowa State University, Ames.

Marsh, S. F., and Payne, G. A. (1984a). Preharvest infection of corn silks and kernels by *Aspergillus flavus*. *Phytopathology* **74**, 1284–1289.

Marsh, S. F., and Payne, G. A. (1984b). Scanning EM studies on the colonization of dent corn by *Aspergillus flavus*. *Phytopathology* **74**, 557–561.

Payne, G. A. (1992). Aflatoxin in maize. *Crit. Rev. Plant Sci.* **10**, 423–440.

Payne, G. A., Cassel, D. K., and Adkins, C. R. (1986). Reduction of aflatoxin contamination by irrigation and tillage. *Phytopathology* **76**, 679–684.

Payne, G. A., Hagler, W. M., and Adkins, C. R. (1988a). Aflatoxin accumulation in inoculated ears of field grown maize. *Plant Dis.* **72**, 422–424.

Payne, G. A., Thompson, D. L., Lillehoj, E. B., Zuber, M. S., and Adkins, C. R. (1988b). Effect of temperature on the preharvest infection of maize kernels by *Aspergillus flavus*. *Phytopathology* **78**, 1376–1380.

Payne, G. A., Kamprath, E. J., and Adkins, C. R. (1989). Increased aflatoxin contamination in nitrogen stressed corn. *Plant Dis.* **73**, 556–559.

Pitt, J. I. (1989). Field studies on *Aspergillus flavus* and aflatoxins in Australian groundnuts. *In* "Aflatoxin Contamination of Groundnuts" (D. McDonald and V. K. Mehan, eds.), pp. 223–236. ICRISAT, Patancheru, India.

Porter, D. M., Wright, F. S., and Steele, J. L. (1986). Relationship of microscopic shell damage to colonization of peanut by *Aspergillus flavus*. *Oleagineux* **41**, 24–30.

Rambo, G. W., Tuite, J., and Caldwell, R. W. (1974). *Aspergillus flavus* and aflatoxin in preharvest corn from Indiana in 1971 and 1972. *Cereal Chem.* **51**, 848–853.

Russell, T. E. (1980). Aflatoxin contamination of cottonseed—Climate, insects, cultural practices and segregation. *Proceedings of the Beltwide Cotton Production Conference, Special Session on Aflatoxin*, St. Louis, pp. 53–55.

Russell, T. E., Watson, T. F., and Ryan, G. F. (1976). Field accumulation of aflatoxin in cottonseed as influenced by irrigation termination dates and pink bollworm infestation. *Appl. Environ. Microbiol.* **31,** 711–713.

Sanders, T. H., Hill, R. A., Cole, R. J., and Blankenship, P. D. (1981). Effect of drought on occurrence of *Aspergillus flavus* in maturing peanuts. *J. Am. Oil Chem. Soc.* **58,** 966A–970A.

Simpson, M. E., and Batra, L. R. (1984). Ecological relations in respect to a boll rot of cotton caused by *Aspergillus flavus. In* "Toxigenic Fungi: Their Toxins and Health Hazard" (H. Kurceta and Y. Uneo, eds.), pp. 24–32. Elsevier, Amsterdam.

Smart, M. G., Wicklow, D. T., and Caldwell, R. W. (1990). Pathogenesis in *Aspergillus* ear rot of maize: Light microscopy of fungal spread from wounds. *Phytopathology* **80,** 1287–1294.

Stephenson, L. W., and Russell, T. E. (1974). The association of *Aspergillus flavus* with hemipterous and other insects infesting cotton bracts and foliage. *Phytopathology* **64,** 1502–1506.

Taubenhaus, J. J. (1920). A study of the black and yellow molds of ear corn. Bulletin 270. Texas Agricultural Experimental Station, College Station.

Thompson, D. L., Payne, G. A., Lillehoj, E. B., and Zuber, M. S. (1983). Early appearance of aflatoxin in developing corn kernels after inoculation with *Aspergillus flavus. Plant Dis.* **67,** 1321–1322.

Wang, H. L., and Hesseltine, C. W. (1982). Oriental fermented foods. *In* "Prescott and Dunn Industrial Microbiology" (G. Reed, ed.), 4th Ed., pp. 492–538. AVI, Westport, Connecticut.

Wells, T. R., and Kreutzer, W. A. (1972). Aerial invasion of peanut flower tissue by *Aspergillus flavus* under gnotobiotic conditions. *Phytopathology* **62,** 797.

Wicklow, D. T. (1983). Taxonomic features and ecological significance of sclerotia. *In* "Aflatoxin and *Aspergillus flavus* in Corn" (U. L. Diener, R. L. Asquith, and J. W. Dickens, eds.), pp. 6–12. Alabama Agricultural Experimental Station, Auburn University, Auburn.

Wicklow, D. T., and Shotwell, O. L. (1983). Intrafungal distribution of aflatoxins among conidia and sclerotia of *Aspergillus flavus* and *Aspergillus parasiticus. Can. J. Microbiol.* **29,** 1–5.

Widstrom, N. W. (1979). The role of insects and other plant pests in aflatoxin contamination of corn, cotton, and peanuts—A review. *J. Environ. Qual.* **8,** 5–11.

Widstrom, N. W., Lillehoj, E. B., Sparks, A. N., and Kwolek, W. F. (1976). Corn earworm damage and aflatoxin B$_1$ on corn ears protected with insecticide. *J. Econ. Entomol.* **69,** 677–679.

Widstrom, N. W., Wilson, D. M., and McMillian, W. W. (1986). Differentiation of maize genotypes for aflatoxin concentration in developing kernels. *Crop Sci.* **26,** 935–937.

Widstrom, N. W., McMillian, W. W., Beaver, R. W., and Wilson, D. M. (1990). Weather associated changes in aflatoxin contamination of preharvest maize. *J. Prod. Agric.* **3,** 196–199.

Wilson, D. M., and Abramson, D. (1992). Mycotoxins. *In* "Storage of Cereal Grains and Their Products" (D. B. Sauer, ed.), 4th ed., pp. 341–391. American Association of Cereal Chemists, St. Paul, Minnesota.

Wilson, D. M., and Stansell, J. R. (1983). Effect of irrigation regimes on aflatoxin contamination of peanut pods. *Peanut Sci.* **10,** 54–56.

Wilson, D. M., McMillian, W. W., and Widstrom, N. W. (1986). Use of *Aspergillus flavus* and *A. parasiticus* color mutants to study aflatoxin contamination of corn. *Biodeterioration* **6,** 284–288.

Wilson, D. M., Walker, M. E., and Gascho, G. J. (1989a). Some effects of mineral nutrition on aflatoxin contamination of corn and peanut. *In* "Soilborne Plant Pathogens: Management with Macro- and Micro-elements" (A. W. Engelhard, ed.), pp. 137–151. APS Press, St. Paul, Minnesota.

Wilson, D. M., Widstrom, N. W., McMillian, W. W., and Beaver, R. W. (1989b). Aflatoxins in corn. *In* "Proceedings of the 44th Annual Corn and Sorghum Research Conference," pp. 1–26. American Seed Trade Association, Washington, D.C.

Wotton, H. R., and Strange, R. N. (1987). Increased susceptibility and reduced phytoalexin accumulation in drought stressed peanut kernels challenged with *Aspergillus flavus. Appl. Environ. Microbiol.* **53,** 270–273.

15

Mycological Aspects
of Aflatoxin Formation

❖

Deepak Bhatnagar, Thomas E. Cleveland, and Peter J. Cotty

INTRODUCTION

Aflatoxins, produced by *Aspergillus flavus* and *Aspergillus parasiticus,* are worldwide contaminants of food and feed (Jelinek *et al.,* 1989). The aflatoxin family of compounds is among the most carcinogenic of naturally occurring substances; therefore extensive efforts are under way to remove or detoxify the toxins in animal and human food chains (Park *et al.,* 1988; Hagler, 1991; Phillips *et al.,* 1991; see also Chapter 18). In addition, methods have been developed to protect animals and humans against aflatoxins by prior chemical treatments that induce protective detoxifying liver enzymes (Kensler *et al.,* 1991; see also Chapter 13). The denoted techniques could have immediate application in aflatoxin-contaminated commodities, since no practical method is currently available for prevention of preharvest toxin contamination of foods and feeds. However, prevention of aflatoxin contamination before harvest is the best long-term approach because the pertinent technology would insure consumer safety and eliminate the need for handling aflatoxin-contaminated commodities by growers and processors.

The aflatoxins that produce the most severe pre- and postharvest contamination of foods and feeds are aflatoxins B_1, B_2, G_1, and G_2 (Figure 1). Special emphasis in research has been placed on aflatoxin B_1, since this compound is the most prevelant and the most toxic and carcinogenic of the four. In addition, a

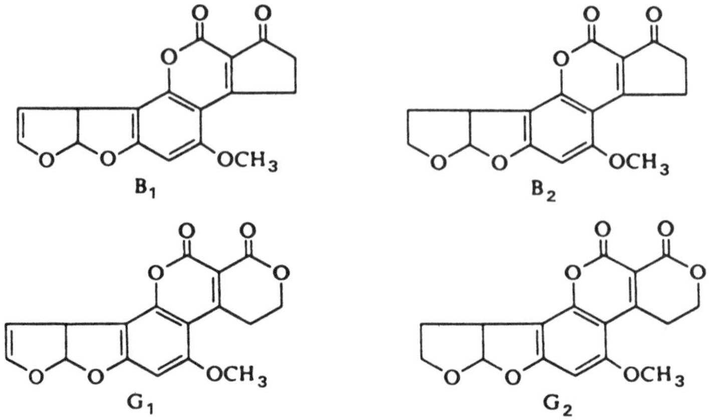

FIGURE 1   Structures of aflatoxins B_1, B_2, G_1, and G_2.

variety of chemical derivatives of the four basic aflatoxins (Detroy *et al.*, 1971; Bhatnagar *et al.*, 1991a) can be produced synthetically by physicochemical methods or after contact with enzymes in living systems, such as the detoxifying enzymes (cytochrome P450) present in human and animal liver tissue.

Several conventional agronomic methods can influence preharvest aflatoxin contamination of crops, including use of pesticides, altered agronomic practices (such as irrigation), and use of resistant varieties (Widstrom, 1987; Scott and Zummo, 1988; Darrah and Barry, 1991). However, the procedures have demonstrated only a limited potential for reducing aflatoxin levels in the field. Domestic growers and food processors are under increased pressure from consumer groups, merchants, and regulatory agencies to eliminate mycotoxins from food and feed. Therefore, the need to develop new technology to reduce and eventually eliminate preharvest aflatoxin contamination is increasing.

A solution to the aflatoxin problem is being sought by acquisition of information on the molecular regulation of aflatoxin biosynthesis within the fungus and on the ecological and biological factors that influence toxin production in the field (Cleveland and Bhatnagar, 1992). This chapter reviews various aspects of aflatoxin formation and the parameters that control fungal growth and aflatoxin contamination of various crops.

FIELD BIOLOGY OF AFLATOXIGENIC FUNGI

Aspergillus flavus is adapted to use a broad assortment of organic resources. In addition to being a saprobe, this organism is an opportunistic pathogen of plants, insects, and vertebrates including humans and domestic animals. In agricultural fields, during hot dry conditions, *A. flavus* populations increase on crop debris, on senescent or dormant tissues, and on damaged or weakened crops (Ashworth *et al.*, 1969; Stephenson and Russell, 1974). Crops grown in these fields become associated with large populations of *A. flavus* that may remain associated with the crop throughout crop maturation, harvest, and storage. Although the source of the initial infecting inoculum can be disputed, the contamination process can be divided into two phases: one occurs during crop maturation and the second after maturation (Diener *et al.*, 1987; Cotty, 1991; Cleveland and Bhatnagar, 1992; see also Chapter 14). Most fungal contamination appears to be associated with damaged crop components (Widstrom, 1979; Hill *et al.*, 1985; Sommer *et al.*, 1986; Cotty and Lee, 1989). Damage can occur either through mechanical damage such as insect activity or through environmental or chemical stress on the host plant (Diener and Davis, 1966). Crop components damaged prior to maturity are, apparently, particularly susceptible to contamination. If damage and infection occur during the appropriate stage of maturity, very high levels of contamination (exceeding 100,000 μg/kg) can result (Lillehoj *et al.*, 1987; Cotty 1989a). Highly contaminated seeds, nuts, or kernels are relatively rare but drastically influence the overall contamination of the crop (Lee *et al.*, 1990).

At maturity, crops susceptible to aflatoxin contamination are generally susceptible to infection by *A. flavus*. When mature crops carry large *A. flavus* spore loads, the extent of infection is determined by the environment to which the standing and harvested crops are exposed as well as by mechanical and insect damage during handling and storage (Bullerman *et al.*, 1975; Cotty, 1991). Extensive colonization of the crops by the fungus does not necessarily lead to significant accumulation of aflatoxin (Diener *et al.*, 1987). Also, strains of peanut have been identified that support extensive colonization by the fungus but accumulate little aflatoxin (Mehan *et al.*, 1986). If the mature crop is remoistened after drying, components previously infected by *A. flavus* may show an increase in aflatoxin content. Further, if *A. flavus* predominates, new crop components can become infected and contaminated (within 24 hr) with aflatoxins (Cotty, 1991). Variation in aflatoxin levels is puzzlingly high, even among equally infected seed. Therefore an understanding of the molecular regulation of aflatoxin biosynthesis and that of intraspecific genetic variation may increase our understanding of, and ability to interfere with, the phenomenon of aflatoxin contamination.

AFLATOXIN BIOSYNTHESIS

Aflatoxins are secondary metabolites with no obvious physiological role in primary growth and metabolism of the organism. However, the metabolic pathways of primary metabolism supply the precursors of secondary metabolism. Therefore, factors that influence primary metabolism also affect secondary metabolism (Drew and Demain, 1977). Biosynthesis of secondary metabolites in fungi shares a common lipid and protein pool with primary metabolites (see Bhatnagar *et al.*, 1991b, for review). A close association also has been established between aflatoxin and lipid biosynthesis (Detroy and Hesseltine, 1968;

Gupta *et al.*, 1970; Shih and Marth, 1976; Townsend *et al.*, 1984). In some cases, protein synthesis is known to decline during the aflatoxin-producing phase (idiophase) (Detroy and Hesseltine, 1970; Maggon *et al.*, 1977).

Chemistry

Studies of the biosynthesis of aflatoxin B_1 (AFB_1) have shown that the basic skeleton of the toxin molecule is derived entirely from acetate units via the polyketide pathway. Current findings support a biosynthetic scheme based on a C_{20} polyketide (decaketide) precursor (see Bhatnagar *et al.*, 1991b, for review).

Success in elucidating early intermediate compounds in the aflatoxin biosynthetic scheme was based on mutants blocked in AFB_1 synthesis. Until 1986, pertinent mutants and their intermediates were used in incorporation and degradation experiments of relevant radiolabeled compounds to develop a biosynthetic scheme (Figure 2) (see Bhatnagar *et al.*, 1991b, for review).

Improved HPLC resolution of the major metabolites of *A. flavus* and *A. parasiticus* cultures (McCormick *et al.*, 1988) has resulted in an expanded aflatoxin biosynthetic scheme to include additional biosynthetic intermediates such as averufanin (Figure 2) (McCormick *et al.*, 1987). Yabe *et al.* (1991b) have, however, suggested that 5-hydroxyaverantin may be an intermediate at this step. Townsend *et al.* (1988) showed that 1-hydroxyversicolorone (HVN) is an aflatoxin precursor between averufin (AVF) and versicolal hemiacetal acetate (VHA). Yabe *et al.* (1989) identified demethylsterigmatocystin (DMST) as an AFB_1 precursor. *O*-Methylsterigmatocystin (OMST) (Bhatnagar *et al.*, 1987) also has been included in the pathway (Figure 2); conversion of ST to OMST involves a simple methylation, but the conversion of OMST to AFB_1 is a very complex reaction (Bhatnagar *et al.*, 1989, 1991b).

Some studies have suggested that AFB_1 is the progenitor of additional aflatoxins (AFB_2, AFG_1, AFG_2), whereas other workers have postulated that AFB_1 and AFB_2 are produced by separate pathways (Bhatnagar *et al.*, 1991b). That AFB_2 is produced from a unique branch in the aflatoxin biosynthetic pathway has now been established. AFB_2 arises from dihydrosterigmatocystin (DHST) and dihydro-*O*-methylsterigmatocystin (DHOMST) (Figure 3), whereas AFB_1 arises from ST and OMST (Cleveland *et al.*, 1987a; Yabe *et al.*, 1988; Bhatnagar *et al.*, 1991c). Additional information on the branch point in the pathway has been acquired by observations that link versicolorin A (Ver A) as a precursor of AFB_1 and Ver B as a precursor of AFB_2 and Ver A (McGuire *et al.*, 1989; Yabe *et al.*, 1991a). However, Hsieh *et al.* (1989) have reported that versiconal hemiacetal alcohol (VAL) is formed by deacetylation of VHA and rapidly converted, with closure of the ring, to either Ver C under acidic conditions or to Ver A enzymatically. The early precursors [norsolorinic acid (NOR), averantin (AVN), averufanin (AVNN), AVF, and VHA] generally are accepted to be common to both pathways. A metabolic grid for interconversion of AFB_1 and AFB_2 precursors has been developed (Anderson *et al.*, 1990; Bhatnagar *et al.*, 1991c).

POLYKETIDE

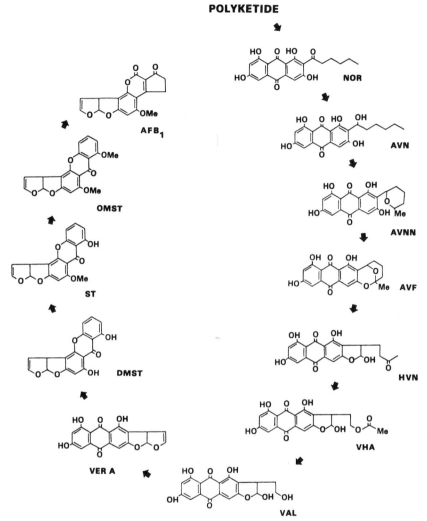

FIGURE 2   The generally accepted scheme of known precursors in the biosynthesis of aflatoxin B_1. NOR, Norsolorinic acid; AVN, averantin; AVNN, averufanin; AVF, averufin; HVN, 1'-hydroxyversicolorone; VHA, versiconal hemiacetal acetate; VAL, versiconal; Ver A, versicolorin A; DMST, demethylsterigmatocystin; ST, sterigmatocystin; OMST, O-methylsterigmatocystin, AFB_1, aflatoxin B_1.

The biosynthesis of AFB_1 and AFB_2 is fairly well understood, but the biosynthetic origins of AFG_1 and AFG_2 have received less attention. AFB_1 generally is assumed to be a precursor of AFG_1 and AFG_2; several reviews include postulated schemes illustrating AFB_1 to AFG_1 conversion. Feeding precursors to pertinent fungal isolates shows that AFG_1 is synthesized from ST/OMST (AFB_1 precursors) (Bhatnagar *et al.,* 1987) and that AFG_2 is synthesized from

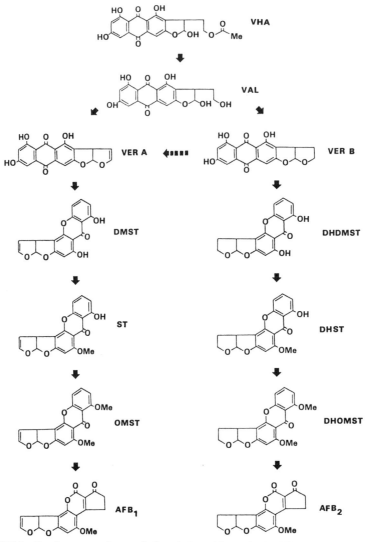

FIGURE 3 The divergent pathways of aflatoxin B_1 and B_2 synthesis. VHA, Versiconal hemiacetal acetate; VAL, versiconal; Ver A, versicolorin A; Ver B, versicolorin B; DMST, demethylsterigmatocystin; DHDMST, dihydro-demethylsterigmatocystin; ST, sterigmatocystin; DHST, dihydrosterigmatocystin; OMST, O-methylsterigmatocystin; DHOMST, dihydro-O-methylsterigmatocystin; AFB_1, aflatoxin B_1; AFB_2, aflatoxin B_2.

DHST/DHOMST (AFB_2 precursors) (Yabe *et al.,* 1988). However, in the presence of 50 μ*M* NADP, NADPH, NAD, or FMN, [^{14}C]-labeled AFB_1 or AFB_2 was not converted to either AFG_1 or AFG_2 by whole cells or by cell-free extracts of *A. parasiticus;* the systems were otherwise capable of producing AFG_1/G_2 (D. Bhatnagar, unpublished observation). Mutant fungal strains that were blocked early in the aflatoxin pathway did not biotransform AFB_1 or AFB_2 into either AFG_1 or AFG_2 (Floyd *et al.*, 1987; Henderberg *et al.*, 1988). These results suggest independent pathways to the B and G aflatoxins. A mechanism for aflatoxin G synthesis has been postulated (Bhatnagar *et al.*, 1991b).

Enzymes

Initial attempts to purify enzymes from *A. parasiticus* mycelial extracts that catalyzed aflatoxin synthesis were unsuccessful because the enzymes were present in relatively low concentrations and were extremely short lived (Dutton, 1988). Subsequently, several techniques were examined for disruption of large quantities of mycelia to obtain active and stable cell-free preparations (Dutton 1988; Bhatnagar *et al.*, 1989). Pertinent enzymes were recovered from cell-free extracts after grinding mycelia under liquid nitrogen (Bhatnagar *et al.*, 1989). The optimum age of mycelial cultures for recovery of aflatoxin pathway enzymes is between 72 and 84 hr (Cleveland *et al.*, 1987b; Chuturgoon and Dutton, 1991).

Several specific enzyme activities have been associated with precursor conversions in the aflatoxin pathway (Dutton, 1988; Bhatnagar *et al.*, 1991b,c; Yabe *et al.*, 1991a,b). Some of these activities have been partially purified (Chuturgoon *et al.*, 1990; Bhatnagar *et al.*, 1991c). Bhatnagar *et al.* (1988) and Keller *et al.* (1993) have purified two methyltransferases (168 kDa and 40 kDa) to homogeneity; both enzymes catalyze ST to OMST conversion. In addition, a 38-kDa reductase that catalyzes the reduction of NOR to AVN has been purified to homogeneity (Bhatnagar and Cleveland, 1990); an isozyme (43 kDa) of the reductase also has been purified to homogeneity (D. Bhatnagar, unpublished observations). Bhatnagar *et al.* (1991c) postulated that alternative pathways may exist at several steps in the aflatoxin pathway; hence, different enzymes with the same catalytic function may be isolated from pertinent fungal cells.

Independent chemical reactions in AFB_1 and AFB_2 syntheses have been demonstrated to be catalyzed by common enzyme systems (Yabe *et al.*, 1988,1989; Bhatnagar *et al.*, 1991c). AFB_1 precursors are the preferred substrates for the relevant enzymes (Bhatnagar *et al.*, 1991c). A desaturase activity also has been demonstrated in cell-free fungal extracts at the branch in the AFB_1/B_2 biosynthetic pathways (Yabe *et al.*, 1991a).

Genetics

Classical Genetic Investigations

Early genetic investigations of *A. flavus* and *A. parasiticus* were hampered by the lack of any known sexual reproduction in these imperfect fungal species. However, marked *A. parasiticus* and *A. flavus* strains have been analyzed genetically using the parasexual cycle (for reviews, see Bennett and Papa, 1988; Keller *et al.*, 1991). The genetics of *A. flavus* are better understood than those of *A.*

parasiticus. Over 30 genes have been mapped to eight linkage groups (Bennett and Papa, 1988). Currently, 24 distinct aflatoxin nonproducing mutants and a number of spore color and auxotropic mutants of *A. flavus* have been derived from the common parental line PC-7. In strain PC-7, 11 aflatoxin loci have been mapped through parasexual recombination; 9 are on linkage group VII; one is on linkage group II; one is on linkage group VIII. The sequence of markers on linkage group VII are (*nor, afl-1*), *leu, afl-15, arg-7, afl-17,* and centromere. The aflatoxin loci are all recessive in diploids, with the exception of the mutant containing the *afl-1* allele.

One technology helpful in resolving karyotypes and defining genetic maps of imperfect fungi has been pulsed-field gel electrophoresis. Karyotype analyses for several *A. flavus* and *A. parasiticus* strains show six to eight chromosomes ranging in size from approximately 3 to ≥ 7 Mb (Keller *et al.,* 1992a). Strains that were used for linkage group studies of *A. flavus* have eight chromosomes; assignment of the eight linkage groups to these chromosomes currently is being determined. In addition, chromosome length polymorphisms and the presence of small chromosomes were observed in several *A. flavus* strains (Keller *et al.,* 1992a).

Molecular Genetics

With strains made available by Papa, Payne and Woloshuk (1989) demonstrated that selectable markers could be transferred from one strain of *A. flavus* to another (perhaps to a more desirable strain for studying aflatoxin biosynthesis) through parasexual recombination. Genes in the aflatoxin biosynthetic pathway might be identified readily by their ability to complement strains of *A. flavus* with specific blocks in the pathway through transformation (Payne and Woloshuk, 1989). Genetic transformation systems also have been developed for *A. parasiticus* in the laboratory of J. Linz (Horng *et al.,* 1990; Skory *et al.,* 1990). Individual conserved structural genes have been cloned from both species (Horng *et al.,* 1989; Seip *et al.,* 1990).

Using complementary DNA hybridization with *A. flavus,* Payne and co-workers (G. A. Payne, personal communication) have isolated a gene that is likely to encode a dehydrogenase (designated CW3) whose expression parallels that of aflatoxin biosynthesis. Another isolated gene (*afl-2*) appears to be involved directly in aflatoxin biosynthesis (Payne *et al.,* 1993); the gene was identified by its ability to restore aflatoxin-producing ability to a non-aflatoxin-producing strain. Recently, in this laboratory, using gene cloning by genetic complementation, a DNA fragment was isolated that resulted in a 2- to 5-fold overproduction of intermediates in the aflatoxin pathway (Chang, *et al.,* 1993). Specific activities of the corresponding enzymes were also enhanced by 2- to 3-fold in the transformed strains. Southern hybridization showed that this DNA fragment (*apa-2*) shared almost complete homology with the *A. flavus* regulatory gene, *afl-2.* Characterization of *apa-2* gene demonstrates that this gene may be responsible for synthesis of a DNA-binding regulatory protein and that it regulates the expression of other genes in the aflatoxin biosynthetic pathway. Linz and co-workers also have identified two genes associated with aflatoxin biosynthesis in *A. parasiticus:* (1) the *nor*-1 gene, associated with the conversion of norsolorinic acid to averantin (Chang *et al.,* 1992), and (2) the *ver*-1 gene, associated with the conversion of versicolorin A to sterigmatocystin (Skory *et al.,* 1992).

An alternative to identifying aflatoxin genes through genetic transformation and complementation is identifying and characterizing catalysts involved in aflatoxin biosynthesis and subsequently isolating the genes encoding the enzymes. The purified proteins are used to obtain antibodies as immunoscreening probes and amino acid sequences for oligonucleotide probes. Contemporary methods are available for screening and cloning genes of interest through pertinent antisera and oligonucleotide gene probes. cDNA libraries have been constructed (Bhatnagar and Cleveland, 1991; Cary *et al.,* 1992) that consist of cloned cDNAs synthesized with *A. parasiticus* mRNAs as templates using various nucleic acid polymerases; mRNAs were isolated from late growth-phase mycelia (Cleveland *et al.,* 1987b) during production of aflatoxin pathway enzymes. A cDNA library was screened with polyclonal antibody raised against a purified 43-kDa protein exhibiting norsolorinic acid reductase activity, an early pathway enzyme. DNA sequencing data from selected recombinant phagemid clones will confirm that the cloned gene encodes the 43-kDa reductase.

A strategy similar to the one used for reductase characterization has been employed to clone the methyltransferase gene (Keller *et al.,* 1992b). A 1.5-kb genomic DNA clone (pF9-1) from *A. flavus* NRRL 3357 was identified with an oligonucleotide prepared from the N terminus of the 40-kDa methyltransferase; the enzyme is active in aflatoxin biosynthesis. Genomic DNA hybridizing to pF9-1 was found to be present in isolates of *A. flavus* and *A. parasiticus* and in some other members of *A. flavus* group, but was not present in more distantly related aspergilli (Keller *et al.,* 1992b). More recently, using polyclonal antibodies prepared against the 40-kDa methyltransferase, a cDNA encoding this protein has been isolated from *A. parasiticus* (Yu, *et al.,* 1993). A 1460-bp full length cDNA of the *omt*-1 gene encodes for a 46-kDa native protein, consisting of a 41-amino acid leader sequence as well as the mature form of the enzyme (377 amino acids) with a molecular weight of 42 kDa. *In vitro* expression of this gene and enzyme activity were demonstrated in *E. coli* cell-free extracts when the *omt*-1 cDNA was cloned into an *E. coli* expression system.

The strategy outlined here could be used to clone the gene for any aflatoxin pathway enzyme that can be purified to homogeneity. The expression library from late growth-phase fungal mRNA could contain cDNA clones for several of the aflatoxin pathway genes. However, detection of pertinent clones awaits the purification of pertinent enzymes and the synthesis of efficient antisera or oligonucleotide probes. Availability of aflatoxin pathway cDNA probes will permit identification of the native pathway genes in a genomic library. This accomplishment will allow elucidation of the molecular regulation of the pathway through characterization of the native aflatoxin pathway genes with their intact regulatory sequences.

INTRASPECIFIC VARIATION

An important influence on the process of contamination of crops by *A. flavus* is variability within populations of the fungus. Individual isolates of *A. flavus* vary widely in their ability to produce aflatoxins, both *in vitro* and during infection of crops (Diener and Davis, 1966; Schroeder and Hein, 1967; Cotty, 1989b). This heterogeneity is an important contributor to the great variability observed in the aflatoxin content of infected seed (Lee *et al.*, 1990). However, aflatoxin production in culture is similar among related isolates, suggesting that production is a genetic trait (Bayman and Cotty, 1991a). Ability of *A. flavus* to produce aflatoxins is independent of its ability to infect and multiply in crops and animals (Cotty, 1989a; Drummond and Pinnock, 1990). The incidence of strains that do not produce aflatoxins can be as high as 20–45% (Diener and Davis, 1966; Harrison *et al.*, 1987). Further, atoxigenic strains can interfere with the contamination process by competitively excluding toxigenic strains and by directly interfering with toxigenesis (Cotty *et al.*, 1990). This observation has resulted in suggestions that atoxigenic strains may be useful as biological control agents directed at preventing aflatoxin contamination (Cotty, 1990).

Aflatoxins occur in high concentrations in both conidia and sclerotia of *A. flavus* (Wicklow and Shotwell, 1983; Thanoboripat, 1988), suggesting that aflatoxins may have a functional relationship with the asexual fruiting bodies. (Wicklow, 1984). Several researchers have looked for an association between morphogenesis and aflatoxin biosynthesis. Aflatoxin production parallels spore production but in some cases convincing evidence linking the two is lacking (Reiss, 1982). Some controversy exists over the relationship of sclerotial formation by fungal strains and the potential of strains to produce aflatoxins. Clearly isolates that produce both very high and very low aflatoxin levels produce sclerotia (Cotty, 1989b). However, some researchers have associated increased aflatoxin production with increased production of sclerotia (see Cotty, 1989b). The observations are complicated by a dichotomy in *A. flavus* that divides isolates into two distinct types based on behavior in culture (Cotty, 1989b): (1) the S strain produces numerous small sclerotia and abundant quantities of aflatoxin and (2) the L strain produces fewer but larger sclerotia and is very erratic in production of aflatoxin. The S strain appears to be a derived, genetically distinct group (Bayman and Cotty, 1991a). Within a strain, the quantity of sclerotia produced by a given isolate is not correlated with the quantity of aflatoxin produced by that strain. Evidence suggesting that aflatoxin biosynthesis and sclerotial morphogenesis are interrelated in strains that produce both includes (1) association of toxicant- or pH-mediated increases in aflatoxin production with inhibition of sclerotial development, (2) correlation of sclerotial maturation with cessation of aflatoxin production, and (3) the high aflatoxin content of sclerotia (Cotty, 1988; Bayman and Cotty, 1991b).

INTERFERENCE WITH AFLATOXIN BIOSYNTHESIS

Aflatoxin biosynthesis is not linked unequivocally to fungal growth and infection of seed. Numerous observations have been made of nonaflatoxigenic strains of *A. flavus* in field environments (Cotty, 1989b). In addition, even known aflatoxigenic strains can discontinue production of aflatoxin on certain media or under certain environmental conditions (Wei and Jong, 1986; Lee, 1989). Fungal strains with the potential to produce high levels of aflatoxins exhibit greatly diminished ability to produce toxins in the presence of certain plant metabolites as well (Bhatnagar and McCormick, 1988; McCormick *et al.*, 1988; Zeringue and McCormick, 1989). Several observations indicate that the aflatoxin biosynthetic "machinery" is very sensitive to certain genetic, biochemical, or environmental influences. The results suggest that field strategies could be devised to perturb or interrupt the aflatoxin biosynthetic process.

Biotic

Numerous microorganisms have the capacity to interfere with aflatoxin production when they are associated with *A. flavus* in monoxenic culture on synthetic media or on various natural substrates (Roy and Chourasia, 1990). Interference occurs either through competition for nutrient resources and space or through elaboration of substances that interfere with toxigenesis. The outcome of competition in the complex microbial community associated with a crop is determined by unique substrate properties and by the overriding influence of the environment. When *A. flavus* dominates, toxin formation proceeds, unless other inhabitants of the crop such as *Bacillus subtilis* (Kimura and Hirano, 1988) and *Lactobacillus* spp. (Karunaratne *et al.*, 1990) elaborate interfering compounds. Acidification of substrates by competing fungi has been suggested as a major mechanism through which nonaflatoxigenic fungi prevent contamination by *A. flavus* (Horn and Wicklow, 1983). Attributing toxin interference to reduced pH is a persistent misconception; lowering the pH of substrates stimulates aflatoxin production in a pH-dependent manner to below pH 3.0 (Cotty, 1988). Production of compounds such as gluconic acid that inhibit aflatoxin production independent of pH probably explains the inhibitory effects of *Aspergillus niger* and *Aspergillus tamarii* (Shantha *et al.*, 1990).

Environment

Environmental factors can influence aflatoxin production dramatically in axenic culture through direct effects on the fungus and through indirect effects on the crop and crop-associated insects and microflora. Temperature, water potential, pH, and specific salts or acids common in foods and feeds can reduce or prevent aflatoxin production through direct effects on *A. flavus*. Toxin production is prevented by temperatures above 40°C, although high temperature and associated conditions can permit *A. flavus* to dominate during colonization and infection of the crop. When temperatures cycle into a range conducive to aflatoxin biosynthesis, dominance of *A. flavus* results in severe contamination (Lutey and Christensen, 1963; Schroeder and Hein, 1967; Mehan and Chohan, 1974; Hill *et al.*, 1985). Environmental conditions in which crops are grown and stored are important determinants of associated microflora (Lutey and Christensen, 1963; Phillips *et al.*, 1979). Low moisture content and high temperatures can favor dominance of aflatoxin-producing aspergilli and, subsequently, aflatoxin contamination. Damaged and weakened crop components are particularly susceptible to infection and degradation by fungi, but temperature and moisture largely determine both the microflora that exploit the natural substrates and the elaboration of specific mycotoxins (Bullerman *et al.*, 1975; Lacey, 1989). If the environment does not favor dominance of *A. flavus*, contamination can be expected to be minor (Bullerman *et al.*, 1975).

The environment also influences fungal contamination through direct and indirect effects on the host. Drought and high temperature can result in corn kernels with small faults that are excellent infection sites for aggressive wound pathogens such as *A. flavus* (Diener *et al.*, 1987). Environmental conditions also can reduce the natural defenses of peanut by shutting down phytoalexin production and making the pods more susceptible (Hill *et al.*, 1985; Cole *et al.*, 1989). Severity of contamination in the field is affected by the timing and extent of crop damage by insects in the field (i.e., European corn borer on maize, lesser corn stalk borer on peanuts, and pink bollworm on cotton) and in storage (i.e., various weevils and beetles) (Widstrom, 1979).

Host Plant Resistance

Conventional plant breeding techniques have identified varieties of corn (Gorman and Kang, 1991; Lisker and Lillehoj, 1991) and peanuts (Mehan *et al.*, 1986) with reduced susceptibility to infection and/or aflatoxin contamination by aflatoxigenic fungi. Attempts are underway to identify chemical compounds that are linked to the lower susceptibility to infection by toxigenic fungi. The host resistance traits could be enhanced in crop plants by plant breeding or through molecular engineering (for review, see Cleveland and Bhatnagar, 1992). An essential requirement for a molecular engineering approach is that specific genes for the trait be identified, cloned, and inserted stably into the plant chromosome. Possible resistance-related substances include enzymes or proteins and low molecular weight inhibitory compounds that have been identified in crop plants. Although potential antifungal enzymes have been identified in seed (such as corn kernels), still no definitive information is available about whether the differences in kernel resistance or susceptibility to *A. flavus* and aflatoxin contamination (Lisker and Lillehoj, 1991) are linked to biochemical traits such as hydrolytic antifungal enzymes. High molecular weight (McCormick *et al.,* 1988) and low molecular weight (Zeringue and McCormick, 1990) compounds that inhibit aflatoxin biosynthesis in *A. flavus* liquid fermentations were detected in developing cottonseed or in cotton leaves, respectively. Cotton and peanut plants can respond to invading aflatoxigenic molds by producing defense-related compounds (phytoalexins) (Bell, 1981; Wotton and Strange, 1987; Cole *et al.,* 1989; Zeringue and McCormick, 1989). Exposure of cotton leaf tissue to the fungus as well as to certain volatile compounds derived from the fungus were shown to elicit phytoalexins (Bell, 1981; Zeringue, 1987).

Genetic engineering strategies to enhance existing plant defenses might include insertion of multiple gene copies and use of pertinent powerful plant gene promoters or enhancers. The strategies assume that native antifungal hydrolases, inhibitory peptides, or other relevant plant substances either are not expressed in effective quantities to inhibit fungal growth or possess the wrong specificity to lyse fungal cell walls efficiently. Future strategies, therefore, could include cloning and amplifying native antifungal genes or incorporating new nonhost inhibitor genes encoding potent antifungal substances into corn, peanut, and cottonseed.

Many aspects of host plant–fungal interactions must be understood before pertinent procedures can be applied. Major technological hurdles must be overcome in the genetic engineering approach, including (1) understanding the processes by which aflatoxin pathway modulating substances in the plant are produced, (2) developing highly efficient procedures to transform the particular aflatoxin-contamination-susceptible varieties of crops with stably expressed foreign genes, and (3) expressing genetically engineered resistance genes, once inserted into the plant genome, at the proper levels, time, and site in plant tissues to prevent infection of plants by toxigenic fungi.

Chemical or Biochemical

Several compounds and substances have been tested that effectively inhibit fungal growth or aflatoxin production or stimulate *Aspergillus* spp. growth with or without enhancing aflatoxin production. An exhaustive review of compounds affecting the biosynthesis or bioregulation of aflatoxins has been presented (Zaika and Buchanan, 1987). Several inhibitors with specific and unknown mechanism of action have been identified (Bhatnagar and McCormick, 1988; Wheeler *et al.,* 1989; Zeringue and McCormick, 1989). These chemicals could be used to develop ecologically safe pesticides based on the chemical structure or biochemical nature of these substances. Certain natural products in crop plants have been found to be both inhibitory and stimulatory to *A. flavus* growth or aflatoxin production (Bhatnagar and McCormick, 1988; Zeringue and McCormick, 1989; Zeringue and Bhatnagar, 1990).

CONCLUSION

Aflatoxin synthesis has no obvious physiological role in primary growth and metabolism of the organism and, therefore, is considered a "secondary" process (Malik, 1982). To date, no biological role of aflatoxin in the ecological survival of the fungal organism has been confirmed. However, since aflatoxins are toxic to certain potential competitor microbes in the ecosystem (Detroy *et al.*, 1971), a survival benefit to the producing fungi is implied. Theories also have been proposed about a possible biological role of aflatoxins or related compounds as deterrents to insect feeding on fungal-resistant structures (sclerotia) (Wicklow and Shotwell, 1983; Dowd, 1991). Note, however, that aflatoxin per se is a poor antibiotic (Ciegler, 1983). Lillehoj (1991) proposed that intensive agricultural practices are responsible for the creation of unique niches that, under certain conditions, select toxin-producing fungi. Contemporary crop production is based on intensive practices that are unlikely to be altered in a significant manner in the near future. Therefore, developing comprehensive understanding of both the mycology of aflatoxin contamination in various ecosystems and the molecular regulation of aflatoxin formation is imperative. This understanding ultimately will provide the tools for developing strategies for effective control of aflatoxigenic fungi and elimination of aflatoxin contamination from animal feed and human food chains.

REFERENCES

Anderson, J. A., Chung, C. H., and Cho, S-H. (1990). Versicolorin A hemiacetal, hydroxydihydrosterigmatocystin, and aflatoxin G_{2a} reductase activity in extracts from *Aspergillus parasiticus*. *Mycopathologia* **111**, 39–45.

Ashworth, L. J., McMeans, J. L., and Brown, C. M. (1969). Infection of cotton by *Aspergillus flavus*: Epidemiology of the disease. *J. Stored Prod. Res.* **5**, 193–202.

Bayman, P., and Cotty, P. J. (1991a). Evolution and DNA polymorphisms in *Aspergillus flavus*; Regulatory aspects of the S and L problem. *Mycol. Soc. Amer. Newslett.* **42**, 5.

Bayman, P., and Cotty, P. J. (1991b). Vegetative compatibility and genetic diversity in the *Aspergillus flavus* population of a single field. *Can. J. Bot.* **69**, 1707–1711.

Bell, A. A. (1981). Biochemical mechanisms of disease resistance. *Ann. Rev. Plant Physiol.* **32**, 21–81.

Bennett, J. W., and Papa, K. E. (1988). The aflatoxigenic *Aspergillus* spp. *In* "Advances in Plant Pathology" (G. S. Sidhu, ed.), Vol. 6, pp. 263–280. Academic Press, New York.

Bhatnagar, D., and Cleveland, T. E. (1990). Purification and characterization of a reductase from *Aspergillus parasiticus* SRRC 2043 involved in aflatoxin biosynthesis. *FASEB J.* **4**, A2164.

Bhatnagar, D., and Cleveland, T. E. (1991). Aflatoxin biosynthesis: Developments in chemistry, biochemistry, and genetics. *In* "Aflatoxin in Corn: New Perspectives" (O. L. Shotwell and C. R. Hurburgh, Jr., eds.), pp. 391–405. Iowa State University Press, Ames.

Bhatnagar, D., and McCormick, S. P. (1988). The inhibitory effect of neem (*Azadirachata indica*) leaf extracts on aflatoxin synthesis in *Aspergillus parasiticus*. *J. Am. Oil Chem. Soc.* **65**, 1166–1168.

Bhatnagar, D., McCormick, S. P., Lee, L. S., and Hill, R. A. (1987). Identification of O-methylsterigmatocystin as an aflatoxin B_1 and G_1 precursor in *Aspergillus parasiticus. Appl. Environ. Microbiol.* **53,** 1028–1033.

Bhatnagar, D., Ullah, A. H. J., and Cleveland, T. E. (1988). Purification and characterization of a methyltransferase from *Aspergillus parasiticus* SRRC 163 involved in aflatoxin biosynthetic pathway. *Prep. Biochem.* **18,** 321–369.

Bhatnagar, D., Cleveland, T. E., and Lillehoj, E. B. (1989). Enzymes in late stages of aflatoxin B_1 biosynthesis: Strategies for identifying pertinent genes. *Mycopathologia* **107,** 75–83.

Bhatnagar, D., Lillehoj, E. B., and Bennett, J. W. (1991a). Biological detoxification of mycotoxins. *In* "Mycotoxins and Animal Foods" (J. E. Smith and R. S. Henderson, eds.), pp. 815–826. CRC Press, Boca Raton, Florida.

Bhatnagar, D., Ehrlich, K. C., and Cleveland, T. E. (1991b). Oxidation-reduction reactions in biosynthesis of secondary metabolites. *In* "Mycotoxins in Ecological Systems" (D. Bhatnagar, E. B. Lillehoj, and D. K. Arora, eds.), pp. 255–286. Marcel Dekker, New York.

Bhatnagar, D., Cleveland, T. E., and Kingston, D. G. I. (1991c). Enzymological evidence for separate pathways for aflatoxin B_1 and B_2 biosynthesis. *Biochemistry* **30,** 4343–4350.

Bullerman, L. B., Baca, J. M., and Scott, W. T. (1975). An evaluation of potential of mycotoxin-production molds in corn meal. *Cereal Foods World* **20,** 248–253.

Cary, J. W., Cleveland, T. E., and Bhatnagar, D. (1992). Regulation by thiamine of expression of a gene from *Aspergillus parasiticus* encoding norsolorinic acid reductase activity. *FASEB J.* **6,** A228.

Chang, P. K., Skory, C. D., and Linz, J. E. (1992). Cloning of a gene associated with aflatoxin B_1 biosynthesis in *Aspergillus parasiticus. Curr. Genet.* **21,** 231–233.

Chang, P.-K., Cary, J. W., Bhatnagar, D., Cleveland, T. E., Bennett, J. W., Linz, J. E., Woloshuk, C. P., and Payne, G. A. (1993). Cloning of the *A. parasiticus apa*-2 gene associated with the regulation of aflatoxin biosynthesis. *Appl. Environ. Microbiol.* Submitted.

Chuturgoon, A. A., and Dutton, M. F. (1991). The appearance of an enzyme activity catalysing the conversion of norsolorinic acid to averantin in *Aspergillus parasiticus* cultures. *Mycopathologia* **113,** 41–44.

Chuturgoon, A. A., Dutton, M. F., and Berry, R. K. (1990). The preparation of an enzyme associated with aflatoxin biosynthesis by affinity chromatography. *Biochem. Biophys. Res. Commun.* **166,** 38–42.

Ciegler, A. (1983). Evolution, ecology, and mycotoxins: Some musings. *In* "Secondary Metabolism and Differentiation in Fungi" (J. W. Bennett and A. Ciegler, eds.), pp. 429–440. Marcel Dekker, New York.

Cleveland, T. E., and Bhatnagar, D. (1992). Molecular strategies for reducing aflatoxin levels in crops before harvest. *In* "Molecular Approaches to Improving Food Quality and Safety" (D. Bhatnagar and T. E. Cleveland, eds.), pp. 205–228. Van Nostrand Reinhold, New York.

Cleveland, T. E., Bhatnagar, D., Foell, C. J., and McCormick, S. P. (1987a). Conversion of a new metabolite to aflatoxin B_2 by *Aspergillus parasiticus. Appl. Environ. Microbiol.* **53,** 2804–2807.

Cleveland, T. E., Lax, A. R., Lee, L. S., and Bhatnagar, D. (1987b). Appearance of enzyme activities catalyzing conversion of sterigmatocystin to aflatoxin B_1 in late-growth-phase *Aspergillus parasiticus* cultures. *Appl. Environ. Microbiol.* **53,** 1711–1713.

Cole, R. J., Sanders, T. H., Dorner, J. W., and Blankenship, P. D. (1989). Environmental conditions required to induce preharvest aflatoxin contamination of groundnuts: Summary of six years research. *In* "Aflatoxin Contamination of Groundnuts," (D. McDonald and V. K. Mehan, eds.), pp. 279–287. ICRISAT, Patancheru, India.

Cotty, P. J. (1988). Aflatoxin and sclerotial production by *Aspergillus flavus:* Influence of pH. *Phytopathology* **78,** 1250–1253.

Cotty, P. J. (1989a). Effects of cultivar and boll age on aflatoxin in cottonseed after inoculation with *Aspergillus flavus* at simulated exit holes of the pink bollworm. *Plant Dis.* **73,** 489–492.

Cotty, P. J. (1989b). Virulence and cultural characteristics of two *Aspergillus flavus* strains pathogenic on cotton. *Phytopathology* **79**, 808–814.

Cotty, P. J. (1990). Effect of atoxigenic strains of *Aspergillus flavus* on aflatoxin contamination of developing cottonseed. *Plant Dis.* **74**, 233–235.

Cotty, P. J. (1991). Effect of harvest date on aflatoxin contamination of cottonseed. *Plant Dis.* **75**, 312–314.

Cotty, P. J., and Lee, L. S. (1989). Aflatoxin contamination of cottonseed: Comparison of pink bollworm damaged and undamaged bolls. *Trop. Sci.* **29**, 273–277.

Cotty, P. J., Bayman, P., and Bhatnagar, D. (1990). Two potential mechanisms by which atoxigenic strains of *Aspergillus flavus* prevent toxigenic strains from contaminating cottonseed. *Phytopathology* **90**, 995.

Darrah, L. L., and Barry, B. D. (1991). Reduction of preharvest aflatoxin in corn. *In* "Pennington Center Nutrition Series" (G. A. Bray and D. H. Ryan, eds.), Vol. 1, pp. 283–310. Louisiana State University Press, Baton Rouge.

Detroy, R. W., and Hesseltine, C. W. (1968). Net synthesis of carbon-14-labeled lipids and aflatoxins in resting cells of *Aspergillus parasiticus*. *Dev. Ind. Microbiol.* **10**, 127–133.

Detroy, R. W., and Hesseltine, C. W. (1970). Secondary biosynthesis of aflatoxin B_1 in *Aspergillus parasiticus*. *Can J. Microbiol.* **16**, 959–963.

Detroy, R. W., Lillehoj, E. B., and Ciegler, A. (1971). Aflatoxin and related compounds. *In* "Microbial Toxins" (A. Ciegler, S. Kadis, and S. J. Ajl, eds.), Vol. 6, pp. 3–178. Academic Press, New York.

Diener, U. L., and Davis, N. D. (1966). Aflatoxin production by isolates of *Aspergillus flavus*. *Phytopathology* **56**, 1390–1393.

Diener, U. L., Cole, R. J., Sanders, T. H., Payne, G. A., Lee, L. S., and Klich, M. A. (1987). Epidemiology of aflatoxin formation by *Aspergillus flavus*. *Ann. Rev. Phytopathol.* **25**, 249–270.

Dowd, P. F. (1991). Insect interactions with mycotoxin producing fungi and their hosts. *In* "Mycotoxins in Ecological Systems" (D. Bhatnagar, E. B. Lillehoj, and D. K. Arora, eds.), pp. 137–156. Marcel Dekker, New York.

Drew, S. W., and Demain, A. L. (1977). Effect of primary metabolites on secondary metabolism. *Ann. Rev. Microbial.* **31**, 343–356.

Drummond, J., and Pinnock, D. E. (1990). Aflatoxin production by entomopathogenic isolates of *Aspergillus parasiticus* and *Aspergillus flavus*. *J. Invert. Pathol.* **55**, 332–336.

Dutton, M. F. (1988). Enzymes and aflatoxin biosynthesis. *Microbiol. Rev.* **52**, 274–295.

Floyd, J. C., Mills, J., and Bennett, J. W. (1987). Biotransformation of sterigmatocystin and absence of aflatoxin biotransformation by blocked mutants of *Aspergillus parasiticus*. *Exp. Mycol.* **11**, 109–114.

Gorman, D. P., and Kang, M. S. (1991). Preharvest aflatoxin contamination in maize: Resistance and genetics. *Plant Breeding* **107**, 1–10.

Gupta, S. K., Viswanathan, L., and Venkitasubramanian, T. A. (1970). A comparative study of lipids of a toxigenic and a non-toxigenic strain of *Aspergillus flavus*. *Indian J. Biochem. Biophys.* **7**, 108–111.

Hagler, W. M., Jr. (1991). Potential for detoxification of mycotoxin-contaminated commodities. *In* "Pennington Center Nutrition Series" (G. A. Bray and D. H. Ryan, eds.), Vol. 1, pp. 253–269. Louisiana State University Press, Baton Rouge.

Harrison, M. A., Silas, J. C., and Carpenter, J. A. (1987). Incidence of aflatoxigenic isolates of *Aspergillus flavus/parasiticus* obtained from Georgia corn processing plants. *J. Food Quality* **10**, 101–105.

Henderberg, A., Bennett, J. W., and Lee, L. S. (1988). Biosynthetic origin of aflatoxin G_1: Confirmation of sterigmatocystin and lack of confirmation of aflatoxin B_1 as precursors. *J. Gen. Microbiol.* **134**, 661–667.

Hill, R. A., Wilson, D. M., McMillian, W. W., Widstrom, N. W., Cole, R. J., Sanders, T. H., and Blankenship, P. D. (1985). Ecology of the *A. flavus* group and aflatoxin formation in maize and groundnut. *In* "Trichothecenes and Other Mycotoxins." (J. Lacey, eds.), pp. 79–95. John Wiley and Sons, Chichester, England.

Horn, B. W., and Wicklow, D. T. (1983). Factors influencing the inhibition of aflatoxin production in corn by *Aspergillus niger. Can. J. Microbiol.* **29,** 1087–1091.

Horng, J. S., Linz, J. E., and Pestka, J. J. (1989). Cloning and characterization of the *trpC* gene from an aflatoxigenic strain of *Aspergillus parasiticus. Appl. Environ. Microbiol.* **55,** 2561–2568.

Horng, J. S., Chang, P-K., Pestka, J. J., and Linz, J. E. (1990). Development of a homologous transformation system for *Aspergillus parasiticus* with the gene encoding nitrate reductase. *Mol. Gen. Genet.* **224,** 294–296.

Hsieh, D. P. H., Wan, C. C., and Billington, J. A. (1989). A versiconal hemiacetal acetate converting enzyme in aflatoxin biosynthesis. *Mycopathologia* **107,** 121–126.

Jelinek, C. F., Pohland, A. E., and Wood, G. E. (1989). Worldwide occurrence of mycotoxins in foods and feeds: An update. *J. Assoc. Off. Anal. Chem.* **72,** 223–230.

Karunaratne, A., Wezenberg, E., and Bullerman, L. B. (1990). Inhibition of mold growth and aflatoxin production by *Lactobacillus* spp. *J. Food Protect.* **53,** 230–236.

Keller, N. P., Cleveland, T. E., and Bhatnagar, D. (1991). A molecular approach towards understanding aflatoxin production. *In* "Mycotoxins in Ecological Systems" (D. Bhatnagar, E. B. Lillehoj, and D. K. Arora, eds.), pp. 287–310. Marcel Dekker, New York.

Keller, N. P., Dischinger, H. C., Bhatnagar, D., Cleveland, T. E., and Ullah, A. H. J. (1993). Purification of a 40-KDa methyltransferase active in the aflatoxin biosynthetic pathway. *Appl. Environ. Microbiol.* **59,** 479–484.

Keller, N. P., Cleveland, T. E., and Bhatnagar, D. (1992a). Variable electrophoretic karyotypes of members of *Aspergillus* section *Flavi. Curr. Genet.* **21,** 371–375.

Keller, N. P., Cary, J. W., Cleveland, T. E., Bhatnagar, D., and Payne, G. A. (1992b). Characterization of a DNA unique to *Aspergillus* section *Flavi. FASEB J.* **6,** A228.

Kensler, T. W., Davidson, N. E., Egner, P. A., Guyton, K. Z., Groopman, J. D., Curphey, T. J., Lin, Y-L., and Roebuck, B. D. (1991). Chemoprotection against aflatoxin-induced hepatocarcinogenesis by dithiolethiones. *In* "Pennington Center Nutrition Series" (G. A. Bray and D. H. Ryan, eds.), Vol. 1, pp. 238–252. Louisiana State University Press, Baton Rouge.

Kimura, N., and Hirano, S. (1988). Inhibitory strains of *Bacillus subtilis* for growth and aflatoxin-production of aflatoxigenic fungi. *Agric. Biol. Chem.* **52,** 1173–1178.

Lacey, J. (1989). Prevention of mould growth and mycotoxin production through control of environmental factors. *In* "Mycotoxins and Phycotoxins '88," (S. Natori, K. Hashimoto and Y. Ueno, eds.), pp. 161–169. Elsevier Science, Amsterdam.

Lee, L. S. (1989). Metabolic precursor regulation of aflatoxin formation in toxigenic and non-toxigenic strains of *Aspergillus flavus. Mycopathologia* **107,** 127–130.

Lee, L. S., Wall, J. H., Cotty, P. J., and Bayman, P. (1990). Integration of enzyme-linked immunosorbent assay with conventional chromatographic procedures for quantification of aflatoxin in individual cotton bolls, seeds, and seed sections. *J. Assoc. Off. Anal. Chem.* **73,** 581–584.

Lillehoj, E. B. (1991). Aflatoxin: Genetic mobilization agent. *In* "Mycotoxins in Ecological Systems" (D. Bhatnagar, E. B. Lillehoj, and D. K. Arora, eds.), pp. 1–22. Marcel Dekker, New York.

Lillehoj, E. B., Wall, J. H., and Bowers, E. J. (1987). Preharvest aflatoxin contamination: Effect of moisture and substrate variation in developing cottonseed and corn kernels. *Appl. Environ. Microbiol.* **53,** 584–586.

Lisker, N., and Lillehoj, E. B. (1991). Prevention of mycotoxin contamination (principally aflatoxins and *Fusarium* toxins) at the preharvest stage. *In* "Mycotoxins and Animal Foods" (J. E. Smith and R. S. Henderson, eds.), pp. 689–719. CRC Press, Boca Raton, Florida.

Lutey, R. W., and Christensen, C. M. (1963). Influence of moisture content, temperature and length of storage upon survival of fungi in barley kernels. *Phytopathology* **53,** 713–717.

McCormick, S. P., Bhatnagar, D., Goynes, W. R., and Lee, L. S. (1988). An inhibitor of aflatoxin biosynthesis in developing cottonseed. *Can. J. Bot.* **66**, 998–1002.

McCormick, S. P., Bhatnagar, D., and Lee, L. S. (1987). Avenufanin is an aflatoxin B_1 precursor between averantin and averufin in the biosynthetic pathway. *Appl. Environ. Microbiol.* **53**, 14–16.

McCormick, S. P., Bowers, E., and Bhatnagar, D. (1988). High-performance liquid chromatographic procedure for determining the profiles of aflatoxin precursors in wildtype and mutant strains of *Aspergillus parasiticus. J. Chromatogr.* **441**, 400–405.

McGuire, S. M., Brobst, S. W., Graybill, T. L., Pal, K., and Townsend, C. A. (1989). Partitioning of tetrahydro- and dihydrobisfuran formation in aflatoxin biosynthesis defined by cell-free and direct incorporation experiments. *J. Am. Chem. Soc.* **111**, 8308–8309.

Maggon, K. K., Gupta, S. K., and Venkitasubramanian, T. A. (1977). Biosynthesis of aflatoxins. *Bacteriol. Rev.* **41**, 822–855.

Malik, V. S. (1982). Genetics and biochemistry of secondary fungal metabolites. *Adv. Appl. Microbiol.* **28**, 27–116.

Mehan, V. K., and Chohan, J. S. (1974). Effect of temperature on growth and aflatoxin production by isolates of *Aspergillus flavus. Indian Phytopathol.* **27**, 160–170.

Mehan, V. K., McDonald, D., and Ramakrishna, N. (1986). Varietal resistance of peanut to aflatoxin production. *Peanut Sci.* **13**, 7–10.

Park, D. L., Lee, L. S., Price, R. L., and Pohland, A. E. (1988). Review of the decontamination of aflatoxins by ammoniation: Current status and regulation. *J. Assoc. Off. Anal. Chem.* **71**, 685–703.

Payne, G. A., and Woloshuk, C. P. (1989). Transformation of *Aspergillus flavus* to study aflatoxin biosynthesis. *Mycopathologia* **107**, 139–144.

Phillips, D. J., Mackey, B., Ellis, W. R., and Hansen, T. N. (1979). Occurrence and interaction of *Aspergillus flavus* with other fungi on almonds. *Phytopathology* **69**, 829–831.

Payne, G. A., Nystorm, G. J., Bhatnagar, D., Cleveland, T. E., and Woloshuk, C. P. (1993). Cloning of the *afl*-2 gene involved in aflatoxin biosynthesis from *Aspergillus flavus. Appl. Environ. Microbiol.* **59**, 156–162.

Phillips, T. D., Sarr, B. A., Clement, B. A., Kubena, L. F., and Harvey, R. B. (1991). Prevention of aflatoxicosis in farm animals via selective chemisorption of aflatoxin. *In* "Pennington Center Nutrition Series" (G. A. Bray and D. H. Ryan, eds.), Vol. 1, pp. 223–237. Louisiana State University Press, Baton Rouge.

Reiss, J. (1982). Development of *Aspergillus parasiticus* and formation of aflatoxin B_1 under the influence of conidiogenesis affecting compounds. *Arch. Microbiol.* **133**, 236–238.

Roy, A. K., and Chourasia, H. K. (1990). Inhibition of aflatoxin production by microbial interaction. *J. Gen. Appl. Microbiol.* **53**, 2250–2527.

Schroeder, H. W., and Hein, H., Jr. (1967). Aflatoxins: Production of the toxins *in vitro* in relation to temperature. *Appl. Microbiol.* **15**, 441–445.

Skory, C. D., Chang, P.-K., Cary, J., and Linz, J. E. (1992). Isolation and characterization of a gene from *Aspergillus parasiticus* associated with the conversion of versicolorin A to sterigmatocystin in aflatoxin biosynthesis. *Appl. Environ. Microbiol.* **58**, 3527–3537.

Scott, G. E., and Zummo, N. (1988). Sources of resistance in maize to kernel infection by *Aspergillus flavus* in the field. *Crop Sci.* **28**, 504–507.

Seip, E. R., Woloshuk, C. P., Payne, G. A., and Curtis, S. E. (1990). Isolation and sequence analysis of a β-tubulin gene from *Aspergillus flavus* and its use as a selectable marker. *Appl. Environ. Microbiol.* **56**, 3686–3692.

Shantha, T., Rati, E. R., and Bhavani Shankar, T. N. (1990). Behaviour of *Aspergillus flavus* in the presence of *Aspergillus niger* during biosynthesis of aflatoxin B_1. *Antonie van Leeuwenhoek* **58**, 121–127.

Shih, C.-N., and Martin, E. H. (1976). Aflatoxin formation: Lipid synthesis and glucose metabolism by *Aspergillus parasiticus* during incubation with and without agitation. *Biochim. Biophys. Acta* **338**, 286–296.

Skory, C. D., Horng, J. S., Pestka, J. J., and Linz, J. E. (1990). Transformation of *Aspergillus parasiticus* with a homologous gene (*pyr G*) involved in pyrimidine biosynthesis. *Appl. Environ. Microbiol.* **56**, 3315–3320.

Somer, N. F., Buchanan, J. R., and Fortlage, R. J. (1986). Relation of early splitting and tattering of pistachio nuts to aflatoxin in the orchard. *Phytopathology* **76**, 692–694.

Stephenson, L. W., and Russell, T. E. (1974). The association of *Aspergillus flavus* with hemipterous and other insects infesting cotton bracts and foliage. *Phytopathology* **64**, 1502–1506.

Thanoboripat, D. (1988). Aflatoxin in spores of *Aspergillus flavus*. *ASEAN Food J.* **4**, 71–72.

Townsend, C. A., Christensen, S. B., and Trautwein, K. (1984). Hexanoate as a starter unit in polyketide synthesis. *J. Am. Chem. Soc.* **106**, 3868–3869.

Townsend, C. A., Plancan, K. A., Pal, K., and Brobst, S. W. (1988). Hydroxyversicolorone: Isolation and characterization of a potential intermediate in aflatoxin biosynthesis. *J. Org. Chem.* **53**, 2472–2477.

Wei, D-L., and Jong, S-C. (1986). Production of aflatoxin by strains of the *Aspergillus flavus* group maintained in ATCC. *Mycopathologia* **93**, 19–24.

Wheeler, M. H., Bhatnagar, D., and Rojas, M. G. (1989). Chlobenthiazone and tricyclazole inhibition of aflatoxin biosynthesis by *Aspergillus flavus*. *Pesticide Biochem. Physiol.* **35**, 315–323.

Wicklow, D. T. (1984). Ecological approaches to the study of mycotoxigenic fungi. *In* "Toxigenic Fungi—Their Toxins and Health Hazard" (H. Kurata and Y. Ueno, eds.), pp. 33–34. Kodansha, Tokyo.

Wicklow, D. T., and Shotwell, O. L. (1983). Intrafungal distribution of aflatoxins among conidia and sclerotia of *Aspergillus flavus* and *A. parasiticus*. *Can. J. Microbiol.* **29**, 1–5.

Widstrom, N. W. (1979). The role of insects and other plant pests in aflatoxin contamination of corn, cotton, and peanuts—A review. *J. Environ. Qual.* **8**, 5–11.

Widstrom, N. W. (1987). Breeding strategies to control aflatoxin contamination of maize through host plant resistance. *In* "Aflatoxin in Maize: A Proceedings of the Workshop" (M. S. Zuber, E. B. Lillehoj, and B. L. Renfro, eds.), pp. 212–220. CIMMYT, Mexico City, Mexico.

Wotton, H. R., and Strange, R. N. (1987). Increased susceptibility and reduced phytoalexin accumulation in drought-stressed peanut kernels challenged with *Aspergillus flavus*. *Appl. Environ. Microbiol.* **53**, 270–273.

Yabe, K., Ando, Y., and Hamasaki, T. (1988). Biosynthetic relationship among aflatoxins B_1, B_2, G_1, G_2. *Appl. Environ. Microbiol.* **54**, 2101–2106.

Yabe, K., Ando, Y., Hashimoto, J., and Hamasaki, T. (1989). Two distinct *O*-methyltransferases in aflatoxin biosynthesis. *Appl. Environ. Microbiol.* **55**, 2172–2177.

Yabe, K., Ando, Y., and Hamasaki, T. (1991a). Desaturase activity in the branching step between aflatoxins B_1 and G_1 and aflatoxins B_2 and G_2. *Agric. Biol. Chem.* **55**, 1907–1911.

Yabe, K., Nakamura, Y., Nakajima, H., Ando, Y., and Hamasaki, T. (1991b). Enzymatic conversion of norsolorinic acid to averufin in aflatoxin biosynthesis. *Appl. Environ. Microbiol.* **57**, 1340–1345.

Yu, J., Cary, J. W., Bhatnagar, D., Cleveland, T. E., Keller, N. P., and Chu, F. S. (1993). Cloning and characterization of a cDNA from *Aspergillus parasiticus* encoding an *O*-methyltransferase involved in aflatoxin biosynthesis. *Appl. Environ. Microbiol.* Submitted.

Zaika, L. L., and Buchanan, R. L. (1987). Review of compounds affecting the biosynthesis or bioregulation of aflatoxins. *J. Food Protect.* **50**, 691–708.

Zeringue, H. J., Jr. (1987). Changes in cotton leaf chemistry induced by volatile elicitors. *Phytochemistry.* **26**, 1357–1360.

Zeringue, H. J., Jr., and Bhatnagar, D. (1990). Inhibition of aflatoxin production in *Aspergillus flavus* infected cotton bolls after treatment with neem (*Azadirachta indica*) leaf extracts. *J. Am. Oil. Chem. Soc.* **67**, 215–216.

Zeringue, H. J., Jr., and McCormick, S. P. (1989). Relationships between cotton leaf-derived volatiles and growth of *Aspergillus flavus*. *J. Am. Oil Chem. Soc.* **66,** 581–585.

Zeringue, H. J., Jr., and McCormick, S. P. (1990). Aflatoxin production in cultures of *Aspergillus flavus* incubated in atmospheres containing selected cotton leaf-derived volatiles. *Toxicon* **28,** 445–448.

16

Veterinary Diseases Related to Aflatoxins

❖

Doris M. Miller and David M. Wilson

INTRODUCTION

Mycotoxins long have been known to cause disease in humans and animals. Aflatoxins are likely to be the best studied of the mycotoxins. Much of the justification for aflatoxin research has been based on its carcinogenic potential in animals. However, the major effects are decrease of animal production because of slow growth and chronic or acute disease induced by aflatoxin consumption. The increased incidence of tumors in humans in certain countries has been postulated to be the result of aflatoxin ingestion.

Research on aflatoxin contamination and aflatoxicosis has been carried out for more than 30 years. The effects of aflatoxin consumption on many species of animals have been studied. The effects on some species, such as the rat and duckling, have been investigated more thoroughly than those on other species such as cattle and horse.

Between 1944 and 1946, primary hepatic carcinomas were observed in swine slaughtered at a Morocco abattoir (Shalkop and Ambroecht, 1974). These pigs had been fed oil cakes made from cottonseed, peanuts, and cocoa. In 1952, hepatomas were described in rats, but not until 1960, when turkey X disease occurred in England, did the feed-based relationship between these similar diseases begin to unfold (Wogan, 1973). All these diseases and others affecting guinea pigs, turkeys, swine, ducklings, and trout later were postulated and usually proven to have a similar etiology (Wilson *et al.*, 1967; Wyllie and Morehouse, 1978). Many laboratories and reports have shown that the common factor was a mycotoxin termed aflatoxin (Wyllie and Morehouse, 1978).

The aflatoxins constitute a group of fungal metabolites that have varied toxic and carcinogenic properties, depending on dose and duration of exposure. These compounds cause serious disease in trout, poultry, livestock, and other animals (Burnside *et al.*, 1957; Sisk *et al.*, 1968; Newberne and Butler, 1969; Wyllie and Morehouse, 1978). Susceptibility to aflatoxin varies with species, age, breed, and dietary protein content (Carnaghan *et al.*, 1967; Newberne, 1973). Several reports have shown the association between aflatoxicosis and decreased humoral and cellular immunity in cattle, guinea pigs, dogs, rabbits, trout, laboratory animals, and swine (Pier and Heddleston, 1970; Pier, 1973b; Pier *et al.*, 1977; Cysewski *et al.*, 1978; Miller *et al.*, 1978; Richard *et al.*, 1978). Aflatoxin in the diet has been associated with increased incidences of infectious disease (Pier, 1973a; Richard *et al.*, 1978) and tumors (Newberne, 1973; Shalkop and Ambroecht, 1974).

CLINICAL SIGNS

Various clinical signs are associated with aflatoxin exposure in animals, depending on the purity of the aflatoxin, the amount consumed, the length of consumption, the age and species of animal, and the quality of the feed. The clinical signs can consist of anorexia, icterus, depression, weight loss, nasal discharge, gastrointestinal signs, ascites, or hydrothorax.

Swine

Several investigators (Sisk *et al.*, 1968; Miller *et al.*, 1978; Harvey *et al.*, 1988; Coppock *et al.*, 1989) have reported observing weight loss, anorexia, icterus, hemorrhages, and acute death in swine naturally or experimentally exposed to aflatoxins.

Equine

In horses, Bortell *et al.* (1983) and Cysewski *et al.* (1982) observed anorexia, depression, tremors, red-brown urine, or bloody feces in experimental cases of aflatoxicosis.

Bovine

Anorexia, depression, and weight loss are consistent clinical signs observed in both natural and experimental cases of aflatoxicosis in cattle (Loosmore and Harding, 1961; Loosmore and Markson, 1961; Osuna *et al.*, 1977; Brown *et al.*, 1981; McKenzie *et al.*, 1981; Colvin *et al.*, 1984; Pugh *et al.*, 1984; Brucato *et al.*, 1986). Decreased milk production and photosensitization also were reported by Colvin *et al.* (1984) and McKenzie *et al.* (1981).

Ovine, Caprine

Anorexia, depression, and icterus were observed in sheep and goats exposed to aflatoxins (Hatch *et al.*, 1971; Samarajeewa *et al.*, 1975; Abdelsalam *et al.*, 1989). The goats also developed a nasal discharge and a dark brown urine was noted in the sheep.

Canine

Dogs exposed to aflatoxins developed the typical anorexia, depression, and icterus but also may have had hemorrhages, melena, and pulmonary edema (Liggett *et al.*, 1986; Bastianello *et al.*, 1987; Thornburg and Raisbeck, 1988).

Poultry (Avian)

Weight loss, decreased weight gains, impaired blood coagulation, poor pigmentation, and decreased bone strength have been reported and discussed by several researchers (Bryden and Cumming, 1980; Huff *et al.*, 1980; Ubosi *et al.*, 1985). Bryden *et al.* (1980) also reported a gender response because of an increased susceptibility to alfatoxin in males. Poultry vary in their recovery from aflatoxin toxicity depending on dose, species, strain, and sex (Bryden *et al.*, 1980).

PATHOLOGY

Aflatoxicosis is the term used to refer to the specific intoxication of animals that generally was characterized by hemorrhage, hepatic necrosis, and bile ductule proliferation (Wilson *et al.*, 1967; Heathcote and Hibbert, 1978). The earliest clinical signs and lesions observed in turkey X disease, hepatitis X of dogs, and similar cases of acute aflatoxicosis were anorexia, lethargy, hemorrhages, hepatic necrosis, and bile ductule proliferation (Wilson *et al.*, 1967; Wyllie and Morehouse, 1978). In swine, acute aflatoxicosis was manifested by icterus, hemorrhages, and centrilobular hepatic necrosis (Wilson *et al.*, 1967; Sisk and Carlton, 1972; Edds, 1973). Chronic aflatoxicosis initially was characterized by decreased feed conversion efficiency and decreased weight gain in animals (Wilson *et al.*, 1967; Newberne, 1973; Wyllie and Morehouse, 1978). Continued consumption of the toxin resulted in anorexia, icterus, yellow livers, and bile ductule proliferation (Sisk *et al.*, 1968; Wyllie and Morehouse, 1978). Table 1 presents a summary of animal organs and systems reported to be affected by aflatoxin poisoning.

Aflatoxins are metabolized in the liver. Therefore, histological changes are observed primarily within this organ. Rats and ducklings exhibit periportal necrosis and bile ductule proliferation (Theron, 1965; Clifford and Rees, 1967). The bile ductule proliferation within the duckling varies with the amount of toxin ingested, which was the basis for using the duckling in biological assays of feeds believed to be contaminated with aflatoxins (Theron, 1965; Newberne, 1974). The pig, cow, guinea pig, and monkey all have centrilobular necrosis whereas

TABLE 1 Organs or Systems Reported to be Affected by Aflatoxin

Animal	Exposure[a]	Aflatoxin dosage[b]	Organs or systems affected[c]	Aflatoxin involved[a]	Reference
Swine	E	1–4 mg/kg of feed for 28 days	Liver	Mixed	Harvey et al. (1988)
Swine	E	0.4–0.8 mg/kg of feed for 10 weeks	Liver	Mixed	Miller et al. (1978, 1981)
Swine	E	0.1 mg/kg b.w. (protein-deficient diet)	Liver, GI, kidney	B_1	Sisk and Carlton (1972)
Swine	E	1.2 mg/kg; single dose	Liver	Mixed	Miller et al. (1982)
Swine	E	0.1–0.3 mg/kg b.w. for 23 days	Liver	Mixed	Sisk et al. (1968)
Swine	E	0.02–0.8 mg/kg	Liver	B_1	Gumbmann and Williams (1969)
Swine	E	2–3.5 mg/kg	Liver	B_1 or mixed	Cysewski et al. (1968)
Swine	E	1.3 mg/day for 25 days	Immune system	B_1	Cysewski et al. (1978)
Swine	E	0.02, 0.03, and 0.04 mg/kg/day for 28–30 months	Hepatic tumors	B_1	Shalkop and Armbrecht (1974)
Swine	E	0.02–0.2 mg/kg for 107 days	Liver, GI, kidney	B_1	Gagne et al. (1968)
Swine	E	0.06–0.8 mg/kg for 117 days	Liver, GI, kidney	B_1	Gagne et al. (1968)
Swine	N	2.3–4.5 mg/kg in corn	Liver, kidney	B_1, B_2, G_1	Coppock et al. (1989)
Bovine	E	1.0 mg/kg; single dose	Liver	B_1	Brucato et al. (1986)
Bovine	E	0.5–1.0 mg/kg; single dose	Liver	B_1	Osuna et al. (1977)

Species	Route	Dose	Tissue affected	Toxin	Reference
Bovine	N	2.23 mg/kg in peanut hay	Liver	Mixed	McKenzie et al. (1981)
Bovine	N	1.5 mg/kg in feed	Liver	B_1, B_2	Colvin et al. (1984)
Bovine	N	0.9 mg/kg for 10 days	Liver	B_1	Hoerr et al. (1986)
Equine	E	0.1–0.4 mg/kg b.w. for 5 days	Liver	B_1, B_2	Aller et al. (1981)
Equine	E	0.02–0.05 mg/kg b.w. for 21 days	None	B_1, B_2	Aller et al. (1981)
Equine	E	0.5–2 mg/kg b.w. liver; single dose	Liver	B_1	Asquith et al. (1980)
Equine	E	0.5–7.4 mg/kg b.w.; single dose	Liver	B_1	Bortell et al. (1983)
Equine	E	0.3–0.75 mg/kg daily for 39 days	Liver	B_1, G_1	Cysewski et al. (1982)
Equine	N	0.13 mg/kg in corn	Liver	B_1, B_2, M_1	Vesonder et al. (1991)
Equine	N	2.5 mg/kg b.w.	Heart, brain, kidney	B_1, B_2	Angsubhakorn et al. (1981)
Caprine	E	4 mg/kg b.w.; single dose	Liver, kidney	B_1	Hatch et al. (1971)
Caprine	E	3 mg/kg b.w.	Liver, lung, kidney	B_1	Hatch et al. (1982)
Caprine	E, N	1.3–5.0 mg/animal/day until death	Liver, kidney	B_1, G_1, M_1	Samarajeewa et al. (1975)
Ovine	N	90–200 mg/kg in feed	Liver, kidney	Mixed	Abdelsalam et al. (1989)
Rabbit	E	0.4 mg/kg for 7 days	Hepatic, renal	B_1	Clark et al. (1982)
Rabbit	E	0.025–0.625 mg/kg b.w. for 24 days	Hepatic	B_1	Clark et al. (1982)
Avian	E	0.5–5.0 mg/kg for 5 weeks	Hepatic	B_1	Bryden et al. (1980)
Avian	E	0.1 mg/kg once per 50 g duckling	Hepatic	B_1	Theron (1965)
Avian	E	0.1–0.8 mg/kg for 35 days	Hepatic	B_1	Hoerr et al. (1986)

(continues)

TABLE 1 (*Continued*)

Animal	Exposure[a]	Aflatoxin dosage[b]	Organs or systems affected[c]	Aflatoxin involved[d]	Reference
Rat	E	7 mg/kg b.w. once	Hepatic carcinoma	B_1	Clifford and Rees (1967)
Rat	E	5 mg/kg for 6 weeks	Hepatic	B_1	Butler and Hempsall (1981)
Rat	E	0.5 mg/kg b.w. for 4 days postnatal	Tumors in digestive tract, urogenital system, hepatic, fetus	B_1	Goettler et al. (1980)
Monkey	E	0.05 mg/kg b.w. twice weekly for 18 weeks	Hepatic	B_1	Singhal et al. (1981)
Monkey	E	1 mg/kg b.w./day for 3 weeks to 0.062 mg/kg once a week for 2 years	Hepatic	B_1	Deo et al. (1970)
Monkey	E	0.01–1.0 mg daily until death	Hepatic	Mixed	Alpert et al. (1970)
Pregnant hamster	E	4 mg or 6 mg/kg once	Hepatic, renal, fetal	B_1	Schmidt and Panciera (1980)
Guinea pigs	E	0.63 mg/kg b.w. once	Hepatic	B_1	Thurston et al. (1980)

[a]E, Experimental exposure; N, natural exposure.
[b]Dosage range reported; low concentrations did not necessarily result in observable effects.
[c]Organs and systems reported.
[d]Aflatoxins reported in reference; mixed usually means $B_1 + B_2 + G_1 + G_2$.

rabbits have midzonal necrosis with acute aflatoxicosis (Heathcote and Hibbert, 1978). Some breeds of monkey have focal hepatic cell necrosis rather than centrilobular necrosis as the primary lesion (Heathcote and Hibbert, 1978). These variations in site of necrosis are the result of the area of primary metabolism and action of the aflatoxins in each species (Patterson, 1973).

The rate of metabolism is one factor in the differences among species in the location of hepatic lesions caused by aflatoxin (Wogan, 1973; Cysewski et al., 1978). Swine differ in their response and lesions to a single estimated LD_{50} dose of aflatoxin compared with rats and guinea pigs (Theron, 1965; Patterson, 1973). Histologically, the centrilobular necrosis observed at 24 hr (Figure 1) in swine continues to resolve over a 72-hr period in acute aflatoxicosis (Miller et al., 1982). Rats and guinea pigs show progressive intensification of cellular necrosis and damage from 24 to 72 hr after ingestion of high levels of aflatoxin. The rat, which is reported to metabolize aflatoxin slowly, has periportal lesions with acute aflatoxicosis (Wogan, 1973). The aflatoxin is believed to act directly on the periportal liver cells (Cysewski et al., 1978). However, in animals that metabolize aflatoxins rapidly, a metabolite is thought to be the primary toxic agent and

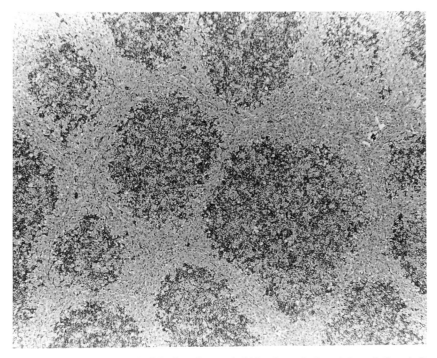

FIGURE 1   Photomicrograph of the liver from a pig 24 hr after a single oral dose of aflatoxin B_1 (1.2 mg/kg b.w.). Note centrilobular necrosis and hemorrhage. Peripheral lobular area of liver is intact. Hematoxylin and eosin stain; x10.

the lesions are centrilobular, as observed in guinea pigs and swine (Wogan, 1973).

The microscopic lesions associated with aflatoxicosis vary with species, duration of exposure (acute or chronic), amount of toxin consumed, and quality of feed. As mentioned earlier, the location of the liver necrosis varies with the rate of metabolism in the species. Centrilobular hepatic necrosis or hepatocellular vacuolar change and bile ductule proliferation are consistent lesions in cow, sheep, goat, and swine (Hatch *et al.*, 1971; Miller *et al.*, 1982; Colvin *et al.*, 1984; Hoerr *et al.*, 1986; Abdelsalam *et al.*, 1989). Chronic obliterating endophlebitis of centrilobular hepatic veins has been reported in some cases of bovine aflatoxicosis but is not observed in other species (Loosmore and Harding, 1961; Loosmore and Markson, 1961; Wyllie and Morehouse, 1978). Fibrosis is reported in all species when the animals do not die from acute aflatoxicosis.

In the horse, various lesions have been described in natural cases of exposure. However, the suspected rations may have contained other unidentified mycotoxins. Centrilobular fatty change, hepatic vacuolization, necrosis, bile duct proliferation, and fibrosis generally were reported in both natural and experimental cases (Cysewski *et al.*, 1978; Angsubhakorn *et al.*, 1981; Bortell *et al.*, 1983; Vesonder *et al.*, 1991). In addition, encephalomalacia of cerebral hemispheres and myocardial degeneration were reported by Angsubhakorn *et al.* (1981) in a case of natural exposure to aflatoxin-contaminated feed. Encephalomalacia has been associated with the mycotoxin fumonisin (Marasas *et al.*, 1988).

In 1961, Loosmore and Harding and later Harding *et al.* (1963) found hepatic lesions in swine fed different batches of the same toxic Brazilian groundnuts. Acute lesions described by Loosmore and Harding (1961) consisted of disruption of normal hepatic architecture, enlarged nuclei, bile stasis, and increased hepatic lipids. Increased fibrous connective tissue was the most pronounced lesion observed in subacute to chronic aflatoxicosis. Gagne *et al.* (1968) reported hepatic lesions of karyomegaly, cytoplasmic degeneration, bile ductule proliferation, necrosis, and hemorrhage in swine fed high levels of aflatoxin (450 μg/kg to 810 μg/kg) for an extended period of time, as shown in Figure 2. Sisk and Carlton (1972) demonstrated similar acute lesions in young pigs fed protein deficient diets and given aflatoxin B_1 (AFB_1).

In poultry, the duckling is the most sensitive species to aflatoxin, followed by turkey poult, pheasant, chicken, and quail (Theron, 1965; Newberne, 1974; Bryden and Cumming, 1980). In ducklings, 100 μg AFB_1 caused severe periportal liver cell necrosis and hemorrhage within 12 hr. Bile ductule hyperplasia developed in 48–72 hr after exposure. Prolonged exposure resulted in marked hepatic nodular hyperplasia, fibrosis, and cholangiocarcinoma or hepatic carcinoma (Theron, 1965; Newberne, 1973). Chickens consuming 1.5 mg/kg aflatoxin developed hepatic fibrosis, nodular regeneration, petechial hemorrhage, and bile duct proliferation (Newberne, 1973). Hoerr *et al.* (1986) reported pleomorphic hepatocytes, necrosis, and disruption of the hepatocellular plates in chickens fed 100–200 μg/kg AFB_1 for 35 days. In addition, occasionally heterophils and

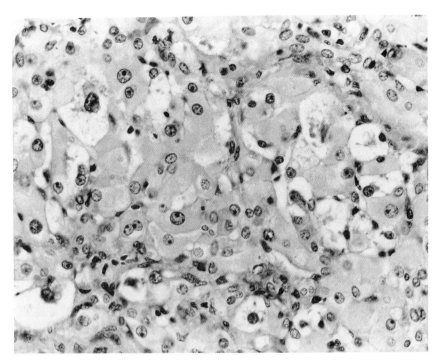

FIGURE 2 Photomicrograph of the liver from a pig fed 0.8 mg aflatoxin/kg feed for 10 weeks. Note karyomegaly, vacuolization of hepatocytes, and pseudolobulation. Hematoxylin and eosin stain; x730.

mononuclear cells were observed in the portal areas in chickens fed 400 and 800 μg/kg AFB$_1$ for 35 days. Turkey poults fed 250–500 μg/kg AFB$_1$ for several weeks developed enlarged gall bladders, fatty hepatic changes, bile duct epithelium hyperplasia, and disorientation of hepatic cords into nodular hyperplasia or tubule-like structures (Pier and Heddleston, 1970; Newberne, 1973).

Long-term consumption of aflatoxin can result in hepatic fibrosis, tumors, and fetal anomalies. Hepatic carcinomas have been reported in birds, ferrets, trout, pigs, sheep, rats, and humans after long-term aflatoxin consumption (Wogan, 1973; Rodricks and Stoloff, 1977). Ryan et al. (1979) discussed the evidence for alfatoxin consumption in the pathogenesis of Reye's syndrome in children. Stillbirths and an increased incidence of tumors were reported by Goerttler et al. (1980) in the F$_1$ generation of adult rats fed aflatoxin diets. The adults also had numerous hepatic tumors. Shalkop and Ambroecht (1974) found nodular adenomas, hepatic cell hyperplasia, focal cystic bile duct proliferation, and extramedullary hepatopoiesis in the livers of sows fed aflatoxin for 30 months. Stunted piglets have been reported from sows fed aflatoxins (Shalkop and Ambroecht, 1974).

CLINICAL PATHOLOGY

The various biological effects of aflatoxin on animals can be summarized by the following statements. Aflatoxin causes an alteration in carbohydrate metabolism and an impairment of lipid transport, resulting in decreased glucose levels and accumulations of lipid within hepatocytes (Wogan 1973; Rodricks and Stoloff, 1977; Heathcote and Hibbert, 1978; Naber and Wallace, 1979). AFB_1 has been shown to alter mitochondrial function by inhibiting the Krebs cycle and the electron transport system and by uncoupling oxidative phosphorylation (Clifford and Rees, 1967; Naber and Wallace, 1979). This compound also has been shown to inhibit RNA polymerase action and to modify the DNA template (Sisk and Carlton, 1972). AFB_1 requires metabolic activation by the microsomal mixed function oxidase system to form the active 8,9-epoxide compound. The epoxide is conjugated readily and thus detoxified by sulfhydryl compounds. The metabolism and detoxification of aflatoxin are influenced by species, sex, age, nutrition, and hormonal status of the animal. These same factors influence the sensitivity of the animals to aflatoxins (Heathcote and Hibbert, 1978).

Alterations in serum enzyme and hematologic parameters have been reported in both natural and experimental exposure to aflatoxins. Table 2 summarizes these reports. The liver is the primary organ affected by aflatoxins. Therefore, enzymes originating from this organ often are used to evaluate toxicity in the animal. However, alterations vary with dose, duration, and species. For example, acute aflatoxicosis in swine is characterized by increases in alkaline phosphatase, isocitric dehydrogenase, aspartate aminotransferase, prothombin time, and sorbitol dehydrogenase (Cysewski et al., 1968; Wyllie and Morehouse, 1978; Miller et al., 1982). Chronic aflatoxicosis also results in increased activity of the liver enzymes. Sisk et al. (1968) also reported changes in the packed cell volume, hemoglobin, and bilirubin with chronic porcine aflatoxicosis but, in general, the common hematologic tests were useful only when the disease was advanced.

IMMUNE SYSTEM

Aflatoxin impairs the cellular and humoral immune system, making animals more susceptible to bacterial, viral, fungal, and parasitic diseases (Rodricks and Stoloff, 1977; Cysewski et al., 1978; Miller et al., 1978; Boulton et al., 1980; Thurston et al., 1986; Harvey et al., 1988). Pier (1973a) pointed out that animals consuming aflatoxin often succumb to infectious diseases. Therefore the initiating factor (aflatoxin) often is overlooked in the final diagnosis. Lanza et al. (1982) reported a variation in the susceptibility to aflatoxin in relation to sex and genetics in chickens. This variation may explain, in part, conflicting reports on the effects of aflatoxin on antibody production in animals (Chang and Hamilton, 1979; Boulton et al., 1980; Ubosi et al., 1985). Low concentrations of aflatoxin have been shown to reduce the resistance of chickens to *Pasteurella, Salmonella,*

TABLE 2 Clinicopathologic Responses Caused by Aflatoxin Exposure in Animals[a]

Parameter tested[b]	Porcine	Bovine	Equine	Caprine	Rabbit	Avian	Monkey	Rat	Guinea pig
Aspartate aminotransferase	I	I	I	I	I		I		
Gamma glutamyltransferase	I	I	I						
Serum alkaline phosphatase	I	I						I	
Isocitrate dehydrogenase	I							I	I
Sorbital dehydrogenase	I	I	I						I
Ornithine dehydrogenase	I	I	I						
Prothrombin time	I		I						
Hemoglobin	I								
Bilirubin	I	I	I	I	I		I	I	
White blood cell count		I							
Packed cell volume	I	I,D[c]	I	D		D			
Iditol dehydrogenase			I						
Arginase			I						
Alanine aminotransferase			I		I				
Creatinine kinase	D								
Blood urea nitrogen	D								
Total protein	D		D	D	D	D			
Albumin	D	D	D	D	D				
Cholesterol			D			D			

[a]References: Aller et al., 1981; Alpert et al., 1970; Asquith et al., 1980; Bortell et al., 1983; Brown et al., 1981; Brucato et al., 1986; Clark et al., 1980, 1982; Clifford and Rees, 1967; Colvin et al., 1984; Cysewski et al., 1968, 1982; Gagne et al., 1968; Gumbmann and Williams, 1969; Harding et al., 1963; Harvey et al., 1988; Hsieh et al., 1971; Lanza et al., 1982; McKenzie et al., 1981; Miller et al., 1981, 1982; Osuna et al., 1977; Sisk et al., 1968; Thurston et al., 1980; Wyllie and Morehouse, 1978.

[b]I, Increased; D, decreased.

[c]Brucato et al., (1986) reported increased and McKenzie et al. (1981) reported decreased PCV with aflatoxin exposure.

Coccidia, and *Candida* and to increase the severity of swine dysentery, *Erysipelothrix rhusiopathiae* infection, and salmonellosis in swine (Pier and Heddleston, 1970; Richard *et al.,* 1975, 1978; Cysewski *et al.,* 1978; Miller *et al.,* 1978; Joens *et al.,* 1981). Bovine and swine peripheral blood lymphocytes showed a decreased response to phytomitogens and specific antigens when exposed to aflatoxin either *in vivo* or *in vitro* (Paul *et al.,* 1977; Miller *et al.,* 1978). Aflatoxin was shown by Miller *et al.* (1978) and Pier *et al.* (1977) to cause decreased cutaneous hypersensitivity in swine and guinea pigs. Lowered antibody response was observed in mice, chickens, and swine fed aflatoxins (Annau *et al.,* 1964; Pier and Heddleston, 1970; Pier, 1973a; Richard *et al.,* 1975; Naber and Wallace, 1979). Several investigators have shown that swine fed aflatoxin-contaminated diets have decreased levels of albumin (alpha and beta) and increased levels of gamma globulins (Annau *et al.,* 1964; Richard *et al.,* 1975; Naber and Wallace, 1979). Complement activity and C_4 were depressed in guinea pigs fed aflatoxin (Thurston *et al.,* 1980). In cows, an effect on complement-dependent and -independent serum bacteriostatic activity was noted by Thurston *et al.* (1986).

The time of consumption of aflatoxin in comparison to the exposure by infectious agents or vaccination is critical. Pier and Heddleston (1970) stated that, to affect acquired resistances, aflatoxin had to be consumed while the body was exposed to the antigen or while the body was producing the immune response. Vaccination response or susceptibility to infectious agents after aflatoxin exposure was variable depending on the recovery rate from the aflatoxin exposure (Pier, 1973b; Osuna *et al.,* 1977; Boulton *et al.,* 1980; Ubosi *et al.,* 1985). Pier *et al.* (1980) concluded that the native and acquired resistance of animals may be impaired at levels of toxin exposure that produce no clinical mycotoxicosis but render the animals predisposed to infectious agents.

OTHER DISEASES

Several researchers have investigated the effects of aflatoxin on various diseases in animals. Because aflatoxin alters the cellular and humoral immune system, animals are generally more susceptible to infectious diseases. Some were shown to be more susceptible to salmonellosis, dysentery, and *Erysipelothrix* (Miller *et al.,* 1978; Joens *et al.,* 1981). An increased number of liver flukes and incidences of pneumonia were associated with aflatoxin consumption in cattle (Osuna *et al.,* 1977). Aflatoxin consumption in chickens reportedly caused a decrease in the breaking strength of bone, which may contribute to bone abnormalities known as "field rickets" (Huff *et al.,* 1980).

However, aflatoxin does not result in increased severity of all diseases. In a study by Pugh *et al.* (1984), the consumption of aflatoxin did not have an influence on the pathogenesis of infectious bovine keratoconjunctivitis although clinical signs of aflatoxicosis were observed in the animals. Brown *et al.* (1981)

also reported no exacerbations of acute clinical mastitis when aflatoxin was consumed by cows that had existing bacterial intramammary infections. Numbers of bacterial populations did increase in the milk, however.

AFLATOXIN INTAKE AND DISEASE RISK

How much aflatoxin can be tolerated in an animal's diet? This question has no easy answer because aflatoxin affects different animals in different ways. Young animals are usually more sensitive than older animals (Newberne, 1973). The question can be separated into two logical questions. First, what is the maximum aflatoxin content that will not affect production economically or result in unacceptable tissue residues? Second, what is the maximum aflatoxin amount that can be fed without inducing clinical symptoms of aflatoxicosis? For broiler chickens, Hamilton (1987) claimed that the economic threshold was below 10 μg/kg aflatoxins. Dairy cows are limited to feed with a maximum of 20 μg/kg total aflatoxin to insure that milk contains below 0.5 μg/kg aflatoxin M_1, the current Food and Drug Administration (FDA) guideline. For other animals, the economic threshold may be ~ 100 μg/kg because demonstrating economic or biological effects in otherwise well-nourished animals has been difficult. However, the nutrient status of the animal may affect susceptibility. For example, Sisk and Carlton (1972) experimentally demonstrated detrimental effects at 100 μg/kg in protein-deficient swine.

Ingestion of aflatoxin may have subtle or easily recognized effects on farm animals. Hamilton (1987) reviewed the literature on the minimum effective dose (safe level) of aflatoxin on the growth of young broiler chicks. When the criterion applied was growth inhibition, a diet with 1250 μg/kg aflatoxin had no effect. However, when bruising, immunosuppression, and physiological factors were considered, the minimum effective dose was below 625 μg/kg. Hamilton (1987) compared two poultry operations with the same diet formula and common sources of feed ingredients in which no disease was detected but a 1% difference existed in feed conversion and a 2% difference existed in growth rate. The highest aflatoxin content in the less efficient operation was 30 μg/kg with a 30% frequency, compared with 6 μg/kg and a 2% frequency in the more efficient one. Using these data as well as surveys of other broiler chicken operations, Hamilton (1975) concluded that the minimum effective dose for aflatoxin in broiler chickens was below 10 μg/kg. Therefore, aflatoxin in feed probably begins to affect chickens somewhere between 10 μg/kg and 625 μg/kg; above that level, histological lesions are likely to become frequent.

Aflatoxin toxicity is documented easily when histopathological changes are observed. Depressed feed intake and low weight gain were observed in swine when total aflatoxin concentrations were 525 μg/kg feed. No gross pathology and only mild hepatocellular changes were seen after the test animals had been on the diet 35 days (Colvin *et al.*, 1989). The ingestion of smaller doses of aflatoxin, between 200 and 500 μg/kg, rarely results in definite pathology, but swine may show reduced rates of growth, occasional liver lesions, and increased susceptibility to infectious disease (Pier, 1973a; Miller *et al.*, 1981). Wilson *et al.* (1984) analyzed feed from 18 of 54 Georgia swine herds diagnosed with aflatoxicosis by gross and histologic examinations. Of the feed samples, 13 contained more than 360 μg/kg total aflatoxins whereas the 5 other feed samples contained less than 100 μg/kg. Acquiring a truly representative sample of feed to confirm diagnosis was difficult because producers had changed feed or inadequate samples were submitted to the Veterinary Diagnostic Laboratory.

Data on aflatoxin levels and initiation of disease in other animals is difficult to extract from the literature. Therefore, the recommendation to keep aflatoxin contamination under 100 μg/kg is probably acceptable for otherwise healthy, well-nourished, growing and finishing animals. However, 100 μg/kg may be excessive at times, especially for breeding stock.

CONCLUSIONS

1. Ingestion of aflatoxins results in varying clinical signs depending on the amount consumed and the species and age of the animal.
2. Aflatoxin consumption or exposure may make an animal more susceptible to infectious diseases by impairing the immune system of the animal.
3. The symptoms of the secondary infections may obscure the symptoms of aflatoxicosis.
4. Aflatoxin exposure during pregnancy may affect the offspring as well as the adults.

ACKNOWLEDGMENTS

We thank Jan Bolt (Athens) and Darlene Morrison (Tifton) for their help in typing the manuscript and Jeff King, Bonnie Evans, and Susan Wilson for their help in proofreading and verifying the references. D. M. Wilson's portion of the research was supported by State and Hatch funds allocated to the Georgia Agriculture Experiment Station.

REFERENCES

Abdelsalam, E. B., El-Tayeb, A. E., Nor Eldin, A. A., and Abdulmagid, A. M. (1989). Aflatoxicosis in fattening sheep. *Vet. Rec.* **124,** 487–488.

Aller, W. W., Edds, G. T., and Asquith, R. L. (1981). Effects of aflatoxins in young ponies. *Am. J. Vet. Res.* **42,** 2162–2164.

Alpert, E., Serck-Hanssen, A., and Rajagopolan, B. (1970). Aflatoxin-induced hepatic injury in the African monkey. *Arch. Environ. Health* **20,** 723–728.

Angsubhakorn, S., Poomvises, P., Romruen, K., and Newberne, P. M. (1981). Aflatoxicosis in horses. *J. Am. Vet. Med. Assoc.* **178,** 274–278.

Annau, E., Corner, A. H., Magwood, S. E., and Jericho, K. (1964). Electrophoretic and chemical studies on sera of swine following the feeding of toxic groundnut meal. *Can. J. Comp. Med. Vet. Sci.* **28,** 264–269.

Asquith, R. L., Edds, G. T., Aller, W. W., and Bortell, R. (1980). Plasma concentrations of iditol dehydrogenase (sorbitol dehydrogenase) in ponies treated with aflatoxin B₁. *Am. J. Vet. Res.* **41,** 925–927.

Bastianello, S. S., Nesbit, J. W., Williams, M. C., and Lange, A. L. (1987). Pathological findings in a natural outbreak of aflatoxicosis in dogs. *Onderstepoort J. Vet. Res.* **54,** 635–640.

Bortell, R., Asquith, R. L., Edds, G. T., Simpson, C. F., and Aller, W. W. (1983). Acute experimentally induced aflatoxicosis in the weanling pony. *Am. J. Vet. Res.* **44,** 2110–2114.

Boulton, S. L., Dick, J. W., and Hughes, B. L. (1980). Effects of dietary aflatoxin and ammonia-inactivated aflatoxin on Newcastle disease antibody titers in layer-breeders. *Avian Dis.* **26,** 1–5.

Brown, R. W., Pier, A. C., Richard, J. L., and Krogstad, R. E. (1981). Effects of dietary aflatoxin on existing bacterial intramammary infections of dairy cows. *Am. J. Vet. Res.* **42,** 927–933.

Brucato, M., Sundlof, S. F., Bell, J. U., and Edds, G. T. (1986). Aflatoxin B₁ toxicosis in dairy calves pretreated with selenium-vitamin E. *Am. J. Vet. Res.* **47,** 179–183.

Bryden, W. L., and Cumming, R. B. (1980). Observations on the liver of the chicken following aflatoxin B₁ ingestion. *Avian Pathol.* **9,** 551–556.

Bryden, W. L., Cumming, R. B., and Lloyd, A. B. (1980). Sex and strain responses to aflatoxin B₁ in the chicken. *Avian Pathol.* **9,** 539–550.

Burnside, J. E., Sippel, W. L., Forgacs, J., Carll, W. T., Atwood, M. A., and Doll, E. R. (1957). A disease of swine and cattle caused by eating moldy corn. II. Experimental production with pure cultures of molds. *Am. J. Vet. Res.* **18,** 817–824.

Butler, W. H., and Hempsall, V. (1981). Histochemical studies of hepatocellular carcinomas in the rat induced by aflatoxin. *J. Pathol.* **134,** 157–170.

Carnaghan, R. B., Herbert, C. N., Patterson, D. S. P., and Sweasey, D. (1967). Comparative biological and biochemical studies in hybrid chicks. II. Susceptibility to aflatoxin and effects on serum protein constituents. *Br. Poult. Sci.* **8,** 279–284.

Chang, C. F., and Hamilton, P. B. (1979). Impairment of phagocytosis in chicken monocytes during aflatoxicosis. *Poultry Sci.* **58,** 562–566.

Clark, J. D., Jain, A. V., Hatch, R. C., and Mahaffey, E. A. (1980). Experimentally induced chronic aflatoxicosis in rabbits. *Am. J. Vet. Res.* **41,** 1841–1845.

Clark, J. D., Hatch, R. C., Jain, A. V., and Weiss, R. (1982). Effect of enzyme inducers and inhibitors and glutathione precursor and depleter on induced acute aflatoxicosis in rabbits. *Am. J. Vet. Res.* **43,** 1027–1033.

Clifford, J. I., and Rees, K. R. (1967). The action of aflatoxin B₁ on the rat liver. *Biochem. J.* **102,** 65–75.

Colvin, B. M., Harrison, L. R., Gosser, H. S., and Hall, R. F. (1984). Aflatoxicosis in feeder cattle. *J. Am. Vet. Med. Assoc.* **184,** 956–958.

Colvin, B. M., Sangster, L. T., Haydon, K. D., Beaver, R. W., and Wilson, D. M. (1989). Effect of a high affinity aluminosilicate sorbent on prevention of aflatoxicosis in growing pigs. *Vet. Hum. Toxicol.* **31,** 46–48.

Coppock, R. W., Reynolds, R. D., Buck, W. B., Jacobsen, B. J., Ross, S. C., and Mostrom, M. S. (1989). Acute aflatoxicosis in feeder pigs, resulting from improper storage of corn. *J. Am. Vet. Med. Assoc.* **195,** 1380–1381.

Cysewski, S. J., Pier, A. C., Engstrom, G. W., Richard, J. L., Dougherty, R. W., and Thurston, J. R. (1968). Clinical pathological features of acute aflatoxicosis of swine. *Am. J. Vet. Res.* **29,** 1577–1590.

Cysewski, S. J., Wood, R. L., Pier, A. C., and Baetz, A. L. (1978). Effects of aflatoxin on the development of acquired immunity to swine erysipelas. *Am. J. Vet. Res.* **39,** 445–448.

Cysewski, S. J., Pier, A. C., Baetz, A. L., and Cheville, N. F. (1982). Experimental equine aflatoxicosis. *Toxicol. Appl. Pharmacol.* **65**, 354–365.

Deo, M. G., Dayal, Y., and Ramalingaswami, V. (1970). Aflatoxins and liver injury in the rhesus monkey. *J. Pathol.* **101**, 47–56.

Edds, G. T. (1973). Acute aflatoxicosis: A review. *J. Am. Vet. Med. Assoc.* **162**, 304–309.

Gagne, W. E., Dungworth, D. L., and Moulton, J. E. (1968). Pathologic effects of aflatoxin in pigs. *Pathol. Vet.* **5**, 370–384.

Goerttler, K., Lohrke, H., Schweizer, H., and Hesse, B. (1980). Effects of aflatoxin B$_1$ on pregnant inbred Sprague–Dawley rats and their F$_1$ generation. A contribution to transplacental carcinogenesis. *J. Natl. Cancer Inst.* **64**, 1349–1354.

Gumbmann, M. R., and Williams, S. N. (1969). Biochemical effects of aflatoxin in pigs. *Toxicol. Appl. Pharmacol.* **15**, 393–404.

Hamilton, P. B. (1975). Determining safe levels of mycotoxins. *J. Food Prot.* **47**, 570–575.

Hamilton, P. B. (1987). Aflatoxicosis in farm animals. *In* "Aflatoxin in Maize: A Proceedings of the Workshop" (M. S. Zuber, E. B. Lillehoj, and B. L. Renfro, eds.), pp. 51–57. CIMMYT, Mexico City, Mexico.

Harding, J. D. J., Done, J. T., Lewis, G., and Allcroft, R. (1963). Experimental groundnut poisoning in pigs. *Res. Vet. Sci.* **4**, 217–229.

Harvey, R. B., Huff, W. E., Jubena, L. F., Corrier, D. E., and Phillips, T. D. (1988). Progression of aflatoxicosis in growing barrows. *Am. J. Vet. Res.* **49**, 482–487.

Hatch, R. C., Clark, J. D., Jain, A. V., and Mahaffey, E. A. (1971). Experimentally induced acute aflatoxicosis in goats treated with ethyl maleate glutathione precursors, or thiosulfate. *Am. J. Vet. Res.* **40**, 505–511.

Hatch, R. C., Clark, J. D., Jain, A. V., and Weiss, R. (1982). Induced acute aflatoxicosis in goats: Treatment with activated charcoal or dual combinations of oxytetracycline, stanozolol, and activated charcoal. *Am. J. Vet. Res.* **43**, 644–648.

Heathcote, J. G., and Hibbert, J. R. (eds.) (1978). "Aflatoxins: Chemical and Biological Aspects." Elsevier Science, New York.

Hoerr, F. J., D'Andrea, G. H., Giambrone, J. J., and Panangala, V. S. (1986). Comparative histopathological changes in aflatoxicosis. *In* "Diagnosis of Mycotoxicoses" (J. L. Richard and J. R. Thurston, eds.), pp. 179–189. Martinus Nijhoff, Boston.

Huff, W. E., Doerr, J. A., Hamilton, P. B., Hamann, D. D., Peterson, R. E., and Ciegler, A. (1980). Evaluation of bone strength during aflatoxicosis and ochratoxicosis. *Appl. Environ. Microbiol.* **40**, 102–107.

Joens, L. A., Pier, A. C., and Cutlip. R. C. (1981). Effects of aflatoxin consumption on the clinical course of swine dysentery. *Am. J. Vet. Res.* **42**, 1170–1172.

Lanza, G. M., Washburn, K. W., Wyatt, R. D., and Marks, H. L. (1982). Genetic variation of physiological response to aflatoxin in *Gallus domesticus. Theor. Appl. Genet.* **63**, 207–212.

Liggett, A. D., Colvin, B. M., Beaver, R. W., and Wilson, D. M. (1986). Canine aflatoxicosis: A continuing problem. *Vet. Hum. Toxicol.* **28**, 428–430.

Loosmore, R. M., and Harding, J. D. J. (1961). A toxic factor in Brazilian groundnut causing liver damage in pigs. *Vet. Rec.* **73**, 1362–1364.

Loosmore, R. M., and Markson, L. M. (1961). Poisoning of cattle by Brazilian ground nut meal. *Vet. Rec.* **73**, 813–814.

McKenzie, R. A., Blaney, B. J., Connole, M. D., and Fitzpatrick, L. A. (1981). Acute aflatoxicosis in calves fed peanut hay. *Aust. Vet. J.* **57**, 284–286.

Marasas, W. F. O., Kellerman, T. S., Gelderblom, W. C. A., Coetzer, J. A. W., Theil, P. G., and Van Derlugt, T. J. (1988). Leukoencephalomalacia in a horse induced by fumonisin B$_1$ isolated from *Fusarium moniliforme. Onderstepoort J. Vet. Res.* **55**, 197–203.

Miller, D. M., Stuart, B. P., Crowell, W. A., Cole, J. R., Goven, A. J., and Brown, J. (1978). Aflatoxicosis in swine: Its effect on immunity and relationship to salmonellosis. *Am. Assoc. Vet. Lab. Diagnost.* **21**, 135–146.

Miller, D. M., Stuart, B. P., and Crowell, W. A. (1981). Experimental aflatoxicosis in swine: Morphological clinical pathological results. *Can. J. Comp. Med.* **45**, 343–351.

Miller, D. M., Crowell, W. A., and Stuart, B. P. (1982). Acute aflatoxicosis in swine: Clinical pathology, histopathology, and electron microscopy. *Am. J. Vet. Res.* **43**, 273–277.

Naber, E. C., and Wallace, H. D. (eds.) (1979). "Interactions of Mycotoxins in Animal Production," pp. 3–76. National Academy of Science, Washington, D.C.

Newberne, P. M. (1973). Chronic aflatoxicosis. *J. Am. Vet. Med. Assoc.* **163**, 1262–1269.

Newberne, P. M. (1974). Mycotoxins: Toxicity, carcinogenicity, and the influence of various nutritional conditions. *Environ. Health Perspect.* **9**, 1–32.

Newberne, P. M., and Butler, W. H. (1969). Acute and chronic effects of aflatoxin on the liver of domestic and laboratory animals. *Cancer Res.* **29**, 236–250.

Osuna, O., Edds, G. T., and Blankespoor, H. D. (1977). Toxic effects of aflatoxin B_1 in male holstein calves with prior infection by flukes (*Fasciola hepatica*). *Am. J. Vet. Res.* **38**, 341–349.

Patterson, D. S. P. (1973). Metabolism as a factor in determining the toxic action of the aflatoxins in different animal species. *Food Cosmet. Toxicol.* **11**, 287–294.

Paul, P. S., Johnson, D. W., Mirocha, C. J., Soper, F. F., Thoen, C. O., Muscoplat, C. C., and Weber, A. F. (1977). *In vitro* stimulation of bovine peripheral blood lymphocytes: Suppression of phytomitogen and specific antigen lymphocyte responses by aflatoxin. *Am. J. Vet. Res.* **38**, 2033–2035.

Pier, A. C. (1973a). An overview of the mycotoxicoses in domestic animals. *J. Am. Vet. Med. Assoc.* **163**, 1259–1261.

Pier, A. C. (1973b). Effects of aflatoxin on immunity. *J. Am. Vet. Med. Assoc.* **163**, 1268–1269.

Pier, A. C., and Heddleston, K. L. (1970). The effect of aflatoxin on immunity in turkeys. I. Impairment of actively acquired resistance to bacterial challenge. *Avian Dis.* **14**, 797–809.

Pier, A. C., Richtner, R. E., and Cysewski, S. J. (1977). Effects of aflatoxin on the cellular immune system. *Ann. Nutr. Alim.* **31**, 781–788.

Pier, A. C., Richard, J. L., and Thurston, J. R. (1980). Effects of aflatoxin on the mechanisms of immunity and native resistance. *In* "Medical Mycology" (H. J. Preusser, ed.), pp. 301–309. Gustav Fischer Verlag, Stuttgart, Germany.

Pugh, G. W., Richard, J. L., Kopecky, K. E., and McDonald, T. J. (1984). Effects of aflatoxin ingestion on the development of *Moraxella bovis* infectious bovine keratoconjunctivitis. *Cornell Vet.* **74**, 96–110.

Richard, J. L., Thurston, J. R., and Pier, A. C. (1975). Mycotoxin-induced alterations of immunity. *In* "Microbiology 1975" (D. Schlessinger, ed.), pp. 388–396. American Society of Microbiologists, Washington, D.C.

Richard, J. L., Thurston, J. R., and Pier, A. C. (1978). Effects of mycotoxins on immunity. *In* "Toxins: Animal, Plant and Microbial" (P. Rosenberg, ed.), pp. 801–817. Pergamon Press, New York.

Rodricks, J. V., and Stoloff, L. (1977). Aflatoxin residues from contaminated feed in edible tissues of food-producing animals. *In* "Mycotoxins in Human and Animal Health" (J. V. Rodricks, C. W. Hesseltine, and M. A. Mehlman, eds.), pp. 67–69. Pathotox, Park Forest South, Illinois.

Ryan, N. J., Hogan, G. R., Hayes, A. W., Unger, P. D., and Siraj, M. Y. (1979). Aflatoxin B_1: Its role in the etiology of Reye's syndrome. *Pediatrics* **64**, 71–75.

Samarajeewa, U., Arseculeratne, S. N., and Tennekoon, G. E. (1975). Spontaneous and experimental aflatoxicosis in goats. *Res. Vet. Sci.* **19**, 269–277.

Schmidt, R. E., and Panciera, R. J. (1980). Effects of aflatoxin on pregnant hamsters and hamster foetuses. *J. Comp. Pathol.* **90**, 339–347.

Shalkop, W. T., and Armbroecht, B. H. (1974). Carcinogenic response of brood sows fed aflatoxin for 28 to 30 months. *Am. J. Vet. Res.* **35**, 623–627.

Singhal, V., Mathur, M., Ramalingaswami, V., and Deo, M. G. (1981). Aflatoxin B_1-induced injury in monkeys. *Arch. Pathol. Lab. Med.* **105**, 105–108.

Sisk, D. B., and Carlton, W. W. (1972). Effects of dietary protein concentration on response of miniature swine to aflatoxins. *Am. J. Vet. Res.* **33**, 107–114.

Sisk, D. B., Carlton, W. W., and Curtin, T. M. (1968). Experimental aflatoxicosis in young swine. *Am. J. Vet. Res.* **29,** 1591–1602.

Theron, J. J. (1965). Acute liver injury in ducklings as a result of aflatoxin poisoning. *Lab. Invest.* **14,** 1586–1603.

Thornburg, L. P., and Raisbeck, M. F. (1988). A study of canine hepatobiliary diseases, Part 9: Hemolytic disease, mycotoxicosis, and pregnancy toxemia in the bitch. *Comp. Anim. Pract.* **2,** 13–17.

Thurston, J. R., Baetz, A. L., Cheville, N. F., and Richard, J. L. (1980). Acute aflatoxicosis in guinea pigs: Sequential changes in serum proteins, complement, C_4, and liver enzymes and histopathologic changes. *Am. J. Vet. Res.* **41,** 1272–1276.

Thurston, J. R., Cook, W., Driftmier, K., Richard, J. L., and Sacks, J. M. (1986). Decreased complement and bacteriostatic activities in the sera of cattle given single or multiple doses of aflatoxin. *Am. J. Vet. Res.* **47,** 846–849.

Ubosi, C. O., Gross, W. B., Hamilton, P. B., Ehrich, M., and Siegel, P. B. (1985). Aflatoxin effects in white leghorn chickens selected for response to sheep erythrocyte antigen. 2. Serological and organ characteristics. *Poultry Sci.* **64,** 1071–1076.

Vesonder, R., Haliburton, J., Stubblefield, R., Gilmore, W., and Peterson, S. (1991). *Aspergillus flavus* and aflatoxins B_1, B_2, and M_1 in corn associated with equine death. *Arch. Environ. Contam. Toxicol.* **20,** 151–153.

Wilson, B. J., Teer, P. A., Barney, G. H., and Blood, F. R. (1967). Relationship of aflatoxin to epizootics of toxic hepatitis among animals in Southern United States. *Am. J. Vet. Res.* **28,** 1217–1230.

Wilson, D. M., Sangster, L. T., and Bedell, D. M. (1984). Recognizing the signs of porcine aflatoxicosis. *Vet. Med. Small An. Clin.* **79,** 974–977.

Wogan, G. N. (1973). Aflatoxin carcinogenesis. *In* "Methods in Cancer Research" (H. Bush, ed.), pp. 309–344. Academic Press, New York.

Wyllie, T. D., and Morehouse, L. G. (eds.) (1978). "Mycotoxic Fungi, Mycotoxins, Mycotoxicoses," pp. 217–277. Marcel Dekker, New York.

17

Aflatoxins in Milk

❖

Hans P. van Egmond

INTRODUCTION

Shortly after the discovery of aflatoxins as feed contaminants, Allcroft and Carnaghan (1963) suggested that aflatoxin residues might occur in milk and other animal products from animals that had ingested aflatoxins with the feedstuff. A toxic factor in the milk of cows fed aflatoxin B_1-contaminated feedstuff produced the same toxic effects in young ducklings as did aflatoxin B_1. The nature of the milk toxin was studied further by De Iongh *et al.* (1964), who showed by thin layer chromatography on silica gel that the toxic factor had a blue fluorescence similar to that of aflatoxin B_1, but had a much lower R_f value. A trivial name, aflatoxin M, was suggested to indicate its original isolation from milk.

The structure of aflatoxin M was elucidated by Holzapfel *et al.* (1966), who found two components that could be separated by paper chromatography that were designated aflatoxins M_1 and M_2 and identified as the 4-hydroxy derivatives of aflatoxins B_1 and B_2, respectively. The chemical structures of aflatoxins M_1 and M_2 are shown in Figure 1.

Considered from the viewpoints of both toxicity and occurrence, aflatoxin B_1 (AFB$_1$) is the most important of the primary aflatoxins (AFB$_1$, AFB$_2$, AFG$_1$, AFG$_2$). Thus, considerable attention was given to aflatoxin M_1 (AFM$_1$) as a principal AFB$_1$ metabolite. A new hydroxy derivative of AFB$_1$ detected in milk (Lafont *et al.*, 1986), aflatoxin M_4 (see Figure 1), seems to co-occur with AFM$_1$ in certain milks in amounts up to $\sim 16\%$ of the AFM$_1$ content. The knowledge

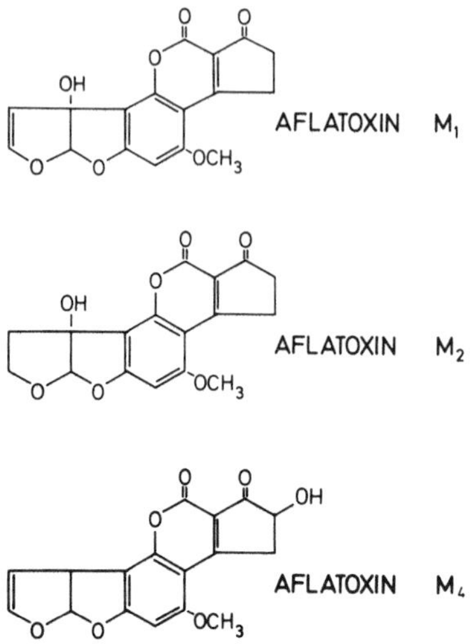

FIGURE 1   Chemical structures of aflatoxins M_1, M_2, and M_4.

about aflatoxin M_4 is so limited, however, that any meaningful assessment of its significance is currently not possible.

The fact that milk, one of the most important and valuable of foods, could harbor aflatoxin(s) initiated a boom of scientific research that continues to this date, and involves various disciplines concerned with the presence of AFM_1 in dairy products. "Carry-over" experiments were done with cows, and attention was focused on toxicity studies of various kinds. Analytical methodologies capable of detecting sub-μg/kg levels of AFM_1 in milk and milk products were developed. Many surveys on the occurrence of the toxin in dairy products were undertaken, and official regulations were established. Studies were done on the stability of AFM_1, the effects of processing, and the possibilities for elimination. In this chapter, these issues are discussed and summarized. More detailed information about AFM_1 in milk and other dairy products can be found in a recently published book on mycotoxins in dairy products (Van Egmond, 1989a).

CONVERSION OF AFLATOXIN B_1 TO AFLATOXIN M_1

Since the early 1960s, various studies have been undertaken to establish quantitatively the extent of conversion of AFB_1 to AFM_1. Most of these studies have been conducted at high (mg range) daily intake levels. Later, some studies were carried out at low (sub-mg range) levels.

At High Aflatoxin Levels

Van der Linde *et al.* (1964) were likely to be the first investigators to study the extent of transmission of AFB_1 into the milk of dairy cows. These researchers

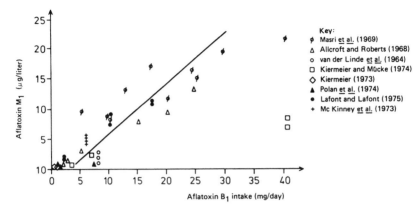

FIGURE 2 Relationship between aflatoxin B_1 intake with feedstuff and aflatoxin M_1 secretion in milk (Sieber and Blanc, 1978). Reprinted with permission of the Editorial Board of the Mitt. Gebiete Lebensm. Hyg.

used cows with high milk yields (28 liters per day) and cows with low milk yields (12 liters per day). Each of the cows was given groundnut meal contaminated with AFB_1 for a period of 18 days. The toxin was detected readily in the milk 12–24 hr after the first AFB_1 ingestion. After a few days, the toxin content in the milk reached a maximum value, although the amount of toxin in the milk was less than 1% of the ingested AFB_1. After cessation of AFB_1 intake, the AFM_1 concentration in the milk declined to an undetectable level after 3 days.

The study by Van der Linde *et al.* (1964) was followed by various other studies on the carry-over of AFB_1 to milk. Reviews are given by Rodricks and Stoloff (1977) and Sieber and Blanc (1978). The data reviewed by Rodricks and Stoloff (1977) showed that the ratios of the concentration of AFB_1 in cattle feed to that of AFM_1 in milk ranged from 34 to 16000, with an average ratio near 300. Sieber and Blanc (1978) presented the conversion data of several studies graphically (Figure 2), and concluded that a linear relationship existed between AFB_1 intake from feedstuff and AFM_1 secretion in milk. These researchers also estimated that the excreted amount of AFM_1 as a percentage of the AFB_1 intake ranged from 0 to 4%, with an average value of 1%.

At Low Aflatoxin Levels

Until 1980, most of the studies were conducted with feedstuffs containing dietary levels of AFB_1 that often were much higher than the existing official limits for aflatoxins in the feed of dairy cattle. Usual legal limits for AFB_1 in dairy rations are 5–20 μg/kg (see Section V,A). Figure 2 is not applicable for these more realistic levels of contamination. In a study by Patterson *et al.* (1980), a group of dairy cows was given a daily diet contaminated with approximately 10 μg/kg AFB_1. Over a period of 7 days, the milk obtained from the herd was analyzed for AFM_1. An average of approximately 2.2% of ingested AFB_1 appeared daily in the milk as AFM_1.

Lafont *et al.* (1980) conducted an experiment with a large group of cows, some in the early and others in the late lactation period. The cows were divided into four groups, each group receiving AFB_1 at 0.09, 0.18, 0.86, and 2.58 mg/day/animal, respectively, compared with 0.15–0.23 mg/day/animal in the study by Patterson *et al.* (1980). Over a period of 12 days, the AFM_1 content in the milk of the various groups of cows was determined. Lafont *et al.* (1980) concluded that the quantity of secreted AFM_1, expressed as a percentage of the parent aflatoxin, averaged 0.22% in the animals in the late lactation period and 0.78% in the cows in the early lactation period.

The differences in results of the studies of Lafont *et al.* (1980) and Patterson *et al.* (1980) make determining a conversion rate for low levels of AFB_1 in feed-stuffs for dairy cattle difficult. The estimated percentages of 0.78% and 2.2% fall within the range of 0–4% carry-over as estimated by Sieber and Blanc (1978).

Given the limited data on conversion at low aflatoxin levels and the fact that milk yields have increased considerably in recent years, Veldman *et al.* (1992) conducted a study in which dairy cows in early lactation (2–4 weeks) and dairy cows in late lactation (34–36 weeks) were given AFB_1 at a level of 34–39 μg/day. After the animals received these diets for 12 days, AFM_1 excretion in the milk was determined. The carry-over rate found was 6.2% and 1.8% for cows in early and late lactation, respectively. This difference was caused not only by milk production level but also by differences in liver AFB_1 metabolism. The experiment was followed by another experiment in which eight high (40 kg milk/day) and eight low (16 kg milk/day) milk-yielding cows were fed compound feed contaminated with AFB_1 at intake levels between 7.3 and 56.9 μg a day. Independent of AFB_1 intake, high-producing dairy cows had a higher carry-over than low producing animals (3.8 versus 2.5%). Since dairy cows in the early to mid-lactation period consume large quantities of compound feed, a maximum tolerable level of AFB_1 in dairy feedstuffs should be adjusted to consider the large feedstuff intake to guarantee that AFM_1 levels in milk do not exceed tolerable levels.

Conclusion

The ingestion of AFB_1 in feedstuffs by the dairy cow leads to the excretion of AFM_1 in the milk. Experimentally established carry-over rates for highly contaminated rations average ∼ 1% at the relatively low levels of AFB_1 contamination of feedstuffs that are common; the percentage of AFB_1 found in milk as AFM_1 tends to be higher, perhaps as high as 6%. The percentages vary from animal to animal, from day to day, and from one milking to the next. In addition, quantities seem to depend on the milk yield and the lactation period.

TOXICOLOGY

Several studies have been undertaken of the toxic effects of AFM_1 in laboratory animals. However, in comparison to AFB_1, relatively little is known about the toxicity of AFM_1, primarily because of the difficulty in obtaining sufficient quantities of the pure compound necessary for extensive toxicity testing.

Short-Term Toxicity Studies

The first detection methods for aflatoxins were bioassays with newly hatched ducklings (Allcroft and Carnaghan, 1963). These animals are extremely sensitive to both AFB_1 and AFM_1, with LD_{50} values ranging from 12 to 16 µg per animal (Holzapfel et al. 1966; Purchase, 1967). Histopathological examinations revealed that AFM_1 caused liver lesions similar to those caused by AFB_1. In addition, necrosis of the kidney tubules occurred.

Frémy et al. (1982) compared the effects of oral administration of milk naturally and artificially contaminated with AFM_1 in newly hatched ducklings. These investigators found that the naturally contaminated milk produced fewer lesions than the artificially contaminated milk, suggesting differences in the bioavailability of naturally and artificially occurring AFM_1.

In addition to the 1-day-old duckling studies, some investigations have been done on the acute toxic effects of synthetically produced (racemic) AFM_1 in Fischer rats (Pong and Wogan, 1971). Synthetic (racemic) and natural AFB_1 also were administered in the same study. Both synthetic aflatoxins were lethal to male rats at a single dose of 1.5 mg/kg body weight, whereas natural (non-racemic) AFB_1 was lethal at a dose of 0.6 mg/kg body weight. Changes in the liver parenchymal cells, such as dissociation of ribosomes from the rough endoplasmic reticulum and proliferation of the smooth endoplasmic reticulum, were observed for the racemic and natural toxins. Quantitatively, the structural changes induced by the synthetic racemic toxins at 1.0 mg/kg were indistinguishable from those caused by natural AFB_1 at a dose of 0.5 mg/kg. The data suggest that AFM_1 acts through the same mechanism as AFB_1 in causing acute toxicity and subcellular alterations, and that only the naturally occurring isomer of each aflatoxin is biologically active.

Chronic Toxicity Studies

Two carcinogenicity studies with rainbow trout have been published. In the first study (Sinnhuber et al., 1974) rainbow trout received various experimental diets, including diets that contained AFB_1 at 4 µg/kg feed and diets that contained AFM_1 at 0, 4, 16, 32, and 64 µg/kg feed respectively. The fish were fed the experimental diets continuously for 12 months, after which they were kept on a control diet. Certain groups were held for 20 months to determine the effect of maturation on tumor development. In addition, the effect of limited oral intake of AFM_1 was determined by feeding groups of trout for 5–30 days at an AFM_1 level of 20 µg/kg. Levels of AFM_1 at 4 and 16 µg/kg and AFB_1 at 4 µg/kg produced 13, 60, and 48% incidences of hepatoma respectively, in 12 months. Significant mortality occurred among female trout with AFM_1-induced hepatomas at the time of maturation (16–20 months), in contrast to males who showed no mortality. Trout that received AFM_1 at levels of 20 µg/kg during 5–30 days developed a 3–12% incidence of hepatoma in 12 months. The investigators concluded that AFM_1 is a potent liver carcinogen, but less potent than AFB_1.

Canton *et al.* (1975) conducted a similar study with groups of rainbow trout fed diets containing AFM_1 at 0, 5.9, and 27.3 μg/kg and AFB_1 at 5.8 μg/kg for 16 months. Necropsy of fish killed after 5, 9, and 12 months revealed ceroid degeneration of the liver in all three groups and in the control group, but no tumors or preneoplastic changes. Autopsy of survivors at month 16, however, revealed 13% with hepatocellular carcinoma and 23% with hyperplastic nodules in the group fed 5.8 μg/kg AFB_1, and 2% with hepatocellular carcinoma and 6% with hyperplastic nodules in the group fed 27.3 μg/kg AFM_1. Comparing their results with those of the study by Sinnhuber *et al.* (1974), these investigators concluded that the different trout strains used may have been responsible for the differences in results, but they confirmed the finding of Sinnhuber *et al.* (1974) that AFM_1 is less carcinogenic than AFB_1.

Probably the most important carcinogenicity studies with AFM_1 are those carried out with rats. In one study (Wogan and Paglialunga, 1974), weanling Fischer rats were given 25 μg synthetic AFM_1 per day by intubation 5 days a week for 8 consecutive weeks. For comparison, a second group of rats was given natural AFB_1 at the same concentration and under similar conditions. A control group also was included in the study. The experiment was terminated when animals died or began to show clinical deterioration. Only one rat (3%) dosed with AFM_1 developed a hepatocellular carcinoma, whereas 28% had liver lesions diagnosed as early or advanced preneoplastic lesions. All rats treated with AFB_1 developed tumors, whereas the controls showed no significant liver pathology. AFM_1 was concluded to have a much lower carcinogenic potency than AFB_1.

Cullen *et al.* (1987) described a carcinogenicity study with natural AFM_1 in Fischer rats. Five test groups were maintained on diets containing natural AFM_1 at 0, 0.5, 5, and 50 μg/kg and AFB_1 at 50 μg/kg. The animals were sacrificed between 18 and 22 months of age. Between 19 and 21 months, hepatocellular carcinomas were detected in 5% of the rats and neoplastic nodules were found in 16% of the rats fed AFM_1 at 50 μg/kg. No nodules or carcinomas were observed in the lower AFM_1 dose groups. Of the rats fed the diet containing AFB_1 at 50 μg/kg, 95% developed hepatocellular carcinomas. In the 50 μg/kg AFM_1 group, a few rats also developed intestinal carcinomas, whereas no intestinal neoplasms were observed in any of the other groups. The authors suggested that the greater polarity of AFM_1 compared with AFB_1 might lead to poor absorption from the digestive tract and that this difference might be associated with the higher incidence of intestinal tumors. AFM_1 was concluded to be a hepatic carcinogen, although with a potency 2–10% of that of AFB_1, and an intestinal carcinogen.

Conclusion

The limited animal studies carried out to determine toxicity and carcinogenicity of AFM_1 tend to come to the same qualitative conclusion: AFM_1 has hepatotoxic and carcinogenic properties. Quantitatively, the (sub)acute toxicity of AFM_1 in ducklings and rats seems to be similar to or slightly less than that of AFB_1. The carcinogenicity is probably one to two orders of magnitude less than that of the highly carcinogenic AFB_1.

ANALYSIS

The analytical methodology for AFM_1 usually follows the general pattern for aflatoxin assays. A uniform sample is obtained easily with milk because AFM_1 is distributed evenly throughout the fluid milk. The initial problem that is encountered in milk analyses is the extraction step. Because milk is a complex natural product, AFM_1 is not easily extracted and purified for final assay. A process is needed that separates AFM_1 from milk easily, efficiently, and economically. Once purified extracts are obtained, several possibilities exist to determine the concentration of AFM_1.

Extraction and Clean-Up

AFM_1 is a semipolar component, extractable with solvents such as methanol, acetone, chloroform, or combinations of one or more of these solvents with water. In practice, the choice of solvents depends on the clean-up and determinative steps. In classical methods, chloroform often is used as an extraction solvent, in combination with adsorption chromatography over SiO_2 columns.

Solid-phase extraction and chromatography cartridges such as SEP-PAK® and Baker® have become popular (Stubblefield and Van Egmond, 1989). Milk usually is added directly, or after dilution, to the cartridge; then the cartridge is washed with one or more solvents to remove impurities. The solvent composition subsequently is changed to elute AFM_1 selectively from the cartridge, and the eluate is collected.

The most recent advance in AFM_1 extraction and subsequent clean-up is the use of immunoaffinity (IA) cartridges. These columns are composed of monoclonal antibodies specific for AFM_1 that are immobilized on Sepharose® and packed into small cartridges. In Figure 3, a schematic diagram for IA chromatography for concentration and purification of AFM_1 is given. A milk sample containing AFM_1 first is loaded on the affinity gel column containing antibody against AFM_1 (A). After washing to remove impurities (B), AFM_1 is eluted from the column with methanol (C). IA columns excellently remove contaminants because the AFM_1 antibody specifically recognizes AFM_1, so the column should not adsorb any other materials. Currently, the provisional results of an international collaborative exercise to study the merits of an analysis procedure for AFM_1 using IA chromatography in combination with high performance liquid chromatography (HPLC) have become available (Tuinstra *et al.*, 1993). These results indicate good performance of the new methodology.

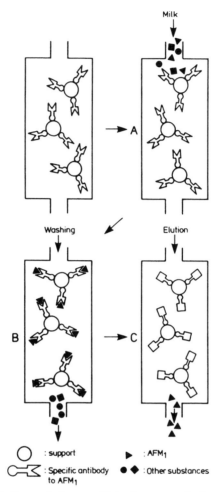

FIGURE 3 Schematic diagram for immunoaffinity chromatography for concentration and purification of aflatoxin M_1. (Frémy and Chu, 1989)

Ultimate Separation and Determination

Extracts of milk that have been cleaned up often are concentrated to lower the limits of quantitation for AFM_1. Matrix components still present in these final extracts can be separated from AFM_1 by chromatographic or immunochemical procedures.

In the 1960s and 1970s, thin layer chromatography (TLC) was used almost exclusively in the determinative step for AFM_1, which readily fluoresces when exposed to UV light. In particular, two-dimensional application of TLC with

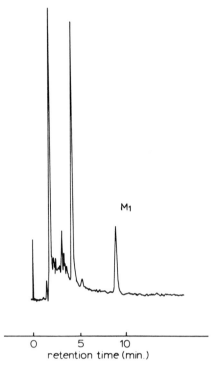

M₁

FIGURE 4 C$_{18}$ reverse phase HPLC chromatogram of an extract of milk powder containing aflatoxin M$_1$ at ~ 0.4 μg/kg. Figure courtesy of E. J. Mulders.

fluorodensitometric detection is a powerful and sensitive separation technique that offers the possibility of very low level determinations of AFM$_1$ (Schuller *et al.*, 1973). In the 1980s, reverse phase HPLC procedures with fluorescence detection (see Figure 4) became more attractive, however, because separations could be accomplished in a matter of minutes and the equipment for HPLC analyses could be automated rather easily. Reported limits of determination for AFM$_1$ in milk with TLC and HPLC range from 0.005 to 0.1 μg/kg. Coincidence of appearance and chromatographic properties in TLC and HPLC does not provide proof that the isolated compound from milk is chemically identical to AFM$_1$. Therefore, several chromatographic methods for determining AFM$_1$ have included a confirmatory step, usually by derivatization with trifluoroacetic acid (Stubblefield and Van Egmond, 1989).

Of growing importance are the immunochemical procedures to determine AFM$_1$, especially enzyme-linked immunosorbent assay (ELISA). Because of their simplicity and specificity, ELISAs represent a great potential for rapid determination of AFM$_1$ in milk products (Frémy and Chu, 1989), although their performance characteristics have not been evaluated fully. ELISAs are partic-

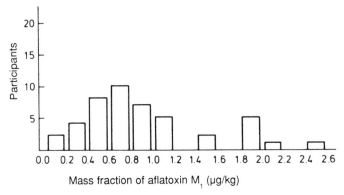

FIGURE 5 Frequency distribution of analytical results for spray-dried milk in the Check Sample Programme for aflatoxin M_1. Reprinted with permission from Friesen and Garren (1982).

ularly useful for rapid screening of milk, and some countries apply commercial test kits for semiquantitative monitoring of milk for AFM_1.

Reference Materials

The availability of reliable methods of analysis is no guarantee of accurate results. "Check sample" programs for AFM_1 in milk powder have shown that wide variability in results should be considered the norm rather than the exception (Friesen and Garren, 1982). The large scatter of analytical results (Figure 5) is of little comfort to those people who must pay for the measurements or must base potentially important decisions on them. This situation prompted the Community Bureau of Reference (BCR) of the European Commission to develop certified reference materials for AFM_1. The activities were undertaken by the BCR in collaboration with a group of European laboratories experienced in analysis for AFM_1. The efforts were successful, and four full-cream milk powder reference materials certified for their AFM_1 content at < 0.05, ~ 0.1, ~ 0.3, and ~ 0.8 µg/kg, respectively, became available (Van Egmond, 1988).

Conclusion

Many methods have become available for the determination of AFM_1 in milk. In particular, solid-phase extraction and IA chromatography cartridges offer good possibilities for efficient clean up. Both TLC and HPLC are adequate techniques to separate and determine AFM_1 in extracts of milk. ELISA is experiencing greater use, since this technique is easy to perform and offers good possibilities for rapid screening and semiquantitative determination.

The reliability of AFM_1 analytical data can be improved by systematically making use of certified milk powder reference materials.

REGULATION AND OCCURRENCE

AFM_1 enters human food by indirect contamination after consumption of AFB_1-contaminated feeds by animals. Many feedstuff ingredients may contain aflatoxins; groundnut meal, cottonseed meal, and maize meal are among the most important. To control AFM_1 in milk and milk products, specific regulations for aflatoxins in feedstuffs for dairy cattle have been established. Surveillance programs are carried out for aflatoxins in feed for dairy cattle and for AFM_1 in milk and milk products. A summary of relevant existing regulations is given here, as is a summary of selected data on occurrence of AFM_1 in milk. Full details have been published elsewhere (Van Egmond, 1989a).

Regulation

An inquiry made in 1987 at the request of the Food and Agriculture Organization (FAO) showed 56 countries with known mycotoxin legislation (Van Egmond, 1989c). Some of these countries had limits for AFB_1 or for the sum of the aflatoxins B_1, B_2, G_1, and G_2. Most of these countries had an acceptable level for AFB_1 at 10 µg/kg (dairy) feedstuff. This group mainly consists of the 12 European Community (EC) countries that must follow a common directive. By the end of 1991, the EC limit was tightened further to 5 µg/kg to concur with the trend in western European countries to establish limits for AFM_1 at a level of 0.05 µg/kg milk or even less (Commission of the European Communities, 1991). The current acceptable levels for AFM_1 in liquid milk (not for specified milk-containing products, such as infant foods) as they existed worldwide in 1990 are presented in the form of a frequency distribution in Figure 6. Two major tolerance peaks occurred—0.05 µg/kg (some western European countries) and 0.5 µg/kg (American and eastern European countries). A study on the rationales for the various limits established (Stoloff *et al.,* 1991) revealed that only the United States and The Netherlands provided some information. The United States rationale was said to be based on the tenuous concept that exposure of (young) individuals to AFM_1 should be kept at a minimum, without jeopardizing the continued supply of milk. The Netherlands based its rationale on a risk determination extrapolated from animal studies. Apparently, in most countries the scientific basis for regulation of AFM_1 is either nonexistent or very weak.

Occurrence of Aflatoxin M_1 in Milk

Surveys Carried Out in the 1970s

The results of various surveys of milk for AFM_1 presence carried out in various countries in the late 1960s and the 1970s have been summarized and published (Brown, 1982). A seasonal trend in milk contamination was noted in a few of these surveys, with lower AFM_1 levels in milk in the summer months. This phenomenon was attributed to the fact that the cows are receiving less

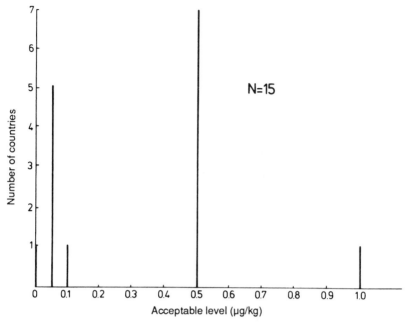

FIGURE 6   Frequency distribution of adopted and proposed "acceptable" levels of aflatoxin M_1 in liquid milk in various countries (circa 1990).

concentrated feeds in the summer when they are grazing. In almost all surveys, positive samples were found with AFM_1 levels exceeding 0.05 μg/kg. In various studies, samples were reported with levels in the range 0.05–0.5 μg/kg, which is in the current range of (proposed) tolerance values for AFM_1 in milk, with the exception of infant milk, for which lower tolerance levels have been mandated.

Surveys Carried Out in the 1980s

The establishment of new regulations in the late 1970s and in the 1980s to control the aflatoxin contents in dairy rations led to the expectation that AFM_1 levels in milk would be reduced in the 1980s. At the same time, newer analytical developments would make detection possible at lower concentrations of AFM_1. A published overview of surveys carried out in 12 countries in the 1980s (Van Egmond, 1989b) showed that the situation for AFM_1 contamination of milk has improved. In general, both the incidence of positive samples and the AFM_1 levels were lower in the 1980s than in the 1970s. In several countries, positive samples could not be found and levels >0.5 μg/kg seldom occurred. Most of the samples of milk taken in the 1980s in various countries had AFM_1 levels in ranges that are acceptable for these products. Note, however, that only a limited number of countries was involved in both the surveys of the 1970s and the 1980s, making drawing firm conclusions difficult.

Conclusion

Several countries have actual or proposed legislation on limits and regulations for aflatoxin(s) in feedstuffs for dairy cattle and AFM_1 in milk and milk products. The differences between legal tolerance levels for AFM_1 in milk vary widely, yet international consistency would be highly desirable. Results of surveys of milk for AFM_1 in the 1970s and those of surveys of milk in the 1980s have shown that the frequency of occurrence and levels of AFM_1 in milk have decreased considerably in recent years. Most of the samples of milk taken in the 1980s in various countries have AFM_1 levels that meet regulatory standards. However, for milk and milk powder intended to be used for infants and infant formulas, some countries adopt very low AFM_1 tolerances that will not be easily enforceable.

STABILITY AND DEGRADATION

Because of the frequent occurrence of AFM_1 in milk, two questions are raised frequently. What happens if such contaminated milk is processed in the regular way by the dairy industry? What can be done to reduce or destroy the AFM_1 in contaminated milk and milk products? Many investigations have been done on these subjects and have been reviewed in detail by Yousef and Marth (1989). Some major studies and conclusions are given in this section.

Fate of AFM_1 during Processing of Milk

The treatments that are common in the dairy industry can be separated into two distinct processes: (1) those that do not involve separation of milk components, such as heat treatment, low temperature storage, and yogurt-preparation, and (2) processes that separate milk components, such as concentration, drying, and cheese and butter production.

Processes that Do Not Separate Milk Components

Several studies have examined the stability of AFM_1 during heat processing treatments such as pasteurization (Allcroft and Carnaghan, 1962) and heating milk directly on a fire for 3–4 hr (Patel *et al.*, 1981). Although the results of the various studies are not highly consistent, most do indicate that heat treatments similar to those used with dairy food do not cause an appreciable change in the amount of AFM_1 in these products.

Studies on the stability of AFM_1 in milk during cool or frozen storage gave variable results, both for cool and for frozen storage (Yousef and Marth, 1989). However, storage of frozen contaminated milk and other dairy products for a few months does not appear to affect AFM_1 content.

The effect of the manufacture of cultured dairy products, such as the preparation of kefir or yogurt, on AFM_1 concentration was studied by Wiseman and Marth (1983). As with the preceding studies, the results varied, but the general trend was that these processes also do not lead to a significant decrease in AFM_1 content.

Processes that Separate Milk Components

Several investigations have been published in which the effects of the removal of water on AFM_1 were studied, including those that involved heat (spray drying, roller drying) as well as freeze-drying. These studies were reviewed by Yousef and Marth (1989). Severe losses of AFM_1 were reported in some studies, whereas other studies revealed that concentrating milk did not affect AFM_1 content substantially.

In the cream manufacturing process, a part of the aqueous phase is separated. Because AFM_1 is a semipolar component, it occurs predominantly in the nonfat fraction, suggesting a lower AFM_1 concentration in the cream than in the milk from which it is made. When making butter, this phenomenon should be even more significant. The few studies that addressed partitioning of AFM_1 during cream and butter processing confirm that a small proportion of AFM_1 is carried over to cream and yet a smaller proportion to butter. No losses of AFM_1 occur, since the remainder of AFM_1 remains in skim milk and buttermilk, respectively.

The manufacture of cheese involves several processes. In the first phase—the conversion of milk into pressed curd—AFM_1 seems not to be degraded since the amount of AFM_1 in whey and curd is approximately the same as in the original milk (Yousef and Marth, 1989). AFM_1 seems to occur predominantly with casein, however, causing the cheese curd to contain a higher concentration than the whey. The association of AFM_1 with casein is also manifested in a higher concentration of AFM_1 in cheese than in the milk from which the cheese is made. Yousef and Marth (1989) expressed the ratio

$$\frac{\text{concentration } AFM_1 \text{ in milk}}{\text{concentration } AFM_1 \text{ in cheese}}$$

as the enrichment factor (EF). From several studies, these researchers conclude that the EF is 2.5–3.3 in many soft cheeses and 3.9–5.8 in hard cheeses. During the second phase of cheese manufacture—the ripening process—some discrepancies in AFM_1 stability were noticed but, in general, AFM_1 did not seem to be degraded during ripening of most cheeses.

Degrading of AFM_1 in Milk

The processes described in the previous section generally do not lead to losses of AFM_1, an observation that is of considerable practical importance. Researchers must determine what can be done to eliminate AFM_1 from milk?

Several possibilities that involve chemical and physical treatments have been investigated to eliminate or inactivate AFM_1 in milk.

Chemical Treatments

Chemicals that have been explored for their ability to degrade AFM_1 are limited to some that are permitted as food additives: sulfites, bisulfites, and hydrogen peroxide (Applebaum and Marth, 1982a,b). Raw milk that was naturally contaminated with AFM_1 was treated with 0.4% potassium bisulfite at 25°C for 5 hr. The concentration of AFM_1 in milk decreased by 45% as a result of this treatment. A higher concentration of bisulfite was less effective in eliminating AFM_1 from milk. AFM_1 in naturally contaminated milk was not affected by the presence of 1% H_2O_2 at 30°C for 30 min. However, use of H_2O_2 plus lactoperoxidase made it possible to reach a $\sim$ 50% reduction of AFM_1 at concentrations of 0.05–0.1%.

Physical Treatments

Physical processes that have been explored to remove AFM_1 from milk include adsorption and radiation. Applebaum and Marth (1982a) investigated the possible adsorption of AFM_1 in milk onto bentonite and found that 5% bentonite in milk adsorbed 89% of AFM_1. Yousef and Marth (1985) studied the effect of ultraviolet radiation with and without addition of H_2O_2. AFM_1 was reduced by 3.6–100%, depending on the length of time the milk was exposed to UV radiation, the volume of treated milk, the presence of H_2O_2 in milk, and other aspects of experimental design.

The chemical and physical treatments described are not readily applicable for the dairy industry, at least not at this time, since little is known about the biological safety, the nutritional value, the functional properties, and the organoleptic qualities of the treated products. Moreover, the costs of the processes may be considerable and prohibitive for large-scale application.

If the destruction or removal of AFM_1 cannot be achieved readily, the toxin can be excluded from milk only by elimination of AFB_1 from the diet of the animals. Because this chapter deals with AFM_1 in milk, these aspects will not be discussed here.

Conclusion

Processing of milk in ways that are common to the dairy industry does not lead to appreciable degradation of AFM_1. Cream and butter contain lower concentrations of AFM_1 than the milk from which these products are made. In contrast, cheese contains much higher concentrations of AFM_1 than the original milk.

Chemical and physical treatments of milk may lead to some degradation of AFM_1. The applicability and cost effectiveness of these techniques require further study.

REFERENCES

Allcroft, R., and Carnaghan, R. B. A. (1962). Groundnut toxicity: *Aspergillus flavus* toxin (aflatoxin) in animal products: Preliminary communication. *Vet Rec.* **74,** 863–864.

Allcroft, R., and Carnaghan, R. B. A. (1963). Groundnut toxicity: An examination for toxin in human food products from animals fed toxic groundnut meal. *Vet. Rec.* **75,** 259–263.

Allcroft and Roberts (1968). Toxic groundmeat meal; The relationship between aflatoxin B_1 intake by cows and excretion of aflatoxin M_1 in milk. *Vet. Rec.* **82,** 116–118.

Applebaum, R. S., and Marth, E. H. (1982a). Use of sulfite or bentonite to eliminate aflatoxin M_1 from naturally contaminated raw whole milk. *Z. Lebensm. Unters. Forsch.* **174,** 303–305.

Applebaum, R. S., and Marth, E. H. (1982b). Inactivation of aflatoxin M_1 in milk using hydrogen peroxide and hydrogen peroxide plus riboflavin or lactoperoxidase. *J. Food Protect.* **45,** 557–560.

Brown, C. A. (1982). Aflatoxin M in milk. *Food Technol. Austral.* **34,** 228–231.

Commission of the European Communities (1991). Commission Directive of 13 February 1991 amending the Annexes to Council Directive 74/63 EEC on undesirable substances and products in animal nutrition (91/126/EEC). *Off. J. Eur. Commun.* **L60,** 16–17.

Canton, J. H., Kroes, R., Van Logten, M. J., Van Schothorst, M., Stavenuiter, J. F. C., and Verhülsdonk, C. A. H. (1975). The carcinogenicity of aflatoxin M_1 in rainbow trout. *Food Cosmet. Toxicol.* **13,** 441–443.

Cullen, J. M., Ruebner, B. H., Hsieh, L. S., Hyde, D. M., and Hsieh, D. S. P. (1987). Carcinogenicity of dietary aflatoxin M_1 in male Fisher rats compared to aflatoxin B_1. *Cancer Res.* **47,** 1913–1917.

De Iongh, H., Vles, R. O., and Van Pelt, J. G. (1964). Investigation of the milk of mammals fed on aflatoxin containing diet. *Nature (London)* **202,** 466–467.

Frémy, J. M., and Chu, F. S. (1989). Immunochemical methods of analysis for aflatoxin M_1. *In* "Mycotoxins in Dairy Products" (H. P. van Egmond, ed.), pp. 97–125. Elsevier Applied Science, London.

Frémy, J. M., Billon, J., Delpech, P., and Meurant, F. (1982). Application of two bioassays for aflatoxin M_1 toxicity evaluation in milk. *Rec. Méd. Vét.* **158,** 461–466. (in French)

Friesen, M. D., and Garren, L. (1982). International Mycotoxin Check Sample Program. Part II. Report on the performance of participating laboratories for the analysis of aflatoxin M_1 in lyophilized milk. *J. Assoc. Off. Anal. Chem.* **65,** 864–868.

Holzapfel, C. W., Steyn, P. S., and Purchase, I. F. H. (1966). Isolation and structure of aflatoxins M_1 and M_2. *Tet. Lett.* **25,** 2799–2803.

Kiermeier, F., and Mücke, W. (1974). Einfluss der Qualität des Futtermittels auf den Aflatoxin-gehalt der Milch. XIX *Int. Milchw. Kongres ID,* 114–115.

Kiermeier, F., Reinhardt, V., and Behringer, G. (1975). Zum Vorkommen von Aflatoxinen in Rohmilch. *Dtsch. Lebensm. Rdsch.* **71,** 35–38.

Lafont, P., and Lafont, J. (1975). Elimination d'aflatoxine par la mamelle chez la vache. *Cah. Nutr. Diét.* **10,** 55–57.

Lafont, P., Lafont, J., Mousset, S., and Frayssinet, C. (1980). Etude de la contamination du lait de vache lors de l'ingestion de faibles quantités d'aflatoxine. *Ann. Nutr. Alim.* **34,** 699–708. (in French)

Lafont, P., Platzer, N., Siriwardana, M. G., Sarfati, J., Mercier, J., and Lafont, J. (1986). Un nouvel hydroxy-dérivé de l'aflatoxine B_1: l'aflatoxine M_4. I. Production *in vitro*—Structure. *Microbiol. Alim. Nutr.* **4,** 65–74. (in French)

McKinney, J. D., Cavanaugh, G. C., Bell, J. T., Hoverland, A. S., Nelson, D. M., Pearson, J., and Selkirk, R. J. (1973). Effects of ammoniation on aflatoxins in rations fed lactating cows. *J. Am. Oil. Chem. Soc.* **50,** 79–84.

Masri, M. S., Garcia, N. C., and Page, J. R. (1969). The aflatoxin M_1 content of milk from cows fed known amounts of aflatoxin. *Vet. Rec.* **84,** 146–147.

Patel, P. M., Netke, S. P., Gupta, D. S., and Dabadghao, A. K. (1981). Note on the effect of processing milk into khoa on aflatoxin M_1 content. *Indian J. Anim. Sci.* **51**, 791–792.

Patterson, D. S. P., Glancy, E. M., and Roberts, B. A. (1980). The carry-over of aflatoxin M_1 into the milk of cows fed rations containing a low concentration of aflatoxin B_1. *Food Cosmet. Toxicol.* **18**, 35–37.

Polan, C. E., Hayes, J. R., and Campbell, T. C. (1974). Consumption and fate of aflatoxin B_1, by lactating cows. *J. Agric. Food Chem.* **22**, 635–638.

Pong, R. S., and Wogan, G. N. (1971). Toxicity and biochemical and fine structural effects of synthetic aflatoxins M_1 and B_1 in rat liver. *J. Natl. Cancer Inst.* **47**, 585–590.

Purchase, I. F. H. (1967). Acute toxicity of aflatoxins M_1 and M_2 in day old ducklings. *Food Cosmet. Toxicol.* **5**, 339–342.

Rodricks, J. V., and Stoloff, L. (1977). Aflatoxin residues from contaminated feed in edible tissues of foodproducing animals. *In* "Mycotoxins in Human and Animal Health" (J. V. Rodricks, C. W. Hesseltine, and M. A. Mehlmann, eds.), pp. 67–79. Pathotox, Park Forest South, Illinois.

Schuller, P. L., Verhülsdonk, C. A. H., and Paulsch, W. E. (1973). Analysis of aflatoxin M_1 in liquid and powdered milk. *Pure Appl. Chem.* **35**, 391–396.

Sieber, R., and Blanc, B. (1978). Zur Ausscheidung von Aflatoxin M_1 in die Milch und dessen Vorkommen in Milch und Milchprodukten—Eine Literaturübersicht. *Mitt. Gebiete Lebensm. Hyg.* **69**, 477–491. (in German)

Sinnhuber, R. O., Lee, D. J., Wales, J. H., Landers, M. K., and Keyl, A. C. (1974). Hepatic carcinogenesis of aflatoxin M_1 in rainbow trout *(Salmo gairdneri)*. *J. Natl. Cancer Inst.* **53**, 1285–1288.

Stoloff, L., Van Egmond, H. P., and Park, D. L. (1991). Rationales for the establishment of limits and regulations for mycotoxins. *Food Addit. Contam.* **8**, 213–222.

Stubblefield, R. D., and Van Egmond, H. P. (1989). Chromatographic methods of analysis for aflatoxin M_1. *In* "Mycotoxins in Dairy Products" (H. P. van Egmond, ed.), pp. 57–95. Elsevier Applied Science, London.

Tuinstra, L. G. M. Th., Roos, A. H., Van Trÿp, J. M. P. (1993). HPLC determination of aflatoxin M_1 in milk powder, using immuno-affinity columns for clean-up: collaborative study. *J. Assoc. Off. Anal Chem.* **76**, in press.

Van der Linde, J. A., Frens, A. M., de Iongh, M., and Vles, R. O. (1964). Inspection of milk from cows fed aflatoxin-containing groundnut meal. *Tijdschr. Diergeneesk.* **89**, 1082–1088. (in Dutch)

Van Egmond, H. P. (1988). Development of mycotoxin reference materials. *Fresenius Z. Anal. Chem.* **332**, 598–601.

Van Egmond, H. P. (1989a). "Mycotoxins in Dairy products." Elsevier Applied Science, London.

Van Egmond, H. P. (1989b). Aflatoxin M_1: Occurrence, toxicity, regulation. *In* "Mycotoxins in Dairy Products" (H. P. van Egmond, ed.), pp. 11–55. Elsevier Applied Science, London.

Van Egmond, H. P. (1989c). Current situation on regulations for mycotoxins. Overview of tolerances and status of standard methods of sampling and analysis. *Food Addit. Cont.* **6**, 139–188.

Veldman, A., Meijs, J. A. C., Borggreve, G. J., and Heeres-van der Tol, J. J. (1992). Carryover of aflatoxin from cows' food to milk. *Animal Prod.* **55**, 163–168.

Wisemann, D. W., and Marth, E. H. (1983). Behaviour of aflatoxin M_1 in yogurt, buttermilk and kefir. *J. Food Protect.* **46**, 115–118.

Wogan, G. N., and Paglialunga, S. (1974). Carcinogenicity of synthetic aflatoxin M_1 in rats. *Food Cosmet. Toxicol.* **12**, 381–384.

Yousef, A. E., and Marth, E. H. (1985). Degradation of aflatoxin M_1 in milk by ultraviolet energy. *J. Food Prot.* **48**, 697–698.

Yousef, A. E., and Marth, E. H. (1989). Stability and degradation of aflatoxin M_1. *In* "Mycotoxins in Dairy Products" (H. P. van Egmond, ed.), pp. 127–161. Elsevier Applied Science, London.

18

Approaches to Reduction
of Aflatoxins in Foods
and Feeds

❖

Timothy D. Phillips, Beverly A. Clement, and Douglas L. Park

INTRODUCTION

Mycotoxins are fungal poisons that occur as frequent contaminants of grains and result in substantial losses to agriculture [Stoloff, 1976; Council for Agricultural Science and Technology (CAST), 1989]. The most thoroughly studied and best understood of the mycotoxins are the aflatoxins. *Aspergillus flavus* and *Aspergillus parasiticus* (the fungi primarily responsible for the production of aflatoxins) are widespread and especially present a problem in corn and peanuts during periods of prolonged drought. Fungal growth on host plants in the field is favored by warm ambient temperatures and plant stress; fungal growth on grain in storage is favored by warm temperatures and high humidity (Diener *et al.,* 1979). Four aflatoxins are elaborated by these fungi as common "invisible poisons" in foods and feeds. Of the aflatoxins, aflatoxin B_1 has invoked the most concern since it is a potent animal carcinogen (Carnaghan, 1965; Dickens and Jones, 1965; Lewis *et al.,* 1967; Wogan and Newberne, 1967; Lancaster, 1968; Wieder *et al.,* 1968; Berry, 1988).

Every year a significant percentage of the world's grain and oilseed crops is contaminated with hazardous mycotoxins such as the aflatoxins. Unfortunately, discontinuing the feeding of aflatoxin-contaminated grain is not always practical, especially when alternate feedstuffs are not readily available or affordable. Thus, these poisons frequently are detected in animal feeds and result in substantial economic losses to the poultry and livestock industries (Nichols, 1987; Shane, 1991).

Also, the prevalence of aflatoxins in a variety of foods destined for human consumption is a major concern. For example, dairy animals can secrete a carcinogenic metabolite of aflatoxin B_1 (aflatoxin M_1) in their milk after ingestion of aflatoxin-contaminated feed. Aflatoxins are heat stable and survive a variety of processing procedures. These compounds have been found as natural contaminants (sometimes at very high concentrations) in food crops and foods such as peanut butter and other peanut products, breakfast cereals, corn and cornmeal, dairy products, and other processed foods (Diener, 1981; Jelinek, 1987; CAST, 1989; Wood, 1989). More importantly, aflatoxin B_1 has been implicated as a factor in human liver cancer and classified as a probable human carcinogen (International Agency for Research on Cancer (IARC), 1987. In light of these important findings, clearly viable strategies to reduce the levels of aflatoxins in foods and feeds are critically needed. A synopsis of selected aflatoxin reduction strategies is provided in this chapter, with a major focus on practical approaches to detoxification or inactivation of aflatoxins. The reader is referred to several excellent reviews on the subject for details (Goldblatt and Dollear, 1977, 1979; Diener *et al.,* 1979; Marth and Doyle, 1979; Anderson, 1983; Palmgren and Hayes, 1987; Park *et al.,* 1988; CAST, 1989; Park and Rua, 1990; Cleveland and Bhatnagar, 1991; Darrah and Barry, 1991; Hagler, 1991; Phillips *et al.,* 1991). Because of size constraints, this chapter is limited to categories of aflatoxin reduction strategies, including (1) food and feed processing, (2) biocontrol and microbial inactivation, (3) structural degradation after chemical treatment, (4) dietary modification of toxicity, and (5) reductions in bioavailable aflatoxin by selective chemisorption.

AFLATOXIN REDUCTION STRATEGIES

Food and Feed Processing

Foods processed from peanuts, corn, and milk have received the most attention concerning the stability of aflatoxins; although corn and peanuts are probably the commodities of greatest worldwide concern since they are staple foods for many countries and usually are grown in climates that are favorable to fungal growth and aflatoxin production (CAST, 1989).

Thermal Inactivation

Aflatoxins are resistant to thermal inactivation and are not destroyed completely by boiling water, autoclaving, or a variety of food and feed processing procedures (Christensen *et al.,* 1977). Aflatoxins may be destroyed partially by conventional processing procedures such as oil- and dry-roasting of peanuts to be used as salted nuts, in confections, or in peanut butter. Lee and co-workers (1969) reported a 45–83% reduction in aflatoxin that was dependent on roasting conditions and initial aflatoxin concentrations in raw peanuts. Comparable results were obtained in an investigation by Waltking (1971). In other studies, roasting conditions resulted in a significant decrease in the aflatoxin content of nuts, oilseed meals (Escher *et al.*, 1973; Marth and Doyle, 1979), and corn (Conway *et al.*, 1978). The degradation of aflatoxins was a direct function of temperature, heating interval, and moisture content (Mann *et al.*, 1967).

Only partial destruction of aflatoxins in contaminated wheat occurs during the various stages of bread making (Jemmali and Lafont, 1972). Baking temperatures do not significantly alter the levels of aflatoxin in dough (Reiss, 1978).

A range of reduction (32–87%) in the level of aflatoxin M_1 in freeze-dried milk (resulting from pasteurization, sterilization, preparation of evaporated milk, roller drying, and spray drying) was reported by Purchase *et al.* (1972). Detoxification was confirmed by the duckling bioassay. Also, aflatoxins were not detected in cottage cheese (prepared from the contaminated milk), but were present in the whey. In other studies, aflatoxin M_1 was apparently stable in raw milk and was resistant to pasteurization and processing (Stoloff *et al.*, 1975; Stoloff, 1980).

A considerable reduction in aflatoxin levels has been associated with the limewater treatment (nixtamalization) of corn to produce tortillas (Ulloa-Sosa and Schroeder, 1969). However, subsequent studies have shown that much of the original aflatoxin is reformed on acidification of the products (Price and Jorgensen, 1985). Comparable results for aflatoxins B_1, B_2, G_1, and G_2 have been reported by Bailey *et al.* (1990,1991). Lime (generating calcium hydroxide in water) was postulated to react with aflatoxin, resulting in (1) a loss of indigenous fluorescence and (2) a major change in the extractability of aflatoxin in solvents such as chloroform. Evidence suggests that aflatoxin may be "masked" chemically in alkaline-processed corn (i.e., tortillas and corn chips) and, thus, may escape analytical detection. More importantly, reformation of parent aflatoxin may occur under the acidic conditions that are found in the stomach (Martinez, 1979; Price and Jorgensen, 1985; Bailey *et al.*, 1990,1991).

Specific criteria have been established for evaluating the acceptance of a given aflatoxin reduction or decontamination procedure. The process must: (1) inactivate, destroy, or remove the toxin; (2) not produce or leave toxic residues in food; (3) retain nutritive value and feed acceptability of the product, (4) not alter significantly the technological properties of the product; and, if possible, (5) destroy fungal spores (Park *et al.*, 1988). Although some destruction of aflatoxins in feed-grade meals has been achieved by conventional processing procedures, heat and moisture alone do not provide a very effective method of detoxification.

Irradiation

Studies have demonstrated that exposure of contaminated peanut oil to shortwave and longwave UV light causes a marked reduction in the concentration of aflatoxins (Shantha and Sreenivasa, 1977). A 14-hr exposure to sunlight destroyed approximately 77–90% of aflatoxin B_1 that was added to groundnut flakes, whereas only 50% of the toxin was destroyed in naturally contaminated product (Shantha *et al.,* 1986). Gamma irradiation did not result in degradation of aflatoxin in contaminated peanut meal, and UV light produced no observable change in fluorescence or toxicity of the treated sample (Feuell, 1966). Exposure of contaminated milk to UV light for 20 min at 25°C decreased the aflatoxin M_1 content by 89.1% in the presence of 0.05% peroxide, compared with 60.7% in peroxide-free milk (Yousef and Marth, 1986). Aflatoxins produce singlet oxygen after exposure to UV light and singlet oxygen further activates these chemicals to mutagens. DNA photobinding and mutagenesis by aflatoxins were enhanced by aflatoxin B_2 in a synergistic manner. Interestingly, these findings imply that aflatoxin B_2 (which often is found in combination with aflatoxin B_1) may amplify the activation of aflatoxin B_1 by sunlight (Stark *et al.,* 1990). Jorgensen and co-workers (1992) found comparative mutagenic activity for equal concentrations of aflatoxin B_1 and mixtures of aflatoxin B_1 and B_2 using the *Salmonella* assay.

Solvent Extraction and Mechanical Separation

Several suitable solvent systems are capable of extracting aflatoxins from different commodities with minimal effects on protein content or nutritional quality (Rayner *et al.,* 1977; Goldblatt and Dollear, 1979). These systems include 95% ethanol, 90% aqueous acetone, 80% isopropanol, hexane–ethanol, hexane–methanol, hexane–acetone–water, and hexane–ethanol–water. However, current extraction technology for the detoxification of aflatoxin-contaminated oilseed meals appears to be impractical and cost prohibitive (Shantha, 1987). Most of the aflatoxin associated with contaminated corn or peanuts can be found in a relatively small number of kernels or seeds, providing an excellent opportunity to reduce the level of aflatoxin contamination by a variety of separation approaches. For example, peanuts and Brazil nuts are sorted (electronically and manually) to reduce the levels of aflatoxins successfully, although complete removal of all contaminated particles is not achieved by these methods (Dickens and Whitaker, 1975; Natarajan *et al.,* 1975).

Conventional methods used to clean corn (e.g., dry cleaning, wet cleaning, density separation, and preferential fragmentation) are somewhat effective in reducing the aflatoxin content (Brekke *et al.,* 1975a). Milling of corn resulted in better results. The distribution of aflatoxins was apparently low in grits and high in the germ, hull, or degermer fines of dry milled corn (Brekke *et al.,* 1975b). Aflatoxin occurred mainly in the steepwater and fiber of wet milled corn, with smaller amounts present in the gluten and germ (Yahl *et al.,* 1971).

Density Segregation

A novel method has been reported for reducing the aflatoxin concentration in corn (Huff, 1980; Huff and Hagler, 1982, 1985) and peanuts (Kirksey *et al.,*

1987,1989; Cole, 1989) by flotation and density segregation of toxic kernels. In a study by Kirksey *et al.* (1989), 95% of the aflatoxin in 21 of 29 samples of peanuts was contained in kernels that floated in tap water. The mean aflatoxin level was decreased from 301 ppb to 20 ppb using this procedure. Flotation for whole-kernel grain and shelled unblanched peanuts should be compatible with current methods for wet milling or alkaline processing of corn (Hagler, 1991).

Adsorption from Solution

Adsorbent materials including activated carbon (Decker, 1980) and clay and zeolitic minerals (Masimanco *et al.,* 1973) have been shown to bind and remove aflatoxins from aqueous solutions such as water, Sorensen buffer, Czapek's medium, Pilsner beer, sorghum beer, whole milk, and skimmed milk. The adsorption of aflatoxins by clay (which is incorporated to remove pigments from crude oils) is a major factor resulting in the significant reduction of aflatoxins in refined peanut and corn oils. Reports indicate that a phyllosilicate clay (HSCAS or NovaSil™) effectively removes aflatoxins from contaminated peanut oil and prevents its mutagenicity and toxicity *in vitro* (Machen *et al.,* 1988,1991). Ellis and co-workers (1990, 1991) reported that HSCAS, at levels as low as 0.5% (w/w), resulted in significant removal of aflatoxin M_1 from naturally contaminated skim milk, with negligible change in nutritional quality (i.e., total milk solids, lactose, and protein) as determined by proximate analysis (Figure 1).

Biocontrol and Microbial Inactivation

Established techniques for the management of aflatoxin contamination during crop production have not adequately insured aflatoxin-free commodities prior to

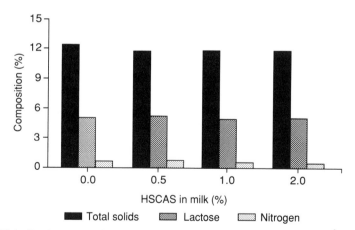

FIGURE 1 Proximate analysis of nutritional components from HSCAS-treated skim milk that was contaminated naturally with aflatoxin M_1 at 0.75 ppb. Milk samples were treated with HSCAS at levels of 0.5, 1.0, and 2.0% w/w ($n = 3$). Data from Ellis *et al.* (1990).

harvest (Cole *et al.,* 1989; Cotty, 1989). A new preharvest strategy for the prevention of aflatoxin contamination of peanuts and cottonseed is the use of nontoxigenic strains of *A. parasiticus* and *A. flavus* fungi to compete with (and exclude) toxin-producing strains. Initial studies have shown that bioprevention can be used to reduce preharvest aflatoxin contamination in peanuts and cottonseed significantly. Additional studies are justified to address human health and environmental impact concerns of this technology (Cole and Cotty, 1990).

A promising new approach to the prevention (or elimination) of preharvest aflatoxin contamination is the molecular regulation of aflatoxin biosynthesis by toxigenic *Aspergillus* fungi. Enzymes that govern multiple steps in the biosynthetic pathway of aflatoxin have been purified (Bhatnagar *et al.,* 1989; Bhatnagar and Cleveland, 1990), providing investigators with the key to characterizing aflatoxin pathway genes and their regulation at the molecular level (Cleveland and Bhatnagar, 1991). A complete understanding of the molecular regulation of aflatoxin biosynthesis may lead to enhanced food safety through selection of crops that minimize expression of aflatoxin pathway genes (Cleveland *et al.,* 1990; Cleveland and Bhatnagar, 1991).

Microorganisms (including yeasts, molds, and bacteria) have been screened for their ability to modify or inactivate aflatoxins. *Flavobacterium aurantiacum* (NRRL B-184) was shown to remove aflatoxin from a liquid medium significantly without the production of toxic by-products (Ciegler *et al.,* 1966). The same investigators also found that certain acid-producing molds could catalyze the hydration of aflatoxin B_1 to B_{2a} (a less toxic product). The applications of microbial detoxification of aflatoxins have been reviewed (Ciegler, 1978; Marth and Doyle, 1979). Hao and co-workers (1987) reported that *F. aurantiacum* removed aflatoxin B_1 from peanut milk. This bacterium grew in both defatted and partially defatted peanut milk and was not inhibited by the presence of aflatoxin.

Fermentation of contaminated grains has been shown to degrade aflatoxins (Dam *et al.,* 1977), but ensiling aflatoxin-contaminated high-moisture corn was apparently not as effective. This difference may be attributed to the production of insufficient acid to catalyze the transformation of aflatoxin B_1 to aflatoxin B_{2a} by this method (Lindenfelser and Ciegler, 1970).

Structural Degradation Following Chemical Treatment

A diverse group of chemicals has been tested for the ability to degrade and inactivate aflatoxins. These chemicals include numerous acids, bases, aldehydes, bisulfite, oxidizing agents, and various gases (Feuell, 1966; Trager and Stoloff, 1967; Mann *et al.,* 1970; Goldblatt and Dollear, 1977, 1979; Anderson, 1983; Park *et al.,* 1988; CAST, 1989; Park and Lee, 1990; Hagler, 1991; Norred, 1990; Park and Rua, 1990; Samarajeewa *et al.,* 1991). A number of these chemicals can react to destroy (or degrade) aflatoxins effectively but most are impractical or potentially unsafe because of the formation of toxic residues or the perturbation of nutrient content, flavor, odor, color, texture, or functional properties of the product. Two chemical approaches to the detoxification of aflatoxins that have received considerable attention are ammoniation and reaction with sodium bisulfite.

Ammoniation

Treatment of grain with ammonia appears to be a viable approach to the detoxification of aflatoxins. Ammoniation (under appropriate conditions) results in a significant reduction in the level of aflatoxins in contaminated peanut and cottonseed meals (Dollear *et al.,* 1968; Masri *et al.,* 1969; Gardner *et al.,* 1971; Park *et al.,* 1984) and corn (Brekke *et al.,* 1977,1979). The ammoniation process, using ammonium hydroxide or gaseous ammonia, has been shown to reduce aflatoxin levels in corn, peanut-meal cakes, and whole cottonseed and cottonseed products by more than 99%. If the reaction is allowed to proceed sufficiently, the process is irreversible. Primarily, two procedures are used: a high pressure and temperature process (HP/HT) at feed mills or an atmospheric pressure and ambient temperature procedure (AP/AT) on the farm (Table 1).

Arizona permits the use of ammonia to reduce aflatoxin levels in whole cottonseed and cottonseed meal and is studying the feasibility of using the process for aflatoxin-contaminated corn. Two processes have been approved: (1) HP/HT for whole cottonseed and cottonseed meal and (2) AP/AT for whole cottonseed. The HP/HT process involves the treatment of the contaminated product with anhydrous ammonia and water in a contained vessel. The amount of ammonia (0.5–2%), moisture (12–16%), pressure (35–50 psi), time (20–60 min), and temperature (80–120°C) varies with respect to the initial levels of aflatoxin in the product. The AP/AT process also uses a 13% ammonia solution, which is sprayed on the seed as it is packed into a plastic silage-type bag (approximately 10 feet diameter by 100 feet long). After this procedure, the bag is sealed and held at ambient temperature (25–45°C) for 14–42 days. The hold-

TABLE 1 Parameters and Application of Ammoniation Aflatoxin Decontamination Procedures

	Process	
Parameter	High pressure/high temperature	Ambient pressure/ atmospheric temperature
Ammonia Level (%)	0.5–2	1–5
Pressure (PSI)	35–50	atmospheric
Temperature (°C)	80–120	ambient
Duration	20–60 min	14–42 d
Moisture (%)	12–16	12–16
Commodities	Whole cottonseed; cottonseed meal; peanut meal; corn	Whole cottonseed; corn
Application	Feed mill	Farm

ing time will vary according to the ambient temperature, that is, a lower ambient temperature will require a longer holding time. The amount of ammonia (1–5%), moisture (12–16%), and time (14–42 days) will vary according to the initial levels of aflatoxin present. With this process, the bag is probed and tested periodically until test results show that aflatoxin levels are equal to or below 20 ppb.

The safety of ammoniated corn has been evaluated in rainbow trout (Brekke *et al.,* 1979), chickens (Hughes *et al.,* 1979), and rats (Norred, 1979,1981; Southern and Clawson, 1980). In a long-term feeding study in rats, Norred and Morrissey (1983) reported that ammoniation of corn containing 750 ppb aflatoxins resulted in significant protection from toxicity and hepatic neoplasia in experimental animals. Frayssinet and Frayssinet-LaFarge (1990) confirmed these results using peanut meal. In this study, rats were fed diets containing ammonia-treated and untreated peanut-meal cakes for 18 months. A satisfactory dose–effect relationship was shown between the residual content of the cakes and the observed incidence of tumors. In another study, Bassuni and co-workers (1990) reported that ammoniated diets significantly decreased the immunotoxicity of aflatoxin in chicks that were vaccinated with Newcastle disease virus. These findings (and others) provide strong evidence that chemical treatment via ammoniation can provide an effective strategy to detoxify aflatoxin-contaminated crops.

The mechanism for this action appears to involve hydrolysis of the lactone ring and chemical conversion of the parent compound aflatoxin B_1 to numerous products that exhibit greatly decreased toxicity. Two major products (Figure 2),

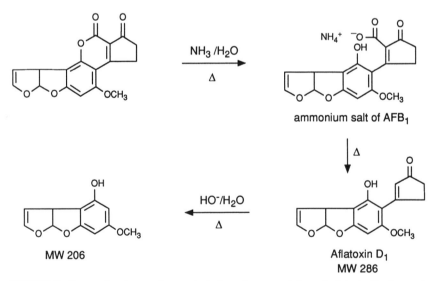

FIGURE 2   Proposed major reaction products from the ammoniation of aflatoxin B_1. Ammonia $(NH_3/H_2O, NH_4OH)$ is proposed to react with parent aflatoxin forming an ammonium salt of the lactone, which is decarboxylated by heat to form numerous products including aflatoxin D_1 (MW 286) and a compound commonly designated MW 206.

TABLE 2   Relative Levels of Conversion of Aflatoxin B_1 to Aflatoxin D_1

Substrate	Conversion (%)	Reference
Pure aflatoxin B_1	30	Lee *et al.* (1974)
Cultured aflatoxin-containing peanut meals	0.31	Lee and Cucullu (1978)
Cultured or natural aflatoxin-containing cottonseed meals	0.10	Lee and Cucullu (1978)
Aflatoxin-spiked peanut and cottonseed meals	0.90	Lee and Cucullu (1978)

identified as compounds with molecular weight 286 (aflatoxin D_1) and molecular weight 206, have been isolated and tested in various biological systems (Park *et al.*, 1988). The first step in the reaction is reversible if the ammoniation process is carried out under mild conditions. However, when the reaction is allowed to proceed past the first step, the products formed do not revert back to aflatoxin B_1. Reaction products of ammoniation are dependent on temperature, pressure, and the source of ammonia. Increased quantities of reaction products are formed in the presence of pure aflatoxin B_1 versus aflatoxin-contaminated meals during ammoniation. Apparently, the meal matrix strongly influences the formation of aflatoxin by-products of ammoniation (Table 2). The mutagenic potential, ability to covalently bind with nucleic acid, and embryotoxicity of pure aflatoxin D_1 and the molecular weight 206 compound from ammoniation have been reported using a variety of bioassays (Table 3). In every case, the toxicity from these products was several orders of magnitude lower than that of aflatoxin B_1. Also, the formation of ammoniation products in the feed matrix is usually < 1% of the

TABLE 3   Relative Toxic Potential of Ammonia/Aflatoxin Reaction Products and Animal Metabolites[a]

Compound	Salmonella assay		Covalent binding index		Chick embryo assay[b]	
	(μg/plate)	Reference	Data	Reference	(μg/egg)	Reference
Aflatoxin B_1	0.005	Lawlor *et al.* (1985)	22000	Schroeder *et al.* (1985)	0.125	Lee *et al.* (1981)
Aflatoxin M_1	0.16	Jorgensen and Price (1981)	2100	Schroeder *et al.* (1985)	ND[c]	
Aflatoxin D_1	2.25[d]	Lee *et al.* (1981)	<70	Schroeder *et al.* (1985)	2.5	Lee *et al.* (1981)
206 MW compound	3.30	Haworth *et al.* (1989)	ND		ND	
Pronase digestion	180	Lawlor *et al.* (1985)	ND		ND	

[a]Modifed with permission from Park *et al.* (1988).
[b]Caused 40% mortality.
[c]Not determined.
[d]Calculations based on 450-fold decrease in mutagenic potential.

original aflatoxin level (Table 2). A high percentage of these products is bound tightly to endogenous components in the feed and is not biologically available.

The use of aqueous or gaseous ammonia, as well as other chemical treatments for the detoxification of aflatoxins, has been reviewed thoroughly (Goldblatt and Dollear, 1979; Anderson, 1983; Palmgren and Hayes, 1987; Park et al., 1988; Park and Lee, 1990). Because of the severity of aflatoxin contamination in selected agricultural commodities from various locales, specific decontamination procedures have been approved and put into use. In the United States, Arizona, Texas, and California permit the ammoniation of cottonseed products and Texas, North Carolina, Georgia, and Alabama have approved the ammoniation procedure for aflatoxin-contaminated corn. Mexico has approved ammoniation for corn. Peanut meal is a frequent component of animal feeds in Europe and elsewhere. Consequently, ammoniation is used routinely in France, South Africa, Senegal, and Brazil and soon will be utilized in the Sudan and India to lower aflatoxin contamination levels. Several member countries of the European Economic Community import ammonia-treated peanut meal on a regular basis (Park and Lee, 1990).

Treatment with Bisulfite

Sodium bisulfite has been shown to react with aflatoxins (B_1, G_1, M_1, and aflatoxicol) under various conditions of temperature, concentration, and time to form water-soluble products (Doyle and Marth, 1978a,b; Moerck et al., 1980; Hagler et al., 1982). The addition of 0.04 g potassium bisulfite per 10 ml milk resulted in a 45% reduction in the level of aflatoxin M_1 after 5 min (Doyle et al., 1982). The aflatoxin B_1–sodium bisulfite reaction products have been proposed (Hagler et al., 1983; Yagen et al., 1989), as have the sulfonate derivatives of aflatoxins G_1, M_1, and aflatoxicol (Lee et al., 1989). In a series of elegant NMR studies, Yagen et al. (1989) established the structure of the sodium bisulfite adduct of aflatoxin B_1 (AFB_1–S) and proposed a mechanism for this addition (Figure 3). This thermally stable adduct is intriguing because the bisulfite addition to the vinyl ether is both stereospecific and anti-Markovnikov. The stereospecificity of such a reaction could be accounted for if the addition were concerted or if a bridged intermediate were involved. However, a concerted addition of bisulfite would lead to the formation of the opposite stereoisomer (cis addition rather than trans). Moreover, the mechanism proposed suggested the transient formation of a carbanion (not a bridged intermediate), which ultimately would lead to racemization. Bisulfite is known to undergo free radical addition (Gilbert, 1965); a mechanism that would explain both the stereochemistry and the anti-Markovnikov orientation of the observed adduct would be the free radical addition of bisulfite to the vinyl ether (Figure 4).

Several studies indicate that peroxide and heat enhance the destruction of aflatoxin B_1 by sodium bisulfite added to dried figs (Altug et al., 1990). The biological activity of AFB–S has not been reported. Clearly more research is warranted in this area to determine the potential of bisulfites to reduce or inactivate aflatoxins in food and feed (Hagler, 1991).

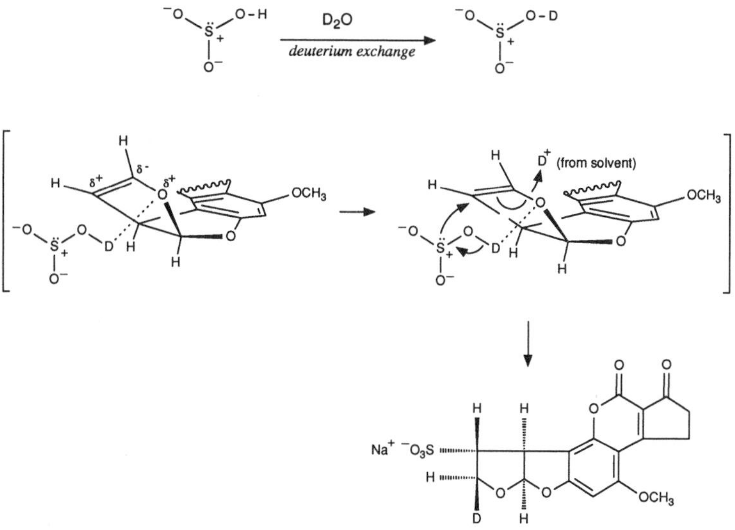

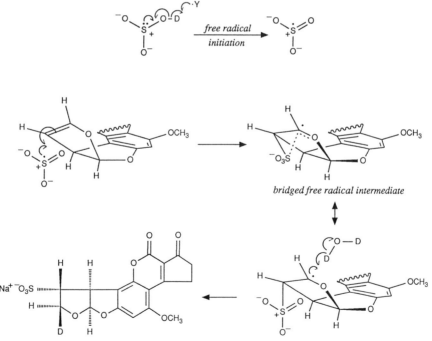

FIGURE 3 Proposed reaction mechanism of sodium bisulfite addition to aflatoxin B$_1$. Modified with permission from Yagen *et al.* (1989).

FIGURE 4 Suggested mechanism of free radical addition of bisulfite to aflatoxin B$_1$, explaining both the stereochemistry and the anti-Markovnikov orientation of the adduct (AFB$_1$–S).

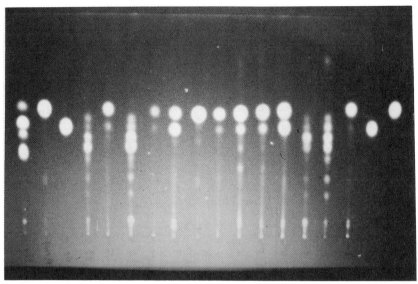

A   B   C   D   E   F   G   H   I   J   K   L   M   N   O   P   Q   R

FIGURE 5   TLC of 18 pellet extracts from aflatoxin-treated samples. Aflatoxin B_1 was reacted for 1 hr with samples of activated charcoal, alumina, zeolite, various phyllosilicates, and modified phyllosilicates in water. The reaction was terminated by centrifugation and sample pellets were extracted vigorously with methanol and chloroform. Extracts were analyzed by thin-layer chromatography. A, mixed aflatoxin standard; B, aflatoxin B_1 control; C, aflatoxin B_2 control; D, modified phyllosilicate; E, HSCAS; F, phyllosilicate; G, zeolite; H, neutral alumina; I, activated charcoal; J, phyllosilicate; K, phyllosilicate; L, phyllosilicate; M, phyllosilicate; N, modified phyllosilicate; O, modified phyllosilicate; P, phyllosilicate; Q, aflatoxin B_2 control; R, aflatoxin B_1 control. Data from Phillips *et al.* (1990b).

Heterogeneous Catalytic Degradation

The formation of aflatoxin B_1 adsorption complexes on the surfaces of certain inorganic materials (including various aluminas, silicas, aluminosilicates, and chemically modified aluminosilicates) may promote heterogeneous catalytic degradation of the parent molecule after desorption. Phillips and co-workers (1988b,1990b) observed that organic extracts from reactions of aflatoxin B_1 with alumina, zeolites, and phyllosilicates contained varying levels of aflatoxin degradation products including aflatoxin B_2, which was identified as the major product in most casts (Figure 5). The parent aflatoxin that was reacted with activated charcoal was recovered from the complex unchanged. Perhaps chemical degradation of aflatoxins by reactive inorganic adsorbents represents another useful chemical approach to detoxification.

Reduction in Bioavailable Aflatoxin by Selective Chemisorption

A phyllosilicate clay (HSCAS or NovaSil™) currently available as an anticaking agent for animal feeds has been reported to (1) tightly bind aflatoxins in

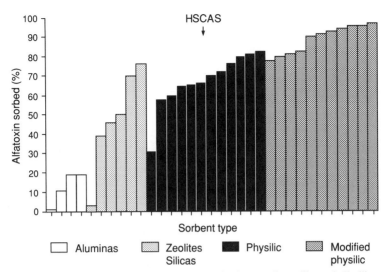

FIGURE 6 Binding of radiolabeled aflatoxin B_1 by aluminas, zeolites, silicas, phyllosilicates, and chemically modified phyllosilicates. Modified with permission from Phillips *et al.* (1988a).

aqueous suspensions (Figure 6); (2) markedly diminish aflatoxin uptake by the blood and distribution to target organs (Figure 7); (3) prevent aflatoxicosis in farm animals, including chickens (Figure 8), turkey poults, goats, pigs, and mink; and (4) decrease the level of aflatoxin M_1 residues in milk from lactating dairy cattle (Davidson *et al.,* 1987; Kubena *et al.,* 1987,1988,1989,1990a,b,c,d, 1991a,b; Phillips 1987; Phillips *et al.,* 1987,1988a,b,1989,1990a,b,1991,1992; Harvey *et al.,* 1988a,b,1989a,b,1990,1991a,b,c; Colvin *et al.,* 1989; Beaver *et al.,* 1990; Chung *et al.,* 1990; Bonna *et al.,* 1991). In other studies, Ellis *et al.* (1990,1991) reported that HSCAS effectively removed aflatoxin M_1 from contaminated skim milk.

Several test compounds with one or more of the functional groups in common with aflatoxin were reacted with HSCAS *in vitro* in an attempt to elucidate the specificity and mechanism of tight binding (or chemisorption). Test compounds that were homologous to AFB_1 and AFG_1 (i.e., AFB_2, AFG_2, AFB_{2a}, AFG_{2a}, AFQ_1, AFP_1, AFM_1, AFM_2, and AFM_{2a}) were bound tightly to HSCAS, whereas those that contained a dihydrofurofuran fused to an aromatic ring did not show any significant interaction and binding. Coumarin, 4-methyl umbelliferone, esculetin, xanthotoxin, aflatoxicol, and tetrahydrodeoxyaflatoxin B_1 were bound initially but were desorbed significantly from the HSCAS pellet. These findings suggest that the β-dicarbonyl system contained in the aflatoxin B and G series is essential for chemisorption by HSCAS.

Phyllosilicates are layer-lattice silicates and chain silicates that are essentially composed of repeating layers (or chains) or (1) divalent or trivalent cations (eg., aluminas) held in octahedral coordination with oxygens and hydroxyls and (2) silicas that are coordinated tetrahedrally with oxygens and

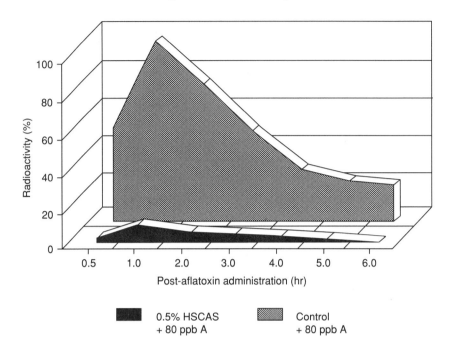

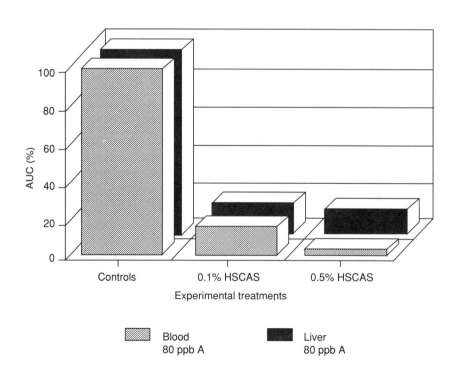

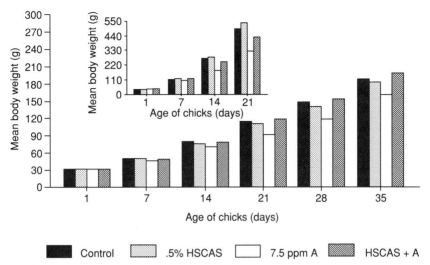

FIGURE 8 Protective effects of HSCAS in broiler (inset) and Leghorn chicks fed AFB$_1$-contaminated diets. Modified with permission from Phillips *et al.* (1988a).

hydroxyls. Condensation of layers in a 1:1 ratio gives rise to dimorphic phyllosilicate clays with the general formula $M_{2-3}Si_2O_5(OH)_4$. Trimorphic phyllosilicates are formed by a 1:2 condensation of layers (with octahedrally coordinated aluminums sandwiched between two layers of tetrahedrally coordinated silicas) with the formula $M_{2-3}Si_4O_{10}(OH)_2$. Chain phyllosilicates are composed of trimorphic layers arranged in chains (or bands) that are joined through oxygen ions (Schulze, 1989). Generally, these materials possess three types of binding sites for aflatoxins: (1) those located within interlayer channels, (2) those located on the surface, and (3) uncoordinated metal ions located at the edges of HSCAS particles.

Diffuse reflectance infrared Fourier transform spectroscopy (DRIFTS) was used to examine sorbed pellets of aflatoxin–HSCAS. DRIFTS clearly showed dramatic shifts of the C=O absorptions that were identical to those known to arise from metal–acetylacetonates. Also, preliminary X-ray diffraction studies indicated a limited penetration of aflatoxin into the interlayer spaces of HSCAS (Sarr *et al.*, 1990,1991). These findings suggest that the molecular mechanism of aflatoxin-selective chemisorption may involve the chelation of the β-dicarbonyl

FIGURE 7 (*Top*) [^{14}C]Aflatoxin B$_1$ in the blood of broiler chicks fed the equivalent of 80 ppb aflatoxin with (*front*) and without (*back*) 0.5% HSCAS is their diet. Data represent percentage of radioactivity relative to peak concentration of the control (100%) over a sampling period of 6 hr. Data from Davidson *et al.*, (1987) and Phillips *et al.* (1990b). (*Bottom*) Relative bioavailability of [^{14}C]aflatoxin B$_1$ in the blood (hatched) and liver (solid) of chicks fed the equivalent of 80 ppb total aflatoxin with and without 0.1% and 0.5% HSCAS in their diet. Data expressed as percentage of AUC (area under the curve) of control (100%). Data from Davidson *et al.* (1987) and Phillips *et al.* (1990b).

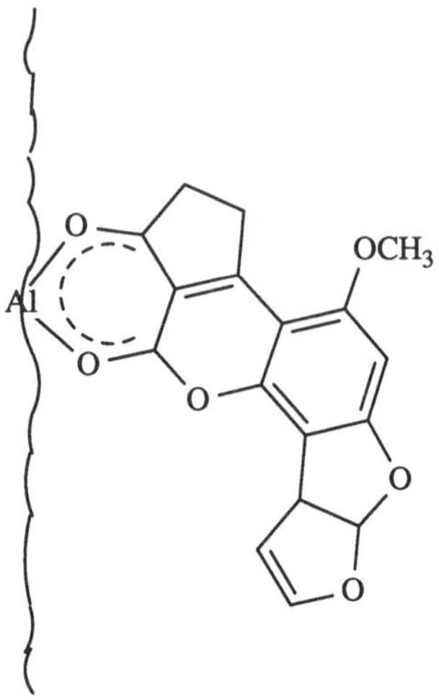

FIGURE 9 Proposed reaction of the β-dicarbonyl moiety of aflatoxin B_1 with uncoordinated "edge site" aluminum ions in HSCAS.

moiety in aflatoxin with uncoordinated metal (i.e., aluminum) ions in HSCAS (Figure 9). A chemisorption index (C_α) was developed allowing direct comparison of various phyllosilicates with HSCAS. C_αs were determined by high performance liquid chromatography (HPLC) analysis of extracts of the supernatants and sorbed pellets (exhaustively extracted with methanol, chloroform, and hexane). Comparison of the C_αs of various classes of compounds with spectral data (DRIFTS) supported the proposed mechanism of chemisorption (Sarr *et al.*, 1990,1991). Additional research is warranted to delineate clearly the sites and chemical mechanisms involved in the action of HSCAS and similar phyllosilicate clays.

CONCLUSIONS

Although good crop management and surveillance programs are essential for the control of aflatoxin contamination, they do not resolve the mycotoxin dilemma. In other words, aflatoxin levels can be lessened in food and feed by these strategies, but not eliminated completely. Since aflatoxins are produced from ubiquitous fungi as unavoidable adulterants of food and feed and are very potent carcinogens, they are regulated explicitly in the United States. Therefore, viable strategies for the effective remediation of aflatoxins are critically needed.

Several potentially feasible strategies for the reduction or inactivation of aflatoxins have been reported in the scientific literature. Some of these methods are clearly more effective and practical than others, but most result in a degree of reduction in levels of parent aflatoxins or in a modification of the toxicity associated with these poisons. These approaches include (1) food and feed processing (e.g., thermal inactivation, irradiation, solvent extraction and mechanical separation, density segregation, and adsorption from solution), (2) biocontrol and microbial inactivation, (3) structural degradation following chemical treatment (e.g., ammoniation, bisulfite, catalytic adsorbents), (4) dietary modification of toxicity, and (5) reduction in bioavailable aflatoxin by selective chemisorption.

The method that has received the most attention is the treatment of aflatoxin-contaminated feeds (primarily cottonseed, corn, and peanut products) with ammonia (i.e., ammoniation). The ammoniation process has been used successfully for many years in the United States, France, and Africa but has not yet been sanctioned by the U.S. Food and Drug Administration. However, the use of ammonia to treat cottonseed for dairy cows has been a significant factor in keeping Arizona's milk supply free from major aflatoxin contamination. Current research supports the use of ammoniation in an effort to reduce markedly the risk posed by aflatoxin contamination of grains and oilseeds.

Presently, no "silver bullet" exists that can be used to eliminate the aflatoxin problem. Newer strategies for the prevention of aflatoxins in grains and the modification of toxicity (and carcinogenicity) in animals show considerable promise and innovation, but more research is warranted to establish their individual efficacy and safety. Perhaps in the future, a combination of various reduction strategies will lead to the elimination of aflatoxins in foods and feeds or to the prevention of aflatoxicoses.

REFERENCES

Altug, T., Yousef, A. E., and Marth, E. H. (1990). Degradation of aflatoxin B_1 in dried figs by sodium bisulfite with or without heat, ultraviolet energy or hydrogen peroxide. *J. Fd. Protect.* **53**, 581–582.

Anderson, R. A. (1983). Detoxification of aflatoxin-contaminated corn. *In* "Aflatoxin and *Aspergillus flavus* in Corn" (U. Diener, R. Asquith, and J. Dickens, eds.), pp. 87–90. Southern Cooperative Series Bulletin 279. Auburn University, Auburn, Alabama.

Bailey, R. H., Clement, B. A., Phillips, J. M., Sarr, A. B., Turner, T. A., and Phillips, T. D. (1990). Fate of aflatoxins in lime processed corn. *Toxicologist* **10**, 163.

Bailey, R. H., Clement, B. A., Chase, T. A., and Phillips, T. D. (1991). The distribution of aflatoxins in the production of corn tortillas. *Toxicologist* **11**, 280.

Bassuni, A. A., Nmer, M. M., El-Nabarawy, A. M., El-Zenaty, I., and Afif, M. A. (1990). Effect of ammoniated aflatoxin contaminated ration on the immune response of chicks vaccinated with Newcastle disease virus. International Symposium on Food Contamination: Mycotoxins and Phycotoxins, Cairo, Egypt.

Beaver, R. W., Wilson, D. M., James, M. A., and Haydon, K. D. (1990). Distribution of aflatoxins in tissues of growing pigs fed an aflatoxin-contaminated diet amended with a high affinity aluminosilicate sorbent. *Vet. Hum. Toxicol.* **32**, 16–18.

Berry, C. (1988). The pathology of mycotoxins. *J. Pathol.* **154**, 301–311.

Bhatnagar, D., and Cleveland, T. E. (1990). Purification and characterization of a reductase from *Aspergillus parasiticus* SRRC 2043 involved in aflatoxin biosynthesis. *FASEB J.* **4**, A2164.

Bhatnagar, D., Cleveland, T. E., and Lillehoj, E. B. (1989). Enzymes in late stages of aflatoxin B_1 biosynthesis: Strategies for identifying pertinent genes. *Mycopathologia.* **107**, 75–83.

Bonna, R. J., Aulerich, R. J., Bursian, S. J., Poppenga, R. H., Braselton, W. E., and Watson, G. L. (1991). Efficacy of hydrated sodium calcium aluminosilicate and activated charcoal in reducing the toxicity of dietary aflatoxin to mink. *Arch. Environ. Contam. Toxicol.* **20,** 441–447.

Brekke, O. L., Peplinski, A. J., and Griffin, E. L., Jr. (1975a). Cleaning trials for corn containing aflatoxin. *Cereal Chem.* **52,** 198–204.

Brekke, O. L., Peplinski, A. F., Nofsinger, G. W., Conway, H. F., Stringfellow, A. C., Montgomery, R. R., Silman, R. W., Sohns, V. E., and Bagley, E. B. (1975b). Pilot-plant dry milling of corn containing aflatoxin. *Cereal Chem.* **52,** 205–211.

Brekke, O. L., Sinnhuber, R. O., Peplinski, Wales, J. H., Putnam, G. B., Lee, D. J., and Ciegler, A. (1977). Aflatoxin in corn: Ammonia inactivation and bioassay with rainbow trout. *Appl. Environ. Microbiol.* **34,** 34–37.

Brekke, O. L., Peplinski, A. J., Nofsinger, G. W., Conway, H. F., Stringfellow, A. C., Montgomery, R. R., Silman, R. W., Sohns, V. E., and Bagley, E. B. (1979). Aflatoxin inactivation in corn by ammonia gas: A field trial. *Trans. Am. Soc. Ag. Engr.* **22,** 425–432.

Carnaghan, R. B. A. (1965). Hepatic tumors in ducks fed a low level of toxic groundnut meal. *Nature (London)* **208,** 308–312.

Christensen, C. M., Mirocha, C. J., and Meronuck, R. A. (1977). "Mold, Mycotoxins and Mycotoxicoses." Agricultural Experiment Station, Report No. 142. University of Minnesota, St. Paul, Minnesota.

Chung, T. K., Erdman, J. W., Jr., and Baker, D. H. (1990). Hydrated sodium calcium aluminosilicate: Effects on zinc, manganese, vitamin A and riboflavin utilization. *Poultry Sci.* **69,** 1364–1370.

Ciegler, A. (1978). Trichothecenes: Occurrence and toxicoses. *J. Fd. Protect.* **41,** 399–403.

Ciegler, A., Lillehoj, E. B., Peterson, R. E., and Hall, H. H. (1966). Microbial detoxification of aflatoxin. *Appl. Microbiol.* **14,** 934–939.

Cleveland, R. E., and Bhatnagar, D. (1991). Molecular regulation of aflatoxin biosynthesis. *In* "Mycotoxins, Cancer, and Health" (G. Bray and D. Ryan), pp. 270–287. Louisiana State University Press, Baton Rouge.

Cleveland, R. E., Bhatnagar, D., and Cotty, P. J. (1990). Control of biosynthesis of aflatoxin in strains of *Aspergillus flavus. In* "Perspectives on Aflatoxins in Field Crops and Animal Food Products in the United States" (ARS-83), pp. 67–73. National Technical Information Services, Springfield, Virginia.

Cole, R. J. (1989). Technology of aflatoxin decontamination. *In* "Mycotoxins and Phycotoxins '88" (S. Natori, K. Hashimoto, and Y. Ueno, eds.). pp. 177–184. Elsevier Scientific, Amsterdam.

Cole, R. J., and Cotty, P. J. (1990). Biocontrol of aflatoxin production by using biocompetitive agents. *In* "Perspectives on Aflatoxin in Field Crops and Animal Food Products in the United States" (ARS-83), pp. 62–66. National Technical Information Services, Springfield, Virginia.

Cole, R. J., Dorner, J. W., and Blankenship, P. D. (1989). Environmental conditions required to induce preharvest aflatoxin contamination of groundnuts: Summary of six years research. *In* "Aflatoxin Contamination of Groundnuts," pp. 279–287. ICRISAT, Patancheru, India.

Colvin, B. M., Sangster, L. T., Hayden, K. D., Bequer, R. W., and Wilson, D. M. (1989). Effect of high affinity aluminosilicate sorbent on prevention of aflatoxicosis in growing pigs. *Vet. Hum. Toxicol.* **31,** 46–48.

Conway, H. F., Anderson, R. A., and Bagley, E. B. (1978). Detoxification of aflatoxin-contaminated corn by roasting. *Cereal Chem.* **55,** 115–117.

Cotty, P. J. (1989). Aflatoxins in cottonseed—What can be done? *Proceedings of the 38th Oilseed Processing Clinic, Southern Regional Research Center,* New Orleans, Louisiana, pp. 30–38.

Council for Agricultural Science and Technology (1989). "Mycotoxins: Economic and Health Risks." CAST, Task Force Report No. 116. Ames, Iowa.

Dam, R., Tam, S. W., and Satterlee, L. D. (1977). Destruction of aflatoxins during fermentation and by-product isolation from artificially contaminated grain. *Cereal Chem.* **54,** 705–714.

Darrah, L. L., and Barry, B. D. (1991). Reduction of preharvest aflatoxin in corn. *In* "Mycotoxins, Cancer, and Health" (G. Bray and D. Ryan), pp. 270–287. Louisiana State University Press, Baton Rouge.

Davidson, J. N., Babish, J. G., Delaney, K. A., Taylor, D. R., and Phillips, T. D. (1987). Hydrated sodium calcium aluminosilicate decreases the bioavailability of aflatoxin in the chicken. *Poultry Sci.* **66**, 89.

Decker, W. J. (1980). Activated charcoal absorbs aflatoxin B_1. *Vet. Human Toxicol.* **22**, 388–389.

Dickens, F., and Jones, H. E. H. (1965). Further studies on the carcinogenic action of certain lactones and related substances in the rat and mouse. *Br. J. Cancer* **19**, 392–403.

Dickens, J. W., and Whitaker, T. B. (1975). Efficacy of electronic color sorting and hand picking to remove aflatoxin contaminated kernels from contaminated lots of shelled peanuts. *Peanut Sci.* **2**, 45–50.

Diener, U. L. (1981). Unwanted biological substances in foods: Aflatoxins. *In* "Impact of Toxicology on Food Processing" (J. C. Ayres and J. C. Kirschman, eds.), pp. 122–150. AVI, Westport, Connecticut.

Diener, U. L., Hesseltine, C. W., Ayres, J. C., Carlton, W. W., Cole, R. J., Goldblatt, L. A., Hamilton, P. B., Hsieh, D. P. H., Marth, E. H., Mirocha, C. J., Pier, A. C., Sinnhuber, R. O., and Wallin, J. R. (1979). "Aflatoxin and Other Mycotoxins: An Agricultural Perspective." CAST Task Force Report No. 80. Council for Agricultural Science and Technology, Ames, Iowa.

Dollear, F. G., Mann, G. E., Codifer, L. P., Gardner, H. K., Koltun, S. P., and Vix, H. L. E. (1968). Elimination of aflatoxins from peanut meal. *J. Am. Oil Chem. Soc.* **45**, 862–865.

Doyle, M. P., and Marth, E. H. (1978a). Bisulfite degrades aflatoxin: Effects of temperature and concentration of bisulfite. *J. Food Protect.* **41**, 774–780.

Doyle, M. P., and Marth, E. H. (1978b). Bisulfite degrades aflatoxin. Effect of citric acid and methanol and possible mechanisms of degradation. *J. Food Protect.* **41**, 891–896.

Doyle, M. P., Applebaum, R. S., Brackett, R. E., and Marth, E. H. (1982). Physical, chemical and biological degradation of mycotoxins in foods and agricultural commodities. *J. Food Protect.* **45**, 964–971.

Ellis, J. A., Harvey, R. B., Kubena, L. F., Bailey, R. H., Clement, B. A., and Phillips, T. D. (1990). Reduction of aflatoxin M_1 residues in milk utilizing hydrated sodium calcium aluminosilicate. *Toxicologist* **10(1)**, 163.

Ellis, J. A., Bailey, R. H., Clement, B. A., and Phillips, T. D. (1991). Chemisorption of aflatoxin M_1 from milk by hydrated sodium calcium aluminosilicate. *Toxicologist* **11**, 96.

Escher, F. E., Koehler, P. E., and Ayres, J. C. (1973). Effect of roasting on aflatoxin content of artificially contaminated pecans. *J. Food Sci.* **38**, 889.

Feuell, A. J. (1966). Aflatoxin in groundnuts. IX. Problems of detoxification. *Trop. Sci.* **8**, 61–70.

Frayssinet, C., and Frayssinet-LaFarge, C. (1990). Effect of ammoniation on the carcinogenicity of aflatoxin-contaminated groundnut oil cakes: Long-term feeding study in the rat. *Food Add. Cont.* **7(1)**, 63–68.

Gardner, H. K., Koltun, S. P., Dollear, F. G., and Rayner, E. T. (1971). Inactivation of aflatoxins in peanut and cottonseed meals by ammoniation. *J. Am. Oil Chem. Soc.* **48**, 70–73.

Gilbert, E. E. (1965). "Sulfonation and Related Reactions," pp. 148–156. Interscience Publishers, New York.

Goldblatt, L. A., and Dollear, F. G. (1977). Detoxification of contaminated crops. *In* "Mycotoxins in Human and Animal Health" (J. V. Rodricks, C. W. Hesseltine, and M. A. Mehlman, eds.), pp. 139–150. Pathotox, Park Forrest South, Illinois.

Goldblatt, L. A., and Dollear, F. G. (1979). Modifying mycotoxin contamination in feeds—Use of mold inhibitors, ammoniation and roasting. *In* "Interactions of Mycotoxins in Animal Production," pp. 167–184. National Academy of Sciences, Washington, D.C.

Hagler, W. M., Jr. (1991). Potential for detoxification of mycotoxin-contaminated commodities. *In* "Mycotoxins, Cancer and Health" (G. Bray and D. Ryan), pp. 253–269. Louisiana State University Press, Baton Rouge.

Hagler, W. M., Jr., Hutchins, J. E., and Hamilton, P. B. (1982). Destruction of aflatoxin in corn with sodium bisulfite. *J. Food Protect.* **45**, 1287–1291.

Hagler, W. M., Jr., Hutchins, J. E., and Hamilton, P. B. (1983). Destruction of aflatoxin B_1 with sodium bisulfite: Isolation of the major product aflatoxin B_1S. *J. Food Prot.* **46**, 295–300.

Hao, D. Y. Y., Brackett, R. E., and Nakayama, T. O. M. (1987). Removal of aflatoxin B₁ from peanut milk by *Flavobacterium aurantiacum*. *In* "Summary and Recommendations of the International Workshop on Aflatoxin Contamination of Groundnut," p. 15. ICRISAT Center, India.

Harvey, R. B., Kubena, L. F., Phillips, T. D., Huff, W. E., and Corrier, D. E. (1988a). Approaches to the prevention of aflatoxicosis. *Proceedings of the Maryland Nutrition Conference*, pp. 102–107. University of Maryland Cooperative Extension Service, Agricultural Experiment Station.

Harvey, R. B., Phillips, T. D., Kubena, L. F., and Huff, W. E. (1988b). Dietary hydrated sodium calcium aluminosilicate and its impact on aflatoxin toxicity in pigs and milk residues in dairy cows. *In* "Dairy Agent Update," Vol. 7, pp. 3–4. Cornell University, Ithaca, New York.

Harvey, R. B., Kubena, L. F., Phillips, T. D., Huff, W. E. (1989a). Mycotoxins: Danger, prevention and control. *Proceedings of the Latin American Poultry Health Symposium*, Guanacaste, Costa Rica.

Harvey, R. B., Kubena, L. F., Phillips, T. D., Huff, W. E., and Corrier, D. E. (1989b). Prevention of clinical signs of aflatoxicosis with hydrated sodium calcium aluminosilicate. *Proceedings of the American Association of Swine Practice*, pp. 99–102.

Harvey, R. B., Kubena, L. F., Phillips, T. D., and Huff, W. E. (1990). Diagnosis, prevention and control of mycotoxins in swine. *Proceedings of the Indiana Veterinary Medical Association*, Indianapolis, pp. 1–5.

Harvey, R. B., Kubena, L. F., Phillips, T. D., Corrier, D. E., Elissalde, M. H., and Huff, W. E. (1991a). Diminution of aflatoxin toxicity to growing lambs by dietary supplementation with hydrated sodium calcium aluminosilicate. *Am. J. Vet. Res.* **52,** 152–156.

Harvey, R. B., Phillips, T. D., Ellis, J. A., Kubena, L. F., Huff, W. E., and Petersen, D. V. (1991b). Effects on aflatoxin M₁ residues in milk by addition of hydrated sodium calcium aluminosilicate to aflatoxin-contaminated diets of dairy cows. *Am. J. Vet. Res.* **52,** 1556–1559.

Harvey, R. B., Kubena, L. F., Phillips, T. D., and Huff, W. E. (1991c). Review of the protective effects of hydrated sodium calcium aluminosilicate (HSCAS) on the toxicity of aflatoxin to chicks and turkey poults. *Proceedings of the 40th Western Poultry Disease Conference*, pp. 115–117.

Haworth, S. R., Lawlor, T. E., Zeiger, E., Lee, L. S., and Park, D. L. (1989). Mutagenic potential of ammonium-related aflatoxin reaction products in model system. *J. Am. Oil Chem. Soc.* **66,** 102.

Huff, W. E. (1980). A physical method for the segregation of aflatoxin contaminated corn. *Cereal Chem.* **57,** 236–238.

Huff, W. E., and Hagler, W. M., Jr. (1982). Evaluation of density segregation as a means to estimate degree of aflatoxin contamination of corn. *Cereal Chem.* **59,** 152–154.

Huff, W. E., and Hagler, W. M., Jr. (1985). Density segregation of corn and wheat naturally contaminated with aflatoxin, deoxynivalenol and zearalenone. *J. Food Protect.* **48,** 416–420.

Hughes, B. L., Barnett, B. D., Jones, J. E., and Dick, J. W. (1979). Safety of feeding aflatoxin-inactivated corn with the Leghorn layer-breeders. *Poultry Sci.* **58,** 1202–1209.

International Agency for Research on Cancer (1987). IARC monograph on the evaluation of carcinogenic risk to humans. IARC, Lyon, France.

Jelinek, C. F. (1987). Distribution of mycotoxins: An analysis of world-wide commodities data, including data from FAO/WHO/UNEP food contamination monitoring programme. Second Joint FAO/WHO/UNEP International Congress on Mycotoxins, Bangkok, Thailand.

Jemmali, M., and Lafont, P. (1972). Evolution de l'aflatoxine B₁ au cours de la panification. *Cah. Nut. Diet* **7,** 319–322.

Jorgensen, K. V., and Price, R. L. (1981). Atmospheric pressure-ambient temperature reduction of aflatoxin B₁ in ammoniated cottonseed. *J. Agric. Food. Chem.* **29,** 555–558.

Jorgensen, K. V., Park, D. L., Rua, S. M., Jr., Mauteiga, R., and Price, R. L. (1992). Effects of glutathione and aflatoxin B₂ on the mutagenic potential of aflatoxin B₁. *J. Environ. Mut.*

Kirksey, J. W., Cole, R. J., Dorner, J. W., and Henning, R. J. (1987). Density segregation of peanuts naturally contaminated with aflatoxin. *Proc. Am. Peanut Res. Ed. Soc.* **19,** 36.

Kirksey, J. W., Cole, R. J., and Dorner, J. W. (1989). Relationship between aflatoxin content and buoyancy in Florunner peanuts. *Peanut Sci.* **16**, 48–51.

Kubena, L. F., Harvey, R. B., Phillips, T. D., and Heidelbaugh, N. D. (1987). Novel approach to the preventive management of aflatoxicosis in poultry. *Proceedings of the U.S. Animal Health Association,* pp. 302–304.

Kubena, L. F., Harvey, R. B., Phillips, T. D., and Huff, W. E. (1988). Modulation of aflatoxicosis in growing chickens by dietary addition of a hydrated sodium calcium aluminosilicate. *Poultry Sci.* **67**, 106.

Kubena, L. F., Harvey, R. B., Huff, W. E., Corrier, D. E., and Phillips, T. D. (1989). Ameliorating properties of a hydrated sodium calcium aluminosilicate on the toxicity of aflatoxin and T-2 toxin. *Poultry. Sci.* **68**, 81.

Kubena, L. F., Harvey, R. B., Huff, W. E., Corrier, D. E., and Phillips, T. D. (1990a). Ameliorating properties of a hydrated sodium calcium aluminosilicate on the toxicity of aflatoxin and T-2 toxin. *Poultry Sci.* **69**, 1078–1086.

Kubena, L. F., Harvey, R. B., Phillips, T. D., Corrier, D. E. and Huff, W. E. (1990b). Diminution of aflatoxicosis in growing chickens by dietary addition of a hydrated sodium calcium aluminosilicate. *Poultry Sci.* **69**, 727–735.

Kubena, L. F., Huff, W. E., Harvey, Yersin, A. G., Elissalde, M. H., Witzel, D. A., Giroir, L. E., and Phillips, T. D. (1990c). Effects of hydrated sodium calcium aluminosilicate on growing turkey poults during aflatoxicosis. *Poultry Sci.* **69**, 175.

Kubena, L. F., Harvey, R. B., Huff, W. E., Yersin, A. G., Elissalde, M. H., and Witzel, D. A. (1990d). Efficacy of a hydrated sodium calcium aluminosilicate to reduce the toxicity of aflatoxin and diacetoxyscirpenol. *Poultry Sci.* **69**, 76.

Kubena, L. F., Huff, W. E., Harvey, R. B., Yersin, A. G., Elissalde, M. H., Witzel, D. A., Giroir, L. E., and Phillips, T. D. (1991a). Effects of a hydrated sodium calcium aluminosilicate on growing turkey poults during aflatoxicosis. *Poultry Sci.* **70**, 1823–1830.

Kubena, L. F., Harvey, R. B., Phillips, T. D., Clement, B. A., and Elissalde, M. H. (1991b). Effects of sodium calcium aluminosilicate on broiler chicks during aflatoxicosis. *Poultry Sci.* **70**, 68.

Lancaster, M. C. (1968). Comparative aspects of aflatoxin-induced hepatic tumors. *Cancer Res.* **28**, 2288–2292.

Lawlor, T. E., Haworth, S. R., Zeiger, E., Park, D. L., and Lee, L. S. (1985). Mutagenic potential of ammonia-related aflatoxin reaction products in cottonseed meal, *J. Am. Oil Chem. Soc.* **62**, 1136–1138.

Lee, L. S., and Cucullu, A. F. (1978). Conversion of aflatoxin B_1 to aflatoxin D_1 in ammoniated peanut and cottonseed meals. *J. Agric. Food Chem.* **26**, 881–884.

Lee, L. S., Cucullu, A. F., Franz, A. O., Jr., and Pons, W. A., Jr. (1969). Destruction of aflatoxins in peanuts during dry and oil roasting. *J. Agric. Fd. Chem.* **17**, 451–453.

Lee, L. S., Stanley, J. B., Cucullu, A. F., Pons, W. A., Jr., and Goldblatt, L. A. (1974). Ammoniation of aflatoxin B_1: Isolation and identification of the major reaction product, *J. Assoc. Off. Anal. Chem.* **57**, 626–631.

Lee, L. S., Dunn, J. J., DeLucca, A. J., and Ciegler, A. (1981). Role of lactone ring of aflatoxin B_1 in toxicity and mutagenicity. *Experientia* **37**, 16–17.

Lee, Y. J., Hagler, W. M., Jr., Hutchins, J. E., and Voyksner, R. D. (1989). Conversion of aflatoxins to bisulfite adducts for quantitation and confirmation. *Biodeterioration Res.* **2**, 313–335.

Lewis, G., Mackson, L. M., and Allcroft, R. (1967). The effect of feeding toxic groundnut meal to sheep over a period of 5 years. *Vet. Rec.* **80**, 312–314.

Lindenfelser, L. A., and Ciegler, A. (1970). Studies on aflatoxin detoxification in shelled corn by ensiling. *J. Agric. Fd. Chem.* **18**, 640–643.

Machen, M. D., Clement, B. A., Shepherd, E. C., Sarr, A. B., Pettit, R. E., and Phillips, T. D. (1988). Sorption of aflatoxins from peanut oil by aluminosilicates. *Toxicologist* **8**, 265.

Machen, M. D., Mayura, K., Clement, B. A., and Phillips, T. D. (1991). Utilization of *in vitro* bioassays to assess the effectiveness of chemisorption of aflatoxin B_1 by hydrated sodium calcium aluminosilicate. *Toxicologist* **11**, 280.

Mann, G. E., Codifer, L. P., Jr., and Dollear, F. G. (1967). Effect of heat on aflatoxins in oilseed meals. *J. Agric. Fd. Chem.* **15**, 1090.

Mann, G. E., Codifer, L. P., Jr., Gardner, H. K., Koltun, S. P., and Dollear, F. G. (1970). Chemical inactivation of aflatoxins in peanut and cotonseed meals. *J. Am. Oil Chem. Soc.* **47**, 173–176.

Marth, E. H., and Doyle, M. P. (1979). Update on molds: Degradation of aflatoxin. *Fd. Technol.* **33**, 81–87.

Martinez, R. R. (1979). Las aflatoxinas en las tortillas. *Veterinario Mexico* **10**, 37.

Masimanco, N., Remacle, J., and Ramaut, J. (1973). Elimination of alfatoxin B_1 by absorbent clays in contaminated substrates. *Ann. Nutr. Alim.* **23**, 137–147.

Masri, M. S., Vix, H. L. E., and Goldblatt, L. A. (1969). Process for detoxifying substances contaminated with aflatoxin. United States Patent 3, 429, 709.

Moerck, K. E., McElfresh, P., Wohlman, A., and Hinton, B. W. (1980). Aflatoxin destruction in corn using sodium bisulfite, sodium hydroxide and aqueous ammonia. *J. Food Protect.* **43**, 571–574.

Natarajan, K. R., Rhee, K. C., Cater, C. M., and Mattil, K. F. (1975). Distribution of aflatoxins in various fractions separated from raw peanuts and defatted peanut meal. *J. Am. Oil Chem. Soc.* **52**, 44–47.

Nichols, T. E., Jr. (1987). Aflatoxin in the southeastern USA. *In* "Aflatoxins in Maize: Proceedings of a Workshop" (M. S. Zuber, E. B. Lillehoj, and B. L. Renfo, eds.), pp. 339–347. CIMMYT, Mexico City, Mexico.

Norred, W. P. (1979). Effect of ammoniation on the toxicity of corn artificially contaminated with aflatoxin B_1. *Toxicol. Appl. Pharmacol.* **51**, 411–416.

Norred, W. P. (1981). Excretion and distribution of ammoniated ^{14}C-labeled aflatoxin B_1-contaminated corn. *Fed. Proc.* **40**, 280.

Norred, W. P. (1990). Animal testing procedures in assessing the efficacy of ammoniation of commodities. *In* "A Perspective on Aflatoxin in Field Crops and Animal Food Products in the United States: A Symposium," pp. 120–126. National Technical Information Service, Springfield, Virginia.

Norred, W. P., and Morrissey, R. (1983). Effects of long-term feeding of ammoniated aflatoxin contaminated corn to Fisher 344 rats. *Toxicol. Appl. Pharmacol.* **70**, 96–104.

Palmgren, M. S., and Hayes, A. W. (1987). Aflatoxin in food. *In* "Mycotoxins in Food" (P. Krogh, ed.), pp. 65–95. Academic Press, New York.

Park, D. L., and Lee, L. S. (1990). New perspectives on the ammonia treatment for decontamination of aflatoxins. *In* "A Perspective on Aflatoxin in Field Crops and Animal Food Products in the United States: A Symposium," pp. 127–137. National Technical Information Service, Springfield, Virginia.

Park, D. L., and Rua, S. M. Jr. (1990). Biological evaluation of aflatoxins and metabolites in animal tissues. *Drug Metab. Rev.* **22**, 871–890.

Park, D. L., Lee, L. S., Price, R. L., and Pohland, A. E. (1988). Review of the decontamination of aflatoxins by ammoniation: Current status and regulation. *J. Assoc. Off. Anal. Chem.* **71**, 685–703.

Park, D. L., Lee, L. S., and Kolton, S. A. (1984). Distribution of ammonia-related aflatoxin reaction products in cottonseed meal. *J. Am. Oil Chem. Soc.* **61**, 1071–1074.

Phillips, T. D. (1987). Recent developments in the study of mycotoxins. *Proceedings of a Kaiser Aluminum and Chemical Company Symposium*, Rosemont, Illinois, pp. C3–C11.

Phillips, T. D., Kubena, L. F., Harvey, R. B., Taylor, D. R., and Heidelbaugh, N. D. (1987). Mycotoxin hazards in agriculture: New approach to control. *J. Am. Vet. Med. Assoc.* **190(12)**, 1617.

Phillips, T. D., Kubena, L. F., Harvey, R. B., Taylor, D. R., and Heidelbaugh, N. D. (1988a). Hydrated sodium calcium aluminosilicate: A high affinity sorbent for aflatoxin. *Poultry Sci.* **67**, 243–247.

Phillips, T. D., Clement, B. A., Kubena, L. F., and Harvey, R. B. (1988b). Mycotoxins: Detection and detoxification. *Proceedings of the Delmarva Poultry Health and Condemnations Meeting,* Ocean City, Maryland, pp. 94–106.

Phillips, T. D., Clement, B. A., Kubena, L. F., and Harvey, R. B. (1989). Prevention of aflatoxicosis in animals and aflatoxin residues in food of animal origin with hydrated sodium calcium aluminosilicate. *Proceedings of the World Association of Veterinary Food Hygienists,* Stockholm, Sweden.

Phillips, T. D., Clement, B. A., Kubena, L. F., and Harvey, R. B. (1990a). Use of dietary chemisorbents to prevent aflatoxicosis in farm animals. *In* "Perspectives on Aflatoxins in Field Crops and Animal Food Products in the United States." pp. 106–114. National Technical Information Service, Springfield, Virginia.

Phillips, T. D., Clement, B. A., Kubena, L. F., and Harvey, R. B. (1990b). Detection and detoxification of aflatoxins: Prevention of aflatoxicosis and aflatoxin residues with hydrated sodium calcium aluminosilicate. *Vet. Human Toxicol.* **32,** 15–19.

Phillips, T. D., Sarr, B. A., Clement, B. A., Kubena, L. F., and Harvey, R. B. (1991). Prevention of aflatoxicosis in farm animals via selective chemisorption of aflatoxin. *In* "Mycotoxins, Cancer and Health" (G. A. Bray and D. H. Ryan, eds.), Vol. 1, pp. 223–237. Louisiana State University Press, Baton Rouge.

Phillips, T. D., Clement, B. A., Sarr, A. B., Harvey, R. B., and Kubena, L. F. (1992). A practical approach to the control of aflatoxins: Selective chemisorption by phyllosilicate clay. *Proceedings of the 3rd World Congress on Foodborne Infections and Intoxications,* Berlin, Germany.

Price, R. L., and Jorgensen, K. V. (1985). Effects of processing on aflatoxin levels and on mutagenic potential of tortillas made from naturally contaminated corn. *J. Food Sci.* **50,** 347–357.

Purchase, I. F. H., Steyn, M., Rinsma, R., and Tustin, R. C. (1972). Reduction in the aflatoxin M content of milk by processing. *Fd. Cosmet. Toxicol.* **10,** 383–387.

Rayner, E. T., Koltun, S. P., and Dollear, F. G. (1977). Solvent extraction of aflatoxins from contaminated agricultural products. *J. Am. Oil. Chem. Soc.* **54,** 242A–244A.

Reiss, J. (1978). Mycotoxins in foodstuffs. XI. Fate of aflatoxin B_1 during preparation and baking of whole wheat bread. *Cereal Chem.* **55,** 421–423.

Samarajeewa, U., Sen, A. C., Fernando, S. Y., Ahmed, E. M., and Wei, C. I. (1991). Inactivation of aflatoxin B_1 in corn meal, copra meal and peanuts by chlorine gas treatment. *Fd. Chem. Toxicol.* **29,** 41–47.

Sarr, A. B., Clement, B. A., and Phillips, T. D. (1990). Effects of molecular structure on the chemisorption of aflatoxin B_1 and related compounds by hydrated sodium calcium aluminosilicate. *Toxicologist* **10,** 163.

Sarr, A. B., Clement, B. A., and Phillips, T. D. (1991). Molecular mechanism of aflatoxin B_1 chemisorption by hydrated sodium calcium aluminosilicate. *Toxicologist* **11,** 97.

Schroeder, T., Zweifel, U., Sagelsdorff, P., Friederich, U., Luthy, J., and Schlatter, C. (1985). Ammoniation of aflatoxin-containing corn: Distribution, *in vivo* covalent deoxyribonucleic acid binding and mutagenicity. *J. Agric. Food Chem.* **33,** 311–316.

Schulze, D. G. (1989). An introduction to soil mineraology. *In* "Minerals in Soil Environments" (J. B. Dixson and S. B. Weed, eds.), 2d Ed., pp. 1–34. Soil Science Society of America, Madison, Wisconsin.

Shane, S. M. (1991). Economic significance of mycotoxicoses. *In* "Mycotoxins, Cancer and Health" (G. Bray and D. Ryan), pp. 53–64. Louisiana State University Press, Baton Rouge.

Shantha, T. (1987). Detoxification of groundnut seed and products in India. *In* "Summary and Recommendations of the International Workshop on Aflatoxin Contamination of Groundnut," pp. 16 ICRISAT Center, India.

Shantha, T., and Sreenivasa, M. (1977). Photo-destruction of aflatoxin in groundnut oil. *Indian J. Technol.* **15,** 453.

Shantha, T., Sreenivasa, M., Rati, E. R., and Prema, V. (1986). Detoxification of groundnut seeds by urea and sunlight. *J. Fd. Safety* **7,** 225–232.

Southern, L. L., and Clawson, A. J. (1980). Ammoniation of corn contaminated with aflatoxin and its effects on growing rats. *J. Anim. Sci.* **50**, 459–466.

Stark, A. A., Gal, Y., and Shaulsky, G. (1990). Involvement of singlet oxygen in photoactivation of aflatoxins B$_1$ and B$_2$ to DNA-binding forms *in vitro. Carcinogenesis* **11**, 529–534.

Stoloff, L. (1976). Occurrence of mycotoxins in food and feeds. *In* "Mycotoxins and Other Fungal Related Food Problems" (J. V. Rodricks, ed.), pp. 23–50. American Chemical Society, Washington, D.C.

Stoloff, L. (1980). Aflatoxin M$_1$ in perspective. *J. Fd. Protect.* **43**, 226–230.

Stoloff, L., Trucksess, M., Harding, N., Francis, O. J., Hayes, J. R., Polan, C. E., and Campbell, T. C. (1975). Stability of aflatoxin M in milk. *J. Dairy Sci.* **58**, 1789–1793.

Trager, W., and Stoloff, L. (1967). Possible reactions for aflatoxin detoxification. *J. Agric. Fd. Chem.* **15**, 679–681.

Ulloa-Sosa, M., and Schroeder, H. W. (1969). Note on aflatoxin decomposition in the process of making tortillas from corn. *Cereal Chem.* **46**, 397–400.

Waltking, A. E. (1971). Fate of aflatoxin during roasting and storage of contaminated peanut products. *J. Assoc. Off. Anal. Chem.* **54**, 533–539.

Wieder, R., Wogan, G. N., and Shimkin, M. B. (1968). Pulmonary tumors in strain A mice given injections of aflatoxin B$_1$. *J. Natl. Cancer Inst.* **40**, 1195–1197.

Wogan, G. N., and Newberne, P. M. (1967). Dose-response characteristics of aflatoxin B$_1$ carcinogenesis in the rat. *Cancer Res.* **27**, 2370–2376.

Wood, G. E. (1989). Aflatoxins in domestic and imported foods and feeds. *J. Assoc. Off. Anal. Chem.* **72**, 543–548.

Yagen, B., Hutchins, J. E., Cox, R. H., Hagler, W. M., Jr., and Hamilton, P. B. (1989). Aflatoxin B$_1$S: Revised structure for the sodium sulfonate formed by destruction of aflatoxin B$_1$ with sodium bisulfite. *J. Fd. Protect.* **52**, 574–577.

Yahl, K. R., Watson, S. A., Smith, R. J., and Barabolak, R. (1971). Laboratory wet-milling of corn containing high levels of aflatoxin and a survey of commercial wet-milling products. *Cereal Chem.* **48**, 385–391.

Yousef, A. E., and Marth, E. H. (1986). Use of ultraviolet energy to degrade aflatoxin M$_1$ in raw or heated milk with and without added peroxide. *J. Dairy Sci.* **69**, 2243–2247.

Part IV

❖

Analytical Identification
of Aflatoxins

19

Recent Methods of Analysis for Aflatoxins in Foods and Feeds

❖

Mary W. Trucksess and Garnett E. Wood

INTRODUCTION

Aflatoxins, a group of structurally related mycotoxins, are well known for their toxic and carcinogenic effects in certain susceptible animal species as well as in humans. Since their discovery in the early 1960s, the relationship between the toxins and diseases of animals and humans has been of considerable interest. The occurrence of aflatoxins in foods and feeds is unavoidable, and is influenced by certain environmental factors; hence, the extent of aflatoxin contamination is unpredictable and varies with geographic location, agricultural and agronomic practices, and susceptibility of commodities to fungal invasion during preharvest, storage, or processing.

The major aflatoxins of concern are B_1, B_2, G_1, G_2, M_1, and M_2. New information on the mode of action of aflatoxin B_1 (AFB_1), the most potent of the six naturally occurring aflatoxins, has been reported (Bressac *et al.*, 1991; Hsu *et al.*, 1991). Specific mutations were identified as the cause of human hepatocellular carcinoma in patients in China, where the hepatitis B virus and aflatoxin-contaminated food are common. These mutations were consistent with mutations caused by AFB_1 in mutagenesis experiments in rat liver. Because of the diversity of toxicological manifestations and the economic losses caused by exposure to aflatoxins, humans and susceptible animals must be protected from undue exposure to these toxins to safeguard their health.

The literature contains many reports of the occurrence of aflatoxins in foods and feeds. In most developed countries, the levels and incidence of aflatoxins are usually much lower in human foods (generally < 20 ng/g) than in feeds. Many countries have attempted to limit exposure to aflatoxins by imposing regulatory limits on foods and feeds in commercial channels. Food and feed industries should monitor their products routinely to ensure that aflatoxin levels are below the regulatory limits. Sensitive, accurate, and precise methods of analysis are needed for any monitoring program to be effective.

Systematic approaches to sampling, sample preparation, and analysis are absolutely necessary to determine aflatoxins at the parts-per-billion level. In this chapter, emphasis is placed on currently available analytical methodology. Sampling and sample preparation are discussed only briefly in the first section. The use of solid-phase extraction for analyte isolation and various aspects of thin layer chromatography, liquid chromatography, and immunochemical methods are described in subsequent sections. Instrumentation is addressed briefly.

Only articles published after 1984 are included in this update, because excellent reviews of modern methods of analysis for mycotoxins have been published (Cole, 1986; Scott, 1991). The exceptions are some of the earlier, widely used techniques and the collaboratively studied and approved methods of the Association of Official Analytical Chemists (AOAC; Eppley, 1966). The last three sections of this chapter focus on automation, confirmation of identity, and safe handling of moldy grains and the pure aflatoxins.

SAMPLING AND SAMPLE PREPARATION

Specific sampling plans have been developed and tested rigorously for some commodities such as corn, peanuts, and tree nuts; sampling plans for some other commodities have been modeled after them. The U.S. Department of Agriculture (USDA) recommends that 48 lb peanuts and 5–10 lb corn, milo, and other grains be collected for aflatoxin analysis (Whitaker *et al.*, 1979). The Inspection Operations Manual of the U.S. Food and Drug Administration (FDA) gives detailed descriptions of sampling sizes for various commodities and processed products (FDA, 1988). The sample from the same lot can be collected from 10–15 sites using different probe patterns or an automatic sampler (Whitaker *et al.*, 1979).

The entire primary sample must be ground and mixed so the analytical test portion has the same concentration of toxin as the original sample. A 2-lb portion is sufficient when coarse or pelleted feed is tested because any toxins in the individual ingredients already have been mixed throughout the feed. Whenever possible, the grain should be analyzed before it is processed into feed because of the many other components of the feed that might interfere in the analysis. In general, 50 g ground material is used for analysis. One study indicates that a 10-g test portion of a sufficiently ground and blended sample would produce an analyte variance statistically comparable with that of a 50-g portion (Francis *et al.,* 1988b). An overview of sampling and analyte purification for the identification and quantitation of natural toxicants in foods and feeds offers information in these areas (Park and Pohland, 1989).

SOLID-PHASE EXTRACTION

All analytical procedures include three steps: extraction, purification, and determination. The most common solvent system for extraction has been a mixture of chlorohydrocarbon and water. This system gradually is being replaced by the methanol–water or acetonitrile–water system. The most significant recent improvement in the purification step is the use of solid-phase extraction (SPE).

Test extracts are cleaned up before instrumental analysis (thin layer or liquid chromatography) to remove coextracted materials that often interfere with the determination of target analytes. Traditionally, the procedures used are column chromatography (silica gel) and liquid–liquid partition involving large volumes of solvent (> 200 ml). Considerable time is required for preparing the adsorbent, packing the chromatographic columns, eluting the toxins from the columns, and evaporating the solvent. SPE, also referred to as liquid–solid extraction, is the latest technique for analyte clean-up or concentration of aflatoxin. The method is quick, solvent efficient, and therefore economical. Many SPE cartridges are available commercially (Sep-Pak®, Bond-Elut®, Aflatest®, Multifunctional). The SPE cartridge is a microcolumn made of plastic tubing containing 100–500 mg 40-μm stationary-phase particles in the middle and plastic frits at both ends. The most commonly used stationary phases are silica gel, bonded-phase C_{18}, Florisil®, and antibody–agarose (immunoaffinity packing).

Usually a multicartridge vacuum manifold is used to pull extract and eluting solvent through the column. However, the antibody–agarose is quite fragile and requires the application of positive pressure with a piston syringe. The elution conditions for the cartridges typically are chosen to retain the analytes on the adsorbent while the coextracted nonaflatoxin materials are washed from the cartridge with the eluant; alternatively, the coextracted materials are retained while the aflatoxins are eluted from the cartridge, SPE requires less solvent than conventional column chromatography or liquid–liquid partition (Trucksess *et al.,* 1984; Hutchins *et al.,* 1989). The volume of eluate containing the analyte is suitable for subsequent injection into a liquid chromatograph, making automation of the analysis possible.

THIN-LAYER CHROMATOGRAPHY

Thin layer chromatography (TLC), also known as flat bed chromatography or planar chromatography, is one of the most widely used separation techniques in aflatoxin analysis. The first analytical method using this technique was published by Eppley (1966). This method was evaluated in a collaborative study conducted by the AOAC and is now an AOAC official method (AOAC, 1990). Since then, TLC has been the method of choice to identify and quantitate aflatoxins at levels as low as 1 ng/g.

The TLC method often is used to verify findings by newer, more rapid methods. For qualitative testing, this simple technique consists of spotting the test extract on an adsorbent-coated plate and developing the plate in a glass or metal tank with an appropriate mobile phase. The developed plate is then observed under longwave ultraviolet (UV) light, and the fluorescent intensities of the presumptive aflatoxin spots from the extract are compared with those of standard aflatoxin spots.

Reliable quantitative tests are now available because of improvements in instrumentation and the availability of a wide variety of adsorbents for use as the stationary phase on the TLC plates. Great advances have been made in the quality of the stationary phases in recent years. Phases with small particle size and narrow particle size distribution have become available. High performance TLC (HPTLC) plates are prepared using these improved phases.

The HPTLC plates are smaller than the conventional TLC plates, usually 10 × 10 cm or 10 × 20 cm. The separation efficiency is typically 5000 theoretical plates for 5-cm migration. Improvements have been made in the instrumentation necessary to accommodate the smaller plate sizes, the small volumes of test solution applied, the extremely compact fluorescence signal of the aflatoxin spot, and the close migration of the toxin spots. HPTLC instruments for application of the test solution, plate development, and densitometry were evaluated (Coker, 1988). Optimum sensitivity, accuracy, and precision were obtained using a fully automated TLC sampler, an unsaturated conventional TLC glass chamber, and a monochromatic fluorodensitometer. Benzene–acetonitrile (98 + 2) was the most suitable spotting solvent. This study used aflatoxin standards only.

The solubility of the residue remaining after extraction and clean-up should be an important consideration in selecting the spotting solvent. In general, most residues dissolve more readily in chloroform than in the solvent described in the study. Compact spots can be obtained by reducing the rate of solvent delivery. A microcomputer has been interfaced to a fluorodensitometer to simplify the data handling procedure (Whitaker *et al.,* 1990). The system computed and recorded the amount of aflatoxin in the extract spots and the concentration of aflatoxin that was in the original extracted test portion.

TLC generally is more popular in Europe than in the United States. The number of publications on TLC has declined, which is not necessarily an indication of the extent to which TLC is being used worldwide. Apparently, TLC methods are used routinely but are not published unless they are being applied to new commodities or are improvements of previously published methods. Many TLC methods for aflatoxins in foods such as corn, peanuts, peanut butter, cottonseed, milk, meat, and eggs are included in the compendium *Official Methods of Analysis* (AOAC, 1990).

Some of the widely used TLC methods for aflatoxins B_1, B_2, G_1, and G_2 in which silica gel is used as the stationary phase are listed in Table 1 in chronological order. Four types of development are included: one solvent, two solvents, bidirectional, and two-dimensional. The one-solvent system is self-explanatory. In the two-solvent development, the plate is first developed with a solvent that removes the interferences; then the plate is dried and developed with another solvent in the same direction to separate the toxins. In the bidirectional development, extracts are spotted in the middle of the plate. After the first development with a nonpolar solvent to remove the nonpolar components, the top of the plate below the solvent front is cut off; the plate is then turned upside down (180°) and developed with a more polar solvent system to separate the toxins. Two-dimensional TLC is the most powerful technique and offers greater resolution than other chromatographic techniques. This method requires two solvents of different selectivity for the two developments. The test extract is spotted in one corner with reference standards on the two adjacent corners. The plate is developed in one direction, then rotated (90°) and developed in a second direction.

In the past decade, TLC plates precoated with bonded-phase silica gel have become available commercially; they are known as reversed-phase (RP) TLC plates. In the RP-TLC system, the mobile phase is more polar than the stationary phase whereas, in the normal-phase (NP) TLC (silica or alumina), the mobile phase is less polar than the coating medium. RP-TLC plates are made of a variety of bonded-phase adsorbents, including C_2, C_8, C_{12}, C_{18}, and diphenyl types. Quantitation of aflatoxins by RP-TLC is still in the developmental stage, although RP-TLC can be used to confirm the identity of aflatoxins found using NP-TLC plates. This method also can be used for screening: 18 mycotoxins, including the aflatoxins, were identified using RP C_{18} or RP diphenyl plates (Abramson *et al.*, 1989). TLC has maintained its analytical status because of the constant improvements of instrumentation and stationary phases.

LIQUID CHROMATOGRAPHY

Liquid chromatography (LC) is similar to TLC in many respects, including analyte application, stationary phase, and mobile phase. LC and TLC complement each other. For an analyst to use TLC for preliminary work to optimize LC separation conditions is not unusual. LC methods for the determination of aflatoxins in foods include normal-phase LC (NPLC), reversed-phase LC (RPLC) with pre- or before-column derivatization (BCD), RPLC followed by postcolumn derivatization (PCD), and RPLC with electrochemical detection. All these techniques, except electrochemical detection, use fluorescence detectors set at E_x 360 nm, $E_m > 420$ nm. Two review articles on LC methodology have been published (Beaver, 1989; Wilson, 1989) but they include only LC methods developed before 1986 for corn and peanuts.

In the early 1980s, most of the fluorescence detectors were not sensitive

TABLE 1 TLC Methods for Aflatoxins B_1, B_2, G_1, and G_2

Commodity	Clean-up	Spotting[a]	TLC type[b]	Mobile phase[c]	Detection limit ($\mu g/g$)	Reference
Vegetable oil	Silica gel column	C	2D	C + A (9 + 1); E + M + W (96 + 3 + 1)	5	Miller et al. (1985)
Spices	Silica gel column	B + N (98 + 2)	MD	C + A (9 + 1); E		Madhyastha and Bhat (1985)
Poultry feed	Silica gel cartridge	B + N (98 + 2)	HPTLC 2D	C + A + I (92 + 5 + 2.5); E + M + W (88 + 9 + 3)	10	Kozloski (1986)
Animal tissue	Silica gel column	B + N (98 + 2)	2D	C + A + I (87 + 10 + 3); E + M + W (95 + 4 + 1)	0.5	Richard and Lyon (1986)
Mixed feeds	Liquid partition	C	BiD	C + A (88 + 12); T + E + F (48 + 40 + 12)	1	Pennington (1986)
Feeds	Silica gel, Florisil, C_{18}	T + N (98 + 2)	2D	E + M + W (94 + 4.5 + 1.5); C + A (9 + 1)	10	van Egmond et al. (1988b)
Feeds	Silica gel cartridge	B + N (98 + 2)	HPTLC 2D	C + A (9 + 1); B + M + A (9 + 5 + 0.5)	4	Simonella et al. (1987)
Sorghum	Florisil cartridge	B + N	HPTLC BiD	E; C + X + A (6 + 3 + 1)	5	Jewers et al. (1989)
Maize	C_{18}, C_8, C_2 cyclohexyl phenyl	B + N (98 + 2)	HPTLC BiD	E; C + X + A (6 + 3 + 1)	0.8	Tomlins et al. (1989)
Olives	Extraction	B + N	1D	T + E + F (6 + 3 + 1)	5.0	Mahjoub and Bullerman (1990)

[a]B, Benzene; C, Chloroform; N, acetonitrile; T, toluene.

[b]1D, 1 dimensional; 2D, 2 dimensional; BiD, bidirectional; MD, multidevelopments.

[c]A, Acetone; B, benzene, C, chloroform; E, ethyl ether; Et, ethyl acetate; F, formic acid; M, methanol; T, toluene; W, water; X, xylene.

enough to detect the native fluorescence of AFB_1 and AFB_2 at < 0.5 ng in the mobile phase eluted from a normal-phase silica gel column. The use of a detector flow cell packed with silica gel can enhance the fluorescence of AFB_1 and AFB_2 (Panalaks and Scott, 1977). In one study, the determination limit was 0.25 ng/g for AFB_1, 0.5 ng/g for AFG_1, and 0.2 ng/g for AFB_2 and AFG_2 in corn when chloroform–cyclohexane–acetonitrile–isopropanol $(75 + 22 + 3 + 0.2)$ was used as the mobile phase (Francis et al., 1982). The detector required frequent repacking because the silica gel adsorbed the matrix irreversibly and caused elevated noise levels in the detector, decreased the resolution, and lowered the fluorescence response of the toxins.

Because RPLC is among the most effective analytical techniques, it frequently is used to overcome the problems of NPLC. The stationary phase is usually a C_{18} chain chemically bonded to the silica gel support; the mobile phase is a mixture of water, methanol, and acetonitrile. The column dimensions are 3.9–4.6 mm $\times$ 15–30 cm; the particle size is 5–10 μm with 9- to 12-nm pore size. One drawback of RPLC is that AFB_1 and AFG_1 do not fluoresce in aqueous mobile phase. Consequently, pre- and postcolumn derivatization techniques are used to increase sensitivity (Beebe, 1978; Shepherd and Gilbert, 1984).

The BCD procedure can be optimized by adding hexane and trifluoroacetic acid (TFA) to the extracted analyte residue (Tarter et al., 1984), allowing the mixture to react for 5 min at room temperature, and adding aqueous acetonitrile to the test solution. After mixing, a portion of the aqueous layer is injected onto the column for separation and quantitation. The determination limit for aflatoxin in peanut butter is about 0.3 ng/g for AFB_1. The method was studied collaboratively and was adopted as an AOAC official method (Park et al., 1990). The disadvantages of this technique are the occurrence of incomplete reaction and the formation of more than one derivative. The average recovery for added total aflatoxin at 10–30 ng/g levels in corn and peanut products was about 70%.

PCD with iodine can determine 0.7 ng AFB_1/g corn (Thiel et al., 1986). In this system, iodine is introduced as an aqueous solution into the eluant stream between the column outlet and the fluorescence detector. A current publication gives a detailed description of the system (Trucksess et al., 1991). The LC column outlet is connected to one arm of a stainless steel, low dead-volume tee by a short piece of 0.01-in. (0.25-mm) id tubing. The outlet of a second LC pump, which delivers the postcolumn reagent, is connected to the second arm of the tee. One end of a 610-cm $\times$ 0.5-mm id coil of Teflon tubing is connected to a third arm of the tee and the other end of the coil is connected to the detector. The reaction coil is maintained at 70°C with an oven or a constant temperature bath. The following flow rates are used: (1) mobile phase [(μBondapak C_{18} column):water–acetonitrile–methanol $(3 + 1 + 1)$], 1.0 ml/min; (2) postcolumn reagent (100 mg iodine in 2 ml methanol and 200 ml water), 0.3 mL/min; and (3) total rate through reaction coil, 1.3 ml/min. The reaction coil volume is 1.2 ml; thus, with a total flow rate of 1.3 ml/min, postcolumn reaction time is about 55 sec. The disadvantage of this procedure is that it requires two pumps and a reaction coil kept at constant temperature.

An alternative postcolumn derivatization method has been developed as an improvement over the procedure just described. This method involves the use of a postcolumn on-line electrochemical cell to produce bromine, which enhances the fluorescence signal of AFB_1 and AFG_1 (Kok *et al.,* 1986). The method has been modified and used to analyze corn naturally contaminated with aflatoxins (D. M. Wilson, M. W. Trucksess, T. Urano, and Y. Kim unpublished observations).

In this method, the bromine is produced from the bromide present in the mobile phase [water–methanol–acetonitrile (6 + 2 + 2) with 1 mM potassium bromide and 1 mM nitric acid] in a postcolumn electrochemical cell (a Kobra-cell). The cell is made of a Teflon block, stacked in the middle with the following components in sequential order: a working electrode, a spacer, a membrane, a spacer, and a counterelectrode. The column outlet is connected to the inlet of the cell next to the working electrode (at the top of the cell). The outlet on top of the cell is connected to a reaction coil (0.3 mm $\times$ 12 cm), which is connected to a fluorescence detector. The detector outlet is connected to the inlet at the bottom of the cell, and the outlet at the bottom of the cell is drained to the waste container. The voltage to the cell is controlled by a variable DC supply. The optimum voltage required to produce the maximum signal is 7 V. Higher voltage would not increase the signal but would generate too much heat in the cell, resulting in destruction of the membrane. This LC PCD procedure is simple to use; however, to avoid damaging the electrochemical cell the following precautions must be taken: (1) plastic nuts and ferrules should be used to couple the cell to the column and detector and (2) the cell must be cleaned with and stored in distilled water to keep the membrane in good condition.

Two other PCD methods are not as well established as those just described. One of these methods uses postcolumn enhancement with cyclodextrin (Francis *et al.,* 1988a), and is similar to PCD with iodine. The other method uses an electrochemical detector, and is capable of reducing the background noise inter-ference associated with other electrochemical methods (Duhart *et al.,* 1988) by preelectrolyzing the mobile phase, switching to a glass-lined column, and using a better oxygen-removal technique. The electrochemical detector is a po-larographic analyzer containing a pressurized dropping mercury electrode with 1-sec drop time. The differential-pulse mode is used, in which each aflatoxin gives a slightly different peak potential. The potential between the AFBs and AFGs is -1.28 V. The limit of determination is about 10 ng/g in peanut butter. The advantage of this procedure is that it does not require a separate derivatiza-tion step, as is common for fluorescence detection. Some applications of the various LC methods are listed in Table 2.

IMMUNOCHEMICAL METHODS

TLC and LC methods for determining aflatoxins in food are laborious and time consuming. Often these techniques require knowledge and experience of

TABLE 2 LC Methods for Aflatoxins B_1, B_2, G_1, and G_2

Commodity	Clean-up	LC mode[a]	Mobile phase[b]	Detector[c]	Detection limit (ng/g)	Reference
Maize, peanuts, sorghum	Silica gel column	RP; PCD-iodine	Buffer + N + M (39 + 9 + 7)	F	6.3	Thiel et al. (1986)
Cattle feed		RP; PCD-bromine	W + M + N (13 + 7 + 4) in 1 mM KBr, 1 mM HNO$_3$	F	0.8	Kok et al. (1986)
Feed	Florisil cartridge	RP; PCD-iodine	W + M + N (13 + 7 + 4)	F	1.0	Paulsch et al. (1988)
Peanut butter		RP; Hg electrode	Buffer + M + N (62.7 + 17.9 + 19.4)	Electrochemical	5.0	Duhart et al. (1988)
Peanuts, maize	Immunoaffinity	RP; PCD-iodine	W + M + N (50 + 20 + 30)	F	0.5	Candlish et al. (1988)
Peanuts	Silica gel	RP; PCD-iodine	W + THF (8 + 2)	F	5.0	Dorner and Cole (1988)
Maize feeds	Bio-beads S-X3	RP; PCD-iodine	W + N + M (6 + 3 + 1)	F	1.0	Hetmanski and Scudamore (1989)
Spices	Pb(AcO$_3$)$_2$	NP-silica gel packed flow cell	E + M + W (95 + 4 + 1) or C + N + M (97.75 + 1.25 + 1.0)	F	10	Adensam et al. (1989)
Peanuts, corn, cottonseed, cereals	Silica gel	NP-silica gel packed flow cell	E + M + W (95 + 4 + 1)	F	0.1	Tutelyan et al. (1989)
Corn	Silica gel cartridge	RP; BCD-TFA	W + THF + N (84 + 6 + 10)	F	2.0	Hutchins et al. (1989)
Grains, oilseeds, feeds	On-line	RP; PCD-iodine	Buffer + M + N (2 + 1 + 1)	F	5.0	Chamkasem et al. (1989)
Corn	Silica gel column	RP; PC-enhancement	M + 0.005 g/ml β-cyclodextrin (1 + 1)	F	1.0	Francis et al. (1988a)
Corn, treenuts, milo feeds	Multifunctional column	RP; BCD-TFA + A	N + W (1 + 4)	F	0.5	Wilson and Romer (1991)

[a]A, Acetic acid; BCD, precolumn derivatization; TFA, trifluoroacetic acid; PC, precolumn; RP, reversed phase; NP, normal phase; PCD, postcolumn derivatization.
[b]C, Chloroform; E, ethyl ether; N, acetonitrile; M, methanol; W, water; THF, tetrahydrofuran.
[c]F, Fluorescence.

chromatographic techniques to solve separation and interference problems. Through advances in biotechnology, highly specific antibody-based tests are now commercially available that can identify and measure aflatoxins in food in less than 10 min. These tests are based on the affinities of the monoclonal or polyclonal antibodies for aflatoxins. The three types of immunochemical methods (Chu, 1990) are radioimmunoassay (RIA), enzyme-linked immunosorbent assay (ELISA), and immunoaffinity column assay (ICA). The first two methods are based on competition between the unlabeled aflatoxin in the test solution and the labeled aflatoxin in the assay system for the specific binding sites of antibody molecules. Radioactive aflatoxin is used as a labeled ligand in the RIA and an aflatoxin–enzyme conjugate is used as ligand in the ELISA. In the ICA procedure, the antibody column traps or binds the aflatoxins, which then can be eluted from the column with methanol for subsequent measurement.

RIA first was developed in 1959 for the determination of insulin. In the competitive RIA for aflatoxins, a specific antibody is incubated with a constant amount of radiolabeled toxin in the presence of varying amounts of toxin standard or unknown commodity. Ammonium sulfate precipitation is used to remove the toxin–antibody complex from the solution. The toxin content of the commodity is related inversely to the amount of unbound radioactive toxin remaining in the supernatant solution. Although RIA is very sensitive, it has several disadvantages. The radioisotopes used in the assays are hazardous, present disposal difficulties, and may have short shelf lives; nonisotopic labels such as enzymes have been used in place of radioisotopes.

ELISA was developed in 1971. Although both direct and indirect competitive ELISAs have been used for aflatoxin determination, the direct assay is preferable for analytical purposes because it is simpler. This technique consists of a two-step process: (1) the reaction between the antibody and the toxin and (2) measurement of the reaction of the substrate with the enzyme attached to the toxin. Analyte isolation for the ELISA is simple. The test portion is extracted with methanol and water, and the filtrate is then diluted and analyzed. The reagents for ELISA consist of aflatoxin–peroxidase conjugate, enzyme substrate, and antibody-coated microtiter wells, cups, or cards. Diluted filtrate and aflatoxin–peroxidase conjugate are added to the antibody-coated apparatus, the toxin–antibody complex is formed, and the apparatus is washed with water. Substrate is added and the color is developed. The color of the test solution is compared with that of the standards and controls.

In the past few years, several immunoassay kits for aflatoxins have been marketed under various trade names (CAST, 1989). No formal or standard criteria have been established for evaluation of the kits. Several organizations such as the AOAC, International Union of Pure and Applied Chemistry, Environmental Protection Agency, USDA, and FDA are engaged actively in developing evaluation guidelines. Some of the criteria that have been used to evaluate immunoassays include sensitivity, specificity, avidity, applicability, stability, procedure, quality control, cost, precision, accuracy, and simplicity (Koeltzow and Tanner, 1990; Trucksess and Page, 1990).

Three of the commercial test kits have been studied collaboratively according to the AOAC guidelines: the Neogen Screen Kit, the Immuno Dot Screen Cup, and the Aflatest P immunoaffinity column (Park *et al.,* 1989a,b; Trucksess *et al.,* 1989,1991). The first two tests are yes/no types whereas the Aflatest P is quantitative. Performance commonly is assessed by examining the ability of the test to classify test samples correctly as either positive or negative at a certain aflatoxin level (20 ng/g). If no overlap in test results occurs between these two categories, the test can identify all test samples correctly, that is, distinguish the two categories perfectly. However, if the test results for the two categories overlap somewhat, the test cannot distinguish them perfectly.

This criterion raises the question of how much deviation is acceptable. A perfect test exhibits both 100% sensitivity (ability to identify a positive test sample as positive, at some target level) and 100% specificity (ability to identify a negative test sample, that is, below target level, as negative). However, tests are rarely perfect. Each test has a particular sensitivity and specificity. When the target level changes, various sensitivities and the corresponding specificities also change.

Defining the acceptable sensitivity or specificity is extremely difficult. The use of operating characteristic (OC) curves is one solution. The OC curve plots the true positive rate (TP; positive at target level) or percentage as a function of concentration. The false negative rate (FN; negative at target level) is a function of the TP $(1 - FN)$. The false positive rate (FP; positive at levels below the target level) is also a function of concentration. FP is usually highest as the concentration approaches the target level. Good performance of a test is characterized by a high TP $> 90\%$) and a low FP $(< 60\%)$.

Quantitative immunochemical methods should have about the same precision and accuracy as the traditional LC and TLC methods. Precision is the ability of an assay to yield the same result when the assay is repeated on the same test sample. The precision of a measurement commonly is represented by the standard deviation or coefficient of variation of replicate measurement. Accuracy is the degree to which the assay approximates the actual concentration of toxin in the test sample. The ability of an assay to recover known standards added to test samples is an important measure of accuracy. With some commodities, for example, cottonseed, a standard curve should be prepared for the commodity to eliminate matrix effects.

Immunochemical methods are quite specific and can, for the most part, be used to screen for aflatoxins in grain and grain products. Some of the immunochemical methods are also capable of giving quantitative results and are recognized as acceptable analytical methods by the American Oil Chemists Society (AOCS) (McKinney, 1989). Beginning with the AOCS 1990–1991 Smalley Aflatoxin Series, immunoassay test kits have been included in the methodology that can be used in the analysis of Smalley aflatoxin samples (peanut, corn, cottonseed, and nuts). Table 3 lists some of these methods and their applications. Although the methods are specific, simple, fast, and cost effective, they cannot be coupled with mass spectrometry to confirm the identity of the aflatoxins being

TABLE 3   Enzyme-Linked Immunosorbent Assay for Determination of Aflatoxins

Commodity	Aflatoxins	Extraction[a]	Format[b]	Detection limit (ng/g)	Reference
Peanut butter	B_1, G_1	N + W (1 + 1)	p, id, mt	0.25	Mortimer et al. (1987b)
Figs	B_1	M	p, d, mt	10	Reichert et al. (1988)
Peanut butter	B_1	M + W (55 + 45)	c, d, mt	8	Mortimer et al. (1988)
Peanut butter	B_1	M + W (6 + 1)	p, d, mt	1	Kawamura et al. (1988)
Corn, cottonseed, peanuts, peanut butter, feed	B_1, B_2, G_1	M + W (8 + 2)	c, d, cup	20	Trucksess et al. (1989)
Cottonseed, feed	B_1	M + W (55 + 45)	c, d, mt	15	Park et al. (1989a)
Corn, peanuts	B_1	M + W (55 + 45)	c, d, mt	15	Park et al. (1989b)
Peanut butter	B_1, B_2, G_1, G_2	N + W (1 + 1)	c, d, mt	2	Ward and Morgan (1991)

[a]N, Acetonitrile; M, methanol; W, water.
[b]c, Commercially available; d, direct competitive; id, indirect competitive; mt, microtiter wells.

measured. ELISA probably will have the greatest value when used in tandem with existing analytical methods.

METHODS FOR AFLATOXIN M_1 DETERMINATION IN MILK AND MILK PRODUCTS

Analytical methods for determining aflatoxin M_1 generally are based on chromatographic procedures (TLC and LC) followed by fluorescence measurement (AOAC, 1990). Since TLC confirmation of identity of M_1 and LC determination of M_1 require the formation of an M_1 derivative in the presence of TFA, the reaction conditions have been optimized (Stubblefield, 1987). The test residue in 200 μl each of hexane and TFA is heated 10 min at 40°C in a closed container. No unreacted M_1 and by-products remain.

No significant difference exists between the results by TLC and LC methods. Van Egmond et al. (1988a) showed that the ratio of average results by TLC to those by LC is 1.08. In this study, the majority of the participants used a fluorodensitometric reading for TLC to quantitate M_1 in powdered milk. Participants who used both TLC and LC methods preferred LC to TLC. Although the faster methods based on immunochemical procedures are being used more and more, no methods of this type have been studied collaboratively. Comprehensive reviews of the performance characteristics of various methods for M_1 have been published (Newsome, 1986; Groopman and Donahue, 1988; Pestka, 1988; Scott, 1989).

Accurate determinations of M_1 in milk at very low levels are difficult to obtain (van Egmond and Wagstaffe, 1987; Wood, 1991). The validity of the data often is based on the credibility of the laboratory and the method used. The use of a reference material containing M_1 could help verify the reliability and comparability of determinations performed in different laboratories. The Netherlands Community Bureau of Reference (BCR), a bureau of the Commission of the European Communities, has certified four full-cream milk powders as reference materials for their M_1 content (van Egmond and Wagstaffe, 1988). These reference materials are available commercially. One report demonstrated their use in method development and quality assurance of laboratory performance (Gilbert, 1988).

Table 4 shows methods currently used for M_1 analysis. Agreement between results obtained by either TLC or LC is good. However, apparent M_1 contamination in raw and reconstituted dried milk increases considerably when the direct ELISA method is used (Jackman, 1985; Kaveri *et al.*, 1987). This increase could be the result of the ability of the ELISA to identify M_1 residues in milk directly without the extraction and purification steps required in conventional LC or TLC methods. These steps could result in lower recoveries. The method for determining aflatoxin M_4, an isomer of M_1, and M_1 metabolites (Lafont *et al.*, 1986a,b) is not included in Table 4. Evidence suggests that M_4 is more mutagenic than M_1 (Scott, 1989). Scott proposed that an M_4 standard should be included in any TLC determination of M_1. He stated that the limit of determination for M_4 was about 5 ng/liter.

CONFIRMATION OF IDENTITIES OF THE AFLATOXINS

Although analytical methods might consist of different extraction, clean-up, and quantitation steps, the results of the analyses by such methods should be similar when the methods are applied properly. This agreement was illustrated by a study involving more than 20 European laboratories (van Egmond and Wagstaffe, 1989,1990) in preparation of reference materials (peanut butter naturally contaminated with aflatoxins) for the validation and quality assurance of methods. Since the reliability of the quantitative data is not in question, the problem still to be solved is the confirmation of identity of the aflatoxins. The confirmation techniques used involve either chemical derivatization or mass spectrometry (MS).

TFA is the most common reagent used for chemical derivatization of the aflatoxins. TFA is used as the catalyst in the addition of water across the double bond of the vinyl ether function of AFB_1 and AFG_1. Normally in the TLC methods, TFA is added to the spots of the extracts and standards; the plate is

TABLE 4 Methods for Aflatoxin M_1 Detection in Milk and Milk Products

Product type	Clean-up	Detection method[a]	Mobile phase[b]	Detection limit (ng/g)	Reference
Milk	Liquid partition	TLC, 1D	E + M + W (95 + 4 + 1)	0.5	Serralheiro and Quinta (1985)
Milk, yogurt	Alkaline extraction	TLC, 1D	C + A + P (85 + 10 + 5)	0.2	Dominguez et al. (1987)
Cheese	C_{18} cartridge	TLC, 2D	E + M + W (95 + 4 + 1); C + A (7 + 3)	0.1	Bijl et al. (1987)
Milk	C_{18}, silica, Bond-Elut cartridge	TLC, BiD	E, T + F + EOH (6 + 1 + 3); C + A + P (87 + 10 + 3) develop 3X	0.005	Kubicek et al. (1988)
Milk	Silica cartridge	LC, RP	W + N (73 + 27)	0.05	Yousef and Marth (1985)
Milk	C_{18} cartridge	LC, RP PCD-TFA	Linear gradient: M + W (10 + 90); M + W (65 + 35)	0.002	Carisano and Torre (1986)
Powdered milk	IAC[c]	LC, RP	W + N + M (68 + 24 + 8)	5×10^{-5}	Mortimer et al. (1987a)
Cheese	IAC	LC, RP	W + N + M (60 + 10 + 30)	0.005	Sharman et al. (1989)
Milk	IAC	LC, RP	W + M (1 + 1)	0.05	Hansen (1990)
Milk	IAC	Bromination fluorometer		0.05	Hansen (1990)
Milk	C_{18}, Bond-Elut	ELISA, mt		0.1	Jackman (1985)
Milk	IAC	ELISA, mt		0.1	Kaveri et al. (1987)

[a]1D, One dimensional; 2D, two dimensional; BiD, bidirectional; RP, reversed phase; PCD, postcolumn derivatization; TFA, trifluoroacetic acid; mt, microtiter wells.

[b]A, Acetone; C, chloroform; E, ethyl ether; EOH, ethanol; F, formic acid; M, methanol; N, acetonitrile; P, 2-propanol; W, water.

[c]IAC, Immunoaffinity column.

dried in a 40°C oven for 10 min and developed with chloroform–acetone–2-propanol (85 + 10 + 5). The fluorescent water-addition products of AFB_1 and AFG_1 are observed at R_f values of 0.2 and 0.15. The identity of M_1 can be confirmed in a similar manner with minor modifications. The spotted plate is covered with a clean glass plate, heated in a 70°C oven for 8 min, and developed in a slightly more polar solvent (7% 2-propanol). In the LC methods, TFA, iodine, or bromine is used to derivatize the aflatoxins before quantitation; thus, no further chemical confirmation of identity is needed. The chemical methods for confirmation of identity are not as definitive as the MS techniques.

Confirming the identity of aflatoxins by MS once required additional clean-up steps such as TLC isolation or solid-phase extraction (Park *et al.,* 1985) because the presence of impurities in the test extract was a problem. Another approach was to interface gas chromatography with mass spectrometry (GC/MS), that is, use GC to separate the impurities in the extract from the aflatoxins and use MS to confirm the identities of the aflatoxins. The first GC/MS method for AFB_1 used on-column injection at 40°C (Trucksess *et al.,* 1984) and a 6-m × 0.2-mm methyl silicone-coated, fused-silica column. Immediately after the test extract was injected onto the column, the column temperature was raised to 250°C in 4 min; the effluent was analyzed by negative ion chemical ionization (NICI)-MS. The NICI mass spectrum of AFB_1 showed major ions at m/z 312 and 297. The gas chromatograph was used to introduce the analyte to the mass spectrometer.

In another study, Goto *et al.* (1988) used GC to analyze mixtures of the four aflatoxins. The initial and final temperatures were set at 50 and 300°C and the rate of heating was set at 15 or 20°C/min. A 5% phenylmethylsilicone column was used to separate AFB_1, AFB_2, AFG_1, and AFG_2 (2, 2, 4, and 4 ng), which were determined by GC with flame ionization detection. This technique coupled with MS may be used for quantitation and confirmation of aflatoxin identity.

A thermospray MS (TSMS) method was developed to characterize the reaction products of aflatoxins B_1 and G_1 with iodine in methanol–water (Holcomb *et al.,* 1991). About 2 µg of each derivative was injected into an LC/TSMS system. The mobile phase was 0.1 M ammonium acetate in water and the flow rate was 1.2 ml/min. The vaporizer was set at 110°C and the jet was set at 220°C. Test solutions were analyzed in the positive ion, filament off, and discharge off modes. The mass spectra showed m/z 471 and 488, which corresponded to the $[M + H]^+$ derivatized AFB_1 and AFG_1. These results indicated that the reaction products were adducts of one iodine atom and a methoxy group to the furan ring.

An LC/TSMS method was developed for confirmation of identity of aflatoxins in peanuts (Hurst *et al.,* 1991). The column used was C_{18}, 5 µm, 4.6 mm × 25 cm, with a mobile phase of 0.1 M ammonium acetate–methanol–acetonitrile (56 + 22 + 22) at a flow rate of 1.0 ml/min. The interface conditions used were T_{aux} 318°C, T_{block} 290°C, and T_{tip} 185°C. Test solutions and standards were injected under "filament on" mode and were scanned under selected ion monitoring conditions. The detection limits (signal-to-noise ratio < 5) were 60, 40, 100, and

100 pg for AFB_1, AFB_2, AFG_1, and AFG_2, respectively. The mass spectra of AFB_1 and AFB_2 showed strong MH^+ peaks at m/z 313 and 315, respectively. Spectra of AFG_1 and AFG_2 showed strong $[MH^+ - 44]$ peaks at m/z 285 and 287, respectively, in addition to the MH^+ peaks at m/z 329 and 331.

Another approach to confirming the identity of aflatoxins was the use of tandem (MS/MS) mass spectrometry. The identity of aflatoxin M_1 isolated from milk after a disposable immunoaffinity column clean-up was confirmed (J. E. Matusik, personal communication). The eluate of 50 ml milk spiked at the 0.5 ng/ml level was subjected to MS analysis. The test solution was introduced into the mass spectrometer via a direct exposure probe. The tandem instrument was operated in the daughter ion mode. The first quadrupole (Q_1), the mass filter, was set to pass the ion of interest at a particular m/z; the second quadrupole (Q_2) acted as a collision cell, and the third quadrupole (Q_3) scanned the daughter ions formed in Q_2. The molecular ion at m/z 328 was selected in Q_1 and the collisionally activated decompositions occurred in Q_2. To increase sensitivity, Q_3 was set to monitor the following selected ions: m/z 328, 313, 270, 257, and 231. This procedure was used to confirm the identity of M_1 at 0.05 ng/ml in 2% low-fat milk.

AUTOMATION

The analytical laboratory of today is under pressure to increase productivity and decrease cost. To achieve these goals, the quantities of reagents and the amount of time required for the analyte purification step of the analysis must be reduced. Preparation and purification of aflatoxin test samples require grinding and mixing, extracting, filtering, adding a nonpolar solvent to separate fat, removing an aliquot, passing it through a clean-up column, and finally introducing the material into the measurement instrument. Analyte purification is the most labor-intensive operation in the TLC and LC analytical methodology.

Since the development of the simple, fast, and selective immunochemical methods for the determination of aflatoxins, some of the difficulties of conventional methods have been alleviated. However, ELISA is used mainly for screening and cannot be used to determine individual aflatoxins (Pohland *et al.*, 1990). The immunoaffinity clean-up columns require manual operation (Candlish *et al.*, 1988; Trucksess *et al.*, 1991). The steps for purification and isolation of aflatoxins need to be improved.

The technology of automation offers the analyst some solutions, but the application of automated systems to aflatoxin analysis is in its infancy. Several automated analytical methods have been reported. M_1 in milk has been determined using an automated dialysis clean-up and test sample concentration system (Tuinstra *et al.*, 1990). The method did not separate M_1 from matrix interferences completely during the LC separation and quantitation. The recovery of added M_1 at 1 ng/g was about 65%. A partially automated robotic system was coupled with disposable solid-phase extraction after dilution and filtration (Gifford *et al.,* 1990). A similar technique was used for total aflatoxins in foods and animal feeds (Sharman and Gilbert, 1991). A method for screening multimycotoxins (including aflatoxins) in foods and feeds that incorporates an on-line analyte clean-up also has been reported (Chamkasem *et al.,* 1989). All these methods use either very expensive LC (four pumps, two switching valves) or robotic systems (> $50,000 U.S.). The solid-phase extraction and the immunoaffinity columns are added expenses.

The latest development has been the use of an on-line LC affinity column clean-up, coupled with column switching to an analytical column and post-column derivatization using electrochemically generated bromine (Urano *et al.,* 1991). The packing in the LC affinity column is a macroporous hydrophilic copolymer of 2-hydroxyethyl methacrylate and ethylene dimethacrylate. Aflatoxins in the extract (methanol–water) are bound to the support, whereas lipids in the extract are not retained on the column; the proteins and carbohydrates are removed with the water wash. The aflatoxins are eluted with the mobile phase and loaded onto the analytical column. After separation, aflatoxins B_1 and G_1 pass through the electrochemical cell, in which bromine is generated from the mobile phase. The fluorescence of the derivatized AFB_1/AFG_1 and the underivatized AFB_2/AFG_2 is then measured with a fluorescence detector. The four aflatoxins are well resolved. The limits of determination are 0.5 ng/g for AFB_1 and AFG_1 and 0.15 ng/g for AFB_2 and AFG_2, based on a signal-to-noise ratio > 5. Recoveries of total aflatoxins from corn and peanuts spiked at 5–30 ng/g are about 80%. The entire automated LC procedure takes about 30 min, and the LC affinity column support can be used for more than 100 injections.

In the modern laboratory automation will play an important role in future research on aflatoxins, and will contribute to rapid turnaround capabilities, labor saving, and reliability of the data. Human errors in the analysis will be minimized.

SAFETY ISSUES IN HANDLING MOLDY GRAINS
AND AFLATOXINS

Safety is a key issue for scientists working in the aflatoxin area. Steps must be taken to minimize exposure to the toxins as well as to the producing microorganisms, *Aspergillus flavus* and *Aspergillus parasiticus.* A safety program should be established that meets the requirements of the Laboratory Standard of the Occupational Safety and Health Administration (1990) and the guidelines of the National Institutes of Health (1981) covering use of chemical carcinogens. One published paper describes safety training and the use of personal protection equipment, and summarizes safety guidelines for handling mold cultures and moldy commodities, analysis of test samples, preparation and administration of contaminated diets for animal studies, and methods for destruction of toxins in waste products, equipment, work surfaces, and laboratory space (Trucksess and Richard, 1992). The authors also discuss procedures to avoid pollution of the work environment and the use of biological safety cabinets and fume hoods. A 0.5% sodium hypochlorite solution was used most commonly to wash glassware, and a solution of 1% sodium perborate and 1% sodium bicarbonate was used to wash body parts. The main emphasis was on following safety guidelines and using common sense to minimize potential exposure to toxic materials.

CONCLUSIONS

The most up-to-date methods for the determination of aflatoxins have been presented here. TLC, LC, and immunochemical methods of analysis provide similar results. However, TLC and LC methods are well established and have been evaluated. The extract remaining after TLC or LC determination is often suitable for MS confirmation, whereas the extracts prepared for immunochemical methods (with the exception of the immunoaffinity column method) are not.

Commercially prepared immunochemical testing kits have become available within this decade. Only a few of these kits have been evaluated by collaborative studies. The data of some of these studies are no longer valid because the suppliers have changed their manufacturing process. The AOAC has established a new research institute to confirm the claims of the test kit manufacturers (AOAC, 1991). Perhaps the results of their studies will help analysts have a clearer picture of which immunochemical methods are suitable for their needs. Automated immunoassays for clinical uses have been developed. The same approach may be used for aflatoxin analysis.

Many methods of analysis for aflatoxin in foods and feeds are published each year. An appropriate method must be selected for each particular need. Factors that should be considered before selecting a method include number of analyses needed, time, location, cost, equipment, safety, waste disposal, and, above all, the experience of the analyst. The simple, specific, and rapid immunoassays will play an increasing role in monitoring foods and feeds for mycotoxins. Currently, these methods appear to have their greatest value when used in conjunction with existing TLC and LC methods.

REFERENCES

Abramson, D., Thorsteinson, T., and Forest, D. (1989). Chromatography of mycotoxins on precoated reverse-phase thin layer plates. *Arch. Environ. Contam. Toxicol.* **18,** 327–330.

Adensam, L., Lebedova, M., and Turek, B. (1989). Method of aflatoxin estimation in spices. *Cesk. Hyg.* **34,** 207–212.

Association of Official Analytical Chemists (1990). "Official Methods of Analysis," 15th Ed. AOAC, Arlington, Virginia.

Association of Official Analytical Chemists (1991). AOAC prepares to expand methods validation programs. *Referee* **15**, 1.

Beaver, R. W. (1989). Determination of aflatoxins in corn and peanuts using high performance liquid chromatography. *Arch. Environ. Contam. Toxicol.* **18**, 315–318.

Beebe, R. M. (1978). Reverse phase high pressure liquid chromatographic determination of aflatoxins in foods. *J. Assoc. Off. Anal. Chem.* **61**, 1347–1352.

Bijl, J. P., van Peteghem, C. H., and Dekeyser, D. A. (1987). Fluorimetric determination of aflatoxin M_1 in cheese. *J. Assoc. Off. Anal. Chem.* **70**, 472–475.

Bressac, B., Kew, M., Wands, J., and Ozturk, M. (1991). Selective G to T mutations of *p53* gene in hepatocellular carcinoma from southern Africa. *Nature (London)* **350**, 429–431.

Candlish, A. A. G., Haynes, C. A., and Stimson, W. H. (1988). Detection and determination of aflatoxins using affinity chromatography. *Int. J. Food Sci. Technol.* **23**, 479–485.

Carisano, A., and Torre, G. D. (1986). Sensitive reversed-phase high performance liquid chromatographic determination of aflatoxin M_1 in dry milk. *J. Chromatogr.* **355**, 340–344.

Chamkasem, N., Cobb, W. Y., Latimer, G. W., Salinas, C., and Clement, B. A. (1989). Liquid chromatographic determination of aflatoxins, ochratoxin A, and zearalenone in grains, oilseeds, and animal feeds by post-column derivatization and on-line sample cleanup. *J. Assoc. Off. Anal. Chem.* **72**, 336–341.

Chu, F. S. (1990). Immunoassays for mycotoxins: Current state of art, commercial and epidemiological applications. *Vet. Hum. Toxicol. (Suppl.)* **32**, 42–50.

Coker, R. D., Jewers, K., Tomlins, K. I., and Blunden, G. (1988). Evaluation of instrumentation used for high performance thin layer chromatography of aflatoxins. *Chromatographia* **25**, 875–880.

Cole, R. J. (ed.) (1986) "Modern Methods in the Analysis and Structural Elucidation of Mycotoxins." Academic Press, New York.

Council of Agricultural Science and Technology (1989). "Mycotoxins, Economic and Health Risks." CAST Task Force Report 116. CAST, Ames, Iowa.

Dominguez, L., Blanco, J. L., Gomez-Lucia, E., Rodriguez, E. F., and Suarez, G. (1987). Determination of aflatoxin M_1 in milk and milk products contaminated at low levels. *J. Assoc. Off. Anal. Chem.* **70**, 470–472.

Dorner, J. W., and Cole, R. J. (1988). Rapid determination of aflatoxin in raw peanuts by liquid chromatography with post column iodination and modified minicolumn cleanup. *J. Assoc. Off. Anal. Chem.* **71**, 43–47.

Duhart, B. T., Shaw, S., Wooley, M., Allen, T., and Grimes, C. (1988). Determination of aflatoxins B_1, B_2, G_1, and G_2 by high-performance liquid chromatography with electrochemical detection. *Anal. Chim. Acta* **208**, 343–346.

Eppley, R. M. (1966). A versatile procedure for assay and preparatory separation of aflatoxin from peanut products. *J. Assoc. Off. Anal. Chem.* **49**, 1218–1223.

Food and Drug Administration (1988). "Food and Drug Inspection Operating Manual." U.S. Department of Health and Human Services, Washington, D.C.

Francis, O. J., Liginski, L. J., Gaul, J. A., and Campbell, A. D. (1982). High pressure liquid chromatographic determination of aflatoxins in peanut butter using a silica gel-packed flow cell for fluorescence detection. *J. Assoc. Off. Anal. Chem.* **65**, 672–676.

Francis, O. J., Kirschenheuter, G. P., Ware, G. M., Carman, A. S., and Kuan, S. S. (1988a). β-Cyclodextrin post-column fluorescence enhancement for reverse-phase liquid chromatographic determination of aflatoxins in corn. *J. Assoc. Off. Anal. Chem.* **71**, 725–728.

Francis, O. J., Ware, G. M., Carman, A. S., Kirschenheuter, G. P., and Kuan, S. S. (1988b). Use of ten gram samples of corn for determination of mycotoxins. *J. Assoc. Off. Anal. Chem.* **71**, 41–43.

Gifford, L. A., Wright, C., and Gilbert, J. (1990). Robotic analysis of aflatoxin M_1 in milk. *Food Add. Contam.* **7**, 829–836.

Gilbert, J. (1988). Application of mycotoxin RM's in method development and quality assurance. *Fresenius Z. Anal. Chem.* **332**, 602–605.

Goto, T., Matsui, M., and Kitsuwa, T. (1988). Determination of aflatoxins by capillary column gas chromatography. *J. Chromatogr.* **447,** 410–414.

Groopman, J. D., and Donahue, K. F. (1988). Aflatoxins, a human carcinogen: Determination in foods and biological samples by monoclonal affinity chromatography. *J. Assoc. Off. Anal. Chem.* **71,** 861–867.

Hansen, T. J. (1990). Affinity column cleanup and direct fluorescence measurement of aflatoxin M_1 in raw milk. *J. Food Protect.* **53,** 75–77.

Hetmanski, M. T., and Scudamore, K. A. (1989). A simple quantitative HPLC method for determination of aflatoxins in cereals and animal feedstuffs using gel permeation chromatography clean-up. *Food Add. Contam.* **6,** 35–48.

Holcomb, M., Korfmacher, W. A., and Thompson, H. C. (1991). Characterization of iodine derivatives of aflatoxins B_1 and G_1 by thermospray mass spectrometry. *J. Anal. Toxicol.* **15,** 289–292.

Hsu, I. C., Metcalf, R. A., Sun, T., Welsh, J. A., Wang, N. J., and Harris, C. C. (1991). Mutational hotspot in the *p53* gene in human hepatocellular carcinomas. *Nature (London)* **350,** 427–428.

Hurst, W. J., Martin, R. A., Jr., and Vestal, C. H. (1991). The use of HPLC/thermospray MS for the confirmation of aflatoxins in peanuts. *J. Liq. Chromatogr.* **14,** 2541–2550.

Hutchins, J. E., Lee, Y. J., Tyczkowska, K., and Hagler, W. M., Jr. (1989). Evaluation of silica cartridge purification and hemiacetal formation for liquid chromatographic determination of aflatoxins in corn. *Arch. Environ. Contam. Toxicol.* **18,** 319–326.

Jackman, R. (1985). Determination of aflatoxins by enzyme-linked immunosorbent assay with special reference to aflatoxin M_1 in raw milk. *J. Sci. Food Agric.* **36,** 685–698.

Jewers, K., John, A. E., and Blunden, G. (1989). Assessment of suitability of the Florisil and CB clean-up procedures for determining aflatoxin levels in sorghum by bidirectional HPTLC. *Chromatographia* **27,** 617–621.

Kaveri, S. V., Fremy, J. M., Lapeyre, C., and Strosberg, A. D. (1987). Immunodetection and immunopurification of aflatoxins using a high affinity monoclonal antibody to aflatoxin B_1. *Lett. Appl. Microbiol.* **4,** 71–75.

Kawamura, O., Nagayama, S., Sato, S., Ohtani, K., Ueno, I., and Ueno, Y. (1988). A monoclonal antibody based enzyme linked immunosorbent assay of aflatoxin B_1 in peanut products. *Mycotoxin Res.* **4,** 75–88.

Koeltzow, D. E., and Tanner, S. N. (1990). Comparative evaluation of commercially available aflatoxin test methods. *J. Assoc. Off. Anal. Chem.* **73,** 584–589.

Kok, W. T., Van Neer, T. Ch., Traag, W. A., and Tuinstra, L. G. M. T. (1986). Determination of aflatoxins in cattle feed by liquid chromatography and post-column derivatization with electrochemically generated bromine. *J. Chromatogr.* **367,** 231–236.

Kozloski, R. P. (1986). High performance thin-layer chromatographic screening for aflatoxins in poultry feed by using silica Sep-Paks. *Bull. Environ. Contam. Toxicol.* **36,** 815–818.

Kubicek, E. M., Vojir, F. K., and Holzer, H. G. (1988). An inproved rapid method for routine determination of aflatoxin M_1 in milk. *Ernaehrung/Nutr.* **12,** 302–305.

Lafont, P., Platzer, N., Siriwardana, M. G., Sarfati, J., Mercier, J., and Lafont, J. (1986a). Un nouvel hydroxy-derive de l'aflatoxine B_1: l'aflatoxine M4: I-Production *in vitro*—Structure. *Microbiol. Aliment. Nutr.* **4,** 65–74.

Lafont, P., Siriwardana, M. G., Sarfati, J., Debeaupuis, J. P., and Lafont, J. (1986b). Un nouvel hydroxy-derive de l'aflatoxine B_1: l'aflatoxine M4: II-Methode de dosage—mise en evidence de contaminations de laits commerciaux. *Microbiol. Aliment. Nutr.* **4,** 141–145.

McKinney, J. (1989). Use of aflatoxin kits proposed. *J. Am. Oil Chem. Soc.* **66,** 1430.

Madhyastha, M. S., and Bhat, R. V. (1985). Aflatoxin-like fluorescent substance in spices. *J. Food Safety* **7,** 101–106.

Mahjoub, A., and Bullerman, L. B. (1990). A method for aflatoxin B_1 determination in olives. *Rev. Fr. Corps Gras* **4,** 245–246.

Miller, N., Pretorius, H. E., and Trinder, D. W. (1985). Determination of aflatoxins in vegetable oil. *J. Assoc. Off. Anal. Chem.* **68,** 136–137.

Mortimer, D. N., Gilbert, J., and Shepherd, M. J. (1987a). Rapid and highly sensitive analysis of aflatoxin M_1 in liquid and powdered milks using an affinity column cleanup. *J. Chromatogr.* **407**, 393–398.

Mortimer, D. N., Shepherd, M. J., Gilbert, J., and Morgan, M. R. A. (1987b). A survey of the occurrence of aflatoxin B_1 in peanut butters by enzyme-linked immunosorbent assay. *Food Add. Contam.* **5**, 127–132.

Mortimer, D. N., Shepherd, M. J., Gilbert, J., and Clark, C. (1988). Enzyme-linked immunosorbent (ELISA) determination of aflatoxin B_1 in peanut butter: Collaborative trial. *Food Addit. Contam.* **5**, 601–608.

National Institutes of Health (1981). "NIH Guidelines for Laboratory Use of Chemical Carcinogens." NIH, Rockville, Maryland.

Newsome, W. H. (1986). Potential and advantages of immunochemical methods for analysis of foods. *J. Assoc. Off. Anal. Chem.* **69**, 919–923.

Occupational Safety and Health Administration (1990). "Occupational Exposure to Hazardous Chemicals in Laboratories," 29 CFR, Part 1910.1450. U.S. Government Printing Office, Washington, D.C.

Panalaks, T., and Scott, P. M. (1977). Sensitive silica-gel packed flow cell for fluorometric detection of aflatoxins by high pressure liquid chromatography. *J. Assoc. Off. Anal. Chem.* **60**, 583–589.

Park, D. L., and Pohland, A. E. (1989). Sampling and sample preparation for detection and quantitation of natural toxicants in food and feed. *J. Assoc. Off. Anal. Chem.* **72**, 399–404.

Park, D. L., Miller, B. M., Hart, P., Yang, G., McVey, J., Page, S. W., Pestka, J., and Brown, L. H. (1989a). Enzyme-linked immunosorbent assay for screening aflatoxin B_1 in cottonseed products ᐧand mixed feed: Collaborative study. *J. Assoc. Off. Anal. Chem.* **72**, 326–332.

Park, D. L., Miller, B. M., Nesheim, S., Trucksess, M. W., Vekich, A., Bidigare, B., McVey, J. L., and Brown, L. H. (1989b). Visual and semiquantitative spectrophotometric ELISA screening methods for aflatoxin B_1 in corn and peanut products: Follow-up collaborative study. *J. Assoc. Off. Anal. Chem.* **72**, 638–643.

Park, D. L., Nesheim, S., Trucksess, M. W., Stack, M. E., and Newell, R. F. (1990). Liquid chromatographic method for determination of aflatoxins B_1, B_2, G_1, and G_2 in corn and peanut products: Collaborative study. *J. Assoc. Off. Anal. Chem.* **73**, 260–266.

Park, L. P., Diprossimo, V., Abdel-Malek, E., Trucksess, M. W., Nesheim, S., Brumley, W. C., Sphon, J. A., Barry, T. L., and Petzinger, G. (1985). Negative ion chemical ionization mass spectrometric method for confirmation of identity of aflatoxin B_1: Collaborative study. *J. Assoc. Off. Anal. Chem.* **68**, 636–640.

Paulsch, W. E., Sizoo, E. A., and Van Egmond, H. P. (1988). Liquid chromatographic determination of aflatoxins in feedstuffs containing citrus pulp. *J. Assoc. Off. Anal. Chem.* **71**, 957–961.

Pennington, L. J. (1986). Thin layer chromatography and densitometric determination of aflatoxins in mixed feeds containing citrus pulp. *J. Assoc. Off. Anal. Chem.* **69**, 690–696.

Pestka, J. J. (1988). Enhanced surveillance of foodborne mycotoxins by immunochemical assay. *J. Assoc. Off. Anal. Chem.* **71**, 1075–1081.

Pohland, A. E., Trucksess, M. W., and Page, S. W. (1990). Immunoassays in food safety applications: Developments and perspectives. *In* "Immunochemical Methods for Environmental Analysis" (J. M. van Emond and P. O. Mumma, eds.), pp. 38–50. American Chemical Society, Washington, D.C.

Reichert, N., Steinmeyer, S., and Weber, R. (1988). Determination of aflatoxin B_1 in dried figs by visual screening, thin layer chromatography and ELISA. *Z. Lebensm. Unters. Forsch.* **186**, 505–508.

Richard, J. L., and Lyon, R. L. (1986). Aflatoxins and their detection in animal tissues and fluids. *J. Toxicol. Toxin Rev.* **5**, 197–215.

Scott, P. M. (1989). Methods for determination of aflatoxin M_1 in milk and milk products—A review of performance characteristics. *Food Add. Contam.* **6**, 283–305.

Scott, P. M. (1991). Methods of analysis for mycotoxins—an overview. *In* "Analysis of Oilseeds, Fats and Fatty Foods" (J. B. Rossell and J. L. R. Pritchard, eds.), pp. 141–184. Elsevier Applied Science, London.

Serralheiro, M. L., and Quinta, M. L. (1985). Rapid thin layer chromatographic determination of aflatoxin M_1 in powdered milk. *J. Assoc. Off. Anal. Chem.* **68**, 952–954.

Sharman, M., and Gilbert, J. (1991). Automated aflatoxin analysis of foods and animal feeds using immunoaffinity column clean-up and high performance liquid chromatographic determination. *J. Chromatogr.* **543**, 220–225.

Sharman, M., Patey, A. L., and Gilbert, J. (1989). Application of an immunoaffinity column sample clean-up to the determination of aflatoxin M_1 in cheese. *J. Chromatogr.* **474**, 457–461.

Shepherd, M. J., and Gilbert, J. (1984). An investigation of HPLC postcolumn iodination conditions for the enhancement of aflatoxin B_1 fluorescence. *Food Add. Contam.* **1**, 325–335.

Simonella, A., Lorreti, L., Filipponi, C., Falgiani, A., and Ambrosil, L. (1987). Simultaneous determination of aflatoxins G_1, B_1, G_2, B_2 in animal feedstuffs by HPTLC and RP-HPLC. *J. High Res. Chromatogr. Chromatogr. Commun.* **10**, 626–628.

Stubblefield, R. D. (1987). Optimum conditions for formation of aflatoxin M_1-derivative. *J. Assoc. Off. Anal. Chem.* **70**, 1047–1049.

Tarter, E. J., Hanchay, J. P., and Scott, P. M. (1984). Improved liquid chromatographic method for the determination of aflatoxins in peanut butter and other commodities. *J. Assoc. Off. Anal. Chem.* **67**, 597–600.

Thiel, P. G., Stockenstrom, S., and Gathercole, P. S. (1986). Aflatoxin analysis by reverse phase HPLC using post-column derivatization for enhancement of fluorescence. *J. Liq. Chromatogr.* **9**, 103–112.

Tomlins, K. I., Jewers, K., Coker, R. D., and Nagler, M. J. (1989). A bi-directional HPTLC development method for the detection of low level of aflatoxin in maize extracts. *Chromatographia* **27**, 49–52.

Trucksess, M. W., and Page, S. W. (1990). Immunochemical methods for aflatoxins. *In* "Biodeterioration Research 3, (G. C. Llewellyn and C. E. O'Rear, Eds.), pp. 161–173. Plenum Press, New York.

Trucksess, M. W., and Richard, J. L. (1992). Laboratory safety considerations in the handling of natural toxins. *In* "Natural Toxins: Toxicology, Chemistry and Safety" (R. F. Keeler, N. B. Mandava, and A. T. Tu, Eds.), pp. 337–345. Alaken, Fort Collins, Colorado.

Trucksess, M. W., Brumley, W. C., and Nesheim, S. (1984). Rapid quantitation and confirmation of aflatoxins in corn and peanut butter, using a disposable silica gel column, thin layer chromatography, and gas chromatography/mass spectrometry. *J. Assoc. Off. Anal. Chem.* **67**, 973–975.

Trucksess, M. W., Stack, M. E., Nesheim, S., Park, D. L., and Pohland, A. E. (1989). Enzyme-linked immunosorbent assay of aflatoxins B_1, B_2, and G_1 in corn, cottonseed, peanuts, peanut butter, and poultry feed: Collaborative study. *J. Assoc. Off. Anal. Chem.* **72**, 957–962.

Trucksess, M. W., Stack, M. E., Page, S. W., and Albert, R. H. (1991). Immunoaffinity column coupled with solution fluorometry or liquid chromatography postcolumn derivatization for determination of aflatoxins in corn, peanuts, and peanut butter: Collaborative study. *J. Assoc. Off. Anal. Chem.* **74**, 81–84.

Tuinstra, L. G. M. Th., Kienhuis, P. G. M., and Dols, P. (1990). Automated liquid chromatographic determination of aflatoxin M_1 in milk using on-line dialysis for sample preparation. *J. Assoc. Off. Anal. Chem.* **73**, 969–973.

Tutelyan, V. A., Eller, K. I., and Sobolev, V. S. (1989). A survey using normal phase high performance liquid chromatography of aflatoxins in domestic and imported foods in the USSR. *Food Add. Contam.* **6**, 459–465.

Urano, T., Trucksess, M. W., and Page, S. W. (1991). LC immunoaffinity column coupled with on-line bromine production for the determination of aflatoxins in grains. Paper presented at the 105th Annual International Meeting of the Association of Official Analytical Chemists, Phoenix, Arizona.

van Egmond, H. P., and Wagstaffe, P. J. (1987). Development of milk powder reference materials certified for aflatoxin M_1 content (part I). *J. Assoc. Off. Anal. Chem.* **70**, 605–610.

van Egmond, H. P., and Wagstaffe, P. J. (1988). Development of milk powder reference materials certified for M_1 content (part II): Certification of milk powder RM 283. *J. Assoc. Off. Anal. Chem.* **71**, 1180–1182.

van Egmond, H. P., and Wagstaffe, P. J. (1989). Aflatoxin B_1 in peanut meal reference material: Intercomparisons of methods. *Food Add. Contam.* **6**, 307–319.

van Egmond, H. P., and Wagstaffe, P. J. (1990). Aflatoxin B_1 in compound-feed reference materials: An intercomparison of methods. *Food Add. Contam.* **7**, 239–251.

van Egmond, H. P., Leussink, A. B., and Paulsch, W. E. (1988a). The determination of aflatoxin M_1 in milk powder. *Int. Dairy Fed. Bull.* **207**, 150–181.

van Egmond, H. P., Paulsch, W. E., and Sizoo, E. A. (1988b). Comparison of six methods of analysis for the determination of aflatoxin B_1 in feeding stuffs containing citrus pulp. *Food Add. Contam.* **3**, 321–332.

Ward, C. M., and Morgan, M. R. A. (1991). Reproducibility of a commercially available kit utilizing enzyme-linked immunosorbent assay for determination of aflatoxin in peanut butter. *Food Add. Contam.* **8**, 9–15.

Whitaker, T. B., Dickens, J. W., and Monroe, R. J. (1979). Variability associated with testing corn for aflatoxin. *J. Am. Oil Chem. Soc.* **56**, 789–794.

Whitaker, T. B., Dickens, J. W., and Slate, A. B. (1990). Computerized system to quantify aflatoxin using thin-layer chromatography. *Peanut Sci.* **17(2)**, 96–100.

Wilson, D. M. (1989). Analytical methods for aflatoxins in corn and peanuts. *Arch. Environ. Contam. Toxicol.* **18**, 308–314.

Wilson, T. J., and Romer, T. R. (1991). Use of the Mycosep multifunctional cleanup column for liquid chromatographic determination of aflatoxins in agricultural products. *J. Assoc. Off. Anal. Chem.* **74**, 951–955.

Wood, G. E. (1991). Aflatoxin M_1. *In* "Mycotoxins and Phytoalexins" (R. P. Sharma and D. K. Salunkhe, Eds.), pp. 145–164. CRC Press, Boca Raton, Florida.

Yousef, A. E., and Marth, E. H. (1985). Rapid reverse phase liquid chromatographic determination of aflatoxin M_1 in milk. *J. Assoc. Off. Anal. Chem.* **68**, 462–465.

20

Problems Associated with Accurately Measuring Aflatoxin in Food and Feeds: Errors Associated with Sampling, Sample Preparation, and Analysis

❖

Thomas B. Whitaker and Douglas L. Park

INTRODUCTION

In research, regulatory, and quality assurance activities, being able to measure accurately and precisely the concentration of a mycotoxin in a commodity so correct decisions can be made is important. However, Whitaker *et al.* (1974,1976,1979), Campbell *et al.* (1986), and Park *et al.* (1991c) have demonstrated how difficult precisely estimating the mycotoxin concentration in a large bulk of material (lot) is because of the large variability associated with the testing procedure. The testing procedure is a complicated process and generally consists of three steps: (1) a sample is taken from the lot, (2) the sample is comminuted to reduce particle size and a subsample is removed from the comminuted sample for analysis, and (3) the mycotoxin is extracted from the subsample and quantified. General recommendations for sampling products for chemical analysis have been published by Kratochivil and Taylor (1981). Dickens and Whitaker (1982), Campbell *et al.* (1986), and Park and Pohland (1989) published a review of accepted procedures of sampling and sample preparation methods for various agricultural commodities for mycotoxin analysis. These authors listed various types of equipment used for sample preparation, as well as sources of supply. Nesheim (1979), Schuller *et al.* (1976), and Park and Pohland (1986) published reviews of accepted analytical procedures to analyze various products for the mycotoxin aflatoxin. Even when using accepted procedures, random variation is associated with each step of the mycotoxin testing procedure. Because of this variability, the true mycotoxin concentration in a commodity cannot be determined with 100% certainty by measuring the concentration in the sample taken from the lot.

In this chapter, we discuss the different sources of variability that are associated with testing a commodity for a mycotoxin. Specifically, we concentrate on testing the commodity for aflatoxin since most published literature is concerned with this mycotoxin. We show how to reduce the variability of test results and how to design testing programs to detect contaminated commodities as precisely as resources will permit.

VARIATION AMONG TEST RESULTS

Table 1 shows 20 replicated aflatoxin test results from each of 10 contaminated cottonseed lots (Velasco *et al.*, 1975). Each test result was made by comminuting a 2.27-kg sample of dehulled kernels in a subsampling mill with a 1-mm screen (developed by Dickens and Satterwhite, 1969; Dickens *et al.*, 1979), extracting aflatoxin from a 100-g subsample using the Velasco method, and quantifying the aflatoxin densitometrically using minicolumns (Velasco, 1972). The 20 test results from each lot are ranked from low to high to demonstrate several important characteristics about replicated aflatoxin test results taken from a contaminated lot.

First, the wide range among replicated test results from the same lot is reflected in the large variances and coefficients of variation (CV) shown in Table 1. The maximum test result can be as much as five times the lot concentration (the average of the 20 test results is the best estimate of the lot concentration). Second, the amount of variation among the 20 test results appears to be a function of the lot concentration. As the lot concentration increases, the variance among test results increases but the relative variance, as measured by the CV, decreases. Third, the distribution of the 20 test results for each lot in Table 1 are not always symmetrical about the lot concentration (Whitaker *et al.*, 1972; Velasco *et al.*, 1975). The distributions are skewed positively, that is, more than half the test results are below the lot concentration. However, the distribution of test results becomes more symmetrical as the lot concentration increases. This skewness can be observed by counting the number of test results above and below the lot concentration in Table 1. If a single sample is tested from a contaminated lot, more than a 50% probability exists that the sample result will be lower than the true lot concentration. The skewness is greater for small sample sizes and the distribution becomes more symmetrical as sample size increases (Remington and Schrok, 1970).

The variability shown in Table 1 is the sum of variance associated with each step of the testing procedure. The total variance (V_T) is equal to the sum of the

TABLE 1 Replicated Test Results for 20 2.27-kg Samples from Each of 10 Contaminated Lots of Cottonseed

Lot number	Aflatoxin test results (ppb)																				Mean (ppb)	Variance	CV[a] (%)
	1	2	3	4	5	6	7	8	9	10	11	12	13	14	15	16	17	18	19	20			
1	0	0	0	0	0	0	0	0	0	1	1	1	1	1	3	5	7	9	10	14	2.7	17	156.0
2	0	0	0	0	1	1	1	1	1	1	4	6	10	10	12	13	16	27	40	44	9.5	174	139.6
3	0	0	0	0	1	1	1	1	1	1	8	9	12	23	24	24	25	30	40	50	12.6	234	122.0
4	0	1	1	1	1	1	1	4	4	11	12	14	19	20	21	22	24	28	38	56	14.0	223	107.0
5	0	4	9	14	16	16	25	27	30	30	31	32	32	32	34	37	40	42	42	100	30.3	819	94.5
6	6	6	7	14	20	22	24	24	31	33	38	40	42	45	54	60	67	68	68	165	41.7	1260	85.1
7	10	10	14	20	25	31	32	34	37	37	55	61	61	65	70	74	83	86	101	117	51.1	959	60.5
8	15	16	20	21	27	30	48	52	57	67	70	80	80	90	111	118	133	136	144	160	73.8	2183	63.4
9	1	16	29	40	53	73	85	89	100	104	113	118	120	121	128	143	157	175	260	266	109.9	4990	64.3
10	70	80	91	114	116	120	127	130	133	150	178	178	192	196	200	201	206	237	252	269	169.8	4741	40.6

[a]CV, Coefficient of variation = $100(\text{variance})^{0.5}/\text{mean}$.

sampling variance (V_S) subsampling variance (V_{SS}), and analytical variance (V_A).

$$V_T = V_S + V_{SS} + V_A \tag{1}$$

Estimates of the magnitude of these three variance components for several commodities, several sample preparation techniques, and several type analytical methods are described next.

SAMPLING VARIABILITY

Studies by Whitaker *et al.* (1974,1976,1979) on three granular products—peanuts (both in the shell and raw kernels), cottonseed, and shelled corn—and by Park *et al.* (1991b,c) on cottonseed indicate that, for small sample sizes, the sampling step is the largest source of variation in the testing procedure. Sampling variance is large because (1) aflatoxin is found only in a small percentage of the kernels in the lot and (2) the concentration in a single kernel may be extremely high (Whitaker and Wiser, 1969). Studies on peanut kernels (Whitaker *et al.*, 1972) indicate that the percentage of contaminated kernels in a contaminated lot at 20 parts per billion (ppb) is 0.095%, which is less than 1 contaminated kernel per 1000 kernels. The same studies also indicate that the percentage of contaminated particles in a lot are a function of the lot aflatoxin concentration and the percentage increases with lot concentration. However, these few contaminated particles can contain extremely high levels of aflatoxin. Cucullu *et al.* (1966,1977) reported aflatoxin concentrations in excess of 1,000,000 ng/g (ppb) for individual peanut kernels and 5,000,000 ng/g for individual cottonseeds. Shotwell *et al.* (1974) reported finding over 400,000 ng/g aflatoxin in individual corn kernels. A 5-kg sample of peanut kernels with a single kernel containing 10^6 ng aflatoxin will have a sample concentration of 200 ppb.

Because of these extremes in both the percentage of contaminated particles and the range in aflatoxin concentrations among individual particles in a contaminated lot, variation among replicated samples tends to be large. The sampling variance associated with peanut pods (T. B. Whitaker, unpublished data), (V_{Spp}), raw shelled peanut kernels (V_{Spk}), cottonseed (V_{Scs}), and shelled corn (V_{Ssc}), for a given sample size was estimated experimentally by Whitaker *et al.* (1974,1976,1979), as shown in Eqs. 2–5, respectively.

$$V_{Spp} = (95.3565M^{0.9576})/W_{Spp} \tag{2}$$

$$V_{Spk} = (49.0546M^{1.3955} - 0.3494M^{1.7867})/W_{Spk} \tag{3}$$

$$V_{Scs} = (16.3447M^{1.3434} - 5.1611M^{1.3508} - 0.1511M^{1.2424})/W_{Scs} \tag{4}$$

$$V_{Ssc} = 3.9539M/W_{Ssc} \tag{5}$$

where M is the aflatoxin concentration in the lot in ppb, W_{Spp} is the mass of peanut pods in kg (pod count per kg, 884.5), W_{Spk} is the mass of peanut kernels in the sample in kg (kernel count per gram, 1.95), W_{Scs} is the mass of cottonseed in the sample in kg (kernel count per gram, 19.03), and W_{Ssc} is the mass of shelled corn in the sample in kg (kernel count per gram, 3.0). Equations 2–5 show that the sampling variance is a function of the lot aflatoxin concentration M and the type of commodity tested.

A comparative study evaluating devices used for sampling cottonseed was developed by Park *et al.* (1991c). The official sampling procedure for the state of Arizona requires that a 30-lb sample be taken from each lot. The 30-lb sample is obtained by probing the lot at least 10 times using a 3-in. diameter $\times$ 50-in. long probe (3 lb per probe site). The official procedure was compared with two other probing procedures using two probes 0.75-in. diameter and 100 and 50 in. long. Adjacent samples were collected by each probing device at 41 different sites throughout a single cottonseed lot (100-ton pile) for a total of 123 samples. Mean sample weights were 35.1, 10.1, and 5.4 lbs., respectively. Aflatoxin levels and CVs (in parentheses) as determined by Association of Official Analytical Chemists (AOAC) thin layer chromatography (TLC) procedures (90.20 A–H; AOAC, 1990) were 49.8 (6.6), 37.1 (5.1), and 39.4 (2.8), respectively. These data suggest the randomness of the probing sites is a significant factor in sampling variation, as is sample size.

The sampling variation measured in these studies assumes that no biases are being introduced by taking the sample from the lot. All particles in the lot are assumed to have an equal chance of being chosen during the sampling procedure, and the variation measured in these studies is assumed to be random.

SUBSAMPLING VARIABILITY

Once the sample has been taken from the lot, the sample must be prepared for aflatoxin extraction. Since extracting the aflatoxin from a large sample is not possible, the aflatoxin usually is extracted from a much smaller portion of product (subsample) taken from the sample. If the commodity is a granular product such as shelled corn, the entire sample must be comminuted in a suitable mill before the subsample is removed from the sample (Dickens and Whitaker, 1982; Park and Pohland, 1989). Removing a subsample from the sample before the comminution process would eliminate the benefits associated with the larger size sample of granular product. After the sample has been comminuted, a subsample is removed for aflatoxin extraction. The distribution of contaminated particles in the comminuted sample is assumed to be similar to the distribution of contaminated kernels found in the lot. As a result, variability also exists among replicated subsamples from the same comminuted sample. However, the subsampling variance should not be as large as the sampling variance because of the larger number of comminuted particles in the subsample.

The subsampling variance has been measured for various commodities using several types of mills. The subsampling variance for peanut kernels comminuted in a Dickens subsampling mill (Whitaker *et al.,* 1974) and a Stephan vertical cutter mill (VCM) (J. W. Dorner, unpublished data) is given in Eqs. 6 and 7, respectively.

$$V_{SSpk}{}^d = (0.0978M^{1.7867} - 0.0178M^{1.9339})/W_{SSpk} \tag{6}$$

$$V_{SSpk}{}^s = (0.01525M^{1.7920} - 0.003755M^{1.7573})/W_{SSpk} \tag{7}$$

where M is the aflatoxin concentration in the sample in ppb and W_{SSpk} is the mass of comminuted peanut kernels in the subsample in kg. Regardless of the type mill used to comminute the sample, the subsampling variation increases with the aflatoxin concentration in the sample. The Dickens mill uses a screen with 3.18-mm diameter holes to comminute peanut kernels. The Dickens mill and the hole diameter in the screen are designed not to create a peanut paste. The Stephan mill comminutes the peanut kernels into a smaller particle size than the Dickens mill and creates a paste material. The effect of particle size reduction on the subsampling variance can be seen by comparing variances in Eqs. 6 and 7. For example, at a sample concentration of 20 ppb, the subsampling variance associated with a 250-g subsample using the Dickens and Stephan mills is 59.2 and 10.2, respectively. The CV is 38.5 and 16.0%, respectively. The subsampling variance for peanut kernels at 20 ppb is reduced by a factor of 6 mostly because of a reduction in particle size.

The subsampling variance for shelled corn (V_{SSsc}), was measured using the Dickens mill with a 3.2-mm screen and a Willey mill using a 1-mm screen (Whitaker *et al.,* 1979). Equations 8 and 9 describe the subsampling variance for the Dickens and Willey mills, respectively.

$$V_{SSsc}{}^d = 0.1196M/W_{SSsc} \tag{8}$$

$$V_{SSsc}{}^w = 0.0125M/W_{SSsc} \tag{9}$$

where M is the aflatoxin concentration in the sample of shelled corn and W_{SSsc} is the sample mass in kg. As in the case of peanuts, the subsampling variance is a function of the sample aflatoxin concentration. The effect of particle size reduction on reducing the subsampling variance for corn can be seen by comparing Eqs. 8 and 9. The subsampling variances for a 50-g subsample taken from a sample at 20 ppb and comminuted with the Dickens and Willey mills is 47.8 and 5.0, respectively. The CV is 34.6 and 11.2%, respectively.

The subsampling variance was measured for cottonseed using the Dickens mill with a 1.6-mm screen (Whitaker *et al.,* 1976) and an ultracentrifugal mill with a 1-mm screen (Park *et al.,* 1991b). Equations 10 and 11 describe the subsampling variance for the Dickens and ultracentrifugal mills, respectively.

$$V_{SScs}{}^d = 0.03596M^{1.3508}/W_{SScs} \tag{10}$$

$$V_{SScs}{}^p = 0.09045M^{1.3298}/W_{SScs} \tag{11}$$

where M is the aflatoxin concentration in the cottonseed sample and W_{SScs} is the sample mass in kg. Regardless of the type of mill used to comminute the sample, the subsampling variance for cottonseed also is a function of the aflatoxin concentration in the sample. The CV associated with the Dickens and ultracentrifugal mills is 49.3 and 32.1%, respectively. The two mills have about the same screen size (1–2 mm) and would be expected to perform about the same.

For a given commodity, the subsampling variance is much lower than sampling variance and also is a function of the aflatoxin concentration in the sample. For a given commodity and subsample size, the subsampling variance is reduced by increasing the degree of comminution or increasing the number of particles per unit mass.

ANALYTICAL VARIABILITY

Once the subsample is removed from the comminuted sample, the aflatoxin usually is extracted by official methods (Schuller *et al.,* 1976; Nesheim, 1979; Park and Pohland, 1986). These methods usually involve several steps such as solvent extraction, centrifugations, drying, dilutions, and quantification. As a result, considerable variation can arise among replicated analyses on the same subsample extract. Whitaker *et al.* (1974,1976,1979) determined the analytical variance (V_{Abf}) associated with AOAC method II extraction procedure along with TLC and densitometric quantification techniques to measure aflatoxin in peanuts (BF method), the analytical variance (V_{Acb}) associated with AOAC method I extraction procedure along with TLC and densitometric quantification techniques to measure aflatoxin in corn (CB method), and the analytical variance (V_{Av}) associated with the Velasco extraction procedure along with minicolumn and densitometric quantification techniques to measure aflatoxin in cottonseed (Velasco method). The variance relationships for the three analytical methods are:

$$V_{Abf} = 0.0637M^{1.9339}/N_{bf} \tag{12}$$

$$V_{Acb} = 0.0699M^2/N_{cb} \tag{13}$$

$$V_{Av} = 0.0666M^{1.2421}/N_v \tag{14}$$

where M is the aflatoxin concentration in the subsample in ppb, N_{bf} is the number of aliquots quantified in the BF method, N_{cb} is the number of aliquots quantified in the CB method, and N_v is the number of aliquots quantified in in the Velasco method. As in the cases of sampling and subsampling variance, the analytical variance also is a function of the aflatoxin concentration of the subsample. For example, at 20 ppb, the CV associated with the BF method is 22.8%.

Studies by Whitaker and Dickens (1981) on the BF method indicate that the TLC quantification step is the major source of variability in the analytical process associated with testing peanuts for aflatoxin. Although similar studies were not conducted for the CB method, TLC is probably the major source of variation in analyzing corn for aflatoxin using the CB method. This observation is supported by the fact that Eqs. 2 and 3 are about the same. The Velasco method has a lower analytical variance than the BF or CB methods, probably because of the use of a minicolumn and densitometric quantification techniques instead of TLC quantification techniques.

If extraction and clean-up contribute only a small portion of the total analytical variance, then the immunoassay and high performance liquid chromatography (HPLC) type analytical methods should have lower variances than methods that use TLC quantification techniques. Hagler and Whitaker (1991) and Dorner and Cole (1988) independently measured the analytical variance associated with HPLC type methods. Although Hagler and Dorner used slightly different extraction and clean-up procedures (Dorner and Cole, 1988; Wilson and Romer, 1991), both groups obtained almost identical results. The results of Hagler's study are

$$V_{Ah} = 0.004828M^{1.7518}/N_h \tag{15}$$

where M is the aflatoxin concentration in the subsample and N_h is the number of aliquots quantified by the HPLC procedure. At 20 ppb, the CV associated with the HPLC method is 4.8%. A CV of 4.8% associated with HPLC is much lower than the 22.8% associated with the BF and CB methods using TLC quantification techniques.

Immunoassay techniques have been applied to the measurement of aflatoxin in commodities such as peanuts, corn, and cottonseed only recently. Food and feed industries as well as regulatory agencies have begun studying the variability associated with immunoassay type analytical methods (Park et al., 1991,c; T. B. Whitaker, unpublished data). The variability one might expect using immunoassay type analytical methods is:

$$V_{Ai} = 0.01327M^{1.5651}/N_i \tag{16}$$

where M is the aflatoxin concentration in the subsample and N_i is the number of aliquots quantified by the immunoassay procedure. Equation 16 specifically reflects the Aflatest® method and a pooling of variance data from corn, cottonseed, and peanuts. At 20 ppb, the CV computed from Eq. 16 is 6.0%. The variability associated with immunoassay type methods appears to be much less than that of TLC methods and slightly more than that of HPLC methods.

All the analytical variance information described here reflects results from single laboratories and does not reflect variances among laboratories. As a result, some laboratories may have higher and some laboratories may have lower variances than those reported in Eqs. 12–16. Analytical variances that reflect among laboratory effects for several analytical methods have been studied by Horwitz et al. (1993). Among-laboratory variance estimates are generally several magnitudes larger than within-laboratory variance estimates.

REDUCING VARIABILITY OF TEST RESULTS

Because of the large variability associated with aflatoxin test results, estimating the true concentration of a lot with a high degree of confidence is difficult. The only way to achieve a more precise estimate of the true lot concentration is to reduce the total variance or the individual variance components associated with each step of the test procedure. The sampling variance can be reduced by increasing the size of the sample. The subsampling variance can be reduced either by increasing the size of the subsample or by increasing the degree of comminution (increasing the number of particles per unit mass in the subsample). The analytical variance can be reduced either by increasing the number of aliquots quantified by the analytical method or by using more precise quantification methods (using HPLC instead of TLC). If one or more of these variance components is reduced, the total variance associated with test results is reduced.

Sampling variance Eqs. 2–5 have a sample size term, subsampling variance Eqs. 6–11 have a subsample size term, and analytical variance Eqs. 12–16 have terms for the number of aliquots quantified by the analytical method. The effect of increasing sample size, subsample size, or the number of aliquots on reducing variance can be determined from these equations. The sampling and subsampling variance Eqs. 2–11 specify mass rather than number of kernels or particles in the sample and subsample, because mass is correlated directly with the number of kernels or particles for a given commodity and because mass is a more convenient measurement than number of particles. However, the number of kernels or particles is the important criterion. Variance reduction is inversely proportional to sample size. For example, if the sample size of any commodity is doubled, then the sampling variance is cut in half (Walpole, 1974). Likewise, if the subsample size and number of aliquots quantified are doubled, the subsampling and analytical variance components are each halved.

Substituting the appropriate variance equation for each step of the testing procedure into Eq. 1 for total variance, one can determine whether the expected variance in aflatoxin test results is acceptable and can determine the most efficient method to reduce variation when necessary. For example, the expected total variance associated with testing a shelled peanut lot at 20 ppb when using a 5.45-kg sample, taking an 1100-g subsample comminuted in the Dickens mill, and using the BF method with TLC quantification can be estimated by summing Eqs. 3, 6, and 12:

$$V_T = 575.1 + 13.5 + 20.9 = 609.5 \qquad (17)$$

The variance, standard deviation, and CV associated with the total testing procedure described above are 609.5, 24.7, and 123%, respectively. The sampling, subsampling, and analytical variances account for 94.4, 2.2, and 3.4% of the total variance, respectively. The major variance component is sampling, which accounts for 94.4% of the total variation. The best use of resources to reduce the total variance appears to be to increase sample size. Increasing the sample size by a factor of 4 to 21.8 kg will cut the sampling variance in Eq. 17 by a factor of 4 to 143.8. The total variance with the 21.8-kg sample now becomes

$$V_T = 143.8 + 13.5 + 20.9 = 178.2 \qquad (18)$$

The variance, standard deviation, and CV associated with the total testing procedure have been reduced to 178.2, 13.3, and 66.7%, respectively.

TABLE 2   Estimated Range of Aflatoxin Test Results for 95% Confidence Limits when Testing a Contaminated Lot of Cottonseed with 100 ppb Using Different Sample Sizes

Sample size (kg)	Standard deviation[a]	Low[b] (ppb)	High[c] (ppb)	Range (high − low) (ppb)
1	87	0	271	271
2	62	0	222	222
4	45	13	187	174
8	32	37	163	126
16	24	53	147	95
32	19	64	136	72

[a]Standard deviation reflects sample sizes shown in the table plus a 100-g subsample comminuted in a Dickens mill with a 1-mm screen and immunoassay analytical method.
[b]Low = 100 − 1.96 (standard deviation). If low was negative, a value of 0 was recorded.
[c]High = 100 + 1.96 (standard deviation).

As indicated earlier, methods other than increasing sample size are available to reduce the total variance associated with testing a commodity for aflatoxin. Different costs are associated with each method and careful study is required to determine the testing procedure that will provide the lowest variance for a given cost. The optimum balance in sample size, degree of comminution, subsample size, and number and type of analysis will vary with the costs involved with each step of the testing procedure. In general, the costs of properly designed aflatoxin testing procedures will increase as the total variance is reduced.

The range of aflatoxin test results associated with any size sample and subsample and any number of analyses can be estimated from the standard deviation (SD), which can be determined by taking the square root of the total variance in Eq. 1. Approximately 95% of all test results will fall between a low of (M − 1.96SD) and a high of (M + 1.96SD). The two expressions are only valid for a normal distribution in which test results are symmetrical about the mean (M). However, aflatoxin test results will approach a symmetrical distribution as sample size becomes large. The effect of increasing sample size on the range of test results when testing a contaminated cottonseed lot that contains 100 ppb aflatoxin is shown in Table 2. We can see that the range does not decrease at a constant rate as sample size increases. For example, doubling sample size has a greater effect on decreasing the range at small sample sizes than at large sample sizes. This characteristic suggests that increasing sample size beyond a certain point may not be the best use of resources and that increasing subsample size or number of analyses may be a better use of resources in reducing the range of test results once sample size has become significantly large.

DESIGNING AFLATOXIN TESTING PROGRAMS

Bulk shipments of a food or feed product often are tested for aflatoxin. If the estimated lot concentration exceeds a defined guideline (tolerance) M_c, then the lot is diverted from human or animal use. Because of the large variability among aflatoxin test results, two types of mistakes are associated with any aflatoxin testing program. First, good lots (lots with a concentration less than or equal to the guideline) will test as bad and be rejected by the testing program. This type of mistake is often called the processor's risk since these lots will be rejected at an unnecessary cost to the processor. Second, bad lots (lots with a concentration greater than the guideline) will test as good and be accepted by the testing program. This type of mistake is often called the consumer's risk since contaminated lots have a potential for making their way into the food chain. To maintain an effective quality control program, these risks associated with a testing program must be evaluated. Based on these evaluations, the costs and benefits (benefits refers to removal of aflatoxin contaminated lots) associated with a testing program can be evaluated.

A lot is termed "bad" when the sample aflatoxin test result X is above some predefined critical level X_c and the lot is termed "good" when X is less than or equal to X_c. Although X_c is usually equal to the guideline M_c, X_c can be greater than or less than M_c. For a given testing design, lots with an aflatoxin concentration M will be accepted with a certain probability $P(M) = \text{prob}(X < X_c \mid M)$ by the testing plan. A plot of $P(M)$ versus M is called an operating characteristic or OC curve. Figure 1 depicts the general shape of an OC curve. As M approaches 0, $P(M)$ approaches 1 and as M becomes large, $P(M)$ approaches 0. The shape of the OC curve is defined uniquely for a particular testing program with designated values of sample size, degree of comminution, subsample size, number of analyses, and the critical level X_c.

For a given testing plan, the OC curve indicates the magnitude of the processor's and consumer's risks. When M_c is defined as the guideline or the maximum lot concentration acceptable, lots with $M > M_c$ are bad and lots with $M \leq M_c$ are good. In Figure 1, the area under the OC curve for $M > M_c$ represents the consumer's risk (bad lots accepted) whereas the area above the OC curve for $M \leq M_c$ represents the processor's risk (good lots rejected) for a particular testing plan.

Because the shape of the OC curve is defined uniquely by the sample size, subsample size, number of analyses, and critical level X_c, these parameters can be used to reduce the processor's and consumer's risks associated with a testing plan. The effect of increasing sample size on the shape of the OC curve when testing cottonseed lots for aflatoxin is shown in Figure 2, where the critical level is equal to the guideline of 40 ppb. As sample size increases from 2.3 to 20.4 kg, the slope of the OC curve about M_c also increases, forcing the two areas associated with each risk to decrease. As a result, increasing sample size decreases both the processor's and the consumer's risk. The same effect can be obtained by increasing the degree of sample comminution, the subsample size, or the number of analyses. In effect, reducing the variability associated with each step of the testing procedure will reduce both the processor's risk and the consumer's risk.

The effect of changing the critical level X_c, on the two risks when testing cottonseed lots for aflatoxin is shown in Figure 3. If the guideline is assumed to

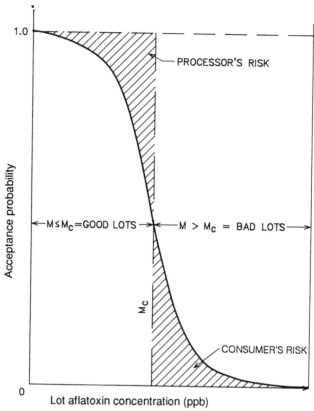

FIGURE 1 Typical operating characteristic curve for evaluating sampling plans. M, Aflatoxin concentration.

be 30 ppb, then changing X_c to a value less than 30 ppb shifts the OC curve to the left. For example, the testing plan in which $X_c = 10$ ppb has a much lower consumer's risk and a much higher processor's risk than the testing plan in which $X_c = 30$ ppb. If X_c becomes larger than 30 ppb, the OC curve shifts to the right. The testing plan in which $X_c = 50$ ppb has a much lower processor's risk and a much higher consumer's risk than the testing plan in which $X_c = 30$ ppb. As a result, only one of the two risks can be reduced by changing X_c relative to the guideline, because reducing one risk will automatically increase the other risk. All these discussions about the effect of X_c on the processor' risk and the consumer's risk assume that the sample size, sample preparation techniques, and analytical methods are the same.

Whitaker (1977) developed methods to evaluate aflatoxin testing designs and compute the OC curve or the acceptance probabilities associated with a given aflatoxin testing plan. He used the negative binomial function (Whitaker and Wiser, 1969) to describe the distribution of sample test results (as seen in Table 1)

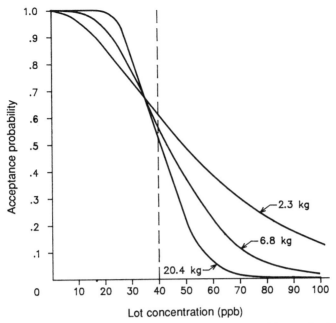

FIGURE 2 Effect of sample size on the processor's and consumer's risks associated with testing cottonseed lots for aflatoxin. Guideline, 40 ppb; accept level, 40 ppb.

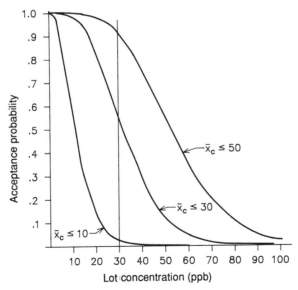

FIGURE 3 Effect of critical level, $\bar{X}_c$, on the processor's and consumer's risks associated with testing cottonseed lots for aflatoxin. Guideline, 30 ppb; sample size, 9.1 kg.

for peanuts, shelled corn, and cottonseed. The evaluation method has been used extensively to design the aflatoxin testing plan used by the peanut industry (Whitaker and Dickens, 1979).

SELECTING SAMPLES

All these discussions assumed that no selection biases are associated with drawing the sample from the lot. If the lot has been blended thoroughly from the various material handling operations, then the contaminated particles probably are distributed uniformly throughout the lot. In this situation, the location in the lot from which the sample is drawn probably is not too important provided the location is selected in a random manner. However, if a commodity has become contaminated because of moisture leaks or for other reasons, then mycotoxin contaminated particles may be located in isolated pockets in the lot (Shotwell *et al.,* 1975). If the sample is drawn from a single location, the contaminated particles may be missed or too many contaminated particles may be collected. Because contaminated particles may not be distributed uniformly throughout the lot, the sample should be an accumulation of small portions taken from many different locations throughout the lot (Bauwin and Ryan, 1982; Hurburgh and Bern, 1983; Park and Pohland, 1989; Park *et al.,* 1991c).

Obtaining a representative sample from a lot at rest (static lot) is generally more difficult than obtaining such a sample from a moving stream of the product. Examples of static lots are commodities contained in storage bins, piles, railcars, or many small containers such as sacks. When drawing a sample from a static lot, a probing pattern should be developed so product can be collected from different locations in the lot. The sampling probe should be driven to the bottom of the container when possible. As a general rule, about 0.5 kg sample should be drawn per 1000 kg feed.

When sampling a static lot contained in separate containers such as sacks, the sample should be taken from many containers dispersed throughout the lot. The recommended number of sacks sampled can vary from one-fourth of the sacks in small lots to the square root of the number of sacks for large lots (U.S. Department of Agriculture, Agricultural Marketing Service, 1975).

If the lot is in a container in which access is limited, the sample should be drawn when the commodity is being removed from or being placed into the container. If the accumulated sample for any sampling method is larger than required, the sample should be blended thoroughly and reduced to the required size with a suitable device such as a riffle divider.

Sampling a moving stream of product often reduces the selection biases associated with sampling a static lot. The sample should be taken as the material leaves the conveyance system, for example, at the end of a conveyor belt. The quantity of material removed each time should be small so not too large a sample

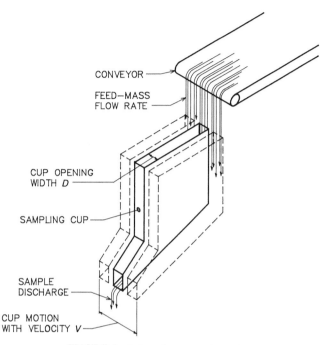

FIGURE 4   Automatic cross-cut sampler.

is accumulated. Commercial samplers are available that cut through the entire commodity stream at predetermined intervals. Cross-cut sampler such as the one shown in Figure 4 should have a sampling cup that (1) moves through the entire stream at a constant velocity, (2) has an opening that is perpendicular to the stream flow, and (3) has an opening that is 1.5–3 times the size of the largest particle in the lot. The sample weight S in kg collected from a lot with a cross-cut sampler is a function of the weight of the lot (W) in kg, the frequency (F) with which the sampling cup moves through the moving stream in seconds, the width (D) of the sampling cup opening in cm, and the velocity (V) of the sampling cup as it moves through the moving stream in cm/sec:

$$S = DW/FV \tag{19}$$

The frequency with which the sampling cup should cut through the moving stream to obtain a given sample weight S may be computed by solving for F in Eq. 19:

$$F = DW/SV \tag{20}$$

If more sample is collected than desired, the sample should be blended thoroughly and the required amount of sample removed.

CONCLUSIONS

Because of the large variability associated with testing granular product for aflatoxin, the true lot concentration cannot be determined with 100% certainty. The total variance is the sum of sampling, subsampling, and analytical variances. For small sample sizes, sampling is usually the largest source of variability. Increasing the sample size, the degree of sample comminution, the subsample size, or the number of analyses will increase the precision of estimating the true lot concentration and also will decrease both the processor's and the consumer's risk associated with a testing program. Decreasing the critical level below the guideline will increase the processor's risk but decrease the consumer's risk. Conversely, increasing the critical level above the guideline will reduce the processor's risk but increase the consumer's risk. The entire sample should be comminuted before a subsample is removed from the sample. Increasing the degree of comminution will reduce the subsampling variability. Care also should be taken in drawing the sample from the lot. The sample should be an accumulation of small portions taken from many different locations throughout the lot.

REFERENCES

Association of Official Analytical Chemists (1990). "Official Methods of Analysis of the Association of Official Analytical Chemists" (Kenneth Helrich, ed.), 15th Ed. Association of Official Analytical Chemists, Arlington, Virginia.

Bauwin, G. R., and Ryan, H. L. (1982). Sampling inspection and grading of grain, *In* "Storage of Cereal Grains and Their Products" (C. M. Christensen, ed.), Vol. 5, pp. 1–35. American Association of Cereal Chemists, St. Paul, Minnesota.

Campbell, A. D., Whitaker, T. B., Pohland, A. E., Dickens, J. W., and Park, D. L. (1986). Sampling sample preparation, and sampling plans for foodstuffs for mycotoxin analysis. *Pure Appl. Chem.* **58**, 305–314.

Cucullu, A. F., Lee, L. S., Mayne, R. Y., and Goldblatt, L. A. (1966). Determination of aflatoxin in individual peanuts and peanut sections. *J. Am. Oil Chem. Soc.* **43**, 89–92.

Cucullu, A. F., Lee, L. S., and Pons, W. A. (1977). Relationship of physical appearance of individual mold damaged cottonseed to aflatoxin content. *J. Am. Oil Chem. Soc.* **54**, 235A–237A.

Dickens, J. W., and Satterwhite, J. B. (1969). Subsampling mill for peanut kernels. *Food Technol.* **23**, 90–92.

Dickens, J. W., and Whitaker, T. B. (1982). Sampling and sampling preparation. *In* "Environmental Carcinogens—Selected Methods of Analysis: Some Mycotoxins" (H. Egan, L. Stoloff, P. Scott, M. Costegnaro, I.K. O'Neill, and H. Bartsch, eds.), Vol. 5, pp. 17–32. IARC, Lyon, France.

Dickens, J. W., Whitaker, T. B., Monroe, R. J., and Weaver, J. N. (1979). Accuracy of subsampling mill for granular material. *J. Am. Oil Chem. Soc.* **56**, 842–844.

Dorner, J. W., and Cole, R. J. (1988). Rapid determination of aflatoxin in raw peanuts by liquid chromatography with post column iodination and modified minicolumn cleanup. *J. Assoc. Off. Anal. Chem.* **71**, 43–47.

Hagler, W. M., and Whitaker, T. B. (1991). One step solid phase extraction cleanup of peanuts and corn extracts for LC quantification of aflatoxin. *Proc. Assoc. Off. Anal. Chem.* 141.

Horwitz, W., Albert, R., and Nesheim, S. (1993). Reliability of mycotoxin assays—An update. *J. Assoc. Anal. Chem.* **76**, 1–30.

Hurburgh, C. R., and Bern, C. J. (1983). Sampling corn and soybeans. 1 Probing method. *Trans. Am. Soc. Agric. Eng.* **26**, 930–934.

Kratochvil, B., and Taylor, J. K. (1981). Sampling for chemical analysis. *J. Anal. Chem.* **53**, 924A–932A.

Nesheim, S. (1979). "Methods of Aflatoxin Analysis" National Bureau of Standards, Gaithersburg, Maryland Special Publication No. 519. 355–372.

Park, D. L., and Pohland, A. E. (1986). Official methods of analysis of foods for mycotoxins. *In* "Foodborne Microorganisms and Their Toxins: Developing Methodology" (H. D. Pierson and N. S. Stern, eds.), pp. 425–438. Marcel Dekker, New York.

Park, D. L., and Pohland, A. E. (1989). Sampling and sample preparation for detection of natural toxicants in food and feed. *J. Assoc. Off. Anal. Chem.* **72**, 399–404.

Park, D. L., Njapau, H., Rua, S. M., Jr., and Jorgensen, K. V. (1991a). Field evaluation of immunoassay for aflatoxin contamination in agricultural commodities. *In* "Immunoassays for Three Chemical Analyses" (M. Vanderlaan, L. H. Stauker, B. E. Watkins, and D. W. Roberts, eds.), pp. 162–169. American Chemical Society, Washington, D.C.

Park, D. L., Rua, S. M., Jr., Paulson, J. H., Harder, D., and Young, A. K. (1991b). Sample collection and sample preparation techniques for aflatoxin determination in whole cottonseed. *J. Assoc. Off. Anal. Chem.* **74**, 73–75.

Park, D. L., Rua, S. M., Jr., Weuc, M. J., Rigsby, E. S., Njapau, H., Whitaker, T. B., and Jorgensen, K. V. (1991c). Determination of sampling variation for aflatoxin concentrations in whole cottonseed. *Proc. Assoc. Offic. Anal. Chem.* 143.

Remington, R. D., and Schrok, M. A. (1970). "Statistics and Applications to the Biological and Health Sciences." Prentice-Hall, Englewood Cliffs, New Jersey.

Schuller, P.O., Horwitz, W., and Stoloff, L. (1976). A review of sampling and collaboratively studied methods of analysis for aflatoxin. *J. Assoc. Off. Anal. Chem.* **59**, 1315–1345.

Shotwell, O. L., Goulden, M. L., and Hesseltine, C. W. (1974). Aflatoxin: Distribution in contaminated corn. *Cereal Chem.* **51**, 492–499.

Shotwell, O. L., Goulden, M. L., Botast, R. J., and Hesseltine, C. W. (1975). Mycotoxins in hot spots in grains. 1. Aflatoxin and zearalenone occurrence in stored corn. *Cereal Chem.* **52**, 687–697.

United States Department of Agriculture (1975). "Inspectors Instructions." Agricultural Marketing Service, Washington, D.C.

Velasco, J. (1972). Modified ferric gel for determining aflatoxin in cottonseed. *J. Assoc. Off. Anal. Chem.* **55**, 1359–1360.

Velasco, J., Whitaker, T. B., and Whitten, M. E. (1975). Sampling cottonseed lots for aflatoxin contamination. *J. Am. Oil Chem. Soc.* **52**, 191–195.

Walpole, R. E. (1974). "Introduction to Statistics." MacMillan, New York.

Whitaker, T. B. (1977). Sampling granular foodstuffs for aflatoxin. *Pure Appl. Chem.* **49**, 1709–1711.

Whitaker, T. B., and Dickens, J. W. (1979). Evaluation of the Peanut Administrative Committee testing program for aflatoxin in shelled peanuts. *Peanut Sci.* **6**, 7–9.

Whitaker, T. .B, and Dickens, J. W. (1981). Errors in aflatoxin analyses of raw peanuts by thin layer chromatography. *Peanut Sci.* **8**, 89–92.

Whitaker, T. B., and Wiser, E. H. (1969). Theoretical investigation into the accuracy of sampling shelled peanuts for aflatoxin. *J. Am. Oil Chem. Soc.* **46**, 377–379.

Whitaker, T. B., Dickens, J. W., and Monroe, R. J. (1972). Comparison of the observed distribution of aflatoxin in shelled peanuts to the negative binomial distribution. *J. Am. Oil Chem. Soc.* **49**, 590–593.

Whitaker, T. B., Dickens, J. W., and Monroe, R. J. (1974). Variability of aflatoxin test results, *J. Am. Oil Chem. Soc.* **51**, 214–218.

Whitaker, T. B., Whitten, M. E., and Monroe R. J. (1976). Variability associated with testing cottonseed for aflatoxin. *J. Am. Oil Chem. Soc.* **53**, 502–505.

Whitaker, T. B., Dickens, J. W., and Monroe, R. J. (1979). Variability associated with testing corn for aflatoxin. *J. Am. Oil Chem. Soc.* **56,** 789–794.

Wilson, T. J., and Romer, T. R. (1991). Utilization of the novel Mycosep #224 cleanup column for the liquid chromatographic determination of aflatoxin in agricultural products. *J. Assoc. Off. Anal. Chem.* **74:(6),** 951–956.

21

Development of Antibodies against Aflatoxins

❖

Fun S. Chu

INTRODUCTION

The toxicology and chemistry of aflatoxins (AFs) have been discussed extensively in various chapters of this book as well as in a number of other reviews and books. Therefore, reiterating the impact of AFs on human and animal health is not necessary. Although removing AFs from foods and feeds completely is difficult with the currently available technology, the most effective measure in controlling the aflatoxin problem depends on a rigorous program for monitoring the toxin in foods and feeds. Thus, extensive research attempting to develop more sensitive, specific, and simple methods for aflatoxin detection has been carried out since the discovery of AFs.

Interest in generating antibodies against AFs originated from the idea that the antibodies could be used as specific reagents for the development of sensitive immunochemical assays for aflatoxins, and also as prophylactic agents. Investigators (Langone and van Vunakis, 1976; Chu and Ueno, 1977) have demonstrated that specific antibodies against aflatoxin could be generated in rabbits after immunization of the animals with aflatoxin–protein conjugates. We also observed some protective effect against aflatoxicosis in rabbits that were immunized with the aflatoxin–protein conjugate (Ueno and Chu, 1978). With advances in hybridoma technology and progress in the medical diagnostic area, rapid progress in the development of immunoassays for aflatoxin and other mycotoxins has been made in the last 10 years. Antibodies against almost all the important mycotoxins are now available. Various immunoassay protocols for mycotoxins also have been developed. Some of these methods have gained wide acceptance in applications for the analysis of aflatoxins in foods, feed, and biological fluids. However, discussing all recent progress on immunoassay of mycotoxins is beyond the scope of this review. Instead, approaches used in the development of antibodies against AFs, properties of various AF antibodies, and the use of these antibodies for monitoring AFs in foods, feed, and body fluid, as well as their role in elucidating the toxic effect of aflatoxin, are addressed here. For immunochemical studies of other mycotoxins, see Chu (1986c, 1991a,b), Fukal and Kas (1989), Morgan and Lee (1990) and Pestka (1988).

METHODS FOR PREPARATION OF AFLATOXIN–PROTEIN CONJUGATES

Preparation of Aflatoxin Derivatives and Site of Attachment of Aflatoxin to Macromolecules

Like most other small molecular weight haptens, AFs are not immunogenic. They must be conjugated be to a protein or polypeptide carrier before subsequent use in immunization. Aflatoxins also do not have a reactive group that can be conjugated readily to a protein carrier; a reactive group first must be introduced into the AF molecular before coupling it to a protein. Most of the earlier efforts in the development of antibodies against AFs were devoted to the investigation of methods for the preparation of aflatoxin derivatives and for conjugation of AF to the protein (Chu, 1984, 1986c). As will be seen from later discussion, the antibody specificity of AF depends greatly on the site in the toxin molecule at which it is conjugated to the protein carriers. Thus, different approaches were tested (Figure 1). The approaches and rationales for selecting an appropriate site for conjugation of AF to protein molecules are summarized here.

Conjugation of Aflatoxin to Protein through the Carboxyl Group in the Cyclopentanone Ring

The carboxyl group of the cyclopentanone ring is one of the most common sites for conjugation of aflatoxin to a protein carrier. The approach involves the introduction of a carboxylic acid group on the AF molecule by converting the carbonyl group in the cyclopentanone ring of aflatoxin B, (AFB$_1$) to an O-carboxymethyl (O-CMO) derivative. The O-CMO-AFB$_1$ then is conjugated to a protein. With the exception of the double bonded oxygen of the cyclopentanone moiety, this approach preserves the whole aflatoxin structure. Thus, the entire AF molecule is exposed for recognition by antibody producing cells. Since its introduction (Langone and van Vunakis, 1976; Chu and Ueno, 1977), this method has

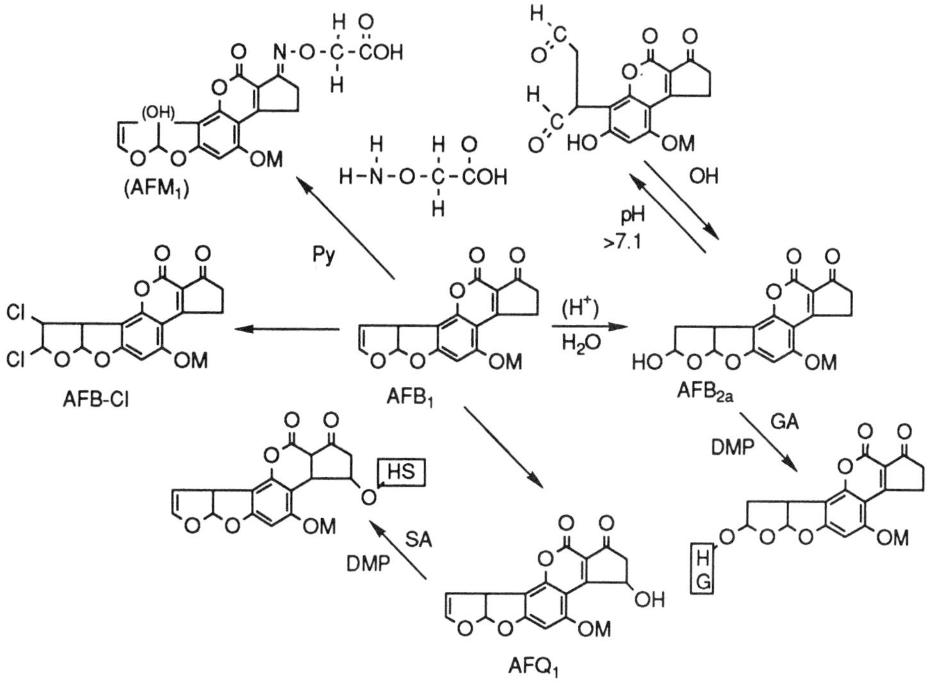

FIGURE 1 Approaches used to prepare various aflatoxin B_1 derivatives. PY, Pyridine; DMP, dimethylaminopyridine; SA, succinic anhydride; HS, hemisuccinate; HG, hemiglutarate; GA, glutaric anhydride.

been used widely by various investigators for the synthesis of immunogens for AFs (Chu, 1991a,b; Fukal *et al.*, 1988a). *O*-Carboxymethyl oximes aflatoxin M_1 (AFM$_1$) (Harder and Chu, 1979) and aflatoxin B_2 (AFB$_2$) (Hastings *et al.*, 1988, 1989) also have been synthesized and conjugated to proteins for the production of antibodies against AFM and AFs. Formation of O-CMO derivatives of AF generally is carried out by refluxing aflatoxins with carboxymethyl hydroxylamine (CMA) with AFB in the presence of NaOH (Langone and van Vunakis, 1976) or under very mild conditions with pyridine (Chu *et al.*, 1977; Chu, 1984, 1986).

Another approach also involving the cyclopentanone ring portion of the molecule was used to generate generic antibodies against aflatoxins (Zhang and Chu, 1989; Hefle and Chu, 1990). Aflatoxin B_3 (AFB$_3$), a minor aflatoxin that has been considered an aflatoxin G_1 (AFG$_1$) degradation product as a result of hydrolysis of the lactone ring and subsequent decarboxylation by *Aspergillus*, was selected for this purpose. Because AFB$_3$ has structural features similar to those of both B and G type aflatoxins, we rationalized that antibody generated against this aflatoxin would be capable of cross-reacting with both types of

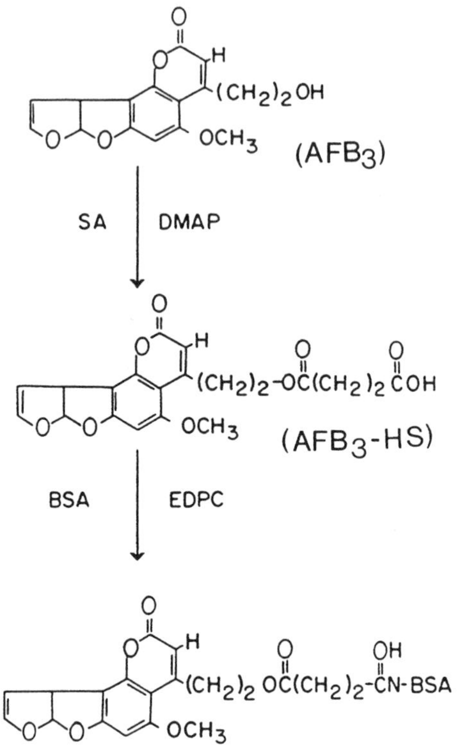

FIGURE 2 Conjugation of aflatoxin B₃ to protein. SA, Succinic anhydride; DMAP, dimethyl amino pyridine; BSA, bovine serum albumin; EDPC, 1-(3-dimethylaminopropyl)-3-ethyl-carbodiimde.

aflatoxin. A hemisuccinate of AFB_3 was prepared and then conjugated to protein for immunization (Figure 2). We found that the antibodies produced by this method, indeed, have a cross-reaction with both AFB_1 and AFG_1 (Zhang and Chu, 1989).

Conjugation of Aflatoxin to Protein through 8,9-Dihydrofuran Position

Because of the reactivity of the α,β unsaturated bond, another very common approach used for the synthesis of immunogens for the production of antibody against aflatoxin is through the 8,9-dihydrofuran (or 2,3-dihydrofuran in older literatures) portion of AFB_1 and AFG_1 molecules. With this approach, the cyclopentanone moeity of the aflatoxin molecule is left intact. The approach involves the conversion of aflatoxins B_1, G_1, and Q_1 to their respective hemiacetals that is, AFB_{2a} (Guar et al., 1981), AFG_{2a} (Chu et al., 1985), and AFQ_{2a} (Fan et al., 1984), under acidic conditions. Thus, a water molecule is added to the α,β unsaturated double bond at the 8,9 position (Ashoor and Chu, 1975). By introducing an –OH group, we were able to conjugate AFB_1 and AFG_1 to a protein carrier by the reductive alkylation method (Ashoor and Chu, 1975) or by reaction with the bifunctional anhydrides such as succinic acid and glutaric acid anhydride in the presence of a catalyst such as pyridine or 4-N,N-dimethylaminopyridine. The corresponding hemisuccinate (HS) and hemiglutarate (HG) than are used for subsequent conjugation. Depending on the stability of these derivatives, HS or HG could be used. For example, the AFB_{2a}-HS was found to be unstable and easily hydrolyzed in a neutral or a slightly alkaline solution (Lau *et al.*, 1981); the HG of AFB_{2a} was prepared and subsequently used as the starting derivative for coupling reactions. However, AFQ_{2a}-HS was stable; thus, it could yield a good immunogen after conjugation to protein for antibody production against AFQ_1 (Fan *et al.*, 1984).

Conjugation of AFB_1 to a protein carrier at this portion of the molecule also was done successfully by converting AFB_1 to its chloride or bromide (Sizaret *et al.*, 1980; Sizaret and Malaveille, 1983; Martin *et al.*, 1984) or aflatoxin diol (Pestka and Chu, 1984), or by activating AFB_1 with m-chloroperoxybenzoic acid (Hertzog *et al.*, 1982; Garner *et al.*, 1985, 1988) and directly reacting the resulting compound with the protein carrier.

Conjugation of Aflatoxins through Other Positions

To preserve the whole aflatoxin molecule, a hemisuccinate derivative of AFQ_1 was prepared in our laboratory to conjugate aflatoxin to the protein carrier. In this approach, an –OH group was introduced at the C22 position of AFB_1 chemically (Figure 1) with good yield (Fan *et al.*, 1984).

Methods of Conjugation

Once a reactive group such as a carboxylic acid is introduced into the aflatoxin molecule, the molecule is ready for conjugation to the proteins by one of the following methods. After reaction, the unreacted reagents were removed by dialysis or Sephadex gel filtration. The amount of aflatoxin conjugated to the protein was determined spectrophotometrically (Chu, 1984, 1986a,c). Conjugates with a ratio of 10–20 mol aflaxtoxin per mol protein generally have been found to be good immunogens. Although bovine serum albumin (BSA) commonly has been used as the carrier protein for the preparation of immunogens, other proteins such as γ-globulin, human serum albumin, ovalbumin, lysozyme, polylysine, and keyhole limpet hemocyanin also have been used successfully (Chu, 1991a). To increase the chain length between aflatoxin molecules and protein, we also prepared a modified BSA as the carrier protein (EDA–BSA). Ethylene diamine modified the side chain carboxyl group in BSA (Chu *et al.*, 1982). Because additional amino groups are made available in the BSA molecules, the efficiency of the coupling reaction also was found to be increased. The modified BSA has been found to be an excellent carrier for aflatoxins and for a number of other mycotoxins and haptens (Chu, 1991a,b). For reasons described in a subsequent section, the method and protein used for the preparation of conjugate for immunization is generally different from those used for preparation of marker (toxin)–enzyme conjugate or marker (toxin)–protein conjugate.

Water-Soluble Carbodiimide Method

Among many conjugation methods, the water-soluble carbodiimide (WSC) method is used most commonly. In this method, aflatoxin derivatives such as CMO-AFB or AFB-HS are reacted with a protein carrier in a molar ratio of 10–100 at pH 5–8 in the presence of a coupling reagent such as 1-(3-dimethylamino-propyl)-3-ethylcarbodiimide (EDPC) (Chu, 1986a, 1991a,b) over a period of 24–48 hr. The reaction is stopped by dialysis. The WSC method is very mild and easy to perform. However, it suffers from the disadvantage of some side reactions, including formation of adducts between the by-product urea and protein carrier and aggregation of carrier protein. When these mixtures are used as the immunogens, antibodies against the products of the side reactions often are generated in the animals. These antibodies may introduce some nonspecific reaction in the subsequent immunoassays. To overcome these problems, an alternative method could be used. One variation of the WSC method is activation of the aflatoxin derivative with N-OH-succinimide (NHS) in the presence of carbodiimide and conjugation to protein directly (Pestka, 1988; Chu, 1991a). Because the WSC method generally is used for the preparation of immunogen for aflatoxin in antibody production, the NHS approach has been used for the preparation of aflatoxin–enzyme marker in the enzyme-linked immunoassay (ELISA) (Chu et al., 1987).

Mixed Anhydride Method

The mixed anhydride method involves activation of aflatoxin derivatives containing a carboxylic acid to their corresponding anhydride by isobutyl chloroformate in dry tetrahydrofuran and triethylamine, followed by direct reaction with protein. The technique has been used to conjugate HS-AFB$_{2a}$ and HG-AFG$_{2a}$ to protein (Lau et al., 1981; Chu et al., 1982). Because several steps of this method are under anhydrous conditions, this method is more difficult to perform; sometimes aggregation of protein occurs. Thus, this method is not recommended for the preparation of the toxin–enzyme conjugate.

Reductive Alkylation Method

The reductive alkylation method is used very commonly for haptens and mycotoxins containing an active aldehyde group, which forms a Schiff base with the amino group in the carrier protein. Stabilization of the bond is achieved by reduction with sodium borohydride or other reagents. Since the furan ring of AFB$_{2a}$ (Ashoor and Chu, 1975; Gaur et al., 1981), AFG$_{2a}$ (Chu et al., 1985), and AFQ$_{2a}$ (Fan et al., 1984) opens up and generates two aldehyde groups available at a pH above 7.0, this method has been used for conjugation of aflatoxins B$_{2a}$, G$_{2a}$, and Q$_{2a}$ to protein.

Cross-linking of Aflatoxin Derivatives to Protein

Several aflatoxin derivatives including AF-dichloride or dibromide (Sizaret and Malaveille, 1983; Martin *et al.*, 1984), AF-diol (Pestka and Chu, 1984), and AF-epoxide formed with *m*-chloroperoxybenzoic acid (Garner *et al.*, 1985, 1988) could be coupled to protein directly. The mechanism of such reactions was proposed by Sabbioni *et al.* (1987, 1990; Sabbioni, 1990). Since the Schiff base formed by an aldehyde with a hydroxyl group at the position of the carbonyl is well known to be susceptible to further prototropic transformation, that is, Amadori rearrangement, Sabbioni *et al.* (1987,1990) found that such a reaction also was involved in the formation of AFB–lysine and AFB_1–serum albumin adducts. Skipper and Tannenbaum (1989) pointed out that this reaction also was involved in the Schiff bases formed by the 8,9-dihydro-8,9-dihydroxyl-related derivatives of aflatoxin B_1 or via the epoxide or the dichloroide (dibromide).

Formaldehyde was used as the condensation reagent (Mannisch reaction) for coupling the phenolic group of AFB_{2a} with the amino group of certain amino acid derivatives, including α-aminohippuric acid, 6-amino nicotinic acid, pterin-6-carboxylic acid, ϵ-amino caproic acid, or 1-amino-4(β-OH-ethyl)piperazine, and methylalbumin (Jackman, 1985). Lawellein *et al.* (1977a), however, used tetrazobenzidine for the condensation of AFB_{2a} to protein.

DNA is well known to be capable of complexing with methylated proteins. This approach was used for the preparation of immunogens for AFB–DNA adducts. Both methylated BSA (Haugen *et al.*, 1981; Groopman *et al.*, 1982) and methylated hemocyanin (Hertzog *et al.*, 1982; Hsieh *et al.*, 1988) have been used as the carrier proteins.

PRODUCTION OF ANTIBODIES AGAINST AFLATOXINS

Production of Polyclonal Antibodies

Two approaches have been used for the production of antibodies. The classic method involves immunizing animals with an immunogen and obtaining antibodies from the sera of immunized animals. Antibodies obtained by this approach are polyclonal antibodies (Pab) because they are derived from different B lymphocyte clones. Antibodies against many mycotoxins, including aflatoxins, have been produced successfully by this procedure. In this method, aflatoxin–protein conjugates are mixed with Freund's complete adjuvant and injected into the backs of rabbits at multiple intradermal sites or subcutaneously (Morgan and Lee, 1990; Chu 1991a,b). Between 200 and 500 μg aflatoxin–protein conjugate generally is used in each rabbit. For production of a large quantity of antiserum, large animals such as goats also have been used for antibody production (Guar *et al.*, 1980). Although production of antibody against aflatoxin has been achieved by immunizing chickens with AFB–BSA conjugate, followed by isolation of antibody from eggs (Hsu and Chu, 1992), the antibody titers were low. Rabbits are still the animal species used most commonly for the production of antibodies against aflatoxin. Although antiserum obtained from animals could be used directly in the immunoassay, a simple purification such as ammonium sulfate precipitation is generally necessary (Chu, 1986c). Further purification of antiserum by various chromatographic methods can improve the efficacy of immunoassays.

If the conjugate is highly immunogenic, the antibody titer as determined by radioimmunoassay (RIA) or ELISA usually starts to increase at 5–7 weeks postinjection. Subsequent booster injections of immunogen generally are performed once each month. Since polyclonal antibodies are heterogeneous, a specific aflatoxin marker must be used in the titer determination to identify the presence of relevant antibodies. Use of inadequate markers may fail to provide unequivocal information on whether the immunogen and immunization protocols are effective. Tritiated aflatoxin B_1 (Chu and Ueno, 1977) and iodinated aflatoxin–tyramine conjugates (Fukal et al., 1987) with high specific radioactivity have been used in the RIA for titer determination. For ELISA, specific AF–enzyme conjugates also have been made available. Note, however, that nonspecific interaction often occurs in the ELISA, which may give a false positive result. If a highly specific radioactive marker is available, it is advisable to use RIA for monitoring the antibody titers.

Production of Monoclonal Antibodies

Although production of polyclonal antibodies is relatively simple and the affinity of polyclonal antibodies is generally high, the antibodies are heterogeneous and their production is limited. With increasing demand for large amounts of antibodies with homogeneous properties for immunoassays as well as with advances in hybridoma technology, specific monoclonal antibodies for aflatoxins and other major mycotoxins have been produced. In monoclonal antibody production, spleen cells from mice that have been immunized with AF–protein conjugates are fused with myeloma cells. The hybrid cells are propagated and selected for clones that produce the specific antibody of interest. Production of antibodies by the hybridoma cell line can be carried out either in tissue cultures or in the ascites fluid of mice. Thus, once a hybridoma cell line is obtained, the potential exists for an unlimited supply of homogeneous antibodies with unique properties. Similar to the production of polyclonal antibodies, the key to success in obtaining a useful hybridoma cell line also lies in the effectiveness of immunogens used and the aflatoxin-marker used in the selection of positive clones.

To eliminate the clones that produce nonspecific antibody, Holladay et al. (1991) have incorporated a second screening step by coating the carrier protein BSA on the ELISA plate after a preliminary screening in which the immunogen (AFM_1–BSA conjugate) was coated on the ELISA plate. By including AFB_1–BSA as the coating antigen, Holladay et al. (1991) were able to obtain clones that

TABLE 1 Antibodies against Aflatoxins

Compound	Type[a]	References
Aflatoxin		
B₁, B₂, M₁	M, P	Lawellen *et al.* (1977a); Harder and Chu (1979); Lubet *et al.* (1983); Sun *et al.* (1983); Woychik *et al.* (1984); Candlish *et al.* (1985); Blankford and Doerr (1986); Kaveri *et al.* (1987); Dixon-Holland *et al.* (1988); Hastings *et al.* (1988); Kawamura *et al.* (1988); Pestka (1988); Fremy and Chu (1989); Zhang and Chu (1989); Hefle and Chu (1990); Morgan and Lee (1990); Chu (1991a,b)
G₁, B₃	M, P	Chu *et al.* (1985); Zhang and Chu (1989); Hefle and Chu (1990); Chu (1991a,b)
Metabolites		
B₂ₐ, Q₁, M₁, Rₒ	M, P	Lawellen *et al.* (1977a); Kawamura *et al.* (1988); Chu (1991a,b)
DNA adducts	M, P	Haugen *et al.* (1981); Groopman *et al.* (1982); Hertzog *et al.* (1982); Garner *et al.* (1985, 1988); Groopman and Kemsler (1987); Groopman and Donahue (1988); Hsieh *et al.* (1988)
Biosynthetic precursors		
Versicolorin A	P	Reynolds and Pestka (1991)
Sterigmatocystin	P	Li and Chu (1984); Morgan *et al.* (1986b)

[a]M, Monoclonal antibody; P, polyclonal antibodies.

produce antibody specific to AFM_1 as well as antibody having cross-reactivity to both AFB_1 and AFM_1. Antibodies against various aflatoxins, including both polyclonal and monoclonal antibodies (Pestka, 1988; Chu 1991,b). Morgan and Lee, 1990; are summarized in Table 1.

CHARACTERIZATION OF ANTIBODIES AGAINST AFLATOXINS

General Considerations

Antibodies obtained by either method just described should be well characterized before they are used in the immunoassay. In general, both the type of immunoglobulin (Ig) and the specificity of the antibody to various analogs of the immunogen or hapten should be determined. Information on the type of Ig is important in selecting adequate markers as well as an appropriate method for separating free antigen (Ag) or antibody (Ab) from the Ag–Ab complex in the immunoassay. The typing of Ig generally is done using commercially available kits. Information on the specificity will provide an accurate assessment of whether or not structurally related compounds will interfere with the immunoassay. This information also could be used to estimate the affinity of the antibody to the immunogen or hapten and its structurally related analogs.

The specificity of the antibody, generally expressed as the cross-reactivity, is determined primarily by the type of aflatoxin that has been used in the antibody production as well as by the site on the aflatoxin molecule where it is linked to the protein carrier. A minor change in the immunogen structure could generate antibodies with different cross-reactivity. Experimentally, the cross-reactivity of antibody is determined by an immunoassay in which various structurally related analogs of aflatoxin at a wide range of concentrations are used to compete with the binding of the marker ligand with the antibody in the assay. The concentration at 50% inhibition (I_{50}) of binding generally is used as the basis for calculating the relative cross-reactivity for each analog. A typical example of such a competitive RIA and ELISA for aflatoxins is shown in Figure 3.

Using various aflatoxin–protein conjugates prepared by different approaches described in the last section, antibodies with different specificity have been produced. Some of these antibodies are very specific to one type of aflatoxin and others have a broad specificity. Although the general trend of the antibody specificity does not change, the "apparent cross-reactivity" of the polyclonal antibody sometimes varies with the type and concentration of the marker and the format used in the immunoassay. Thus, in addition to knowing the specificity of the antibodies generated by a specific immunogen, one also should be aware of the conditions under which that specificity was determined. Preferably, the "apparent cross-reactivity" of the polyclonal antibodies should be re-evaluated under the conditions used for the assay. The specificity of the polyclonal antibodies can be improved by removing nonspecific antibodies through immunoabsorption.

Specificity of Polyclonal Antibodies Obtained from Immunogens Prepared by Conjugation of the Cyclopentanone Ring Portion of Aflatoxin Molecule to Carrier Protein

The relative cross-reactivity with five major aflatoxins of antibodies obtained from rabbits that have been immunized with aflatoxin–protein conjugates that were prepared from the cyclopentanone ring moiety of aflatoxin molecules is shown in Figure 4. The importance of the site of conjugation for the antibody specificity is well demonstrated in this figure. When rabbits were immunized with conjugates prepared by linking the CMO of AFB_1 (lanes 1–7 of Figure 4) and AFM_1 (lanes 11–13 of Figure 4), the antibodies generally recognized the dihydrofuran ring of the molecule. Any structural variation in the dihydrofuran portion of aflatoxin molecules could be recognized by the antibody. Thus, antibodies highly specific against AFB_1 and AFM_1 were produced in this manner (Chu and Ueno, 1977; Harder and Chu, 1979). These two types of antibody currently are used for most of the immunoassays for both aflatoxins. Whereas antibodies for AFB_1 cannot recognize AFM_1, antibodies for AFM_1 bind AFB_1 effectively. This property has been used as an approach for the immunoassay of AFM_1 so radioactive AFB_1 or enzyme-linked AFB_1 could be used as the marker in the assays (Martbauer and Terplan, 1985; Chu, 1986c, 1991a,b; Fremy and

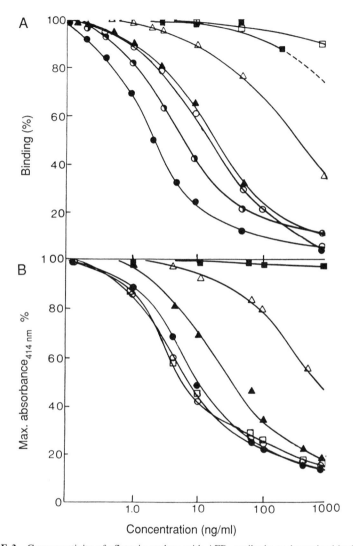

FIGURE 3 Cross-reactivity of aflatoxin analogs with AFB_1 antibody as determined by RIA (A) and competitive direct ELISA (B). In the RIA, an equilibrium dialysis method was used (Chu and Ueno, 1977). [^{3}H]AFB_1, 0.3–0.5 ml (15,000–20,000 cpm/ml), with a constant amount of IgG (30 μg/ml), was dialyzed against an equal volume of different concentrations of unlabeled AF analogs. The extent of binding of [^{3}H]AFB_1 with IgG in the absence of unlabeled toxin (i.e., dialyzed against buffer alone) was considered 100%. In the ELISA, an AFB_1–horseradish peroxidase conjugate was used as an enzyme indicator (Pestka *et al.,* 1980). Antiserum at 1:400 dilution was coated to a microplate for the assay. The figure is in semilog scale, but the *x* axis is in log scale. In the ELISA, 25 μl sample was used in each analysis. Therefore, the amount in each assay was 40 times less on the ng/ml basis. B_1 (●), B_1 oxime (◖), B_2 (○), G_1 (▲), G_2 (△), M_1 (■), and R_o (aflatoxicol, ◲) represent different aflatoxins; ST (▨), sterigmatocystin.

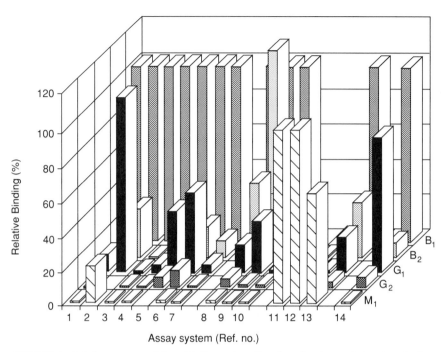

FIGURE 4   Relative cross-reactivity with five major aflatoxins of various polyclonal antibodies obtained from rabbits that have been immunized with aflatoxin–protein conjugates that were prepared from the cyclopentanone ring moiety of aflatoxin molecules. Except where otherwise stated, O-CMO-AFB$_1$ was used as the starting derivative for conjugation and RIAs were used in all the analyses. Reference numbers shown are (1) Chu and Ueno (1977); (2) Morgan *et al.* (1986a); (3) Fukal *et al.* (1988a); (4) Sizaret *et al.* (1980); (5) Fukal *et al.* (1987); (6) Fukal *et al.* (1986a); (7) Fukal *et al.* (1988a); (8) Langone and van Vunakis (1976); (9) Pestka *et al.* (1980)—ELISA, same antisera as 1; (10) Fan and Chu (1984)—indirect ELISA; (11) Harder and Chu (1979)—O-CMO-BSA as the immunogen; (12) Pestka *et al.* (1981)—direct ELISA, same antiserum as 11; (13) Fukal *et al.* (1988a); (14) Zhang and Chu (1989)—AFB$_3$-HS–BSA as the immunogen.

Chu, 1989). From these studies, the side-chain hydroxyl group in AFM$_1$ apparently plays an important role in eliciting antibodies that can recognize AFM$_1$.

Since the majority of the structure of AFB$_1$ and AFG$_1$ is identical to AFB$_3$, a hemisuccinate of AFB$_3$ conjugated to BSA was used as the immunogen for antibody production. We found that the antibodies indeed showed good cross-reaction with both AFB$_1$ and AFG$_1$ (lane 14, Figure 4). Thus, these types of antibody are very useful for immunoassays of both aflatoxins (Zhang and Chu, 1989; Hefle and Chu, 1990).

The role of the marker ligand in the apparent relative cross-reactivity of the antibody is well illustrated in Figure 4. For example, data shown in lanes 1 and 9 were obtained from two separate experiments using identical antibodies, which were obtained from rabbits after immunizing them with AFB$_1$-CMO–BSA conjugate. However, the cross-reactivity of the antibodies with AFB$_2$ was considerably higher in the ELISA when AFB$_1$–enzyme was used as the marker (lane 9) than in the RIA in which tritiated AFB_1 was used as the marker (lane 1). A more drastic difference was apparent from the antibodies against AFM_1. When radioactive AFB_1 was used as the marker ligand, antibodies against AFM_1 showed almost similar cross-reactivity with AFB_1 and AFM_1 (lanes 11,13). Nevertheless, the assay system showed high specificity with AFM_1 when an AFM_1–enzyme marker was used in the ELISA (lane 12, Figure 4).

Specificity of Monoclonal Antibodies Obtained from Immunogens Prepared by Conjugation of the Cyclopentanone Ring Portion of the Aflatoxin Molecule

Several monoclonal antibodies against AFs have been obtained in the last few years (Hertzog *et al.*, 1982; Groopman *et al.*, 1984; Woychik *et al.*, 1984; Candlish *et al.*, 1985,1987,1989; Garner *et al.*, 1985,1988; Blankford and Doerr, 1986; Kaveri *et al.*, 1987; Dixon-Holland *et al.*, 1988; Hastings *et al.*, 1988; Kawamura *et al.*, 1988; Hefle and Chu, 1990). In general, the specificity of different monoclonal antibodies for aflatoxins is also dependent on the type of conjugate used in the immunization. When the same immunogens are used for the preparation of either polyclonal or monoclonal antibody, most of the high affinity monoclonal antibodies have almost the same specificity as that observed for the polyclonal antibodies. This trend is shown in Figure 5, in which the cross-reactivity of some monoclonal antibodies obtained from immunogens prepared by conjugation from the clyopentanone ring portion of the AF molecule is presented. For example, antibodies shown in lanes 1–6 of Figure 5 are highly specific to AFB_1 because CMO-AFB_1–BSA conjugates were used as the immunogen. In contrast, antibodies in lanes 7 and 8 were highly specific for AFB_2 and AFM_1 because BSA conjugates of CMO-AFB_2 and CMO-AFM_1 were used as the immunogens, respectively. However, when AFB_3-HS-BSA was used as the immunogen, the monoclonal antibodies had good cross-reactivity toward both AFB_1 and AFG_1 (lanes 10–12; Hefle and Chu, 1990). Kawamura *et al.*, (1988) have found that some of the monoclonal antibodies for AFB_1 had higher cross-reactivity with aflatoxicol (AFL), one of the major aflatoxin metabolites. Nevertheless, the affinity of these antibodies to aflatoxicol was still lower than the affinity of the best monoclonal antibody to AFB_1.

Specificity of Antibodies Obtained from Immunogens Prepared by Conjugation of the Dihydrofuran Moiety of the Aflatoxin Molecule

Data for the cross-reactivity of both polyclonal and monoclonal antibodies by immunization of animals with aflatoxin derivatives from the dihydrofuran moiety of aflatoxin molecules are shown in Figure 6. Clearly these antibodies had a specificity directed toward the cyclopentanone ring portion of the AF molecule. Antibodies with similar specificity toward B types aflatoxins were obtained from rabbits when conjugates prepared by coupling AFB_{2a} (designated B_{2a}; lanes 1–4, Figure 6), AFB diol (designated AFB-diol; Pestka and Chu, 1984), and AFB

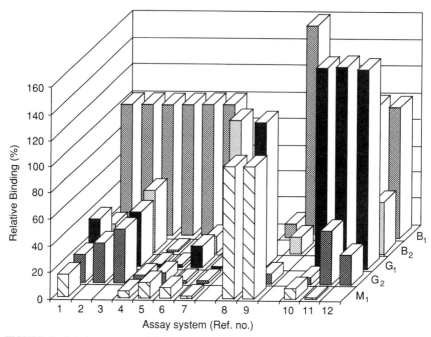

FIGURE 5 Relative cross-reactivity with five major aflatoxins of various monoclonal antibodies obtained from rabbits that have been immunized with aflatoxin–protein conjugates that were prepared from the cyclopentanone ring moiety of aflatoxin molecules. Except where otherwise stated, O-CMO-AFB$_1$ was used as the starting derivative and ELISAs were used in all the analyses. Reference numbers shown are (1) Dixon-Holland *et al.* (1988); (2) Kaveri *et al.* (1987); (3–5) Kawamura *et al.* (1988); (6) Candlish *et al.* (1985); (7) Hastings *et al.* (1988)—O-CMO-AFB$_2$; (8) Woychik *et al.* (1984)—O-CMO-AFM$_1$; (9) Dixon-Holland *et al.* (1988)—O-CMO-AFM$_1$; (10–12) Hefle and Chu (1990)—AFB$_3$-HS–BSA, RIA.

chloride (designated AFB-Cl; lane 6) to protein were used for immunization. These antibodies also have been shown to cross-react with AFB–serum albumin adduct as well as AFB–DNA adducts and other related compounds. These antibodies only have weak cross-reactivity with AFG$_1$ and AFG$_2$. In contrast, antibodies from rabbits that were immunized with AFG$_{2a}$–BSA conjugate were highly specific for AFG$_{2a}$ (lane 9). Figure 6 also demonstrates the role of marker ligand used in the assay. When tritiated AFB$_1$ was used in the analysis, the apparent cross-reaction with AFB$_{2a}$ was considerably weak (lanes 1, 6). However, strong reactions with AFB$_{2a}$ were apparent in the ELISA when AFB$_{2a}$ derivatives were coated to the microtiter plate as the competitor (lanes 2–4, 8) for the antibody bindings.

Cross-Reactivity of Aflatoxin Antibodies with Macromolecule Adducts

Two types of antibody against AFB$_1$–DNA adducts have been prepared (Hertzog *et al.*, 1982; Groopman *et al.*, 1984; Garner *et al.*, 1985,1988; Groop-

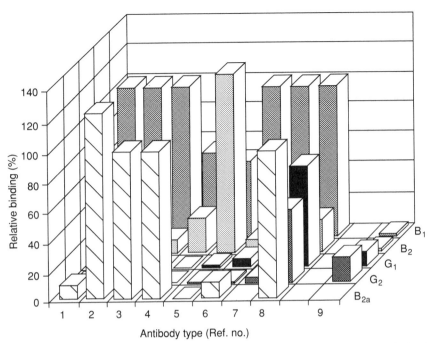

FIGURE 6   Relative cross-reactivity with five major aflatoxins of various antibodies (polyclonal and monoclonal) obtained from rabbits that have been immunized with aflatoxin–protein conjugates that were prepared from the dihydrofuran portion of aflatoxin molecules. The starting derivative and methods of analysis are given for each reference. Reference numbers shown are (1) Guar *et al.* (1981)—AFB$_{2a}$, RIA; (2) Guar *et al.* (1981)—AFB$_{2a}$, ELISA; (3) Lawellen *et al.* (1977a)—AFB$_{2a}$, ELISA; (4) Biermann and Terplan (1980)—AFB$_{2a}$, ELISA; (5) Sizaret *et al.* (1980)—AF-dichloride, RIA; (6) Lau *et al.* (1981)—AFB$_{2a}$-HG, RIA; (7) Groopman *et al.* (1984)—AFB-epoxide, RIA; (8) Dragsted *et al.* (1988)—AFB-epoxide, ELISA; (9) Chu *et al.* (1985)—AFG$_{2a}$, ELISA; AFG$_{2a}$ as 100%.

man and Kemsler, 1987; Groopman and Donahue, 1988). One type of antibody was obtained from animals after immunization with the AFB$_1$–DNA adducts complexed with methylated BSA (Haugen *et al.*, 1981; Groopman *et al.*, 1982) or methylated hemocyanin (Hertzog *et al.*, 1982; Hsieh *et al.*, 1988). Another type of antibody involved immunization with aflatoxin–protein conjugates that were coupled through the 8,9-dihydrofuran (or 2,3-dihydrofuran) portion of the molecules. Monoclonal and polyclonal antibodies of both types have been obtained. With AFB$_1$–DNA immunogens, the monoclonal antibodies prepared against AFB$_1$–guanine-modified DNA and AFB$_1$-ring-opened-modified DNA (Hertzog *et al.*, 1982; Hsieh *et al.*, 1988) showed virtually no cross-reactivity with aflatoxins B$_1$, B$_{2a}$, or diol, and AFB$_1$–guanine; the antibodies exhibited a degree of recognition for epitopes in AFB$_1$–DNA adducts regardless of whether they were AFB$_1$–DNA or AFG$_1$–DNA (Garner *et al.*, 1985). However, the AF–protein immunogen yielded antibodies that cross-reacted with a number of aflatoxins in

TABLE 2 Percentage of Cross-Reactivity of Antibodies with Aflatoxin–Macromolecule Adducts

Conjugate[a]	Aflatoxin–macromolecule adducts							References
	B_1	B_{2a}	Diol	B-DNA	B-Gua	B-FRY-DNA	B-FaYr	
BSA-B_{2a} (P)	80	100	80	50	18	40	22	Pestka et al. (1982)
BSA-B_1-diol (P)	200	100	100	32	0.6	—	—	Pestka and Chu (1984)
BSA-B_1-Cl (P)	100	—	—	—	2		4	Garner et al. (1985)
BGG-AFB (M)	100	—	—	—	3.0		33	Groopman et al. (1984)
MBSA-B_1-DNA (M)	<0.1	<0.1	<1	100	<1		<1	Groopman et al. (1982)
MHB$_1$-FA-Pyr-DNA (M)	—	—	<0.1	53	<0.1	100	<0.1	Hertzog et al. (1982)
				10	<0.1	100		Hsieh et al. (1988)

[a]P, Polyclonal; M, monoclonal.

addition to AF–DNA adducts and the hydrolyzed products of AF–macromolecule adducts, including hydrolyzed products of both AF–protein and AF–DNA adducts. The cross-reactivity of various antibodies to AF–macromolecule adducts is summarized in Table 2.

USE OF AFLATOXIN ANTIBODIES IN VARIOUS INVESTIGATIONS

Development of Immunoassay Protocols and Their Application for the Analysis of Aflatoxins in Foods and Feeds

General Considerations

With the availability of different types of antibody against aflatoxins, many types of immunoassays including RIA and ELISA, as well as several novel immunochemical screening tests, have been developed for aflatoxin analysis. Most of these methods are very sensitive, specific, and simple to perform. Specific antibodies also have been used as immunohistochemical reagents and in affinity columns that are used as clean-up tools for analysis of aflatoxins by other methods. Within the last two years, wide application of immunoassays for aflatoxins as well as other mycotoxins has been noted (Chu, 1990,1991a,b). Thus, a new dimension of methodology for mycotoxin analysis as well as a new tool for diagnosis of mycotoxicoses in humans and animals has emerged (Pestka, 1988; Chu, 1989, 1991b; Morgan and Lee, 1990).

For immunoassays of aflatoxins, several general important criteria must be considered (Chu, 1990,1991b). (1) Since most immunoassays are based on the competition of binding between the unlabeled toxin in the sample and the labeled toxin in the assay system for the specific binding sites of antibody molecules, a good marker aflatoxin derivative is needed in addition to the specific antibody in the assay system. (2) For accurate quantitation, a good method for the separation of free and bound forms of toxin is important. (3) Depending on the approaches that have been used for raising antibodies, the degree of cross-reactivity (specificity) of these antibodies with their respective structural analogs varies considerably. Thus, one must be familiar with the specificity of the antibody to be used in the assay system. (4) For practical application of the immunoassay, the sample matrix interference problems should never be overlooked. This problem is different with different matrices as well as with immunochemical reagents used in the assay. Thus, whereas this problem could be overcome by sample dilution in one assay system because either high affinity antibody or specific marker ligand was used, this approach may not be applicable for another system, and sample clean-up may be necessary (Fukal et al., 1986b; Ramakrishna et al., 1990). Thus, rigorous tests should be carried out (Chu, 1990,1991b). For example, Rauch et al. (1988,1989) have pointed out that caffeic acid and chlorogenic acid, two of most common natural plant constituents, interfere with the RIA of aflatoxin.

With the availability of various immunochemical reagents and development of efficient immunoassay protocols, more than 10 types of immunoassay kit for aflatoxin analysis have been made commercially available within the last few years (CAST, 1989; Lee, 1989; Chu, 1990,1991b). Whereas most of these kits were designed as screening tests for aflatoxins, some could be used for quantitative analysis. The efficacy of some of the commercially prepared kits has been studied (Goto and Manabe, 1988; Dorner and Cole, 1989; Koeltzow and Tanner, 1990; Patey *et al.*, 1990b). The criteria for selection of appropriate kits for mycotoxin analysis have been suggested by Chu (1990) and Pestka (1988).

Radioimmunoassay

The RIA procedure involves incubation of specific antibody simultaneously with a solution of an unknown sample or a known standard and a constant amount of labeled toxin. After separation of the free and bound toxin, the radioactivity in those fractions is determined. The toxin concentration of the unknown sample is determined by comparing the results to a standard curve, which is established by plotting the ratio of radioactivities in the bound fraction and free fraction against the log concentration of unlabeled standard toxin. Several methods, including ammonium sulfate precipitation (Chu and Ueno, 1977), double antibody technique (Langone and van Vunakis, 1976; Chu, 1986a), a solid phase RIA method in which the immunoglobulin G (IgG) was conjugated to CNBr-activated Sepharose gel (Sun and Chu, 1977), a dextran-coated charcoal column (Fukal *et al.*, 1986a) and albumin-coated charcoal (Fontel *et al.*, 1983), and polyethylene glycol 6000 (Fukal *et al.*, 1986a; Rauch *et al.*, 1987) have been used for the separation of free and bound toxins in the RIA. With the introduction of controlled-size fine particle magnetic gels to which specific antibody can be conjugated, RIA is now simpler to perform (R. D. Wei and F. S. Chu, unpublished observations).

The specificity of the radioactive marker plays an important role in determining the sensitivity of the assay. Whereas high specific activity of tritiated AFB_1 commonly was used in the RIA for aflatoxins (Chu and Ueno, 1977), 3′ − [^{125}I]tyramine−AFB_1-1-O-carboxymethyl oxime (2300 Ci/nmol) was used in the RIA of both AFB_1 and AFM_1 (Bludovsky, 1986; Fukal *et al.*, 1986,b, 1988b; Rauch *et al.*, 1987a,b; Fukal, 1988). As little as 10 pg AFB_1 could be measured in each assay when the [^{125}I]tyramine−AFB_1 derivative was used. An RIA commercial kit for AFB_1 and AFM_1 analysis, based on the use of ^{125}I-labeled AFB_1 derivative, was also available (Fukal, 1988).

The sensitivity of RIA for aflatoxin detection in various agricultural commodities is summarized in Table 3. In general, RIA can detect 0.25−0.5 ng purified aflatoxin in each analysis. In the presence of sample matrix, the lower limit for aflatoxin detection in food or feed samples is 2−5 ppb. The sensitivity of RIA can be improved by a simple clean-up procedure after extraction of the sample, in addition to the use of high specific activity radioactive markers (Chu, 1991b). For example, as little as 10 ppt AFB_1 in liver tissue could be detected after treatment of the sample extract twice with a Sep-Pak C_{18} reversed-phase cartridge (Qian

TABLE 3 Sensitivity of Radioimmunoassay for Aflatoxins in Different Agricultural Commodities

Aflatoxin[a]	Commodities[b]	Extraction solvents[c]	Clean-up[d]	Standard range (pg/assay)	Detection limits (μg/kg)	References
B_1-H	C, W	Chl	N	250–5000	5	Sun and Chu (1977); El-Nakib et al. (1981)
	PB	Hx, 55% MeOH	N	250–5000	5	El-Nakib et al. (1981)
B_1-I	E	Chl	N	4	0.1	Rauch et al. (1987a)
M_1-H	M	—	N	5000–50000	5	Pestka et al. (1981)
	M		Y		0.003	Qian et al. (1984)
M_1-I	M, E	Chl	N	0.1–1 ng/mL 100 (pg 50%)	1	Rauch et al. (1987b)

[a]H, Tritiated; I, iodinated.

[b]C, Corn; W, wheat; PB, peanut butter; E, eggs; M, milk.

[c]Chl, Chloroform; Hx, hexane; MeOH, methane.

[d]N, None; Y, yes.

and Yang, 1984; Qian *et al.,* 1984). Although RIA is a less popular method than ELISA, by using the high specific activity radioactive marker ligand, this method is still an efficient and accurate method for the laboratory.

Competitive Enzyme-Linked Immunosorbent Assays

Two types of ELISA have been used for the analysis of aflatoxins. Both types are heterogeneous competitive assays. One type, direct ELISA, involves the use of aflatoxin–enzyme conjugate; the other system, indirect ELISA, involves the use of an AF–protein conjugate and a secondary antibody to which an enzyme has been conjugated. Although horseradish peroxidase (HRP) is used most commonly as the enzyme for conjugation, other enzymes such as alkaline phosphatase and beta-galactosidase also have been used (Chu, 1986c, 1991b; Pestka, 1988; Morgan and Lee, 1990). In most of the analyses, aflatoxins were extracted from the commodities with a solvent containing a mixture of water and methanol or acetonitrile. The samples then were used directly in the immunoassay after dilution in buffer. Interference problems generally were overcome by dilution of the sample either in buffer or in buffer containing sample matrix. Several protocols also included a very simple clean-up procedure. In the ELISA, both direct and indirect types could detect AFB_1 accurately at $\sim$ 10 ppb (μg/kg). For AFB_1 levels less than 10 ppb, protocols for concentration of the toxin and sample clean-up should be included.

Direct Competitive ELISA   In the direct competitive assay, specific antibodies are first coated to a solid phase such as polystyrene beads, polystyrene tubes, nylon beads, and Terisaki plates and most commonly to a microtiter plate (Chu, 1991b). The sample solution or standard toxin generally is incubated simultaneously with enzyme conjugate or incubated separately in two steps. After appropriate washings, the amount of enzyme bound to the plate is determined by incubation with a substrate solution containing hydrogen peroxide and appropriate oxidizable chromogen. The resulting color is measured spectrophotometrically or by visual comparison with the standards. In this assay, toxin in the sample and toxin–enzyme conjugate compete for the same binding site with the antibody coated to the solid phase. Since the toxin–enzyme and antibody concentrations are constant, the color intensity as a result of enzyme reaction is inversely proportional to the toxin concentration in the testing sample.

In general, ELISA is approximately 10–100 times more sensitive than RIA when purified mycotoxins are used. As little as 2.5 pg aflatoxin can be measured. The direct competitive ELISA for aflatoxin generally can be completed in less than 1 hr after extraction of the toxin from the sample (Chu *et al.,* 1987,1988; Chu, 1991b). Samples such as milk generally can be used directly in the assay after dilution in the assay buffer. Like that of RIA, the sensitivity of ELISA is improved when a clean-up treatment is included. For example, the sensitivity of ELISA for AFM_1 in milk increased to 10–25 ppt when milk samples were subjected to a clean-up treatment with a C_{18} reversed-phase Sep-Pak cartridge

(Hu *et al.*, 1983) or an affinity column (Chu, 1986c; Kaveri *et al.*, 1987; Mortimer *et al.*, 1987; Fremy and Chu, 1989; Stubblefield *et al.*, 1991). A clean-up treatment of the sample extract was necessary for the analysis of AFM in cheese (Fremy and Chu, 1984). The sensitivity of ELISA for the determination of aflatoxin in various foods and feeds is summarized in Table 4.

Indirect Competitive ELISA In the indirect ELISA, an AF–protein (or polypeptide) conjugate is prepared and coated onto the microtiter plate before assay. The plate is incubated with specific rabbit (or other type) antibody in the presence or absence of the homologous aflatoxin. The amount of antibody bound to the plate coated with aflatoxin–protein conjugate is determined by reaction with goat anti-rabbit (or anti-other type) IgG–enzyme complex, which is generally commercially available, and by subsequent reaction with the substrate. Thus, toxin in the samples and toxin in the solid phase compete for the same binding site with the specific antibody in the solution. The indirect ELISA also has been used for the analysis of a number of mycotoxins (Chu, 1991b) with a sensitivity that is comparable to or slightly better than that of the direct ELISA. This type of ELISA requires less antibody (100 times less) and does not require preparation of a toxin–enzyme conjugate. However, it requires more analytical time (2–3 hr).

To shorten the assay time for the indirect ELISA, two modifications were made by several investigators. One involved the conjugation of enzyme to the antibody, especially monoclonal antibody, which then is used directly in the ELISA instead of using a second antibody–enzyme conjugate. This modification converts the indirect ELISA format to a direct assay, and some studies have used the "direct ELISA" instead of the indirect ELISA (Canlish *et al.*, 1985; Goodbrand *et al.*, 1987; Itoh *et al.*, 1988). The second approach involves premixing the antibody with the second antibody–enzyme conjugate (Sidwell *et al.*, 1989). Both modifications have shortened the assay time without sacrificing sensitivity. Because only small amounts of antibody are needed for the indirect ELISA, the method also has been used extensively for monitoring the antibody titer of hybridoma culture fluids for the screening of monoclonal antibody-producing cells (Chu, 1991a,b).

Quick Immunoscreening Tests By shortening the incubation time and adjusting the antibody and enzyme concentrations in the direct/modified indirect competitive microtiter plate ELISA assay system, a positive or negative test can be obtained at certain toxin levels (such as 20 ppb) as a quick screening test for aflatoxin (Chu *et al.*, 1987; CAST, 1989; Park *et al.*, 1989a,b; Patey *et al.*, 1989; Lee *et al.*, 1990). Based on the same principle as the competitive ELISA, several other types of immunoscreening tests that have sensitivity similar to that of ELISA were developed also. Rather than coating the antibody onto the microtiter plate, the antibody is immobilized on a paper disk or other membrane, which is mounted either on a plastic card (card screen test), in a plastic cup (cup test and Cite), in a syringe (Cite probe) (Dorner and Cole, 1989; Trucksess *et al.*, 1989b,

TABLE 4 Sensitivity of Enzyme-Linked Immunosorbent Assays for Aflatoxins in Different Agricultural Commodities

Aflatoxin	Commodities[a]	Extraction solvents[b]	Clean-up[c]	Standard range (pg/assay)	Detection limits (μg/kg)	References
D-ELISA[d]						
AFB$_1$/AFs	C, W	Chl (CB)	N	5–1000	5	El-Nakib et al. (1981)
	PB	Hx, 55% MeOH	N	5–1000	3–5	El-Nakib et al. (1981)
	P, PB	Hx, 55% MeOH	N	12.5–300	5–10	Ram et al. (1986)
	PB	Chl/W	Y	10–1000	1	Ueno (1986)
	C, PB	70% MeOH	N	5–50	5–10	Chu et al. (1987)
	P, PB	60% MeOH	N	2.5–1000	1	Kawamura et al. (1988)
M$_1$	M	—	N	25–1000	0.10	Pestka et al. (1981); Kaveri et al. (1987)
	M	—	Y	25–1000	0.01	Hu et al. (1983)
	M	—	Y	5–1000	0.10	Jackman (1985)
	Ch	MeCl	Y	1.25–12.5	0.01	Fremy and Chu (1984)
	F	Chl	Y	1.25–12.5	5	Hu et al. (1983)
ID-ELISA[e]						
AFs	C, P	Hx, 55% MeOH	N	20–1000	5	Fan and Chu (1984)
	PB	50% MeCN	N	0.2–10	0.25	Morgan et al. (1986a)
Md-ID-ELISA[f]						
B$_1$	P, PB, C	Hx, 55% MeOH	N	10–1500	10	Candlish et al. (1987)
B$_1$	P, C	55% MeOH	N	50–1000	10	Itoh et al. (1988)
B$_1$	B	MeCN/KCl/H$_2$SO$_4$	N	50–1000	10	Ramakrishna et al. (1990)

[a]C, Corn; W, wheat; PB, peanut butter; P, raw peanuts; M, milk; Ch, cheese; F, feed; B, barley.
[b]Chl, Chloroform; Hx, hexane; MeOH, methanol; W, water; MeCl, methylene chloride; MeCN, acetonitrile.
[c]N, none, Y, yes.
[d]Direct ELISA.
[e]Indirect ELISA.
[f]Modified indirect ELISA.

1990; Chu, 1990; Stubblefield *et al.*, 1991), or in polystyrene beads (Steimer *et al.*, 1988). The reaction is carried out on the wetted membrane disk. Thus, after reaction, the absence of color (or decrease in color), generally blue, at the sample spot indicates the presence of toxin in the sample. The reaction is generally very rapid and requires less than 10–15 min to complete.

Another screening test is the immunoaffinity method, which is applicable for mycotoxins such as AFs and ochratoxins that have fluorescence (Candlish *et al.*, 1988; Groopman and Donahue, 1988; Fremy and Chu, 1989). In the assay for AFs, AFs extracted from the sample first are diluted with buffer at pH 7.0 and then subjected to purification through a disposable affinity column containing anti-AF antibody coupled to Sepharose gel. Samples such as milk and urine can be applied to the column directly after adjustment of the pH and dilution. After washing, AF is removed from the column with methanol, subjected to treatment with iodine/bromine solution to enhance the fluorescent intensity, and the fluorescence is determined (Groopman and Donahue, 1988; Trucksess *et al.*, 1991). The affinity column serves as a specific clean-up and concentration tool for the analysis.

The application of various screening tests for aflatoxins is summarized in Table 5; most of the screening tests are commercially available as kits (CAST, 1989; Lee, 1989; Chu, 1990,1991b). All the rapid screening test kits permit monitoring of mycotoxins semiquantitatively in the field (CAST, 1989; Chu, 1990). In a comparative study, Koeltzow and Tanner (1990) found that the performance of five commercially available immunoscreening kits for the screening of AF in corn was not statistically different from that of the minicolumn method. Data obtained from some kits were similar to those obtained from TLC methods. Other evaluations of the commercially available kits also arrived at the conclusion that such kits could be used for screening tests (Dorner and Cole, 1989; Patey *et al.*, 1989; Chu, 1991b). Collaborative studies for some of these immunoscreening tests have been made, and some have been adapted by the Association of Official Analytical Chemists (AOAC) as first action for screening for AF in different commodities (Scott, 1989,1990,1991).

Complementing Chemical Analyses by Immunochemical Methods
Immunoaffinity Chromatography as a Clean-up Tool for Chemical Analysis
Immunoaffinity columns (IAC) first were used in the RIA (Sun and Chu, 1977) and later for recovery of AFM from urine and milk samples (Wu *et al.*, 1983) for subsequent analyses. Although earlier application of this method was aimed primarily at biological fluids (Groopman *et al.*, 1984,1986; Groopman and Donahue, 1988; Fremy and Chu, 1989), the IAC has gained wide application as a clean-up tool for a number of mycotoxins (Table 6) and is not limited to fluid samples. Several collaborative studies indicate that this method is an efficient method for clean-up of aflatoxins (Patey *et al.*, 1990a,b,1991; Chu, 1991b). A combination of IAC and postcolumn derivatization in high performance liquid chromatography (HPLC) has gained acceptance as first action of the official

TABLE 5 Sensitivity of Immunoscreening Tests for Aflatoxins in Different Agricultural Commodities

Aflatoxin[a]	Commodities[b]	Extraction solvents[c]	Clean-up[d]	Detection limits (μg/kg)	References
AFC-AFs	C, P, PB	70% MeOH	N	>10	Trucksess et al. (1989a, 1991)
AFC-M₁	M	—	N	0.1	Kaveri et al. (1987); Groopman and Donahue (1988)
Card	C, P	MeOH	N	20	Park et al. (1987); Lee (1989)
	P	80% MeOH	N	10	CAST (1989)
Cite-M₁	M	—	N	0.25	Donner and Cole (1989)
Cite-Probe	C, PB	80% MeOH	N	20	Donner and Cole (1989)
Cup-AFs	PB	Hx, 55% MeOH	N	>20	Trucksess et al. (1989b)
	CS	80% MeOH	N	>20	Trucksess et al. (1989b)
	C, P	80% MeOH	N	>30	Trucksess et al. (1989b)
	P	80% MeOH	N	>10	CAST (1989)
MTP-AFs	C, P, F	70% MEOH	N	1–3	Chu et al. (1987)
	C, P, PB	MeOH	N	>20	Park et al. (1989b)
	CS, MF	Hx/MeOH/W	N	>15	Park et al. (1989a)
MTP-M₁	M	—	N	0.2	Lee (1989)
	M	—	Y	0.01	Lee (1989)

[a]AFC, Affinity column; MTP, microtiter plate.

[b]C, Corn; P, raw peanuts; PB, peanut butter; M, milk; CS, cottonseed; F, feed; MF, mixed feed.

[c]MeOH, Methanol; Hx, hexane; W, water.

[d]N, none; Y, yes.

TABLE 6   Immunoaffinity Chromatography of Aflatoxins

Mycotoxins[a]	Extraction[b]	Analysis[c]	References
AFM (M)	None	F	Fremy and Chu (1989); Hansen (1990)
AFB (F, P)	MeCN/W;MeOH/W	TLC	Candlish *et al.* (1988, 1989); van Egmond and Wagstaffe (1990)
AFB (P)	MeCN/W (60/40)	HPLC	Patey *et al.* (1990a)
AFs (C,P,Ct,F)	AOAC	F/Br; HPLC/PCD	Groopman and Donahue (1988); Trucksess *et al.* (1990)
AF-adducts (T,S,U)	Hydrolysis	HPLC	Sabbioni *et al.* (1990); Wild *et al.* (1990a); Harrison and Garner (1991)
AFs (P)	MeCN/W (60/40)	HPLC/PCD-I	Patey *et al.* (1991); Sharmon and Gilbert (1991)
AFs (C,F,P,PB)	MeOH/W (7/3)	F/Br; HPLC/PCD	Holcomb and Thompson (1991); Trucksess *et al.* (1991)
AFB (P)	MeCN/W (6/2)	TLC	Carvajal *et al.* (1990)

[a]M, Milk; F, feed; P, peanuts; C, corn; CT, cotton seeds; T, tissues; S, serum; U, urine; PB, peanut butter.
[b]MeCN, Acetonitrile; W, water; MeOH, methanol; AOAC, assoc. official analytical chemists official method.
[c]F, Fluorescence detection; PCD, post column derivatization.

method for analysis of AF in several commodities (Trucksess *et al.*, 1991) by the AOAC. Automation involving the use of an affinity column and HPLC was developed for routine analysis of AFM in milk (Carmen *et al.*, 1989) and AFB in peanut butter (Sharman and Gilbert, 1991).

Immunochromatography   A new approach involving the use of ELISA as a postcolumn monitoring system for HPLC has been developed for the analysis of aflatoxin metabolites (Dragsted *et al.*, 1988) and several *Fusarium* mycotoxins (Chu and Lee, 1989; Yu and Chu, 1991). The method involves separation of mycotoxins in HPLC or thin layer chromatography (TLC), and subsequent analysis of each individual fraction eluted from the column or TLC plate by ELISA. This approach can not only identify each individual mycotoxin, but also determine their concentration quantitatively. A combination of HPLC and ELISA technology has proved to be an efficient, sensitive, and specific method for the analysis of aflatoxins as well as other mycotoxins.

Other Immunochemical Methods

Several new immunochemical approaches for aflatoxin analysis have been developed. An approach called HPTLC–ELISA gram method (Pestka, 1991) involves separation of aflatoxins in HPTLC, followed by blotting the chromatogram to nitrocellulose membrane coated with antibody, incubation with AF–enzyme conjugate, and finally incubation with substrate to develop the color. Although this method has good sensitivity, the use of a large amount of antibody limits its wide application. A time-resolved fluoroimmunoassay for AF analysis was developed by Degan *et al.* (1989). This method involves the use of Eu-labeled antibodies and has a sensitivity similar to most ELISA methods with an I_{50} concentration of 0.2 ng AFB_1/ml.

Use of Immunochemical Methods to Monitor Aflatoxin Exposure in Humans and Animals

Monitoring the presence of AFs in human and animal body fluids, tissues, and organs would provide direct evidence for exposure to AFs. Two immunochemical approaches have been used for this purpose. One approach, which has been used more extensively, involves analysis of residual aflatoxin and AF metabolites in biological body fluids, organs, and tissues by RIA or ELISA. Another approach involves analysis of AFs and their metabolites in target organs immunohistochemically. Both methods have been used for samples obtained from humans and experimental animals exposed to aflatoxins (Chu, 1986b,1991b; Groopman and Donahue, 1988).

Monitoring Aflatoxin Metabolites in Human Urine and Milk

Various immunoassay protocols for detection of AF in human urine, milk, and serum are summarized in Table 7. Several types of antibodies have been used for monitoring AF and its metabolites in body fluids. Aflatoxin M_1 long has been considered an important AFB_1 metabolite and has been found both in milk and in urine samples after human and animal ingestion of AFB_1. The ELISA method developed for the analysis of AFM_1 in cow's milk (Fremy and Chu, 1989) has been used successfully in an epidemiological study in China by Zhu *et al.*, (1987) as well as by others (Yu *et al.*, 1989). We found a direct correlation between total

TABLE 7   Immunoassay of Aflatoxins in Human Body Fluids

Body fluids	Aflatoxins	Methods used[a]	Detect limits (ng/ml)
Serum	AFB_1	RIA	0.03
	AFB_1	ID-ELISA	0.02
	AFB_{2a}	DC-ELISA	0.02
Urine	AFM_1	AFC, ELISA	0.01
	AFB_{2a}	AFC, ELISA	0.01
	AFB–DNA	AFC, ELISA	—
Milk	AFM_1	DC-ELISA	0.01

[a]ID, DC, AFC represent indirect, direct, and affinity column, respectively.

dietary AFB_1 intake and total AFM_1 excretion in urine collected from people in high liver cancer incidence regions in China. Between 1.23 and 2.18% of dietary AFB_1 was found to be present as AFM_1 in human urine. A total of 0.04–4.84 μg AFM_1 (0.03–3.2 ng/ml) was excreted by each person per day in the endemic regions, compared with less than 0.04 μg (0.03 ng/ml) in a normal region. Results from this study also support a previous investigation reported by Wu *et al.* (1983), who used an IAC and HPLC method or ELISA method to monitor AFM_1 in human urine (detection limit around 3.0 pg/ml) obtained from a different high cancer incidence district. The results suggested that analysis of AFM_1 in urine by ELISA could be used as an index for human exposure of AFB_1 in an extensive epidemiological study. Aflatoxin M_1 also has been found in human breast milk obtained from women in Zimbabwe and Sudan at levels of 10–50 ppt by ELISA (Wild *et al.*, 1987).

Rather than analyzing specifically for AFM_1 in human urine, antibodies such as anti-AFB_{2a} or anti-AFB_1-dichloride, which have a wider spectrum of cross-reactivities with different aflatoxins, have been used in the ELISA (Garner *et al.*, 1985,1988; Wild *et al.*, 1986a,1988; Garner 1989) to monitor aflatoxin exposure in humans. Although the lower detection limit in estimating the "overall" aflatoxin metabolite levels in urine was ~ 10 ppt, undiluted human urine from normal subjects sometimes gave false positive data. For example, urine samples obtained from people with hepatocellular carcinoma and their immediate contacts in Gambia had a mean "total aflatoxin" concentration of 312 ng/ml, whereas the control group had a level of 81.5 ng/ml. The nonspecific inhibition could be prevented by dilution or by a clean-up procedure using C_{18} reversed-phase treatment. In another study, ELISA analysis of 29 urine samples from the Philippines revealed AFB_1 levels in the range of 0–4.25 ng/ml, with a mean of 0.88 ng/ml, compared with a mean of 0.066 ng/ml in samples obtained from France (Wild *et al.*, 1986a). Thus, the detection limit for aflatoxin in human urine in those ELISA methods was between 50 and 100 ppt when no sample clean-up treatment step was included.

Monitoring Aflatoxin–DNA Adducts in Urine and Tissues

Although immunochemical analysis of AF metabolites such as AFM_1 in human urine and milk has proven to be an effective approach for monitoring human exposure to aflatoxin, pharmacokinetically this method suffers from the disadvantage of the short half-life of such metabolites in urine. Except in the case of continuous exposure, short-term exposure is difficult to determine. In contrast, the macromolecule adducts of AF generally have a longer half-life but occur at a lower level in body fluids (Henderson *et al.*, 1989). Based on the differential effects of ethoxyquin on the kinetics of AF–DNA adduct and gamma-glutamyl transpeptidase-positive foci formation, Groopman and his colleagues (Groopman and Kemsler, 1987; Groopman and Donahue, 1988; Groopman and Zarba, 1991) found that measurement of the major rapidly excised $AFB–N^7$-guanine adduct in tissues and fluids is an appropriate dosimeter for estimating exposure status and risk in individuals consuming AF. Monoclonal antibodies against AFB_1–DNA as well as its major urinary degraded products such as AFB_1–ring-opened-modified DNA [formadopyrimidine (AF–FAPY), 2,3-dihydro-2(N-formyl)-2′,5′,6′-triamino-4′-oxy-N-pyrimidyl-3-hydroxy-AFB_1] have been used as an immunoaffinity agent or used in immunoassay (Haugen *et al.*, 1981; Groopman *et al.*, 1982,1984,1986; Groopman and Kemsler, 1987; Groopman and Donahue, 1988; Garner, 1989). Using immunoaffinity chromatography in combination with other analytical methods (Garner, 1989; Groopman and Zarbra, 1991), such products have been demonstrated in urine samples obtained from high liver cancer incidence regions (Groopman *et al.*, 1986) and in liver samples from hepatocellular carcinoma patients (Hsieh *et al.*, 1988; Garner, 1989; Santella *et al.*, 1991). Between 1.2 and 3.5 AFB_1–FAPY adducts per 10^6 nucleotides were found in 2 of 8 tumor-adjacent normal tissues and 7 of 7 tumor tissues obtained from liver cancer patients (Santella *et al.*, 1991). Harrison and Garner (1991) also found AFB–DNA adduct in human tissues including liver, kidney, and lung. The purified DNA, extracted from formalin-fixed human tissues from persons acutely exposed to aflatoxins during a poisoning incident, was subjected to acid hydrolysis and immunoaffinity column clean-up before indirect ELISA and HPLC analysis.

Monitoring Aflatoxin in Human Serum Aflatoxin

In addition to monitoring aflatoxins in urine and milk, immunoassays have been used to detect AFB_1 in human serum (Tsuboi *et al.*, 1984; Ueno, 1986; Morgan *et al.*, 1989). Using RIA, Tsuboi *et al.* (1984) found that 29 of 80 samples obtained from subjects 2–3 hr after eating had an AFB_1 level of 218.1 ± 268.3 ppt whereas 5 of 20 samples in a fasting group had 33.4 ± 14.6 (SD) ppt. RIA also was used by Czechoslovakian scientists for analysis of AFB_1 in human serum (Fukal and Reisnerova, 1990). In one survey, they found that 1.8% of 227 human sera collected in Czechoslovakia had AFB_1 at levels of 30–100 ppt (ng/liter). A double-antibody indirect ELISA was used by a British research team (Denning *et al.*, 1988; Wilkinson *et al.*, 1988a,b; Morgan *et al.*, 1989), who found that serum samples obtained from Nigeria had a mean AFB_1 level of 665 pg/ml with a standard error of 95 pg/ml (range from less than 20 pg/ml to 3.1 ng/ml), in contrast with a level of AFB_1 less than 20 pg/ml in serum from 78 men from England. Thus, a background of 20–40 ppt, either because of interference, that is, false positives (more likely), or because of a true exposure (questionable) in both studies appears to be normal. Using ELISA, these investigators also demonstrated that serum AFB_1 of neonates immediately following birth was significantly higher than that of the mothers in some Thai subjects (Morgan *et al.*, 1989; Denning *et al.*, 1990). Aflatoxins in concentrations of 0.064–13.6 nmol/l (mean of 3.1 nmol/l) were found in 17 samples of cord sera. Only two samples of maternal sera contained aflatoxins (0.62 nmol/l). These data indicate that transplacental transfer and concentration of aflatoxins did occur.

Aflatoxin–Albumin Adducts Recognizing that aflatoxin–serum albumin (AF–SA) adducts have a longer half-life in animals (Henderson *et al.,* 1989) and possibly in humans, Gan *et al.* (1988) analyzed the aflatoxin–albumin adducts in the blood of 42 residents of Guangxi Province, People's Republic of China. Blood specimens were obtained during the same period that urine was collected for the study of correlation of intake of AFB_1 and excretion of AFM_1 in urine (Zhu *et al.,* 1987). Analysis of AF–SA adducts was achieved by affinity purification of the albumin, enzymatic proteolysis, and immunoaffinity isolation of the digested adducts, followed by RIA. A highly significant correlation of adduct level with AFM_1 excretion was observed. About 1.4–2.3% of ingested AFB_1 was found to be bound covalently to serum albumin. Note that this range is very similar (0.98–2.15%) to that observed by Wild *et al.* (1986b) when rats were fed AFB_1. The level is also similar to the levels of AFM_1 (1.23–2.18% of total AFB_1 converted to AFM_1) found in human urine samples that were collected in the same region during the same period (Zhu *et al.,* 1987). Thus, a large portion of the AFs found in human serum in other studies discussed earlier probably exists in adduct form. A combination of immunoaffinity chromatograph and HPLC has been used very extensively in recent years for the detection of the aflatoxin–albumin and aflatoxin–lysine adducts (Sabbioni *et al.,* 1987,1990; Sabbioni, 1990; Wild *et al.,* 1990a; Wild and Montesano, 1991).

In another study, Wild *et al.* (1990a) compared three approaches, all involving immunochemical methods for quantitation of AFB–albumin adducts. The first two approaches involved the use of ELISA to determine the intact albumin directly (D-ELISA) and the hydrolysate of albumin (H-ELISA); the third approach involved the HPLC and fluorometric analysis of the AFB–lysine adduct after hydrolysis of albumin and affinity purification. The last approach was the most sensitive (5.0 pg AFB/mg human albumin) compared with the other two approaches (5.0 pg for H-ELISA and ~ 100 pg D-ELISA). Analysis of samples obtained from people in Thailand, Gambia, and Kenya revealed a level of 7–338 pg AFB/mg albumin, compared with nondetectable levels in human albumin obtained from France. The average levels of AF–albumin adducts in the human serum samples from Kenya, Thailand, and Gambia were 45 ± 70, 10.7 ± 11.5, and 44.1 ± 51.6 pg AF–lysine/mg of albumin, respectively (Wild *et al.,* 1990a; Wild and Montesano, 1991). Wild *et al.* (1990a) suggested using the H-ELISA for large-scale screening and using the third approach as a confirmatory test.

Antibodies In view of the formation of AF–macromolecule adducts in humans and animals, an intriguing question has been raised regarding whether antibodies against aflatoxin are formed after animal and human exposure. Autrup *et al.* (1990) have provided evidence that such a possibility is highly likely. In an indirect ELISA in which AFB–BSA was coated on the microtiter plate and alkaline phosphate conjugated to anti-human IgA, IgG, or IgM was used as the marker, these investigators were able to demonstrate antibody in human serum.

These researchers also found that the mean antibody titers in human serum of high liver cancer regions (Kenya) were about 3.5-fold higher than those in low liver cancer regions (Denmark). The highest antibody titers were found in individuals recently exposed to aflatoxins. Although the significance of the presence of these antibodies (i.e., protective effect) in humans is not known, this approach may provide another means for determining human exposure to aflatoxins.

Immunohistochemical Methods

Several immunohistochemical methods have been used to monitor aflatoxins in biological samples (Chu, 1986b,c). As early as 1977, Lawellin *et al.* (1977b) used the anti-AFB_{2a} type antibodies to visualize the deposition of AFB_1 in the hyphae of *Aspergillus parasiticus*. Not until 1983 was indirect immunostaining (Pestka *et al.*, 1983) tested for monitoring aflatoxin in organs and tissues of rats receiving aflatoxins. In this system, organs or tissues, after fixation and sectioning, were incubated with the antibody and reacted with respective secondary antibody conjugated to an enzyme such as peroxidase. After washing, the bound peroxidase was localized by incubation with a chromagenic substrate such as H_2O_2 and diaminobenzidine to yield a dense brown-colored water-insoluble product that could be visualized under a light microscope. Using anti-AFB_{2a} polyclonal antibodies (Guar *et al.*, 1981), positive stain was observed, particularly in the nuclear region of hepatocytes in AFB_1-treated rats (Pestka *et al.*, 1983).

Immunohistochemical methods involving the use of different types of antibodies have been used since by several other investigators to localize the presence of aflatoxins in tissues and organs. For example, the presence of aflatoxin adduct in the nuclei and mitochondria was demonstrated immunohistochemically by Shamsuddin *et al.* (1987). Using this approach, Lee *et al.* (1989) found AFB_1-formamidopyrimidine adducts in 4 of 14 nontumorous specimens but in none of the 14 tumor tissues of the livers of patients who died of nonhepatic diseases. These adducts were not found in the 3 control livers. Wild *et al.* (1990b) observed a marked difference in intercellular distribution of AFB_1-DNA and AFG_1-DNA adducts and AFG_1 adducts in the liver, kidney, and lung by a sensitive immunocytochemical approach. Whereas adduct levels in the liver cells were more homogeneous, marked intercellular variations were observed in kidney and lung. No adducts were detected in the esophagus, forestomach, colon, spleen, or testes.

An indirect immunofluorescence method was developed by Zhang *et al.* (1991) for the analysis the accumulation of aflatoxin–DNA adducts in woodchuck hepatocytes and rat liver tissue. The method involves the use of a monoclonal antibody recognizing the imidazole ring-open form of the major N^7-guanine adduct of aflatoxin B_1. In a manner similar to the indirect method described earlier, these investigators used the goat anti-mouse IgG conjugated with fluorescein isothiocyanate (FITC) instead of an enzyme-tagged secondary antibody. Quantitation of AFB–DNA in the tissue was done by measuring the fluorescence intensity of the photographic slide densitometrically. A good correlation of the measurement of adduct formation was observed between the ELISA and this method. These authors observed specific staining in the nuclei of rat livers and woodchuck hepatocytes. In addition, a dose response for the amount of AFB_1 used and the formation of AFB–DNA adducts, that is, the intensity of fluorescence stain, was observed.

SUMMARY AND CONCLUDING REMARKS

Aflatoxins are small molecular weight heptans and, thus, are not immunogenic. However, aflatoxins can be conjugated to a protein or a polypeptide carrier and subsequently used for immunization. Many approaches for coupling of aflatoxins to protein carriers have been developed over the last two decades. Using these conjugates as the immunogens, monoclonal and polyclonal antibodies with diverse specificity against aflatoxins have been generated. Depending on the aflatoxin derivatives used and the site of coupling to the protein carrier, the antibody specificity varies considerably. In general, the antibodies exhibit specificity toward the dihydrofuran portion of the AF molecule when the conjugation is made through the cyclopentanone moiety. On the other hand, immunization with AF–protein conjugates prepared through the dihydrofuran portion of AFs results in antibodies having specificity toward the cyclopentanone ring moiety. Any structural variations in the side-chain structures or in these rings result in antibodies of different specificity.

With the availability of these antibodies, good and reliable immunoassay protocols for AFs in different commodities and body fluids have been established. These methods are very simple, sensitive, and specific. Several immunoassay kits for aflatoxin and other mycotoxins are currently available. However, those who perform the immunoassay must know the specificity of the antibody used in each assay. Depending on the purpose of research or routine analysis, selection of appropriate antibody plays an important role in determining the success of a typical assay. Antibodies highly specific to certain aflatoxins as well as antibodies that cross-react with all the major AFs (generic type) are important; both should be considered for immunoassays.

Immunochemical methods, including affinity chromatography as a concentration step and clean-up tool and immunoassay as a monitoring system for HPLC and TLC, have been used in conjunction with other analytical methodologies. Immunohistochemical methods also have been used to localize aflatoxins in human and animal tissues. Such innovative approaches have led to a new dimension in the immunochemical method for aflatoxins. Within the last 5 years, different immunochemical methods have received wider application in monitoring for AFs in foods and feeds as well as in body fluids and tissues. The versatility of these methods already has provided considerable information regarding the AF contamination of foods and feeds as well as human exposure to AFs. Although I foresee that the immunochemical method will become one of the routine methods in the future for monitoring aflatoxins, I also hope that research on improving present immunoassay protocols and searching for new immunochemical methods will be continued. In addition, the role of antibody in the pathogenesis of aflatoxicosis is another interesting area that must be explored. Early studies already have shown some protective effect against aflatoxicosis in the rabbits that have been immunized with AFB–BSA conjugate; another study also presents evidence that antibodies against aflatoxins can be generated in humans in a result of the formation of aflatoxin–macromolecule adducts. Thus, the possible role and possible mechanism of formation of antibodies in humans after exposure to the toxin remain a mystery.

ACKNOWLEDGMENTS

This work was supported by the College of Agricultural and Life Sciences at the University of Wisconsin, Madison. Part of the work described in this contribution was supported by a Public Health Service grant (CA 15064) from the National Cancer Institute, a contract (DAMD17-86-C-6173) from the U.S. Army Medical Research and Development Command of the Department of Defense, and a USDA North Central Regional Project (NC-129). The author thanks Carole Ayres for her help in the preparation of the manuscript.

REFERENCES

Ashoor, S. H., and Chu, F. S. (1975). Reduction of aflatoxin B_{2a} with sodium borohydride. *Agric. Food Chem.* **23**, 445–447.

Autrup, H., and Wakhisi, J. (1988). Detection of exposure to aflatoxin in an African population. *IARC Sci. Publ.* **89**, 63–66.

Autrup, H., Seremet, T., and Wakhisi, J. (1990). Evidence for human antibodies that recognize an aflatoxin epitope in groups with high and low exposure to aflatoxins. *Arch. Environ. Health* **45**, 31–34.

Biermann, V. A., and Terplan, G. (1980). Nachweis von Aflatoxin B_1 mittels ELISA. *Arch. Lebensmitelhyg.* **31**, 51–57.

Blankford, M. B., and Doerr, J. A. (1986). The development and characterization of monoclonal antibodies against aflatoxin. *Hybridoma* **5**, 57.

Bludovsky, R. (1986). Determination of aflatoxin B_1 in foods by RIA method. *Radioisotopy* **27**, 339–351.

Candlish, A. A. G., Stimson, W. H., and Smith, J. E. (1985). A monoclonal antibody to aflatoxin B_1: Detection of the mycotoxin by enzyme immunoassay. *Lett. Appl. Microbiol.* **1**, 57–61.

Candlish, A. A. G., Stimson, W. H., and Smith, J. E. (1987). The detection of aflatoxin B_1 in peanut kernels, peanut butter and maize using a monoclonal antibody based enzyme immunoassay. *Food Microbiol.* **4**, 147–153.

Candlish, A. A. G., Haynes, C. A., and Stimson, W. H. (1988). Detection and determination of aflatoxins using affinity chromatography. *Intl. J. Food Sci. Technol.* **23**, 479–485.

Canlish, A. A. G., Smith, J. E., and Stimson, W. H. (1989). Monoclonal antibody technology for mycotoxins. *Biotech. Adv.* **7**, 401–418.

Carman, A. S., Kuan, S. S., Ware, G. M., Francis, O. J., and Miller, K. V. (1989). Automated assay of aflatoxin M_1 in fluid milk using robotics. *Abstracts of the 103rd AOAC Annual International Meeting*, St. Louis, Missouri, September 25–28.

Carvajal, M., Muljolland, F., and Garner, R. C. (1990). Comparison of the EASI-EXTRACT immunoaffinity concentration procedure with the AOAC CB method for the extraction and quantitation of aflatoxin B_1 in raw ground unskinned peanuts. *J. Chromatogr.* **511**, 379.

Chu, F. S. (1984). Immunoassays for analysis of mycotoxins. *J. Fd. Protect.* **47**, 562–569.

Chu, F. S. (1986a). *Aspergillus flavus* toxins: Aflatoxin B_1 and M_1. *In* "Methods of Enzymatic Analysis" (H. U. Bergmeyer, ed.), 3d Ed., Vol. XI, pp. 145–156. VCH Publishers, Weinheim, Germany.

Chu, F. S. (1986b). Immunochemical methods for diagnosis of mycotoxicoses. *In* "Diagnosis of Mycotoxicoses" (J. L. Richard and J. R. Thurston, eds.), pp. 163–176. Martinus Nijhoff, Dordrecht, The Netherlands.

Chu, F. S. (1986c). Immunoassay for mycotoxins. *In* "Modern Methods in the Analysis and Structural Elucidation of Mycotoxins" (R. J. Cole, ed.), pp. 207–237. Academic Press, New York.

Chu, F. S. (1989). Current immunochemical methods for analysis of aflatoxin in groundnut and groundnut products. *In* "Aflatoxin Contamination of Groundnuts" (D. MacDonald, V. K. Mehan, and S. D. Hall, eds.), pp. 161–172. ICRISAT, Hyderabad, India.

Chu, F. S. (1990). Immunoassay for mycotoxins, current state of the art, commercial and epidemiological applications. *Vet. Hum. Toxicol. (Suppl)* **32**, 42–50.

Chu, F. S. (1991a). Current immunochemical methods for mycotoxin analysis. *In* "Immunoassays for Trace Chemical Analysis: Monitoring Toxic Chemicals in Humans, Foods and the Environment" (M. Vanderlaan, L. H. Stanker, B. E. Watkins, and D. W. Roberts, eds.), pp. 140–157. American Chemical Society, Washington, D.C.

Chu, F. S. (1991b). Development and use of immunoassays in detection of the ecologically important mycotoxins. *In* "Handbook of Applied Mycology: Mycotoxins" (D. Bhatnagar, E. B. Lillehoj, and D. K. Arora, eds.), Vol. V, pp. 87–136. Marcel Dekker, New York.

Chu, F. S., and Lee, R. C. (1989). Immunochromatography of group A trichothecene mycotoxins. *Food Agric. Immunol.* **1**, 127–136.

Chu, F. S., and Ueno, I. (1977). Production of antibody against aflatoxin B_1. *Appl. Environ. Microbiol.* **33**, 1125–1128.

Chu, F. S., Hsia, M.-T.S., and Sun, P. S. (1977). Preparation and characterization of aflatoxin B_1-1-(O-carboxymethyl) oxime. *J. Assoc. Off. Anal. Chem.* **60**, 791–794.

Chu, F. S., Lau, H. P., Fan, T. S., and Zhang, G. S. (1982). Ethylenediamine modified bovine serum albumin as protein carrier in the production of antibody against mycotoxins. *J. Immunol. Methods* **55**, 73–78.

Chu, F. S., Steinert, B. W., and Guar, P. K. (1985). Production and characterization of antibody against aflatoxin G_1. *J. Food Safety* **7**, 161–170.

Chu, F. S., Fan, T. S. L., Zhang, G. S., Xu, Y. C., Faust, S., and McMahon, P. L. (1987). Improved enzyme-linked immunoassay for aflatoxin B_1 in agricultural commodities. *J. Assoc. Off. Anal. Chem.* **70**, 854–857.

Chu, F. S., Lee, R. C., Trucksess, M. W., and Park, D. L. (1988). Evaluation by enzyme-linked immunosorbent assay of cleanup for thin-layer chromatography of aflatoxin B_1 in corn, peanuts, and peanut butter. *J. Assoc. Off. Anal. Chem.* **71**, 126–129.

Council of Agricultural Science and Technology (1989). "Mycotoxins, Economic and Health Risks." Council of Agricultural Science and Technology CAST Task Force Report No. 116. CAST, Ames, Iowa.

Degan, P., Montagnoli, G., and Wild, C. P. (1989). Time-resolved fluoroimmunoassay of aflatoxins. *Clin. Chem.* **35**, 2308–2310.

Denning, D. W., Onwubalili, J. K., Wilkinson, A. P., and Morgan M. R. A. (1988). Measurement of aflatoxin in Nigerian sera by enzyme-linked immunosorbent assay. *Trans R. Soc. Trop. Med. Hyg.* **82**, 169–171.

Denning, D. W., Allen, R., Wilkinson, A. P., and Morgan, M. R. A. (1990). Transplacental transfer of aflatoxin in humans. *Carcinogenesis* **11**, 1033–1035.

Dixon-Holland, D. E., Pestkan, J. J., Bridigare, B. A., Casale, W. L., Warner, R. L., Ram, B. P., and

Hart, L. P. (1988). Production and sensitive monoclonal antibodies to aflatoxin B_1 and aflatoxin M_1 and their application to ELISA of naturally contaminated foods. *J. Food Protect.* **51,** 201–204.

Dorner, J. W., and Cole, R. J. (1989). Comparison of two ELISA screening tests with liquid chromatography for determination of aflatoxins in raw peanuts. *J. Assoc. Off. Anal. Chem.* **72,** 962–964.

Dragsted, L. O., Bull, I., and Autrup, H. (1988). Substances with affinity to a monoclonal aflatoxin B_1 antibody in Danish urine samples. *Fd. Chem. Toxic.* **26,** 233–242.

El-Nakib, O., Pestka, J. J., and Chu, F. S. (1981). Determination of aflatoxin B_1 in corn, wheat, and peanut butter by enzyme-linked immunosorbent assay and solid phase radioimmunoassay. *J. Assoc. Off. Anal. Chem.* **64,** 1077–1082.

Fan, T. S., and Chu, F. S. (1984). An indirect ELISA for detection of aflatoxin B_1 in corn and peanut butter. *J. Food Protect.* **47,** 263–266.

Fan, T. S., Zhang, G. S., and Chu, F. S. (1984). Production and characterization of antibody against aflatoxin Q_1. *Appl. Environ. Microbiol.* **47,** 526–532.

Fontel, P. A., Beheler, J., Bunner, D. L., and Chu, F. S. (1983). Detection of T-2 toxin by an improved radioimmunoassay. *Appl. Environ. Microbiol.* **45,** 922–928.

Fremy, J. M., and Chu, F. S. (1984). A direct ELISA for determining aflatoxin M_1 at ppt levels in various dairy products. *J. Assoc. Off. Anal. Chem.* **67,** 1098–1101.

Fremy, J. M., and Chu, F. S. (1989). Immunochemical methods of analysis for aflatoxin M_1. *In* "Mycotoxin in Dairy Products" (H. P. van Egmond, ed.), pp. 97–125. Elsevier Science, London.

Fukal, L. (1988). Utilization of commercial RIA-set for milk contamination with M_1 aflatoxin. *Melkarske Listy* **14,** 196.36–196.38.

Fukal, L., and Kas, J. (1989). The advantages of immunoassay in food analysis. *Trends Anal. Chem.* **8,** 112–116.

Fukal, L., and Reisnerova, H. (1990). Monitoring of aflatoxins and ochratoxin A in Czechoslovak human sera by immunoassay. *Bull. Environ. Contam. Toxicol.* **44,** 345–349.

Fukal, L., Prosek, J., Rauch, P., Sova, Z., and Kas, J. (1986a). Choice of procedure conditions for radioimmunoassay of aflatoxin B_1 using ^{125}I as marker and dextran-coated charcoal as a separation matrix. *J. Radioanal. Nucl. Chem. Lett.* **108(5),** 259–268.

Fukal, L., Sova, Z., and Reisnerova, H. (1986b). Some interferences in radioimmunoassay of aflatoxin B_1. *J. Radioanal. Nucl. Chem. Lett.* **108(5),** 297–303.

Fukal, L., Prosek, J., and Rauch, P., Sova, Z., and Kas, J. (1987). Selection of the separation step in the radioimmunoassay for aflatoxin B_1 using ^{125}I as marker. *J. Radioanal. Nucl. Chem. Art.* **109(2),** 383–391.

Fukal, L., Rauch, P., and Kas, J. (1988a). Utilization of immunochemical methods in food production analytical chemistry. I. Determination of non-immunogenic-low molecular compounds. *Chem. Listy* **82,** 959–977.

Fukal, L., Reisnerova, H., and Rauch, P. (1988b). Application of radioimmunoassay with ^{125}-Iodine for determination of aflatoxin B_1 in foods. *Sci. Alim.* **8,** 397–403.

Gan, L. S., Skipper, P. L., Peng, X. C., Groopman, J. D., Chen, J. S., Wogan, G. N., and Tannenbaum, S. R. (1988). Serum albumin adducts in the molecular epidemiology of aflatoxin carcinogenesis, correlation with aflatoxin B_1 intake and urinary excretion of aflatoxin M_1. *Carcinogenesis* **9,**1323–1325.

Garner, R. C. (1989). Monitoring aflatoxin exposure at a macromolecular level in human with immunological methods. *In* "Mycotoxins and Phycotoxins" (S. Natori, K. Hashimoto, and Y. Ueno, eds.), pp. 29–35. Elsevier, Amsterdam.

Garner, R. C., Ryder, R., and Montesano, R. (1985). WHO–IARC Meeting report: Monitoring of aflatoxins in human body fluids and application of field studies. *Cancer Res.* **45,** 922–928.

Garner, R. C., Dvorackova, I., and Tursi, F. (1988). Immunoassay procedures to detect exposure to aflatoxin B_1 and benzo(*a*)pyrene in animals and man at the DNA level. *Int. Arch. Occup. Environ. Health* **60,** 145–150.

Gaur, P. K., El-Nakib, O., and Chu, F. S. (1980). Comparison of antibody production against aflatoxin B_1 in goats and in rabbits. *Appl. Environ. Microbiol.* **40,** 678–680.

Gaur, P. K., Lau, H. P., Pestka, J. J., and Chu F. S. (1981). Production and characterization of aflatoxin B_{2a} antiserum. *Appl. Environ. Microbiol.* **41,** 478–482.

Goodbrand, I. A., Stimson, W. H., and Smith, J. E. (1987). A monoclonal antibody to T-2 toxin. *Lett. Appl. Microbiol.* **5,** 97–99.

Goto, T., and Manabe, M. (1988). Application of ELISA to aflatoxin analysis: Results of preliminary tests of analysis kits on the market. *Rep. Natl. Food Res. Inst.* **52,** 53–59.

Groopman, J. D., and Donahue, K. F. (1988). Aflatoxin, a human carcinogen: Determination in foods and biological samples by monoclonal antibody affinity chromatography. *J. Assoc. Off. Anal. Chem.* **71,** 861–867.

Groopman, J. D., and Kensler, T. W. (1987). The use of monoclonal antibody affinity column for assessing DNA damage and repair following exposure to aflatoxins. *Pharmacol. Ther.* **34,** 321–334.

Groopman, J. D., and Zarba, A. (1991). Immunoaffinity-based monitoring of human exposure to aflatoxins in China and Gambia. *In* "Immunoassays for Trace Chemical Analysis: Monitoring Toxic Chemicals in Humans, Foods and the Environment" (M. Vanderlaan, L. H. Stanker, B. E. Watkins, and D. W. Roberts, eds.), pp. 207–214. American Chemical Society, Washington, D. C.

Groopman, J. D., Haugen, A., Goodrich, G. R., Wogan, G. N., and Harris, C. C. (1982). Quantitation of aflatoxin B_1-modified DNA using monoclonal antibodies. *Cancer Res.* **42,** 3120–3124.

Groopman, J. D., Trudel, L. J., Donahue, P. R., Marshak-Rothstein, A., and Wogan, G. N. (1984). High-affinity monoclonal antibodies for aflatoxins and their application to solid-phase immunoassays. *Proc. Natl. Acad. Sci. U.S.A.* **81,** 7728–7731.

Groopman, J. D., Donahue, P. R., Zhu, J. Q., Chen, J. S., and Wogan, G. N. (1986). Aflatoxin metabolism in humans: Detection of metabolites and nucleic acid adduct in urine by affinity chromatography. *Proc. Natl. Acad. Sci. U.S.A.* **82,** 6492–6496.

Hansen, T. J. (1990). Affinity column cleanup and direct fluorescence measurement of aflatoxin M_1 in raw milk. *J. Food Protect.* **53,** 75–77.

Harder, W. O., and Chu, F. S. (1979). Production and characterization of antibody against aflatoxin M_1. *Experientia* **35,** 104–106.

Harrison, J. C., and Garner, R. C. (1991). Immunochemical and HPLC detection of aflatoxin adducts in human tissues after an acute poisoning incident in S. E. Asia. *Carcinogenesis* **12,** 741–743.

Hastings, K. L., Tulis, J. J., and Dean, J. H. (1988). Production and characterization of a monoclonal antibody to aflatoxin B_2. *J. Agric. Food Chem.* **36,** 404–408.

Hastings, K. L., Hagler, W. M., Harris, T. M., Voyksner, R. D., and Dean, J. H. (1989). Production and characterization of aflatoxin B_2 oximinoacetate. *J. Agric. Food Chem.* **37,** 393–400.

Haugen, A., Groopman, J. D., Hsu, I. C., Goodrich, G. R., Wogan, G. N., and Harris, C. C. (1981). Monoclonal antibody to aflatoxin B_1-modified DNA detected by enzyme immunoassay. *Proc. Natl. Acad. Sci.* **78,** 4124–4127.

Hefle, S., and Chu, F. S. (1990). Production and characterization of monoclonal antibodies cross-reactive with major aflatoxins. *Food Agric. Immunol.* **4,** 181–188.

Henderson, R. F., Bechtold, W. E., Bond, J. A., and Sun, J. D. (1989). The use of biological markers in toxicology. *CRC Crit. Rev. Toxicol.* **20,** 65–82.

Hertzog, P. J., Lindsay Smith, J. R., and Garner, R. C. (1982). Production of monoclonal antibodies to guanine imidazole ring opened aflatoxin B_1-DNA, the persistent DNA adduct *in vivo*. *Carcinogenesis* **38,** 825–828.

Holcomb, M., and Thompson, H. C., Jr. (1991). Analysis of aflatoxins (B_1, B_2, G_1 and G_2) in rodent feed by HPLC using postcolumn derivatization and fluorescence detection. *J. Agric. Food Chem.* **39,** 137.

Holladay, S. D., Brownie, C. F., Corbett, W. T., and Talley, D. D. (1991). Evaluation of an indirect enzyme-linked immunosorbent assay for screening antibody against aflatoxins. *Am. J. Vet. Res.* **52,** 222–223.

Hsieh L. L., Hsu, S. W., Chen, D. S., and Santella, R. M. (1988). Immunological detection of aflatoxin B_1-DNA adducts formed *in vivo*. *Cancer Res.* **48,** 6328–6331.

Hu, W. J., Woychik, N., and Chu, F. S. (1983). ELISA of picogram quantities of aflatoxin M_1 in urine and milk. *J. Food Prot.* **47,** 126–127.

Hsu, K. H. and Chu, F. S. (1992). Production and characterization of antibodies against aflatoxin in laying hens. *Food & Agric. Immunol.* **4,** 83–91.

Itoh, Y., Hifumi, E., Sudoh, K. Uda, T., Ohtano, K., Kawaura, O., Nagayama, S., Satoh, S., and Ueno, Y. (1988). Determination of aflatoxin B_1 by direct ELISA. *Proc. Jpn. Assoc. Mycotoxicol.* **26,** 31–35.

Jackman, R. (1985). Determination of aflatoxins by ELISA with special reference to aflatoxin M_1 in raw milk. *J. Sci. Food Agric.* **36,** 685–698.

Kaveri, S. V., Fremy, J. M., Lapeyre, C., and Strosberg, A. D. (1987). Immunodetection and immunopurification of aflatoxins using a high affinity monoclonal antibody to AFB_1. *Lett. Appl. Microbiol.* **4,** 71–75.

Kawamura, O., Nagayama, S., Sato, S., Ohtani, K., Ueno, I., and Ueno, Y. (1988). A monoclonal antibody-based enzyme-linked immunosorbent assay for aflatoxin B_1 in peanut products. *Mycotoxin Res.* **4,** 75–27.

Koeltzow, D. E., and Tanner, S. N. (1990). Comparative evaluation of commercially available aflatoxin test methods. *J. Assoc. Off. Anal. Chem.* **73,** 584–589.

Langone, J. J., and van Vunakis, H. (1976). Aflatoxin B_1: Specific antibodies and their use in radioimmunoassay. *J. Natl. Cancer Inst.* **56,** 591–595.

Lau, H. P., Gaur, P. K., and Chu, F. S. (1981). Preparation and characterization of aflatoxin B_{2a}-hemiglutarate and its use for the production of antibody against aflatoxin B_1. *J. Food Safety* **3,** 1–13.

Lawellen, D. W., Gant, D. W., and Joyce, B. K. (1997a). Enzyme-linked immunosorbent analysis of aflatoxin B_1. *Appl. Environ. Microbiol.* **34,** 94–96.

Lawellen, D. W., Grant, D. W., and Joyce, B. K. (1977b). Aflatoxin localization by enzyme-linked immunocytochemical techniques. *Appl. Environ. Microbiol.* **34,** 88–93.

Lee, H. S., Sarosi, I., and Vyas, G. N. (1989). Aflatoxin B_1 formamidopyrimidine adducts in human heptocarcinogenesis: A preliminary report. *Gastroenterology* **97,** 1281–1287.

Lee, L. S. (1989). Aflatoxin. *J. Am. Oil Chem. Soc.* **66,** 1398–1413.

Lee, L. S., Wall, J. H., Cotty, P. J., and Bayman, P. (1990). Integration of enzyme-linked immunosorbent assay with conventional chromatographic procedures for quantitation of aflatoxin in individual cotton bolls, seeds, and seed sections. *J. Assoc. Off. Anal. Chem.* **73,** 581–584.

Li, Y. K., and Chu, F. S. (1984). Production and characterization of antibody against sterigmatocystin. *J. Food Safety* **6,** 119–126.

Lubet, M. T., Olson, D. F., Yang, G., Ting, R., and Steuer, A. (1983). Use of a monoclonal antibody to detect aflatoxin B_1 and M_1 in enzyme immunoassay. *Abstracts of the 97th AOAC Annual Meeting,* p. 71.

Martin, C. N., Garner, R. C., Tursi, F., Garner, J. V., Whittle, H. C., Ryder, R. W., Sizaret, P., and Montesano, R. (1984). An ELISA procedure for assaying aflatoxin B_1. *In* "Monitoring Human Exposure to Carcinogenic and Mutagenic Agents" (A. Berlin, M. Draper, K. Hemminki, and H. Vainio, eds.). pp. 313–321. IARC, Lyon, France.

Martlbauer, E., and Terplan, G. (1985). Ein hochempfindlicher hetrologer enzymimmunologischer Nachweis von Aflatoxin M_1 in Milch und Milchpulver. *Arch. Lebensmittelhyg.* **36,** 53–55.

Martlbauer, E., and Terplan, G. (1986). Development of enzyme immunoassays for analysis of mycotoxins. *Proceedings of the Second World Congress of Foodborne Infections and Intoxications,* Vol. II, 920–923.

Morgan, M. R. A. (1985). Newer techniques in food analysis-immunoassay and their application to small molecules. *J. Assoc. Pub. Anal.* **23,** 59–63.

Morgan, M. R. A., and Lee, H. A. (1990). Mycotoxins and natural food toxicants. *In* "Development and Application of Immunoassay for Food Analysis (J. H., Rittenburg, ed.). Elsevier Applied Science, pp. 143–170. London.

Morgan, M. R. A., Kang, A. S., and Chan, H. W. S. (1986a). Aflatoxin determination in peanut butter by enzyme-linked immunosorbent assay. *J. Sci. Food Agric.* **37**, 908–914.

Morgan, M. R. A., Kang, A. S., and Chan, H. W. S. (1986b). Production of antisera against steigmatocystin hemiacetal and its potential for use in an enzyme-linked immunosorbent assay for steigmatocystin in barley. *J. Sci. Food Agric.* **37**, 837–880.

Morgan, M. R. A., Wilkinson, A. P., and Denning, D. W. (1989). Human exposure to mycotoxins monitored by immunoassay. *In* "Mycotoxins and Phycotoxins '88" (S. Natori, K. Hashimoto, and Y. Ueno, eds.), pp. 45–50. Elsevier, Amsterdam.

Mortimer, D. N., Gilbert, J., and Sheperd, M. J. (1987). Rapid and highly sensitive analysis of aflatoxin M_1 in liquid and powdered milks using an affinity column cleanup. *J. Chromatogr.* **407**, 393–398.

Mortimer, D. N., Shepherd, M. J., & Gilbert, J. (1988a). Enzyme-linked immunosorbent (ELISA) determination of aflatoxin B_1 in peanut butter: Collaborative trial. *Food Add. Contam.* **5**, 601–608.

Mortimer, D. N., Sheperd, M. J., & Gilbert, J. (1988b). A survey of the occurrence of aflatoxin B_1 in peanut butters by enzyme-linked immunosorbent assay. *Food Add. Contam.* **5**, 127–132.

Park, D. L., Nesheim, S., Trucksess, N. W., Stack, M., Morris, D., Lewis, E., and Romer, T. (1987). ELISA "Quick Card" screening method for aflatoxins B_1, B_2, G_1, and G_2: Collaborative study. *Abstracts of 101st AOAC Annual International Meeting, San Francisco,* September 14–17.

Park, D. L., Miller, B. M., Hart, P., Yang, G., McVey, J., Page, S. W., Pestka, J., and Brown, L. H. (1989a). Enzyme-linked immunosorbent assay for screening aflatoxin in cottonseed products and mixed feed: Collaborative study. *J. Assoc. Off. Anal. Chem.* **72**, 326–332.

Park, D. L., Miller, B. M., Nesheim, S., Trucksess, M. W., Vekich, A., Bidigare, B., McVey, J. L., and Brown, L. H. (1989b). Visual and semiquantitative spectrophotometriac screening method for aflatoxin B in corn and peanut products: Follow-up study. *J. Assoc. Off. Anal. Chem.* **72**, 638–643.

Patey, A. L., Sharman, M., Wood, R., and Gilbert, J. (1989). Determination of aflatoxin concentrations in peanut butter by enzyme-linked immunosorbent assay (ELISA)—Study of three commercial ELISA kits. *J. Assoc. Off. Anal. Chem.* **72**, 965–969.

Patey, A. L., Sharman, M., and Gilbert, J. (1990). Determination of aflatoxin B_1 levels in peanut butter using an immunoaffinity column clean-up procedure: Inter-laboratory study. *Food Add. Contam.* **7**, 515–520.

Patey, A. L., Sharman, M., and Gilbert, J. (1991). Liquid chromatographic determination of aflatoxin levels in peanut butters using an immunoaffinity column cleanup method: international collaborative Trial. *J. Assoc. Off. Anal. Chem.* **74**, 76.

Pestka, J. J. (1988). Enhanced surveillance of foodborne mycotoxins by immunochemical assay. *J. Assoc. Off. Anal. Chem.* **71**, 1075–1081.

Pestka, J. J. (1991). High performance thin layer chromatography ELISAGRAM: Application of a multi-hapten immunoassay to analysis of the zearalenone and aflatoxin mycotoxin families. *J. Immunol. Meth.* **136**, 177.

Pestka, J. J., and Chu, F. S. (1984). Aflatoxin B_1 dihydrodiol antibody: Production and specificity. *Appl. Environ. Microbiol.* **47**, 472–477.

Pestka, J. J., Gaur, P. K., and Chu, F. S. (1980). Quantitation of aflatoxin B_1 and aflatoxin B_1 antibody by an enzyme-linked immunosorbent microassay. *Appl. Environ. Microbiol.* **40**, 1027–1031.

Pestka, J. J., Lee, Y. K., Harder, W. O., and Chu, F. S. (1981). Comparison of a radioimmunoassay and an enzyme-linked immunosorbent assay for the analysis of aflatoxin M_1 in milk. *J. Assoc. Off. Anal. Chem.* **64**, 294–301.

Pestka, J. J., Li, Y. K., and Chu, F. S. (1982). Reactivity of aflatoxin B_{2a} antibody with aflatoxin B_1-modified DNA and related metabolites. *Appl. Environ. Microbiol.* **44**, 1159–1165.

Pestka, J. J., Beery, J. T., and Chu, F. S. (1983). Indirect immunoperoxidase localization of aflatoxin B_1 in rat liver. *Food Chem. Toxicol.* **21**, 41–48.

Qian, G. S., and Yang, G. C. (1984). Rapid extraction and detection of aflatoxin B_1 and M_1 in beef liver. *Anal. Chem.* **56**, 1071–1073.

❖

Qian, G. S., Yasei, P., and Yang, G. C. (1984). Rapid extraction and detection of aflatoxin M₁ in cow's milk by high-performance liquid chromatography and radioimmunoassay. *Anal. Chem.* **56**, 2079–2080.

Ram, B. P., Hart, L. P., Shotwell, O. L., and Pestka, J. J. (1986). Enzyme-linked immunosorbent assay of aflatoxin B₁ in naturally contaminated corn and cottonseed. *J. Assoc. Off. Anal. Chem.* **69**, 904–907.

Ramakrishna, N., Lacey, J., Candlish, A. A. G., Smith, J. E., and Goodbrand, I. A. (1990). Monoclonal antibody-based enzyme linked immunosorbent assay of aflatoxin-B₁, T-2-toxin, and ochratoxin-A in barley. *J. Assoc. Off. Anal. Chem.* **73**, 71–76.

Rauch, P., Fukal, L., Prosek, J., Bresina, P., and Kas, J. (1987). Radioimmunoassay of Aflatoxin M₁. *J. Radioanal. Nucl. Chem. Lett.* **117**, 163–169.

Rauch, P., Fukal, L., Brezina, P., and Kas, J. (1988). Interferences in radioimmunoassay of aflatoxins in foods and fodder samples of plant origin. *J. Assoc. Off. Anal. Chem.* **71**, 491–493.

Rauch, P., Viden, I., Davidek, T., Velisek, J., and Fukal, L. (1989). Caffeine as main interfering compound in radioimmunoassay of aflatoxin-B₁ in coffee samples. *J. Assoc. Off. Anal. Chem.* **72**, 1015–1017.

Reynolds, G., and Pestka, J. J. (1991). Enzyme-linked immunosorbent assay of versicolorin A and related aflatoxin biosynthetic presurors. *J. Food Protect.* **54**, 105–108.

Sabbioni, G. (1990). Chemical and physical properties of the major serum albumin adduct of aflatoxin B₁ and their implications for the quantification in biological samples. *Chem. Biol. Interact.* **75**, 1–15.

Sabbioni, G., Skipper, P. L., Büchi, G., and Tannenbaum, S. R. (1987). Isolation and characterization of the major serum albumin adduct formed by aflatoxin B₁ in rats. *Carcinogenesis* **8**, 819–824.

Sabbioni, G., Ambs, S., Wogan, G. N., and Groopman, J. D. (1990). The aflatoxin–lysine adduct quantified by high-performance liquid chromatography from human albumin samples. *Carcinogenesis* **11**, 2063–2066.

Santalla, R. M., Zhang, Y. J., Hsieh, L. L., Young, T. L., Lu, X. Q., Lee, B. M., Yang, G. Y., and Perera, F. P. (1991). Immunological methods for monitoring human exposure to bezo(a)pyrene and aflatoxin B₁: Measurement of carcinogen adducts. *In* "Immunoassays for Trace Chemical Analysis: Monitoring Toxic Chemicals in Humans, Foods and the Environment" (M. Vanderlaan, L. H. Stanker, B. E. Watkins, and D. W. Roberts, eds.), pp. 229–245. American Chemical Society, Washington, D.C.

Scott, P. M. (1989). Mycotoxins—General referee reports. *J. Assoc. Off. Anal. Chem.* **72**, 75.

Scott, P. M. (1990a). Mycotoxins—General referee reports. *J. Assoc. Off. Anal. Chem.* **73**, 98.

Scott, P. M. (1990b). A mycotoxin miscellany. *J. Assoc. Off. Anal. Chem.* **73**, 14–16.

Scott, P. M. (1991). Mycotoxins—General referee reports. *J. Assoc. Off. Anal. Chem.* **74**, 120.

Shamsuddin, A. M., Harris, C. C., and Hinzman, M. J. (1987). Localization of aflatoxin B₁-nucleic acid adducts in mitochondria and nuclei. *Carcinogenesis* **8**, 109–114.

Sharman, M., and Gilbert, J. (1991). Automated aflatoxin analysis of foods and animal feeds using immunoaffinity column cleanup and high-performance liquid chromatographic determination. *J. Chromatogr.* **543**, 220.

Sidwell, W. J., Chan, H. W. S., and Morgan, M. R. A. (1989). Reduction in assay time for a double antibody, indirect enzyme-linked immunosorbent assay applied to ochractoxin A. *Food Agric. Immunol.* **1**, 111–118.

Sizaret, P., and Malaveille, C. (1983). Preparation of aflatoxin B₁–BSA conjugate with high hapten/carrier molar ratio. *J. Immunol. Meth.* **63**, 159–162.

Sizaret, P., Malaveille, C., Montesano, R., and Frayssinet, C. (1980). Detection of aflatoxins and related metabolites by R. I. A. *J. Natl. Cancer Inst.* **69**, 1375–1380.

Skipper, P., and Tannenbaum, S. R. (1989). Comments on production and characterization of aflatoxin B₂ oximinoacetate. *J. Agric. Food Chem.* **37**, 1548–1548.

Steimer, J., Hahn, G., Heeschen, W., and Bluthgen, A. (1988). On the enzyme–immunological detection of aflatoxin M_1 in milk: Development of a rapid aflatoxin M screening test using polystyrol beads as solid-phase. *Milchwiss.* **43**, 772–776.

Stubblefield, R. D., Greer, J. I., Shotwell, O. L., and Aikens, A. M. (1991). Rapid immunochemical screening method for aflatoxin B_1 in human and animal urine. *J. Assoc. Off. Anal. Chem.* **74**, 530–532.

Sun, P., and Chu, F. S. (1977). A simple solid-phase radioimmunoassay for aflatoxin B_1. *J. Food Safety* **1**, 67–75.

Sun, T., Wu, Y., and Wu, S. (1983). Monoclonal antibody against aflatoxin B_1 and its potential applications. *Chin. J. Oncol.* **5**, 401–405.

Trucksess, M. W., Stack, M. E., Nesheim, S., Page, S. W., Albert, R. H., Hansen, T. J., and Donahue, K. F. (1989a). Immunoaffinity column coupled with solution fluorometry or LC postcolumn derivatization for aflatoxins in corn, peanuts, and peanut butter: Collaborative study. *Abstracts of The 103rd AOAC Annual International Meeting,* St. Louis, Missouri, September 25–28.

Trucksess, M. W., Stack, M. E., Nesheim, S. D., Park, L., and Pohland, A. E. (1989b). Enzyme-linked immunosorbent assay of aflatoxins B_1, B_2, and G_1 in corn, cottonseed, peanuts, peanut butter, and poultry feed: Collaborative study. *J. Assoc. Off. Anal. Chem.* **72**, 957–962.

Trucksess, M. W., Young, K., Donahue, K. F., Morris, D. K., and Lewis, E. (1990). Comparison of two immunochemical methods with thin-layer chromatographic methods for determination of aflatoxins. *J. Assoc. Off. Anal. Chem.* **73**, 425–428.

Trucksess, M. W., Stack, M. E., Nesheim, S., Page, S. W., Albert, R. H., Hansen, J. T., and Donahue, K. F. (1991). Immunoaffinity column coupled with solution fluorometry or LC postcolumn derivatization for aflatoxins in corn, peanuts, and peanut butter: Collaborative study. *J. Assoc. Off. Anal. Chem.* **74**, 81.

Tsuboi, S., Nakagawa, T., Tomita, M., Seo, T., Ono, H., Kawamura, K., and Iwamura, N. (1984). Detection of aflatoxin B_1 in serum samples of male Japanese subjects by radioimmunoassay and high performance liquid chromatography. *Cancer Res.* **44**, 1231–1234.

Ueno, I. (1986). A simple and improved enzyme-linked immunosorbent assay method for microquantitation of aflatoxin B_1 in peanuts and blood plasma. *Proc. Jpn. Assoc. Mycotoxicol.* 24–27.

Ueno, I., and Chu, F. S. (1978). Modification of hepatotoxic effects of aflatoxin B_1 in rabbits by immunization. *Experientia* **34**, 85–86.

van Egmond, H. P., and Wagstaffe, P. J. (1990). Aflatoxin B_1 in compound-feed reference materials: An intercomparison of methods. *Food Add. Contam.* **7**, 239.

Wild, C. P., and Montesano, R. (1991). Immunological quantitation of human exposure to aflatoxins and N-nitrosamines. *In* "Immunoassays for Trace Chemical Analysis: Monitoring Toxic Chemicals in Humans, Foods and the Environment" (M. Vanderlaan, L. H. Stanker, B. E. Watkins, and D. W. Roberts, eds.), pp. 215–228. American Chemical Society, Washington, D.C.

Wild, C. P., Umbenhauer, D., Chapot, B., and Montesano, R. (1986a). Monitoring of individual human exposure to aflatoxins and *N*-nitrosamines by immunoassay. *J. Cell. Biochem.* **30**, 171–179.

Wild, C. P., Garner, R. C., Montesano, R., and Tursi, F. (1986b). Aflatoxin B_1 binding to plasma albumin and liver DNA upon chronic administration to rats. *Carcinogenesis (London)* **7**, 853–858.

Wild, C. P., Pionneau, F. A., Montesano, R., Mutiro, C. F., and Chetsanga, C. J. (1987). Aflatoxin detected in human breast milk by immunoassay. *Int. J. Cancer* **40**, 328–398.

Wild, C. P., Chapot, B., Scherer, E., Den-Engelse, L., and Montesano, R. (1988). Application of antibody methods to the detection of aflatoxin in human body fluids. *IARC Sci. Publ.* **89**, 67–74.

Wild, C. P., Jiang, Y. Z., Sabbioni, G., Chapot, B., and Montesano, R. (1990a). Evaluation of methods for quantitation of aflatoxin–albumin adducts and their application to human exposure assessment. *Cancer Res.* **50**, 245–251.

Wild, C. P., Montesano, R., Van Benthem, J., Scherer, E., and Den Engelse, L. (1990b). Intercellular variation in levels of adducts of aflatoxin B_1 and G_1 in DNA from rat tissues: A quantitative immunocytochemical study. *J. Cancer Res. Clin. Oncol.* **116**, 134–140.

Wilkinson, A. P., Denning, D. W., and Morgan, M. R. A. (1988a). Analysis of UK sera for aflatoxin by enzyme-linked immunosorbent assay. *Human Toxicol.* **7**, 353–356.

Wilkinson, A. P., Denning, D. W., and Morgan, M. R. A. (1988b). An ELISA method for the rapid and simple determination of aflatoxin in human serum. *Food Add. Contam.* **5**, 609–620.

Woychik, N. A., Hinsdill, R. D., and Chu, F. S. (1984). Production and characterization of monoclonal antibody against aflatoxin M_1. *Appl. Environ. Microbiol.* **48**, 1096–1099.

Wu, S., Yang, G., and Sun, T. (1983). Studies on immuno-concentration and immunoassay of aflatoxins. *Chin. J. Oncol.* **5**, 81–84.

Xu, Y. C., Zhang, G. S., and Chu, F. S. (1986). Radioimmunoassay of deoxynivalenol in wheat and corn. *J. Assoc. Off. Anal. Chem.* **69**, 967–969.

Yu, J., and Chu, F. S. (1991). Immunochromatography of fusarochromanone mycotoxins. *J. Assoc. Off. Anal. Chem.* **74**, 655–660.

Yu, S. Z., Cheng, Z. Q., Liu, X. K., Zhang, L. S., and Zhao, Y. F. (1989). The aflatoxins and water contaminated in the etiological study of primary liver cancer. *In* "Mycotoxins and Phycotoxins '88" (S. Natori, K., Hashimoto, and Y. Ueno, eds.), pp. 37–44. Elsevier, Amsterdam.

Zhang, G. S., and Chu, F. S. (1989). Production and characterization of antibodies cross-reactive with major aflatoxins. *Experientia* **45**, 182–184.

Zhang, Y. J., Chen, C. J., Haghighi, B., Yang, G. Y., Hsieh, L. L., Wang, L. W., and Santella, R. M. (1991). Quantitation of aflatoxin B_1–DNA adducts in woodchuck hepatocytes and rat tissues by indirect immunofluorescence analysis. *Cancer Res.* **51**, 1720–1725.

Zhu, J. Q., Zhang, L. S., Hu, X., Xiao, Y., Chen, J. S., Xu, Y. C., Fremy, J., and Chu, F. S. (1987). Correlation of dietary aflatoxin B_1 levels with excretion of aflatoxin M_1 in human urine. *Cancer Res.* **47**, 1848–1852.

Part V

❖

Economic and Regulatory Aspects of Aflatoxins

22

Human Risk Assessment Based on Animal Data: Inconsistencies and Alternatives

Nancy J. Gorelick, Robert D. Bruce, and Mohammad S. Hoseyni

INTRODUCTION

The qualitative association between aflatoxin B_1 (AFB_1) exposure and cancer risk in different species including humans has been discussed by many authors (e.g., Bruce, 1990). Human exposure to aflatoxin in the diet requires estimation of the potential risk to humans to establish acceptable exposure limits. Approaches to calculate potency estimates and critiques of these approaches are presented in this chapter. The first approach is the classical approach based on tumor data from the most sensitive response in the most sensitive species. However, the potency estimate so derived is not consistent with observed liver cancer incidence in United States' populations. Reasons for this apparent discrepancy are discussed here. Alternative approaches to calculate potency estimates based on epidemiological observations in the presence or absence of confounders are proposed.

COMPARISON OF RODENT AND HUMAN DATA
FOR RISK ASSESSMENT

Human safety assessments traditionally have been calculated from studies in which rodents are exposed over their adult lifetime to the test chemical. High dosages of chemical, relative to typical human exposure levels, are tested to maximize the assay sensitivity with a minimum number of study animals. Consequently, human safety assessments require an extrapolation of the observed effects between species, as well as an extrapolation of the observed effects from high dose levels to the low dose levels characteristic of human exposure. Typically, tumor data from the most sensitive species and strain are used for human safety assessment.

Risk Assessment Based on Rat Data

At least three authors have employed this classical approach to derive an acceptable AFB_1 exposure level for humans: Carlborg (1979), Dichter (1984), and Kuiper-Goodman (1990). The latter is, perhaps, most representative of current regulatory practice. Based on the rat study of Wogan *et al.* (1974), Kuiper-Goodman (1990) derives a virtually safe dose (VSD) of 0.016 ng/kg/day corresponding to a lifetime risk of 1×10^{-5}; the author describes this as a lower confidence limit but does not mention the confidence level. Comparable results may be found in Carlborg (1979) and Dichter (1984), but these authors also report the results of applying a variety of models for low-dose extrapolation. For later use, re-expressing the VSD of 0.16 ng/kg/day as

Excess lifetime risk $= 6.25 \times 10^{-4}$ [aflatoxin intake (ng/kg/day)]

will be convenient.

Liver Cancer in Humans in the United States

Perspective on this estimate may be gained by using it with an estimate of AFB_1 ingestion for the southeast United States calculated by Bruce (1990) using the data of Stoloff (1983). In this region, consumption of aflatoxin-containing foods, particularly full-fat corn meal products, was strikingly high during the period of 1910–1959. For the period of 1910–1979, Bruce calculated a time-weighted average exposure of 110 ng/kg/day. Combining this estimated exposure level with the expression just presented gives

Excess lifetime risk $= 6.88 \times 10^{-2}$

or, assuming a 70-year life-span,

Risk/100,000/yr $= 98$

However, United States liver cancer rates, whatever the cause, are currently around 3.4/100,000/yr for males and 1.3/100,000/yr for females (Yeh *et al.*, 1989), with no indication that rates in the southeast are much higher (Stoloff, 1983) if at all. Thus, the rat data predict a liver cancer rate as a result of aflatoxin exposure alone that is far in excess of the actual rate as a result of all causes.

Possible Explanation for the Differences between Predictions from Rodents and Observations in Humans

Previous chapters have described current knowledge about aflatoxin metabolism and carcinogenesis. Clearly, the metabolic processing and tumorigenic potency of aflatoxin vary widely among species. The methods used for species extrapolation and dose extrapolation are critical to developing an accurate human safety assessment for a chemical, such as aflatoxin, that requires metabolic activation to exert its biological effects. Thus, understanding the biochemistry underlying the carcinogenic response is critical to assessing the appropriateness of data from any particular laboratory animal model as a basis for extrapolations for human safety assessment.

Critical Differences between Species in Aflatoxin Metabolism

Strong evidence indicates that AFB_1 initiates its mutagenic and carcinogenic effects in laboratory animals by cytochrome P450-mediated metabolic activation to an epoxide (Figure 1) (Swenson *et al.,* 1977). AFB-8,9-epoxide reacts with DNA, forming adducts that persist in DNA as well as adducts that are excreted in urine (Essigmann *et al.,* 1983). However, the formation of AFB_1–DNA adducts may be modulated by enzymatic conjugation of the epoxide with glutathione as mediated by cytosolic glutathione *S*-transferases (Kensler *et al.,* 1986). Comparison of the formation and detoxification of the active AFB-8,9-epoxide in different species has shown that the balance between these metabolic processes can explain in part why animal species vary widely in sensitivity to AFB_1 carcinogenicity (reviewed by Gorelick, 1990).

Most of the studies comparing AFB_1 metabolism in different species have been conducted *in vitro,* with subcellular fractions or purified components of these fractions. The loss of cellular compartmentalization and the isolation of only some components of competing pathways for metabolic activation and detoxification limit the conclusions that can be drawn about species differences to qualitative comparisons.

Similar aflatoxin metabolites may be found in humans and in laboratory animals. Currently available data on metabolic activation and detoxification indicate that these reactions are not mediated by the same isozymes in different species (reviewed by Gorelick, 1990; Guengerich, 1990). In fact, one metabolic reaction, for example, epoxidation, may be catalyzed by different isozymes in different species. Consequently, this metabolic reaction may vary in relative rate, dose dependence, sex specificity, and tissue specificity among species. Extrapolation of metabolism-dependent processes in rodents to humans, as well as dose effects among species, therefore, must be done with caution (Gonzalez, 1990; Gonzalez and Nebert, 1990). The next sections of this chapter discuss the species differences in metabolic activation and detoxification of AFB_1. These differences may explain the discrepancy between observed human liver cancer incidence and

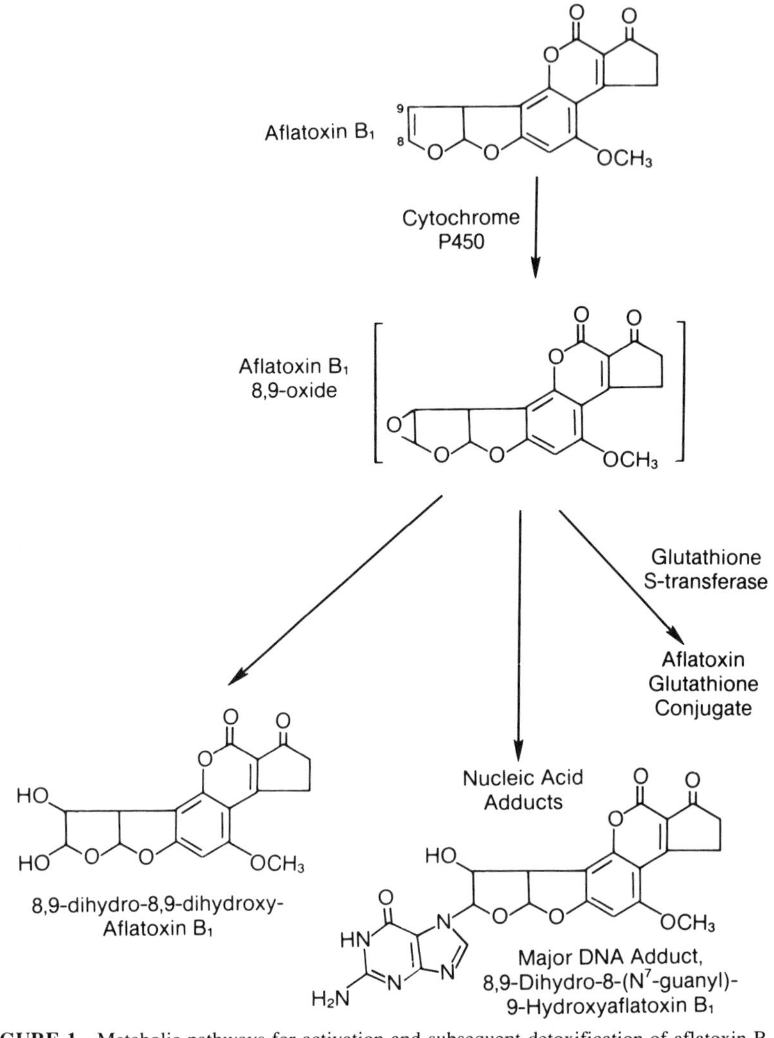

FIGURE 1   Metabolic pathways for activation and subsequent detoxification of aflatoxin B_1. Reprinted with permission from Gorelick (1990).

the liver cancer incidence predicted from rodent studies. The reader is referred to the review by Gorelick (1990) for a more detailed discussion of this topic.

Metabolic Activation   Multiple P450 isozymes from rodents, fish, and humans are capable of catalyzing the epoxidation of AFB_1. The methods used to determine the activity of specific isozymes vary in their selectivity; therefore, the role of specific P450 isozymes in AFB_1 metabolism is known with varying degrees of certainty. A summary of the P450 isozymes currently known to activate AFB_1, and the biological system in which this activity was determined, is presented in Table 1.

The role of particular P450 isozymes in AFB_1 activation has been determined in five kinds of studies. In all five approaches, metabolic activation is monitored by DNA damage or mutagenesis, either in cultured cells or with microsomes or S9 *in vitro*. In one approach, cell lines are created in which cloned genes for individual P450 isozymes are expressed (Gonzalez *et al.*, 1991). Results from these studies are definitive for the ability of a particular isozyme to metabolize AFB_1. However, the relative contribution of an individual isozyme to AFB_1 activation *in vivo* is not defined. Similarly, purified enzymes may be reconstituted to demonstrate specific activities. Difficulties in purification to homogeneity and in reconstitution have been discussed (Guengerich and Kim, 1991). Immunoinhibition is a third approach to study the role of specific P450 isozymes in AFB_1 activation. Immunoinhibition studies are limited by the cross-reactivity of antibodies with related isozymes, but are usually specific for activity within P450 families. In a fourth approach, the effect of P450 inducers or inhibitors on AFB_1 metabolism *in vivo* or *in vitro* has been determined. Results from these studies are less definitive because of the multiplicity of P450 isozymes as well as other phase I and phase II enzymes that are affected by many of the agents studied. For example, a discrepancy exists between results from *in vivo* and *in vitro* studies with P450 inducers or inhibitors. Microsomes from rats pretreated with P450 inducers such as phenobarbital enhance AFB_1 activation (measured by hepatic DNA binding; Gurtoo and Bejba, 1974; Guengerich, 1979; Metcalfe *et al.*, 1981) whereas DNA binding *in vivo* (Garner, 1975; Swenson *et al.*, 1977) as well as tumorigenesis (McLean and Marshall, 1971) is suppressed by pretreatment of rats with phenobarbital. Thus, induction studies *in vitro* may be useful to characterize P450 activities partially, but those capabilities might not have the predicted biological outcome *in vivo*. Finally, the correlation between the formation of AFB_1 metabolites and characteristic P450-specific activities has been determined (Shimada and Guengerich, 1989). A high correlation is observed when the same isozyme catalyzes AFB_1 metabolism and isozyme-specific reactions.

These kinds of studies have shown that the human isozyme P450 3A4 can catalyze AFB_1 activation (Shimada and Guengerich, 1989; Aoyama *et al.*, 1990; Forrester *et al.*, 1990; Crespi *et al.*, 1991). This isozyme may be the predominant human liver P450 isozyme responsible for AFB_1 activation (Shimada and Guengerich, 1989; Forrester *et al.*, 1990; Guengerich and Kim, 1990,1991). Studies referred to in Table 1 indicate that several other isozymes are active *in vitro*, but the relative contribution of each *in vivo* is not certain. The activity level *in vivo* of each protein that can activate AFB_1 may determine the relative contribution of particular isozymes. Low protein levels or low levels of gene expression in adult human liver have been documented for P450 1A2, P450 2A6, P450 3A3, and P450 3A7—some of the human isozymes that have been shown to

TABLE 1 P450 Isozymes that Catalyze the Metabolic Activation of AFB₁

P450[a]	Biological endpoint				Approach[f]	P450 source	Reference
	Mutagenesis in Salmonella[b]	DNA damage[c]	DNA binding[d]	Endogenous mutagenesis[e]			
Human							
CYP1A2			+	+	Dose–response	Expressed cell line[g]	Crespi et al. (1990a)
			+		Dose–response	Expressed cell line	Aoyama et al. (1990)
	+				Ab inhibition[h]	Liver S9	Aoyama et al. (1990)
	+				Ab inhibition	Liver microsomes	Forrester et al. (1990)
CYP2A6				+	Dose–response	Expressed cell line	Crespi et al. (1990b, 1991)
			+		Dose–response	Expressed cell line	Aoyama et al. (1990)
	+				Ab inhibition	Liver S9	Aoyama et al. (1990)
CYP2B7			+		Dose–response	Expressed cell line	Aoyama et al. (1990)
	+				Ab inhibition	Liver S9	Aoyama et al. (1990)
CYP3A[i]	+				Ab inhibition	Liver microsomes	Forrester et al. (1990)
	+				P4503A protein level	Liver microsomes	Forrester et al. (1990)
	+				Ab inhibition	Liver S9	Aoyama et al. (1990)
		+	+		Ab inhibition	Liver microsomes	Shimada and Guengerich (1989)
		+	+		Correlation[j]	Liver microsomes	Shimada and Guengerich (1989)
CYP3A3	+		+		Dose–response	Expressed cell line	Aoyama et al. (1990)
CYP3A4			+	+	Dose–response	Expressed cell line	Crespi et al. (1991)

CYP3A7	+			+	Dose–response	Expressed cell line	Aoyama et al. (1990)
		+			Ab inhibition	Purified from fetal liver	Kitada et al. (1990)
Rat							
CYP2B1				+	Dose–response	Expressed cell line	Doehmer et al. (1988)
mt3					Purified	Liver mitochondria	Shayiq and Avadhani (1989)
CYP2C11			+		Purified	Liver microsomes	Shimada et al. (1987)
CYP2C12			+		Purified	Liver microsomes	Shimada et al. (1987)
Hamster							
CYP2A8	+			+	Purified	Liver microsomes	Fukuhara et al. (1990)
Trout				+	Purified	Liver microsomes	Williams and Buhler (1983)
				+	Purified	Kidney microsomes	Williams et al. (1986)

[a] Nomenclature according to Nebert et al. (1991).
[b] Detected by Ames test.
[c] DNA damage in *Salmonella* detected by induction of *umuC* gene expression (Shimada and Nakamura, 1987) in response to SOS repair. *umuC* gene expression is correlated with mutagenesis in the Ames test (Shimada et al., 1987).
[d] [^{3}H] AFB–DNA binding in cell line or *in vitro*.
[e] Mutagenesis at an endogenous locus.
[f] Experimental approaches to identify isozymes with particular activities are discussed in the text.
[g] A cell line carries and expresses the particular P450 gene.
[h] Antibodies against specific rat isozymes (P4501A2, P4502A6, P4502B2, P4503A1, respectively) inhibited the biological response.
[i] Activity of individual human isozymes in the P4503A family was not distinguishable.
[j] Correlation between characteristic isozyme activities and AFB$_1$ activation.

activate AFB_1. Therefore, many of these P450s may not contribute substantially to AFB_1 activation *in vivo* (Guengerich and Kim, 1990,1991). In addition, AFB_1 detoxification may compete with AFB_1 activation by isozymes that can catalyze both kinds of reactions. For example, the conversion of AFB_1 to AFQ_1 is catalyzed by P450 3A4 (Forrester *et al.*, 1990; Raney *et al.*, 1992a) and by the orthologous rat P450 3A enzymes (Halvorson *et al.*, 1988); P450 1A2 may catalyze the conversion of AFB_1 to AFM_1 in both mouse (Faletto *et al.*, 1988) and human (Guengerich and Kim, 1990, 1991; Raney *et al.*, 1992a); and P450 2A6 may yield both AFP_1 and AFM_1 (Crespi *et al.*, 1990b). Additional studies in human liver microsomes and in cell lines that carry multiple human P450s may reveal the effect of competition between isozymes.

Less information is available about the specific P450 isozymes that catalyze the metabolic activation of AFB_1 in laboratory animals. P450s that activate AFB_1 have been identified in rodents and fish, and are not from the same P450 families as those in humans. Multiple constitutive rat liver P450s in the P450 2C family, including sex-specific isozymes, catalyze AFB_1 activation (Ishii *et al.*, 1986; Shimada *et al.*, 1987). In addition, rat liver microsomal studies indicate that a phenobarbital-inducible P450 activates AFB_1 *in vitro* (reviewed by Gorelick, 1990). Phenobarbital induces multiple P450 isozymes, probably by multiple mechanisms (Kocarek *et al.*, 1990), and only a limited number of purified isozymes actually have been tested for aflatoxin metabolism. Although phenobarbital does induce rat P450 3A proteins, such proteins have not been shown to activate AFB_1, whereas AFB_1 activation by phenobarbital-inducible proteins in the P450 2B family has been demonstrated (Doehmer *et al.*, 1988). Hamster is the only other rodent in which P450s capable of AFB_1 activation have been identified (Mizokami *et al.*, 1986; Fukuhara *et al.*, 1990). Again, this P450 is from a different family (P450 2A; Fukuhara *et al.*, 1989). Studies in cell lines carrying either of two mouse P450 genes (*Cyp1a-1* or *Cyp1a-2*) failed to activate AFB_1 metabolically (Aoyama *et al.*, 1989). Finally, a trout liver P450 (Williams and Buhler, 1983) and a constitutive sex-specific P450 in trout kidney (Williams *et al.*, 1986) are capable of AFB_1 activation, but further characterization of these proteins is not available.

In cases in which orthologous P450s have been identified, the regulation of gene expression for those orthologs is substantially different (Imaoka *et al.*, 1990). Thus, factors other than the protein sequence similarity may be important for extrapolation of function among species. Despite some overlapping isozyme specificity, the regulation of AFB_1 activation in different species is likely to be different.

Overall, metabolic activation of AFB_1 appears to be accomplished in different species by different P450 isozymes. Regulation of gene expression for many of these isozymes is probably also different. A comparison of the complete profile of P450s that metabolically activate AFB_1 in different species is still unavailable.

Metabolic Detoxification

Detoxification of AFB₁ Hydroxylation of AFB_1 on both the furan and the lactone rings [at C10 (AFM_1) or C3 (AFQ_1)] and oxidative demethylation (AFP_1) may be considered detoxification reactions. Species differences in the profile of hydroxylated metabolites have been discussed elsewhere (Busby and Wogan, 1984; Gorelick, 1990; Ramsdell and Eaton, 1990a). For the purposes of risk assessment, hydroxylation is important to the extent that it may mitigate the potency of the original AFB_1 dose. However, AFM_1 and AFP_1 are still susceptible to metabolic activation via epoxidation and also can form DNA adducts, and thus should be considered independently for risk assessment.

Detoxification of AFB-8,9-Epoxide Hydrolysis of AFB-8,9-epoxide, producing AFB_1 dihydrodiol, may occur via an enzymatic or a spontaneous reaction. Studies *in vitro* (Lin *et al.*, 1978; Lotlikar *et al.*, 1984) and *in vivo* (Kensler *et al.*, 1985) on the role of epoxide hydrolase in this reaction indicate that this minor reaction is probably not enzymatic.

An alternative fate of AFB-8,9-epoxide is conjugation with glutathione, which is an enzymatic reaction. Glutathione conjugation of AFB-8,9-epoxide protects against DNA binding *in vivo* in rats, mice, and hamsters and against the development of preneoplastic lesions, as well as tumors, in rats (reviewed by Gorelick, 1990). The competition between glutathione conjugation and DNA binding depends on the rate of epoxide generation and, therefore, on the P450 isozymes, the affinity of glutathione S-transferases for AFB-8,9-epoxide, and the availability of glutathione. Thus, the appropriateness of an animal model for human safety assessment of AFB_1 depends on the interspecies similarity of P450s as well as on detoxification reactions such as glutathione conjugation.

Most studies of glutathione S-transferase affinity for AFB-8,9-epoxide have been conducted *in vitro* with cytosolic fractions from livers of various species. In these studies, the inhibition of AFB_1–DNA binding by AFB_1 conjugation with glutathione *in vitro* has been correlated with low susceptability to AFB_1 tumorigenesis *in vivo*. For example, low glutathione S-transferase activity toward AFB_1 is found in rats, who are highly sensitive to AFB_1, whereas mice have high glutathione S-transferase activity and are resistant to the tumorigenic effects of AFB_1. Preliminary studies of glutathione S-transferase activity toward AFB_1 in human liver cytosol *in vitro* indicate low levels of activity (Raney *et al.*, 1992b), whereas others have not detected any activity (O'Brien *et al.*, 1983; Moss and Neal, 1985; Ramsdell and Eaton, 1989).

Rodent glutathione S-transferase isozymes that show activity toward AFB-8,9-epoxide have been identified. The θ class (Meyer *et al.*, 1991) and the μ class in rats contain enzymes with activity toward the synthetic AFB_1-8,9-epoxide (Raney *et al.*, 1992b). Transferases from the α class are active in both rat (Coles *et al.*, 1985; Raney *et al.*, 1992b) and mouse liver (Quinn *et al.*, 1990; Ramsdell and Eaton, 1990b). The mouse isozyme, however, has a high affinity for AFB-8,9-epoxide whereas the rat isozyme apparently has a low affinity for AFB-8,9-epoxide, since the rat isozyme constitutes a substantial amount of the total rat liver glutathione S-transferase yet shows little activity. Mouse and rat glutathione S-transferases from the α class are highly related immunologically and in amino acid sequence (Ramsdell and Eaton, 1990b). The amino acid sequence difference must be significant, however, because of the difference in affinity for AFB-8,9-epoxide. Preliminary studies of human glutathione S-transferase demonstrate activity toward AFB-8,9-epoxide in isozymes from the μ class (Liu *et al.*, 1990; Raney *et al.*, 1992b), but neither the μ class isozyme M3-3 nor isozymes from the π class generated AFB-glutathione conjugates *in vitro* (Raney *et al.*, 1992b).

Comparison of activity toward AFB-8,9-epoxide by classes of glutathione S-transferase from different species has not been extensive, yet the data indicate that mouse liver is unique in its high affinity constitutive glutathione S-transferase. Depending on the liver sample, human liver glutathione S-transferases have lower (Raney *et al.*, 1992b) or higher (Quinn *et al.*, 1990) activity toward AFB-8,9-epoxide than do rat liver enzymes.

Quantitative Comparison of Metabolism in Different Species   The preceding discussion indicates that the susceptibility of different species to AFB₁ carcinogenesis will be determined by the relative contribution of factors for metabolic activation and detoxification. The balance of these factors, in addition to others such as DNA repair, may be reflected in DNA adduct levels. Studies of species differences in DNA adduct levels *in vivo* have been reviewed (Gorelick, 1990). In brief, relatively higher levels of DNA binding *in vivo* correlate with tissue susceptibility to AFB₁ carcinogenesis in rodents. However, a quantitative relationship between adducts and tumors has not been established rigorously. Ongoing molecular epidemiology studies may indicate the relative susceptibility of humans. Progress in this area is discussed elsewhere.

Conclusion

In summary, the similarity of individual P450 or glutathione S-transferase isozymes from different species can be compared at the molecular level. The data on the molecular differences in AFB₁ metabolism in rat and humans reviewed here indicate that the rat may not be an appropriate animal model for human risk assessment.

USE OF EPIDEMIOLOGICAL DATA AS AN ALTERNATIVE TO ANIMAL DATA FOR ASSESSING HUMAN RISK

Review of Available Human Data

Not only has the potency of AFB₁ in the rat been shown to overestimate greatly the human liver cancer rate in the United States, but use of the rat data ignores the large and growing literature on the epidemiology of liver cancer as it relates to AFB_1 and other factors. This literature was reviewed by Bruce (1990) in an attempt to identify studies suitable for assessing the potential risk of aflatoxin under conditions in the United States. When consideration is limited to those studies that provide quantitative estimates of both aflatoxin exposure levels and liver cancer rates, one is left with just a few studies, all conducted either in sub-Saharan Africa or in southeast Asia. Each of these remaining studies is compromised, to some degree, by the difficulty of measuring aflatoxin exposure levels and of ascertaining liver cancer rates as well as by failure to control adequately for the confounding effects of known causes of liver cancer including alcoholism and chronic infection with hepatitis B virus. In no case was aflatoxin exposure estimated at the level of individual subjects, largely because of the lack of markers for past exposure. Invariably, aflatoxin exposures were presented only for geographical subdivisions of the studied population, so some of the more sophisticated methods often applied to modern epidemiology studies (for example, conditional logistic regression; Breslow and Day, 1989) could not be applied to separate the effect of aflatoxin from that of the confounding variables.

Perhaps the most serious confounding variable was chronic infection with hepatitis B virus, because the virus itself now is identified as an important cause of liver cancer. This relationship has been established most convincingly by the work of Beasley and his co-workers, as reviewed by Bruce (1990). The study by Campbell *et al.* (1990) confirms this relationship, finding a relatively strong correlation between hepatitis B serum antigen (HBsAg) positivity and liver cancer incidence across 48 counties in China. The likelihood of a synergistic relationship between the effects of hepatitis B virus infection and aflatoxin exposure also exists (Armstrong, 1980; Bulatao-Jayme *et al.,* 1982; Yeh *et al.,* 1985). In the studies that attempted to separate the effects of hepatitis B virus and aflatoxin (Peers *et al.,* 1987; Yeh *et al.,* 1989), very little variation in hepatitis carrier rates across the geographic regions was seen; thus, that analyses showed that hepatitis B virus seemed to have little effect on the liver cancer rate was not surprising.

Potency Estimates Based on Human Data

In the studied areas, HBsAg carrier rates typically range from 5 to 15%, whereas in North America this rate is below 1% (Hoofnagle and Alter, 1984). This fact, coupled with the possibility of synergism, makes suspect any attempt to use the African and Asian data to estimate possible effects of aflatoxin in the United States. Nevertheless, if we estimate the potency of aflatoxin from the African and southeast Asian data, we obtain:

$$\text{Excess lifetime risk} = 4.76 \times 10^{-5} \times [\text{aflatoxin intake (ng/kg/day)}]$$

from the data for Africa and Thailand and:

$$\text{Excess lifetime risk} = 8.25 \times 10^{-5} \times [\text{aflatoxin intake (ng/kg/day)}]$$

from the Chinese data (see Bruce, 1990, for details). If these estimates are combined with the aflatoxin consumption value for the southeast United States (100 ng/kg/day), as was done earlier, predicted rates of liver cancer as a result of aflatoxin exposure alone are 7.7/100,000/year and 13.0/100,000/year respectively. As noted earlier, these rates are much higher than the United States liver cancer rates for all causes.

Although no formal human epidemiology studies of the possible relationship between aflatoxin and liver cancer have been conducted in the United States, Bruce (1990) showed how the data of Stoloff (1983) may be used to estimate the potency of aflatoxin. Stoloff estimated that rural white males in the Southeast might have as much as 10% more liver cancer compared with individuals in the North and West (other demographic groups had less excess cancer, or no excess at all). Aflatoxin intake in the Southeast was estimated at 100 ng/kg/day whereas in the North and West it was estimated at 0.34 ng/kg/day. These values may be used with the male liver cancer rate cited earlier to estimate the potency of aflatoxin as

$$\text{Potency} = \frac{\text{Excess risk}/100,000/\text{yr}}{\text{Extra aflatoxin intake (ng/kg/day)}} = \frac{10\% \times 3.4}{(110 - 0.34)} = 0.00310$$

or, assuming a life-span of 70 years,

$$\text{Excess lifetime risk} = 2.17 \times 10^{-6} \times [\text{aflatoxin intake (ng/kg/day)}]$$

This value, although still incorporating some conservative assumptions, is thought to be a more realistic reflection of the potency of aflatoxin for areas such as North America and northern Europe. For an excess lifetime risk of 1×10^{-5} and body weight of 70 kg, this value corresponds to an exposure rate of 323 ng/day. This value is also in agreement with the results of a more formal modeling approach to the data of Yeh et al. (1989) that adjusts for differences in hepatitis B carrier rates.

Relative Risk Model

For projection of lifetime excess risks among the United States population, data from the study of liver cancer in China by Yeh et al. (1989) were used to assess the effect of aflatoxin exposure on the risk of liver cancer. This study was chosen for the relatively high quality of its data and because both liver cancer incidence and HBsAg status were ascertained for the same subjects, allowing for estimation of the carcinogenic potency of aflatoxin. However, the prevalence of hepatitis B infection in China is known to be considerably higher than in United States. Therefore these risk estimates are not directly applicable to the United States population. Precise modeling of the effects of aflatoxin exposure and hepatitis B infection on the risk of liver cancer is essential for valid extrapolation of cancer potency across populations. Two widely used risk models in cancer epidemiology are (1) the relative risk model and (2) the additive (or "absolute risk") model:

$$\lambda(t,\underline{z}) = \lambda_0(t)\, \rho(t,\underline{z};\, \underline{\beta}) \tag{1}$$

$$\lambda(t,\underline{z}) = \lambda_0(t) + \eta(t,\underline{z};\, \underline{\beta}) \tag{2}$$

where $\lambda(t,\underline{z})$ is the hazard rate in the cohort at time t given covariates (risk factors) z, and λ_0 is the baseline hazard rate, that is, in the absence of exposure. ρ and η are relative and excess risk functions, involving vectors of covariates z and unknown parameters $\underline{\beta}$, that describe the effect of additional risk due to carcinogen exposure on the baseline rate. Selection of an appropriate model is based on several criteria.

1. The nature of the interaction between the carcinogens, as defined by ρ in model 1 or η in model 2, is important. Thomas (1981) and Barlow (1985) have discussed general forms of relative risk models in detail. The most commonly used risk function in analytical epidemiology is the exponential-multiplicative form. Given two exposure variables z_1 and z_2, this model is defined by

$$\rho(z_1, z_2) = \exp\,(\beta_1 z_1 + \beta_2 z_2) \tag{3}$$

In this equation, the regression parameters have the usual interpretation of the log-change in relative risk for a unit change in z. Alternative risk functions include

$$\rho(z_1, z_2) = (1 + \beta_1 z_1)(1 + \beta_2 z_2) \tag{4}$$

$$\rho(z_1, z_2) = (1 + \beta_1 z_1 + \beta_2 z_2) \tag{5}$$

or a power function describing the dose–response relationship

$$\rho(z_1, z_2) = \exp\,[\beta_1 \log(1 + z_1) + \beta_2 z_2] \tag{6}$$

2. Interaction between the baseline rates and exposure to carcinogens is important. The absolute risk model is fundamentally different from the relative risk model because the additional risk due to carcinogen exposure is added to the baseline rate rather than multiplied. For assessment and projection of cancer-specific risks, the suitability of the relative risk must be examined in contrast to the additive risk model, since it is highly dependent on the type of tumor (Muirhead and Darby, 1987). Direct comparison of these two models is complicated by the need for a test of separate families of hypotheses. Muirhead and Darby (1987) have proposed a simpler method to compare these two models by a generalized risk function, including both the additive and relative risk models as special cases:

$$\lambda(\underline{z}, \underline{\beta}) = \{\lambda_0^\gamma + [1 + \exp\,(\underline{\beta}'\,\underline{z})]^\gamma - 1\}^{1/\gamma} \tag{7}$$

A value of $\gamma = 1$ corresponds to the absolute risk model, whereas the relative risk model is obtained as $\gamma \to 0$. In prospective studies, such as the one by Yeh *et al.*

(1989), data usually are presented in grouped form; observations (number of deaths) and the corresponding person years are stratified into groups according to the level of exposure. In terms of grouped data, we can define the predicted number of deaths μ_k in group k by

$$\mu_k = Y_k \lambda (\underline{z}_k, \underline{\beta}) \tag{8}$$

where Y_k is the number of person years in group k and $\lambda (\underline{z}_k, \underline{\beta})$ denotes the hazard rate for group k. If we assume that the observed number of deaths (d_k) in group k follows a Poisson process with parameter μ_k, then the likelihood of this outcome would be

$$L_k = \frac{e^{-\mu_k} \mu_k^{d_k}}{d_k!} \tag{9}$$

Inferences about the parameters of interest (β) based on the likelihood function (9) for grouped data, can be shown to be identical to those from the likelihood function based on survivorship data (Holford, 1980), justifying the use of the Poisson regression as the basis for inference. The "iteratively reweighted least squares" method is used to fit model 8. For a broad class of models, this method leads to identical inferences, as does the method of maximum likelihood (Breslow *et al.*, 1983; Frome, 1983). To examine the goodness-of-fit with Poisson log-likelihoods, Muirhead and Darby (1987) propose an alternative to Pearson's χ^2 statistic based on the deviance

$$\sum_{k=1}^{n} D(d_k, \hat{\mu}_k) = 2 \sum_{k=1}^{n} [d_k \log(d_k/\hat{\mu}_k) - d_k + \hat{\mu}_k] \tag{10}$$

With large expected number of deaths (and p free parameters), the deviance is distributed asymptotically as χ^2 with $n - p$ degrees of freedom; when the expected number of deaths per group is small (less than 5), this value tends to be more stable than Pearson's χ^2.

The generalized risk model 7 was fitted to the Chinese data to examine the suitability of the relative risk model in contrast to the additive model. The minimum deviance was attained at $\hat{\gamma} = 0$, indicating the consistency of the data with the relative risk model. The difference in deviance (10) between $\gamma = 1$ (corresponding to the additive model) and $\hat{\gamma}$ is 6.65 (χ^2 with 1 degree of freedom). Next, risk functions 3–6 were fitted to the data to estimate the effect of aflatoxin exposure and HBsAg status on the risk of liver cancer, as well as their effect on each other. Based on the weighted sum of squares (WSS = 5.49 ~ χ_5^2), model 3 provided the best fit; examination of the observed and predicted number of cases also indicated the overall superiority of this model (Figure 2). The effects of AFB_1 and HBsAg both were found to be statistically significant. For those exposed to AFB_1, relative to the unexposed population, risk of liver cancer increases by 0.05% per ng/kg/day of exposure ($p < 0.001$). The results also

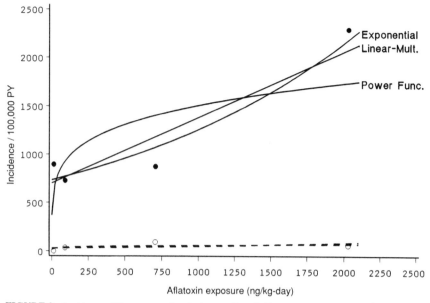

FIGURE 2 Incidence of liver cancer in relation to aflatoxin B₁ exposure: Observed (●, HbsAg +; ○, HbsAg−) vs. predicted (——, HbsAg +; −−, HbsAg−). Reprinted with permission from Hoseyni (1992).

indicated a 25-fold increase in the risk of liver cancer among those infected with hepatitis B virus relative to noncarriers ($p < 0.0001$). For projection of lifetime liver cancer risks in the United States, the AFB_1 and HBsAg parameter estimates based on model 3 were applied to the United States age-adjusted liver cancer mortality rates. The level of aflatoxin intake associated with a lifetime excess risk of 1×10^{-5} in the United States, assuming an average life-span of 70 years and body weight of 70 kg, was estimated as 253 ng/day (based on the upper 95% confidence limit of AFB_1 parameter estimate). Using the best estimate for AFB_1, the corresponding intake level is estimated as 441 ng/day. The reader is encouraged to refer to Hoseyni (1992) for details.

SUMMARY

In summary, the human liver cancer risk predicted by extrapolation of data from AFB_1-exposed rats has been compared with the liver cancer incidence observed in human populations. Predictions based on the rat data significantly overestimate human liver cancer incidence in the United States. This discrepancy may be explained by metabolic differences among species. Analyses of epidemiological data that control for confounding factors, such as infection with hepatitis B virus, may provide better estimates of human potency values for AFB_1.

REFERENCES

Aoyama, T., Gonzalez, F. J., and Gelboin, H. V. (1989). Mutagen activation by cDNA-expressed P_1450, P_3450, and P450a. *Mol. Carcinogen.* **1**, 253–259.

Aoyama, T., Yamano, S., Guzelian, P. S., Gelboin, H. V., and Gonzalez, F. J. (1990). Five of 12 forms of vaccinia virus-expressed human hepatic cytochrome P450 metabolically activate aflatoxin B_1. *Proc. Natl. Acad. Sci. U.S.A.* **87**, 4790–4793.

Armstrong, B. (1980). The epidemiology of cancer in the People's Republic of China. *Int. J. Epid.* **9**, 305–315.

Barlow, W. E. (1985). General relative risk models in stratified epidemiologic studies. *Appl. Stat.* **34**, 246–257.

Breslow, N. E., and Day, N. E. (1980). "Statistical Methods in Cancer Research," Vol. 1. International Agency for Research on Cancer, Lyon, France.

Breslow, N. E., Lubin, J. H., Marek, P., and Langholz, B. (1983). Multiplicative Models and Cohort Analysis. *J. Am. Statist. Assoc.* **78**, 1–12.

Bruce, R. D. (1990). Risk assessment for aflatoxin: II. Implications of human epidemiology data. *Risk Analysis* **10**, 561–569.

Bulatao-Jayme, J., Almero, E. M., Castro, M. C. A., Jardeleza, M. T. R., and Salamat, L. A. (1982). A case-control dietary study of primary liver cancer risk from aflatoxin exposure. *Int. J. Epidemiol.* **11**, 112–119.

Busby, W. F., Jr., and Wogan, G. N. (1984). Aflatoxins. *In* " Chemical Carcinogens" (C. E. Searle, ed.), 2d Ed., pp. 945–1094. American Chemical Society, Washington, D.C.

Campbell, T. C., Chen, J., Liu, C., Li, J., and Parpia, B. (1990). Nonassociation of aflatoxin with primary liver cancer in a cross-sectional ecological survey in the People's Republic of China. *Cancer Res.* **50**, 6882–6893.

Carlborg, F. W. (1979). Cancer, mathematical models and aflatoxin. *Food Cosmet. Toxicol.* **17**, 159–166.

Coles, B., Meyer, D. J., Ketterer, B., Stanton, C. A., and Garner, R. C. (1985). Studies on the detoxication of microsomally-activated aflatoxin B_1 by glutathione and glutathione transferases *in vitro*. *Carcinogenesis* **5**, 693–697.

Crespi, C. L., Steimel, D. T., Aoyama, T., Gelboin, H. V., and Gonzalez, F. J. (1990a). Stable expression of human cytochrome P450IA2 cDNA in a human lymphoblastoid cell line: Role of the enzyme in the metabolic activation of aflatoxin B_1. *Mol. Carcinogen.* **3**, 5–8.

Crespi, C. L., Penman, B. W., Leakey, J. A. E., Arlotto, M. P., Stark, A., Parkinson, A., Turner, T., Steimel, D. T., Rudo, K., Davies, R. L., and Langenbach, R. (1990b). Human cytochrome P450IIA3: cDNA sequence, role of the enzyme in the metabolic activation of promutagens, comparison to nitrosamine activation by human cytochrome P450IIE1. *Carcinogenesis* **11**, 1293–1300.

Crespi, C. L., Penman, B. W., Steimel, D. T., Gelboin, H. V., and Gonzalez, F. J. (1991). The development of a human cell line stably expressing human CYP3A4: Role in the metabolic activation of aflatoxin B_1 and comparison to CYP1A2 and CYP2A3. *Carcinogenesis* **12**, 355–359.

Dichter, C. R. (1984). Risk estimates of liver cancer due to aflatoxin exposure from peanuts and peanut products. *Food Chem. Toxicol.* **22**, 431–437.

Doehmer, J., Dogra, S., Freidberg, T., Monier, S., Adesnik, M., Glatt, H., and Oesch, F. (1988). Stable expression of rat cytochrome P-450IIB$_1$ cDNA in Chinese hamster cells (V79) and metabolic activation of alfatoxin B_1. *Proc. Natl. Acad. Sci. U.S.A.* **85**, 5769–5773.

Essigmann, J. M., Green, C. L., Croy, R. G., Fowler, K. W., Büchi, G. H., and Wogan, G. N. (1983). Interactions of aflatoxin B_1 and alkylating agents with DNA: Structural and functional studies. *Cold Spring Harbor Symp. Quant. Biol.* **47**, 327–337.

Faletto, M. B., Koser, P. L., Battula, N., Townsend, G. K., Maccubbin, A. E., Gelboin, H. V., and

Gurtoo, H. L. (1988). Cytochrome P_3-450 cDNA encodes aflatoxin B_1-4-hydroxylase. *J. Biol. Chem.* **263(25)**, 12187-12189.

Forrester, L. M., Neal, G. E., Judah, D. J., Glancey, M. J., and Wolf, C. R. (1990). Evidence for involvement of multiple forms of cytochrome P-450 in aflatoxin B_1 metabolism in human liver. *Proc. Natl. Acad. Sci. U.S.A.* **87**, 8306-8310.

Frome, E. L. (1983). The analysis of rates using Poisson regression models. *Biomedicine* **39**, 665-674.

Fukuhara, M., Nagata, K., Mizokami, K., Yamazoe, Y., Takanaka, A., and Kato, R. (1989). Complete cDNA sequence of a major 3-methylcholanthrene-inducible cytochrome P-450 isozyme (P-450AFB) of Syrian hamsters with high activity toward aflatoxin B_1. *Biochem. Biophys. Res. Commun.* **162(1)**, 265-272.

Fukuhara, M., Mizokami, K., Sakaguchi, M., Niimura, Y., Kato, K., Inouye, S., and Takanaka, A. (1990). Aflatoxin B_1-specific cytochrome P-450 isozyme (P-450-AFB) inducible by 3-methylcholanthrene in golden hamsters. *Biochem. Pharmacol.* **39(3)**, 463-469.

Garner, R. C. (1975). Reduction in binding of [^{14}-C]aflatoxin B_1 to rat liver macromolecules by phenobarbitone pretreatment. *Biochem. Pharmacol.* **24**, 1553-1556.

Gonzalez, F. J. (1990). Molecular genetics of the P-450 superfamily. *Pharmacol. Ther.* **45**, 1-38.

Gonzalez, F. J., and Nebert, D. W. (1990). Evolution of the P450 gene superfamily: Animal–plant "warfare", molecular drive and human genetic differences in drug oxidation. *Trends Genet.* **6**, 182-6. ˙

Gonzalez, F. J., Crespi, C. L., and Gelboin, H. V. (1991). cDNA-expressed human cytochrome P450s: A new age of molecular toxicology and human risk assessment. *Mutat. Res.* **247**, 113-127.

Gorelick, N. J. (1990). Risk assessment for aflatoxin: I. Metabolism of aflatoxin B_1 by different species. *Risk Analysis* **10**, 539-549.

Guengerich, F. P. (1979). Similarity of nuclear and microsomal cytochromes P-450 in the *in vitro* activation of aflatoxin B_1. *Biochem. Pharmacol.* **28**, 2883-2890.

Guengerich, F. P. (1990). Characterization of roles of human cytochrome P-450 enzymes in carcinogen metabolism. *Asia Pacific J. Pharmacol.* **5**, 327-345.

Guengerich, F. P., and Kim, T. (1990). *In vitro* inhibition of dihydropyridine oxidation and aflatoxin B_1 activation in human liver microsomes by naringenin and other flavonoids. *Carcinogenesis* **11**, 2275-2279.

Guengerich, F. P., and Kim, T. (1991). Oxidation of toxic and carcinogenic chemicals by human cytochrome P450 enzymes. *Chem. Res. Toxicol.* **4**, 391-407.

Gurtoo, H. L., and Bejba, N. (1974). Hepatic microsomal mixed function oxygenase: Enzyme multiplicity for the metabolism of carcinogens to DNA-binding metabolites. *Biochem. Biophys. Res. Commun.* **61**, 735-742.

Halvorson, M. R., Safe, S. H., Parkinson, A., and Philips, T. D. (1988). Aflatoxin B_1 hydroxylation by the pregnenolone-16-alpha-carbonitrile-inducible form of rat liver microsomal cytochrome P-450. *Carcinogenesis* **9**, 2103-2108.

Holford, T. R. (1980). The analysis of rates and of survivorship using log-linear models. *Biomedicine* **36**, 299-305.

Hoofnagle, J. H., and Alter, H. J. (1984). Chronic viral hepatitis. *In* "Viral Hepatitis and Liver Disease" (G. N. Vyas, J. L. Dienstag, and J. H. Hoofnagle, eds.), pp. 97-113. Grune & Stratton, Orlando, Florida.

Hoseyni, M. S. (1992). Risk assessment for aflatoxin: III. Modelling the relative risk of hepatocellular carcinoma. *Risk Analysis.* **12**, 123-128.

Imaoka, S., Enomoto, K., Oda, Y., Asada, A., Fujimori, M., Shimada, T., Fujita, S., Guengerich, F. P., and Funae, Y. (1990). Lidocaine metabolism by human cytochrome P-450s purified from hepatic microsomes: Comparison of those with rat hepatic cytochrome P-450s. *J. Pharmacol. Exp. Ther.* **255**, 1385-1391.

Ishii, K., Meada, K., Kamataki, T., and Kato, R. (1986). Mutagenic activation of aflatoxin B_1 by several forms of purified cytochrome P450. *Mutat. Res.* **174**, 85-88.

Kensler, T. W., Egner, P. A., Trush, M. A., Bueding, E., and Groopman, J. D. (1985). Modification of aflatoxin B$_1$ binding to DNA *in vivo* in rats fed phenolic antioxidants, ethyoxyquin and a dithiothione. *Carcinogenesis* **6(5)**, 759–763.

Kensler, T. W., Egner, P. A., Davidson, N. E., Roebuck, B. D., Pikul, A., and Groopman, J. D. (1986). Modulation of aflatoxin metabolism, aflatoxin-N^7-guanine formation, and hepatic tumorigenesis in rats fed ethyoxyquin: Role of induction of glutathione S-transferases. *Cancer Res.* **46**, 3924–3931.

Kitada, M., Taneda, M., Ohta, K., Nagashima, K., Itahashi, K., and Kamataki, T. (1990). Metabolic activation of aflatoxin B$_1$ and 2-amino-3-methylimidazo[4,5-f]-quinoline by human adult and fetal livers. *Cancer Res.* **50**, 2641–2645.

Kocarek, T. A., Schuetz, E. G., and Guzelian, P. S. (1990). Differentiated induction of cytochrome P450b/e and P450p mRNAs by dose of phenobarbital in primary cultures of adult rat hepatocytes. *Mol. Pharmacol.* **38**, 440–444.

Kuiper-Goodman, T. (1990). Uncertainties in the risk assessment of three mycotoxins: Aflatoxin, ochratoxin, and zearalenone. *Can. J. Physiol. Pharmacol.* **68**, 1017–1024.

Lin, J. K., Kennan, K., Miller, E. C., and Miller, J. A. (1978). Reduced nicotinamide adenine dinucleotide phosphate-dependent formation of 2,3,-dihydro-2,3-dihydroxyaflatoxin B$_1$ from aflatoxin B$_1$ by hepatic microsomes. *Cancer Res.* **38**, 2424–2428.

Liu, Y. H., Taylor, J., Linko, P., Lucier, G. W., and Thompson, C. L. (1991). Glutathione-S-transferase μ in human lymphocyte and liver: role in modulating formation of carcinogen-derived DNA adducts. *Carcinogenesis* **12**, 2269–2275.

Lotlikar, P. D., Jhee, E. C., Insetta, S. M., and Clearfield, M. S. (1984). Modulation of microsome-mediated aflatoxin B$_1$ binding to exogenous and endogenous DNA by cytosolic glutathione S-transferases in rat and hamster livers. *Carcinogenesis* **5(2)**, 269–276.

McLean, A. E. M., and Marshall, A. (1971). Reduced carcinogenic effects of aflatoxin in rat given phenobarbitone. *Br. J. Exp. Pathol.* **52**, 322–329.

Metcalfe, S. A., Colley, P. J., and Neal, G. E. (1981). A comparison of the effects of pretreatment with phenobarbitone and 3-methylcholanthrene on the metabolism of aflatoxin B$_1$ by rat liver microsomes and isolated hepatocytes *in vitro*. *Chem. Biol. Interact.* **35**, 145–157.

Meyer, D. J., Coles, B., Pemble, S. E., Gilmore, K. S., Fraser, G. M., and Ketterer, B. (1991). Theta, a new class of glutathione transferases purified from rat and man. *Biochem. J.* **274**, 409–414.

Mizokami, K., Nohmi, T., Fukuhara, M., Takanaka, A., and Omori, Y. (1986). Purification and characterization of a form of cytochrome P450 with high specificity for aflatoxin B$_1$ from 3-methylcholanthrene-treated hamster liver. *Biochem. Biophys. Res. Commun.* **139(2)**, 466–472.

Moss, E. J., and Neal, G. E. (1985). The metabolism of aflatoxin B$_1$ by human liver. *Biochem. Pharmacol.* **34(17)**, 3193–3197.

Muirhead, C. R., and Darby, S. C. (1987). Modelling the relative and absolute risks of radiation-induced cancers. *Journal of Royal Statistics Society (JRSS)* **150**, 83–118.

Nebert, D. W., Nelson, D. R., Coon, M. J., Estabrook, R. W., Feyereisen, R., Fujii-Kuriyama, Y., Gonzalez, F. J., Guengerich, F. P., Gunsalus, I. C., Johnson, E. F., Loper, J. C., Sato, R., Waterman, M. R., and Waxman, D. J. (1991). The P450 gene superfamily: Update on new sequences, gene mapping, and recommended nomenclature. *DNA Cell Biol.* **10(1)**, 1–14.

O'Brien, K., Moss, E., Judah, D., and Neal, G. (1983). Metabolic basis of the species differences to aflatoxin B$_1$-induced hepatotoxicity. *Biochem. Biophys. Res. Commun.* **114(2)**, 813–821.

Peers, F., Bosch, X., Kaldor, J., Linsell, A., and Pluijmen, M. (1987). Aflatoxin exposure, hepatitis B virus infection and liver cancer in Swaziland. *Int. Cancer* **39**, 545–553.

Quinn, B. A., Crane, T. L., Kocal, T. E., Best, S. J., Cameron, R. G., Rushmore, T. H., Farber, E., and Hayes, M. A. (1990). Protective activity of different hepatic cytosolic glutathione S-transferases against DNA-binding metabolites of aflatoxin B$_1$. *Toxicol. Appl. Pharmacol.* **105**, 351–363.

Ramsdell, H. S., and Eaton, D. L. (1989). Kinetics of aflatoxin B$_1$ biotransformation *in vitro* by hepatic microsomes from the rat, mouse, monkey and human. *Proc. Am. Assoc. Cancer Res.* **30,** 621.

Ramsdell, H. S., and Eaton, D. L. (1990a). Species susceptibility to aflatoxin B_1 carcinogenesis: Comparative kinetics of microsomal biotransformation. *Cancer Res.* **50,** 615–620.

Ramsdell, H. S., and Eaton, D. L. (1990b). Mouse liver glutathione S-transferase isoenzyme activity toward aflatoxin B_1-8,9-epoxide and benzo[*a*]pyrene-7,8-dihydrodiol-9,10-epoxide. *Toxicol. Appl. Pharmacol.* **105,** 216-225.

Raney, K. D., Shimada, T., Kim, D.-H., Groopman, J. D., Harris, T. M., and Guengerich, F. P. (1992a). Oxidation of aflatoxins and sterigmatocystin by human liver microsomes: significance of aflatoxin Q_1 as a detoxification product of aflatoxin B_1. *Chem. Res. Toxicol.* **5,** 202–210.

Raney, K. D., Meyer, D. J., Ketterer, B., Harris, T. M., and Guengerich, F. P. (1992b). Glutathione conjugation of aflatoxin B_1 exo- and endo-epoxides by rat and human glutathione S-transferases. *Chem. Res. Toxicol.* **5,** 470–478.

Shayiq, R. M., and Avadhani, N. G. (1989). Purification and characterization of a hepatic mitochondrial cytochrome P-450 active in aflatoxin B_1 metabolism. *Biochem.* **28,** 7546–7554.

Shimada, T., and Guengerich, F. P. (1989). Evidence for cytochrome P450-NF, the nifedipine oxidase, being the principal enzyme involved in the bioactivation of aflatoxins in human liver. *Proc. Natl. Acad. Sci. U.S.A.* **86,** 462–465.

Shimada, T., and Nakamura, S. (1987). Cytochrome P-450-mediated activation of procarcinogens and promutagens to DNA-damaging products by measuring expression of *umu* gene in *Salmonella typhimurium* TA1535/pSK1002. *Biochem. Pharmacol.* **36,** 1979–1987.

Shimada, T., Nakamura, S., Imaoka, S., and Funae, Y. (1987). Genotoxic and mutagenic activation of aflatoxin B_1 by constitutive forms of cytochrome P450 in rat liver microsomes. *Toxicol. Appl. Pharmacol.* **91,** 13–21.

Stoloff, L. (1983). Aflatoxin as a cause of primary liver-cell cancer in the United States: A probability study. *Nutr. Cancer* **5,** 165–185.

Swenson, D. H., Lin, J.-K., Miller, E. C., and Miller, J. A. (1977). Aflatoxin B_1-2,3-oxide as a probable intermediate in the covalent binding of aflatoxins B_1 and B_2 to rat liver DNA and ribosomal RNA *in vivo*. *Cancer Res.* **37,** 172–181.

Thomas, D. C. (1981). General relative-risk models for survival time and matched case-control analysis. *Biomedicine* **37,** 673–686.

Williams, D. E., and Buhler, D. R. (1983). Purified form of cytochrome P-450 from rainbow trout with high activity toward conversion of aflatoxin B_1 to aflatoxin B_1-2,3-epoxide. *Cancer Res.* **43,** 4752–4756.

Williams, D. E., Masters, B. S. S., Lech, J. J., and Buhler, D. R. (1986). Sex differences in cytochrome P-450 isozyme composition and activity in kidney microsomes of mature rainbow trout. *Biochem. Pharmacol.* **35(12),** 2017–2023.

Wogan, G. N., Paglialunga, S., and Newberne, P. M. (1974). Carcinogenic effects of low dietary levels of aflatoxin B_1 in rats. *Food Cosmet. Toxicol.* **12,** 681–685.

Yeh, F. S., Mo, C. C., and Yen, R. C. (1985). Risk factors for hepatocellular carcinoma in Guangxi, People's Republic of China. *Natl. Cancer Inst. Monogr.* **69,** 47–48.

Yeh, F. S., Yu, M. C., Mo, C. C., Luo, S., Tong, M. J., and Henderson, B. E. (1989). Hepatitis B virus, aflatoxins, and hepatocellular carcinoma in Southern Guangxi, China. *Cancer Res.* **49,** 2506–2509.

23

Economic Issues Associated with Aflatoxins

❖

Simon M. Shane

INTRODUCTION

Mycotoxic fungi are responsible for significant financial losses encompassing a broad spectrum of crop and animal agriculture, and extending through the food chain to the consumer. Crop farmers are impacted by fieldborne, intermediate, and storage fungi. Contamination at successive stages in growing and storage results in destruction and downgrading of grains and oilseeds as well as depression in nutritional value, and generates costs associated with prevention and control of mycotoxic fungi.

Mycotoxins depress the profitability of animal production through decreased growth, feed conversion efficiency, liveability, and reproductive potential in herds, flocks, and ponds. Mycotoxin residues are responsible for destruction of contaminated milk and dairy products. Considerable expense is incurred by producers through programs to detect and ameliorate fungal metabolites in ingredients, rations, and the production environment.

The food industry, consumers, and regulatory agencies incur costs relating to quality control programs, insurance, and destruction of products. Consumers in rural areas and tropical countries are especially subject to the acute and long-term effects of mycotoxins, which may act synergistically with primary infectious agents, malnutrition, or alcoholism. In addition, chronic exposure or ingestion of mycotoxins by human populations represents a risk for carcinogenesis. The purpose of this chapter is to identify the diverse range of costs attributable to mycotoxins that must be borne by consumers, food processors, and producers of commodities and domestic animals.

AGRONOMIC ASPECTS OF AFLATOXINS

Table 1 categorizes the range of costs associated with contamination of cereals and other ingredients such as pulses and legumes that are consumed by livestock and humans. Unfortunately, these costs cannot be quantified readily on a regional or national basis.

Since the mid-1970s the potential for extensive preharvest mycotoxin contamination of corn has been recognized (Lisker and Lillehoj, 1991). This recognition has led to the selective development of cultivars that are resistant to insect damage and drought, and produce phytoalexins. Significant costs are associated with breeding programs, evaluation of genetic resistance to fungal contamination, and selection of cultivars for specific areas and climates. Development of aflatoxin-resistant peanuts and of wheat varieties refractory to kernel blight (*Fusarium graminearum*) occupy the attention of geneticists and plant pathologists. Unfavorable yield of selected fungal-resistant cultivars compared with conventional varieties also should be considered as a real cost of mycotoxicosis.

TABLE 1   Cost of Mycotoxins in the Agronomic Sector

Development of fungus-resistant cultivars
   breeding, testing, and distribution of new strains
   reduced yield from fungus-resistant strains
Altered farming practices requiring incremental production costs
   irrigation to prevent desiccation
   application of insecticides to avert damage
   use of fungicides to prevent contamination
   additional fertilizers to reduce environmental stress
   crop rotation
   modified harvesting to avoid damage
Improved postharvest handling requiring additional facilities and resources
   modified transport and mechanization to reduce damage
   intensified drying of grain to achieve desirable moisture levels
   water-resistant, aerated or heated, capital-intensive storage
   in-storage fumigation to destroy insects
   appropriate quality control procedures to monitor for moisture and toxins
   additional storage capacity and separate bins for assayed consignments
Cost of amelioration
   additives, including fungistats, anticaking agents, and binders of toxins
   physical and chemical treatment
Financial impact
   condemnation with destruction of heavily contaminated consignments
   downgrading and reduced unit revenue for producer
   lowered turnover for elevators, dealers, and exporters

Farming practices to reduce mycotoxin contamination of ingredients incorporated into human and animal diets impose real, but poorly defined, costs. Irrigation to reduce stress; application of insecticides, fungicides, and specific fertilizer combinations; crop rotation; and special postharvest handling all contribute to increased yield but may be required to suppress the development of mycotoxins.

Costs are incurred through improved postharvest handling of grains and other ingredients. Preventing damage to kernels during threshing and on-farm loading, storage, and transport is critical to reducing mold contamination. Grain-drying installations on farms and at receiving elevators are required to reduce moisture content to values ranging from 11% for soybeans to 13.5% for wheat and corn (assuming less than 30°C ambient temperature and 70% relative humidity). In addition to capital and operating costs of grain drying, which is energy intensive, complementary programs of quality control, including sampling and monitoring, are required. Excessive heating of cereals, including corn and wheat, can result in cracking and increased susceptibility to fungal contamination. Fumigation to eliminate insects and the need to segregate types, grades, or consignments of ingredients during storage impose additional costs that ultimately are borne by consumers or the national economy.

The presence of mycotoxins in grains may result in downgrading or condemnation and destruction, depending on the contaminant, level, and regulatory infrastructure. Nichols (1983) has estimated that corn producers in eight southeastern states harvesting a total of 277 million bushels incurred losses of $97 million in 1980 because of inability to market contaminated grain, restricted markets, increased transport and drying costs, inability to obtain loans, and lowered revenue. Many elevators refuse to purchase corn with aflatoxin levels exceeding the Food and Drug Administration (FDA) guideline of 20 ppb, or reduce the price paid by 50¢/bushel for 50 ppb corn and 1¢/ppb over this amount. The grain industry experienced $14 million additional costs in 1980 from extra handling, drying, loss of revenue from operation at reduced storage capacity (30¢/bushel), and lowered turnover.

A review of mycotoxins in food products has shown extensive contamination (Jelinek *et al.,* 1989). The Food Contamination Monitoring Program in the United Nations FAO/WHO/UNEP demonstrated widespread distribution of aflatoxin ranging from 5 to 20 ppb among 5400 assays on corn produced in seven countries. Nearly all (95%) samples of peanuts imported into the United States from Brazil exceeded 25 ppb, contrasted with only a 5% rejection rate in product from Australia. *Fusarium* mycotoxins frequently are detected in cereals from producer countries in temperate (wheat) and tropical (corn) zones. The value of cereals and oilseeds is reduced sharply by mycotoxin contamination. Markets in developed countries usually impose some control over quality of products; consignments of ingredients that contain mycotoxins are either rejected or downgraded, resulting in extensive loss to producer countries.

The various programs of monitoring and analysis imposed by regulatory agencies and by the commercial sector represent a significant component of the cost of mycotoxins. In addition to the field, laboratory, and administrative personnel required, costs are incurred by analyses, maintaining the identity of consignments, and insuring the integrity of batches of ingredients conforming to acceptable standards.

Subjecting contaminated cereals to chemical or physical treatment may reduce levels of mycotoxins. Ultraviolet light or gamma radiation has been shown to degrade toxins of fungal origin (Samarajeewa, 1991). Chemical treatment with acids, oxidizing agents, and alkalis can reduce mycotoxin levels in feed ingredients under laboratory and pilot plant conditions. Commercial application of processes such as reaction with ammonium hydroxide under pressure (Pemberton and Simpson, 1991) has been limited to small-scale plants to treat peanut meal in Senegal, France, and India. Quantifying the financial benefits of treating ingredients to reduce levels of mycotoxins is required. These methods must be evaluated on the basis of contribution to live animal efficiency and optimization of production cost of milk, poultry meat, pork, and other human food products. In tropical countries, where extensive mycotoxin contamination of a specific product such as peanut meal occurs, chemical treatment may be justified. The cost of amelioration should be compared with alternatives such as importation of a substitute or feeding the contaminated ingredient with consequential elevation in livestock mortality and a depression in growth rate and feed conversion efficiency. Mycotoxin-related cancer is emerging as a recognized occupational hazard of workers involved in grain handling and processing (Shotwell, 1991). The costs of reducing dust levels in oil-expressing plants and feed-mixing facilities, providing worker protection, medical insurance, legal liability, regular physical examinations, health surveillance, and records all are related to the risk of carcinogenesis associated with mycotoxins.

ECONOMIC EFFECTS OF AFLATOXINS IN DOMESTIC ANIMALS

Mycotoxins are responsible for suboptimal efficiency in reproduction growth and liveability in a wide range of domestic animals including ruminant, monogastric, and aquatic species (Nelson and Christensen, 1978). Losses to producers usually can be quantified in the case of mycotoxin-induced death, growth suppression, or inferior feed conversion. Subtle effects of mycotoxins such as immunosuppression (Chang and Hamilton, 1982), reproductive dysfunction, or synergistic action with pathogens are difficult to identify and evaluate.

The financial impact of mycotoxins on intensive animal production can be illustrated by studies conducted in the poultry industry (Dalvi, 1986). Because of market competition and relatively large volumes of broiler and egg production in most industrialized countries, a high standard of efficiency is required to maintain profitability. Deviation from predetermined production standards at either the parent, commercial egg, or broiler grow-out level will impact both revenue and return on investment.

The critical determinants of parent-level performance include persistence of egg production, fertility, hatchability, and liveability. A significant intake of aflatoxins can induce profound reduction in egg production. Hamilton (1971) attributed a case of peracute toxicity in laying hens to consumption of moldy corn containing in excess of 100 ppm aflatoxin B_1. Egg production of the affected flock dropped to 5% of normal and 50% of the hens died within 48 hrs. Shlosberg *et at.* (1984) documented a similar decline in production, but without mortality, in laying hens fed a diet containing 3.5 ppm T-2 *Fusarium* toxin. Inclusion of 5% *Diplodia maydis* corn culture material in a layer ration resulted in a 43% decline in egg production (Rabie *et al.*, 1987). At the 0.5% level, hens showed an 8% reduction in daily feed intake and a 16% decrease in egg production compared with controls. Significant differences in feed consumption (6%) and egg production (3%) may be caused by inclusion of corn infected with *Gibberella zeae* in layer rations (Adams and Tuite, 1976). Ochratoxin A included in rations at 1 ppm induces a 10% drop in egg production accompanied by diarrhea, hyperuricemia, and staining of egg shells (Page *et al.*, 1981). The toxicities of *Fusarium roseum* culture and diacetoxyscirpenol (DAS) have been evaluated in mature hens (Allen *et. al.*, 1982). Dietary inclusion of 3% culture reduced hatchability by 99% after 1 week compared with controls. Purified DAS at the 0.5-ppm level lowered hatchability by 24% after 4 weeks. These severe effects were attributed to embryo mortality during the first week of incubation. Significantly, no effects on egg production, egg weight, or clinical abnormalities were noted in hens receiving either *F. roseum* cultures or DAS. Similar effects on hatchability of turkey eggs have been documented (Allen *et al.*, 1983), with 49% reduction in poult yield following inclusion of 2% *F. roseum* culture in rations fed to hens. This study also demonstrated a depression in body weight, egg production, and feed intake after consumption of rations containing 100 ppm zearalenone.

The financial impact of mycotoxins on a parent broiler breeding enterprise can be calculated by applying the methods described by Gifford *et al.* (1987). In this study, a microcomputer spreadsheet program was used to evaluate the financial effect of disease on the liveability and reproductive efficiency of broiler parent flocks. The costs associated with adverse reproductive effects of mycotoxicosis include lowered egg production, decreased hatchability, deaths, and loss in future chick production as a result of mortality. Any factor detracting from optimal production efficiency will reduce the number of viable chicks delivered to farms. In the context of an integrated broiler enterprise, a reduction in number of chicks will influence total production cost of finished product since fixed components represent at least 20% of the total cost of poultry meat. Integrators are obliged to purchase broiler chicks (at 15¢/unit) or fertile eggs ($1.25/dozen) to maintain throughput in the event of shortfalls from their breeder flocks subjected to mycotoxicosis. A 10% drop in egg production concurrent with a 5% depression in hatchability for 20 days in a flock of 25,000 40-week-old broiler breeders is estimated to result in a 19% reduction in chicks produced. The value of the projected loss of 45,625 1-day-old broilers would be $8,193, assuming purchase at 15¢/unit (Table 2).

Several mycotoxins have been identified as responsible for reducing weight

**TABLE 2   Projection of Replacement Cost of Chicks
Associated with Depressed Reproductive Efficiency**

Parameter	Unaffected flock	Affected flock	Difference
Hens	25,000	25,000	—
Average hen week egg production	70%	60%	10%
Allowance for rejected eggs	5%	5%	—
Hatchable eggs produced			
per day	16,625	14,250	2375
per 20 days	332,500	285,000	47,500
Hatchability	85%	80%	5%
Number of broiler chicks produced	282,625	228,000	54,625
Value of chicks at 15¢/unit ($)	42,393	34,200	8,193

gain in broilers (Table 3). Overt effects are relatively simple to quantify under controlled laboratory conditions, but associating low-level suppression of weight gain in commercial flocks to mycotoxicosis may be impossible. In evaluating the significance of mycotoxins, producers rely on screening and quantitative assays of feed, recognition of nonspecific clinical abnormalities such as erosion of the oral mucosa and gizzard (ventriculus), impaired feathering, and eliminating management, disease, and parasite-related causes of growth depression. Field studies have shown that the mean daily weight gain of broilers (standard value of 40 g/day over a 45-day growing period) can be depressed at levels of aflatoxin intake as low as 60 ppb (Hamilton, 1978). A study on broiler flocks in North Carolina showed a statistical relationship between commercial performance and aflatoxin contamination of diets. The parameters of live weight, feed conversion efficiency, liveability, and carcass quality for all flocks produced in a complex were assessed using the standard formula for contractor remuneration. Feed from

**TABLE 3   Growth-Depressing Effect
of Common Mycotoxins in Broilers**

Mycotoxin	Dietary level producing significant effect on weight gain	Reference
Triticale ergot	0.8%	Bragg *et al.* (1970)
Aflatoxin	0.06 ppm	Jones *et al.* (1982)
Aflatoxin	0.8 ppm	Giambrone *et al.* (1985)
Aflatoxin	0.7–2.1 ppm	Arafa *et al.* (1981)
Cyclopiazonic acid	100 ppm	Dorner *et al.* (1983)
Ochratoxin	3.0 ppm	Kubena *et al.* (1983)
Rubratoxin	0.5 ppb	Wyatt and Hamilton (1972)
Zearelenone	4.0 ppm	Wyatt *et al.* (1973)
Citrinin	500 ppm	Ames *et al.* (1976)

only 18% of the flocks achieving superior performance demonstrated aflatoxin at levels exceeding 6 ppm. In comparison, a mean aflatoxin value of 14 ppm was determined in 31% of the flocks regarded as below the acceptable standard of efficiency (Jones *et al.*, 1982).

Immunosuppression is an important component of the costs attributed to mycotoxicosis. Aflatoxins have been shown to affect cell-mediated immunity (Giambrone *et al.*, 1985). Although most immature commercial poultry species show atrophy of the bursa of Fabricius and thymus, no studies have been conducted to show a specific immune response to aflatoxins at dietary levels normally encountered in commercial situations (Hoerr, 1991). Ochratoxin suppresses both humoral and cell-mediated immunity (Dwivedi and Burns, 1985); as with aflatoxicosis, this agent is synergistic with coccidiosis in increasing the severity of cecal and intestinal infection (Huff and Ruff, 1982).

Aflatoxin, *Fusarium* T-2 toxin, and ochratoxin are responsible for coagulopathy at growth-suppressing dietary levels. Aflatoxin at 1.25 ppm in the diet extends whole-blood clotting time, prothrombin time, and recalcification time (Doerr *et al.*, 1974) and, at dietary levels as low as 0.6 ppm (Tung *et al.*, 1971), is responsible for bruising and plant condemnation. Aflatoxin and ochratoxin function in synergy to exacerbate the occurrence and severity of bruising under conditions simulating commercial handling and transport (Huff *et al.*, 1983). A field study initiated in 1988 evaluated the effect of constant low-level aflatoxin (20 ppb) and zearalenone (100 ppb) on downgrading of broilers and turkeys. Data relating to eight categories of carcass quality were monitored in flocks known to have consumed mycotoxins. Results were compared with controls fed noncontaminated rations. Losses associated with mycotoxicosis included bruising, increased mortality during transport, septicemia (following immunosuppression), and lowered feed conversion efficiency. Revenue was lowered by $16/1000 in 2-kg broilers and $400/1000 in 8-kg turkeys (Muirhead, 1989).

The financial impact of mycotoxin-induced depression in broiler growth, feed conversion efficiency, and liveability can be approximated given prevailing costs and performance parameters. Stimulation studies can be performed to quantify losses caused by aflatoxicosis, applying realistic performance and cost data as shown in Table 4. Aflatoxicosis is assumed, in this example, to produce a 100-g (6%) decrease in live weight, a 7.5% widening of the feed conversion ratio, an

TABLE 4   Broiler Performance Parameters:
Aflatoxin Simulation

Parameter	Unaffected	Aflatoxicosis	Difference
Live mass (g)	1800	1700	100
Feed conversion ratio	2.00	2.15	0.15
Mortality (%)	7.0	10.0	3.00
Downgrades (%)	1.0	3.0	2.00
Processing yield (%)	78	78	—

TABLE 5   Broiler Production Volume: Aflatoxin Simulation

Parameter	Unaffected	Aflatoxicosis	Difference
Weekly placement (chicks)	1,000,000	1,000,000	—
Live mass delivered (kg)	1,674,000	1,530,000	144,000
Processed product (kg)	1,292,663	1,157,598	135,065
Feed consumed (kg)	3,348,000	3,289,500	58,500

increase in flock mortality from 7 to 10%, and a significant escalation in down-grades from 1 to 3% of all birds processed. This example presumes equality in plant yield of 78%. If marked variability in size of carcasses occurs or if live weight is depressed, yield generally is lowered.

Table 5 shows the reduction in weekly live weight (8.6%) and processed product (10%) associated with aflatoxin-induced depression in growth, liveability, feed conversion, and carcass quality. The contribution to overhead generated by the broiler-growing and -processing operations of the enterprise are shown in Table 6 for unaffected broilers and flocks consuming rations contaminated with aflatoxins. These calculations are based on the purchase of rations at $200/ton and partitioning of feed and nonfeed costs (chicks, other variable and fixed components) in the ratio of 70:30%. In the case of flocks affected with aflatoxicosis, applying the same nonfeed value of $286,971 is realistic since the operation would purchase (or transfer) 1 million chicks per week irrespective of growth rate and liveability. Most broiler production and processing costs other than feed and chicks should be regarded as fixed, especially if relatively short-term deviations from normal performance occur. In this instance, the enterprise would be unlikely to reduce labor complement in houses and plants and fuel; other operational costs (other than purchase of packaging material and vaccines) also would be unaffected by the 10% reduction in saleable poultry meat. In this example, the unit cost of product is increased by 10.2% as a result of aflatoxicosis, and the 81.6¢/kg cost approaches the selling price of 90¢/kg, reducing contribution to the overhead costs of the enterprise.

TABLE 6   Calculation of Contribution
from Broiler Production

	Unaffected	Aflatoxicosis
Weekly sales at 90¢/kg ($)	1,163,397	1,041,838
Less cost of sales		
Feed at $200/ton (70%) ($)	(669,600)	(657,900)
Chicks and other variable	(286,971)	(286,971)
and fixed costs (30%) ($)		
Subtotal cost ($)	(956,571)	(944,871)
Weekly contribution ($)	206,826	96,967
Unit cost (¢/kg)	74.0	81.6

In the event of aflatoxicosis in a vertically integrated broiler production company using independent contractors, the losses would be borne in part by the growers since they are remunerated at the rate of approximately 13¢/bird, based on live mass of acceptable quality delivered to plants, with downward adjustments for mortality and inferior feed conversion. In totally integrated broiler production companies with ownership of grow-out housing, the entire cost of aflatoxicosis would be absorbed.

This example did not include the deleterious effect of aflatoxicosis on chick production, as previously discussed. If unassayed consignments of aflatoxin-contaminated corn were incorporated into both breeder and broiler diets, the financial impact would be increased by the need to purchase chicks to maintain the planned production level of 1,000,000 broilers per week. If chicks are not commercially available at short notice, as occurs in many developing areas of the world, production from the enterprise would be reduced further. Although feed costs would be lower, fixed and semivariable costs of production would remain constant despite the 10–20% decline in broiler mass. This condition effectively would place the enterprise in a loss situation unless market factors favor a compensatory increase in selling price. In many tropical areas, mycotoxicoses of even short duration or moderate effect can impose severe restraints on cash generation, especially in a competitive environment.

Comparable studies could be conducted to determine the financial impact of mycotoxins on the production of aquatic species, especially salmonids, in which aflatoxin levels of 0.5 ppb can induce hepatoma. Lower levels of contamination can inhibit growth (Sinnhuber and Wales, 1978). Significant problems in the intensive animal industry that are attributable to mycotoxins include facial eczema in sheep (Van Warmelo et al., 1970), fusariotoxicosis (Chang et al., 1979) and ochratoxicosis (porcine nephropathy) in swine (Krogh, 1991), and leukoencephalomalacia in horses (Kellerman et al., 1972). Hog producers in 10 southeastern states suffered losses in 1980 estimated at $100 million in 13.6 million animals because of aflatoxin contamination of locally produced corn. Costs were incurred through additional mortality ($24 million) and inferior feed conversion efficiency ($76 million) compared with unaffected herds (Nichols, 1983). The total $100 million 1980 cost of aflatoxicosis to swine producers increased from $58 million in 1977 because of escalation in value of production costs and a higher prevalence of aflatoxicosis. The losses caused by aflatoxin represented $7.40 per hog marketed.

Although ruminants generally are affected less by aflatoxin contamination of feed than swine and poultry are, dairy farmers have a specific problem of residues. The 1977 FDA action level of 0.5 ppb for AFM_1 in milk may be attained or exceeded with a dietary intake of 30 ppb. As surveillance programs are intensified, farmers and processors will be obliged to improve quality control procedures to avert losses to condemnation. Of 302 samples of fluid milk examined in North Carolina during 1977–1980, 7 exceeded the 0.5 ppb limit and 15 were within the 0.2–0.5 ppb range (Nichols, 1983).

Studies extending the knowledge of the deleterious effects of mycotoxins from laboratory trials to field conditions are necessary to evaluate and select appropriate remedial action. The application of screening programs, addition of mold inhibitors to rations (Williams, 1991), and chemical treatment to detoxify ingredients are all valid techniques to reduce the cost of mycotoxins to animal production systems. A simulation study conducted on the addition of a commercial hydrated sodium calcium aluminosilicate to broiler rations showed a 5:1 benefit-to-cost ratio (Shane, 1991), confirming the extensive laboratory studies on zeolite-binding of aflatoxins in feed (Philips *et al.,* 1988). Modifying management practices, including reducing feed residence time (Good and Hamilton, 1981) and physical decontamination of feed bins and broiler feeding systems (Hamilton, 1975), is cost effective in reversing the deleterious impact of mycotoxins on the animal production industry. Economic analyses to select specific strategies to reduce the effect of mycotoxicoses should be developed and adopted for herds, flocks, and aquatic units.

ECONOMIC IMPACT OF AFLATOXINS IN HUMAN POPULATIONS

Since the first reports of aflatoxin intoxication in humans in Taiwan (Ling *et al.,* 1967), the pathology and toxicity of human mycotoxicosis have been reviewed extensively (Shank, 1978; Denning, 1987). The epidemiology of primary hepatic carcinoma and its relationship to diet and intake of mycotoxins in developing countries has been the subject of extensive research (Krogh, 1984). Aflatoxins are considered etiologic factors in the occurrence of acute hepatitis (Krishnamachari *et al.,* 1975), juvenile hepatic cirrhosis in India (Amla *et al.,* 1971), and hepatocellular carcinoma in southeastern Africa (Van Rensburg *et al.,* 1985). Aflatoxicosis has been implicated as a contributory or exacerbatory factor in kwashiorkor (Hendrickse, 1984) and immunosuppression (Denning, 1987). Despite advances in understanding aspects of detection and metabolism of mycotoxins and their teratogenic and mutagenic effects, no comprehensive studies have been done on their economic and financial impact on specific populations, either in industrialized or in developing nations.

A longitudinal retrospective cohort study was conducted in Holland on a population of workers in an oilseed processing plant (Hayes *et al.,* 1984). Vital records for 1961–1980 reflecting exposure of 71 workers in 1961–1969 were compared with a cohort group in food processing plants. Standard mortality ratios were calculated for the plant workers and controls for all causes of death, cardiovascular disease, and malignant neoplasia. Workers in oilseed processing plants, presumed to be exposed to aflatoxins for an average of 105 weeks, showed a significantly higher mortality rate from cancer of the respiratory tract and all other organ systems than did cohorts. Also significant was that onset of neoplasia was diagnosed within 10 years of exposure in half of the 14 cases. This study serves as a model for the evaluation of mycotoxin-related occupational effects. Conventional methods for monitoring incidence of neoplasia and/or toxicity and projecting the long-term risks and costs associated with environmental exposure should be applied to workers in plants processing ingredients and preparing animal feed. Projecting the cost of death is dependent on the selection of the "human capital" or "willingness-to-pay" approach (Landefeld and Seskin, 1982) and will be influenced by prevailing government policy and the socioeconomic level of the individual exposed. The value of life extends from $50,000 as accepted by the U.S. Consumer Product Safety Commission for accident protection, through $100,000 (Department of Health and Human Services for immunization studies) and $2.6 million (Environmental Protection Agency), to $12 million (Occupational Safety and Health Administration) for exposure to carcinogens.

The potential public health significance of aflatoxicosis was recognized as early as 1961 (Fischbach, 1984). Cooperation among the food industry, federal regulatory agencies, and academic institutions resulted in rapid progress in analytical procedures to identify aflatoxin contamination in peanuts (Baur and Parker, 1984) and in quality control procedures to limit consumption of potentially toxic peanuts and derived products (Tosch et al., 1984).

The selection of an upper limit for aflatoxin in peanuts products in the range of 5–20 ppb has been subjected to a cost-effectiveness analysis (Dichter and Weinstein, 1984). Costs accruing from more intensive analysis and selection of product relating to 5, 10, and 15 ppb levels of contamination were calculated from data collected in a survey of 12 peanut processors. Annual incremental costs compared with the 1969 interim action level of 20 ppb ranged from $300,000 at the 15 ppb level proposed in 1974 to $18.1 million with a more stringent 5 ppb standard. The incremental number of cases of hepatic carcinoma were estimated for the four levels of aflatoxin content by extrapolation from the Fischer rat model. The predicted prevalence ranged from 151 cases per year at the 20 ppb tolerance level to 38 cases at 5 ppb. Cost-effectiveness analysis was employed to determine the most appropriate level, taking into account a 20-year latency period for development of carcinoma and a 5% inflation-corrected discount factor for costs relating to the specific tolerance levels. Average future values were $0.8 million and $48 million for the 15 ppb and 5 ppb levels, respectively. The central estimates for the marginal cost-effectiveness ratios for the levels were 0.056 (15 ppb), 1.7 (10 ppb), and 1.6 (5 ppb), meaning $56,000 would be expended for each case of hepatic carcinoma prevented by lowering the standard from 20 ppb to 15 ppb. At the 10 ppb and 5 ppb levels, expenditures would be $1.7 million and $1.6 million, respectively. The analysis demonstrated that the 15 ppb tolerance level was the most cost effective. The author's conclusion is, however, based on specific assumptions of risk, length of latency period, and the selection of a discount rate. Increasing the last two values to 25 years and 10%, respectively, would increase the ratio at the 15 ppb level from $56,000 to $3 million. Decisions by regulators concerning an appropriate tolerance for aflatoxin in a food product would be based on an appraisal of realistic costs and their perception of consumer preference. The authors estimated that imposing a 5 ppb level would increase the cost of a 500-g container of peanut butter by only 5¢. Since this is a relatively inconsequential cost, lowering the tolerance level in the United States would not cause a significant depression in consumption of peanut butter, which only contributes approximately 1% to the normal United States dietary protein intake.

In contrast, estimating the economic impact of mycotoxicosis in developing countries is much more difficult. Interactions between aflatoxin and chronic alcohol consumption (Bulatao-Jayme *et al.,* 1982) and viral hepatitis (Prince *et al.,* 1975) detract from an understanding of the epidemiology of hepatic carcinoma. Intercurrent malnutrition, parasitism, interactions between mycotoxins, and immunosuppression all complicate risk-assessment in the context of tropical countries. Tolerance levels for mycotoxins may be unrealistic in these areas, where peanuts may be the principal source of protein and analytical resources may not be available. Diversion of contaminated commodities into animal rations may be impractical because of deficiencies in storage and transport infrastructure. In addition, the absence of an organized intensive animal production industry or grain exchange may preclude the sale of contaminated ingredients to allow the purchase of substitute food items. More appropriate cost-effective strategies are required, including selection of fungus-resistant cultivars, improved crop husbandry and postharvest storage, and the use of zeolite feed additives.

In contrast to the extensive knowledge base concerning the biochemistry and pathophysiology of mycotoxicoses in domestic animals and humans, considerable work is necessary on the epidemiology and economic aspects of control. Cost-effectiveness studies are required to evaluate alternative strategies for prevention and remediation of mycotoxicosis. Realistic tolerance levels also are required (Edds and Osuna, 1976), which should be based on economic and health considerations.

REFERENCES

Adams, R. L., and Tuite, J. (1976). Feeding *Gibberella zeae* damaged corn to laying hens. *Poultry Sci.* **55,** 1991–1993.

Allen, N. K., Jevne, R. L., Mirocha, C. J., and Lee, Y. W. (1982). The effect of a *Fusarium roseum* culture and diacetoxyscirpenol on reproduction of white leghorn females. *Poultry Sci.* **61,** 2172–2175.

Allen, N. K., Peguri, A., Mirocha, C. J., and Newman, J. A. (1983). Effects of *Fusarium* cultures, T-2 toxin, and zearalenone on reproduction of turkey females. *Poultry Sci.* **62,** 282–289.

Ames, D. D., Wyatt, R. D., Marks, H. L., and Washburn, K. W. (1976). Effect of citrinin, a mycotoxin produced by *Penicillium citrinum,* on laying hens and young broilers chicks. *Poultry Sci.* **55,** 1294–1301.

Amla, I., Kamala, C. S., Gopalakrishna, G. S., Jayaraj, A. P., Sreenivasamurthy, V., and Parpia, H. A. B. (1971). Cirrhosis in children from peanut meal contaminated by aflatoxin. *Am. J. Clin. Nut.* **24,** 609–614.

Arafa, A. S., Bloomer, R. J., Wilson, H. R., Simpson, C. F., and Harms, R. H. (1981). Susceptibility of various poultry species to dietary aflatoxin. *Br. Poultry Sci.* **22**, 431–436.

Baur, F. J., and Parker, W. A. (1984). The aflatoxin problem: Industry–FDA–USDA cooperation. *J. Assoc. Off. Anal. Chem.* **67**, 3–7.

Bragg, D. B., Salem, H. A., and Devlin, T. J. (1970). Effect of dietary *Triticale* ergot on the performance and survival of broiler chicks. *Can. J. Anim. Sci.* **50**, 259–264.

Bulatao-Jayme, J., Alermo, E. M., Castro, C. A., Jurdeleza, M. T. R., and Salamat, L. A. (1982). A case-control dietary study of primary liver cancer risk from aflatoxin exposure. *Int. J. Epidemiol.* **11**, 112–119.

Chang, C-F., and Hamilton, P. B. (1982). Increased severity and new symptoms of infectious bursal disease during aflatoxicosis in broiler chickens. *Poultry Sci.* **61**, 1061–1068.

Chang, K., Kurtz, H., and Mirocha, C. J. (1979). Effects of the mycotoxin zearalenone on swine. *Am. J. Vet. Res.* **40**, 1260–1267.

Dalvi, R. R. (1986). An overview of aflatoxicosis of poultry: Its characteristics, prevention and reduction. *Vet. Res. Commun.* **10**, 429–443.

Denning, D. W. (1987). Aflatoxin and human disease. *Adv. Drug React. Ac. Pois. Rev.* **4**, 175–209.

Dichter, C. R., and Weinstein, M. C. (1984). Cost-effectiveness of lowering the aflatoxin tolerance level. *Food Chem. Toxicol.* **22**, 439–445.

Doerr, J. A., Huff, W. E., Tung, H. T., Wyatt, R. D., and Hamilton, P. B. (1974). A survey of T-2 toxin, ochratoxin, and aflatoxin for their effects on the coagulation of blood in young broiler chickens. *Poultry Sci.* **53**, 1728–1734.

Dorner, J. W., Cole, R. J., Lomax, L. G., Gosser, H. S., and Diener, U. L. (1983). Cyclopiazonic acid production by *Aspergillus flavus* and its effects on broiler chickens. *Appl. Environ. Microbiol.* **46**, 698–703.

Dwivedi, P., and Burns, R. B. (1985). Immunosuppressive effects of ochratoxin A in young turkeys. *Avian Pathol.* **14**, 213–225.

Edds, G. T., and Osuna, O. (1976). Aflatoxin B$_1$ increases infectious disease losses in food animals. *Proc. U.S. Animal Health Assoc.* **80**, 434–441.

Fischbach, H. (1984). Coping with the aflatoxin problem in the early years. *J. Assoc. Off. Anal. Chem.* **67**, 1–3.

Giambrone, J. J., Diener, U. L., Davis, N. D., Panangala, V. S., and Hoerr, F. J. (1985). Effects of aflatoxin on young turkeys and broiler chickens. *Poultry Sci.* **64**, 1678–1684.

Gifford, D. H., Shane, S. M., Hugh-Jones, M., and Weigler, B. J. (1987). Evaluation of biosecurity in broiler breeders. *Avian Dis.* **31**, 339–344.

Good, R. E., and Hamilton, P. B. (1981). Beneficial effect of reducing the feed residence time in a field problem of suspected moldy feed. *Poultry Sci.* **60**, 1403–1405.

Hamilton, P. B. (1971). A natural and extremely severe occurrence of aflatoxicosis in laying hens. *Poultry Sci.* **50**, 1880–1882.

Hamilton, P. B. (1975). Proof of mycotoxicoses being a field problem and a simple method for their control. *Poultry Sci.* **54**, 1706–1708.

Hamilton, P. B. (1978). Fallacies in our understanding of mycotoxins. *J. Food Protect.* **41**, 404–408.

Hayes, R. B., Van Nieuwenhuize, J. P., Raatgever, J. W., and ten Kate, F. J. W. (1984). Aflatoxin exposures in the industrial setting: An epidemiological study of mortality. *Food Chem. Toxicol.* **22**, 39–43.

Hendrickse, R. G. (1984). The influence of aflatoxins on child health in the tropics with particular reference to kwashiorkor. *Trans. R. Soc. Trop. Med. Hyg.* **78**, 427–435.

Hoerr, F. J. (1991). Mycotoxicoses. *In* "Diseases of Poultry" (B. W. Calnek, H. J. Barnes, C. W. Beard, W. M. Reid, and H. W. Yoder, eds.), 9th Ed., pp. 884–915. Iowa State University Press, Ames.

Huff, W. E., and Ruff, M. D. (1982). *Eimeria acervulina* and *Eimeria tenella* infections in ochratoxin A-compromised broiler chickens. *Poultry Sci.* **61**, 685–692.

Huff, W. E., Doerr, J. A., Wabeck, C. J., Chaloupka, G. W., May, J. D., and Merkley, J. W. (1983).

Individual and combined effects of aflatoxin and ochratoxin A on bruising in broiler chickens. *Poultry Sci.* **62,** 1764–1771.

Jelinek, C. F., Pohland, A. E., and Wood, G. E. (1989). Worldwide occurrence of mycotoxins in foods and feeds—An update. *J. Assoc. Off. Anal. Chem.* **72,** 223–230.

Jones, F. T., Hagler, W. H., and Hamilton, P. B. (1982). Association of low levels of aflatoxin in feed with productivity losses in commercial broiler operations. *Poultry Sci.* **61,** 861–868.

Kellerman, T. S., Marasas, W. F. O., Pienaar, J. G., and Naude, T. W. (1972). A mycotoxicosis of Equidae caused by *Fusarium moniliforme* Sheldon: A preliminary communication. *Onderstepoort J. Vet. Res.* **39,** 205–208.

Krishnamachari, K. A. V. R., Bhat, R. V., Nagarajan, V., and Tilak, T. B. G. (1975). Hepatitis due to aflatoxicosis. *Lancet* **i,** 1061–1063.

Krogh, P. (1984). A review of epidemiological studies of mycotoxin-related diseases in man and animals. *In* "Toxigenic Fungi—Their Toxins and Health Hazard" (H. Kurata and Y. Ueno, eds.), pp. 327–331. Elsevier, Amsterdam.

Krogh, P. (1991). Porcine nephropathy associated with ochratoxin-A. *In* "Mycotoxins and Animal Foods" (J. E. Smith and R. S. Henderson, eds.), pp. 628–640. CRC Press, Boca Raton, Florida.

Kubena, L. F., Phillips, T. D., Creger, C. R., Witzel, D. A., and Heidelbaugh, N. D. (1983). Toxicity of ochratoxin A and tannic acid to growing chicks. *Poultry Sci.* **62,** 1786–1792.

Landefeld, J. S., and Seskin, E. P. (1982). The economic value of life: Linking theory to practice. *Am. J. Public Health* **72,** 555–566.

Ling, K-H., Wang, J-J., Wu, R., Tung, T-C., Lin, C-K, Lin, S-S., and Lin, T-M. (1967). Intoxication possibly caused by aflatoxin B₁ in the mouldy rice in Shaung-Chih township. *J. Formosan Med. Assoc.* **66,** 517–525.

Lisker, N., and Lillehoj, E. B. (1991). Prevention of mycotoxin contamination (principally aflatoxins and *Fusarium* toxins) at the preharvest stage. *In* "Mycotoxins and Animal Foods" (J. E. Smith and R. S. Henderson, eds.), pp. 690–712. CRC Press, Boca Raton, Florida.

Muirhead, S. (1989). Studies show cost of mycotoxin contamination to poultry firms. *Feedstuffs* **20,** 10.

Nelson, G. H., and Christensen, C. M. (1978). Diagnosis of mycotoxicoses in animals and poultry: An overview. *In* "Mycotoxic Fungi, Mycotoxins, and Mycotoxicoses" (T. D. Wyllie and L. G. Moorhouse, eds.), Vol. 2, pp. 3–8. Marcel Dekker, New York.

Nichols, T. E., Jr. (1983). Economic impact of aflatoxin in corn. *In* "Aflatoxin and *Aspergillus flavus* in Corn" (U L. Diener, ed.), pp. 67–71. Southern Cooperative Series Bulletin 279. Auburn University, Auburn, Alabama.

Page, R. K., Stewart, G., Wyatt, R., Bush, P., Fletcher, O. J., and Brown, J. (1981). Influence of low levels of ochratoxin A on egg production, egg-shell stains, and serum uric-acid levels in leghorn-type hens. *Avian Dis.* **24,** 777–787.

Pemberton, A. D., and Simpson, T. J. (1991). The chemical degradation of mycotoxins. *In* "Mycotoxins and Animal Foods" (J. E. Smith and R. S. Henderson, eds.), pp. 798–810. CRC Press, Boca Raton, Florida.

Phillips, T. D., Kubena, L. F., Harvey, R. B., Taylor, D. R., and Heidelbaugh, N. O. (1988). Hydrated sodium calcium aluminosilicate: A high affinity sorbent for aflatoxin. *Poultry Sci.* **67,** 243–247.

Prince, A. M., Szmuness, W., Michon, J., Demaille, J., Diebolt, G., Linhard, J., Quenum, C., and Sankale, M. (1975). A case/control study of the association between primary liver cancer and hepatitis B infection in Senegal. *Int. J. Cancer* **16,** 376–383.

Rabie, C. J., DuPreez, J. J., and Hayes, J. P. (1987). Toxicity of *Diplodia maydis* to broilers, ducklings, and laying chicken hens. *Poultry Sci.* **66,** 1123–1128.

Samarajeewa, U. (1991). *In situ* degradation of mycotoxins by physical methods. *In* "Mycotoxins and Animal Foods" (J. E. Smith and R. S. Henderson, eds.), pp. 786–794. CRC Press, Boca Raton, Florida.

Shane, S. M. (1991). Economic significance of mycotoxicoses. *In* "Mycotoxins, Cancer and Health" (G. A. Bray and D. H. Ryan, eds.), pp. 53–64. Louisiana State University Press, Baton Rouge.

Shank, R. C. (1978). Mycotoxicoses of man: Dietary and epidemiological considerations. *In* "Mycotoxic Fungi, Mycotoxins, and Mycotoxicoses" (T. D. Wyllie and L. G. Moorhouse, eds.), Vol. 3, pp. 1–19. Marcel Dekker, New York.

Shlosberg, A., Weisman, Y., and Handji, V. (1984). A severe reduction in egg laying in a flock of hens associated with trichothecene mycotoxins in the feed. *Vet. Hum. Toxicol.* **26**, 384–386.

Shotwell, O. L. (1991). Mycotoxins in grain dusts: Health implications. *In* "Mycotoxins and Animal Foods" (J. E. Smith and R. S. Henderson, eds.), pp. 416–436. CRC Press, Boca Raton, Florida.

Sinnhuber, R. O., and Wales, J. H. (1978). The effect of mycotoxins in aquatic animals. *In* "Mycotoxic Fungi, Mycotoxins, and Mycotoxicoses" (T. D. Wyllie and L. G. Moorhouse, eds.), pp. 489–503. Marcel Dekker, New York.

Tosch, D., Waltking, A. E., and Schlesier, J. F. (1984). Past and present research on aflatoxin in peanut products. *J. Assoc. Off. Anal. Chem.* **67**, 8–9.

Tung, H-T., Smith, J. W., and Hamilton, P. B. (1971). Aflatoxicosis and bruising in the chicken. *Poultry Sci.* **50**, 795–800.

Van Rensburg, S. J., Cook-Mozaffari, P., Van Schlalkwyk, D. J., Van Der Watt, J. J., Vincent, T. J., and Purchase, I. F. (1985). Hepatocellular carcinoma and dietary aflatoxin in Mozambique and Transkei. *Br. J. Cancer* **51**, 713–726.

van Warmelo, K. T., Marasas, W. F. O., Adelaar, T. F., Kellerman, T. S., Van Rensburg, I. B. J., and Minne, J. A. (1970). Experimental evidence that lupinosis of sheep is a mycotoxicosis caused by the fungus *Phomopsis leptostromi formis* (Kuhn). *J. S. Afr. Vet. Assoc.* **41**, 235–247.

Williams, P. C. (1991). Storage of grains and seeds. *In* "Mycotoxins and Animal Foods" (J. E. Smith and R. S. Henderson, eds.), pp. 722–745. CRC Press, Boca Raton, Florida.

Wyatt, R. D., and Hamilton, P. B. (1972). The effect of rubratoxin in broiler chickens. *Poultry Sci.* **51**, 1383–1387.

Wyatt, R. D., Hamilton, P. B., and Burmeister, H. R. (1973). The effects of T-2 toxin in broiler chickens. *Poultry Sci.* **52**, 1853–1895.

Index

❖